Fundamentals of waves and oscillations

Fundamentals of

WAVES & OSCILLATIONS

K.U. INGARD
Massachusetts Institute of Technology

The right of the
University of Cambridge
to print and sell
all manner of books
was granted by
Henry VIII in 1534.
The University has printed
and published continuously
since 1584.

CAMBRIDGE UNIVERSITY PRESS
Cambridge
New York   Port Chester
Melbourne   Sydney

Published by the Press Syndicate of the University of Cambridge
The Pitt Building, Trumpington Street, Cambridge CB2 1RP
40 West 20th Street, New York, NY 10011, USA
10 Stamford Road, Oakleigh, Melbourne 3166, Australia

First published 1988
Reprinted 1990

Printed in Great Britain at the University Press, Cambridge

British Library cataloguing in publication data

Ingard, K.U.
Fundamentals of waves and oscillations
1. Waves 2. Oscillations
I. Title
531'.1133 QC157

Library of Congress cataloguing in publication data

Ingard, K. Uno.
Fundamentals of waves and oscillations
Includes index.

1. Waves. 2. Oscillations. I. Title.
QC157.I54 1988 531'.1133 86-17116

ISBN 0 521 32734 2 hardback
ISBN 0 521 33957 X paperback

TM

Contents

Preface

This book is based on the notes which I prepared for a one-term undergraduate physics course on waves and oscillations for science and engineering students at M.I.T. (about 40 hours of lectures and 28 hours of recitations). In the process of working over these notes for publication, some material was added, and the book in its present form contains more than was covered in the course.

The book can be divided into three major parts: Oscillations (Chapters 1–5), Waves (Chapters 6–16), and Special Topics (Chapters 17–21). The essentials of the first two parts can be incorporated in a one-term course. The third part has been presented during M.I.T.'s Independent Activity Period (IAP) as a continuation of the regular course, and it has been used also as preparatory reading material for students involved in undergraduate research or independent supervised study of special problems. Even a more accomplished reader may find these topics of interest together with other aspects of the presentation, such as the analogies between various types of waves and the examples and problems used to illustrate basic ideas.

There exist several texts on waves and oscillations, but on the introductory level a consistent use of complex variables generally is avoided. This is unfortunate for several reasons. One is the loss of algebraic simplicity, so that many interesting problems become intractable (examples are given below), and the extension of quantitative studies of idealized systems to 'real' ones (for example inclusion of the spring mass in a mass–spring oscillator and damping in coupled oscillators and electrical transmission lines, etc.), usually becomes quite cumbersome.

From a more general view point, an even more important reason is the failure of such treatments to use the conceptually simple subject of waves and oscillations as a training ground for the student to gain working knowledge and insight in the use of complex amplitudes, a valuable preparation for quantum mechanics and advanced topics in classical fields (plasma physics, geophysics, etc.). Such a training is valuable also for the understanding and handling of modern instrumentation (and the reading of the related instruction manuals), where free use is made of the concepts and terminology based on the application of complex variables.

Furthermore, it is instructive in an introductory treatment to stress the analogies between different types of waves and the field variables involved, and to demonstrate in a simple manner the important transition from the motion of a lattice of discrete elements to waves on a continuum.

In this context, a comment on notation is in order, since it is not as insignificant a matter as it may seem. The understanding of the transition from the study of coupled oscillators and lattices to waves and oscillations of a continuum is greatly aided by a consistent notation. Thus, rather than to use the notation $\xi_1(t)$, $\xi_2(t)$ for the displacements of two coupled oscillators, the present text uses the notation $\xi(x_1, t)$, $\xi(x_2, t)$ (or the shorter versions $\xi(1,t)$, $\xi(2,t)$) so that the transition to the continuum description of a displacement field $\xi(x,t)$ becomes natural. Furthermore, in the discussion of normal modes, the subscripts are reserved for the labelling of the different modes.

These observations gave the initial impetus for this book and dictated the approach used. After a review of elementary concepts and examples, the objective in Part 1 (Oscillations) is to provide a thorough treatment of (a) the frequency response to forced harmonic excitation of single and coupled oscillators, and, through Fourier analysis, the response to an arbitrary driving force, and (b) the 'free' motion, resulting from given initial conditions, expressed as a superposition of normal modes. As an example, the impulse response of the system is derived and used to express the response to an arbitrary driving force.

In Part 2 (Waves), after simple illustrations of the wave concept and a study of the kinematics of waves, the field equations involved in the dynamical analysis are obtained from the equations of a lattice of discrete coupled oscillators (mechanical or electrical) through transition to a continuum.

A 'unified' description of the dynamics of waves is stressed, in which the 'field equations', which lead to the wave equation, are put in the same form for longitudinal and transverse mechanical waves, electromagnetic waves, surface waves on a liquid, plasma oscillations, and hydromagnetic waves, so that analogies between the various waves readily can be established.

Following the approach in Part 1 for oscillators, the analysis of waves on finite systems starts with a study of the frequency response to forced harmonic excitation and the identification of resonances. This is followed by a study of free motions and normal modes, including a discussion of the density of modes for one-, two-, and three-dimensional waves.

One chapter is devoted to the general characteristics of electromagnetic waves and one to acoustic waves in fluids, and phenomena common to all waves are discussed in chapters on wave interference and diffraction, refraction, and the Doppler effect. Wave dispersion and the concepts of group and phase velocity are analyzed in detail in a study of wave propagation on a periodic lattice. Dispersion is discussed also in sections on electromagnetic wave propagation in matter, in a chapter on wave guides and a chapter on matter waves, in which the problem of barrier penetration is compared with similar phenomena in mechanical and EM wave transmission. (Further examples of dispersive waves are given in Part 3: surface waves on a liquid, plasma oscillations, and bending waves on a plate).

The topic of polarization sometimes causes problems for the student at first exposure. It has proved to be of considerable help in this regard to illustrate the phenomenon by means of the conceptually simple model of a wave on a string which runs through a slot in a rigid plane. In this case, calculations of reflected and transmitted waves is straight forward for both linearly and circularly polarized string waves and may serve as a lucid illustration of similar analyses of the EM wave.

Topics ordinarily not included in introductory treatments, such as wave propagation on a periodic lattice with damping, electromagnetic wave propagation in a magnetized plasma, miscellaneous examples of forced and free oscillations of continuous systems, wave reflection and refraction at boundaries in relative motion, etc., are shown to be tractable on an introductory level. Although previous exposure to complex variables obviously is useful in the study of this book, it is not necessary; the required background can be obtained from the review in Chapter 2.

In Part 3, dealing with special topics, a unique feature of the book is the chapter on surface waves, in which discussions of the complete dispersion relation (including viscosity), light scattering from thermal surface fluctuations, and the related determination of surface tension and viscosity are presented for the first time as text book material. The chapters on plasma oscillations and on feedback oscillations and instabilities also contain topics of current interest.

Another feature are the 'Displays', summarizing important ideas and equations throughout the book. They frequently contain results of numerical calculations, where special attention has been paid to the problem of presenting mathematical results in a form suitable for computer programming. The displays have proved useful to the students in reviewing the material and to the teacher in summarizing lectures through the projection of transparencies of the displays.

Solutions to specific problems (often drawn from lecture demonstrations) are discussed throughout the text, and a set of problems is included at the end of each chapter. Some chapters also contain example sections, where a few engineering applications are presented.

Numerious discussions with colleagues and students at M.I.T. of the material in this book and the problems of teaching a course of this kind are gratefully acknowledged. Special thanks go to Professor Adnan Akay for his generous and valuable help in proofreading.

The book is dedicated to the memory of the late Professor Philip M. Morse, mentor, colleague, friend.

K.U.I.

PART 1

Oscillations

1

Review of elementary concepts and examples

It is assumed that you have had introductory courses in mechanics and electromagnetism, in which you have encountered periodic phenomena and harmonic motion. Therefore, this chapter, devoted to elementary concepts and examples, can be regarded as an introductory review.

1.1 Introduction

Among the endless possible motions of matter, periodic motion plays a particularly important role, not only because it occurs so widely in nature and everyday life, in man as well as machine, from the beat of the pulse to the motion of the planets, but also because of its basic role in physics; the concept and measurement of time are intimately linked to it.

A periodic motion of a particle is one which repeats itself over and over again. The motion is bounded, which means that it is confined to a finite region of space. This confinement is the result of the interaction of the particle with other particles, including matter in bulk. Thus, in planetary motion the interaction force is provided by the Sun, and for a body moving up and down on the surface of wavy water, the interaction force on it is the collective effect of all the water 'particles'.

The periodic motions of the Moon and the planets have played important roles in the development of physics. Kepler's discovery that the planetary orbits are elliptical with the Sun at the focus, that the sector velocity of the planet is constant, and that the period is proportional to the 3/2 power of the major axis of the orbit, supported the hypothesis that the gravitational force is central and varies as the inverse square of the Sun–planet distance.

Another important example is the motion of a charged particle in a uniform magnetic field. In a plane normal to the magnetic field, the orbit will be circular with constant speed, and if the particle has a magnetic moment, it will also precess with a frequency proportional to the magnetic field. When applied to the nucleus, the measurement of this precession or gyro frequency is often used as a means of measuring the magnetic field.

1.2 Harmonic motion defined

Let us describe the circular motion in more detail. The round trip time T is called the **period** of the motion. The angular displacement of the particle in one period is 2π radians, and the corresponding **angular velocity** is

$$\omega = 2\pi/T \tag{1}$$

The number of round trips or cycles per second is $1/T$, and it is called the **frequency** of the motion

$$\nu = 1/T = \omega/2\pi \tag{2}$$

The frequency is expressed in cycles per second, cps, frequently called Hz, (Hertz) after the scientist Heinrich Hertz.

If we choose time $t = 0$ when the particle crosses the positive x-axis, the angular displacement in radians at time t will be ωt. The corresponding x- and y-coordinates will be

$$\begin{aligned} x &= A\cos(\omega t) \\ y &= A\sin(\omega t) \end{aligned} \tag{3}$$

where A is the radius of the circular orbit, as shown in Display 1.1. These motions are called **harmonic**. The corresponding velocity and acceleration components are obtained, of course, by differentiation with respect to time, $u_x = \mathrm{d}x/\mathrm{d}t = (-A\omega)\sin(\omega t)$ and $a_x = \mathrm{d}^2x/\mathrm{d}t^2 = (-A\omega^2)\cos(\omega t)$, with similar expressions for the y-components.

The maximum value A of x is obtained when the particle in the circular orbit crosses the positive x-axis, and this occurs when $t = 0$ or generally when $t = nT$, where n is an integer. This maximum value is called the *displacement amplitude* of the harmonic motion. For $t = T/2$ the displacement is $-A$, i.e. the magnitude equals the displacement amplitude.

The maximum value of u_x is obtained when the particle in the circular orbit crosses the negative y-axis, and it equals $A\omega$; it is called the **velocity amplitude** of the harmonic motion. For $t = T/4$, i.e. when the particle in the orbit crosses the positive y-axis, the velocity has the same magnitude but opposite sign. Finally, the maximum value of the acceleration is obtained at the crossing of the negative x-axis, and the value is $A\omega^2$, the **acceleration amplitude** of the harmonic motion. At $t = 0$ or nT the magnitude of the acceleration is the same, but the sign is negative.

Display 1.1.

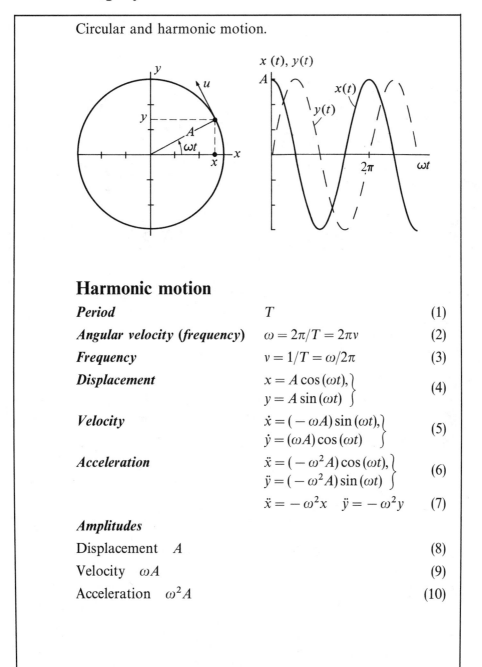

Circular and harmonic motion.

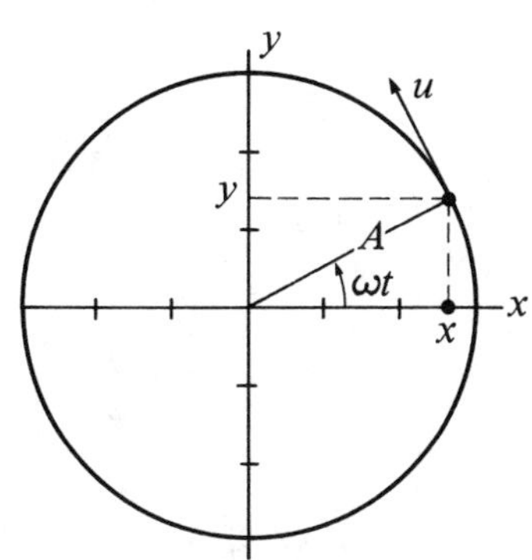

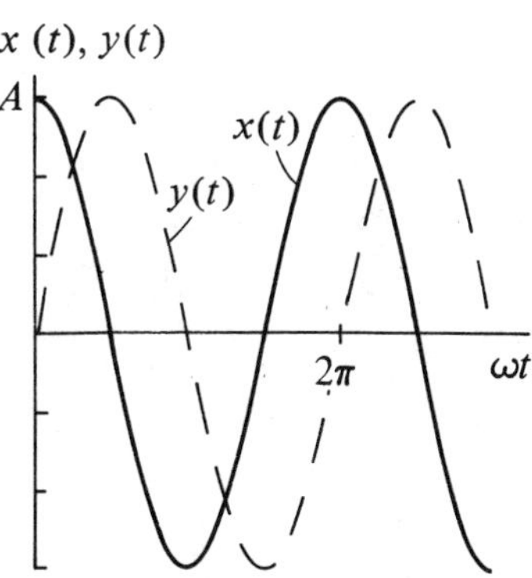

Harmonic motion

Period	T	(1)
Angular velocity (frequency)	$\omega = 2\pi/T = 2\pi\nu$	(2)
Frequency	$\nu = 1/T = \omega/2\pi$	(3)
Displacement	$\left.\begin{array}{l} x = A\cos(\omega t), \\ y = A\sin(\omega t) \end{array}\right\}$	(4)
Velocity	$\left.\begin{array}{l} \dot{x} = (-\omega A)\sin(\omega t), \\ \dot{y} = (\omega A)\cos(\omega t) \end{array}\right\}$	(5)
Acceleration	$\left.\begin{array}{l} \ddot{x} = (-\omega^2 A)\cos(\omega t), \\ \ddot{y} = (-\omega^2 A)\sin(\omega t) \end{array}\right\}$	(6)
	$\ddot{x} = -\omega^2 x \quad \ddot{y} = -\omega^2 y$	(7)

Amplitudes

Displacement A (8)

Velocity ωA (9)

Acceleration $\omega^2 A$ (10)

Display 1.2.

Phase and phase angle.

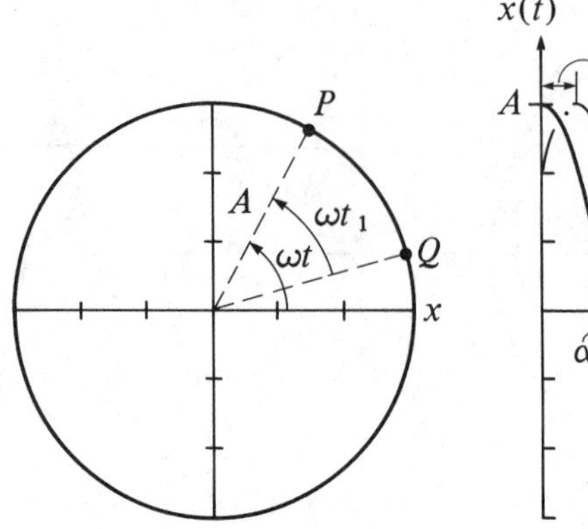

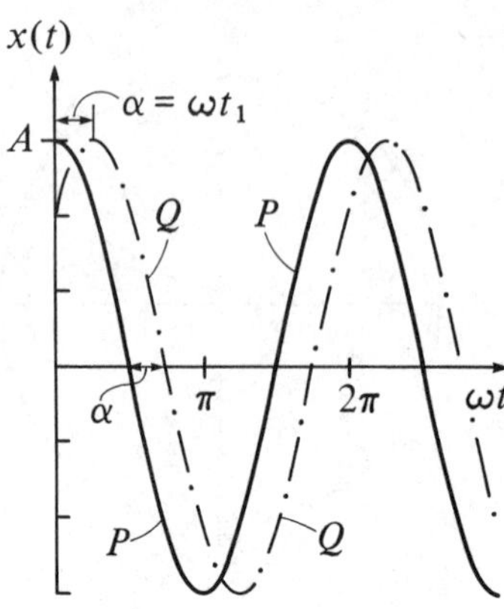

Harmonic motion $x = A\cos[\omega(t - t_1)] = A\cos(\omega t - \alpha)$ (1)

Phase $[\omega(t - t_1)] = (\omega t - \alpha)$ (2)

Phase angle (lag) $\alpha = \omega t_1$ (3)

Example (See Figure)

Curve P $x_p = A\cos(\omega t)$, Phase angle, $\alpha = 0$ (4)

Curve Q $x_q = A\cos[\omega(t - t_1)] = A\cos(\omega t - \alpha)$ (5)

In this case, $t_1 = T/8$

Corresponding phase angle, $\alpha = \omega t_1 = 2\pi t_1/T = \pi/4$ (6)

Note $x_q(0) = A\cos(\omega t_1) = A\cos(\pi/4) = A/\sqrt{2}$ (7)

$x_q(t) = 0$ when $\omega(t - t_1) = \pi/2$, $\omega t = \pi/2 + \pi/4 = 3\pi 4$
(first zero) (8)

Differential equation for harmonic motion

From Eq. D.1.1(7) $\ddot{x} + \omega^2 x = 0$ $(\ddot{x} = d^2x/dt^2)$ (9)

Phase and phase angle. In Display 1.2 are shown particles, P and Q, moving in the same circular orbit with the same angular veocity ω. We choose the origin of the time scale so that $t = 0$ when P crosses the positive x-axis. The second particle crosses at time $t = t_1$, which in this example has been chosen $t_1 = T/8$. The time dependence of the x-coordinate for P is then $x_p = A\cos(\omega t)$ as before, and for Q it is $x_q = A\cos[\omega(t - t_1)]$. The harmonic motion x_q is said to lag behind the motion x_p by the time t_1. A negative value of the lag time t_1 means that Q runs ahead of P.

The argument $\omega(t - t_1)$ is called the **phase** of the harmonic motion. In general it includes a time lag t_1 different from zero, corresponding to the angular displacement $\alpha = \omega t_1$, which is called the **phase angle** or **phase lag**. A negative value of t_1 and of α means that Q runs 'ahead' of P. We shall use

$$x = A\cos[\omega(t - t_1)] = A\cos(\omega t - \alpha) \tag{4}$$

as the **definition of a harmonic motion** and the corresponding phase angle. According to this definition the function $A\sin(\omega t) = A\cos(\omega t - \pi/2)$ is a harmonic motion with the phase angle $\pi/2$.

Differential equation for harmonic motion. Starting with the harmonic motion $x = A\cos[\omega(t - t_1)]$, we obtain for the acceleration $\mathrm{d}^2x/\mathrm{d}t^2 = -(A\omega^2)\cos[\omega(t - t_1)] = -\omega^2 x$. In other words, the second derivative of x is proportional to x itself, with the constant of proportionality $-\omega^2$. This relation can be written as a differential equation

$$\mathrm{d}^2x/\mathrm{d}t^2 = -\omega^2 x \quad \mathrm{d}^2x/\mathrm{d}t^2 + \omega^2 x = 0 \tag{5}$$

Thus, if an equation of this form is encountered in the study of motion, we know that the solution is a harmonic motion. For example, a mass M acted on by a 'restoring' force $F = -Kx$ will have the equation of motion, $M\mathrm{d}^2x/\mathrm{d}t^2 = -Kx$, with $\omega^2 = K/M$, and the motion will be harmonic.

1.3 The mass–spring oscillator

One of the reasons for the unique importance of the harmonic motion is that a small displacement of a particle from its equilibrium position generally results in a restoring (reaction) force proportional to the displacement. If the particle is released from a displaced position, the only force acting on it will be the reaction force and the subsequent motion of the particle will be harmonic.

In our continued discussion, this motion will be illustrated by means of a mass M connected to a coil spring, as indicated in Display 1.3. As an idealization, we assume the mass of the spring itself to be negligible, so that the only function of the spring is to provide a restoring force. (We shall later

Display 1.3.

The mass–spring oscillator. Initial value problem.

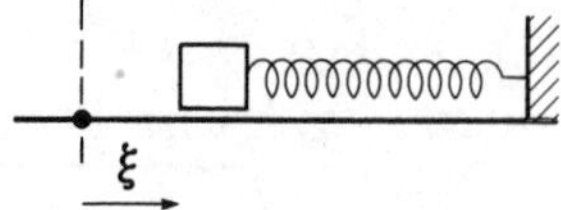

Displacement from equilibrium ξ (1)

Spring constant, K (compliance $C = 1/K$) (2)

Restoring force on M $\quad -K\xi$ (3)

Equation of motion for M

$$M\ddot{\xi} = -K\xi \tag{4}$$

$$\ddot{\xi} + \omega_0^2\xi = 0 \tag{5}$$

$$\omega_0 = \sqrt{K/M} \tag{6}$$

General solution

$$\xi = A\cos(\omega_0 t - \alpha) \tag{7}$$

Frequency $\quad \nu_0 = \omega_0/2\pi = (1/2\pi)\sqrt{K/M}$ (8)

Period $\quad T_0 = 1/\nu_0 = 2\pi\sqrt{M/K}$ (9)

Initial value problem

$$\xi = A\cos(\omega_0 t - \alpha) \tag{10}$$

At $t = 0$, $\quad \xi = \xi_0, \quad u = \dot{\xi} = u_0$ (initial conditions) (11)

From Eq. (10)

$$\xi_0 = A\cos(\alpha), \quad u_0 = \omega_0 A\sin(\alpha) \tag{12}$$

$$A = \sqrt{\xi_0^2 + (u_0/\omega_0)^2}, \quad \tan(\alpha) = u_0/\omega_0\xi_0 \tag{13}$$

be able to account for the mass of the spring). The mass is located on a frictionless table, which can be closely approximated experimentally by means of an air-suspended cart on a track. The equilibrium position of M is then determined by the relaxed length of the spring.

We shall denote the displacement of M from the equilibrium position by ξ rather than x to avoid conflict in notation in later chapters, where x will be used to designate the equilibrium position of a particle.

It is found experimentally that the force required to change the length of the spring by an amount ξ is $K\xi$ (at least for small ξ), where K is a constant, called the **spring constant**. The reaction force produced by the spring is equal but opposite in direction, and this reaction force, $-K\xi$, will act on M. Thus, if M is displaced from equilibrium and then released, only the spring reaction force will be involved, so that Newton's equation of motion for M becomes

$$M\ddot{\xi} = -K\xi \quad (\ddot{\xi} = \mathrm{d}^2\xi/\mathrm{d}t^2) \tag{1}$$

$$\ddot{\xi} + \omega_0^2\xi = 0 \tag{2}$$

$$\omega_0 = \sqrt{K/M} \tag{3}$$

General solution:

$$\xi = A\cos(\omega_0 t - \alpha)$$

The general solution $\xi = A\cos(\omega_0 t - \alpha)$ contains the amplitude A and the phase angle α, which are unspecified. By assigning values to these constants, we select from the infinite set of possible harmonic motions the particular solution, which refers to a particular situation. The constants can be determined if the 'state' of the motion (displacement and velocity) is known at a given time. We then have two equations for the determination of A and α. In Display 1.3 this is illustrated in terms of the known initial conditions of the oscillator (at $t = 0$).

Discussion. The spring constant depends not only on the elastic properties of the material in the spring, but also on its length and shape. In an ordinary uniform coil spring, for example, the pitch angle of the coil (spiral) plays a role, and another relevant factor is the thickness of the material. The deformation of the spring material is complicated; it involves both torsion and bending of the coil wire, and the calculation of the spring constant from first principles is not simple. Therefore, the spring constant, generally, should be regarded as an experimentally determined quantity.

The linear relation between force and deformation is valid only for small deformations. For example, for very large elongations of a coil spring it will ultimately end up a straight wire, and, conversely, a large compression will make it into a tube-like configuration, corresponding to zero pitch angle of the coil. In both these limits, the stiffness of the spring is much larger than for the relaxed spring.

The inverse of the spring constant K is called the **compliance**, $C = 1/K$. The compliance is proportional to the length of the spring. Later, in the study of wave motion on a spring, we shall introduce the compliance per unit length of the spring, which is a quantity independent of the length of the spring.

The fact that we have neglected the mass of the spring in comparison with the mass M attached to it, shows up in the expression for the frequency oscillation of the mass–spring oscillator, $\omega_0 = \sqrt{K/M}$. According to this formula, the frequency goes to infinity when M goes to zero. This, clearly, cannot be correct, since if the mass is removed from the spring, the spring itself can be brought into oscillations with a certain finite frequency. Thus, in using the results obtained for the mass–spring oscillator, the conditions for the validity of the analysis should be borne in mind. The question of the role of the spring mass will be discussed in a subsequent chapter.

Frequently, several springs are combined in order to obtain a desired resulting spring constant. If the springs are in 'parallel', the deformations of both springs will be the same, and the restoring forces by the springs will add. Consequently, the resulting spring constant of the combined springs will be the sum of the individual spring constants. If the springs are in 'series', the force in each spring will be the same, and the deformations will add. In this case the resulting compliance will be the sum of the individual compliances.

When we indicated above, that the spring constant should be regarded as an experimentally determined quantity, it was tacitly assumed that the experiment involved a static deformation of the spring, in which the elongation (or compression) d, produced by a given force F, is determined, yielding the 'static' spring constant $K = F/d$.

In deriving the equation of motion for the mass–spring oscillator, we assumed that this static spring constant could be used even though the spring is in motion, subject to periodic compressions and extensions. Although this assumption is a good approximation in most cases, there are some materials, such as rubber and plastic, for which it is not. For example, there exist substances, which are plastic for slow but elastic for rapid deformations. This is related to the molecular structure of the material, and the effect is often strongly dependent on temperature. A cold tennis ball, for example, does not bounce very well.

1.4 Examples of harmonic motion

In addition to the mass–spring oscillation, there are many other familiar motions, which belong to the class of harmonic oscillations. We shall devote this section to a discussion of a few examples.

The simple pendulum. The simple or 'mathematical' pendulum is an idealization, consisting of a point mass M at the end of a massless string or bar, which oscillates in a plane about a fixed horizontal axis through the upper end of the bar. A playground swing, for example, can be approximately described as a simple pendulum.

A displacement of the pendulum away from its vertical equilibrium position is described in terms of the angle ϕ, as shown in Display 1.4. It can be described also in terms of the displacements ξ and η in the x- and y-directions, as shown.

With the length of the pendulum denoted by L, the relations between these variables are $\xi = L\sin(\phi)$ and $\eta = L(1-\cos\phi)$. For small values of the angle, $\phi \ll 1$, these relations reduce to $\xi \simeq L\phi$ and $\eta \simeq L\phi^2/2$. Under these conditions, terms of second and higher order in ϕ will be neglected so that $\eta \simeq 0$.

In free motion of the pendulum, the only external force on the mass M is the gravitational force. In addition, there is a tension force S in the bar acting on M. Since we neglect the displacement and acceleration of M in the y-direction, the net force $S\cos\phi - Mg$ in this direction will be zero, so that $S\cos\phi = Mg$.

The net force in the x-direction is $-S\sin\phi \simeq -S\phi \simeq -S\xi/L \simeq (Mg/L)\xi$, where we have put $S \simeq Mg$. The equation of motion in the x-direction then becomes $M\ddot{\xi} = -(Mg/L)\xi$, which has the same form as the mass–spring oscillator equation

$$\ddot{\xi} + \omega_0^2\xi = 0 \quad \omega_0 = \sqrt{g/L} \tag{1}$$

with the general solution

$$\xi = \xi_0\cos(\omega_0 t - \alpha) \tag{2}$$

and with the period

$$T_0 = 2\pi/\omega_0 = 2\pi\sqrt{L/g} \tag{3}$$

There is a corresponding solution for the angle ϕ.

The physical pendulum. As already indicated, a pendulum is simple or 'mathematical' only if the pendulum mass is concentrated in a point at the

Display 1.4.

The pendulum

The mathematical pendulum

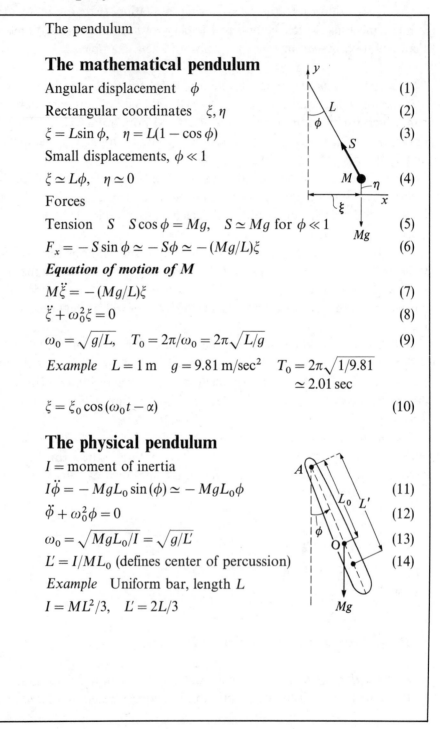

Angular displacement ϕ (1)

Rectangular coordinates ξ, η (2)

$\xi = L\sin\phi, \quad \eta = L(1-\cos\phi)$ (3)

Small displacements, $\phi \ll 1$

$\xi \simeq L\phi, \quad \eta \simeq 0$ (4)

Forces

Tension S $\quad S\cos\phi = Mg, \quad S \simeq Mg$ for $\phi \ll 1$ (5)

$F_x = -S\sin\phi \simeq -S\phi \simeq -(Mg/L)\xi$ (6)

Equation of motion of M

$M\ddot{\xi} = -(Mg/L)\xi$ (7)

$\ddot{\xi} + \omega_0^2\xi = 0$ (8)

$\omega_0 = \sqrt{g/L}, \quad T_0 = 2\pi/\omega_0 = 2\pi\sqrt{L/g}$ (9)

Example $L = 1\,\text{m} \quad g = 9.81\,\text{m/sec}^2 \quad T_0 = 2\pi\sqrt{1/9.81} \simeq 2.01\,\text{sec}$

$\xi = \xi_0\cos(\omega_0 t - \alpha)$ (10)

The physical pendulum

I = moment of inertia

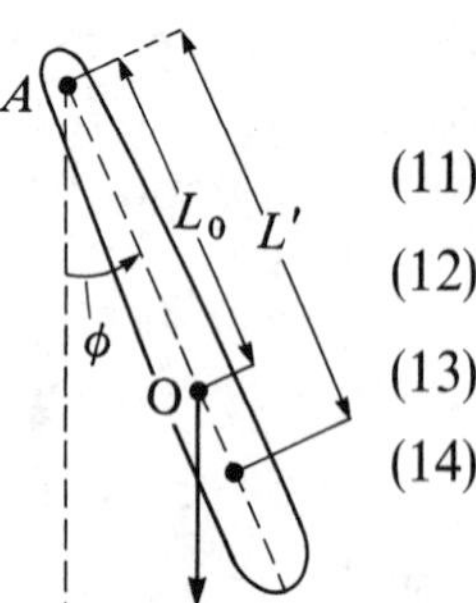

$I\ddot{\phi} = -MgL_0\sin(\phi) \simeq -MgL_0\phi$ (11)

$\ddot{\phi} + \omega_0^2\phi = 0$ (12)

$\omega_0 = \sqrt{MgL_0/I} = \sqrt{g/L'}$ (13)

$L' = I/ML_0$ (defines center of percussion) (14)

Example Uniform bar, length L

$I = ML^2/3, \quad L' = 2L/3$

end of the pendulum. This is idealization, which can be realized only approximately. For a real or 'physical' pendulum, the mass is distributed in some manner over the volume of the pendulum. The **moment of inertia** is $I = I_0 + ML_0^2$, where I_0 is the moment of inertia about a center of mass axis, which is parallel with the axis of rotation, and L_0 the distance from the center of mass to the axis of rotation.

As for the simple pendulum, the torque of the force of gravity with respect to the axis of rotation is $-MgL_0 \sin\phi$, where ϕ is the angle of deflection of the pendulum. With the angular momentum being $I\dot{\phi}$, the equation of motion is $I\ddot{\phi} = -MgL_0 \sin\phi$, which for small deflections becomes the harmonic oscillator equation $I\ddot{\phi} = -MgL_0\phi$ or $\ddot{\phi} + \omega_0^2\phi = 0$, where

$$\begin{aligned} \omega_0 &= \sqrt{MgL_0/I} = \sqrt{g/L'} \\ L' &= I/ML_0 \end{aligned} \tag{4}$$

We have here introduced the equivalent length L' of a mathematical pendulum with the same frequency of oscillation as the physical pendulum.

As an example, let us consider a pendulum consisting of a **uniform bar** of length L and total mass M. The moment of inertia of this bar with respect to an axis perpendicular to the bar through the center of mass is $ML^2/12$. With the center of mass distance $L_0 = L/2$, the corresponding moment of inertia with respect to an axis through the end of the bar is $I = ML^2/12 + M(L/2)^2 = ML^2/3$. The length of the equivalent mathematical pendulum is $2L/3$.

With the bar replaced by a **circular disc** of radius R and mass M, oscillating in the plane of the disc, we have $I_0 = MR^2/2$, $L_0 = R$, $I = I_0 + MR^2 = 3MR^2/2$, and $L' = 3R/2$.

The length L' defines a location on the line AO through the axis of rotation and the center of mass of the physical pendulum, called the **center of percussion**, which has an interesting property. If an impulse goes through the center of percussion perpendicular to AO, the reaction force on the axis of rotation will be zero. This means that if a ball is hit by a baseball bat (or a tennis racket) at the center of percussion of the bat, there will be no 'sting' in the hands of the hitter.

If the body does not have rotational symmetry about the AO-axis, the moment of inertia will be different for different planes of oscillation, and the same is then true for the period of oscillation and the location of the center of percussion.

With reference to the example about the baseball bat, it may be mentioned, that the location of the center of percussion is one of the dynamical parameters which is likely to influence the 'feel' of the bat (or tennis racket). The mass and the location of the center of mass are also involved; these may be referred to as 'static' parameters. The location of the center of percussion can be determined from measurement of the period of pendulum oscillations.

Cyclotron motion. As we have seen, the harmonic motion is defined as the components of circular motion and any circular motion can be considered to be an example of harmonic motion. An important case is the motion of a charged particle in a constant magnetic field (cyclotron motion). We recall that the angular velocity of the motion is

$$\omega_c = qB/m$$

and if the magnitude of particle momentum is p, the orbit radius becomes $r = p/Bq$. We note that for an electron we obtain $\nu_c = \omega_c/2\pi = 2.8\,B$ MHz and for a proton $\nu_c = 1.5B$ kHz (B in Gauss).

Particle in a potential well. The previous specific examples can be considered to be special cases of a particle oscillating in a potential well. To treat this general problem, we let the potential energy of the particle be $V(\xi)$, where ξ is the displacement in the x-direction from the equilibrium position at the bottom of the well, which is placed at $x = 0$. At this position we let $V(0) = 0$. The force on the particle in the x-direction is $-\mathrm{d}V/\mathrm{d}\xi = -V'(\xi)$. At the equilibrium position this force is zero, i.e. $V'(0) = 0$. At the bottom of the well $V''(0)$ is positive.

The Taylor expansion of $V(\xi)$ then becomes

$$V(\xi) = V(0) + \xi V'(0) + (\xi^2/2)V''(0) + \cdots = (\xi^2/2)V''(0) + \cdots \tag{5}$$

and the force on the particle

$$F = -V'(\xi) = -V''(0)\xi - \cdots \tag{6}$$

In other words, for small displacements, the potential energy is proportional to the square of the displacement, and the force is a restoring force proportional to ξ. The equivalent 'spring constant' is $K = V''(0)$. Checking this for a spring, we recall that $V = K\xi^2/2$ with $V'' = K$, as it should.

The equation of motion for small oscillations becomes $M\ddot{\xi} = -V''(0)\xi$ or

$$\begin{aligned} &\ddot{\xi} + \omega_0^2\xi = 0 \\ &\omega_0^2 = V''(0)/M \end{aligned} \tag{7}$$

The discussion above is summarized in Display 1.5. As an example is considered a particle oscillating in a one-dimensional potential $V(\xi) = 1 - \cos(k\xi)$, where $k = 2\pi/\lambda$, where λ is the wavelength of the periodic potential. This type of motion occurs, when electrons are trapped in the

Display 1.5.

Particle in a one-dimensional potential 'well'

Displacement ξ (1)

Potential energy $V(\xi)$ (2)

At equilibrium $V(0) = 0, \quad V'(0) = 0 \quad (V' = \mathrm{d}V/\mathrm{d}\xi)$ (3)

Taylor expansion

$$V(\xi) = V(0) + \xi V'(0) + (\xi^2/2)V''(0) + \cdots = (\xi^2/2)V''(0) + \cdots \quad (4)$$

Force $F = -V'(\xi) = -\xi V''(0) - \cdots$ (5)

Small amplitudes $F = -V''(0)\xi$ (6)

Equation of motion

$$M\ddot{\xi} = -V''(0)\xi \quad (7)$$

$$\ddot{\xi} + \omega_0^2\xi = 0 \quad (8)$$

$$\omega_0^2 = V''(0)/M \quad (9)$$

Example

$$V(\xi) = V_0[1 - \cos(k\xi)] \quad (10)$$

Small displacements: $V(\xi) \simeq V_0(k\xi)^2/2$ (11)

$$V''(0) = V_0k^2 \quad \omega_0^2 = V_0k^2/M \quad (12)$$

potential well created by an electrostatic harmonic wave, encountered in plasma physics. For a sufficiently small energy of the electrons it will perform harmonic oscillations about the equilibrium position at the 'bottom' of the potential. With a kinetic energy larger than the potential barrier height V_0, the electrons will escape and move from one well to the next.

This motion of an electron is equivalent to the motion of a simple pendulum of length L and with $V_0 = 2MgL$. With a kinetic energy larger than $2MgL$, the pendulum will go over the top.

Torsional oscillation. Consider a rod or wire clamped at one end. A torque applied at the other end about the axis of the rod produces an angular displacement proportional to the torque, at least for sufficiently small displacements. The ratio of the torque and the angular displacement at the point of an application of the torque is called the **torsion constant *Q***. It is the analogue of the spring constant and generally should be regarded as an experimentally determined quantity, although it can be calculated in terms of the **shear modulus *G*** of the rod. Thus, for a uniform rod with circular cross section, it can be shown that

$$Q = (\pi a^4/2L)G \tag{8}$$

$a =$ radius, $L =$ length

$G =$ shear modulus. For steel: $G = 8.1 \times 10^{10}\,\mathrm{Nm}^{-2}$.

The dimension of the spring constant K is force divided by length, and for the torsion constant it is torque, i.e. force multiplied by length.

As an example, consider a rod or wire clamped at the upper end and supporting a body at the lower end, such as a circular disc or a dumbbell, with a **moment of inertia *I*** about the axis of the rod much larger than that of the rod itself. After having been released from an angular displacement, the body is acted on only by the restoring torque from the rod and the equation for the **angle or rotation *θ*** of the body is

$$I\ddot{\theta} = -Q\theta \tag{9}$$

$$\ddot{\theta} + \omega_0^2\theta = 0 \tag{10}$$

$$\omega_0 = \sqrt{Q/I} \tag{11}$$

which is a harmonic oscillator equation.

With the 'rod' consisting of a thin wire or filament, the torsion constant can be made extremely small, and minute torques and corresponding forces can be measured from the angular deflection of the torsional pendulum. The deflection can be 'amplified' by means of a light beam reflected from a mirror attached to pendulum. This technique has been used in sensitive galvanometers and for the measurement of light pressure and the gravitational constant. In the light pressure experiment it is advantageous to pulse the light at a frequency equal to the frequency of oscillation of the pendulum.

Oscillator with nonlinear spring and a bias force. In a vertical mass–spring oscillator, with the mass M supported by a **linear** spring with a spring constant K, the weight Mg will produce a **static deflection** of the spring equal to $\boldsymbol{y_0 = Mg/K}$. If M is displaced in the vertical direction from this equilibrium position and then released, the subsequent motion will be harmonic with the same angular frequency $\sqrt{K/M}$ as for the corresponding horizontal oscillator.

If the spring is **nonlinear**, however, this is not true in general. The force required to produce a displacement y is no longer proportional to y but is expressed by a nonlinear relation $F(y)$. The **static displacement** $\boldsymbol{y_0}$, produced by the weight Mg, has to be found by solving the equation

$$F(y_0) = Mg \tag{12}$$

We now consider a small displacement dy from the equilibrium position y_0. The force required to produce this displacement is $dF = F'(y_0)\,dy$, where F' is the derivative of F with respect to y at $y = y_0$. The corresponding **'local' spring constant** $\boldsymbol{K(y_0)}$ and frequency of small oscillations about the equilibrium position y_0 are

$$\begin{aligned} K(y_0) &= dF/dy = F'(y_0) \\ \omega_0(y_0) &= \sqrt{F'(y_0)/M} \end{aligned} \tag{13}$$

In the first example in Display 1.6, the relation $F(y)$ is of the form by^3. Note that as y changes sign, so does F. (Can $F(y)$ be a quadratic function of y?) In this case we find that the frequency of oscillation will be $\sqrt{3}$ times larger than for a linear oscillator with the same static deflection, as shown in the display.

The second example deals with a body in water oscillating under the influence of gravity and the buoyancy force. The magnitude of the latter is equal to the weight of the displaced water. In equilibrium, the buoyancy force equals the weight of the body.

To determine the equilibrium position, we introduce the *area function* $\boldsymbol{A(y)}$ which gives the area of the body in a horizontal plane as a function of the height y of the plane above the lowest point of the body. Then, if the water line falls at $y = y_0$ in equilibrium, and if the **density** of the water is $\boldsymbol{\rho}$, the mass of the displaced water is $\rho\int A(y)\,dy$, and the equation for the

Display 1.6.

Nonlinear spring with bias force. Small oscillations about equilibrium position.

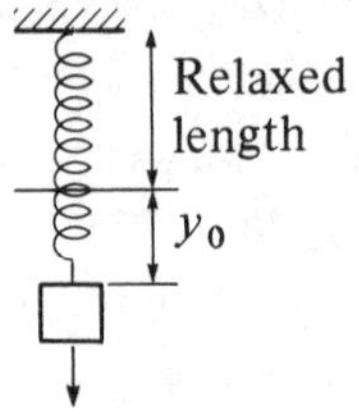

Displacement y Applied force F (1)

Spring characteristic $F = F(y)$ (2)

Bias force Mg (3)

Static displacement y_0 obtained from

$F(y_0) = Mg$ (4)

Small displacement from equil. $\mathrm{d}y$ (5)

Corresponding force $\mathrm{d}F$ (6)

$\mathrm{d}F = F'(y_0)\,\mathrm{d}y$ (7)

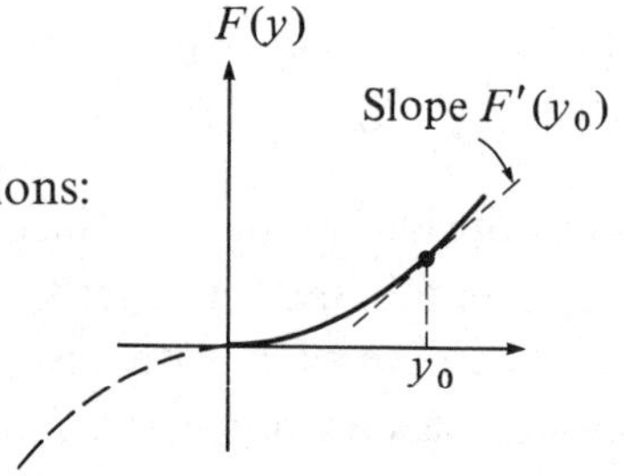

Local spring constant:

$K(y_0) = \mathrm{d}F/\mathrm{d}y = F'(y_0)$ (8)

Frequency of small oscillations:

$\omega_0 = \sqrt{F'(y_0)/M}$ (9)

Example:

Let $F(y) = by^3$ (10)

Static displacement $y_0 = \sqrt[3]{Mg/b}$ (11)

Local spring constant $K(y_0) = F'(y_0) = 3by_0^2 = 3Mg/y_0$ (12)

Angular frequency $\omega_0 = \sqrt{F'(y_0)/M} = \sqrt{3g/y_0}$ (13)

For linear spring with spring constant K:

$y_0 = Mg/K \quad \omega_0 = \sqrt{K/M} = \sqrt{g/y_0}$ (14)

Example:

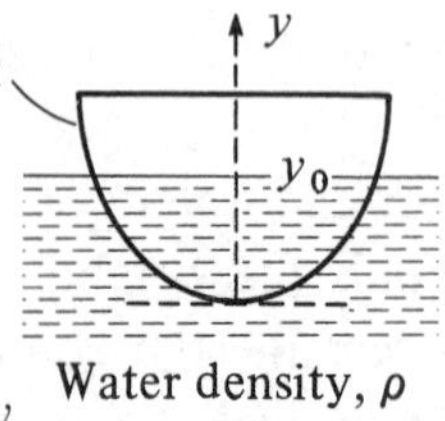

Area in horizontal plane at y $A(y)$. (15)

Calc. of y_0 $\rho \int_0^{y_0} A(y)\,\mathrm{d}y = M$ (16)

Buoyant force in displ. $\mathrm{d}y$ $\rho g A(y)\,\mathrm{d}y$ (17)

Local 'spring' constant' $F'(y_0) = \rho g A(y_0)$ (18)

Angular frequency $\omega_0 = \sqrt{F'(y_0)/M} = \sqrt{\rho g A(y_0)/M}$ (19)

determination of y_0 is

$$\rho \int_0^{y_0} A(y)\,\mathrm{d}y = M \tag{14}$$

where M is the mass of the body.

A displacement $\mathrm{d}y$ from the equilibrium position results in a restoring force with the magnitude $\mathrm{d}F = \rho g A(y)\,\mathrm{d}y$ with the corresponding local 'spring' constant and angular frequency of oscillation given by

$$\begin{aligned} K(y_0) &= \rho g A(y_0) \\ \omega_0 &= \sqrt{\rho g A(y_0)/M} \end{aligned} \tag{15}$$

It should be remarked, that in this equation, M should include an 'induced' mass contribution, resulting from the oscillatory motion of the water, but we shall neglect it in this context.

1.5 The two-body oscillator. Reduced mass

In the study of the vibration of a diatomic molecule, for example, we deal with two interacting bodies, as illustrated schematically in Display 1.7. Here two masses M_1 and M_2 are connected with the spring with a force constant K. If the spring is compressed or extended and then released, the bodies will oscillate to and fro. Since there are no external forces acting on the system, the center of mass will remain at rest if it was at rest initially. If the center of mass has an initial velocity, it will continue to move with the same constant velocity, and it is advantageous to study the motion in a coordinate system, which moves with the center of mass.

With the displacements of the two particles denoted by $\xi(1)$ and $\xi(2)$, the extension of the spring will be $\xi = \xi(2) - \xi(1)$ and the corresponding interaction force between the two particles will be $K\xi$. The equations of motion for the two particles then become

$$M_1 \ddot{\xi}(1) = K\xi \tag{1}$$

$$M_2 \ddot{\xi}(2) = -K\xi \tag{2}$$

If we multiply the first of these equations by M_2 and the second by M_1 and subtract the equations, we obtain

$$\mu\ddot{\xi} = -K\xi \tag{3}$$

$$\mu = M_1 M_2/(M_1 + M_2) \tag{4}$$

$$\ddot{\xi} + \omega_0^2 \xi = 0, \quad \omega_0^2 = K/\mu \tag{5}$$

In other words, the extension ξ of the spring satisfies a harmonic oscillator equation, in which the mass of the particle is equal to $\boldsymbol{\mu}$, often called the

Display 1.7.

The two-body oscillator.

Spring extension

$\xi = \xi(2) - \xi(1)$ (1)

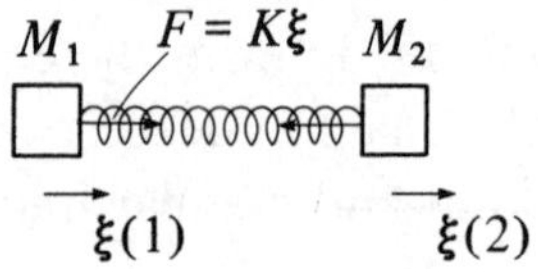

M_1 $F = K\xi$ M_2

$\xi(1)$ $\xi(2)$

Equations of motion

$M_1\ddot{\xi}(1) = K\xi$ (2)

$M_2\ddot{\xi}(2) = -K\xi$ (3)

Multiply Eqs. (2) and (3) by M_2 and M_1, respectively, and then subtract Eq. (3) from Eq. (2):

$\mu\ddot{\xi} = -K\xi$ or $\ddot{\xi} + \omega_0^2\xi = 0$ $\xi = \xi_0 \cos(\omega_0 t - \alpha)$ (4)

$\mu = M_1 M_2/(M_1 + M_2)$ Reduced mass (5)

$\omega_0^2 = K/\mu$ (6)

Energy

With respect to CM $E_0 = K\xi_0^2/2 = p_0^2/2\mu$ (7)

Associated with motion of CM $MU^2/2$ $(M = M_1 + M_2)$ (8)

Total energy $E = E_0 + MU^2/2$ (9)

Example

Impulse J delivered to M_1 at $t = 0$. (10)

Velocity of CM after impulse $U = J/M$ (11)

Velocity of M_1 immediately after impulse J/M_1 (12)

Total kinetic energy transfer $J^2/2M_1 = E$ (13)

Energy with respect to CM

$E_0 = (J^2/2)(1/M_1 - 1/M) = (\mu/M_1)E$ (14)

Maximum extension of spring $\xi_0 = \sqrt{2E_0/K}$ (15)

Energy of M_1, M_2 w.r.t. CM $E_0(1) = (\mu/M_1)E_0$
$E_0(2) = (\mu/M_2)E_0$ (16)

reduced mass. The frequency of oscillation is the same for both mass elements.

The **total energy** of oscillation with respect to the center of mass can be expressed either as the maximum potential energy $K\xi_0^2/2$, where ξ_0 is the amplitude of the spring extension, or as the sum of the kinetic energies of the two particles when the spring is relaxed. The amplitude of the momenta of the particles with respect to the center of mass are the same, $p_0(1) = p_0(2) = p_0$, and the total energy can be expressed as

$$E_0 = K\xi_0^2/2 = (p_0^2/2)(1/M_1 + 1/M_2) = p_0^2/2\mu \tag{6}$$

If the displacement amplitudes of the two bodies are $\xi_0(1)$ and $\xi_0(2)$, we have $\xi_0(1)/\xi_0(2) = M_2/M_1$. Furthermore, $\xi_0(1) + \xi_0(2) = \xi_0$. From these relations we can determine $\xi_0(1)$ and $\xi_0(2)$ in terms of ξ_0.

If the center of mass is not at rest, but has a velocity U, the total energy of the system is the sum of the energy relative to the center of mass and the energy $MU^2/2$ associated with the center of mass motion ($M = M_1 + M_2$). By subtracting $MU^2/2$ from the total energy, which often is known, we obtain the energy of oscillation with respect to the center of mass. This is demonstrated explicitly in the example in Display 1.7.

Molecular vibrations and related matters. As indicated earlier, the linear relationship between restoring force and displacement in an oscillator applies only for sufficiently small amplitudes. The resulting motion is then harmonic and can be described in terms of simple formulas. At larger amplitudes, a quantitative description, in general, is considerably more complicated.

The qualitative features of the motion, however, can be understood from the potential energy curve of the oscillator. As an example, let us consider the interaction potential energy between two atoms or molecules, as illustrated schematically in Display 1.8. The coordinate x represents the separation of the atoms.

The potential energy curve has a minimum value $-D$ for a separation x_0, and it is equal to zero when the atom are infinitely far apart. The corresponding force is attractive for $x > x_0$ and repulsive for $x < x_0$. The potential energy and the corresponding repulsive force increase to very large values as x approaches zero. In the vicinity of equilibrium the interaction force is of the form $-K\xi$, where ξ is the increase of the separation from equilibrium.

If the kinetic energy of the atoms is zero in the center of mass system, they are in equilibrium at a separation x_0. This corresponds to a relaxed

Display 1.8.

Thermal vibration of diatomic molecule.

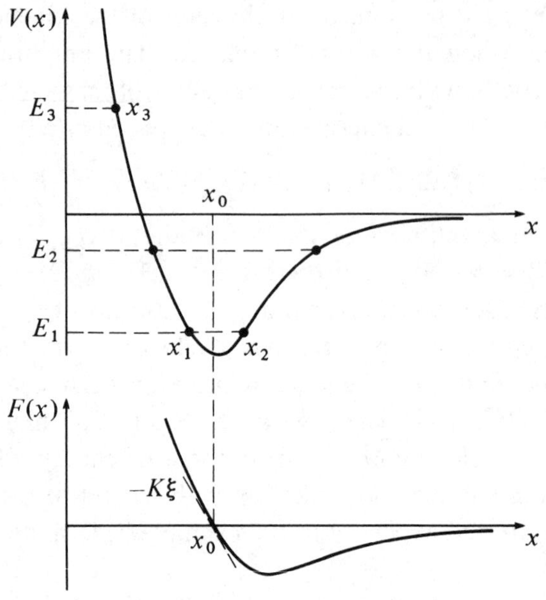

Example

The Morse potential

Let $\xi = x - x_0$ (1)

$V(\xi) = D[\mathrm{e}^{-2b\xi} - 2\mathrm{e}^{-b\xi}]$ (2)

$V(x_0) = -D. \quad V(\infty) = 0$ (3)

Interaction force $F(\xi) = -\mathrm{d}V/\mathrm{d}\xi = 2bD[\mathrm{e}^{-2b\xi} - \mathrm{e}^{-b\xi}]$ (4)

For small ξ $F(\xi) \simeq -2b^2 D\xi = -K\xi$ (5)

Angular frequency of oscillation $\omega_0^2 = V''(0)/M = 2b^2 D/M$ (6)

spring in the two body oscillator in Display 1.7. At a temperature different from absolute zero, the atoms are in thermal motion, and as a result they oscillate to and fro with the smallest and largest values of the separation being the 'turning points' x_1 and x_2. If the thermal energy is small, corresponding to the total energy E_1, these turning points are symmetrically located with respect to x_0 so that the average separation of the atoms will still be x_0.

As the thermal energy increases, say to a value E_2, the average separation will increase, however, since the potential energy curve is asymmetrical; the restoring force being weaker when $x > x_0$ then when $x < x_0$, (the analogous spring being softer in extension then in compression) and the atoms reach further on the outside than on the inside. This is related to the **thermal expansion** of a solid.

If the thermal energy is increased until it exceeds D, the total energy will be positive, and the interatomic distance goes to infinity, i.e. the atoms separate. This separation, in essence, is the 'atomic' explanation of the *thermal dissociation* of molecules and the **melting** of a solid.

Conversely, in a head-on collision of two atoms (the energy being greater than zero) the distance of closest approach will be x_3, but if the collision is elastic, the atoms will bounce back and escape from each other again. On the other hand, in an inelastic collision, it is possible that during the collision energy is removed to make the remaining energy of the atoms less than zero so that the atoms will be trapped in the potential well. This is the process of **recombination** of atoms in the formation of a molecule.

For interacting nuclei, the potential energy starts out with the Coulomb repulsion, but for sufficiently small separations, the attractive nuclear forces will dominate, and the potential will again have a well. However, in order for two nuclei to fuse in an inelastic collision, they must first overcome the Coulomb barrier. If they do fuse, the related energy release can be quite large. It is the source of the Sun's radiation. Several **fusion** processes are possible in the Sun because of the very large particle density in the Sun, and these fusion reactions typically require a thermal energy of the nuclei corresponding to a temperature of about 20 million Kelvin.

Under laboratory conditions, with lower particle densities, a most likely process for fusion involves deuterons (one proton and two neutrons, heavy hydrogen). The required thermal energy for fusion now corresponds to a temperature of about 350 million Kelvin. The generation of such high temperature under controlled conditions is one of the basic challenging problems in today's fusion research (in the hydrogen bomb the required temperature is produced by an atomic bomb).

The example given in Display 1.8 refers to a simple empirical mathematical expression for a potential (called the Morse potential), which has the same general form as the interaction potential between two atoms. It contains three parameters x_0, D, and b.

The quantity x_0 is the equilibrium separation, as before, and D is the depth of the potential well. The value of the potential at $x = x_0$ is $V(x_0) = -D$, which represents the energy of dissociation. The constant b measures the steepness with which $V(x)$ rises on both sides of the minimum and thus determines the equivalent 'spring constant' of the potential. In Section 1.4 we treated some of the more quantitative aspects of motion in a potential well.

1.6 Two-dimensional oscillator

So far, we have considered motion only in one dimension, but the extension to two and three dimensions is straight forward. A simple demonstration of a two-dimensional oscillator involves an air-suspended puck of mass M, which is attached to two springs perpendicular to each other, as indicated in Display 1.9.

The spring constants of the individual springs in the x- and y-directions are $K_x/2$ and $K_y/2$ so that the resulting spring constant in these directions are K_x and K_y, respectively. The equilibrium position of the puck is at the origin, $x = 0$, $y = 0$. A displacement with the coordinates ξ and η changes the lengths of the springs in the x- and y-directions by these amounts, respectively (to first order in the displacements), and the components of the force on M become

$$F_x = -K_x\xi, \quad F_y = -K_y\eta \tag{1}$$

It should be noted that the force vector generally is not along the same line as the vector displacement.

The corresponding equations of motion are

$$M\ddot{\xi} = -K_x\xi, \quad \ddot{\xi} + \omega_x^2\xi = 0, \quad \omega_x = \sqrt{K_x/M} \tag{2}$$

$$M\ddot{\eta} = -K_y\eta, \quad \ddot{\eta} + \omega_y^2\eta = 0, \quad \omega_y = \sqrt{K_y/M} \tag{3}$$

with the solutions

$$\xi = \xi_0 \cos(\omega_x t) \tag{4}$$

$$\eta = \eta_0 \cos(\omega_y t - \beta) \tag{5}$$

where we have assigned a phase angle of zero to the motion in the x-direction.

The resulting trajectory of M in the xy-plane will be closed if the ratio between the two frequencies ω_x and ω_y is a rational fraction, and the shape

Display 1.9.

Two-dimensional oscillator.

Displacement components ξ, η (1)

Length of deformed spring A

$$[(L-\xi)^2+\eta^2]^{\frac{1}{2}} = [L^2 - 2\xi L + \xi^2 + \eta^2]^{\frac{1}{2}}$$
$$= L[1 - 2(\xi/L) + (\xi^2+\eta^2)/L^2]^{\frac{1}{2}} = L - \xi \quad (2)$$

Restoring force

$$F_x = -K_x\xi, \quad F_y = -K_y\eta \quad (3)$$

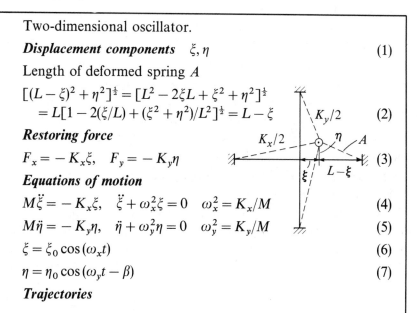

Equations of motion

$$M\ddot{\xi} = -K_x\xi, \quad \ddot{\xi} + \omega_x^2\xi = 0 \quad \omega_x^2 = K_x/M \quad (4)$$

$$M\ddot{\eta} = -K_y\eta, \quad \ddot{\eta} + \omega_y^2\eta = 0 \quad \omega_y^2 = K_y/M \quad (5)$$

$$\xi = \xi_0 \cos(\omega_x t) \quad (6)$$

$$\eta = \eta_0 \cos(\omega_y t - \beta) \quad (7)$$

Trajectories

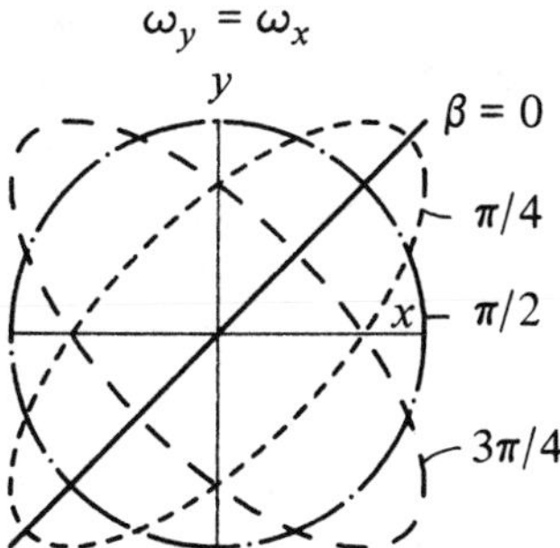

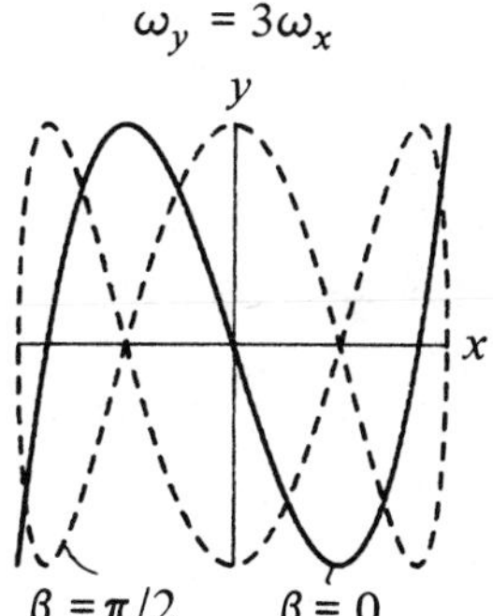

of the trajectory depends, of course, on the relative amplitudes of the two motions as well as on the phase difference between them.

We consider first the simple case when the frequencies as well as the amplitudes are the same. Then, if the phase difference is zero, the trajectory becomes a straight line inclined an angle of 45° with the x-axis. For a phase difference less than 90°, the trajectory will be an ellipse with the major axis inclined 45° with the x-axis. With increasing phase difference β, the trajectory opens up to an ellipse with the major axis inclined 45° with the x-axis and becomes a circle when the phase difference is 90°. A further increase in β in the range between 90° and 180°, turns the trajectory into an ellipse inclined 135° with respect to the x-axis.

We have shown also the trajectories obtained when $\omega_y = 3\omega_x$. With the phase difference being zero the particle moves back and forth along the solid line trajectory in the figure; one period in the x-direction for every three in the y-direction. As the phase difference increases, the trajectory 'opens', as illustrated by the dashed curve in the figure, for which the phase difference is 90°.

Trajectories of this kind are often called **Lissajou figures**. They can be used as a simple means of comparing two electrical signals on an oscilloscope. One signal is applied to the horizontal plates and the other to the vertical plates of the oscilloscope. If the signals are harmonic the phase difference and the ratio between the frequencies can be inferred from the resulting Lissajou figure.

1.7 Nonlinear oscillations

Although this chapter focuses on the harmonic oscillator, we shall include a brief discussion of oscillations in general, which can be considered to be a supplement to the qualitative considerations about molecular vibrations in the last section.

Thus, let us consider the motion in a one-dimensional potential well as illustrated in Display 1.10. The potential energy is $V(x)$ with $x = 0$ placed at the bottom of the well. The displacement of the particle from the equilibrium position $x = 0$ is denoted by ξ and the total energy by E. There is no friction, and conservation of energy is then expressed by

$$(M/2)(\mathrm{d}\xi/\mathrm{d}t)^2 + V(\xi) = E \qquad (1)$$

The particle oscillates back and forth with the velocity being a maximum at the bottom of the well and zero at the 'turning points' ξ_1 and ξ_2, which are determined by $V(\xi) = E$.

From conservation of energy follows $\mathrm{d}\xi/\mathrm{d}t = \{(2/M)[E - V(x)]\}^{\frac{1}{2}}$, which

Display 1.10.

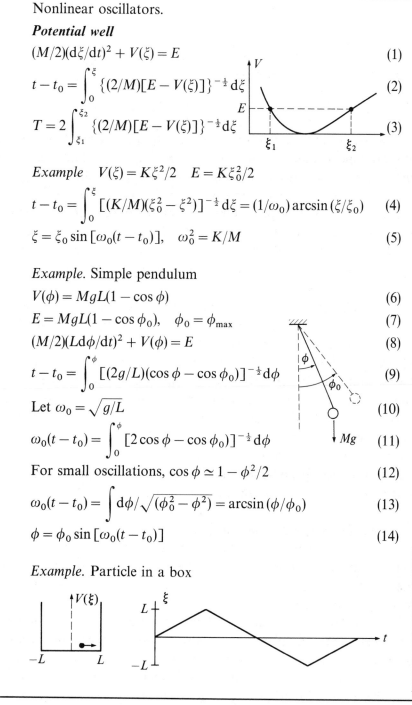

Nonlinear oscillators.

Potential well

$$(M/2)(\mathrm{d}\xi/\mathrm{d}t)^2 + V(\xi) = E \quad (1)$$

$$t - t_0 = \int_0^{\xi} \{(2/M)[E - V(\xi)]\}^{-\frac{1}{2}}\,\mathrm{d}\xi \quad (2)$$

$$T = 2\int_{\xi_1}^{\xi_2} \{(2/M)[E - V(\xi)]\}^{-\frac{1}{2}}\,\mathrm{d}\xi \quad (3)$$

Example $V(\xi) = K\xi^2/2 \quad E = K\xi_0^2/2$

$$t - t_0 = \int_0^{\xi} [(K/M)(\xi_0^2 - \xi^2)]^{-\frac{1}{2}}\,\mathrm{d}\xi = (1/\omega_0)\arcsin(\xi/\xi_0) \quad (4)$$

$$\xi = \xi_0 \sin[\omega_0(t - t_0)], \quad \omega_0^2 = K/M \quad (5)$$

Example. Simple pendulum

$$V(\phi) = MgL(1 - \cos\phi) \quad (6)$$

$$E = MgL(1 - \cos\phi_0), \quad \phi_0 = \phi_{\max} \quad (7)$$

$$(M/2)(L\mathrm{d}\phi/\mathrm{d}t)^2 + V(\phi) = E \quad (8)$$

$$t - t_0 = \int_0^{\phi} [(2g/L)(\cos\phi - \cos\phi_0)]^{-\frac{1}{2}}\,\mathrm{d}\phi \quad (9)$$

Let $\omega_0 = \sqrt{g/L}$ (10)

$$\omega_0(t - t_0) = \int_0^{\phi} [2\cos\phi - \cos\phi_0)]^{-\frac{1}{2}}\,\mathrm{d}\phi \quad (11)$$

For small oscillations, $\cos\phi \simeq 1 - \phi^2/2$ (12)

$$\omega_0(t - t_0) = \int \mathrm{d}\phi/\sqrt{(\phi_0^2 - \phi^2)} = \arcsin(\phi/\phi_0) \quad (13)$$

$$\phi = \phi_0 \sin[\omega_0(t - t_0)] \quad (14)$$

Example. Particle in a box

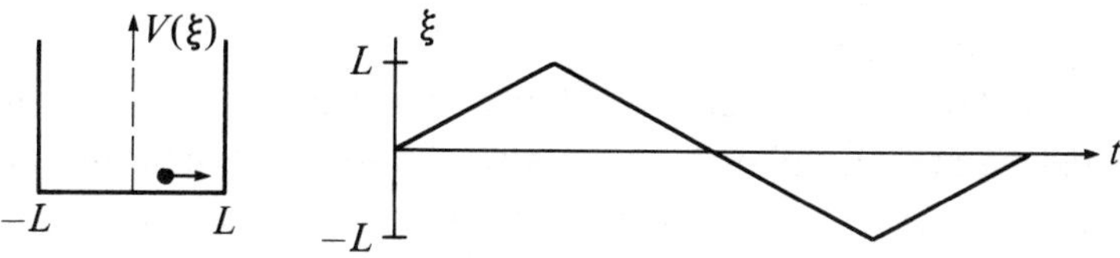

can be integrated, with t expressed as a function of ξ,

$$t - t_0 = \int_0^{\xi} \{(2/M)[E - V(\xi)]\}^{-\frac{1}{2}} \mathrm{d}\xi \tag{2}$$

where $t = t_0$ for $\xi = 0$.

The period of oscillation is the time required for the particle to go back and forth between the turning points,

$$T = 2\int_{\xi_1}^{\xi_2} \{(2/M)[E - V(\xi)]\}^{-\frac{1}{2}} \mathrm{d}\xi \tag{3}$$

As a first example in the display we check the general result by considering the harmonic oscillator for which $V(\xi) = K\xi^2/2$, and we find indeed that the integral for t leads to the harmonic motion, as it should.

As a second example we apply the general result to the simple pendulum. The displacement of the pendulum is expressed by the angle ϕ, as before, and in the expression $MgL(1 - \cos\phi)$ for the potential energy, L is the length of the pendulum and $L(1 - \cos\phi)$ is the height of the mass M above the equilibrium position. If the angle corresponding to the turning point is ϕ_0, the total energy is $E = MgL(1 - \cos\phi_0)$.

The integral for t in terms of ϕ can then be expressed as shown, and for small displacements, the harmonic motion is obtained, as it should.

As a third example, we consider the motion of a particle in a 'box', i.e. a square well potential with infinitely high walls. The motion then involves elastic collisions at the walls, and the particle will move back and forth between the walls with constant speed. The time dependence of the displacement will be represented by straight line segments with slopes equal to the particle velocity. This function should be compared with the sine function, which is obtained for a square law potential. The period of motion will be inversely proportional to the speed of the particle; in harmonic motion the period is independent of the speed (energy).

1.8 Forced harmonic motion

A detailed study of the forced harmonic oscillator will be carried out in Chapter 3 in terms of the complex amplitude approach. The analysis presented here is given for comparison and may serve as an introduction to subsequent studies.

The oscillators considered so far were idealized, since damping was neglected. In a real oscillator there is always some form of damping (except in a superconductor and a superfluid); in an electric circuit we have the resistance and in a mechanical oscillator there is friction.

In the electric circuit the voltage across the resistance is RI, where I is the current. In the mechanical oscillator we shall assume that the friction force is $-R\dot{\xi}$, i.e. proportional to the velocity $\dot{\xi}$. The constant of proportionality R is a friction constant, which generally has to be experimentally determined.

The friction force on the mass element in a mass–spring oscillator can be the result of contact friction between the element and the surroundings. It can be produced also by means of a device called a **dashpot**, which is connected in parallel with the spring. The dashpot consists of a piston, which slides inside a 'leaky' cylinder, and the force required to move the piston will be proportional to the velocity of the piston relative to the cylinder, at least for small velocities. With the cylinder held fixed, the force will be proportional to the velocity of the piston and is expressed as $-R\dot{\xi}$, where the force constant R generally has to be determined experimentally.

With reference to Display 1.11, we consider here a mass–spring oscillator with a dashpot damper, in which the mass element is driven by an external harmonic force $F = F_0 \cos(\omega t)$, with the phase angle chosen to be zero. Accounting also for the spring force and the friction force, the equation of motion of the mass element is $M\ddot{\xi} = F_0 \cos(\omega t) - R\dot{\xi} - K\xi$, or

$$M\ddot{\xi} + R\dot{\xi} + K\xi = F_0 \cos(\omega t) \tag{1}$$

The oscillator is linear (the force terms on the left hand side of the equation contain only first order terms in ξ and its derivatives), and in order to satisfy the equation, the displacement must be harmonic with the same frequency as the driving force. Thus, the displacement will be of the form

$$\xi = \xi_0 \cos(\omega t - \alpha) \tag{2}$$

where the amplitude ξ_0 and the phase angle α have to be chosen so as to satisfy the equation of motion. Clearly, the displacement cannot be a $\cos(\omega t)$ function, since the friction force then would be proportional to $\sin(\omega t)$, which makes it impossible to satisfy the equation at all times. Thus, we have to introduce a certain phase angle α, which is to be determined together with the amplitude.

To determine these two quantities, we insert $\xi = \xi_0 \cos(\omega t - \alpha)$ into the equation of motion. With the expressions for the velocity and acceleration being $-\omega\xi_0 \sin(\omega t - \alpha)$ and $-\omega^2\xi_0 \cos(\omega t - \alpha) = -\omega^2\xi$, the equation of motion becomes

$$(K - \omega^2 M)\xi_0 \cos(\omega t - \alpha) - R\omega\xi_0 \sin(\omega t - \alpha) = F_0 \cos(\omega t) \tag{3}$$

We now use the relations $\cos(\omega t - \alpha) = \cos(\omega t)\cos(\alpha) + \sin(\omega t)\sin(\alpha)$ and $\sin(\omega t - \alpha) = \sin(\omega t)\cos(\alpha) - \cos(\omega t)\sin(\alpha)$ to obtain

$$[(K - \omega^2 M)\cos(\alpha) + R\omega \sin(\alpha)]\xi_0 \cos(\omega t) + [(K - \omega^2 M)\sin(\alpha) - R\omega \cos(\alpha)]\xi_0 \sin(\omega t) = F_0 \cos(\omega t) \tag{4}$$

Display 1.11.

Forced and free motion of damped oscillator.

Forced harmonic motion

$$M\ddot{\xi} + R\dot{\xi} + K\xi = F_0 \cos(\omega t) \quad (1)$$

$$\xi = \xi_0 \cos(\omega t - \alpha) \quad (2)$$

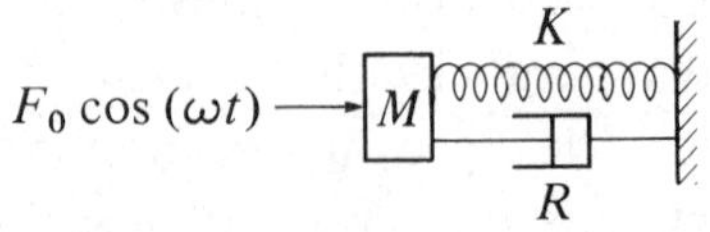

$$\dot{\xi} = -\xi_0 \omega \sin(\omega t - \alpha) \quad (3)$$

$$\ddot{\xi} = -\omega^2 \xi \quad (4)$$

Eq. (1) becomes

$$(K - \omega^2 M)\xi_0 \cos(\omega t - \alpha) - \omega R \xi_0 \sin(\omega t - \alpha) = F_0 \cos(\omega t) \quad (5)$$

Use $\cos(\omega t - \alpha) = \cos(\omega t)\cos(\alpha) + \sin(\omega t)\sin(\alpha)$ (6)

$\sin(\omega t - \alpha) = \sin(\omega t)\cos(\alpha) - \cos(\omega t)\sin(\alpha)$ (7)

Eq. (5) becomes

$$\xi_0[(K - \omega^2 M)\cos(\alpha) + \omega R \sin(\alpha)]\cos(\omega t) + \xi_0[(K - \omega^2 M)\sin(\alpha) - \omega R \cos(\alpha)]\sin(\omega t) = F_0 \cos(\omega t) \quad (8)$$

$$(K - \omega^2 M)\cos(\alpha) + \omega R \sin(\alpha) = F_0/\xi_0 \quad (9)$$

$$(K - \omega^2 M)\sin(\alpha) - \omega R \cos(\alpha) = 0 \quad (10)$$

$$\tan(\alpha) = \omega R/(K - \omega^2 M) \quad (11)$$

$$\xi_0 = F_0/\sqrt{(K - \omega^2 M)^2 + (\omega R)^2} \quad (12)$$

Free damped motion

$$M\ddot{\xi} + R\dot{\xi} + K\xi = 0 \quad (13)$$

Assume $\xi(t) = \xi_0 e^{-\gamma t} \cos(\omega' t - \alpha)$ (14)

$$\dot{\xi} = \xi_0 e^{-\gamma t}[-\gamma \cos(\cdots) - \omega' \sin(\cdots)] \quad (15)$$

$$\ddot{\xi} = \xi_0 e^{-\gamma t}[\gamma^2 \cos(\cdots) + 2\gamma\omega' \sin(\cdots) - \omega'^2 \cos(\cdots)] \quad (16)$$

Insertion into Eq. (13):

$$[M(\gamma^2 - \omega'^2) + K - \omega' R]\cos(\cdots) + [2M\gamma\omega' - \gamma R]\sin(\cdots) = 0 \quad (17)$$

i.e. $[M(\gamma^2 - \omega'^2) + K - \gamma R] = 0$ and $[2M\gamma\omega' - \omega' R] = 0$ (18)

$$\gamma = R/2M, \quad \omega' = \sqrt{\omega_0^2 - \gamma^2}, \quad \omega_0^2 = K/M \quad (19)$$

In order for this equation to be satisfied identically (i.e. at all times) the coefficient for $\sin(\omega)$ must be zero and the coefficient for $\cos(\omega t)$ must be F_0, i.e.

$$(K - \omega^2 M)\sin(\alpha) - R\cos(\alpha) = 0 \tag{5}$$

$$(K - \omega^2 M)\cos(\alpha) + R\omega\sin(\alpha) = F_0 \tag{6}$$

From these equations we get [recall $\cos(\alpha) = 1/\sqrt{1 + \tan^2(\alpha)}$],

$$\tan(\alpha) = R\omega/(K - \omega^2 M) \tag{7}$$

$$\xi_0 = F_0/\sqrt{(K - \omega^2 M)^2 + (R\omega)^2} \tag{8}$$

which express the 'frequency response' of the oscillator

In the limit of zero frequency, the displacement amplitude is $\xi' = F_0/K$, and it is convenient to express the amplitude at some other frequency in terms of it. Similarly, we shall express the frequency in terms of the resonance frequency given by $\omega_0 = \sqrt{K/M}$, and the corresponding 'normalized' frequency will be denoted by $\Omega = \omega/\omega_0$. Furthermore, we normalize the resistance by introducing the damping parameter $D = R/\omega_0 M = R\omega_0/K$. This parameter can be interpreted as the ratio between the friction force and the inertia force in the oscillator (this follows by multiplying both numerator and denominator by the velocity). In terms of these quantities the frequency dependence of the displacement amplitude and the phase angle can be written

$$\xi_0/\xi' = 1/\sqrt{(1 - \Omega^2)^2 + (D\Omega)^2} \qquad \xi' = F/K \tag{9}$$

$$\tan(\alpha) = D\Omega/(1 - \Omega^2) \qquad \Omega = \omega/\omega_0 \quad D = R/\omega_0 M \tag{10}$$

These relations will be discussed in detail in Chapter 3. It is sufficient to point out now that when $\omega = \omega_0$, i.e. $\Omega = 1$, the amplitude will be $1/D$ times the displacement amplitude at zero frequency. This amplification, $Q = 1/D$, is usually called the **'quality factor'** or simply the quality of the oscillator.

The calculation we have carried out for determining the frequency dependence of the amplitude and the phase angle of the oscillator in forced harmonic motion is straight forward but algebraically quite cumbersome. In more complicated problems involving coupled oscillators and waves, a similar analytical approach becomes even more intractable and awkward. In the method described in Chapter 3, however, we in essence operate with both the $\cos(\omega t)$ and $\sin(\omega t)$ terms simultaneously, and the algebra is simplified considerably.

1.9 Free damped motion

If the driving force in the previous section is 'turned off' the oscillator motion will decay because of the friction damping. We wish to determine the characteristics of this decaying motion, and to do so, we have to solve the equation for free damped motion

$$M\ddot{\xi} + R\dot{\xi} + K\xi = 0 \qquad (1)$$

A detailed analysis of this motion will be given in Chapter 4, and, as in the case of forced harmonic motion, we discuss it here for future reference in our comparison of different methods of analysis.

In our present approach we have to make a guess as to the form of the solution and try

$$\xi = \xi_0 e^{-\gamma t} \cos(\omega' t - \alpha) \qquad (2)$$

where the **decay constant** γ and the frequency ω' are to be determined. The corresponding velocity is

$$\dot{\xi} = \xi_0 e^{-\gamma t}[-\gamma \cos(\cdots) - \omega' \sin(\cdots)]$$

and for the acceleration we get

$$\ddot{\xi} = \xi_0 e^{-\gamma t}[\gamma^2 \cos(\cdots) + 2\gamma\omega' \sin(\cdots) - \omega'^2 \cos(\cdots)]$$

where $(\cdots) = (\omega' t - \alpha)$.

Inserting these expressions into the equation of motion and collecting the terms involving $\cos(\cdots)$ and $\sin(\cdots)$, we note that the equation can be satisfied only if the coefficients for these terms vanish separately. This leads to the equations $M(\gamma^2 - \omega'^2) + K - R\gamma = 0$ and $2M\gamma\omega' - R\omega' = 0$ which yield

$$\gamma = R/2M \qquad (3)$$

$$\omega' = \sqrt{\omega_0^2 - \gamma^2} \qquad (4)$$

In other words, the displacement $\xi = \xi_0 \exp(-\gamma t) \cos(\omega' t - \alpha)$ satisfies the equation of motion with the values of the decay constant and the frequency given above. Furthermore, the solution contains two arbitrary constants ξ_0 and α, which can be determined from known values of displacement and velocity at a given time. A solution which fulfills these requirements is the unique solution to the differential equation of motion.

This motion is reconsidered in Chapter 4, where a detailed discussion of the results is given.

Problems

1.1 *Period of a satelite.* Regard Earth as a homogeneous sphere with a density $\rho = 5.5\,\mathrm{g/cm^3}$. Calculate the period (round trip time) of a satelite moving in a circular orbit about the Earth. Approximate the orbit radius with the Earth radius. The gravitational constant is $G = 6.67 \times 10^{-11}$ MKS.

1.2 *Cyclotron motion.* Consider the cyclotron motions of an electron and a proton in a uniform magnetic field $B = 10\,000$ gauss $= 1$ Weber/m^2 $= 1$ volt sec/m^2. Electron mass $m = 9.11 \times 10^{-31}$ kg. Electron charge $e = 1.6 \times 10^{-19}$ Coulomb. Proton mass $M = 1836\,m$.

(a) Calculate the cyclotron frequencies of these motions.
(b) Determine the orbit radii if the energy of each of the particles is 1 keV (1000 electron volts $= 1.6 \times 10^{-16}$ joules).
(c) What is the maximum value of the velocity and the acceleration of the x-coordinate of the electron?

1.3 *Harmonic motion.* The harmonic displacements of two particles are $A\cos(\omega t)$ and $A\cos[\omega t - (\pi/6)]$.

(a) The latter motion lags behind the former in time. Determine this time lag in terms of the period T.
(b) At what times do the particles have their positive maxima of velocity and acceleration?
(c) If the amplitude A is 1 cm, at what frequency (in Hz) will the acceleration equal $g = 981$ cm/sec^2.

1.4 *Average speed.* Show that the average speed of a particle in harmonic motion is $(2/\pi)$ times the maximum speed.

1.5 *Initial value problem.* A mass–spring oscillator with $\omega_0 = 400\,\text{sec}^{-1}$ has a displacement 10 cm and a velocity 20 cm/sec at $t = 0$. Determine the subsequent motion.

1.6 *Initial value problem.* Consider a horizontal mass–spring oscillator on a frictionless table. (Mass M, spring constant K). The end of the spring is attached to a rigid wall. At $t = 0$ a bullet of mass m is shot at a speed v into the block in the direction of the spring. The collision takes place in a very short time, so that the spring can be considered to be in the relaxed position during the collision.

(a) Determine the subsequent time dependence of the displacement.
(b) What is the total mechanical energy of oscillation?
(c) When should a second bullet be shot into the block in order to increase (decrease) the amplitude as much as possible? In each case determine the new expression for the time dependence of the motion (both amplitude and phase) using the same origin of time t as before.
(d) With $M = 1$ kg, $m = 5$ g, $v = 300$ m/sec, and $K = 400$ N/m determine the numerical answers in (a), (b), and (c).

1.7 *Mass–spring oscillator.* A body of mass M, on a horizontal frictionless plane, is attached to the junction of two horizontal springs, with spring constants K_1 and

K_2. The relaxed lengths of the springs are the same and equal to L. The free ends of the springs are pulled apart and fastened to fixed supports a distance $3L$ apart.

(a) Determine the equilibrium position of the body.

(b) What is the frequency of oscillation of the body about the equilibrium position?

1.8 *Pendulum.* A pendulum and a mass–spring oscillator have identical periods on Each. What is the ratio of the periods of oscillation on a planet where bodies weigh 8 times more than they do on Earth?

1.9 *Floating body.* The cross section of a uniform body perpendicular to its length has the shape defined in the figure. The length of the body is L and the mass M. The body floats on water.

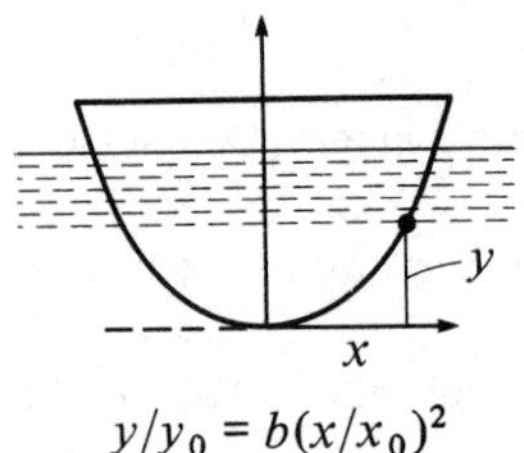

$y/y_0 = b(x/x_0)^2$

(a) Determine the equilibrium position of the body.

(b) What will be the frequency of small vertical oscillations of the body about the equilibrium position?

1.10 *Springs in series and in parallel.* Two springs with the same relaxed lengths have the spring constants $K_1 = K$ and $K_2 = 1.5K$, where $K = 10^5$ N/m. With these springs and a mass $M = 10$ kg, we can construct four oscillators each with its upper end of the spring(s) held fixed. In each case determine the static deflection and the frequency of oscillation.

1.11 *Pendulum.* A body hung at the end of a vertical spring stretches the spring statically to twice its initial length. This system can be set into oscillation either as a simple pendulum or as a mass–spring oscillator. Determine the ratio between the periods of these motions. In the pendulum mode of motion, assume the length of the spring to be constant.

1.12 *Two-body oscillator.* Two air-suspended carts on a track of masses M_1 and M_2 are connected with a spring of negligible mass and spring constant K. With M_2 backed by a rigid wall, the system is set in motion by compressing the spring a distance d (by moving M_1 toward M_2) and then releasing it at $t = 0$.

(a) How far does M_1 move before M_2 starts moving?

(b) After M_2 has lost contact with the wall, what is the velocity of the center of mass of the system?

(c) Determine the expression for the time dependence of the separation of the two bodies in the subsequent motion.

1.13 *Two-body oscillator.* Two particles of equal mass M are connected with a spring. The spring is linear with a spring constant K when the change of its length is less than ξ_m, but will break when it is stretched beyond this value.

The two-body oscillator is struck in a head-on collision (along the spring) by a body of mass M with a velocity U. If the collision is elastic, for what value of U will the collision result in a 'dissociation' of the two-body oscillator?

1.14 *Two-body oscillator.* In a two-body oscillator $M_1 = 1$ kg and $M_2 = 3$ kg. The spring between them is compressed a distance $d = 20$ cm from the relaxed length and the bodies are then released in such a way that the center of mass is at rest.

(a) What is the reduced mass of the system?
(b) What is the frequency of the ensuing motion?
(c) What is the maximum kinetic energy of M_1 and M_2?

1.15 *Physical pendulum.* A tennis racket weighs 13 ounces and its center of mass is 13 inches from the top of the handle. The period of oscillation about an axis through the top of the handle and in the plane of the racket is 1.3 sec.

(a) Determine the location of the center of percussion.
(b) Determine the radius of gyration $R_g = \sqrt{I/M}$ of the racket (I = moment of inertia, M = mass) with respect to the axis of rotation.

1.16 *Physical pendulum.* Calculate the period of small oscillations of a square plate about an axis through one of the corners perpendicular to the plate.

1.17 *Thermal vibration.* The frequency of oscillation and the mass of an atom in a diatomic molecule are of the order of 10^{13} Hz and 10^{-22} g. What is the equivalent spring constant in N/m?

The energy of oscillation is $kT/2$, where k is the Boltzmann constant and T the absolute temperature ($k = 1.38 \times 10^{-23}$ joule/K). What then is the amplitude of oscillation of the atoms at a temperature $T = 300$ K (27 °C)?

1.18 *Nonlinear oscillator.* In most mechanical systems, a small displacement of a particle from its equilibrium position will result in a restoring force proportional to the dislacement. There are exceptions, however, and a simple example is indicated in the figure.

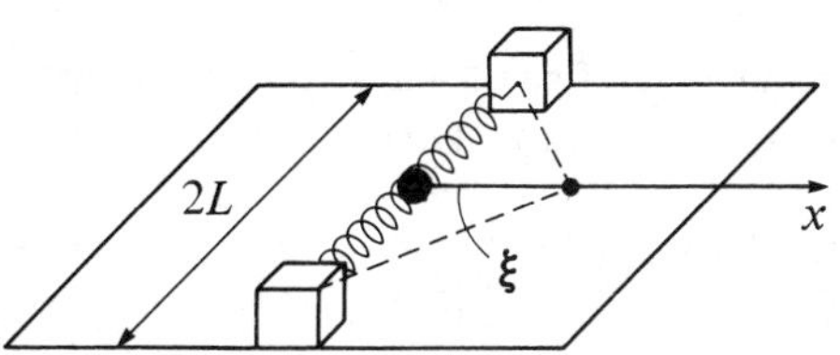

This figure illustrates a mass M attached to the center of a horizontal spring, which is clamped at both ends. The spring with a length $2L$ and a spring constant K, is initially slack. The body is set in motion perpendicular to the spring, i.e. along the direction of the x-axis, as shown.

(a) What is the potential energy $V(\xi)$ of the particle?
(b) What is the corresponding restoring force for small oscillations, $\xi \ll L$? Will the motion be harmonic?
(c) Qualitatively, will the period of oscillation increase or decrease with increasing amplitude of oscillation?
(d) What is the answer to (c) in the case of a simple pendulum?

1.19 *Motion in the gravitational field inside a sphere.* Imagine a straight tunnel through the center of the Earth, which we regard as a sphere with a uniform mass density ρ. A particle is dropped into the tunnel from the surface at $t = 0$.

(a) Show that the subsequent motion will be harmonic.
(b) When will the particle reach the other end of the tunnel?
(c) As a generalization of (b), show that the motion will be harmonic even if the tunnel does not go through the center of the Earth, and that the period will be the same as before. (Neglect friction).

1.20 *The Thomson model of the atom.* J.J. Thomson (British physicist, 1856–1940) imagined the atom as a swarm of electrons contained within a uniform spherical distribution of positive charge equal in magnitude to the total charge of the electrons. In this atomic model, consider the case of one electron within the sphere under the influence of the electric field of the uniform positive charge distribution (Hydrogen atom).

(a) Show that the motion is harmonic.
(b) If the diameter of the sphere is taken to be 1 Ångström $= 10^{-8}$ cm, what then would be the frequency of oscillation of the electron?

It is interesting to note that J.J. Thomson's son, G.P. Thomson, was one of the pioneers in establishing the wave nature of the electron. His father's work dealt with the particle nature of the electron.

1.21 *Thermal oscillations.* The Young's modulus of a solid is defined by the relation $\sigma = E(\Delta l/l)$, where σ is the stress, i.e. the force per unit area of a uniform rod, and $\Delta l/l$ the strain, i.e. the relative change in the length l of the rod produced by the stress. For steel $E = 2 \times 10^{11}$ N/m^2.

(a) What is the corresponding 'spring' constant of a bar of length 1 m and a cross sectional area of 1 cm^2?
(b) Imagine the bar composed of parallel atomic 'strings' consisting of mass elements (atoms) connected with springs. The atoms are assumed to be arranged in a cubical lattice with an interatomic distance $d = 10^{-8}$ cm. What then is the spring constant of one of the strings?
(c) The density of steel is $\rho = 7.8$ g/cm^3. From the result obtained in (b), estimate the frequency of thermal atomic oscillations.

2

The complex amplitude

As a preparation for the analysis of forced and free motion of a damped oscillator in the next two chapters and the subsequent treatment of coupled oscillations and waves, we introduce in this chapter the complex amplitude approach in the analysis of harmonic motion. As part of this preparation, we present a brief review of complex numbers and show how a harmonic motion can be described in terms of a complex amplitude.

2.1 A review of complex numbers

As we have seen in the previous chapter, a harmonic motion with a given frequency is uniquely defined by the amplitude and the phase angle. A complex number can be described in a similar manner, and this suggests the possibility of using complex numbers and the mathematics of complex numbers in the description and analysis of harmonic motion. This, indeed, turns out to be a most useful approach. Before applying it, however, we shall review some of the essentials of complex numbers for the benefit of those who have not been exposed to them before.

It is instructive in this context to start with some reminders about the real number system. In particular, we recall the role of the basic operations of addition and multiplication in the process of building up the set of real numbers from the set of positive integers.

Consider, for example, the relation $A + B = C$. For any two positive integers A, B, the number C can always be found amongst the set of positive integers. On the other hand, for any two positive integers B, C, say 5 and 3, we cannot always find A amongst the set to satisfy the relation. Then, to make the relation meaningful, we use it to define negative numbers, thus extending the number system. In the particular example of $B = 5$ and $C = 3$, the number $A = -2$ is defined as the number, which, when added to 5 gives 3 as a result.

In analogous manner, rational and irrational numbers are introduced to make the operation $AB = C$ of multiplication meaningful for arbitrary integers A and B. For example, with $B = 3$ and $C = 2$, the fraction $A = 2/3$ is defined as the number, which, when multiplied by 3 gives 2 as a result. If $A = B$ and $C = 3$, so that $A^2 = 3$, the irrational number $A = \sqrt{3}$ (or $A = -\sqrt{3}$) is defined as the number, which when squared, gives 3 as a result. In this manner, the set of **real numbers** can be generated, and these numbers are represented geometrically as points on the real number line.

Imaginary numbers. Further extension of the number system is required, however. For example, there is no real number A such that $A^2 = C$, if C is negative. For this reason, a number, i, called the **imaginary unit number** is defined, such that $\mathrm{i}^2 = -1$. Similarly, with y a real number, the imaginary number $A = \mathrm{i}y$ has the property $A^2 = -y^2$. The imaginary number $\mathrm{i}y$ is represented geometrically as a point on a new number line, **the imaginary axis**, perpendicular to the real number line, and the point is a distance y from the origin, as indicated in Display 2.1.

We require the associative law of algebra $(AB)C = A(BC)$ [and $(A + B) + C = A + (B + C)$] to be valid also for imaginary numbers and ABC is defined as $(AB)C$ or $A(BC)$. Thus with $\mathrm{i}^2 = -1$ we have $\mathrm{i}^3 = (\mathrm{i}^2)\mathrm{i} = -\mathrm{i}$ and $\mathrm{i}^4 = 1$. From these relations and with $b = \mathrm{i}$ it follows, as an example, that $z = b^4 + b = 1 + \mathrm{i}$, which contains both a real and imaginary part.

Complex numbers. In general, a number $z = x + \mathrm{i}y$ with a real part x and an imaginary part y defines a complex number. It is represented geometrically by a point in the **complex plane**, which is spanned by the real and imaginary axes, as shown.

With the number system extended to include complex numbers, an algebraic equation of nth order always has n roots (Theorem of Algebra). For example, the second order equation $z^2 + 2z + 5 = 0$ has the two roots $z = -1 \pm \mathrm{i}2$, and the fourth order equation $z^4 = -1$ has the four roots $(\pm 1 \pm \mathrm{i})/\sqrt{2}$, located symmetrically in the complex plane, as will be shown later.

The **sum** of two complex numbers $z_1 = x_1 + \mathrm{i}y_1$ and $z_2 = x_2 + \mathrm{i}y_2$ is

$$z = z_1 + z_2 = (x_1 + x_2) + \mathrm{i}(y_1 + y_2) = x + \mathrm{i}y \tag{1}$$

i.e. it is a complex number with the real part $x = x_1 + x_2$ and the imaginary part $y = y_1 + y_2$. The **difference** between two complex numbers is defined in a similar manner.

Display 2.1.

Complex number algebra.

Imaginary unit number i, $i^2 = -1$ $(i^3 = -i, \quad i^4 = 1)$ (1)

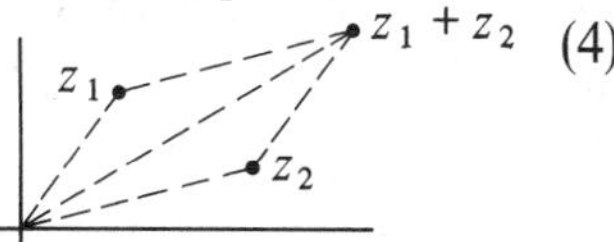

Complex number $z = x + iy$ (2)

Addition

$z_1 = z_1 + iy_1, \quad z_2 = x_2 + iy_2$ (3)

Complex plane

$z_1 + z_2 = (x_1 + x_2) + i(y_1 + y_2)$ (4)

(Geometrically, addition is similar to addition of vectors)

Multiplication

$z_1 z_2 = (x_1 + iy_1)(x_2 + iy_2) = (x_1 x_2 - y_1 y_2) + i(x_2 y_1 + x_1 y_2)$ (5)

Complex conjugate $z^* = x - iy$ (6)

$x = (z + z^*)/2 \quad y = (z - z^*)/2i$ (7)

Division

$z_1/z_2 = z_1 z_2^*/z_2^* z_2 = (x_1 x_2 + y_1 y_2)/A_2^2 + i(x_2 y_1 - x_1 y_2)/A_2^2$ (8)

$A_2^2 = x_2^2 + y_2^2$ (9)

Amplitude and phase angle

$A = \sqrt{x^2 + y^2}$ (10)

$\tan(\alpha) = y/x, \quad x = A\cos(\alpha), \quad y = A\sin(\alpha)$ (11)

Example

Determine amplitudes and phase angles of the complex numbers

(a) $z_1 = 1 + i\sqrt{3}$, (b) $z_2 = \sqrt{3} + i$, (c) $z_1 + z_2$, (d) $z_1 z_2$, (e) z_1/z_2 (12)

(a) $A_1 = \sqrt{1 + (\sqrt{3})^2} = 2 \quad \tan(\alpha_1) = \sqrt{3}/1 \quad \alpha = \pi/3$ (13)

(b) $A_2 = \sqrt{(\sqrt{3})^2 + 1} = 2 \quad \tan(\alpha_2) = 1/\sqrt{3} \quad \alpha = \pi/6$ (14)

(c) $z_1 + z_2 = 1 + \sqrt{3} + i(\sqrt{3} + 1) \quad A = \sqrt{2(1 + \sqrt{3})^2}$
$= (1 + \sqrt{3})\sqrt{2} \quad \tan(\alpha) = 1 \quad \alpha = \pi/4$ (15)

(d) $z_1 z_2 = (1 + i\sqrt{3})(\sqrt{3} + i) = i2\sqrt{3} \quad \tan(\alpha) = \infty, \quad \alpha = \pi/2$ (16)

(e) $z_1/z_2 = (1 + i\sqrt{3})/(\sqrt{3} + i) = (1 + i\sqrt{3})(\sqrt{3} - i)/4$
$= (\sqrt{3} + i)/2 \quad A = 1 \quad \alpha = \pi/6$ (17)

We require the distributive law of multiplication, $A(B+C)=AB+AC$ to be valid also for complex numbers. Thus, the **product** of the two complex numbers z_1 and z_2 is obtained by using the distributive law for the components, accounting for $\mathrm{i}^2=-1$,

$$w=z_1z_2=(x_1+\mathrm{i}y_1)(x_2+\mathrm{i}y_2)=(x_1x_2-y_1y_2)+\mathrm{i}(-x_1y_2+x_2y_1)=u+\mathrm{i}v \qquad (2)$$

With $z=x+\mathrm{i}y$ the number $z^*=x-\mathrm{i}y$ is called the **complex conjugate** of z. It is obtained simply by changing the sign of the imaginary part of z, and can be thought of as the mirror image of z in the complex plane with respect to the real axis. It follows that

$$x=\tfrac{1}{2}(z+z^*) \quad y=\tfrac{1}{2}(z-z^*)/\mathrm{i}$$
$$zz^*=(x+\mathrm{i}y)(x-\mathrm{i}y)=x^2+y^2=A^2=|z|^2 \qquad (3)$$

where we have introduced the **amplitude (magnitude)** A of z. It is sometimes denoted $|z|$. The geometrical meaning of A is simply the distance from the orgin to z in the complex plane, the 'radius vector' to z. The angle between this radius and the x-axis, denoted by α, will be called the **phase angle** of the complex number. It follows that

$$x=A\cos(\alpha),\ y=A\sin(\alpha) \qquad (4)$$
$$\tan(\alpha)=y/x$$
$$A=\sqrt{x^2+y^2}$$

The ratio between two complex numbers z_1 and z_2 follows from the definition of multiplication. Thus, by multiplying both the numerator and the denominator by the complex conjugate z_2^*, we obtain

$$z_1/z_2=z_1z_2^*/z_2z_2^*=(1/A_2^2)[x_1x_2+y_1y_2+\mathrm{i}(x_2y_1-x_1y_2)] \qquad (5)$$
$$z_2z_2^*=x_2^2+y_2^2=A_2^2$$

where we have introduced the magnitude (amplitude) A_2. The operations of multiplication and division are made simpler by the use of a remarkable relation, to be discussed next.

2.2 Euler's formula

By repeated use of multiplication of complex numbers, we can determine the complex number z^n, where n is a positive or negative integer. The meaning of a function of a complex number, $f(z)$, can then be defined in terms of the power series expansion of the function.

In the particular case of the exponential function $f(z)=\exp(z)$, the power series is

$$\exp(z)=1+z+z^2/2!+z^3/3!+z^4/4!+\cdots \qquad (1)$$

This series has the familiar and remarkable property that it remains unchanged after differentiation. It has another remarkable property, when the argument is purely imaginary, $z = \mathrm{i}\alpha$. The power series then takes the form

$$\exp(\mathrm{i}\alpha) = 1 + (\mathrm{i}\alpha) + (\mathrm{i}\alpha)^2/2! + (\mathrm{i}\alpha)^3/3! + \cdots \tag{2}$$

The terms of even powers are real and those of odd power imaginary. Collecting the real and imaginary terms, we obtain two series which we recognize as those of $\cos(\alpha)$ and $\sin(\alpha)$, respectively,

$$\exp(\mathrm{i}\alpha) = (1 - \alpha^2/2! + \alpha^4/4! + \cdots) + \mathrm{i}(\alpha - \alpha^3/3! + \alpha^5/5! - \cdots)$$

$$\exp(\mathrm{i}\alpha) = \cos(\alpha) + \sin(\alpha) \tag{3}$$

In other words, $\exp(\mathrm{i}\alpha)$ is a complex number with the real and imaginary parts $\cos(\alpha)$ and $\sin(\alpha)$. This is **Euler's formula**. The magnitude of $\exp(\mathrm{i}\alpha)$ is unity, and the phase angle is α, as illustrated in Display 2.2.

It follows that a complex number with the magnitude A rather than unity and with a phase angle α can be written $A\exp(\mathrm{i}\alpha)$. The real and imaginary parts are $x = A\cos(\alpha)$ and $y = A\sin(\alpha)$.

Expressing two complex numbers z_1 and z_2 as $A_1 \exp(\mathrm{i}\alpha_1)$ and $A\exp(\mathrm{i}\alpha_2)$ we obtain the product and the ratio simply as

$$z_1 z_2 = A_1 A_2 \exp[\mathrm{i}(\alpha_1 + \alpha_2)] \tag{4}$$

$$z_1/z_2 = (A_1/A_2)\exp[\mathrm{i}(\alpha_1 - \alpha_2)] \tag{5}$$

Thus, the magnitude of the product (ratio) is the product (ratio) of the magnitude, and the phase angle of the product (ratio) is the sum (difference) of the individual phase angles.

From Euler's formula we have $\exp(-\mathrm{i}\alpha) = \cos(\alpha) - \mathrm{i}\sin(\alpha)$ and it follows that

$$\cos(\alpha) = [\exp(\mathrm{i}\alpha) + \exp(-\mathrm{i}\alpha)]/2 \tag{6}$$

$$\sin(\alpha) = [\exp(\mathrm{i}\alpha) - \exp(-\mathrm{i}\alpha)]/2\mathrm{i} \tag{7}$$

Furthermore, if Re stands for 'real part of' we have

$$\cos(\alpha) = \mathrm{Re}\{\exp(\mathrm{i}\alpha)\} = \mathrm{Re}\{\exp(-\mathrm{i}\alpha)\} \tag{8}$$

The exponential function $\exp(\mathrm{i}\alpha)$ is periodic in α with the period 2π. In other words, $\exp(\mathrm{i}\alpha) = \exp[\mathrm{i}(\alpha + q2\pi)]$, where q is an integer. This extended form of the exponential must be used if we wish to find all the roots to an equation of the form $z^n = w$, where w is a complex number, $w = A\exp[\mathrm{i}(\alpha + q2\pi)]$ and n an integer. The n roots to the equation are then

$$z = A^{(1/n)} \exp[\mathrm{i}(\alpha + q2\pi)/n], \quad q = 0, 1, 2, \ldots n - 1 \tag{9}$$

As a simple example consider the equation $z^4 = -1$. We rewrite -1 as

Display 2.2.

Euler's formula and applications.

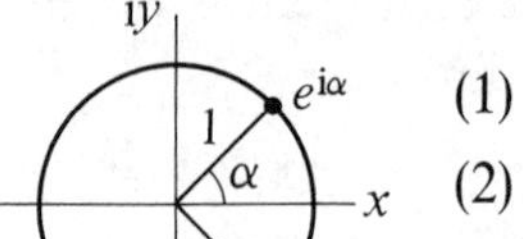

$$\exp(i\alpha) = \cos(\alpha) + i\sin(\alpha) \tag{1}$$

$$\cos(\alpha) = [\exp(i\alpha) + \exp(-i\alpha)]/2 \tag{2}$$

$$\sin(\alpha) = [\exp(i\alpha) - \exp(-i\alpha)]/2i \tag{3}$$

$$\cos(\alpha) = \mathrm{Re}\{\exp(i\alpha)\} = \mathrm{Re}\{\exp(-i\alpha)\} \tag{4}$$

Amplitude A and phase angle α (Magnitude and argument)

$$z = x + iy = A\exp(i\alpha) = A[\cos(\alpha) + i\sin(\alpha)] \tag{5}$$

$$x = A\cos(\alpha), \quad y = A\sin(\alpha) \tag{6}$$

$$A = \sqrt{x^2 + y^2} \tag{7}$$

$$\tan(\alpha) = y/x \tag{8}$$

Product, ratio

$$\text{Let } z_1 = A_1\exp(i\alpha_1), \quad z_2 = A_2\exp(i\alpha_2) \tag{9}$$

Complex plane

$$z_1 z_2 = A_1 A_2 \exp[i(\alpha_1 + \alpha_2)] \tag{10}$$

$$z_1/z_2 = (A_1/A_2)\exp[i(\alpha_1 - \alpha_2)] \tag{11}$$

Roots

$$\text{If } z^n = A\exp(i\alpha) = A\exp[i(\alpha + q2\pi)] \tag{12}$$

$$z = A^{(1/n)}\exp[i(\alpha/n + q2\pi/n)], \quad q = 0, 1, 2, \ldots (n-1) \tag{13}$$

Example

Calculate the amplitude and phase angle of $\int_0^b \exp(i\beta)\,d\beta$ (14)

$$\int_0^b \exp(i\beta)\,d\beta = (1/i)[e^{ib} - 1] = (1/i)e^{ib/2}[e^{ib/2} - e^{-ib/2}]$$

$$= be^{ib/2}\sin(b/2)/(b/2) \tag{15}$$

Amplitude $b\sin(b/2)/(b/2)$

Phase angle $b/2$

the complex number $\exp[i(\pi + q2\pi)]$, where q is an integer. The four solutions are $z = \exp[i(\pi/4 + q\pi/2)]$ with $q = 0$, 1, 2, and 3, i.e. $z = (\pm 1 \pm i)/\sqrt{2}$.

The two roots to the equation $z^2 = A + iB$, of course, can be obtained from the general expression above. Thus, we express $A + iB$ in terms of the amplitude $\sqrt{A^2 + B^2}$ and the phase factor $\exp(i\alpha)$, where $\tan\alpha = B/A$. It follows then that

$$z = (A^2 + B^2)^{1/4}\exp(i\alpha/2 + q2\pi/2) = \pm (A^2 + B^2)^{1/4}\exp(i\alpha/2) \quad (10)$$

where the plus sign corresponds to $q = 0$ and the minus sign to $q = 1$ $[\exp(i\pi) = -1]$. It is usually quite apparent from the physical context of the problem which of the roots is relevant in a particular application.

It is sometimes convenient to express an arbitrary complex number $A\exp(i\alpha)$ in terms of a single exponential, $A\exp(i\alpha) = \exp(a + i\alpha)$, where $a = \ln(A)$. This can be written $\exp[i(\alpha - ia)] = \exp(i\Psi)$, where $\Psi = \alpha - ia$ can be regarded as a complex phase angle. The meaning of the quantity z^w, where both z and w are complex numbers, can then be defined. With $z = \exp(a + i\alpha)$, we obtain $z^w = \exp[w(a + i\alpha)]$. As an example, let $z = 2i$ and $w = i$. Express 2i as $\exp(a + i\pi/2)$, where $a = \ln(2)$, and obtain $(2i)^i = \exp(ia - \pi/2) = \exp(-\pi/2)\exp(ia)$.

2.3 The complex amplitude of a harmonic function

The usefulness of complex numbers in the description and analysis of oscillations and waves is linked to Euler's formula, which makes it possible to express a harmonic function as the real part of a complex exponential function.

The harmonic function of interest in this context is the displacement of a mass–spring oscillator $\xi(t) = \xi_0(\omega)\cos(\omega t - \alpha(\omega))$, where we have indicated that the amplitude $\xi_0(\omega)$ and the phase angle $\alpha(\omega)$ generally are functions of the frequency. In the following, for simplicity, we shall generally use the shorter notations ξ_0 and α for $\xi_0(\omega)$ and $\alpha(\omega)$. We can express this function as the real part of the complex number $z = \xi_0\exp[i(\omega t - \alpha)]$ or its complex conjugate $z^* = \xi_0\exp[-i(\omega t - \alpha)]$. Using the latter option we have

$$\begin{aligned}\xi(t) &= \xi_0\cos(\omega t - \alpha) = \mathrm{Re}\{\xi_0\exp[-i(\omega t - \alpha)]\}\\ &= \mathrm{Re}\{\xi(\omega)e^{-i\omega t}\} \quad (1)\end{aligned}$$

$$\xi(\omega) = \xi_0\exp(i\alpha) \quad (2)$$

where we have introduced the quantity $\xi(\omega) = \xi_0(\omega)\exp(i\alpha(\omega))$, which we shall call the **complex displacement amplitude** of the harmonic motion. It is a complex number with the magnitude equal to the displacement amplitude ξ_0 and a phase angle equal to the phase angle α of the harmonic motion. As we said before, at a given frequency the harmonic motion is uniquely determined by the amplitude and the phase angle, and these two quantities are contained in the complex amplitude. This complex number, therefore, describes the harmonic motion uniquely.

The velocity and the acceleration are also harmonic functions and have their corresponding complex amplitudes. They are found by differentiation of $\xi(t) = \mathrm{Re}\{\xi(\omega)\exp(-\mathrm{i}\omega t)\}$ with respect to time. Thus, the velocity is

$$u(t) = \mathrm{d}\xi/\mathrm{d}t = \dot{\xi} = \mathrm{Re}\{(-\mathrm{i}\omega)\xi(\omega)\mathrm{e}^{-\mathrm{i}\omega t}\} = \mathrm{Re}\{u(\omega)\mathrm{e}^{-\mathrm{i}\omega t}\} \tag{3}$$

$$u(\omega) = -\mathrm{i}\omega\xi(\omega) \tag{4}$$

where $u(\omega)$ is the **complex velocity amplitude**. Differentiation with respect to time corresponds to multiplying the complex amplitude by $(-\mathrm{i}\omega)$. Thus, the **complex acceleration amplitude** is

$$a(\omega) = (-\mathrm{i}\omega)u(\omega) = (-\mathrm{i}\omega)^2\xi(\omega) = -\omega^2\xi(\omega) \tag{5}$$

We are now prepared to demonstrate how the complex amplitude description is used, and for this purpose let us consider as an example the equation of motion for a mass–spring oscillator, which is driven by a harmonic force $F_0(\omega)\cos(\omega t)$, amplitude F_0 generally being frequency dependent. We have assigned a phase angle zero to this force. For the sake of generality, we shall include also a friction force $R\mathrm{d}\xi/\mathrm{d}t$ proportional to the velocity. The equation of motion is then

$$M\mathrm{d}^2\xi/\mathrm{d}t^2 + R\mathrm{d}\xi/\mathrm{d}t + K\xi = F_0(\omega)\cos(\omega t) \tag{6}$$

For simplicity, we usually shall write F_0 for $F_0(\omega)$.

This is a linear equation in the displacement variable $\xi(t)$, and in order for this equation to be satisfied at all times, it is necessary that $\xi(t)$ be harmonic with the same frequency as the frequency of the driving force. Specifically, the sum of the terms on the left hand side of the equation must be a $\cos(\omega t)$ function, the same as on the right hand side.

The displacement function $\xi(t)$ itself cannot be a $\cos(\omega t)$ function because if it were, the term $R\mathrm{d}\xi/\mathrm{d}t$ would be a $\sin(\omega t)$ function, which would prevent the left hand side from being equal to the right hand side at all times. Consequently, we have to assume that the displacement is of the form $\xi_0\cos(\omega t - \alpha)$, where both the amplitude ξ_0 and the phase angle α are to be determined.

Without the use of complex amplitudes the procedure to determine these two unknowns is to write $\cos(\omega t - \alpha) = \cos(\omega t)\cos(\alpha) + \sin(\omega t)\sin(\alpha)$

so that the left hand side of the equation of motion will be of the form $C\cos(\omega t) + S\sin(\omega t)$, where C and S contain the constants M, R, K, and ω. Then, the equation is satisfied if we have $C = F_0$ and $S = 0$. From these two conditions (equations) the two unknown ξ_0 and α can be determined, as was demonstrated explicitly in Section 1.8. Although this procedure is straight forward, it becomes quite cumbersome and awkward, particularly in more complicated problems.

In the **complex amplitude approach**, the two equations $C = F_0$ and $S = 0$ for ξ_0 and α are replaced by **one** equation for the complex displacement amplitude $\boldsymbol{\xi}(\omega)$ which, as we have seen, contains both ξ_0 and α. The reason for this simplification is that by using complex numbers, we operate with **two** real numbers simultaneously in much the same way as in vector analysis, where a vector contains two or more components. A single vector equation then replaces two or more equations for the components.

Thus, in the complex amplitude approach in solving the equation of motion for the amplitude ξ_0 and the phase angle α, we insert $\xi(t) = \mathrm{Re}\{\xi(\omega)\exp(-\mathrm{i}\omega t)\}$, where $\xi(\omega) = \xi_0\exp(\mathrm{i}\alpha)$. Similarly, the driving force is expressed as $F(t) = \mathrm{Re}\{F(\omega)\exp(-\mathrm{i}\omega t)\}$, with a complex amplitude, which in this case is real, $F(\omega) = F_0$, since the phase angle is zero.

In order for the equation of motion to be satisfied at all times, it is necessary that the complex amplitude on both sides of the equation be the same. On the left hand side of the equation, the complex amplitude is the sum of three contributions. In accordance with the discussion above, the complex amplitude for the spring force is simply $K\xi(\omega)$, the friction force $R\mathrm{d}\xi/\mathrm{d}t$ yields $R(-\mathrm{i}\omega)\xi(\omega)$, and $M\mathrm{d}^2\xi/\mathrm{d}t^2$ corresponds to $-\omega^2 M\xi(\omega)$.

The equation for determination of the unknown complex displacement amplitude $\xi(\omega)$ is then

$$(-\omega^2 M - \mathrm{i}\omega R + K)\xi(\omega) = F(\omega) \tag{7}$$

The solution to this equation is the ratio between the two complex numbers $F(\omega)$ and $(K - \omega^2 M - \mathrm{i}\omega R)$. In this case the first is simply F_0, the amplitude of the given driving force. The second contains the properties of the oscillator with a magnitude $\sqrt{(K-\omega^2 M)^2 + (\omega R)^2}$ and a phase angle given by $\tan(\delta) = -\omega R/(K - \omega^2 M)$. The amplitude of the ratio between two complex numbers is the ratio between their amplitudes and the phase angle is the difference between the individual phase angles. Consequently, the amplitude ξ_0 and phase angle α of the displacement are given by

$$\xi(\omega) = \xi_0\exp(\mathrm{i}\alpha) = F(\omega)/(K - \omega^2 M - \mathrm{i}\omega R) \tag{8}$$

$$\xi_0 = F_0/\sqrt{(K-\omega^2 M)^2 + (\omega R)^2} \tag{9}$$

$$\tan(\alpha) = \omega R/(K - \omega^2 M) \quad (\alpha = -\delta) \tag{10}$$

Display 2.3.

The complex amplitude of a harmonic function. Forced motion of a mass-spring oscillator.

Complex amplitudes

$$\xi(t) = \xi_0 \cos(\omega t - \alpha) = \mathrm{Re}\{\xi_0 e^{-i(\omega t - \alpha)}\} = \mathrm{Re}\{\xi(\omega) e^{-i\omega t}\} \tag{1}$$

$$\xi = \xi_0 e^{i\alpha} \tag{2}$$

$$u(\omega) = \dot{\xi}(\omega) = (-i\omega)\xi(\omega) \tag{3}$$

$$a(\omega) = \ddot{\xi}(\omega) = (-i\omega)^2 \xi(\omega) = -\omega^2 \xi(\omega) \tag{4}$$

Forced harmonic motion of oscillator

Differential equation

$$M\ddot{\xi} + R\dot{\xi} + K\xi = F_0 \cos(\omega t - \phi) \tag{5}$$

Complex amplitude equation

$$\text{Let } \xi(\omega) = \xi_0 \exp(i\alpha), \quad F = F_0 \exp(i\phi) \tag{6}$$

$$[-\omega^2 M - i\omega R + K]\xi(\omega) = F(\omega) \tag{7}$$

$$\xi(\omega) = F(\omega)/(K - \omega^2 M - i\omega R) \tag{8}$$

$$\xi_0 = F_0/\sqrt{(K - \omega^2 M)^2 + (\omega R)^2} \tag{9}$$

$$\alpha = \phi - \delta \tag{10}$$

$$\tan(\delta) = -\omega R/(K - \omega^2 M) \quad \tan(\alpha - \phi) = \omega R/(K - \omega^2 M) \tag{11}$$

$$\text{If } \phi = 0, \quad \alpha = \arctan[\omega R/(K - \omega^2 M)] \tag{12}$$

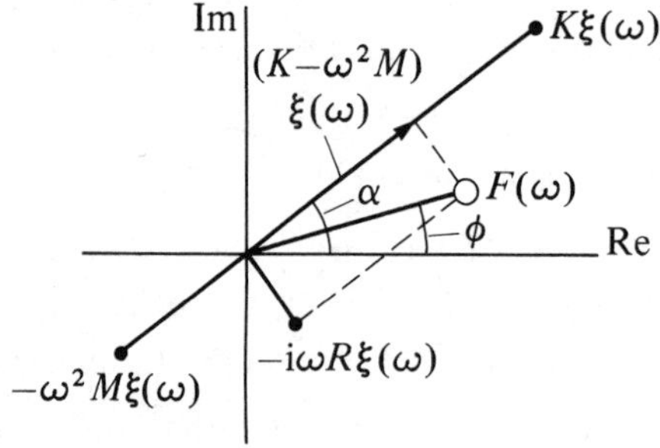

The complex amplitudes for velocity and acceleration are obtained by multiplying $\xi(\omega)$ by $(-\mathrm{i}\omega)$ and $(-\mathrm{i}\omega)^2 = -\omega^2$,

$$u(\omega) = u_0 \exp(\mathrm{i}\beta) = -\mathrm{i}\omega\xi(\omega) = (\omega\xi_0)\exp \mathrm{i}(\alpha - \pi/2) \quad (11)$$

$$(-\mathrm{i} = \mathrm{e}^{-\mathrm{i}\pi/2})$$

$$u_0 = \omega\xi_0$$

$$\beta = \alpha - (\pi/2)$$

$$a(\omega) = a_0 \exp(\mathrm{i}\gamma) = -\omega^2\xi(\omega) = (\omega^2\xi_0)\exp \mathrm{i}(\alpha - \pi) \quad (12)$$

$$(-1 = \mathrm{e}^{-\mathrm{i}\pi})$$

$$a_0 = \omega^2\xi_0$$

$$\gamma = \alpha - \pi$$

Having determined ξ_0 and α we obtain the real displacement function

$$\xi(t) = \xi_0 \cos(\omega t - \alpha) \quad (13)$$

with similar expressions for the velocity and acceleration. These relations as illustrated further in Display 2.3.

2.4 Discussion

As we proceed in our study of oscillations and waves, it will become apparent that the complex amplitude description of harmonic motion simplifies the analysis considerably. It is used almost exclusively in modern treatments of the subject, and familiarity with it is a prerequisite for the use of modern instrumentation in the field of oscillations (and the understanding of the related instruction manuals).

Furthermore, the complex amplitude approach in the study of classical waves and oscillations serves as a valuable preparation for the conceptually more difficult subject of quantum mechanics, in which a complex amplitude is an intrinsic component of the theory.

In regard to notation, we use the argument (t) to indicate the time dependence of a real quantity, for example $\xi(t)$, and the corresponding complex amplitude is denoted by $\xi(\omega)$. Usually, there is little risk for confusion if the arguments (t) and (ω) are left out, and this simplification in notation will be used whenever possible.

Sign convention. It should be pointed out that in defining the complex amplitude we could equally well have expressed $\xi_0 \cos(\omega t - \alpha)$ as the real part of $z = \xi_0 \exp[\mathrm{i}(\omega t - \alpha)]$ rather than $z^* = \xi_0 \exp[-\mathrm{i}(\omega t - \alpha)]$, which was our choice. The use of z leads to the definition $\xi(\omega) = \xi_0 \exp(-\mathrm{i}\alpha)$ for the complex amplitude rather than $\xi(\omega) = \xi_0 \exp(\mathrm{i}\alpha)$, i.e. with $-\mathrm{i}$ rather than $+\mathrm{i}$ in the phase factor.

In electrical engineering, particularly in dealing with electrical circuits, the use of the complex amplitude $\xi(\omega) = \xi_0 \exp(-i\alpha)$ is common practice, and the symbol j is often used instead of i.

In physics and in wave propagation studies, our choice of $\xi(\omega) = \xi_0 \exp(i\alpha)$ is common, and we shall use it throughout in our analysis of oscillations and waves. If it is desired to convert the results thus obtained to the alternate form, we merely have to change $+i$ to $-i$ (or $-j$). We shall comment further on this question of sign convention in connection with our discussion of harmonic waves.

Linearity. The equation of motion for the mass–spring oscillator, which we have considered, is linear; it contains only first order terms of the displacement and its derivatives. As a result, if the driving forces F_1 and F_2 individually produce the displacements $\xi_1(t)$ and $\xi_2(t)$, the force $F_1 + F_2$ will produce the displacement $\xi_1(t) + \xi_2(t)$. This property can be taken as a definition or 'test' of linearity.

An important consequence of linearity is that a harmonic driving force with an angular frequency ω produces a harmonic displacement of the same frequency. With the displacement being harmonic, the same holds true also for the velocity and acceleration, and by proper choice of the displacement amplitude and the phase angle, the oscillator equation can be satisfied, as we have seen.

If, on the other hand, the equation had contained a nonlinear term, say ξ^3, this term would have produced not only a harmonic term with a frequency ω, but also a term with a frequency 3ω, if a solution of the form $\xi = \xi_0 \cos(\omega t - \alpha)$ had been assumed. This means that a driving force (the right hand side of the equation) which contains only the frequency ω will **not** yield a harmonic displacement as a solution if the oscillator is nonlinear.

Examples

E2.1 *Complex amplitude.* What are the complex amplitudes of

(a) $\xi(t) = A\cos(\omega t - \pi/4)$
(b) $A\cos(\omega t + \pi/4)$
(c) $A\sin(\omega t)$
(d) $A\cos(\omega t) + A\sin(\omega t)$

Solution. The complex amplitude ξ of $A\cos(\omega t - \alpha)$ is defined by $\xi = A\exp(i\alpha)$, which follows from $A\cos(\omega t - \alpha) = \mathrm{Re}\{A\exp[-i(\omega t - \alpha)]\} = \mathrm{Re}\{A\exp(i\alpha)\exp(-i\omega t)\} = \mathrm{Re}\{\xi\exp(-i\omega t)\}$.

In (a) we have $\alpha = \pi/4$, and it follows that the complex amplitude becomes

$\xi = A\exp(i\pi/4) = (A/\sqrt{2})(1+i)$. Similarly, in (b) we have $\alpha = -\pi/4$ so that $\xi = A\exp(-i\pi/4) = (A/\sqrt{2})(1-i)$

In (c) we note that $\sin(\omega t) = \cos(\omega t - \pi/2)$, i.e. $\alpha = \pi/2$ so that $\xi = A\exp(i\pi/2) = iA$. It follows then that the complex amplitude in (d) will be $\xi = A(1+i)$

It is important to keep in mind that multiplication of a complex amplitude by i is equivalent to an additional phase lag of $\pi/2$ and hence a change from a cosine- to a sine-function.

E2.2 *Forced harmonic motion.* A mass–spring oscillator is driven by the force $F(t) = F_0\cos(\omega t)$. Determine the amplitude and phase angle of the displacement and the velocity if (a) $\omega = \omega_0 = \sqrt{K/M}$, (b) $\omega \ll \omega_0$, (c) $\omega \gg \omega_\sigma$.

Solution. With the phase angle of the force being zero, the complex force amplitude is simply $F = F_0$. The complex amplitude equation which corresponds to the equation of motion of the mass element, $M\mathrm{d}^2\xi/\mathrm{d}t^2 + R\mathrm{d}\xi/\mathrm{d}t + K\xi = F_0\cos(\omega t)$ is then

$$-\omega^2 M\xi - i\omega R\xi + K\xi = F_0 \tag{1}$$

from which follows

$$\xi = F_0/[K - \omega^2 M - i\omega R] \tag{2}$$

(a) At the resonance frequency, Eq. (2) reduces to

$$\xi = F_0/(-i\omega R) \quad \text{or} \quad u = i\omega\xi = F_0/R \quad \omega = \omega_0 \tag{3}$$

In other words, at resonance, the velocity u is in phase with the driving force and the velocity amplitude is simply $u_0 = F_0/R$. The complex displacement amplitude is $\xi = u/-i\omega = i(u/\omega)$, which means that the displacement lags behind the velocity by a phase angle $\pi/2$. Since the real velocity function is $u = (F_0/R)\cos(\omega t)$, we get $\xi(t) = (F_0/\omega R)\cos(\omega t - \pi/2) = (F_0/\omega R)\sin(\omega t)$ at $\omega = \omega_0$.

(b) When the frequency is much smaller than the resonance frequency, the terms containing ω in the denominator of Eq. (2) can be neglected compared to K, and we obtain

$$\xi \simeq F_0/K \quad \omega \ll \omega_0 \tag{4}$$

which is the same as the 'static' displacement.

(c) When the frequency is much higher than the resonance frequency, the response of the oscillator is dominated by the inertial mass, and we obtain

$$\xi \simeq -F_0/\omega^2 M \quad \omega \gg \omega_0 \tag{5}$$

The corresponding real displacement is $\xi(t) = -(F_0/\omega^2 M)\cos(\omega t)$.

To recapitulate: With the complex amplitudes of displacement and velocity being $\xi = \xi_0\exp(i\alpha)$ and $u = u_0\exp(i\beta)$, the corresponding real functions are $\xi = \xi_0\cos(\omega t - \alpha)$ and $u = u_0\cos(\omega t - \beta)$. At resonance the oscillator is **resistance controlled** and we have $u_0 = F_0/R$, $\beta = 0$ and $\xi = F_0/\omega R$ and $\alpha = \pi/2$, so that $\xi(t) = \xi_0\sin(\omega t)$.

In the stiffness controlled regime, $\omega \ll \omega_0$, we have $\xi_0 = F_0/K$, $\alpha = 0$, and $u_0 = \omega F_0/K$ and $\beta = -\pi/2$ so that $u(t) = -u_0\sin(\omega t)$.

Finally, in the **mass-controlled** regime, $\omega \gg \omega_0$, we have $\xi_0 = F_0/\omega^2 M$, $\alpha = \pi$, and $u_0 = F_0/\omega M$, $\beta = \pi/2$ so that

$\xi(t) = -(F_0/\omega^2 M)\cos(\omega t)$ and $u = (F_0/\omega M)\sin(\omega t)$

In this particular example, the force amplitude F_0 is assumed to be independent of frequency. The displacement and velocity amplitudes, however, are frequency dependent, $\xi_0 = \xi_0(\omega)$ and $u_0 = u_0(\omega)$. Actually, it is this frequency dependence, together with the corresponding frequency dependence of the phase angle, which describe the frequency response of the system, as will be discussed in detail in the next chapter.

Problems

2.1 *Complex numbers.* With $z_1 = 1 - \mathrm{i}\sqrt{3}$ and $z_2 = \sqrt{3} + \mathrm{i}$, calculate (a) $z_1 + z_2$. (b) $z_1 - z_2$. (c) $z_1 z_2$. (d) z_1/z_2. (e) $z_1 z_2^*$. In each case, indicate the locations of these quantities in the complex plane.

2.2 *Amplitude and phase angle.* Express the following complex numbers in the form $A\exp(\mathrm{i}\alpha)$, and determine A and α.

(a) $3 + \mathrm{i}4$. (b) $(3 + \mathrm{i}4)/(4 + \mathrm{i}3)$. (c) $(1 + \mathrm{i})$. (d) $\sqrt{1 + \mathrm{i}}$. (e) i.

2.3 *Euler's formula.* (a) From Eq. 2.2(3), express $\cos(\alpha)$ in terms of exponentials and show that $2\cos(\alpha)\cos(\beta) = \cos(\alpha + \beta) + \cos(\alpha - \beta)$. (b) Show that $1 + \exp(\mathrm{i}\alpha) = 2\mathrm{e}^{\mathrm{i}\alpha/2}\cos(\alpha/2)$.

2.4 *Integral.* Determine the amplitude and phase angle of $\int_0^b \cos(\beta)\exp(\mathrm{i}\beta)\,\mathrm{d}\beta$.

2.5 *Complex amplitude.* A harmonic displacement is of the form $\xi = A\cos[\omega(t - t_1)]$, where $A = 10$ cm, $t_1 = T/10$, and the period $T = 0.1$ sec. What are the complex amplitudes of displacement, velocity, and acceleration?

2.6 *Complex conjugate.* Consider the complex numbers $z_1 = 1 + \mathrm{i}3$ and $z_2 = 4\exp(\mathrm{i}\pi/6)$. Determine

(a) z_1^2. (b) $z_1 z_1^*$. (c) $z_2 z_2^*$. (d) $z_2 + z_2^*$.

2.7 *Complex roots.* Determine the roots of the equations

(a) $z^2 - 4z + 8 = 0$,
(b) $z^4 = 1 + \mathrm{i}$,

and indicate the location of the roots in the complex plane.

2.8 *Hyperbolic functions.* Prove that

(a) $\sin(\mathrm{i}x) = \mathrm{i}\sinh(x)$
(b) $\cos(\mathrm{i}x) = \cosh(x)$

3

Forced oscillations and frequency response

In the last chapter, we introduced the complex amplitude and illustrated its use in the analysis of forced harmonic motion. In this chapter we shall discuss in more quantitative detail the characteristics of the forced harmonic motion of an oscillator. In the process we discuss the power transfer to the oscillator and introduce the concepts of impedance and admittance.

It should be realized from the very beginning, that the analysis can be applied not only to mechanical but also to electrical circuits, as will be demonstrated in the next section on electro-mechanical analogies.

3.1 Electro-mechanical analogies

The mechanical oscillator under consideration is a mass–spring oscillator (mass M and spring constant K). A friction force Ru on M, proportional to the velocity u, is provided by a 'dashpot', which is in parallel with the spring, as indicated in Display 3.1. The dashpot is simply a piston, which moves inside a 'leaky' cylinder, and the force required to move the piston is proportional to the velocity of the piston relative to the cylinder. In this case, the cylinder is attached to the same rigid support as the spring. (As an alternative to the dashpot, the contact between M and the surrounding medium can be considered responsible for the friction force.)

The driving force $F(t)$ is applied to M, and with $\xi(t)$ being the displacement of M from the equilibrium position and $u = \dot{\xi}$ the velocity, the equation of motion is

$$M\,\mathrm{d}u/\mathrm{d}t + Ru + K\xi = F \tag{1}$$

or

$$M\ddot{\xi} + R\dot{\xi} + K\xi = F \tag{2}$$

The analogous electrical system consists of an inductance L, a resistance

Display 3.1.

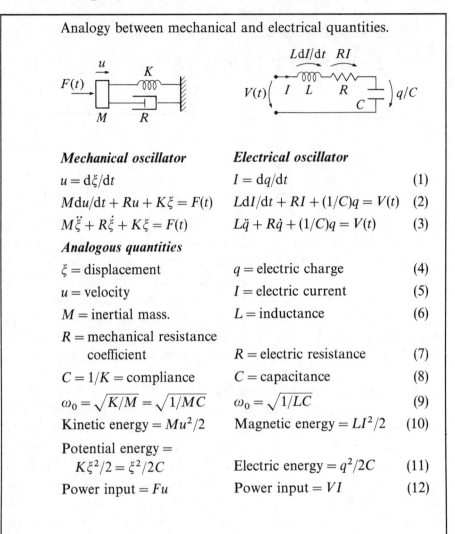

Analogy between mechanical and electrical quantities.

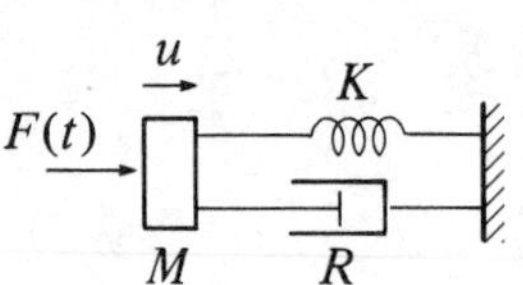

Mechanical oscillator	*Electrical oscillator*	
$u = \mathrm{d}\xi/\mathrm{d}t$	$I = \mathrm{d}q/\mathrm{d}t$	(1)
$M\mathrm{d}u/\mathrm{d}t + Ru + K\xi = F(t)$	$L\mathrm{d}I/\mathrm{d}t + RI + (1/C)q = V(t)$	(2)
$M\ddot{\xi} + R\dot{\xi} + K\xi = F(t)$	$L\ddot{q} + R\dot{q} + (1/C)q = V(t)$	(3)
Analogous quantities		
ξ = displacement	q = electric charge	(4)
u = velocity	I = electric current	(5)
M = inertial mass.	L = inductance	(6)
R = mechanical resistance coefficient	R = electric resistance	(7)
$C = 1/K$ = compliance	C = capacitance	(8)
$\omega_0 = \sqrt{K/M} = \sqrt{1/MC}$	$\omega_0 = \sqrt{1/LC}$	(9)
Kinetic energy $= Mu^2/2$	Magnetic energy $= LI^2/2$	(10)
Potential energy $= K\xi^2/2 = \xi^2/2C$	Electric energy $= q^2/2C$	(11)
Power input $= Fu$	Power input $= VI$	(12)

R, and a capacitance C in series. A voltage $V(t)$ is applied across the circuit terminals. With the current through the circuit denoted by $I(t)$ and the charge on the capacitor by $q(t)$, we have $I = \mathrm{d}q/\mathrm{d}t$, and the voltages across the inductance, resistance, and the capacitance are $L\mathrm{d}I/\mathrm{d}t$, RI, and q/C, respectively. The sum of these must equal the applied voltage $V(t)$,

$$L\mathrm{d}I/\mathrm{d}t + RI + q/C = V \tag{3}$$

or

$$L\ddot{q} + R\dot{q} + (1/C)q = V \tag{4}$$

The differential equations for the mechanical and electrical 'circuits' have the same form, and they indicate the following correspondence between mechanical and electrical quantities: Displacement ↔ electrical charge, velocity ↔ current, mass ↔ inductance, mechanical resistance ↔ electrical resistance, compliance ↔ capacitance, and force ↔ voltage.

These electro-mechanical analogies, summarized in the display, apply also in more complex systems. Because of the simplicity and small size of an electrical circuit and the ease with which circuit parameters can be varied, an electrical analogue of a mechanical system is often used in the study of the dynamical characteristics of a mechanical system.

3.2 Frequency response

With reference to Display 3.2, we shall now study in more quantitative detail than in the last chapter the 'response' of an oscillator. Specifically, we consider now the electrical circuit, which is analogous to the mass–spring oscillator. The circuit is driven by a harmonic voltage $V(t) = V_0 \cos(\omega t - \gamma)$ with the corresponding complex amplitude $V(\omega) = V_0 \exp(\mathrm{i}\gamma)$. For simplicity, we shall choose the phase angle γ to be zero, so that the complex amplitude is simply $V(\omega) = V_0$. In general, both V_0 and γ are frequency dependent.

The driving voltage is assumed to have constant amplitude but variable frequency, and by the 'frequency response' of the circuit we mean the frequency dependence of the amplitude and phase angle of the current through the circuit. We could determine also the frequency dependence of the charge on the capacitor, and this would lead to a different frequency response. Ordinarily, we are interested only in the current.

For a mechanical oscillator, the velocity corresponds to the current, as we have seen, and the frequency response for the current in the electrical circuit applies directly to the velocity in the mass–spring oscillator. The displacement of the oscillator, which corresponds to the charge on the capacitor, is frequently of interest, and we shall consider it separately.

Display 3.2.

Frequency dependence of current amplitude and phase angle in forced harmonic motion of electric resonator circuit.

Differential equation

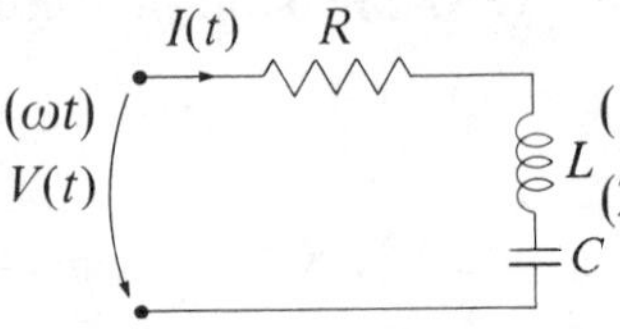

$$L\mathrm{d}I/\mathrm{d}t + RI + q/C = V(t) = V_0 \cos(\omega t) \tag{1}$$

$$I = \mathrm{d}q/\mathrm{d}t \tag{2}$$

I(t)
R
V(t)
L
C

Complex amplitude equations

$$I(\omega) = -\,\mathrm{i}\omega q(\omega) \quad q(\omega) = \mathrm{i}I(\omega)/\omega \tag{3}$$

$$(R - \mathrm{i}\omega L + \mathrm{i}/\omega C)I = V \tag{4}$$

$$I = V/Z \tag{5}$$

where

$$Z = R - \mathrm{i}\omega L + \mathrm{i}/\omega C = R + (\mathrm{i}/\omega C)(1 - \omega^2 LC) \tag{6}$$

Let $\omega_0^2 = 1/LC$ and $\Omega = \omega/\omega_0$ (7)

Impedance becomes

$$Z = R[1 + (\mathrm{i}/D\Omega)(1 - \Omega^2)] \tag{8}$$

$$I = I_0 \exp(\mathrm{i}\beta) = I_\mathrm{m}(-\,\mathrm{i}D\Omega)/(1 - \Omega^2 - \mathrm{i}D\Omega) \tag{9}$$

where

$I_\mathrm{m} = V_0/R$ and $D = R/\omega_0 M$ (10)

Frequency response function

$$I_0/I_\mathrm{m} = D\Omega/\sqrt{[(1 - \Omega^2)^2 + (D\Omega)^2} \tag{11}$$

$$\beta = \arctan[(\Omega^2 - 1)/D\Omega] \tag{12}$$

$$\Omega \ll 1, \quad I_0/I_\mathrm{m} \simeq D\Omega \tag{13}$$

$$\Omega \gg 1, \quad I_0/I_\mathrm{m} \simeq D/\Omega, \tag{14}$$

$$\begin{aligned} \Omega \simeq 1, \quad I_0/I_\mathrm{m} &\simeq D/\sqrt{4(1 - \Omega)^2 + D^2} \\ &= (D/2)/\sqrt{(1 - \Omega)^2 + (D/2)^2} \\ &[(1 - \Omega^2) = (1 + \Omega)(1 - \Omega) \simeq 2(1 - \Omega)] \end{aligned} \tag{15}$$

Width of resonance curve

From Eq. (11) or (15):

$$I_0/I_\mathrm{m} = \sqrt{\tfrac{1}{2}}$$

when

$$\Omega_\pm \simeq 1 \pm (D/2), \quad \Delta\Omega = \Omega_+ - \Omega_- = \Delta\omega/\omega_0 \simeq D = 1/Q = R/\omega_0 L \quad (D \ll 1)$$

Current or velocity. As shown in the last section, the differential equation for the electrical oscillator is $L\mathrm{d}I/\mathrm{d}t + RI + q/C = V_0 \cos(\omega t)$, where q is the charge on the capacitor. The current is the time rate of change of the charge, $I = \mathrm{d}q/\mathrm{d}t$, and the corresponding relation between the complex amplitudes is $I(\omega) = -\mathrm{i}\omega q(\omega)$ or $q(\omega) = \mathrm{i}I(\omega)/\omega$. Making use of $q = \mathrm{i}I/\omega$, we obtain the complex amplitude equation

$$-\mathrm{i}\omega LI + RI + \mathrm{i}(1/\omega \mathrm{C})I = V \tag{1}$$

where $V = V_0$.

This equation states that the product of the complex current amplitude and a certain complex number $Z = R - \mathrm{i}(\omega L - 1/\omega C)$ equals the complex amplitude of the driving voltage. Thus, we can express the current as

$$I(\omega) = V(\omega)/Z(\omega) \tag{2}$$

$$Z = R - \mathrm{i}[\omega L - (1/\omega C)] \tag{3}$$

The quantity $Z(\omega)$ is called the **impedance** of the electrical circuit. The inverse, $Y(\omega) = 1/Z(\omega)$, is called the **admittance**, in terms of which

$$I = YV \tag{4}$$

We are interested in the amplitude and phase angle of the current, $I = I_0 \exp(\mathrm{i}\beta)$. With the impedance expressed as $Z = Z_0 \exp(\mathrm{i}\phi)$, where $Z_0 = \sqrt{R^2 + (\omega L - 1/\omega C)^2}$ and $\phi = \arctan[-(\omega L - 1/\omega C)/R]$, we obtain

$$I_0 = V_0/Z_0 = V_0/\sqrt{R^2 + (\omega L - 1/\omega C)^2} \tag{5}$$

$$\beta = -\phi = \arctan[(\omega L - 1/\omega C)/R] \tag{6}$$

In the limit of zero frequency (corresponding to a DC voltage on the circuit), the capacitor prevents current from flowing and $I_0 = 0$. In the other limit of infinite frequency, the current is again zero; now it is the inductance that blocks the current. The **maximum** current amplitude, $I_\mathrm{m} = V_0/R$, is obtained at the angular frequency

$$\omega_0 = \sqrt{1/LC} \tag{7}$$

and the corresponding frequency $\nu_0 = \omega_0/2\pi$ is called the **resonance frequency** of the oscillator. It is the same as the frequency of free undamped oscillations.

The current lags behind the voltage by the phase angle β. At resonance, this angle is zero, which means that the current and the driving voltage are 'in phase'. In the limits of zero and infinite frequency we have $\beta = -\pi/2$ and $\beta = \pi/2$, respectively; the velocity runs ahead of the voltage for $\omega < \omega_0$ and lags behind the voltage for $\omega > \omega_0$.

In expressing the complete frequency dependence of the current amplitude and phase angle, we shall use the normalized frequency $\Omega = \omega/\omega_0$ and the normalized resistance defined by $D = R\omega_0 C = R/\omega_0 L$. It can be interpreted as the ratio between the amplitudes of the voltage over the resistance and the voltage over the inductance (or capacitor) at resonance. (At resonance the voltage over the inductance is the same as the voltage over the capacitor except for the sign). The inverse of the normalized resistance coefficient, $Q = 1/D = \omega_0 L/R$ is usually called the **Q-value** or **quality factor** of the circuit.

In terms of these normalized quantities, the impedance can be expressed as $Z = R(\mathrm{i}/D\Omega)[(1-\Omega^2) - \mathrm{i}D\Omega]$, and with $I_m = V/R$, the current amplitude at $\omega = \omega_0$, we obtain

$$I = I_0 e^{\mathrm{i}\beta} = I_m[-\mathrm{i}D\Omega/(1-\Omega^2-\mathrm{i}D\Omega)] \tag{8}$$

$$I_0/I_m = D\Omega/\sqrt{(1-\Omega^2)^2 + (D\Omega)^2} \tag{9}$$

$$\beta = \arctan[(\Omega^2 - 1)/D\Omega] \tag{10}$$

The frequency response thus obtained can be applied to a mass–spring oscillator driven by the force $F_0 \cos(\omega t)$ if we replace V by F, I by u, L by M, and C by $1/K$. The resonance frequency is now given by $\omega_0 = \sqrt{K/M}$, and the normalized resistance is $D = R/\omega_0 M$, sometimes called the friction factor.

In Display 3.3 we have plotted the amplitude ratio u_0/u_m (or I_0/I_m) versus the normalized frequency Ω for some different values of the friction factor D from 0.125 to 4 (with corresponding Q-values from 8 to 0.25).

At frequencies much lower than the resonance frequency, $\Omega \ll 1$, the amplitude is proportional to frequency. The spring stiffness then provides the dominant reaction force. In the other limit, $\Omega \gg 1$, the inertial mass in the oscillator controls the response, and the amplitude is inversely proportional to frequency.

At resonance the reaction forces of the mass and the spring cancel each other, and only the resistance R limits the velocity (to the value $u_m = F_0/R$). In the vicinity of resonance, $\Omega \simeq 1$, we have $u_0/u_m \simeq D/\sqrt{(1-\Omega^2)^2 + D^2}$. The amplitude ratio is then reduced from the maximum of 1 to the value $1/\sqrt{2}$ (the energy is reduced by a factor of 2) when $(1-\Omega^2) = \pm D$. For $D \ll 1$, the corresponding values of Ω are $\Omega \simeq 1 \pm (D/2)$. Under these conditions, the width of the resonance curve at the 'half power points' is given by $\Delta\Omega \simeq D$, or $\Delta\omega/\omega_0 \simeq D = 1/Q$.

In the case of an electrical circuit, with fixed values of L and R, the resonance frequency can be varied with a variable capacitance. Then, for a given driving frequency ω, the maximum current is obtained if the circuit is 'tuned' to the driving frequency, i.e. $\omega_0 = \omega$.

If the driving voltage consists of two harmonic components of different frequencies, the corresponding current maxima can be determined by

Display 3.3.

Frequency dependence of the amplitude ratio (u_0/u_m for mass–spring oscillator or I_0/I_m for analogous electrical oscillator) and phase angle, as given by Eqs. D.3.2.(11)–(12), in forced harmonic motion.

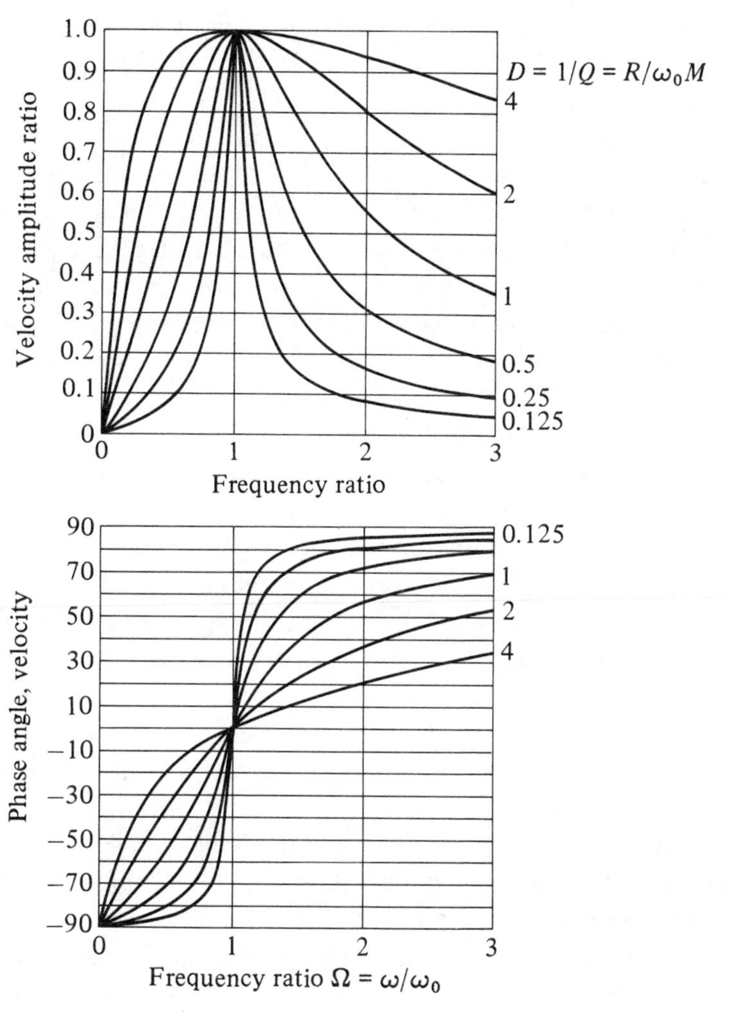

varying the resonance frequency, and the maxima can be 'resolved' if the width $\Delta\Omega$ of the response curve is small compared to the separation of the two frequencies. The smaller the resistance factor D, i.e. the higher the Q-value or 'quality' of the circuit, the better is the resolution or the selectivity.

The phase angle β varies from $-\pi/2$ to $+\pi/2$ as the frequency goes from zero to infinity, and is zero at resonance, $\Omega = 1$. In the limit of zero resistance, the transition of the phase angle from $-\pi/2$ to $+\pi/2$ is sudden. As the resistance increases, the transition becomes gradual, and in the limit of infinite resistance, the phase angle will be zero at all frequencies different from zero and infinity.

Returning to the frequency dependence of the amplitude ratio (Eq. D.3.2.(11)), we note that it is symmetrical with respect to $\Omega = 1$ in the sense that, if Ω is replaced by $1/\Omega$, the amplitude ratio remains the same. This is not immediately apparent from the curves in Display 3.3, however. On the other hand, in a log–log plot, as in Display 3.4, the symmetry is made quite clear. At frequencies much lower than the resonance frequency, the slope of the curves is $+1$, and at frequencies much higher than the resonance frequency, the slope is -1, corresponding to $u_0/u_m \simeq D\Omega$ and $u_0/u_m \simeq D/\Omega$, respectively (see Eqs. D3.2.13–14).

Rather than to plot the frequency dependence of amplitude ratio and the phase angle separately, we can show the frequency dependence of the **complex amplitude ratio** u/u_m in the complex plane. This complex number describes a path which starts at the origin at $\omega = 0$ in the direction of the negative imaginary axis (since the phase angle initially is $-\pi/2$) and continues in the fourth quadrant toward the real axis (Display 3.5).

At the resonance frequency, where the phase angle is zero, the path crosses the real axis at $u/u_m = 1$. As the frequency increases further, the path moves into the first quadrant and goes back to the origin as the frequency goes to infinity.

As shown in the display, the path turns out to be a circle with unit diameter with the center on the real axis, and it has the remarkable property that it is **independent of the resistance factor *D***. The frequency scale along the path does depend on D, however; i.e. the location of u/u_m at a given frequency depends on D.

Displacement (charge). In an electrical oscillator the current generally is the quantity of interest; the charge on the capacitor is rarely used in discussions of the frequency response. In a mechanical oscillator, however, the displacement is of considerable importance, and the frequency functions of the amplitude and the phase angle are of interest.

Display 3.4.

Frequency dependence of the amplitude ratio (u_0/u_m for mass–spring oscillator or I_0/I_m for analogous electrical oscillator), as given by Eq. D.3.2.(11), in forced harmonic motion.

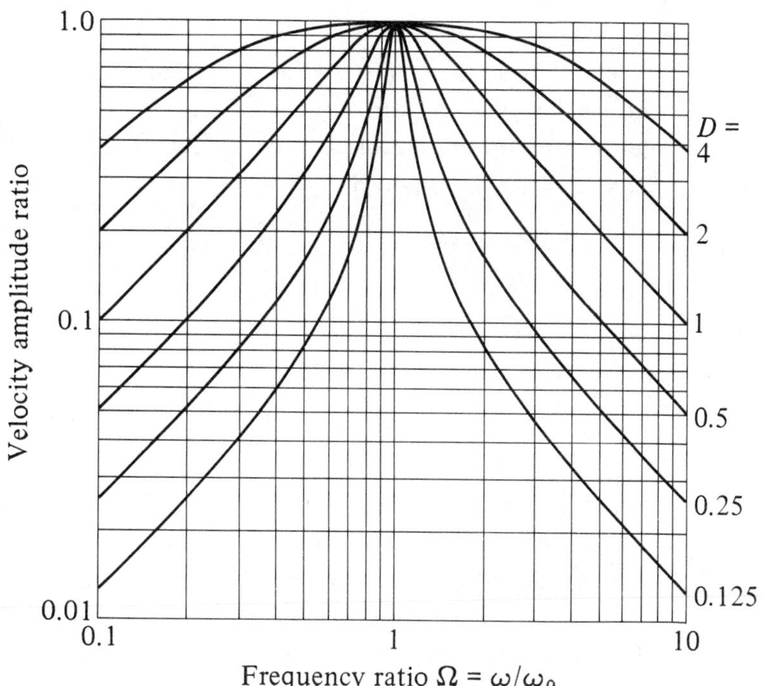

$D = 1/Q = R/\omega_0 M$

Display 3.5.

The frequency dependence of the complex amplitude ratio (velocity or current) in forced harmonic excitation of an oscillator is described by a circle in the complex plane.

Example

Prove that as the frequency of forced harmonic excitation of an oscillator goes from zero to infinity, the complex velocity (current) amplitude moves along a circle in the complex plane.

$$u(\omega)/u_{\mathrm{m}} = (u_0/u_{\mathrm{m}}) \exp(\mathrm{i}\beta) \tag{1}$$

From Eqs. D.3.2.(11)–(12)

$$u_0/u_{\mathrm{m}} = D\Omega/\sqrt{(1-\Omega^2)^2 + (D\Omega)^2} \tag{2}$$

$$\tan(\beta) = (\Omega^2 - 1)/D\Omega \tag{3}$$

Eq. (2) can be rewritten as

$$u_0/u_{\mathrm{m}} = 1/\sqrt{1 + (1-\Omega^2)^2/(D\Omega)^2} = 1/\sqrt{1+\tan^2(\beta)} = \cos(\beta) \tag{4}$$

The radius vector from the origin to a point of a circle with a diameter 1 and with the center at 0.5 on the real axis has the magnitude $\cos(\beta)$, a characteristic of the circle.

Complex velocity amplitude ratio u/u_{m}

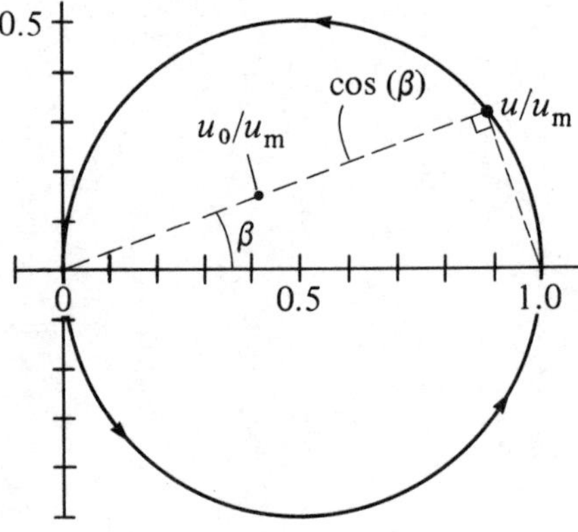

The complex displacement amplitude can be obtained simply by dividing the complex velocity amplitude in Eq. D.3.2(9) by $(-\mathrm{i}\omega)$ (with I replaced by u). On the other hand, it is equally simple to start from the equation of motion $M\ddot{\xi} + R\dot{\xi} + K\xi = F_0 \cos(\omega t)$ with the corresponding complex amplitude equation $(-\omega^2 M - \mathrm{i}\omega R + K)\xi = F$ from which follows

$$\xi = \xi_0 \exp(\mathrm{i}\alpha) = F/[K - \omega^2 M - \mathrm{i}\omega R] = \xi'/[1 - \Omega^2 - \mathrm{i}D\Omega] \quad (11)$$

$$\xi'(\omega) = F/K = F_0/K, \quad \Omega = \omega/\omega_0 \quad \omega_0 = \sqrt{K/M} \quad (12)$$

We have here introduced the quantity $\xi' = F_0/K$ for normalization, which is the displacement amplitude in the limit of zero frequency. The corresponding amplitude ξ_0 and phase angle α can then be obtained from the relations

$$|\xi_0/\xi'| = 1/\sqrt{(1 - \Omega^2)^2 + (D\Omega)^2} \quad (13)$$

$$\tan(\alpha) = D\Omega/(1 - \Omega^2) \quad (14)$$

The displacement response is summarized in Display 3.6, where we have shown also the path of the complex amplitude ratio ξ/ξ' in the complex plane as the frequency goes from zero to infinity. The paths are no longer circles, as for the complex velocity ratio, and there will be one path for each value of the resistance factor D. The curves shown correspond to the values $D = 0.125, 0.25, 0.50, 1, 2$, and 4. On each curve is indicated also the location of the complex amplitude at various values of the normalized frequency from 0.6 to 1.4 at intervals of 0.1. The displacement amplitude is the length of the radius vector from the origin to each point on the path. We note, particularly for large values of D, that maximum amplitude is obtained for values of Ω less than 1.

At frequencies much below the resonance frequency, i.e. for $\Omega \ll 1$, the amplitude is approximately equal to the static value, and the phase angle is close to zero. Thus, the curves for the complex displacement amplitude ratio start from 1 on the real axis in the direction along the real axis. As the frequency increases the phase angle increases to the value $\pi/2$ at resonance, where the curves cross the imaginary axis. As the frequency is increased further, the curves move into the second quadrant and go toward the origin (zero amplitude) and a phase angle of π.

Conventionally, the frequency functions for the amplitude ξ_0 and the phase angle α are presented in separate graphs as in Display 3.7.

The amplitude is $\xi' = F_0/K$ at $\Omega = 0$ and reaches the value $Q = \omega_0 M/R = 1/D$ times ξ' at resonance, $\Omega = 1$. For values of $D \ll 1$, this displacement differs but little from the maximum displacement amplitude, which occurs at a value of Ω less than 1, but as D increases the difference increases. Actually, for a value of the friction factor D equal to or larger than

Display 3.6.

Complex displacement amplitude in forced harmonic motion of a mass–spring oscillator.

Complex force amplitude

$F = F_0 \exp(\mathrm{i}\phi) = F_0$ (ϕ chosen to be zero) (1)

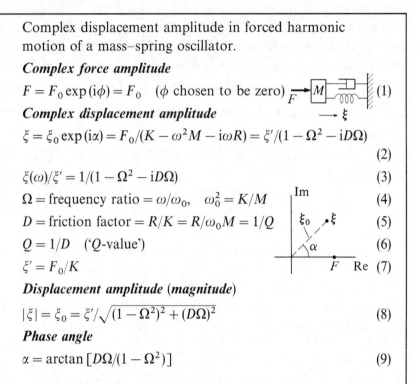

Complex displacement amplitude

$\xi = \xi_0 \exp(\mathrm{i}\alpha) = F_0/(K - \omega^2 M - \mathrm{i}\omega R) = \xi'/(1 - \Omega^2 - \mathrm{i}D\Omega)$ (2)

$\xi(\omega)/\xi' = 1/(1 - \Omega^2 - \mathrm{i}D\Omega)$ (3)

$\Omega =$ frequency ratio $= \omega/\omega_0, \quad \omega_0^2 = K/M$ (4)

$D =$ friction factor $= R/K = R/\omega_0 M = 1/Q$ (5)

$Q = 1/D$ ('Q-value') (6)

$\xi' = F_0/K$ (7)

Displacement amplitude (magnitude)

$|\xi| = \xi_0 = \xi'/\sqrt{(1 - \Omega^2)^2 + (D\Omega)^2}$ (8)

Phase angle

$\alpha = \arctan[D\Omega/(1 - \Omega^2)]$ (9)

Complex displacement amplitude ratio ξ/ξ'

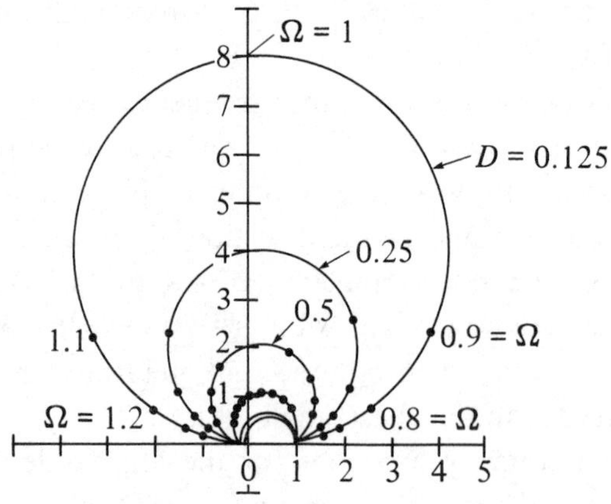

Display 3.7.

Displacement amplitude ratio $\xi_0/\xi' = 1\sqrt{(1-\Omega^2)^2 + (D\Omega)^2}$, (Eq. D.3.6.(8)) Phase angle $\alpha = \arctan[D\Omega/(1-\Omega^2)]$, (Eq. D.3.6.(9))

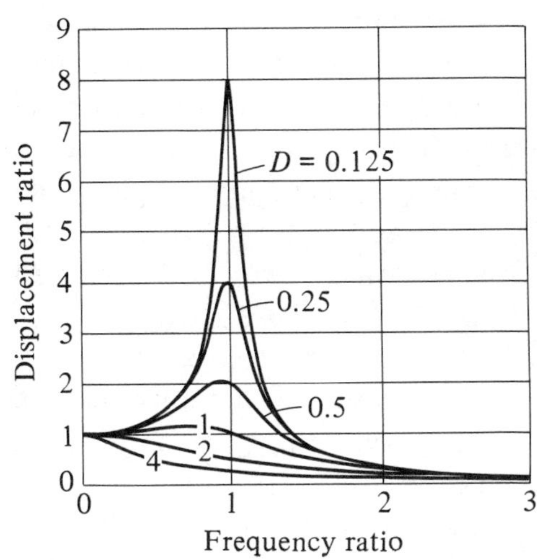

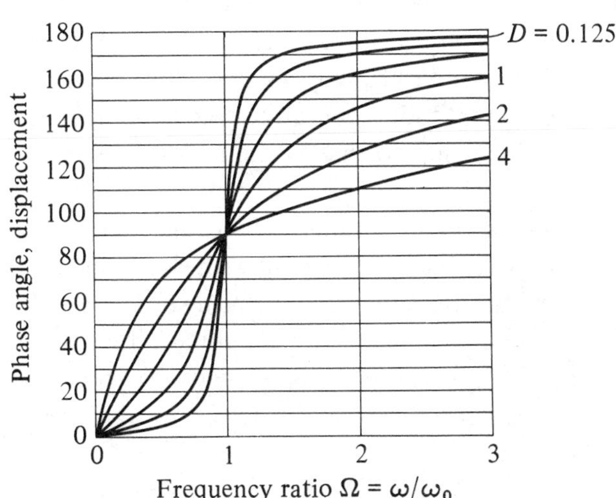

$\sqrt{2}$, the maximum occurs at $\Omega = 0$; we leave the proof as a problem for the reader.

3.3 Impedance and admittance

The concepts of impedance and admittance were discussed briefly in the previous section. They are commonly used in electric circuit analysis. The impedance is defined as the ratio between the complex amplitude of the voltage applied to the terminals of a circuit and the complex amplitude of the current entering the circuit. By analogy, the mechanical impedance is the ratio between the complex amplitude of a driving force and the complex amplitude of the velocity at the point of application of the force.

The impedance $Z = R + \mathrm{i}X$ is a complex number; the real part R, is called the resistance and the imaginary part, X, the reactance. For the mass–spring oscillator, we have $Z = R - \mathrm{i}\omega M + \mathrm{i}K/\omega$.

It should be noted that with the sign convention we have adopted (corresponding to the time factor $\exp(-\mathrm{i}\omega t)$ rather than $\exp(\mathrm{i}\omega t)$), the reactance of a mass (inductance) will be negative and the reactance of a spring (capacitance) will be positive.

The admittance is the inverse of the impedance, $Y = 1/Z$. It is more convenient to use than the impedance in the description of circuits consisting of a number of elements coupled in parallel, as will be discussed shortly.

The ratio between the complex force (voltage) amplitude and the complex displacement (charge) amplitude generally is not given a special name, although it plays a role similar to that of the impedance.

Circuit elements in series and in parallel. The concepts of impedance and admittance are useful particularly in dealing with electrical and mechanical circuits more complex than the simple mass–spring or L–C oscillator. When a large number of circuit elements are combined, the analysis can be facilitated by means of simple rules for the calculation of the impedance and the admittance of the entire system in terms of the impedance and admittances of the elements.

When the elements are in series, the current through each element is the same, and the sum of the voltages across the elements equals the applied voltage. Accordingly, the total impedance is the sum of the individual impedances.

With the elements in parallel, the voltage across each element is the same, and the total current through the circuit is the sum of the individual currents. It follows that the total admittance of the circuit is the sum of the individual admittances. These observations are summarized in Display 3.8.

Display 3.8.

Impedance and admittance. Circuit elements in series and in parallel.

Impedance and admittance

$$Z = R + iX \tag{1}$$

$$Y = G + iH = 1/Z = (R - iX)/(R^2 + X^2) \tag{2}$$

$$G = R/(R^2 + X^2), \quad H = -X/(R^2 + X^2) \tag{3}$$

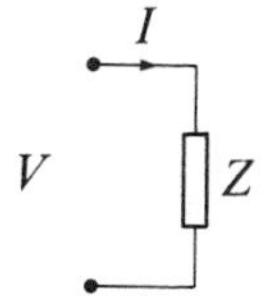

Circuit elements in series

$$V = V_1 + V_2 = Z_1 I + Z_2 I \tag{4}$$

$$Z = V/I = Z_1 + Z_2 \tag{5}$$

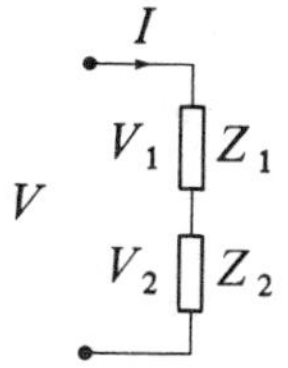

Circuit elements in parallel

$$I = I_1 + I_2 = Y_1 V + Y_2 V \tag{6}$$

$$Y = I/V = Y_1 + Y_2 \tag{7}$$

$$1/Z = 1/Z_1 + 1/Z_2 \tag{8}$$

$$Z = Z_1 Z_2/(Z_1 + Z_2) \tag{9}$$

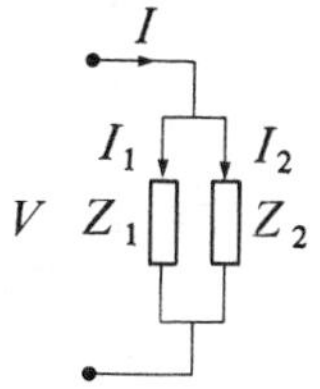

Example

The circuit shown is driven by the voltage $V = V_0 \cos(\omega t)$. Find the amplitude and phase angle of the current I.

$$Y_1 = 1/R \quad Y_2 = -i\omega C \tag{10}$$

$$I = V(Y_1 + Y_2) = V[(1/R) - i(\omega C)] = I_0 \exp(i\beta) \tag{11}$$

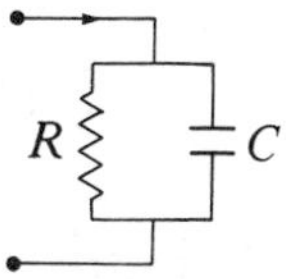

$$I_0 = V_0\sqrt{(1/R)^2 + (\omega C)^2} \tag{12}$$

$$\tan(\beta) = -\omega CR \tag{13}$$

Display 3.9.

Examples of electro-mechanical analogues.

Electrical circuit ***Mechanical analogue***

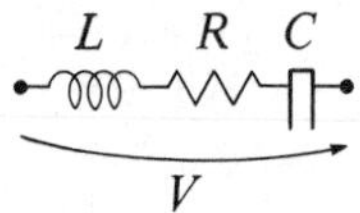

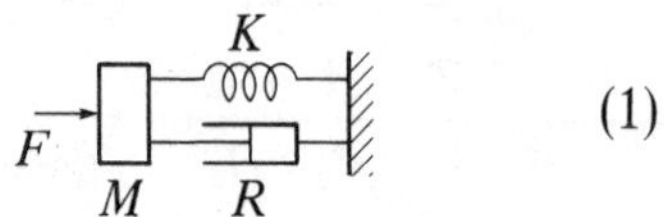

(1)

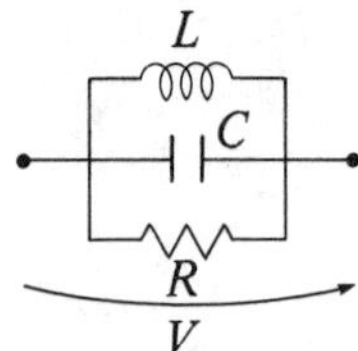

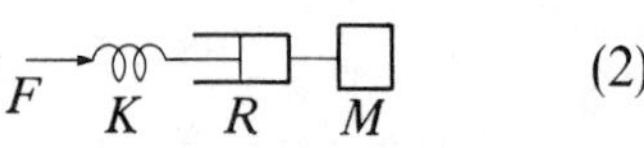

(2)

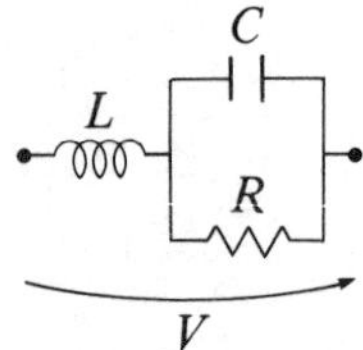

F M R K (3)

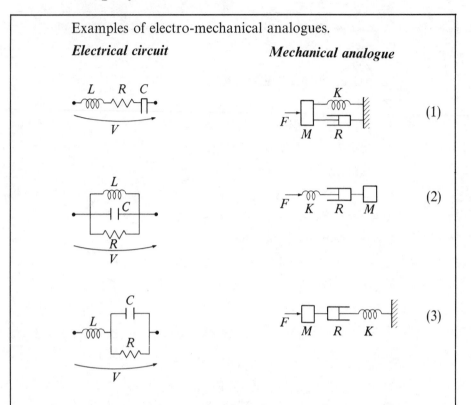

Example

Derive an expression for the complex velocity amplitude of M in the mechanical system in the third example above.

Force on dashpot cylinder (and on spring) F_1 (4)

Velocity of spring (and dashpot piston) $u_1 = (-\text{i}\omega)F_1/K$ (5)

Velocity of M (and of dashpot cylinder) u (6)

$$F_1 = R(u - u_1) \quad (7)$$

$$(-\text{i}\omega)\,Mu = F - F_1 \quad (8)$$

From Eqs. (5), (7), and (8)

$$u = F/[-\text{i}\omega M + R/(1 + \text{i}\omega R/K)]$$
$$= F/[Z + Z_1 Z_2/(Z_1 + Z_2)] \quad (9)$$

$$Z = -\text{i}\omega M, \quad Z_1 = R, \quad Z_2 = \text{i}K/\omega \quad (10)$$

Electro-mechanical analogies. In Section 3.1. we established the equivalence between an electrical and a mechanical oscillator, corresponding to a resistance, inductance, and capacitance in series. The current through each of these elements is the same.

In the analogous mechanical oscillator, the mass is attached to a spring and a dashpot, as shown in Display 3.9, and, being connected, these elements have the same velocity. The total force on the mass, required to drive the system, is the sum of the force associated with the acceleration of the mass, the spring force, and the friction force from the dashpot.

In the electrical circuit with a resistance, inductance, and capacitance in parallel, the voltage across each element is the same, and in the corresponding mechanical analogous system, the force acting on each mechanical element must be the same. This holds true for the mechanical circuit shown in the second example in the display. Here a spring, a dashpot, and a mass are arranged in 'series', with the driving force applied to the end of the spring. Since the mass of the spring is neglected, the force is transmitted through the spring to the dashpot piston, which in turns transmits the force to the dashpot cylinder, which is connected to the mass. Thus, the force on each element is the same, as it should be. Furthermore, the velocity of the driven end of the spring is the sum of the velocity of the mass (which is the same as the velocity of the cylinder of the dashpot), the velocity of the piston in the dashpot, and the velocity associated with the compression of the spring. In other words, the requirements for equivalence between the mechanical and electrical circuits are fulfilled.

In the third example, we drive the mechanical system in example 2 from the opposite side with the end of the spring 'anchored'. The force is applied to the mass, and the force transmitted to the cylinder in the dashpot is reduced by an amount equal to the force associated with the acceleration of the mass. The force on the dashpot, however, is transmitted without reduction to the spring.

The corresponding electrical circuit is an inductance in series with a parallel combination of a capacitance and a resistance, as shown.

3.4 Power transfer

A harmonic function, strictly speaking, has no beginning and no end. In dealing with physical phenomena, however, we generally are concerned with finite time intervals, and the mathematical description has to be chosen accordingly. For example, in an experiment with forced motion, the force is applied at a certain time, say $t=0$, and may be described mathematically as $F = F_0 \cos(\omega t)$ for $t > 0$ and $F(t) = 0$ for $t < 0$, for example.

As a result, the motion starts at $t = 0$, but the oscillator does not settle down in a harmonic motion with frequency ω until after a certain time of adjustment. After this 'transient' interval, the motion, for all practical purposes, is the same as that which would have been obtained if the force had been acting at all times. This motion is called the **steady state motion** or response of the oscillator, and it is this motion we have studied so far. In the next chapter, we shall analyze the transition to this steady state.

The energy of the oscillator, the sum of the kinetic and potential energy, is built up during the initial transition interval, increasing from zero to the steady state value. When steady state has been reached, the energy of the oscillator is independent of time. To maintain the steady state motion, the power dissipated in the resistance of the oscillator has to be delivered by the driving force; if the force is turned off, the motion will decay.

To express the steady state power delivered by the force $F(t) = F_0 \cos(\omega t - \phi)$, we express the velocity at the point of application of the force as $u(t) = u_0 \cos(\omega t - \beta)$. The instantaneous power transfer is then

$$\begin{aligned} F(t)u(t) &= F_0 u_0 \cos(\omega t - \phi)\cos(\omega t - \beta) \\ &= (F_0 u_0/2)[\cos(\phi - \beta) + \cos(2\omega t - \phi - \beta)] \end{aligned} \tag{1}$$

The power varies with time, but usually we are interested only in the time average over one period. The contribution from the fluctuating component is zero, and we obtain for the time average power

$$P = \langle F(t)u(t) \rangle = \tfrac{1}{2} F_0 u_0 \cos(\phi - \beta) \tag{2}$$

which is proportional to the cosine of the phase difference between the force and the velocity.

Since the analysis of harmonic oscillations generally is carried out in terms of complex amplitudes, it is useful to express the time average power directly in terms of these quantities. The complex amplitudes of the force and the velocity are $F(\omega) = F_0 \exp(\mathrm{i}\phi)$ and $u(\omega) = u_0 \exp(\mathrm{i}\beta)$ with the corresponding complex conjugates $F^* = F_0 \exp(-\mathrm{i}\phi)$ and $u^* = u_0 \exp(-\mathrm{i}\beta)$. We have $Fu^* = F_0 u_0 \exp[\mathrm{i}(\phi - \beta)]$ and

$$P = \tfrac{1}{2}\mathrm{Re}\{Fu^*\} = \tfrac{1}{2}\mathrm{Re}\{F^*u\} = \tfrac{1}{2} F_0 u_0 \cos(\phi - \beta) \tag{3}$$

If the impedance or the admittance of the system is known, it is sometimes useful to express the power in terms of the square velocity or force amplitude. Thus, with the impedance being $Z = R + \mathrm{i}X$, we get $F = Zu$ and

$$P = \tfrac{1}{2}\mathrm{Re}\{Fu^*\} = \tfrac{1}{2}\mathrm{Re}\{Zuu^*\} = \tfrac{1}{2}u_0^2\,\mathrm{Re}\{Z\} = \tfrac{1}{2}Ru_0^2 \tag{4}$$

Similarly, if we express the admittance as $Y = G + \mathrm{i}H$, so that $u = YF$ it

Display 3.10.

Power transfer to a circuit.

Voltage, current

$$V = V_0 \cos(\omega t) \quad \text{(phase angle chosen to be zero)} \tag{1}$$

$$I = I_0 \cos(\omega t - \beta) \tag{2}$$

$$V(\omega) = V_0, \quad I(\omega) = I_0 \exp(\mathrm{i}\beta) \tag{3}$$

Power

$$V(t)I(t) = V_0 I_0 \cos(\omega t)\cos(\omega t - \beta)$$
$$= \tfrac{1}{2} V_0 I_0 [\cos(\beta) + \cos(2\omega t - \beta)] \tag{4}$$

Time average:

$$P = \langle V(t)I(t) \rangle = \tfrac{1}{2} V_0 I_0 \cos(\beta) = \tfrac{1}{2}\mathrm{Re}\{VI^*\} = \tfrac{1}{2}\mathrm{Re}\{V^*I\} \tag{5}$$

Example 1

With reference to Eq. 4, calculate the time average power in the interval from $t = 0$ to $t = \tau$.

$$(1/\tau)\int_0^\tau V(t)I(t)\,\mathrm{d}t = \tfrac{1}{2} V_0 I_0 \cos(\beta)$$
$$+ (\tfrac{1}{2} V_0)(1/\tau)\int_0^\tau \cos(2\omega t - \beta)\,\mathrm{d}t$$
$$= P + \tfrac{1}{2} V_0 I_0 [\sin(2\omega\tau - \beta) + \sin(\beta)]/2\omega\tau \tag{6}$$

where $P = \tfrac{1}{2} V_0 I_0 \cos(\beta)$.

We see that this average goes to P when $\omega\tau$ goes to infinity and is equal to P when τ is an integer number of periods.

Example 2

An electric circuit consists of an inductance L, a capacitance C, and a resistance R in series. It is driven by a voltage $V = V_0 \cos(\omega t)$. What is the frequency dependence of the power delivered to the circuit.

$$Z = R - \mathrm{i}\omega L + \mathrm{i}/\omega C = R + \mathrm{i}X \tag{7}$$

$$Y = G + \mathrm{i}H = 1/Z = R/(R^2 + X^2) - \mathrm{i}X/(R^2 + X^2). \tag{8}$$

From Eq. (5) we obtain

$$P = \tfrac{1}{2} V_0^2 G = \tfrac{1}{2} V_0^2 R/(R^2 + X^2) = \tfrac{1}{2} V_0^2 R/[R^2 + (\omega L - 1/\omega C)^2]$$
$$= \tfrac{1}{2}(V_0^2/R)/[1 + Q^2(\Omega - 1/\Omega)^2] \tag{9}$$

where $Q = \omega_0 M/R$ and $\Omega = \omega/\omega_0 \quad \omega_0 = 1/\sqrt{LC}$ (10)

follows that

$$P = \tfrac{1}{2}\mathrm{Re}\{YFF^*\} = \tfrac{1}{2}F_0^2 G \tag{5}$$

The expression $P = \frac{1}{2}Ru_0^2$ is of particular interest, since it shows that in the steady state motion the power delivered by the force is dissipated by the friction force.

It should be realized, however, that during the transient interval, this energy balance is not fulfilled; more power is transferred than dissipated, and the excess goes into the building up of kinetic and potential energy of the oscillator, until steady state is reached. This buildup will be discussed in the next chapter, where we consider the motion produced by an arbitrary driving force.

3.5 Sources of vibration

In our analysis of **forced harmonic motion** of a mass–spring oscillator, it was implied that the mass element was driven with a harmonic force with constant amplitude, independent of the frequency. This can be approximately achieved by means of an electromagnetic vibrator, for which the frequency can be varied without a significant change in amplitude over a wide range of frequencies. The displacement amplitude produced in this manner usually is quite small, however, and a vibration transducer is required so that the displacement can be made visible on an oscilloscope.

A variable speed motor with an unbalanced rotor is a commonly used force generator; it does not have any limitation in regard to the resulting amplitude of oscillation. The force amplitude is not constant but is proportional to the square of the frequency, however. If the width of the frequency response curve is small, it may still be acceptable for demonstrating the response curve in the vicinity of resonance.

If such a motor is hung from the end of a spring, the pendulum mode of motion can be excited also, and this may interfere with the demonstration of the axial response of the spring. This problem exists also if the motor is placed on a platform supported by springs; a lateral mode of oscillation of the platform can be excited.

Instead of using a driving force applied to the mass element, we can drive the end of the spring with a harmonic **displacement**, as in a hand-held vertical mass–spring oscillator. An air-suspended cart on a horizontal track can be driven in similar manner, with the cart placed between two springs. The end of one spring is anchored and the other is driven with a harmonic displacement by means of a string, which runs over a pulley and is connected eccentrically to a disc, driven by a variable speed motor.

The intrinsic damping in this and other similar mechanical oscillators often is quite small, and additional damping might be desirable. A simple method of providing additional damping is the use of the air drag on cardboard mounted on the mass element. The air drag can be varied (and made quite large) by air flow from a fan. For an analysis of this damping mechanism, we refer to one of the problems.

The mass-spring oscillator, driven by a force applied to the mass element, is analogous to an L–C electric series circuit driven by a voltage between the circuit terminals. The current through the circuit can be monitored by the voltage across a (small) resistor in the circuit. This voltage corresponds to the velocity in the mechanical oscillator. Similarly the voltage across the capacitor corresponds to the displacement of the mass element.

When a mass–spring oscillator is driven by a harmonic velocity of given amplitude at the end of the spring (second example in Display 3.9), the electric analogue is a parallel L–C circuit driven by a harmonic current of constant amplitude. Again, the voltage over a (small) resistance in series with the inductance can be used to represent the velocity of the mass element, and the voltage across the capacitor corresponds to the compression of the spring.

3.6 Example

An acoustic resonator, consisting simply of bottle with a pronounced neck, is often called a Helmholtz resonator. At sufficiently low frequencies, this bottle acts like a mass–spring oscillator; the air plug in the neck corresponds to the mass, and the air in the rest of the bottle represents the spring. The tone produced by blowing over the opening of the bottle is produced by the resonant oscillations of this 'mass–spring' oscillator.

Oscillations can be produced also by using a closely fitting (metal) ball in the neck of the bottle. After having been displaced from its equilibrium position, the ball will oscillate in harmonic motion in the neck as a result of the restoring force produced by the air cushion in the bottle.

The compressibility of the gas in the bottle is known to be $\kappa = -(1/V)\,\mathrm{d}V/\mathrm{d}P = 1/\rho v^2$ (see Chapters 6 and 12), where $\mathrm{d}P$ is the pressure change caused by the volume change $\mathrm{d}V$, ρ is the mass density of the gas and v the sound speed.

If the air plug in the neck is displaced inwards a distance ξ from equilibrium, the corresponding change in volume of the gas in the bottle will be $\mathrm{d}V = -A\xi$, where A is the area of the neck. The change in pressure is then $\mathrm{d}P = (A\xi/V)\rho v^2$ and the restoring force on the mass plug will be $A\mathrm{d}P = (A^2\xi/V)\rho v^2 = K\xi$, where K is the equivalent spring constant. The mass of the plug is $M = A\rho L$, where L is the length of the bottle neck, and if follows that the resonance frequency of the acoustic (**Helmholtz**) resonator is given by

$$\omega_0^2 = K/M = (A^2\rho v^2/V)/(A\rho L) = v^2 A/VL \quad (1)$$

$$\nu_0 = \omega_0/2\pi = (v/2\pi)\sqrt{A/VL} \quad (2)$$

If the resonator is exposed to a harmonic sound pressure, the mass plug in the neck will be forced to oscillate in harmonic motion and the air in the bottle will be subject to harmonic compressions and rarefactions. A microphone inside the bottle will record the corresponding pressure fluctuations, and the frequency dependence of this pressure represents the frequency dependence of the displacement in the analogous mass–spring oscillator or the voltage over the capacitor in the analogous electrical circuit.

Damping of the acoustic resonator can easily be provided and varied by means of a porous plug placed in the bottle neck. A demonstration of the acoustic resonator is particularly illuminating, when the resonator is exposed to a pulse modulated sound pressure, as will be discussed in the next chapter.

3.7 Fourier analysis

The special importance of harmonic motion and the frequency response of a system is to a great extent due to the fact that an arbitrary motion can be expressed as a superposition of harmonic motions and that the response to an arbitrary driving force can be derived from the frequency response. We consider first a periodic function and discuss the decomposition of it into an infinite series of harmonic functions and then proceed to the general case of an arbitrary time dependence.

Fourier series. Consider a sum of harmonic functions

$$F(t) = a_1 \cos(\omega_1 t - \alpha_1) + a_2 \cos(\omega_2 t - \alpha_2) + \cdots \quad (1)$$

in which the fundamental angular frequency is ω_1, and the other frequencies are multiples thereof, $\omega_n = n\omega_1 = n(2\pi/T_1)$, where T_1 is the fundamental period. An increase of t by T_1 returns each term in the series to its value at t and the same applies to the sum so that

$$F(t + T_1) = F(t) \quad (2)$$

which expresses the fact that the function is periodic.

By varying the amplitudes a_n and the phase angles α_n, we can produce an infinite number of different functions of time with a given period.

It seems plausible that an arbitrary periodic function $F(t)$ can be expressed as an infinite series of harmonic functions

$$F(t) = \sum_0^\infty a_n \cos(\omega_n t - \alpha_n), \tag{3}$$

$$\omega_n = n\omega_1 = n2\pi/T_1$$

and, indeed, this Fourier series expansion is an important theorem of mathematics which includes the method of determining a_n and α_n, as described below. These quantities are contained in the complex amplitude $A_n = a_n \exp(i\alpha_n)$, and it is useful to express the Fourier series in terms of these amplitudes.

Thus, we have

$$a_n \cos(\omega_n t - \alpha_n) = \mathrm{Re}\{A_n e^{-\omega_n t}\} = \mathrm{Re}\{z_n\} = \tfrac{1}{2}(z_n + z_n^*) \tag{4}$$
$$z_n = A_n \exp(-i\omega_n t), \quad A_n = a_n \exp(i\alpha_n)$$

where we have introduced the temporary notation z_n (with its complex conjugate $z_n^* = A_n^* \exp(i\omega_n t)$), to simplify writing somewhat. Furthermore, if we let A_{-n} be defined to mean A_n^*, the sum in Eq. (3) can be rewritten to include negative values of n,

$$F(t) = \frac{1}{2}\sum_0^\infty (z_n + z_n^*) = \frac{1}{2}\sum_0^\infty z_n + \frac{1}{2}\sum_{-\infty}^0 z_n$$

$$= \sum_{-\infty}^\infty f_n \exp(-i\omega_n t)$$

$$f_n = A_n/2 \quad |n| \neq 0, \quad f_0 = A_0 \tag{5}$$

$$A_{-n} = A_n^* \text{ (by definition)}, \quad \omega_n = n\omega_1,$$

In other words, the function $F(t)$ can be expanded in a ('double-sided') series of harmonic functions $\exp(-in\omega_1 t)$ with n extending from minus to plus infinity. The expansion coefficients f_n, apart from a factor $\frac{1}{2}$, are the complex amplitudes of the harmonic functions in the original expansion in Eq. (3).

To obtain the coefficient f_m, we multiply both sides of Eq. (5) by $\exp(i\omega_m t)$ and integrate both sides over the fundamental period T_1, using the range from $-\frac{1}{2}T_1$ to $\frac{1}{2}T_1$. The only surviving term in the series on the right hand side is obtained for $n = m$. It is simply $f_m T_1$, and we obtain

$$f_m = (1/T_1)\int_{-\frac{1}{2}T_1}^{\frac{1}{2}T_1} F(t)\exp(i\omega_m t)\,dt \tag{6}$$

To check the result let us choose $F(t)$ simply as $F(t) = A\cos(\omega_1 t) = \frac{1}{2}A[\exp(i\omega_1 t) + \exp(-i\omega_1 t)]$. Then, the only terms which contribute to the integral correspond to $m=1$ and $m=-1$, so that $f_1=f_{-1}=A/2$, and insertion into Eq. (5) yields $F(t)=A\cos(\omega_1 t)$, as it should.

Example. As another example we consider a periodic square wave function, as shown in Display 3.11. The height of each pulse is H and the width τ, and we place the origin of t in the middle of one of the pulses, as shown. The function is then symmetrical with respect to $t=0$, i.e. $F(t)=F(-t)$, and in calculating the complex amplitude coefficients from Eq. (6), we note that only the $\cos(\omega_m t)$ term of the exponential $\exp(\mathrm{i}\omega_m t)$ will contribute to the integral. In the range of integration $-\frac{1}{2}T_1$ to $\frac{1}{2}T_1$ the function is different from zero only between $-\tau/2$ and $\tau/2$. Therefore, the integral for f_m becomes $f_m=(1/T_1)\int F(t)\cos(\omega_m t)\,\mathrm{d}t=(2H/\omega_m T_1)\sin(\omega_m\tau/2)$ or

$$f_m=H(\tau/T_1)[\sin(X_m)/X_m] \qquad (7)$$
$$X_m=\omega_m\tau/2=m\pi\tau/T_1 .$$

We could have obtained this result also by direct integration of the complex exponential

$$f_m=(H/T_1)\int\exp(\mathrm{i}\omega_m t)\,\mathrm{d}t=(H/\mathrm{i}\omega_m T_1)[\exp(\mathrm{i}\omega_m\tau/2)$$
$$-\exp(-\mathrm{i}\omega_m\tau/2)]=(H/\mathrm{i}\omega_m T_1)2\mathrm{i}\sin(\omega_m\tau/2)$$

which is the same as before.

We note that f_m is real, which means that the phase angles α_m in Eq. (3) are all zero so that $F(t)=f_0+\sum_1^\infty 2f_m\cos(\omega_m t)$, (see Eqs. 3 and 5).

The fact that all the phase angles are zero is a consequence of the choice of the origin of the coordinate system so that $F(t)$, like $\cos(\omega_m t)$, is an 'even' function of t. For any other choice of the origin of the coordinate system, f_m will be complex, and the magnitude a_m and phase angle α_m have to be determined accordingly.

The term corresponding to $m=0$ gives the time average value of the function. We note that $\sin(X_0)/X_0=1$, and it follows from Eq. (7) that in this example the average value is $f_0=H(\tau/T_1)$, as it should.

The amplitude of the fundamental frequency component is $2H(\tau/T_1)\sin(X_1)/X_1$, which in the special case $\tau=T_1/2$ becomes $f_1=2H/\pi$.

Fourier Transform. With reference to the example above, it seems plausible that a single 'pulse' in the periodic function in Display 3.11 may be expressible as a Fourier series by letting T_1 go to infinity. Then, by linear superposition, the same would be true for a more general function.

To explore this idea, we let T_1 go to infinity in Eq. (5), which, with

Display 3.11.

Fourier series.

$$F(t)=\sum_0^\infty a_n\cos(\omega_n t-\alpha_n)\quad \omega_n=n\omega_1=n2\pi\nu_1=n2\pi/T_1 \tag{1}$$

Complex amplitudes:

$$A_n=a_n\exp(\mathrm{i}\alpha_n)=2f_n \quad \text{for } n\neq 0 \text{ and } A_0=f_0 \text{ yields} \tag{2}$$

$$F(t)=\sum_{-\infty}^{\infty} f_n\exp(-\mathrm{i}\omega_n t) \tag{3}$$

$$f_n=(1/T_1)\int_{-\frac{1}{2}T_1}^{\frac{1}{2}T_1} F(t)\exp(\mathrm{i}\omega_n t)\,\mathrm{d}t \tag{4}$$

Example

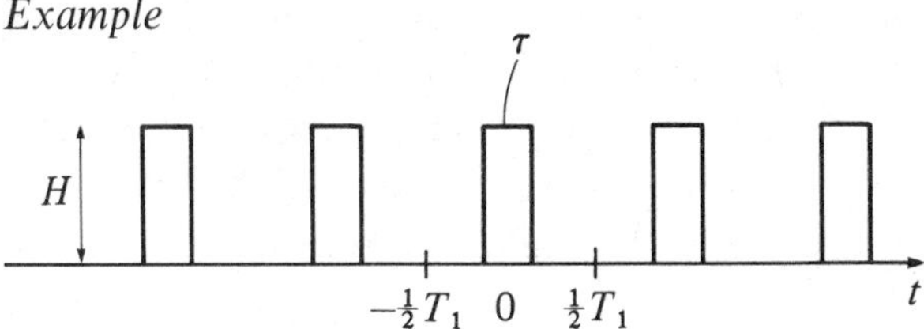

$$f_n=(H\tau/T_1)\sin(X_n)/X_n,\quad X_n=\omega_n\tau/2=n\pi\nu_1\tau \tag{5}$$

$$a_n=2(H\tau/T_1)\sin(X_n)/X_n \tag{6}$$

$$\alpha_n=0 \tag{7}$$

$$F(t)=\sum_0^\infty a_n\cos(\omega_n t) \tag{8}$$

Example

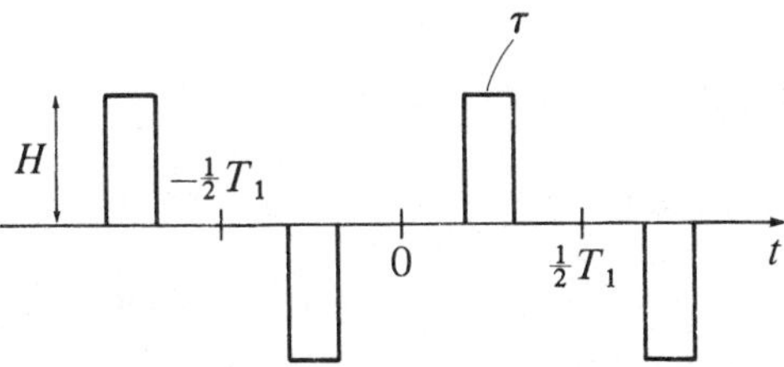

$$f_n=\mathrm{i}2(H\tau/T_1)\sin(n\pi/2)\sin(X_n)/X_n=\frac{a_n}{2}\mathrm{e}^{\mathrm{i}\alpha_n} \tag{9}$$

$$a_n=4(H\tau/T_1)\sin(n\pi/2)\sin(X_n)/X_n \tag{10}$$

$$\alpha_n=\pi/2 \tag{11}$$

$$F(t)=\sum_0^\infty a_n\sin(\omega_n t) \tag{12}$$

$\omega_n = n2\pi\nu_1$, we express as

$$F(t) = \sum_{-\infty}^{\infty} f_n \exp(-i2\pi n\nu_1 t) \tag{8}$$

$$f_n = (1/T_1)\int_{-\frac{1}{2}T_1}^{\frac{1}{2}T_1} F(t)\exp(i2\pi n\nu_1 t)\,dt \tag{9}$$

$$\nu_1 = 1/T_1$$

The separation $\nu_1 = 1/T_1$ of the frequencies of two adjacent terms can be made as small as we wish by making T_1 large, and it is meaningful to introduce the average number of terms $\Delta\nu_s/\nu_1 = T_1\Delta\nu_s$, contained in a (small) frequency interval $\Delta\nu_s$. The sum over n in Eq. (8) can then be replaced by a sum over the frequency intervals $\Delta\nu_s$.

With T_1 going to infinity, the complex amplitude f_n in Eq. (9) goes to zero in such a way that f_nT_1 is finite, which is demonstrated explicitly in the example in Eq. (7), where $f_nT_1 = H\tau$ in the limit $T_1 \to \infty$. We shall denote f_nT_1 by F_n; actually, with the average frequency in the interval $\Delta\nu_s$ denoted by ν_s, we replace F_n by the average value $F(\nu_s)$ in this interval. The sum over n in Eq. (8) can then be replaced by a sum over the frequency intervals $\Delta\nu_s$

$$F(t) = \sum_n f_n \exp(-i2\pi n\nu_1 t) = \sum_s F(\nu_s)\exp(-i2\pi\nu_s t)\Delta\nu_s \tag{10}$$

where we have used $T_1\Delta\nu_s$ as the number of terms in the frequency interval $\Delta\nu_s$ and expressed $F_n = f_nT_1$ as $F(\nu_s)$, as mentioned above.

In the limit $\Delta\nu_s \to 0$, this sum (and Eq. (8)) can be expressed as the integral

$$F(t) = \int_{-\infty}^{\infty} F(\nu)\exp(-i2\pi\nu t)\,d\nu \tag{11}$$

and Eq. (6) becomes

$$F(\nu) = \int_{-\infty}^{\infty} F(t)\exp(i2\pi\nu t)\,dt \tag{12}$$

These equations are often called the Fourier Transform pair.

Uncertainty relation. As an example, we consider a single period of the square wave function considered in the previous subsection. As before, the height of the 'pulse' is H and the width τ with the center at $t = 0$. According to Eq. 12, the complex Fourier amplitude becomes

$$F(\nu) = H\int_{-\frac{1}{2}\tau}^{\frac{1}{2}\tau} \exp(i2\pi\nu t)\,dt = (H\tau)\sin(X)/X \quad X = \pi\tau\nu \tag{13}$$

The quantity $H\tau$ is the area under the pulse. If this area is kept constant and equal to 1 as the width τ of the pulse decreases toward zero (and correspondingly, H goes to infinity), the factor $\sin(X)/X$ goes to 1 so that $F(\nu)=1$. The corresponding time function $F(t)$ is called the **delta function**

$$\delta(t)=\int_{-\infty}^{\infty}\exp(-\mathrm{i}2\pi\nu t)\,\mathrm{d}\nu \tag{14}$$

$$F(\nu)=1=\int_{-\infty}^{\infty}\delta(t)\exp(\mathrm{i}2\pi\nu t)\,\mathrm{d}t$$

The delta function is of considerable importance in physics and applied mathematics, and some illustrations of its use will be given below.

It is important to notice, that the infinitely narrow 'spike', which represents the delta function, has a complex Fourier amplitude which has the same value, $F(\nu)=1$, at all frequencies. In other words, to build a delta function from harmonic components, all frequencies from zero to infinity have to be included with equal 'weight'.

For a duration τ of the pulse different from zero, the Fourier amplitude $F(\nu)$ decreases in an oscillatory manner toward zero as the frequency goes to infinity. Actually, we can define a characteristic frequency range $\Delta\nu$ of $F(\nu)$ as the frequency, at which $F(\nu)$ first becomes zero. According to Eq. (13), this occurs when $X=\pi\tau\Delta\nu=\pi$, or $\tau\Delta\nu=1$. If we denote the duration τ of the pulse by Δt, we obtain the relation

$$\Delta t\Delta\nu=1 \tag{15}$$

which is sometimes called the **uncertainty relation**. It states, that the duration Δt of a pulse is inversely proportional to the width of $\Delta\nu$ of the corresponding Fourier amplitude function in the frequency domain. This has important practical consequences in physics and engineering, in measurements, communication, signal processing, etc.

If the pulse in Display 3.12 is centered at $t=t'$ rather than at $t=0$, the corresponding delta function will be $\delta(t-t')$. Since this function is zero for every value of t different from $t=t'$, it follows that

$$\int_{-\infty}^{\infty}f(t)\delta(t-t')\,\mathrm{d}t=f(t')\int\delta(t-t')\,\mathrm{d}t=f(t') \tag{16}$$

a result which will be applied shortly.

Energy and power spectrum. A finite 'pulse' of an electric current, described as a function of time by $I(t)$, goes through a resistance R. The time dependence of the power dissipated in the resistance is then $RI^2(t)$, and the total energy absorption is $R\int I^2(t)\,\mathrm{d}t$, where the integration extends from $-\infty$ to $+\infty$.

Through the Fourier Transform, Eqs. (11)–(12), we can express $I(t)$ in

Display 3.12.

Fourier Transform.

$$F(t) = \int_{-\infty}^{\infty} F(\nu) \exp(-i2\pi\nu t)\, d\nu \qquad (1)$$

$$F(\nu) = \int_{-\infty}^{\infty} F(t) \exp(i2\pi\nu t)\, dt \qquad (2)$$

Example

$F(t)$: See Fig.

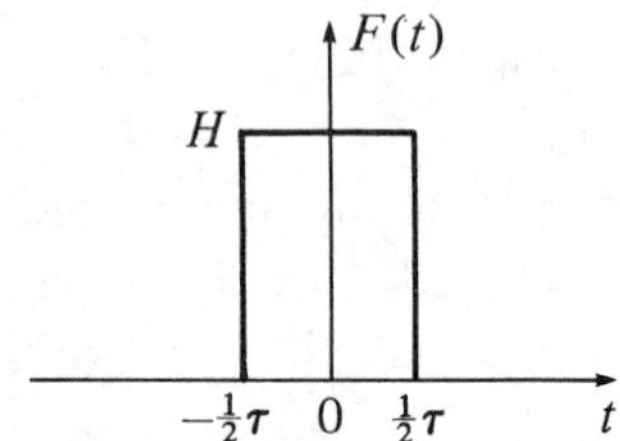

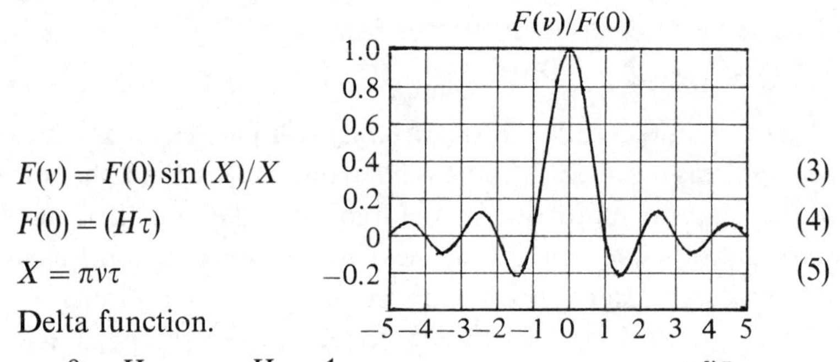

$$F(\nu) = F(0) \sin(X)/X \qquad (3)$$

$$F(0) = (H\tau) \qquad (4)$$

$$X = \pi\nu\tau \qquad (5)$$

Delta function.

$\tau \to 0, \quad H \to \infty, \quad H\tau \to 1$

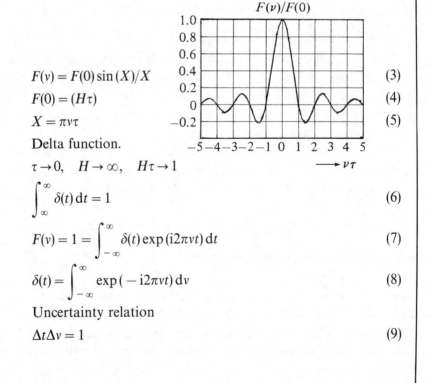

$$\int_{\infty}^{\infty} \delta(t)\, dt = 1 \qquad (6)$$

$$F(\nu) = 1 = \int_{-\infty}^{\infty} \delta(t) \exp(i2\pi\nu t)\, dt \qquad (7)$$

$$\delta(t) = \int_{-\infty}^{\infty} \exp(-i2\pi\nu t)\, d\nu \qquad (8)$$

Uncertainty relation

$$\Delta t \Delta\nu = 1 \qquad (9)$$

terms of its complex Fourier amplitude $I(\nu)$

$$I(t)=\int_{-\infty}^{\infty} I(\nu)\exp(-i2\pi\nu t)\,d\nu \tag{17}$$

It is of interest to determine also how the total energy is distributed in frequency, and to do that we introduce the energy spectrum density $E(\nu)$ defined by

$$\int_{-\infty}^{\infty} I^2(t)\,dt=\int_0^{\infty} E(\nu)\,d\nu \tag{18}$$

To calculate $E(\nu)$, we start with the Fourier Transform of the current in Eq. (1), from which follows

$$\int_{-\infty}^{\infty} I^2(t)\,dt = \int_{-\infty}^{\infty}\left(\int_{-\infty}^{\infty} I(\nu)e^{-i2\pi\nu t}\,d\nu\right)\left(\int_{-\infty}^{\infty} I(\nu')e^{-i2\pi\nu' t}\,d\nu'\right)dt \tag{19}$$

With reference to the definition of the delta function in Eq. (14), we note that the integration over t in Eq. (19) can be expressed as

$$\int_{-\infty}^{\infty} \exp[-i2\pi(\nu+\nu')t]\,dt=\delta(\nu+\nu') \tag{20}$$

Application of Eq. (16) to the integration over ν' reduces the integral in Eq. (19) to $\int I(\nu)I(-\nu)\,d\nu$. The Fourier amplitude $I(\nu)$ has the property that $I(-\nu)=I^*(\nu)$, in complete analogy with the result for the Fourier series. Thus, with $I(\nu)I^*(\nu)=|I(\nu)|^2$, we obtain

$$\int_{-\infty}^{\infty} I^2(t)\,dt=\int_{-\infty}^{\infty} |I(\nu)|^2\,d\nu=2\int_0^{\infty} |I(\nu)|^2\,d\nu$$
$$=\int_0^{\infty} E(\nu)\,d\nu \tag{21}$$
$$E(\nu)=2|I(\nu)|^2$$

where $E(\nu)$ is the **energy spectrum density.**

If the current $I(t)$ represents a sample of a random function of length T, we obtain for the mean square value of the current

$$(1/T)\int_0^T I^2(t)\,dt=\langle I^2(t)\rangle=\int_0^{\infty} W(\nu)\,d\nu \tag{22}$$
$$W(\nu)=(2/T)|I(\nu)|^2$$

where $W(\nu)$ is called the **power spectrum density** of the current.

Example. As an example we consider a decaying oscillatory current

$$I(t)=I(0)\exp(-\gamma t)\cos(\omega_1 t) \quad t>0, \quad I(t)=0 \quad t<0 \tag{23}$$
$$\omega_1=\sqrt{\omega_0^2-\gamma^2}$$

Display 3.13.

Energy spectrum of decaying oscillation.

$$\int_{-\infty}^{\infty} I^2(t)\,\mathrm{d}t = \int_0^{\infty} E(\nu)\,\mathrm{d}\nu \tag{1}$$

Current

$$I(t) = I(0)\mathrm{e}^{-\gamma t}\cos(t\sqrt{\omega^2-\gamma^2}) \quad \text{for } t>0 \quad I(t)=0 \quad t<0 \tag{2}$$

$$\text{Let } \Gamma = \gamma/\omega_0 \tag{3}$$

$$I(t)/I(0) = \mathrm{e}^{-2\pi\Gamma t}\cos(2\pi t\sqrt{1-\Gamma^2}/T_0) \tag{4}$$

Energy spectrum density

$$E(\nu) = 2I^2(0)(\omega^2+\gamma^2)/[(\omega_0^2-\omega^2)^2+4\omega^2\gamma^2] \tag{5}$$

$$\text{Let } \Omega = \omega/\omega_0 \quad \Gamma = \gamma/\omega_0 \tag{6}$$

$$E(\Omega)/E(1) = [4\Gamma^2/(1+\Gamma^2)](\Omega^2+\Gamma^2)/[(1-\Omega^2)^2+4\Omega^2\Gamma^2] \tag{7}$$

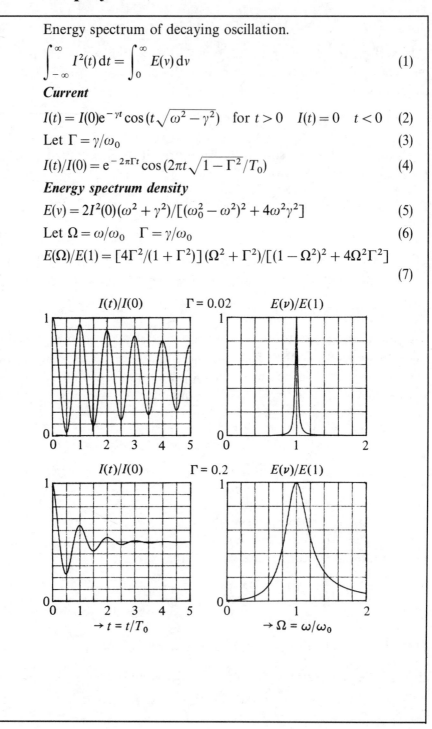

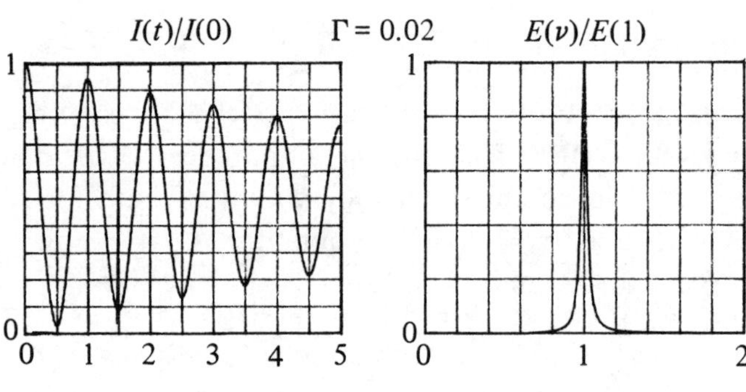

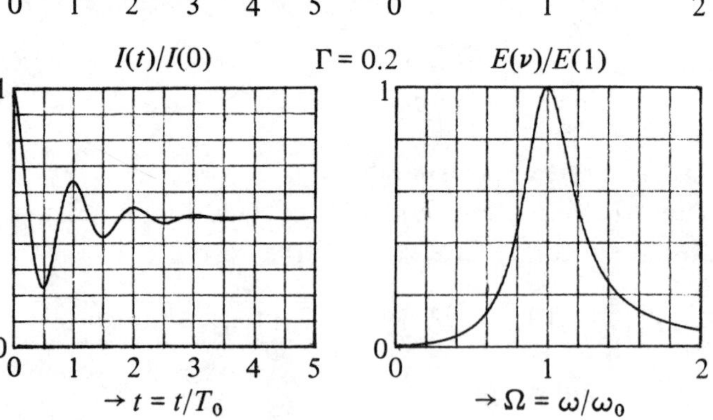

in a resonator circuit with a resonance frequency $\omega_0 = 1/\sqrt{LC}$ and a resistance R, so that the decay constant is $\gamma = R/2L$.

Since $I(t) = 0$ for negative value of t, it follows that the limits of integration in the Fourier Transform can be set equal to zero and infinity, and we obtain

$$F(\nu) = I_0 \int \exp(-\gamma t)\cos(\omega_1 t)\exp(-i\omega t)\,dt \tag{24}$$

$$= \tfrac{1}{2}I_0 \int_0^\infty [e^{i(\omega_1 - \omega + i\gamma)t} + e^{-i(\omega_1 + \omega - i\gamma)t}]\,dt$$

$$= I_0(\gamma - i\omega)/(\omega_0^2 - \omega^2 + i2\gamma\omega)$$

The calculation is then straight forward, and we leave it for one of the problems to show that (see Eq. 21)

$$E(\nu) = 2|F(\nu)|^2 = 2(I_0/\omega_0)^2(\Omega^2 + \Gamma^2)/[(1 - \Omega^2)^2 + 4\Omega^2\Gamma^2] \tag{25}$$

$$\Omega = \omega/\omega_0, \quad \Gamma = \gamma/\omega_0$$

For $\Gamma \ll 1$, the maximum of $E(\nu)$ is $(I_0^2/\omega_0^2)/2\Gamma^2$, which occurs at $\Omega = 1$. The value of $E(\nu)$ is then reduced by a factor of 2 when $\Omega = 1 \pm \Gamma$, which means that the corresponding width of the curve is $2\Gamma = 2\gamma/\omega_0 = 2R/2\omega_0 L = 1/Q$, inversely proportional to the characteristic decay time $\tau = 1/\gamma$ and to the Q-value of the circuit. (See Display 3.13).

3.8 Arbitrary driving force

After the discussion of Fourier analysis in the previous section, we are now in a position to determine the response of a system to an arbitrary driving force $F(t)$.

Since we assume the system to be linear, its response to a collection of several harmonic driving forces acting simultaneously is the sum of the responses to the individual forces. To make use of the linearity, we express the driving force in terms of the Fourier Transform

$$F(t) = \int_{-\infty}^{\infty} F(\omega)\exp(-i\omega t)(d\omega/2\pi) \tag{1}$$

where we have put $d\nu = d\omega/2\pi$.

Thus, in terms of the impedance $Z(\omega)$ of the system at the point of application of the force, the contribution $u(\omega)\,d\omega$ to the complex velocity amplitude at this point is $u(\omega)\,d\omega = F(\omega)\,d\omega/Z(\omega)$, and for the complex displacement amplitude we have $\xi(\omega) = u(\omega)/-i\omega$. The corresponding expressions for the time dependence of the velocity and displacement are then

$$u(t) = \int_{-\infty}^{\infty} u(\omega) \exp(-i\omega t)(d\omega/2\pi)$$

$$= \int [F(\omega)/Z(\omega)] \exp(-i\omega t)(d\omega/2\pi) \tag{2}$$

$$\xi(t) = \int_{-\infty}^{\infty} [F(\omega)/(-i\omega)Z(\omega)] \exp(-i\omega t)(d\omega/2\pi) \tag{3}$$

Through these relations, the response problem for a one-dimensional system is formally solved, and it remains to evaluate the integrals involved. The general systematic integration procedures will not be discussed here, however.

Examples

E3.1 *Vibration isolation.* As an application of the material in Chapter 3, we shall consider here the important engineering problem of vibration isolation.

A variety of sources of vibration are encountered in industry, in transportation, and even in the home. Examples of sources are various reciprocating engines, rotating devices (always with some dynamic imbalance), punch presses, pile drivers, etc.

Vibration not only can be annoying, both directly, as in an automobile, for example, or indirectly, as a source of noise. It can also be responsible for structural damage and can interfere with the operation of sensitive equipment and instruments such as electron microscopes, lithographic tools for computer chip manufacturing, optical equipment, etc. Human beings are sensitive to vibration as well, and if the amplitude is high enough it can cause annoyance and damage.

It is highly desirable, therefore, to isolate sources of vibration as well as sensitive equipment. As an example, we shall consider the vibration isolation characteristics of a spring with an associated dashpot damper, as shown schematically in Display 3.14.

The mass of the machine to be isolated is M, which includes the (concrete) mounting slab, often called the inertia block. The slab is placed on springs, and the resulting spring constant and friction constant are K and R. The driving force in most cases is due to dynamical imbalance of rotors and reciprocating engine components. The driving force is assumed to be known, and the problem is to determine the amplitude of the force transmitted to the floor under the machine and to establish the conditions, under which a reduction of the vibration amplitude is achieved by insertion of the spring isolator.

With reference to the display, we denote the complex amplitudes of force and displacement by F and ξ and use the subscript 1 to refer to the quantities associated with the machine and the subscript 2 for the floor. Furthermore, we introduce the static deflection of the spring

Display 3.14.

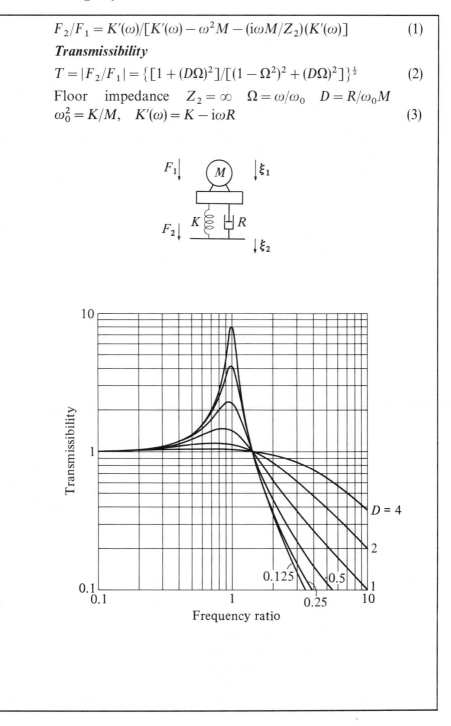

$$F_2/F_1 = K'(\omega)/[K'(\omega) - \omega^2 M - (i\omega M/Z_2)(K'(\omega)] \quad (1)$$

Transmissibility

$$T = |F_2/F_1| = \{[1 + (D\Omega)^2]/[(1 - \Omega^2)^2 + (D\Omega)^2]\}^{\frac{1}{2}} \quad (2)$$

Floor impedance $Z_2 = \infty$ $\Omega = \omega/\omega_0$ $D = R/\omega_0 M$

$$\omega_0^2 = K/M, \quad K'(\omega) = K - i\omega R \quad (3)$$

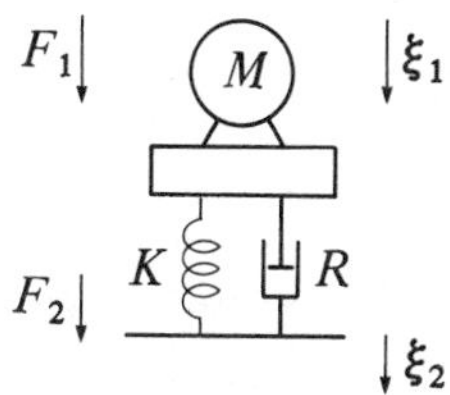

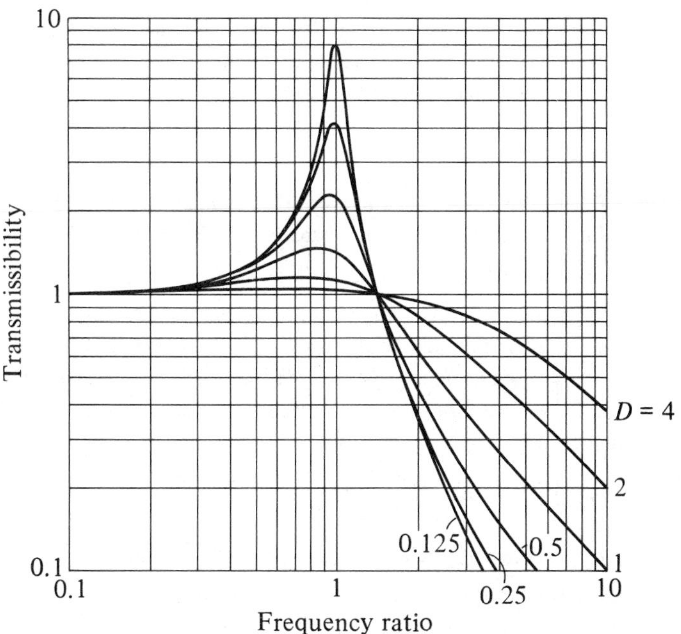

$$\xi_s = Mg/K = g/\omega_0^2 \quad \omega_0^2 = K/M \tag{1}$$

$$\omega_0 = \sqrt{g/\xi_s} \tag{2}$$

It should be noted that the resonance frequency of the system can be expressed in terms of the static deflection.

The force transmitted to the floor consists of two parts, the contribution from the compression of the spring and the friction force from the dashpot. The first is simply $K(\xi_1 - \xi_2)$ and the second, $R(\dot{\xi}_1 - \dot{\xi}_2)$, is proportional to the relative velocity of the piston and the cylinder of the dashpot and hence to the relative velocity of the top and the bottom of the spring. Thus, the complex force amplitude on the floor is

$$F_2(\omega) = K(\xi_1 - \xi_2) - \mathrm{i}\omega R(\xi_1 - \xi_2) = K'(\xi_1 - \xi_2)$$

where

$$K'(\omega) = K - \mathrm{i}\omega R \tag{3}$$

By introducing the **complex spring constant $K'(\omega)$** we can account for both the spring and the dashpot.

The floor impedance Z_2 is assumed to be known, and the relation between the complex amplitudes of force and velocity can be written

$$F_2 = (-\mathrm{i}\omega\xi_2)Z_2 \tag{4}$$

Since we neglect the inertia of the spring and the dashpot, the reaction force on M from the spring and the dashpot is equal in magnitude to the force on the floor. Thus, the net force on M is $F_1 - F_2$, and the equation of motion $M\ddot{\xi}_1 = F_1 - F_2$ yields the complex amplitude equation

$$-\omega^2 M\xi_1 = F_1 - F_2 \tag{5}$$

From the three equations thus obtained, we can determine the three unknowns F_2, ξ_1, and ξ_2 in terms of F_1. For example, we find

$$F_2/F_1 = K'/[K' - \omega^2 M - (\mathrm{i}\omega M/Z_2)K'] \tag{6}$$

Of particular interest is the magnitude of this force ratio, $T = |F_2/F_1|$, usually called the **transmissibility** of the spring isolator. Generally, the floor impedance Z_2 is quite large, and it is instructive to consider the case of an infinite impedance. Then, if we introduce the normalized frequency Ω and the friction factor D, we obtain

$$T = |\dot{F}_2/F_1| = \{[1 + (D\Omega)^2]/[(1 - \Omega^2)^2 + (D\Omega)^2]\}^{\frac{1}{2}}$$

$$\Omega = \omega/\omega_0 \quad D = 1/Q = R/\omega_0 M \quad \omega_0^2 = K/M \quad Z_2 = \infty \tag{7}$$

In Display 3.14 we have plotted the transmissibility T as a function of the normalized frequency $\Omega = \omega/\omega_0$ for some different values of the friction factor D. We note that T is less than 1 at frequencies $\omega > \sqrt{2}\omega_0$. In this frequency range the value of T decreases (isolation increases) with decreasing D. It should be remembered, however, that this result refers to the steady state motion of the system. In order to dampen out transients in the system some friction is desired. This question will be discussed further in the next chapter.

If the machine is placed directly on the floor, assumed to have infinite impedance, the force amplitude on the floor will be the same as the driving force, and if the isolator is to yield a reduction of the force amplitude, the transmissibility T must be less than unity. This condition is fulfilled if the normalized frequency ω/ω_0 is larger than $\sqrt{2}$. If we express the resonance frequency in terms of the static deflection, $\omega_0 = \sqrt{g/\xi_s}$, the condition $T < 1$ can be written $\xi_s > 2g/\omega^2$, where ω is the angular frequency of the driving force.

Although the transmissibility generally is a good indicator of the performance of an isolator, the resulting displacement or velocity amplitude of the floor is of more direct interest. We leave it for one of the problems to determine the displacement amplitude of the floor in terms of the driving force. If the impedance of the floor is small, as is the case if the driving frequency happens to coincide with the floor resonance, we find that the displacement amplitude ξ_2 of the floor will be approximately the same as the amplitude ξ_1 of the machine, independent of the characteristics of the spring.

E3.2 *Complex mass (inductance) and compliance (capacitance).* As indicated in the text, friction in a harmonic oscillator can be provided by a dashpot in parallel with the spring or by a contact force acting on the mass element. In Chapter 3 we dealt mainly with the dashpot damper.

The complex amplitude of the total reaction force $R\mathrm{d}\xi/\mathrm{d}t + K\xi$ from a spring and a dashpot in parallel is $(K - \mathrm{i}\omega R)\xi$, which can be expressed as $K'(\omega)\xi$, where

$$K'(\omega) = K - \mathrm{i}\omega R \tag{1}$$

is a complex spring constant and $C'(\omega) = 1/K'(\omega)$ a complex compliance.

If the contact friction force on the mass element in the oscillator is denoted by $R_1 u$, the equation of motion for the mass element is $M\mathrm{d}u/\mathrm{d}t + R_1 u = F$, where F is the force on M, and the corresponding complex amplitude equation is $(-\mathrm{i}\omega M + R_1)u = F$, which can be written as $-\mathrm{i}\omega M'(\omega)u = F$, where

$$M'(\omega) = M + \mathrm{i}(R_1/\omega) \tag{2}$$

is the complex mass, which incorporates the friction constant.

In analogous manner a capacitance C in series with a resistance R can be replaced by a single capacitor with the complex capacitance C' given by

$$1/C' = 1/C - \mathrm{i}\omega R \tag{3}$$

and an inductance in series with a resistance by the single complex inductance

$$L'(\omega) = L + \mathrm{i}(R/\omega) \tag{4}$$

Similarly, a capacitance C in parallel with a resistance R has the total admittance $-\mathrm{i}\omega C + 1/R = -\mathrm{i}\omega C'$, where the equivalent complex single capacitance is

$$C' = C + \mathrm{i}(1/\omega R) \tag{5}$$

The use of these complex quantities is often quite convenient in the analysis of the frequency response of electrical and mechanical circuits; the results obtained for a loss-free system can be used directly to determine the effect of losses by substituting the parameters K, M, C, and L by the corresponding complex quantities.

E3.3 *Charged particle in electric field.* A water droplet with a radius a and an electric charge Q is acted on by an electric field $E = E_0 \cos(\omega t)$. The droplet moves in air and the viscous drag force on it is known to be Ru, where $R = 6\pi\eta a$, where η is the coefficient of shear viscosity of air. (This force, the Stokes' force, is familiar to the physicist from the Millikan oil drop experiment).

(a) Determine the amplitude and phase angle of the displacement and of the velocity of the particle.
(b) Determine the numerical values in (a) if $a = 5$ microns, $E_0 = 500$ V/cm, $\omega/2\pi = 1000$ Hz, and $Q = 10^5 e$ where e is the electron charge $= 1.6 \times 10^{-19}$ C. $\eta = 1.8 \times 10^{-4}$ CGS.

Solution. (a) The force on the droplet is $F(t) = QE(t) = QE_0 \cos(\omega t)$ with the corresponding complex amplitude $F = QE_0$.

From the equation of motion $M du/dt + Ru = F(t)$ follows the complex amplitude equation $-i\omega Mu + Ru = F$, and the complex velocity amplitude

$$u = u_0 \exp(i\beta) = QE_0/(R - i\omega M) = (iQE_0/\omega M)/(1 + i\delta) \qquad (1)$$

$$\delta = R/\omega M$$

The corresponding complex displacement amplitude is

$$\xi = \xi_0 \exp(i\alpha) = u/-i\omega = (-QE_0/\omega^2 M)/(1 + i\delta) \qquad (2)$$

It follows from Eqs. (1) and (2) that the velocity and displacement amplitudes are

$$u_0 = QE_0/\omega M\sqrt{1 + \delta^2}, \quad \xi_0 = u_0/\omega \qquad (3)$$

The phase angle of the velocity in Eq. (1) has one contribution $\pi/2$ from the factor i and a contribution $-\phi = -\arctan(\delta)$ from the denominator so that $\beta = \pi/2 - \phi$. Consequently, the time dependence is $u = u_0 \cos(\omega t - \beta) = u_0 \sin(\omega t + \phi)$.

Similarly, for the displacement the contribution from the minus sign to the phase angle in Eq. (2) is π, and the denominator contributes $-\phi$, as before, so that $\alpha = \pi - \phi$. This result could have been obtained directly from $\xi = u/-i\omega = iu/\omega$, which shows that $\alpha = \beta + \pi/2 = \pi - \phi$ and $\xi(t) = \xi_0 \cos(\omega t - \pi + \phi) = -\xi_0 \cos(\omega t + \phi)$.

Solution. (b) Let us compute first the dimensionless quantity δ, which can be considered to be the ratio between the viscous drag force and the inertial force.

$$\delta = R/\omega M = 6\pi\eta a/(4\pi a^3/3)\omega\rho = 4.5 \times 1.8 \times 10^{-4}/2\pi \times 1000 \times 25 \times 10^{-8} = 0.52$$

where we have used $a = 10^{-4}$ cm and $\rho = 1$ g/cm^3.

The velocity amplitude becomes

$$u_0 = QE_0/\omega M\sqrt{1 + \delta^2} = QE_0/\omega(4\pi a^3/3)\rho\sqrt{1 + \delta^2} = 0.22 \text{ m/s}$$

where we have used $\rho = 1000$ kg/m^3, $a = 10^{-6}$ m, and $E_0 = 50\,000$ V/m.

The corresponding displacement amplitude becomes $\xi_0 = 35 \times 10^{-6}$ m, i.e. 7 sphere radii.

Problems

3.1 *Frequency response for current (velocity).* With reference to Display 3.2, show that the frequency dependence of the current amplitude is symmetrical with respect to $\Omega = 1$ in the sense that if Ω is replaced by $\Omega' = 1/\Omega$, the response curve remains the same.

3.2 *Frequency response for displacement.* Consider the frequency dependence of the displacement amplitude of a mass–spring oscillator, driven by a harmonic force, as shown in Display 3.6.

(a) Determine the maximum value of the displacement amplitude and the corresponding value of the normalized frequency.
(b) For what value of the resistance factor D will a maximum occur only at zero frequency?

3.3 *Forced motion.* A centrifugal fan with its motor and concrete slab (total mass of system is 1000 kg) is mounted on springs at the corners of the slab. The static compression of the springs is 1 cm. Due to an imbalance in the fan, the system is oscillating in the vertical direction. The rotation speed of the fan is 1200 rpm (revolutions per minute) and the friction factor D is 0.15. The amplitude of the vertical force component due to the imbalance is 0.001 of the total weight of the system.

(a) What is the steady state displacement amplitude of the system?
(b) What will be the amplitude if the number of springs under the concrete slab is doubled?

3.4 *Complex velocity components.* A damped harmonic oscillator is driven by the force $F_0 \cos(\omega t)$. The complex velocity amplitude is $u = F/Z$, where $Z = R + iX = R + i(K/\omega) - i\omega M$ is the mechanical impedance of the oscillator.

(a) With $u(\omega) = u_r + iu_i$, determine the frequency dependence of the real and imaginary parts of the velocity, u_r and u_i, and plot them versus the normalized frequency ω/ω_0, where $\omega_0 = \sqrt{K/M}$.
(b) At what value of ω/ω_0 is u_r a maximum, and determine the maximum value.
(c) Show that the maximum value of the imaginary part is half of the maximum of the real part and that it occurs at the frequency given by $\omega/\omega_0 = D/2 + \sqrt{1 + (D/2)^2}$, where $D = R/\omega_0 M$. Determine also the minimum value of the imaginary part.

3.5 *A seismometer.* A seismometer, in principle, consists of a mass M supported by a spring (spring constant K), which is mounted on a rigid housing, as shown schematically in the figure. The housing stands firmly on the ground surface, the vibration of which is to be measured. We assume this vibration to be harmonic, and denote the displacement by $\eta(t)$. The friction force on M is proportional to the velocity of M with respect to the housing.

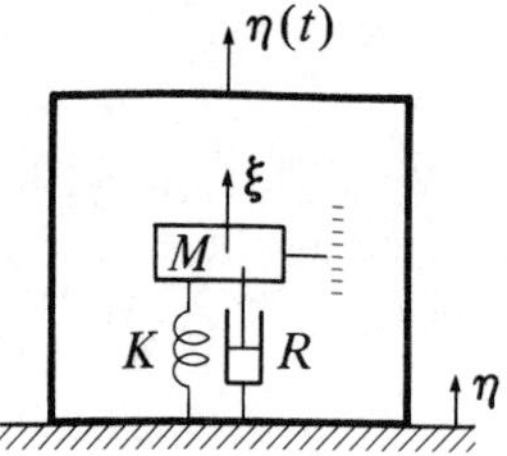

(a) Derive the equation for the displacement of M with respect to the laboratory (inertial) frame of reference, and determine also the displacement with respect to the housing.
(b) Derive the equation of motion of M with respect to a coordinate system S' attached to the housing. Use the appropriate inertial force, and determine the displacement of M with respect to S'. Compare the result with that in (a).
(c) Suppose that M contains a coil, which moves in the magnetic field of a magnet attached to the housing. What is the frequency dependence of the voltage amplitude of the signal from the coil?

3.6 *Integrator circuit.* In the seismometer discussed in Problem 3.5, the output voltage is proportional to the velocity of M with respect to the housing. To obtain a signal proportional to the relative displacement, the output signal needs to be integrated.

Show that if the voltage is applied over a circuit consisting of a resistance R in series with a capacitance C, the voltage over the capacitor will be proportional to the desired integral if $R \gg 1/\omega C$, where ω is the angular frequency of oscillation.

3.7 *The dashpot damper.* The dashpot is a damping device, which consists of a piston, which can move in a cylinder filled with air or some other fluid. The cylinder contains small holes so that the fluid is forced through the holes when the piston moves. This produces a resistive force on the piston, which is proportional to the velocity of the piston relative to the cylinder (at least at sufficiently small velocities). With reference to the figure, the force on the pistion is $-R(u_1 - u_2)$.

In the first arrangement in this figure, the dashpot is in 'series' with the spring and in the second arrangement it is in 'parallel'. In each case determine the displacement amplitude of the mass element as a function of frequency as the end of the spring is driven with a harmonic displacement of constant amplitude.

3.8 *Impedance.* An inductance L_2 and a capacitance C_2 are combined in parallel, and the combination is in series with an inductance L_1. Calculate the impedance of this circuit and show (at least qualitatively) the frequency dependence of the impedance.

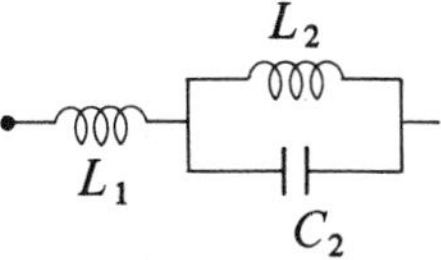

3.9 *Vibration damper.* An engine component of mass M_1 is acted on by a force $F_0 \cos(\omega t)$. As a result it vibrates with the steady state displacement amplitude $F_0/M\omega^2$.

(a) An 'anti-vibration' device, consisting simply of a mass M_2 at the end of a spring (spring constant K_2), is connected to M_1 as indicated. Calculate the new displacement amplitude of M_1, and suggest an appropriate choice of the combination of values of M_2 and K_2 to reduce the amplitude of M_1 as much as possible.
(b) Show that the electrical analogue of this mechanical system is that described in Problem 3.8.

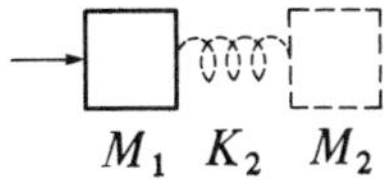

3.10 *Power transfer.* The plug in a control valve can be considered to be the mass element in a mechanical oscillator. In a particular case, the resonance frequency of this oscillator was 30 Hz, the mass $M = 25$ kg and the Q-value of the oscillator $Q = \omega_0 M/R = 10$.

The valve plug was observed to have a steady state displacement amplitude of 0.1 cm at a frequency equal to the resonance frequency. What was the average power transfer to the valve plug in watts?

3.11 *Vibration isolation.* To isolate a sensitive piece of equipment from floor vibration, a spring isolator is inserted between the floor and the equipment. The displacement amplitude of the floor vibration is η_1 and the corresponding amplitude of this equipment is denoted by η_2. Show that the ratio η_2/η_1 is the same as the transmissibility given in Eq. E3.1.7.

3.12 *Vibration isolation.* With reference to the discussion in Example 3.1, show that the displacement amplitude of the floor can be expressed as $\xi_2 = (F_1/-\omega^2 M)\omega_0^2/[\omega_0^2 - (\omega_0^2 - \omega^2)(Z_2/i\omega M)]$.

3.13 *Aerodynamic damping.* The drag force on a body in a fluid stream usually is expressed as $F = C(R)A\rho U^2/2$, where U is the fluid speed relative to the body, A the cross sectional area of the body, ρ the fluid density, and $C(R)$ an experimentally determined quantity, called the 'drag coefficient', which is a function of the dimensionless parameter $R = UL/\nu$, the 'Reynolds number'. (L = characteristic dimension of the body, ν = kinematic viscosity, $\nu = \mu/\rho$, where μ is the shear viscosity).

Consider now a body in an air stream with a velocity U_0. The body oscillates in the direction of the flow with the velocity $u(t)$, so that the relative flow velocity is $U = U_0 - u(t)$.

If the flow velocity is sufficiently large, the drag coefficient is approximately equal to 1, independent of R. In that case, show that the time dependent force on the body represents a friction force proportional to U_0.

3.14 *Forced harmonic motion.* A simple demonstration of forced harmonic motion of an oscillator is illustrated in the figure. An air-suspended cart, mass M, is mounted between two springs, as shown. The end of one of the springs is driven in harmonic motion with constant displacement amplitude by means of a string, that runs over a pulley and is connected eccentrically to a disc on a motor shaft. The damping in such a system is usually quite small, but can be increased by mounting a cardboard on the cart, and, if necessary blowing on it with a fan (see problem 3.13).

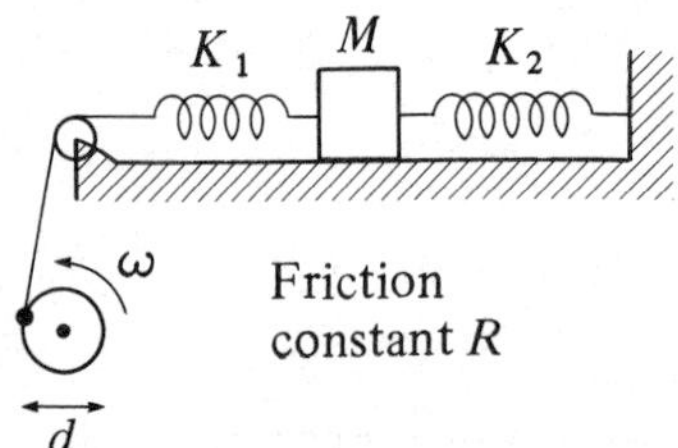

Use the parameters given in the figure together with the resistance factor $D = R/\omega_0 M$, and express the frequency dependence of the displacement amplitude of M.

3.15 *Helmholtz resonator.* A bottle with a diameter $D = 10$ cm and a height $H =$ 20 cm, has a neck with a diameter $d = 1$ cm and a length $h = 4$ cm. What is the resonance frequency of the air in this bottle? Speed of sound in air, $v = 340$ m/sec.

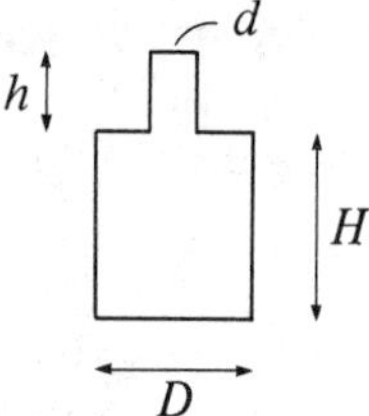

3.16 *Power.* Consider an electric circuit driven by a voltage $V = V_0 \cos(\omega t)$. The current into the circuit is found to be $I = I_0 \cos(\omega t - \pi/4)$.

Calculate the time average of the power delivered to the circuit in the time interval between 0 and τ, and make a graph of the power versus τ/T, where T is the period.

3.17 *Effect of vibration on instrument.* Some instruments and processes can be adversely affected by vibration, because misalignment of various components can result.

For the purpose of illustration we consider the vibration induced misalignment of two components mounted on a platform, as indicated. The two components are modelled as mass–spring oscillators. If the maximum permissible amplitude of misalignment between the mass element is Δ determine the frequency dependence of the maximum permissible displacement amplitude of the platform.

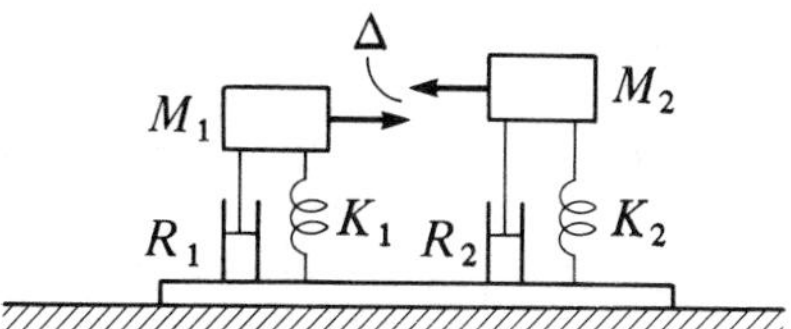

3.18 *Fourier series.* Prove Eqs. 9–12 in Display 3.11 for the Fourier decomposition of the periodic function shown in the display.

3.19 *Fourier series.* Make a Fourier decomposition of the periodic function $F(t) = A|\cos(\Omega t)|$.

3.20 *Fourier Transform.* The displacement function for an overdamped harmonic oscillator of mass M is $\xi(t) = A\exp(-\gamma t)$. What is the energy density spectrum of the kinetic energy in this motion?

4

Free oscillations and impulse response

Having discussed forced harmonic oscillations, we now turn to the free oscillations and the response of an oscillator to other than harmonic excitation. In particular, we shall consider an impulse excitation of unit strength and show how the corresponding response can be used in the analysis involving an excitation with arbitrary time dependence. In the process, such matters as the physical significance of a complex frequency, oscillatory and non-oscillatory decay, and the transition to steady state will be discussed.

4.1 Free oscillations

In the discussion of free oscillations we shall consider a mass–spring oscillator. As discussed in Section 3.2, the complex amplitude equation for the displacement in forced harmonic motion is

$$(-\omega^2 M - i\omega R + K)\xi = F \tag{1}$$

For simplicity we have used the notations ξ and F rather than $\xi(\omega)$ and $F(\omega)$ for the complex amplitudes of displacement and driving force.

Complex frequency. In free motion, the driving force is zero, and the complex amplitude equation above reduces to a product of two factors, one being the complex displacement amplitude. The equation is satisfied if one of these factors is zero. The possibility that the amplitude is zero is of no interest, and it remains to explore the consequences of the second factor being zero, i.e. $-\omega^2 M - i\omega R + K = 0$ or

$$\omega^2 + 2\gamma\omega - \omega_0^2 = 0 \quad \text{where } \gamma = R/2M, \quad \omega_0^2 = K/M \tag{2}$$

This equation is of the second order in the frequency, and the two solutions are

$$\omega_{\pm} = -\mathrm{i}\gamma \pm \sqrt{\omega_0^2 - \gamma^2} = -\mathrm{i}\gamma \pm \omega', \tag{3}$$
$$\text{where } \omega' = \sqrt{\omega_0^2 - \gamma^2},\ \gamma = R/2M$$

These solutions are complex numbers, and in order to determine their physical meaning, we recall the definition of the complex amplitude, $\xi(t) = \xi_0 \cos(\omega t - \alpha) = \mathrm{Re}\{\xi \exp(-\mathrm{i}\omega t)\}$. The general solution to the equation is a linear combination of the functions $\xi(t)$ obtained when the two frequencies are inserted in the exponential factor $\exp(-\mathrm{i}\omega t)$. This factor takes the form

$$\exp(-\mathrm{i}\omega_{\pm} t) = \exp(-\gamma t)\exp(\pm \mathrm{i}\omega' t) \tag{4}$$

In other words, the imaginary part of the frequency leads to an exponential decay of the amplitude, as expressed by the factor $\exp(-\gamma t)$, the same for both solutions.

We express the two solutions as $\xi_+ = \mathrm{Re}\{\xi_+ \exp(-\mathrm{i}\omega_+ t)\} = A_+ \exp(-\gamma t)\cos(\omega' t - \alpha_+)$ and $\xi_-(t) = A_- \exp(-\gamma t)\cos(\omega' t - \alpha_-)$. The general solution is then $\xi(t) = a\xi_+ + b\xi_-$, where a and b are real constants. This solution is the sum of two cosine functions with different phase and amplitude, and this sum can be expressed as a single cosine function

$$\xi(t) = a\xi_+ + b\xi_- = A\exp(-\gamma t)\cos(\omega' t - \alpha) \tag{5}$$

The quantity $\gamma = R/2M$ is called the **decay constant** and its inverse is the **decay time** τ, in terms of which the time dependence of the amplitude becomes $\exp(-t/\tau)$; the amplitude is reduced by the factor 1/e in the time τ, also called the 'life time' of the oscillation.

Sometimes the relative change of the amplitude in one period, called the **logarithmic decrement** $\boldsymbol{\delta}$, is used to describe the rate of decay. It is $\delta = \gamma T'$ and with $T' = 2\pi/\omega'$, it can be expressed as

$$\delta = \gamma T' = \gamma 2\pi/\omega' = (\pi R/\omega_0 M)(\omega_0/\omega') = \pi D/\sqrt{[1-(D/2)^2]} \tag{6}$$

For small values of the resistance factor $D = R/\omega_0 M = 1/Q$, we have $\delta \simeq \pi D = \pi/Q$.

The resistance in the oscillator not only produces a decay of the amplitude but also a reduction in the frequency of oscillation and a corresponding increase in the period T'. The effect is determined by the ratio γ/ω_0, which is proportional to τ/T_0.

As the resistance parameter γ/ω_0 increases from zero to 1, the frequency decreases from ω_0 to zero. A further increase of the resistance makes ω' imaginary, and both roots to the frequency equation will be purely imaginary corresponding to non-oscillatory decaying motions. Actually,

Display 4.1

Complex frequencies

Differential equation

$$M\ddot{\xi} + R\dot{\xi} + K\xi = 0 \tag{1}$$

Complex amplitude equation

$$(-\omega^2 M - i\omega R + K)\xi = 0 \tag{2}$$

Frequency equation

$$-\omega^2 M - i\omega R + K = 0 \tag{3}$$

$$\omega^2 + i2\gamma\omega - \omega_0^2 = 0 \quad \gamma = R/2M \quad \omega_0^2 = K/M \tag{4}$$

Complex frequency solutions

$$\omega_\pm = -i\gamma \pm \sqrt{\omega_0^2 - \gamma^2} = -i\gamma \pm \omega' \quad \omega' = \sqrt{\omega_0^2 - \gamma^2} \tag{5}$$

Displacement, oscillatory decay

$$\xi = A\exp(-\gamma t)\cos(\omega' t - \alpha) \tag{6}$$

Decay time: $\tau = 1/\gamma$ (7)

Logarithmic decrement: $\delta = \gamma T' = T'/\tau$
$= (\pi/Q)(\omega_0/\omega') \simeq \pi D$ for $D \ll 1$ (8)

Complex ω-plane

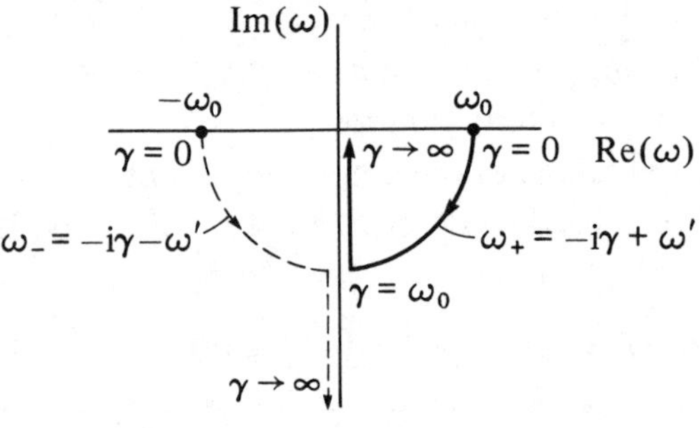

depending on the value of γ/ω_0, we can distinguish between three different types of free motion, as described below. The corresponding frequency dependence of the complex frequencies is shown in Display 4.1.

Oscillatory decay, $\boldsymbol{\gamma < \omega_0}$**.** In this regime we have $D < 2$ ($Q > \frac{1}{2}$) and the frequency $\omega' = \sqrt{\omega_0^2 - \gamma^2}$ is real. The corresponding displacement function is

$$\xi(t) = A \exp(-\gamma t) \cos(\omega' t - \alpha) \tag{7}$$

The constants A and α are determined from a knowledge of the displacement and the velocity at a given time.

Overdamped motion, $\boldsymbol{\gamma > \omega_0}$**.** The frequency ω' is now purely imaginary, $\omega' = \mathrm{i}\sqrt{\gamma^2 - \omega_0^2}$, and the two solutions to the frequency equation become

$$\left.\begin{aligned} \omega_+ &= -\mathrm{i}(\gamma - \sqrt{\gamma^2 - \omega_0^2}) = -\mathrm{i}\gamma_1 \\ \omega_- &= -\mathrm{i}(\gamma + \sqrt{\gamma^2 - \omega_0^2}) = -\mathrm{i}\gamma_2 \end{aligned}\right\} \tag{8}$$

The corresponding general motion will be the sum of two exponential functions with different decay constants,

$$\xi = A_1 \exp(-\gamma_1 t) + A_2 \exp(-\gamma_2 t) \tag{9}$$

Again, the two constants A_1 and A_2 are determined from the initial conditions of displacement and velocity.

It is of interest to determine the values of the decay constants γ_1 and γ_2 when $\gamma \gg \omega_0$,

$$\left.\begin{aligned} \gamma_1 &= \gamma - \sqrt{\gamma^2 - \omega_0^2} = \gamma[1 - \sqrt{1 - (\omega_0^2/\gamma)^2}] \\ &\qquad \simeq \tfrac{1}{2}\omega_0^2/\gamma = K/R \quad \gamma \gg \omega_0 \\ \gamma_2 &= \gamma + \sqrt{\gamma^2 - \omega_0^2} \simeq R/M \quad \gamma \gg \omega_0 \end{aligned}\right\} \tag{10}$$

The first of these corresponds to an oscillator, in which R is so large that the effect of the inertia of the oscillator can be neglected. The equation of motion can then be approximated by $R\dot{\xi} = -K\xi$, which has the solution $A \exp[-(K/R)t]$.

Similarly, the second root corresponds to an oscillator, in which R is much larger than the stiffness reactance of the spring. The approximate equation of motion is then $M\dot{u} = -Ru$ with the solution $B \exp[-(R/M)t]$. The motion of a particle through a viscous fluid belongs to this class of motions.

Critically damped motion, $\gamma = \omega_0$. A special mention should be made of this 'degenerate' case (with $D = 2$, i.e. $Q = 1/2$), in which the two roots of the frequency equation are equal, $\omega_+ = \omega_- = -i\omega_0$. To obtain the general solution in this case we cannot merely put $\gamma_1 = \gamma_2 = \omega_0$ in the expression for the overdamped motion since this leads to a particular solution with only one adjustable constant. The general solution must contain two constants so that the two conditions of initial displacement and velocity can be satisfied (the general solution to a second-order differential equation has two constants of integration). To obtain this solution we can proceed as follows.

We start from the overdamped motion $\xi = A_1 \exp(-\gamma_1 t) + A_2 \exp(-\gamma_2 t)$. Let $\gamma_2 = \gamma_1 + \Delta$ and denote $\exp(-\gamma_2 t)$ by $f(\gamma_2, t)$. Expansion of this function to first order in Δ yields $f(\gamma_2, t) = f(\gamma_1, t) + (\partial f/\partial\Delta)_0 \Delta = \exp(-\gamma t) - t\Delta \exp(-\gamma_1 t)$. The expression for the displacement then becomes $\xi = (A_1 + A_2)\exp(-\gamma_1 t) - t(A_2\Delta)\exp(-\gamma_1 t)$ or

$$\xi = (A + Bt)\exp(-\omega_0 t) \quad \gamma_1 = \gamma_2 = \omega_0 \tag{11}$$

where we have put $A = A_1 + A_2$ and $B = -A_2\Delta$, going to the limit in such a way that B is finite. Direct insertion into the differential equation $\ddot{\xi} + 2\gamma\dot{\xi} + \omega_0^2\xi = 0$ shows that this indeed is the solution when $\gamma = \omega_0$.

Discussion. With the use of a complex frequency, the complex amplitude approach is applicable not only for forced harmonic motion but also for free motion. For the sake of completeness and to make a comparison, we shall review briefly the procedure of analyzing forced and free motion without the use of complex amplitudes.

In the case of forced harmonic motion with the driving force $F = F_0 \cos(\omega t - \phi)$ the frequency is given, and a solution for the displacement of the form $\xi = \xi_0 \cos(\omega t - \alpha)$ is assumed. The amplitude ξ_0 and the phase angle α are to be determined, and to do so we insert this expression for ξ into the differential equation of motion, $M\ddot{\xi} + R\dot{\xi} + K\xi = F_0 \cos(\omega t - \phi)$. The expressions for velocity and acceleration are $-\omega\xi_0 \sin(\omega t - \phi)$ and $-\omega^2\xi_0 \cos(\omega t - \alpha)$, respectively. Then, applying the trigonometric relations $\sin(a + b) = \sin(a)\cos(b) + \cos(a)\sin(b)$ and $\cos(a + b) = \cos(a)\cos(b) - \sin(a)\sin(b)$, and collecting terms, we can express the left hand side of the equation in the form $C\cos(\omega t) + S\sin(\omega t)$, where C and S contain the unknown quantities ξ_0 and α.

With the right hand side, $F_0 \cos(\omega t - \phi)$, written as $F_0 \cos(\omega t)\cos(\phi) + F_0 \sin(\omega t)\sin(\phi)$, the equation is satisfied if $C = F_0 \cos(\phi)$ and $S = F_0 \sin(\phi)$. From these two relations we can determine the amplitude ξ_0 and the phase angle α.

In the case of free motion, with $F = 0$, the procedure described above is no longer applicable. The frequency is not known, *a priori*, and the assumption of a constant amplitude is unrealistic. Rather, we have to guess a solution with an exponential decay of the form $\exp(-\gamma t)\cos(\omega' t - \alpha)$. After insertion of this expression into the differential equation $M\ddot{\xi} + R\dot{\xi} + K\xi = 0$, one finds, after some algebra, that the equation can be satisfied if the damping factor is chosen to be $R/2M$ and the frequency $\omega' = \sqrt{\omega_0^2 - \gamma^2}$.

The approach thus outlined is not only algebraically cumbersome, but it has the disadvantage, that the forced and the free motion cannot be treated in a unified manner. These shortcomings become even more apparent later in our studies of coupled oscillators, periodic lattices, feed-back oscillations and waves.

Another reason, which speaks strongly in favour of the complex amplitude approach and the concepts and techniques that it entails, is their wide use in modern electronic equipment for vibration analysis and data processing.

Furthermore, the complex amplitude approach in the study of classical oscillators and waves serves as a valuable preparation for the study of the conceptually more difficult subject of quantum mechanics, in which the complex amplitude is an intrinsic component of the theory.

(Actually, in mathematics the commonly used procedure to solve the homogenous differential equation describing free oscillations is to make a transformation of coordinates from ξ to $\exp(zt)$, which leads to an algebraic equation for z. This transformation, in essence, is contained in the complex amplitude approach).

The discussion of the free motion of an oscillator is summarized in Display 4.2. As an example we have considered the motion which results after the oscillator is released at $t = 0$ from rest at $\xi = \xi(0)$. Curves for several values of γ/ω_0 have been shown, $\gamma/\omega_0 = 0, 0.25, 1$, and 4. The corresponding values of $D = R/\omega_0 M = 2\gamma/\omega_0$ are $0, 0.5, 2$, and 8.

For values of D less than 2, the motion will be an oscillatory decay as given by $\xi = \xi_0 \cos(\omega' t - \alpha)$. The constants ξ_0 and α are determined from the initial conditions of $\xi = \xi(0)$ and $\dot{\xi} = 0$ at $t = 0$, which yield $\xi_0/\xi(0) = 1/\cos(\alpha)$ and $\tan(\alpha) = \gamma/\omega'$.

If the oscillator is critically damped, i.e. $\gamma = \omega_0$ or $D = 2$, the motion will be of the form $\xi = (A + Bt)\exp(-\omega_0 t)$. In this case the initial conditions require that $A = \xi(0)$ and $B = \omega_0 \xi(0)$.

In the overdamped regime, corresponding to $\gamma > \omega_0$ or $D > 2$, the displacement is $\xi = A_1 \exp(-\gamma_1 t) + A_2 \exp(-\gamma_2 t)$. From the initial conditions we now find $A_1/\xi(0) = \gamma_2/(\gamma_2 - \gamma_1)$ and $A_2/\xi(0) = \gamma_1/(\gamma_2 - \gamma_1)$.

Display 4.2.

Different types of free motion of an oscillator.

Oscillatory decay $\gamma = R/2M < \omega_0 \quad (D < 2, Q > 1/2)$ (1)

$\xi = A\exp(-\gamma t)\cos(\omega' t - \alpha), \quad \gamma = R/2M \quad \omega' = \sqrt{\omega_0^2 - \gamma^2}$ (2)

Overdamped oscillator $\gamma > \omega_0$ (3)

$\xi = A_1\exp(-\gamma_1 t) + A_2\exp(-\gamma_2 t), \quad \gamma_1 = \gamma - \sqrt{\gamma^2 - \omega_0^2},$
$\gamma_2 = \gamma + \sqrt{\gamma^2 - \omega_0^2}$ (4)

Critically damped oscillator $\gamma = \omega_0$ (5)

$\xi = (A + Bt)\exp(-\omega_0 t)$ (6)

Example (Note: $D = 2\gamma/\omega_0 = 1/Q$)

At $t = 0, \quad \xi = \xi(0), \quad \dot{\xi} = 0$ (7)

$\gamma < \omega_0 \quad \xi/\xi(0) = [1/\cos(\alpha)]\exp(-\gamma t)\cos(\omega' t - \alpha)$ (8)

$\alpha = \arctan(\gamma/\omega'), \quad \omega' = \sqrt{\omega^2 - \gamma^2}$ (9)

$\gamma = \omega_0 \quad \xi/\xi(0) = (1 + \gamma t)\exp(-\gamma t)$ (10)

$\gamma > \omega_0 \quad \xi/\xi(0) = [1/(\gamma_2 - \gamma_1)](\gamma_2 e^{-\gamma_1 t} - \gamma_1 e^{-\gamma_2 t})$ (11)

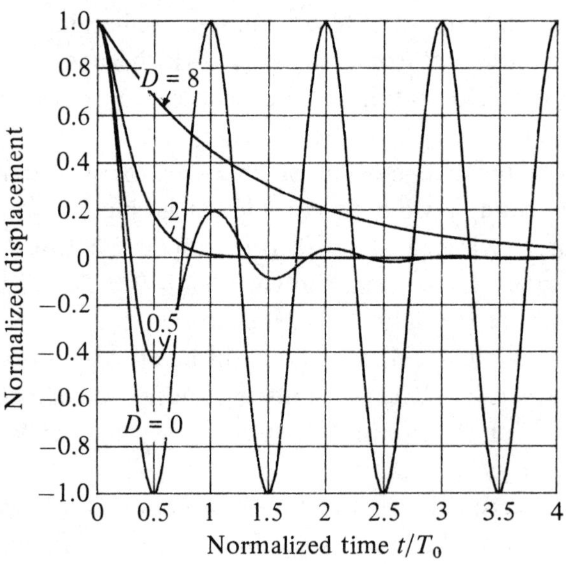

We note that the frequency of oscillation decreases with increasing D and goes to zero when the oscillator is critically damped. The corresponding decay curve separates the oscillatory decay curves from the non-oscillatory. As the damping increases above the critical value, the decay time increases as the return to equilibrium is impeded by the resistance in the system.

Instead of starting the oscillator from a displaced position with zero velocity, we could have started it at $t = t'$ with a certain velocity. The resulting motion is of particular significance when the motion is generated by a unit impulse, as will be discussed in the next section.

4.2 Impulse response

An instantaneous impulse I on a mass M changes the velocity of M by an amount I/M. We now start a mass–spring oscillator from rest by means of an impulse of unit strength.

The impulse is applied at time $t = t'$, and since it is instantaneous, the displacement immediately after the impulse will be $\xi = 0$ and the velocity $\dot{\xi} = 1/M$. In the subsequent motion, the oscillator is free from external forces, and in the case of oscillatory decay, the subsequent general expression for the free motion will be $\xi = \xi_0 \exp[-\gamma(t - t')] \cos[\omega'(t - t') - \alpha]$, where $\gamma = R/2M$ and $\omega' = \sqrt{\omega_0^2 - \gamma^2}$. The constants ξ_0 and α are determined from the displacement $\xi = 0$ and the velocity $\dot{\xi} = 1/M$ at $t = t'$.

In order to make the displacement zero at $t = t'$ we must have $\alpha = \pi/2$, which means that the displacement will be of the form $\xi = \xi_0 \exp[-\gamma(t - t')] \sin[\omega'(t - t')]$. The corresponding velocity is $\dot{\xi} = \xi_0 \exp[-\gamma(t - t')]\{\omega' \cos[\omega'(t - t') - \gamma \sin[\omega'(t - t')]\}$, and to make this equal to $1/M$ at $t = t'$ we must have $\xi_0 = 1/M\omega'$. In other words, the *impulse response function* for displacement is

$$h(t, t') = (1/\omega' M) e^{-\gamma(t - t')} \sin[\omega'(t - t')] \tag{1}$$

In the case of a critically damped oscillator, this expression takes the indeterminate form 0/0 (since $\omega' = 0$), which can readily be evaluated by replacing the sine-function by its argument, which is valid for small arguments. The impulse response function then becomes $h(t, t') = (1/M)(t - t') \exp[-\gamma(t - t')]$.

In the overdamped regime $\gamma > \omega_0$, the frequency ω' is imaginary and $h(t, t')$ can then be expressed as a hyperbolic sine function, since $\sin(\mathrm{i}x) = \mathrm{i} \sinh(x)$. This makes the impulse response function a sum of two decaying exponentials, as it should.

The argument of $h(t, t')$ contains the two times t and t' to indicate that if a unit impulse is applied at t' the impulse response function yields the

Display 4.3.

Impulse response function.
Normalized displacement $= \xi/(1/M\omega_0)$
Normalized time $= t/T_0$, where $T_0 = 2\pi/\omega_0$, $\omega_0^2 = K/M$.

General displacement function

$$\xi(t) = \xi_0 e^{-\gamma(t-t')} \cos[\omega'(t-t') - \alpha], \quad \gamma = R/2M, \quad \omega' = \sqrt{\omega_0^2 - \gamma^2} \tag{1}$$

Initial conditions

Unit impulse delivered (and completed) at $t = t'$;

$$t = t', \quad \xi(t') = 0 \qquad \text{Yields } \alpha = \pi/2 \tag{2}$$
$$t = t' \quad \dot{\xi}(t') = 1/M \quad \text{Yields } \xi_0 = 1/\omega' M \tag{3}$$

Impulse-response function

$$h(t, t') = (1/\omega' M) e^{-\gamma(t-t')} \sin[\omega'(t-t')] \quad t > t' \quad \gamma = R/2m \tag{4}$$
$$h(t, t') = 0 \qquad t < t' \quad \omega' = \sqrt{\omega_0^2 - \gamma^2} \tag{5}$$

$$h(t, t') = (1/M)(t - t') e^{-\gamma(t-t')} \qquad \gamma = \omega_0 \quad \omega' = 0 \tag{6}$$
$$h(t, t') = (1/2\varepsilon M)(e^{-\gamma_1(t-t')} - e^{-\gamma_2(t-t')}) \quad \gamma > \omega_0 \quad \omega' = i\varepsilon \tag{7}$$

$$\varepsilon = \sqrt{\gamma^2 - \omega_0^2} \quad \gamma_1 = \gamma - \varepsilon \quad \gamma_2 = \gamma + \varepsilon \tag{8}$$

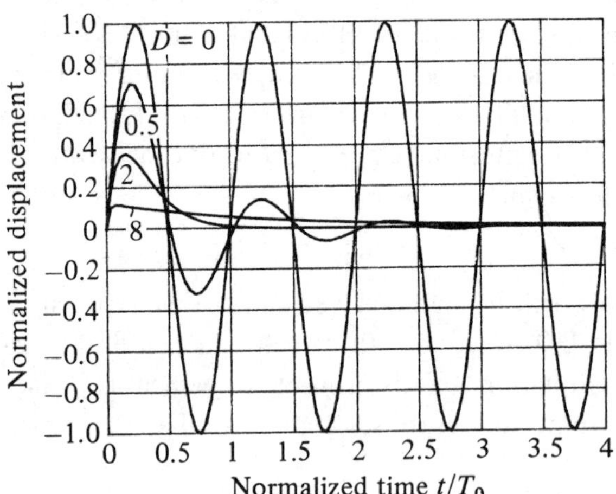

resulting displacement at t. The dependence on t and t' is expressed through the combination $(t-t')$ only, i.e. the time difference between the 'cause' and the 'effect'. If we accept the **causality principle**, that the effect cannot occur before the cause, we have to include $h(t,t')=0$ for $t<t'$ in the definition of the impulse response function, as indicated in Display 4.3. We have here plotted the impulse response function for some different values of D from 0 to 8 in terms of the normalized time t/T_0, where $T_0=2\pi/\omega_0$.

4.3 Response to an arbitrary driving force

The reason for the particular importance of the impulse response function is that the response to an arbitrary driving force can be constructed from it.

To prove this, we consider first the displacement resulting from two unit impulses delivered at t' and t''. Although not necessary, we assume for simplicity that the displacement and the velocity of the oscillator are zero when the first impulse is delivered at $t=t'$. Then, by definition, the displacement produced at time t is the impulse response function $h(t,t')$.

The time t'' is greater than t', and when the second impulse is delivered, the displacement and the velocity are both different from zero. It should be realized, however, that the increase in the displacement produced at t by the second impulse does not depend on the state of motion when the impulse is delivered (because of linearity), and the total displacement at t will be $\xi(t)=h(t,t')+h(t,t'')$. (To be more explicit, let $\xi(t)$ denote the total displacement function. The function $\xi(t)-h(t,t')$ then is zero when the second impulse is delivered and the same holds true for the velocity. Thus, this function satisfies the condition under which the impulse response function was defined, and it follows that the value of this function at t, after the second impulse, will be $\xi(t)-h(t,t')=h(t,t'')$, or $\xi(t)=h(t,t')+h(t,t'')$).

If the impulses at t' and t'' have the values I' and I'' rather than unity, the displacement at t will be $I'h(t,t')+I''h(t,t'')$, since the oscillator is linear.

We can now proceed to the general case of the displacement produced by an arbitrary force $F(t)$. The effect of this force is the same as that of a succession of impulses of magnitude $F(t')\Delta t'$ over the entire time of action of the force up to the time t, as indicated in Display 4.4. The displacement produced by one of these impulses is $h(t,t')F(t')\Delta t'$, and the sum of the contributions from all the elementary impulses can be expressed as the integral

$$\xi(t)=\int_{-\infty}^{t} h(t,t')F(t')\,\mathrm{d}t' \tag{1}$$

Display 4.4.

Oscillator response to an arbitrary driving force.

Impulse-response function

$$h(t, t') = (1/M\omega')e^{-\gamma(t-t')} \sin[\omega'(t-t')] \quad t > t' \tag{1}$$

$$h(t, t') = 0 \quad t < t' \tag{2}$$

$$\gamma = R/2M, \quad \omega' = \sqrt{\omega_0^2 - \gamma^2}, \quad \omega_0^2 = K/M \tag{3}$$

Driving force $F(t)$

Elementary impulse at t'

$F(t')\Delta t'$ (see figure)

yields displacement at t: $\Delta\xi(t) = h(t, t')F(t')\Delta t'$ (4)

Total displacement

$$\xi(t) = \int_{-\infty}^{t} F(t')h(t, t')\,dt' \tag{5}$$

Alternative form: Let $\tau = t - t'$ (6)

$$\xi(t) = \int_{0}^{\infty} F(t-\tau)h(\tau)\,d\tau \tag{7}$$

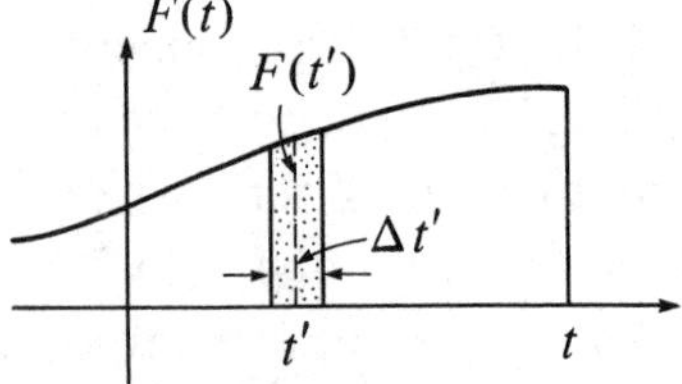

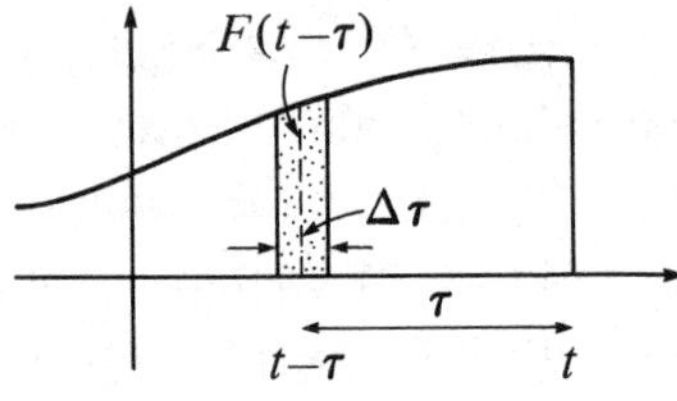

where the range of integration goes from $-\infty$ to t to cover the entire past time period.

It should be emphasized that the validity of this result relies on the linearity of the system, so that the **incremental change** of the displacement will be the same for a given impulse independent of the state of motion when the impulse is delivered.

Frequently, it is convenient to introduce a new variable, $\tau = t - t'$, in which case the integral for the displacement is transformed into

$$\xi(t) = \int_0^\infty F(t-\tau)h(\tau)\,d\tau. \tag{2}$$

We now integrate backwards from the time of observation to infinity to account for all displacement contributions by the force.

4.4 Transition to steady state

A harmonic force is an idealization, since it has no beginning and no end. The same applies to the steady state motion it produces.

To consider a more realistic situation, we study the motion produced by a force, defined by

$$\left.\begin{aligned} F(t) &= F_0 \cos(\omega t) \quad & t > 0 \\ F(t) &= 0 & t < 0 \end{aligned}\right\} \tag{1}$$

For large values of t, we expect the motion produced by this force to approach the steady state motion of the harmonic force $F_0 \cos(\omega t)$. This steady state motion cannot be the complete solution in the present case, however, since both the displacement and the velocity are zero at $t = 0$. An additional motion is required to satisfy these conditions; it is called the **transient motion** or simply the transient.

Both the steady state and the transient motion are contained in the integral expression

$$\xi(t) = \int_{-\infty}^{t} h(t, t')F(t')\,dt \tag{2}$$

where

$$h(t, t') = (1/\omega' M)e^{-\gamma(t-t')} \sin[\omega'(t-t')]$$

With the particular driving force given above and in Eq. D.4.5(1) in Display 4.5, the integral takes the form of Eq. D.4.5(5) and it can be expressed in terms of elementary functions. The integration can be carried out in several different ways; it is facilitated by expressing the sine- and cosine functions in terms of exponentials (Euler's formula). The result can be written in the form of Eq. D.4.5(7). The first term is the steady state

Display 4.5.

Transition to steady state.
Normalized displacement $= \xi/A$.
Normalized time $= t/T_0$, where $T_0 = 2\pi/\omega_0$

Driving force

$$F(t) = F_0 \cos(\omega t) \quad t > 0 \tag{1}$$
$$F(t) = 0 \quad t < 0 \tag{2}$$

Impulse-response function

$$h(t, t') = (1/\omega' M)e^{-\gamma(t-t')} \sin[\omega'(t-t')] \tag{3}$$

$$\gamma = R/2M \quad \omega' = \sqrt{\omega_0^2 - \gamma^2} \quad \omega_0^2 = K/M$$
$$\tau = 1/\gamma = 2/D\omega_0 = (Q/\pi)T_0 \tag{4}$$

Displacement produced by F(t)

$$\xi(t) = \int_{-\infty}^{t} F(t')h(t, t')\,\mathrm{d}t'$$
$$= \xi_0 e^{-\gamma t} \int_0^t e^{\gamma t'} \sin[\omega'(t-t')] \cos(\omega t')\,\mathrm{d}t' \tag{5}$$

$\xi_0 = F_0/\omega' M$ Let $\Omega = \omega/\omega_0$, $D = R/\omega_0 M$, $A = F_0/K$.

$$(F_0/\omega' M = A\omega_0^2/\omega') \tag{6}$$

Eq. (5) becomes:

$$\xi/A = [1/\sqrt{(1-\Omega^2)^2 + (D\Omega)^2}][\cos(\omega t - \alpha) - (\omega_0/\omega')e^{-\gamma t}\cos(\omega' t - \beta)] \tag{7}$$

$$\tan(\alpha) = D\Omega/(1-\Omega^2), \quad \tan(\beta) = \gamma(1+\Omega^2)/\omega'(1-\Omega^2) \tag{8}$$

Special case

$$\omega = \omega_0, \quad \gamma = 0 \quad \xi/A = (\omega_0 t/2)\sin(\omega_0 t) \tag{9}$$

Example $\omega = \omega_0$

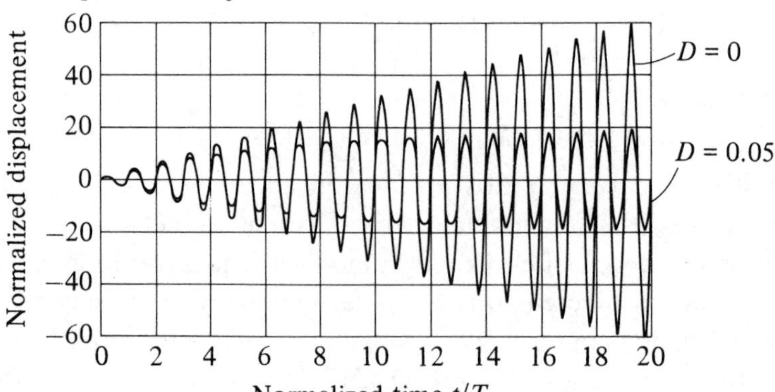

part of the solution, which remains after the transient, represented by the second term, has died out. Notice that the steady state motion has the frequency ω of the driving force, whereas the transient has the frequency ω' of the free motion of the oscillator.

There are several aspects of the solution, which deserve special notice. One concerns the response of the oscillator when it is driven at the resonance frequency, $\omega = \omega_0$. The steady state response is contained in the first term. If the damping is zero, $D = \gamma = 0$, the steady state amplitude is infinite. The present analysis is meaningful even in the 'anomalous' case of an undamped oscillator driven at resonance, $\omega = \omega_0$. It shows how the amplitude grows with time toward the infinite value. In this particular case, $\gamma = 0$ ($\omega' = \omega_0$) and $\omega = \omega_0$, the integration of Eq. D.4.5(5) is straight forward, and we obtain

$$\begin{aligned}\xi(t) &= (F_0/\omega' M)\int_0 \sin[\omega_0(t-t')]\cos(\omega_0 t')\,dt' \\ &= (F_0/K)(\omega_0 t/2)\sin(\omega_0 t) \qquad (3)\end{aligned}$$

The general solution in Eq. D.4.5(7) reduces to this expression for $\gamma = 0$ and $\omega = \omega_0$ (i.e. $\Omega = 1$). Actually, for these values the expression becomes indeterminate, of the form 0/0, and we have to determine the limit value as γ goes to zero. To do this we express the exponential function in terms of the first two terms in its power series, $\exp(-\gamma t) \simeq 1 - \gamma t$. Then, with $\omega = \omega_0$, i.e. $\mathbf{\Omega} = 1$, the steady state term is cancelled out, and, realizing that $\alpha = \beta = \pi/2$ and $D = 2\gamma/\omega_0$, we note that the remaining term indeed is the result given above and in Eq. D.4.5(9).

The amplitude of this motion grows linearly with time (toward the steady state value of infinity); the growth over the first twenty periods is shown in the graph in the display.

For comparison, we have shown also the transition to steady state in the case when the damping parameter $D = 0.05$. The normalized steady state amplitude of 20 is reached approximately after about 20 periods.

Beats. When the frequency of the driving force is not equal to the resonance frequency, and if the damping is sufficiently small, the transition to steady state exhibits 'beats', i.e. variations in the amplitude, as shown in Display 4.6. These beats result from the interference between the steady state motion, with the frequency ω of the driving force, and the transient motion, with the frequency ω' of the free motion of the oscillator.

The curves shown in the display refer to a driving frequency 1.1 times the resonance frequency ω_0 and the values 0.01, 0.02, and 0.04 of the damping parameter D (corresponding Q-values are 100, 50 and 25). The

Display 4.6.

Approach to steady state. Driving force: $F(t) = F_0 \cos(\omega t)$ for $t > 0$, $F(t) = 0$ for $t < 0$, with $\omega = 1.1\ \omega_0$, $D = R/\omega_0$, $M = 0.01, 0.02$, and 0.04. Normalized displacement $= \xi/(F_0/K)$. $T_0 = 2\pi/\omega_0$

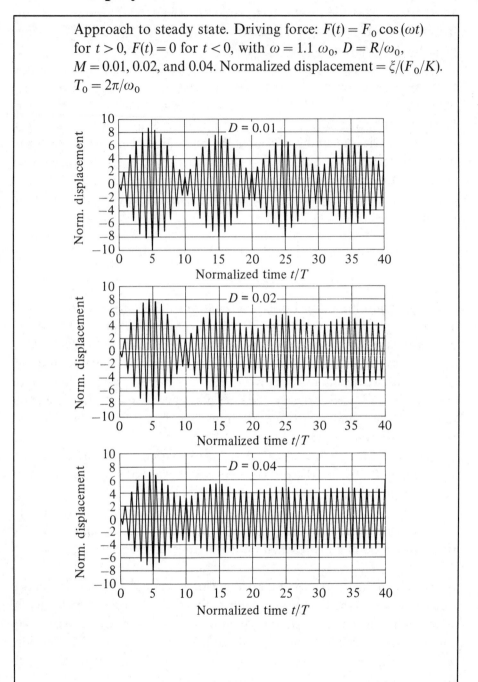

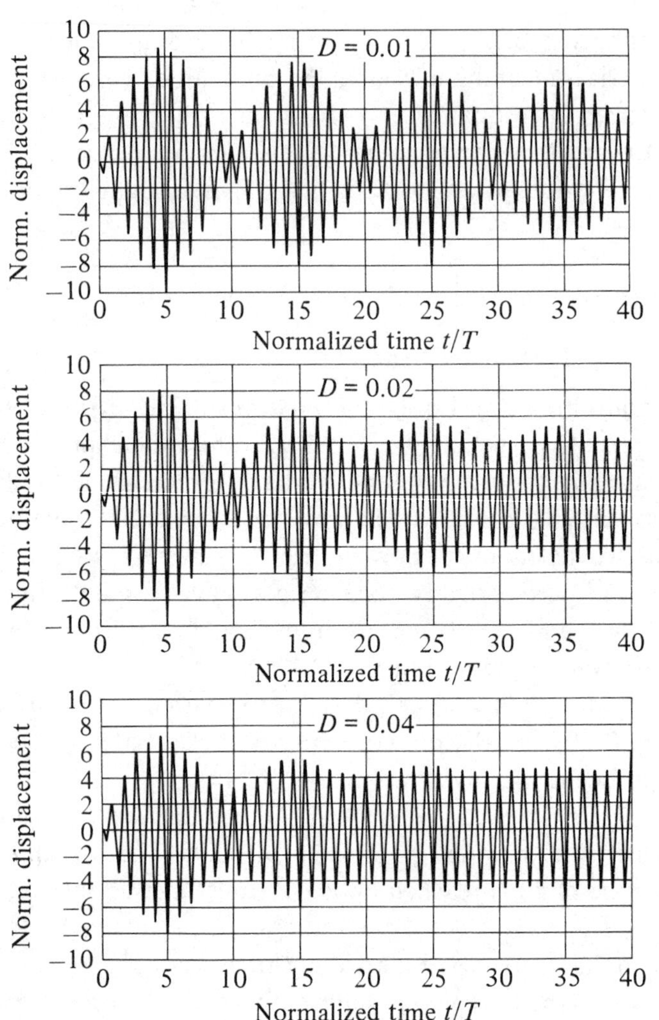

frequency ω' of the transient motion is approximately equal to the resonance frequency ω_0, and the interference between the two motions periodically goes from destructive to constructive as the phase difference $(\omega - \omega')t \simeq (\omega - \omega_0)t$ between the two motions increases with time.

With $\Delta\omega = \omega - \omega'$, the phase difference is increased by 2π in a time interval Δt given by $\Delta\omega\Delta t = 2\pi$, which in the present case yields $0.1\omega'\Delta t \simeq 2\pi$ or $\Delta t/T' \simeq 10$, where $T' \simeq T_0$ is the period of free motion. This will be the time interval between two successive maxima or minima in the resulting displacement function, which is consistent with the result shown in the display.

At a driving frequency below the resonance frequency a similar result is obtained. For example, with $\omega/\omega_0 = 0.9$, the curves are much like those in the display except that they start out in the positive rather than the negative direction.

If the driving frequency is brought sufficiently close to the resonance frequency, the time interval between beats will be so large that the amplitude of the transient will be reduced so much in this interval that the beats will be less pronounced. For a more detailed discussion of this question we refer to the problem section.

4.5 Discussion

The various oscillators mentioned in connection with demonstration of the frequency response in the last chapter can be used, of course, also for demonstration of impulse response and free motion.

For example, the transition from oscillatory to non-oscillatory decay of a free oscillator can be demonstrated particularly well by means of pulse excitation of an electrical resonator circuit in which the resistance can be varied. The repetition rate of the pulse generator is chosen small enough so that the time between pulses is large compared to the characteristic decay time τ of the circuit. Displaying the voltage over the capacitor on an oscilloscope as the resistance is varied, we can readily demonstrate oscillatory, overdamped, and critically damped decay.

Another instructive demonstration is the excitation of an acoustic cavity (Helmholtz) resonator with repeated harmonic wave trains (pulse modulated sound wave), as shown in Display 4.7.

A microphone inside the cavity of the resonator measures the sound pressure, and the corresponding signal from the microphone is displayed on one channel of a dual beam oscilloscope. On the other channel is shown the input voltage to the loudspeaker, which produces the incident sound. As can be seen, the amplitude of this sound is constant during the duration of each 'pulse'.

Display 4.7.

Demonstration of beats. Pressure response of an acoustic resonator to a pulse modulated incident tone with (carrier) frequency ν. Resonance frequency $= \nu_0$. (a) $\nu = \nu_0$. (b) $\nu < \nu_0$.

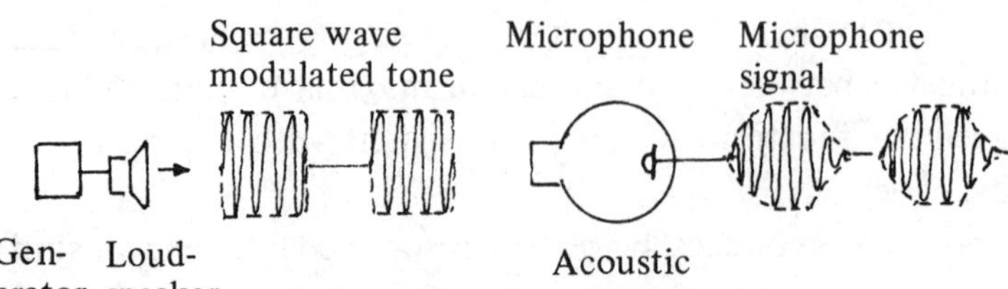

(a) $\nu = \nu_0 = 88$ Hz

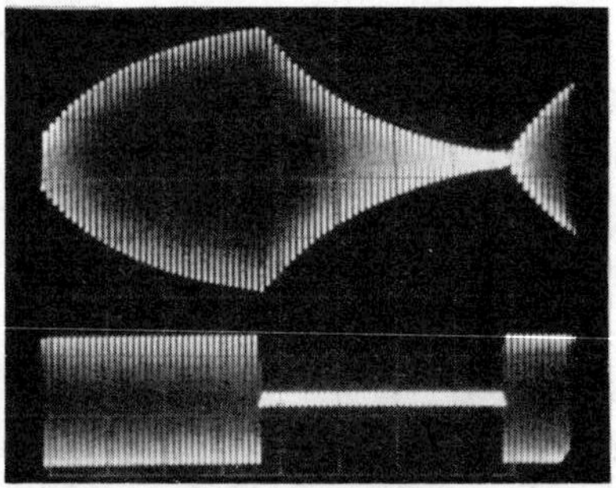

(b) $\nu = 82$ Hz

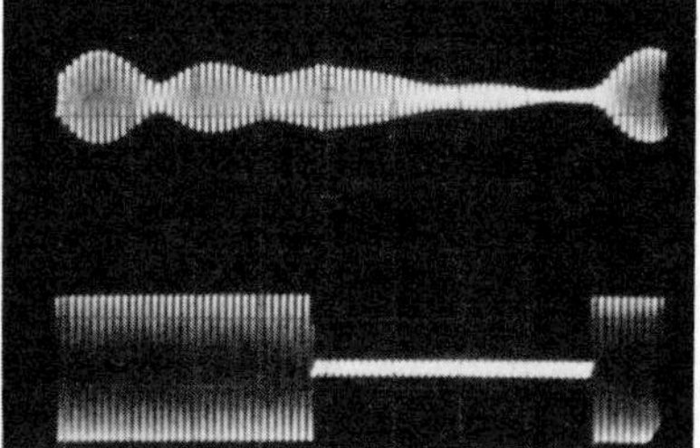

The time dependence of the pressure in the resonator is quite different from that of the incident pressure. (The sound pressure in the cavity, incidentally, corresponds to the compression of the equivalent mass–spring oscillator or the voltage over the capacitor in an electrical circuit). At the beginning of each pulse, it shows the characteristic growth of the amplitude toward the steady state value, similar to the curves in Displays 4.5 and 4.6. Before the steady state is reached, however, the pulse is terminated, and the pressure in the cavity decays exponentially. This process is repeated for each pulse.

Photographs of the response curves obtained for a resonator with a resonance frequency of 88 Hz are shown in the display. In the first, the frequency (the 'carrier' frequency) of the incident sound is equal to the resonance frequency, and the growth and (exponential) decay are monotonic. During the decay of the pressure, the resonator re-radiates sound and can be heard as a 'reverberation' after the incident sound has been shut off. (Resonators of this kind were built into the walls under the seats in some Greek open air amphitheaters).

In the second example, the frequency of the incident sound is somewhat lower than the resonance frequency, and the beats resulting from the interference of the free and forced oscillations occur, as expected from the results in Display 4.6. A similar result is obtained if the carrier frequency is somewhat higher than the resonance frequency. For further details, we refer to the problem section.

Examples

E4.1 *Radiation damping of electric oscillator.* According to Eq. 9.5 (3) the instantaneous power radiated from an accelerated charged particle is $(1/4\pi\varepsilon_0)2q^2a^2/3c^3$, where a is the acceleration and q the electric charge.

(a) Apply this result to the case of harmonic motion of the particle and calculate the average power radiated in one period.

(b) Assume the radiated energy in one period to be small compared to the energy of oscillation, determine the decay constant of the free motion of an elastically bound charged particle and the corresponding Q-value of the oscillator.

Solution. (a) With the harmonic displacement $\xi = \xi_0 \cos(\omega t)$, the acceleration becomes $a = -(\omega^2\xi_0)\cos(\omega t)$ and the time average of a^2 over one period becomes $\frac{1}{2}\omega^4\xi_0^2$. The corresponding average radiated power in one period is then

$$\langle P \rangle = (1/4\pi\varepsilon_0)(2/3)q^2\omega^4\xi_0^2/2c^3 \qquad (1)$$

(b) If the radiation loss is small, the frequency of the free oscillations can be approximated by $\omega_0 = \sqrt{K/M}$, where K is the spring constant and M the mass of the particle.

The energy of oscillation and the amplitude of oscillation will vary with time and we put $\xi(t) = \xi_0(t)\cos(\omega_0 t)$, where the variation of the amplitude will be slow in comparison to ω_0 for small losses.

Then, in calculating the average energy of oscillation over one period, we can neglect the variation in $\xi_0(t)$ so that the time average of the potential energy becomes $\frac{1}{4}K\xi_0^2(t)$. The time average of the kinetic energy has the same value and the average of the total energy is

$$\langle E(t)\rangle = \tfrac{1}{2}K\xi_0^2(t) = \tfrac{1}{2}M\omega_0^2\xi_0^2(t) \tag{2}$$

The radiated power in Eq. (1) can be expressed as

$$\langle P\rangle = (1/4\pi\varepsilon_0)(2/3)(q^2\omega_0^2/Mc^3)\langle E(t)\rangle \tag{3}$$

where the bracket indicates time average over one period.

From the energy balance equation $\mathrm{d}\langle E\rangle/\mathrm{d}t = -\langle P\rangle$ follows then

$$\mathrm{d}\langle E\rangle/\mathrm{d}t = -(1/\tau)\langle E\rangle \quad \langle E\rangle = A\exp(-t/\tau) \tag{4}$$

$$1/\tau = (1/4\pi\varepsilon_0)(2/3)(q^2\omega_0^2/Mc^3) \tag{5}$$

where $1/\tau$ is the decay constant and τ the corresponding 'life' time of the oscillator.

The Q-value of an oscillator is defined as $Q = \omega_0 M/R = 2\pi M/RT_0$, where R is the friction constant. It can be written also as $Q = 2\pi M\langle u^2\rangle/T_0 R\langle u^2\rangle$, where we have multiplied both denominator and numerator by the average of u^2 over one period, where u is the velocity. For small damping, $M\langle u^2\rangle$ is the average oscillator energy and $T_0 R\langle u^2\rangle$ is the energy dissipated in one period,

$$Q = 2\pi\langle E\rangle/T_0\langle P\rangle = \omega_0\langle E\rangle/\langle P\rangle = \omega_0\tau \tag{6}$$

where we have used Eq. (3).

If we let the charged particle represent an atomic electron, we have $|q| = 1.6 \times 10^{-19}$ C, $M = 9.1 \times 10^{-31}$ kg, and the frequency of oscillation typically is $\omega_0/2\pi = 10^{15}$ Hz. Then, with $1/4\pi\varepsilon_0 = 9 \times 10^9$ and $c = 3 \times 10^8$ m/s, Eq. (5) gives $\tau \simeq 0.4 \times 10^{-8}$ sec and from Eq. (6) follows $Q \simeq 2.5 \times 10^7$.

The result obtained here is consistent with the generally accepted order of magnitude of 10^{-8} sec for the life time of the oscillation which is associated with an electronic transition in an atom. The corresponding length of the emitted wave train will be $c\tau \simeq 3$ m.

E4.2 *Effect of mass on pendulum decay.* A uniform sphere at the end a string oscillates as a simple pendulum in air. The amplitude decay is due to the friction force (drag force) of the air. It is found that the amplitude is reduced by a factor of e in N periods of oscillation. Assuming the friction force to be proportional to the radius of the sphere (Stokes' law), how many periods of oscillation are required to reduce the amplitude by the same factor if the radius of the sphere is doubled?

Solution. The decay of the displacement of the pendulum is expressed by the factor $\exp(-\gamma t)$, where $\gamma = R/2M$. Since the mass M increases as the cube of the radius and the friction constant R as the first power, it follows that the characteristic decay time ('life' time) $\tau = 1/\gamma$ will be proportional to the square of the radius. Thus a doubling of the radius increases the decay time by a factor of 4, so that $4N$ periods are required to reduce the amplitude by the factor e (assuming that the damping is small so that the period can be approximated by the undamped value).

Problems

4.1 *Complex frequencies.* Solve the frequency equation D.4.1(3) for the free oscillation of a mass–spring oscillator with $M = 1$ kg and $K = 100$ N/m for the following values of the damping parameter $D = R/\omega_0 M$:

(a) 1. (b) 2. (c) 4.

In each case indicate the location of the complex frequencies in the complex plane and determine the decay rate and period of oscillation.

4.2 *Overdamped oscillator.* (a) The mass element in a mass–spring oscillator is so small that the corresponding inertia force can be neglected in comparison with the friction force and the spring force. At $t = 0$ the mass element is released from zero velocity and a displacement A from the equilibrium position. What is the subsequent motion of the oscillator?
(b) A resistance R is connected in parallel with a capacitance C. At $t = 0$ the charge on the capacitor is Q. Determine the time dependence of the charge on the capacitor. At what time is the charge 1/e of the initial value?
(c) A small sphere is started at $t = 0$ with a velocity U in a viscous fluid. The friction force is $-Ru$. Determine the time dependence of the velocity u. How far does the sphere travel before it comes to rest?

4.3 *Critically damped oscillator.* Show by direct insertion into the differential equation $\ddot{\xi} + 2\gamma\dot{\xi} + \omega_0^2\xi = 0$ that when $\gamma = \omega_0$, the solution is $\xi = (A + Bt)\exp(-\omega_0 t)$.

4.4 *Impulse response.* A critically damped oscillator is given a unit impulse at $t = 0$. Determine the subsequent motion. Show how this result can be obtained from the general impulse response function in Eq. D.4.3 (4) in the limit when $\gamma = \omega_0 (\omega' = 0)$.

4.5 *Forced motion. Non-harmonic force.* A damped oscillator is driven by the force $F(t) = A\exp(-\gamma t)$ for $t > 0$ and $F(t) = 0$ for $t < 0$, where $\gamma = R/2M$. Calculate the resulting displacement as a function of time.

4.6 *Kinetic energy in decaying motion.* With reference to Display 4.2, determine the maximum kinetic energy of the various motions and the corresponding times of occurrence.

4.7 *Impulse response. Energy considerations.* With reference to Display 4.3, consider an oscillator with $M = 2$ kg and a resonance period of one second. It is given an impulse of 10 Newton-sec at $t = 0$.

(a) For each of the types of decaying motion in the display, determine the maximum excursion of the mass element and the corresponding time of occurrence.

(b) In each case determine the amount of mechanical energy lost in the motion to the point of maximum excursion.

4.8 *Secular growth.* An undamped oscillator is driven at its resonance frequency (i.e. $\omega = \omega_0$) by a force defined by $F(t) = F_0 \cos(\omega t)$ for $t > 0$ and $F(t) = 0$ for $t < 0$. Using Eq. D.4.5 (5), show that the displacement is given by Eq. D.4.5 (9). Show also that this result follows from the general result in Eq. D.4.5 (7) in the limit when $\mathbf{\Omega} = 1$ ($D = 0$, $\omega' = \omega_0$).

4.9 *Beats between steady state and transient response.* Accepting Eq. D.4.5 (7) as the displacement of a damped oscillator driven by the force described in Eqs. D.4.5. (1)–(2), we note that beats can result from the interference between the steady state motion (frequency ω) and the transient motion (frequency ω'). For small values of the decay constant $\gamma = R/2M$, the time T_b between successive maxima (minima) is given approximately by $|(\omega - \omega_0)| T_b = 2\pi$ or $T_b \simeq T_0/|\Omega - 1|$, where $\Omega = \omega/\omega_0$ and $T_0 = 2\pi/\omega_0$.

The closer the driving frequency is to the resonance frequency, the larger is T_b and the greater will be the amplitude decay of the amplitude between beats. For a given value of Ω, determine the value of the damping constant D above which the difference between the maximum and the subsequent minimum is less than, say, 20%.

4.10 *Demonstration of beats. Pulse excitation of resonator.* In a lecture demonstration, an acoustic cavity resonator was excited by a pulse modulated wave train as indicated in Display 4.7.

(a) The first curve shows the measured sound pressure in the resonator when the 'carrier' frequency of the incident sound equals the resonance frequency. From the decay portion of this response curve, determine the Q-value of the resonator. ($Q = \omega_0 M/R = \pi/\gamma T_0$, where T_0 is the period of undamped oscillation and $\gamma = R/2M$. For small damping, $T_0 \cong T'$).

(b) In the second curve, the carrier frequency is 82 Hz, i.e. lower than the resonance frequency 88 Hz, and through the interference of the steady state and the transient responses, beats occur. Determine the time (beat period) between two successive minima and compare the result with the predicted.

5

Coupled oscillators

We start this chapter with a study of the forced harmonic motion of two coupled oscillators. From the resulting frequency response, the two characteristic or normal modes of the system are identified, which are studied further in the analysis of the free motion of the system. The motion resulting from given initial conditions is determined, with particular emphasis on the impulse response. In terms of this, the motion produced by an arbitrary driving force is obtained. As matters of general interest, the concepts of configuration space and normal coordinates are discussed.

5.1 **Forced harmonic motion**

As for the single oscillator, we begin our study of coupled oscillators with forced harmonic motion. This is simpler than the free motion in as much as the frequency is known *a priori* (the frequency of the driving force). To illustrate the basic procedure involved in as simple a manner as possible, we consider two equal oscillators, as indicated in Display 5.1.

Each oscillator has a mass M and a spring constant K, and they are coupled by means of a spring with a spring constant K_{12}. Damping is neglected. The harmonic force is applied to mass element #1, as shown, and the object is to determine the frequency dependence of the displacement amplitude and the phase angle for each oscillator. From the result thus obtained, the displacement produced by a driving force applied to mass element #2 follows by analogy. With both forces acting simultaneously, the resulting motion is obtained by linear superposition of the individual displacements.

The equilibrium coordinates of the mass elements are denoted by x_1 and x_2 and the displacements from equilibrium by $\xi(x_1, t)$ and $\xi(x_2, t)$, or simply $\xi(1, t)$ and $\xi(2, t)$.

Display 5.1.

Forced harmonic motion of two coupled oscillators

Differential equations

Equilibrium positions x_1, x_2 (1)

Displacement $\xi(x_1, t)$, $\xi(x_2, t)$ or simply $\xi(1, t)$, $\xi(2, t)$ (2)

Driving force $F(x_1, t)$ or simply $F(1, t)$

$M\ddot{\xi}(1,t) = -K\xi(1,t) - K_{12}[\xi(1,t) - \xi(2,t)] + F(1,t)$ (3)

$M\ddot{\xi}(2,t) = -K\xi(2,t) - K_{12}[\xi(2,t) - \xi(1,t)]$ (4)

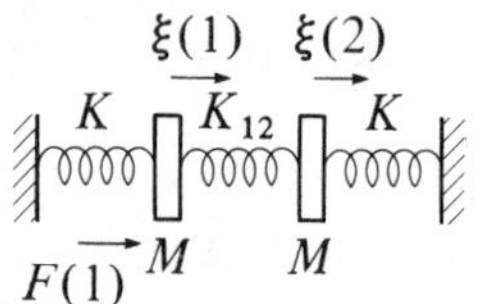

Complex amplitude equations

Let $a = (K + K_{12})/M$, $c = K_{12}/M$ (5)

Complex amplitudes $\xi(1), \xi(2)$ and $F(1)$

$(\omega^2 - a)\xi(1) + c\xi(2) = -f(1)$, $f(1) = F(1)/M$ (6)

$c\xi(1) + (\omega^2 - a)\xi(2) = 0$ (7)

Solution

$\xi(1) = -f(1)(\omega^2 - a)/N$ (8)

$\xi(2) = f(1)c/N$ (9)

$$N = (\omega^2 - a)^2 - c^2 = [(\omega^2 - a) - c][(\omega^2 - a) + c] = (\omega^2 - \omega_1^2)(\omega^2 - \omega_2^2) \quad (10)$$

Resonance frequencies $[N = 0 \quad (\omega^2 - a) = \pm c]$

$\omega_1^2 = a - c = K/M$ (11)

$\omega_2^2 = a + c = (K + 2K_{12})/M$ (12)

Displacement ratio

$\xi(2)/\xi(1) = -c/(\omega^2 - a)$ (13)

Mode 1 $\omega_1^2 = a - c$ $\xi_1(2)/\xi_1(1) = 1$ (14)

Mode 2 $\omega_2^2 = a + c$ $\xi_2(2)/\xi_2(1) = -1$ (15)

As a result of the displacements, the first spring will be stretched by an amount $\xi(1,t)$, the coupling spring compressed by $\xi(1,t)-\xi(2,t)$, and the remaining spring compressed by $\xi(2,t)$. The corresponding total spring force acting on the first mass element becomes $-K\xi(1,t)-K_{12}[\xi(1,t)-\xi(2,t)] = -(K+K_{12})\xi(1,t)+K_{12}\xi(2,t)$; the spring force on the second mass element has the same form with $\xi(1,t)$ and $\xi(2,t)$ interchanged.

The equations of motion for the two mass elements are then

$$M\ddot{\xi}(1,t) = -(K+K_{12})\xi(1,t)+K_{12}\xi(2,t)+F(1,t) \tag{1}$$

$$M\ddot{\xi}(2,t) = -(K+K_{12})\xi(2,t)+K_{12}\xi(1,t) \tag{2}$$

The two displacement variables occur in each of these equations, and, like the oscillators, the equations are said to be coupled. To make them somewhat more compact, we introduce the quantities $a=(K+K_{12})/M$ and the 'coupling constant' $c=K_{12}/M$ and obtain

$$\ddot{\xi}(1,t) = -a\xi(1,t)+c\xi(2,t)+F(1)/M \tag{3}$$

$$\ddot{\xi}(2,t) = -a\xi(2,t)+c\xi(1,t) \tag{4}$$

In steady state, the displacements will be harmonic functions with the same angular frequency as the driving frequency ω because of the linearity of the system (no products of the field variables or their derivatives occur). Each of these harmonic displacements is described by an amplitude and a phase angle and our task is to determine the frequency dependence of these quantities in terms of the amplitude and phase of the driving force.

As in the case of the forced motion of a single oscillator, the analysis is facilitated by introducing the complex amplitudes of the variables involved. We shall denote the complex displacement amplitudes by $\xi(1,\omega)$ and $\xi(2,\omega)$, or simply by $\xi(1)$ and $\xi(2)$. Similarly, the complex force amplitude is $F(1,\omega)$. To avoid any possible confusion, we shall denote the time dependence of the real quantities by $\xi(1,t)$, $\xi(2,t)$, and $F(1,t)$. Actually, since we generally are interested only in the amplitude and the phase angle, there is seldom need to deal with the complete time dependence, unless specifically desired.

With reference to previous chapters we obtain the equations for the complex amplitudes simply by replacing $\mathrm{d}/\mathrm{d}t$ by the operator $(-\mathrm{i}\omega)$, so that $\dot{\xi}(1,t)$ turns into $(-\mathrm{i}\omega)\xi(1)$ and $\ddot{\xi}(1,t)$ becomes $(-\mathrm{i}\omega)^2\xi(1) = -\omega^2\xi(1)$, with similar expressions for the other displacement variable. Since resistive forces have been omitted, we have no velocity terms $\dot{\xi}(1,t)$ and $\dot{\xi}(2,t)$, but they will be included later in this section, when the effect of damping is analyzed.

From the two previous equations, we then obtain the corresponding equations for the complex amplitudes

$$\left.\begin{aligned}&(\omega^2-a)\xi(1)+c\xi(2)=-f(1)\quad f(1)=F(1)/M\\&c\xi(1)+(\omega^2-a)\xi(2)=0\end{aligned}\right\}\tag{5}$$

The two unknown complex displacement amplitudes then can be expressed as follows

$$\left.\begin{aligned}&\xi(1)=-f(1)(\omega^2-a)/[(\omega^2-a)^2-c^2]\\&\xi(2)=f(1)c/[(\omega^2-a)^2-c^2]\end{aligned}\right\}\tag{6}$$

The amplitudes are proportional to the force amplitude. We recall that the force is applied to mass element #1, and it is interesting to note that the amplitude of mass element #2 is proportional to the coupling constant c.

Resonances. It is significant that both amplitudes can become infinite, since the denominator can be zero. Such resonance is familiar from the forced motion of the undamped single oscillator. The essential difference is that instead of having one resonance frequency we now have two, being solutions to the equation

$$(\omega^2-a)^2-c^2=0\quad\text{or}\quad\omega^2-a=\pm c\tag{7}$$

$$\omega_1^2=a-c=K/M\tag{8}$$

$$\omega_2^2=a+c=(K+2K_{12})/M\tag{9}$$

The ratio between the displacement amplitudes is frequency dependent, and it follows from the equations above that

$$\xi(2)/\xi(1)=-c/(\omega^2-a)\tag{10}$$

It is of particular interest to determine this ratio at the resonance frequencies. Thus, by introducing these frequencies we obtain

$$\text{Mode 1:}\quad\omega_1^2=a-c\quad\xi_1(2)/\xi_1(1)=1\tag{11}$$

$$\text{Mode 2:}\quad\omega_2^2=a+c\quad\xi_2(2)/\xi_2(1)=-1\tag{12}$$

In other words, at the first resonance, the amplitudes of the two mass elements are the same; they move in the same direction and the length of the coupling spring between them remains unchanged. At the second resonance, the two displacements are equal in amplitude but opposite in direction. As we shall see shortly, these are the characteristics of the two normal modes of free motion of the system.

In Display 5.2 we have considered an example in which the spring constants are all the same and equal to K. As before, damping is neglected and the harmonic driving force is acting on mass element #1, as shown.

In the limit of zero frequency the displacements of the two elements are

Display 5.2.

Example of forced harmonic motion of two coupled oscillators.

$\xi(1)/\xi'(1)$: Solid lines. $\xi(2)/\xi'(1)$: Dashed lines.

Example (*Reference: Display 5.1*)

$K_{12} = K \quad a = (K + K_{12})/M = 2K/M \quad c = K_{12}/M = K/M$ (1)

$\omega_1^2 = a - c = K/M \quad \omega_2^2 = (a + c) = 3K/M \quad \omega_2 = \sqrt{3}\omega_1$ (2)

For $\omega = 0 \quad \xi'(1) = f(1)a/\omega_1^2\omega_2^2 = (2/3)F(1)/K$

$\xi'(2) = \frac{1}{2}\xi'(1)$ (3)

Let $\Omega = \omega/\omega_1$ (4)

$\xi(1)/\xi'(1) = -(3/2)(\Omega^2 - 2)/(\Omega^2 - 1)(\Omega^2 - 3)$ (5)

$\xi(2)/\xi'(1) = (3/2)/(\Omega^2 - 1)(\Omega^2 - 3)$ (6)

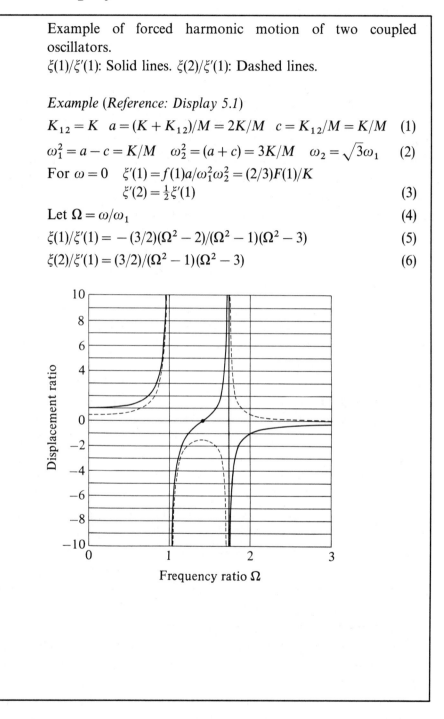

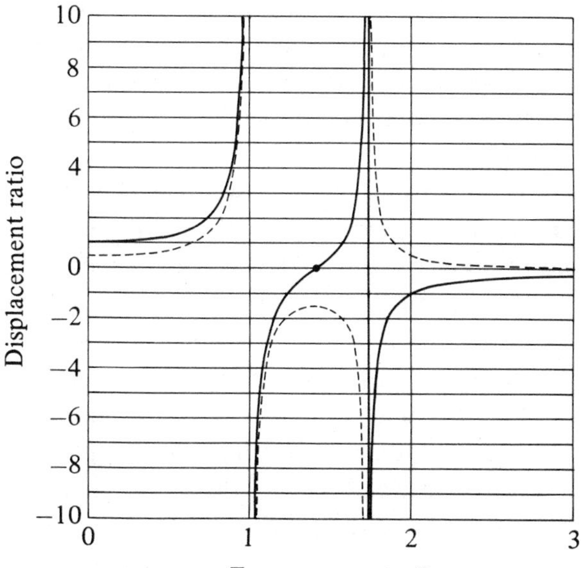

the same as if a static force $F(1)$ had been applied, and it is instructive to check Eq. D.5.2(3) from statical considerations.

If we choose the driving force to have zero phase angle, the force parameter $f(1)$ is real. In this case of no damping, the complex displacement amplitudes (Eqs. D.5.1(8)–(9)) are real also; if a displacement amplitude is positive, it is in phase with the driving force, and if it is negative, it has a phase angle π; the direction of the displacement being opposite that of the force.

In presenting the frequency dependence of the displacement amplitudes graphically, we have normalized the displacements with respect to the displacement of element #1 at zero frequency, and the frequency is normalized with respect to the resonance frequency ω_1. Thus, the solid curve, which refers to element #1, starts out from the value 1 at zero frequency. The corresponding displacement for the second element (dashed curve) is 1/2.

As the frequency approaches the first resonance, both amplitudes approach infinity, the ratio between them approaching 1, both being in phase with the force. As we pass through the first resonance, the direction of both displacements become opposite that of the force, the ratio between the displacements still being 1 at the resonance frequency. This shift in phase of 180 degrees is the same as for a single undamped oscillator.

As the second resonance is approached, the displacement of element #1 (the driven element) goes to plus infinity, whereas the amplitude of the second element goes to minus infinity, their ratio approaching -1. Going through the second resonance yields a 180 degree change of phase of both displacements. As the frequency is increased further, both displacement amplitudes approach zero monotonically. The relationship between the displacements in the vicinity of the resonances is characteristic of the two modes of motion of the system, to be discussed further in the next section.

It is interesting to notice that in the frequency region between the resonances, the displacement amplitude of the driven element goes from minus to plus infinity and in the process goes to zero at the angular frequency $\omega = \sqrt{a} = \sqrt{2}\omega_1$. The fact that the displacement of the driven element is zero can be used in practice for vibration damping of machines, as discussed in one of the problems.

To demonstrate this response experimentally, we can use air suspended carts on a track, as in the study of the one-dimensional oscillator. It is difficult, however, to generate a harmonic driving force with constant amplitude. For this reason it is simpler to consider an equivalent system, in which the left spring is driven with a harmonic displacement $A\cos(\omega t)$, rather than being held fixed, as in Display 5.1. The equations of motion will be the same as before, with the complex force amplitude $F(1)$ replaced by KA.

With no damping in the system, the complex displacement amplitudes are real (assuming the complex force amplitude to be real). The phase angle, 0 or 180 degrees, is accounted for by the positive or negative sign. Thus, in Display 5.2, both the amplitude and the phase angle are contained in one and the same curve. In the presence of damping, however, this presentation is no longer possible, and we have to plot the amplitude (magnitude of the complex amplitude) and the phase angle separately, as we did in Display 3.7 for the single damped oscillator. This is demonstrated in the following example.

Effect of damping. The analysis of forced motion of coupled oscillators readily can be extended to include damping (for free motion it is more complicated), and we shall demonstrate the effect of damping in the case when the coupled oscillators are equal. The resistance is introduced by means of dashpot dampers in parallel with the springs, as indicated in Display 5.3. The corresponding friction constants are denoted by R and R_{12}, the latter referring to the coupling spring.

It follows, as in Chapter 3, that we can account for the damping simply by replacing each spring constant K by $K - i\omega R$, where R is the associated dashpot resistance coefficient. Apart from the fact that the parameters $a = (K + K_{12})/M$ and $c = K_{12}/M$ are now complex numbers, the complex amplitude equations and the expressions for the complex displacement amplitudes obtained for the undamped system are still valid,

$$\left.\begin{aligned} \xi(1) &= -f(1)(\omega^2 - a)/[(\omega^2 - a)^2 - c^2] \\ \xi(2) &= f(1)c/[(\omega^2 - a)^2 - c^2] \\ a &= (K' + K'_{12})/M \quad c = K'_{12}/M \\ K' &= K - \mathrm{i}\omega R \quad K'_{12} = K_{12} - \mathrm{i}\omega R_{12} \end{aligned}\right\} \tag{13}$$

To illustrate quantitatively the effect of damping we consider again the special case of Display 5.2 in which $K_{12} = K$, and we let $R_{12} = R$. For computational purposes, as before, we introduce the normalized frequency $\Omega = \omega/\omega_0$, and the damping parameter $D = R/\omega_0 M$, where $\omega_0 = \sqrt{K/M}$. In terms of these quantities we obtain $a = (K' + K'_{12})/M = 2K'/M = 2\omega_0^2(1 - \mathrm{i}\Omega D)$ and $c = \omega_0^2(1 - \mathrm{i}\Omega D)$.

The solutions for the complex displacement amplitudes in terms of $f(1)$ now will be ratios between complex numbers, in which the imaginary parts contain the quantity ΩD. The magnitude (amplitude) of these ratios and the phase angles are obtained in the usual manner, and in Display 5.3 we

Display 5.3.

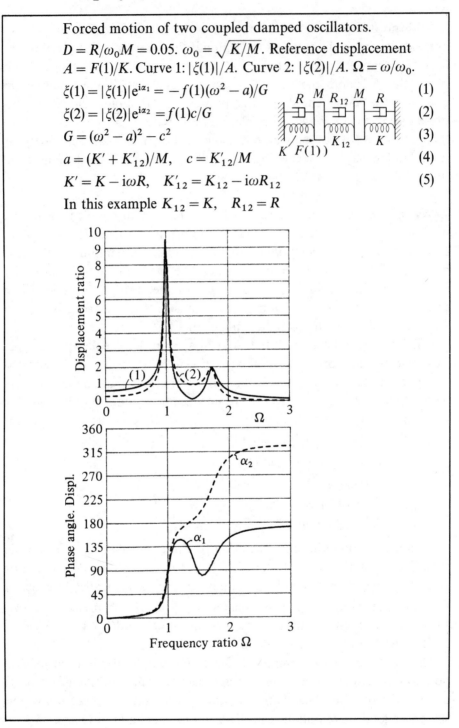

Forced motion of two coupled damped oscillators.
$D = R/\omega_0 M = 0.05$. $\omega_0 = \sqrt{K/M}$. Reference displacement
$A = F(1)/K$. Curve 1: $|\xi(1)|/A$. Curve 2: $|\xi(2)|/A$. $\Omega = \omega/\omega_0$.

$$\xi(1) = |\xi(1)|e^{i\alpha_1} = -f(1)(\omega^2 - a)/G \quad (1)$$

$$\xi(2) = |\xi(2)|e^{i\alpha_2} = f(1)c/G \quad (2)$$

$$G = (\omega^2 - a)^2 - c^2 \quad (3)$$

$$a = (K' + K'_{12})/M, \quad c = K'_{12}/M \quad (4)$$

$$K' = K - i\omega R, \quad K'_{12} = K_{12} - i\omega R_{12} \quad (5)$$

In this example $K_{12} = K$, $R_{12} = R$

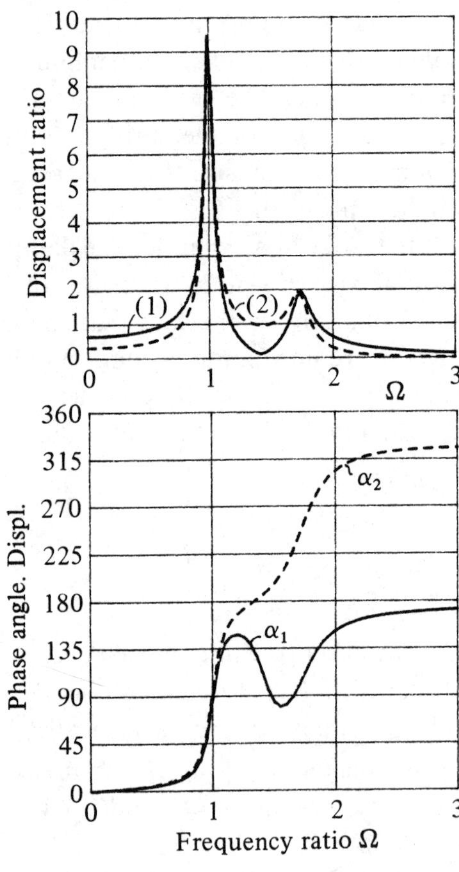

have plotted the amplitude and phase angle as functions of the normalized frequency for a particular value of the damping parameter. The maxima in the two amplitudes occur very nearly at the two resonance frequencies ω_1 and ω_2 familiar from the undamped oscillator.

The forced harmonic motion of two unequal coupled oscillators is obtained in a similar manner.

5.2 Free motion. Normal modes

We shall continue to consider the two identical coupled oscillators, now in free motion. As was demonstrated in the forced harmonic motion, the displacement amplitudes become infinite at the two resonance frequencies of the system, and the ratio between the complex displacement amplitudes at these frequencies are 1 and -1, respectively. This result suggests that, in the absence of damping, the system can oscillate in harmonic motion at these frequencies without a driving force.

In other words, if the two oscillators initially are given the same displacements from their equilibrium positions and then released, we expect the subsequent free motion of each of the mass elements to be harmonic with the same frequency $\omega_1 = \sqrt{K/M}$. That this is indeed so follows from the fact that in this **symmetric** mode of motion, the length of the coupling spring remains unchanged, and the oscillators move as if the coupling spring were not present. Therefore, the frequency of oscillation of each of the elements will be the same as for the uncoupled oscillators, $\omega_1 = \sqrt{K/M}$.

Similarly, if we start the oscillators from equal but opposite displacements, the subsequent motion of each element will be harmonic with the frequency $\omega_2 = \sqrt{(K + 2K_{12}/M}$. This mode of motion is **anti-symmetric** in the sense that the displacements of the two elements are equal but opposite. The midpoint of the coupling spring remains at rest, and the motion of each element is as if it were connected to two springs with spring constants K and $2K_{12}$ (half of the coupling spring) which are clamped at the ends.

These two motions, called the **normal modes**, are illustrated in Display 5.4. The important characteristic of a normal mode is that the motion is harmonic and the frequency is the same for the two mass elements. This property can be taken as the definition of a normal mode of motion.

A comment should be made about the notation which we have used. As before, the number in an argument for the displacement is a shorthand notation for the equilibrium coordinate. The subscript identifies the mode.

In the particular example of identical oscillators, we were able to determine the normal modes from symmetry considerations. If the mass elements and the spring constants are different, however, the determination

Display 5.4.

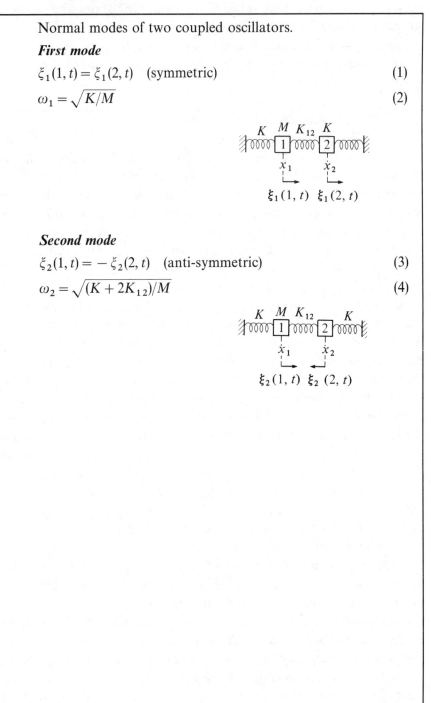

Normal modes of two coupled oscillators.

First mode

$\xi_1(1, t) = \xi_1(2, t)$ (symmetric) (1)

$\omega_1 = \sqrt{K/M}$ (2)

Second mode

$\xi_2(1, t) = -\xi_2(2, t)$ (anti-symmetric) (3)

$\omega_2 = \sqrt{(K + 2K_{12})/M}$ (4)

of the normal mode displacements by trial and error can be a lengthy process. We then have to turn to the systematic mathematical modal analysis of free motion, which will be discussed later in this section, or extract the normal modes from a study of the forced harmonic motion of the system, as illustrated in the special case in the previous section.

Free motion determined by initial conditions. We recall that the general free motion of the undamped one-dimensional oscillator is given by the displacement function $\xi = \xi_0 \cos(\omega_0 t - \alpha)$, with the two constants ξ_0 and α determined by the known displacement and velocity of the oscillator at a given time. A motion is uniquely determined by the displacement and the velocity, and since the displacement function satisfies both these conditions and the equation of motion, it is the unique solution for all t.

Similarly, for two coupled oscillators, the motion is uniquely determined by the displacement and the velocity of each of the two oscillators at a given time. For each oscillator, a displacement function, which satisfies these conditions as well as the equations of motion, is obtained as a linear superposition of the normal mode motions, which, for the case of two identical coupled oscillators, are

$$\xi_1(1,t) = \xi_1(2,t) = A_1 \cos(\omega_1 t - \alpha_1) \tag{1}$$

$$\xi_2(1,t) = -\xi_2(2,t) = A_2 \cos(\omega_2 t - \alpha_2) \tag{2}$$

Thus, superimposing these solutions, we obtain the general expression for the displacement of the two mass elements

$$\xi(1,t) = \xi_1(1,t) + \xi_2(1,t) = A_1 \cos(\omega_1 t - \alpha_1) + A_2 \cos(\omega_2 t - \alpha_2) \tag{3}$$

$$\xi(2,t) = \xi_1(2,t) + \xi_2(2,t) = A_1 \cos(\omega_1 t - \alpha_1) - A_2 \cos(\omega_2 t - \alpha_2) \tag{4}$$

These expressions contain four unknown constants, A_1, A_2, α_1, and α_2, which can be determined from known values of displacements and velocities at a given time. It follows then from the uniqueness of the motion that these expressions, satisfying both the equations of motion as well as the 'initial' conditions, represent the displacements at all times.

Let us assume that the system is started at $t = 0$ from the displacements $\xi(1,0)$, $\xi(2,0)$ and the velocities $\dot{\xi}(1,0)$, $\dot{\xi}(2,0)$. The equations for the determination of the four unknown constants are then

$$\left.\begin{aligned}
\xi(1,0) &= A_1 \cos(\alpha_1) + A_2 \cos(\alpha_2)\\
\xi(2,0) &= A_1 \cos(\alpha_1) - A_2 \cos(\alpha_2)\\
\dot{\xi}(1,0) &= \omega_1 A_1 \sin(\alpha_1) + \omega_2 A_2 \sin(\alpha_2)\\
\dot{\xi}(2,0) &= \omega_1 A_1 \sin(\alpha_1) - \omega_2 A_2 \sin(\alpha_2)
\end{aligned}\right\} \tag{5}$$

Display 5.5.

Free motion of two coupled equal oscillators.
Ref. Eqs. 5.1.(1)–(2) with $F=0$

$$M\ddot{\xi}(1,t)=-K\xi(1,t)-K_{12}[\xi(1,t)-\xi(2,t)] \quad (1)$$

$$M\ddot{\xi}(2,t)=-K\xi(2,t)-K_{12}[\xi(2,t)-\xi(1,t)] \quad (2)$$

Complex amplitude equations

Denote complex amplitudes by $\xi(1)$, and $\xi(2)$

Let $a=(K+K_{12})/M, \quad c=K_{12}/\mathrm{M}$ (3)

From Eqs. 1 and 2

$$(\omega^2-a)\xi(1)+c\xi(2)=0 \quad (4)$$

$$c\xi(1)+(\omega^2-a)\xi(2)=0 \quad (5)$$

Frequency equation $(\omega^2-a)^2-c^2=0 \quad \omega^2-a=\pm c$ (6)

Mode #1 $\omega_1^2=a-c=K/M,$
$\xi_1(2)/\xi_1(1)=-(\omega^2-a)/c=1$ (7)

Mode #2 $\omega_2^2=a+c=(K+2K_{12})/M,$
$\xi_2(2)/\xi_2(1)=-(\omega^2-a)/c=-1$ (8)

$$\xi_1(1,t)=\xi_1(2,t)=A_1\cos(\omega_1 t-\alpha_1) \quad (9)$$

$$\xi_2(1,t)=-\xi_2(2,t)=A_2\cos(\omega_2 t-\alpha_2) \quad (10)$$

$$\xi(1,t)=\xi_1(1,t)+\xi_2(1,t)=A_1\cos(\omega_1 t-\alpha_1)+A_2\cos(\omega_2 t-\alpha_2) \quad (11)$$

$$\xi(2,t)=\xi_1(2,t)+\xi_2(2,t)=A_1\cos(\omega_1 t-\alpha_1)-A_2\cos(\omega_2 t-\alpha_2) \quad (12)$$

Example

With $\xi(1,0)$, $\xi(2,0)$ and $\dot{\xi}(1,0)$, $\dot{\xi}(2,0)$ given, determine the subsequent motion of the system

Let $X(t)=\xi(1,t)+\xi(2,t), \quad Y(t)=\xi(1,t)-\xi(2,t)$ (13)

From Eqs. 9–12 follows:

$$4A_1^2=X^2(0)+(1/\omega_1)^2\dot{X}^2(0) \quad (14)$$

$$\tan(\alpha_1)=\dot{X}(0)/\omega_1 X(0) \quad (15)$$

$$4A_2^2=Y^2(0)+(1/\omega_2)^2\dot{Y}^2(0) \quad (16)$$

$$\tan(\alpha_2)=\dot{Y}/\omega_2 Y(0) \quad (17)$$

Addition of the first two equations yields $2A_1\cos(\alpha_1) = \xi(1,0) + \xi(2,0)$ and from the last two we get $2A_1\omega_1\sin(\alpha_1) = \dot{\xi}(1,0) + \dot{\xi}(2,0)$. The corresponding expressions for A_1 and α_1 are given by

$$\left.\begin{aligned} 4A_1^2 &= [\xi(1,0) + \xi(2,0)]^2 + (1/\omega_1)^2[\dot{\xi}(1,0) + \dot{\xi}(2,0)]^2 \\ \tan(\alpha_1) &= [\dot{\xi}(1,0) + \dot{\xi}(2,0)]/\omega_1[\xi(1,0) + \xi(2,0)] \end{aligned}\right\} \quad (6)$$

By subtracting the equations, we obtain the corresponding expressions for A_2 and α_2, which are the same as for A_1 and α_1 if the plus signs are replaced by minus signs in both numerators and denominators and with ω_1 replaced by ω_2 (see Display 5.5).

As a numerical example we study the case when the system is started from rest with $\xi(1,0) = 1$ and $\xi(2,0) = 0$. The resulting normal modes are then $\xi_1(1,t) = \xi_2(2,t) = 0.5\cos(\omega_1 t)$ and $\xi_2(1,t) = -\xi_2(2,t) = 0.5\cos(\omega_2 t)$. Since the spring constants are all the same we have $\omega_2 = \sqrt{3}\omega_1$. The initial displacement, the decomposition into normal modes, and the time dependence of the subsequent displacements of the mass elements are illustrated in Display 5.6.

The displacement of each of the mass elements is a superposition of the normal mode contributions. Note that for element #1 the displacement starts from unity at $t = 0$, as it should; the corresponding displacement of element #2 is zero.

The ratio between the two normal mode frequencies in this case is $\sqrt{3}$, an irrational number (the frequencies are incommensurable), and no matter how long we wait, the sum of the normal mode displacements will not repeat in a periodic manner. Therefore, the motion of the system gives the appearance of being quite complicated. Only after decomposition into normal modes is the hidden simplicity of the motion revealed. (If the ratio had been a commensurable number, a rational fraction, say 2/3, the resulting motion would have been periodic with a period equal to six times the shorter of the two normal mode periods).

This simple example illustrates the significance of the normal modes in the analysis of motion of coupled oscillators. By decomposition of the displacement into normal modes, the problem in essence is reduced to a study of single harmonic oscillations as will be discussed in more detail later.

5.3 Weakly coupled oscillators. Beats

As was illustrated in the previous section, the displacement of each of the elements in two coupled oscillators is the sum of contributions from the two normal modes of the system. As in Display 5.6, the resulting motion can be quite irregular, and only if the normal mode frequencies are commensurable, will the motion be periodic.

Display 5.6.

Decomposition of a displacement into normal modes.
$T_1 = 2\pi/\omega_1$

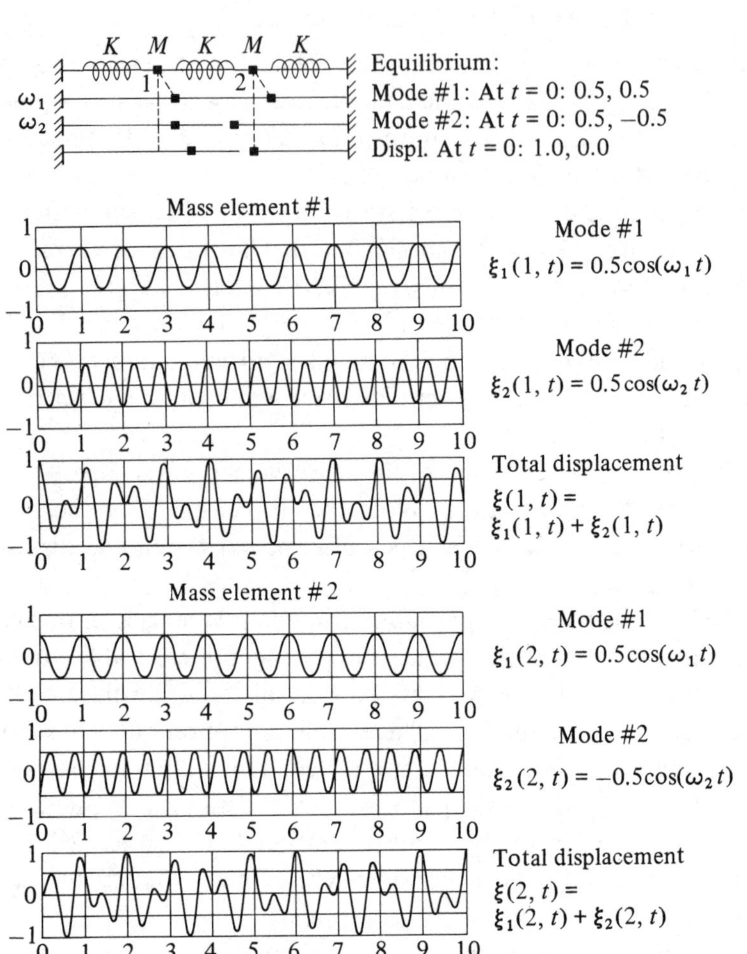

We shall now consider a special and interesting motion which occurs when the coupling between the oscillators in Display 5.5 is weak, i.e. when $\varepsilon = K_{12}/K = \ll 1$. The normal mode frequencies are then nearly the same

$$\omega_1 = \sqrt{K/M} \tag{1}$$

$$\omega_2 = \sqrt{(K + 2K_{12})/M} = \sqrt{K/M}\sqrt{1 + 2(K_{12}/K)}$$
$$\simeq \omega_1(1 + K_{12}/K) = \omega_1(1 + \varepsilon) \tag{2}$$

In the last two steps we have approximated the square root by the first two terms in the power expansion ($\sqrt{1+x} \simeq 1 + x/2$) and under these conditions $(\omega_2 - \omega_1)/\omega_1 \simeq \varepsilon$.

We shall assume the system started from rest with the displacements $\xi(1, 0) = 2A$, $\xi(2, 0) = 0$. According to the analysis in the previous section, the subsequent displacement will be

$$\xi(1, t) = A\cos(\omega_1 t) + A\cos(\omega_2 t) \tag{3}$$
$$\xi(2, t) = A\cos(\omega_1 t) - A\cos(\omega_2 t) \tag{4}$$

With the trigonometric relations

$$\cos(\alpha) + \cos(\beta) = 2\cos[(\alpha - \beta)/2]\cos[(\alpha + \beta)/2]$$
$$\cos(\alpha) - \cos(\beta) = -2\sin[(\alpha - \beta)/2]\sin[(\alpha + \beta)/2]$$

in which $\alpha = \omega_1 t$ and $\beta = \omega_2 t$, and with $(\omega_1 + \omega_2)/2 = \omega_1 + \varepsilon/2 \simeq \omega_1$ and $(\omega_1 - \omega_2)t \simeq -\varepsilon\omega_1 t$, we obtain

$$\xi(1, t) = A[\cos(\omega_1 t) + \cos(\omega_2 t)] \simeq [2A\cos(\varepsilon\omega_1 t/2)]\cdot\cos(\omega_1 t) \tag{5}$$
$$\xi(2, t) = A[\cos(\omega_1 t) - \cos(\omega_2 t)] \simeq [2A\sin(\varepsilon\omega_1 t/2)]\cdot\sin(\omega_1 t) \tag{6}$$

The first factor in each case has the argument $\varepsilon\omega_1 t/2 = \varepsilon\pi(t/T_1)$. Since $\varepsilon \ll 1$, this factor varies slowly with time and can be regarded as a **time-dependent amplitude** of oscillation. The maximum values of the amplitude (magnitude) for element #1 occurs when $\varepsilon\pi(t/T_1) = n\pi$ i.e. $t/T_1 = n/\varepsilon$. The time between maxima is T_1/ε, i.e. inversely proportional to the coupling spring constant. Similarly, the amplitude maxima for element #2 occur when $\varepsilon\pi(t/T_1) = (2n + 1)\pi/2$ i.e. $t/T_1 = (2n + 1)/2\varepsilon$.

It is important to note that when the amplitude of oscillator #1 is a maximum, the amplitude of #2 is a minimum. In other words, the energy of oscillation, being proportional to the square of the amplitude, is exchanged between the oscillators periodically with the period T_1/ε.

This behaviour is shown explicitly in Display 5.7 when $\varepsilon = K_{12}/K = 0.1$. Oscillator #1 starts out with all the energy, being displaced by the amount $2A$ at $t = 0$. At that time the second oscillator starts out from the origin with zero velocity, but after 5 periods of oscillation, the energy has been

Display 5.7.

Weakly coupled oscillators and beats.

Initial conditions

$$\xi(1,0) = 2A, \quad \xi(2,0) = 0 \quad \text{and} \quad \dot{\xi}(1,0) = \dot{\xi}(2,0) = 0 \tag{1}$$

Displacement functions

$$\xi(1,t) = A\cos(\omega_1 t) + A\cos(\omega_2 t) \tag{2}$$
$$\xi(2,t) = A\cos(\omega_1 t) - A\cos(\omega_2 t) \tag{3}$$

Normal mode frequencies

$$\omega_1 = \sqrt{K/M} \tag{4}$$
$$\omega_2 = \sqrt{(K + 2K_{12})/M} \tag{5}$$

Weak coupling

$$\varepsilon = K_{12}/K \ll 1 \tag{6}$$
$$\omega_2 \simeq \omega_1(1 + \varepsilon) \tag{7}$$
$$\omega_2 - \omega_1 \simeq \varepsilon\omega_1, \quad \omega_1 + \omega_2 \simeq 2\omega_1 \tag{8}$$
$$\xi(1,t) \simeq [2A\cos(\varepsilon\omega_1 t)]\cos(\omega_1 t) \tag{9}$$
$$\xi(2,t) \simeq [2A\sin(\varepsilon\omega_1 t/2)]\sin(\omega_1 t) \tag{10}$$

Amplitude function

$|2A\cos(\varepsilon\omega_1 t/2)|$ is max. for $\varepsilon\omega_1 t/2 = n\pi\,(n = 0, 1, \ldots)$ (11)

Amplitude function

$|2A\sin(\varepsilon\omega_1 t/2)|$ is max. for $\varepsilon\omega_1 t/2 = (2n+1)\pi/2$ (12)

Period between maxima $T \simeq 2\pi/\varepsilon\omega_1 = T_1/\varepsilon$ (13)

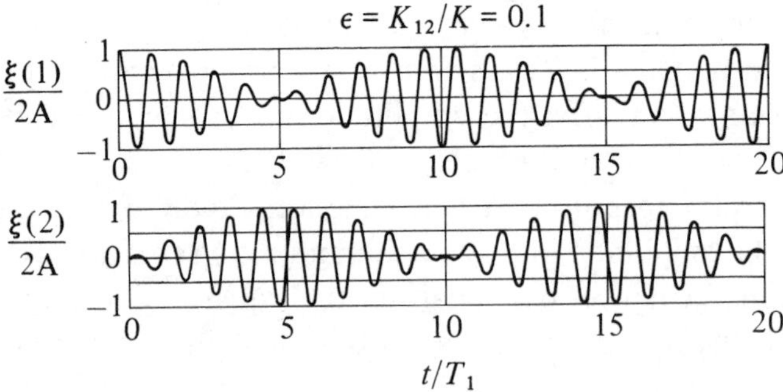

transferred from #1 to #2. After another 5 periods it has been returned to #1.

The time dependence of the amplitude for each of the oscillators is an example of the 'beats' produced when two harmonic motions of slightly different frequencies are superimposed.

A striking demonstration of the periodic energy change between two modes of motion involves the weakly coupled longitudinal and torsional modes of a vertical mass–coil spring oscillator. A coil spring has the property that an elongation induces a small torsional motion and conversely a torsion results in a small change in the length of the spring. The frequency of the torsional mode is adjusted to be nearly the same as the longitudinal, which can be facilitated by having a mass element in the form of a dumbbell with variable moment of inertia.

5.4 Impulse response

An initial-value problem of more general importance than the previous example involves the motion produced by an instantaneous unit impulse applied to one of the two coupled oscillators, and we have chosen mass element #1 to receive the impulse at time $t = t'$, as indicated in Display 5.8. The velocity of this element immediately after the impulse then will be $1/M$ and the velocity of the other element is zero.

Assuming the oscillators to be identical, we can apply the result from the general initial value problem summarized in Display 5.5, but it is instructive to go through the calculation once again.

Thus, starting from the general expressions (Eqs. D.5.5(11)–(12))

$$\xi(1, t) = A_1 \cos(\omega_1 t - \alpha_1) + A_2 \cos(\omega_2 t - \alpha_2)$$
$$\xi(2, t) = A_1 \cos(\omega_1 t - \alpha_1) - A_2 \cos(\omega_2 t - \alpha_2)$$

we note that in order for these displacements to be zero at $t = t'$ we must have $\omega_1 t' - \alpha_1 = \pi/2$, i.e. $\alpha_1 = \omega_1 t' - \pi/2$ with analogous result for α_2. Consequently, in this case the displacements will be

$$\xi(1, t) = A_1 \sin[\omega_1(t - t')] + A_2 \sin[\omega_2(t - t')] \tag{1}$$
$$\xi(2, t) = A_1 \sin[\omega_1(t - t')] - A_2 \sin[\omega_2(t - t')] \tag{2}$$

The corresponding velocities are

$$\dot{\xi}(1, t) = A_1\omega_1 \cos[\omega_1(t - t')] + A_2\omega_2 \cos[\omega_2(t - t')] \tag{3}$$
$$\dot{\xi}(2, t) = A_1\omega_1 \cos[\omega_1(t - t')] - A_2\omega_2 \cos[\omega_2(t - t')] \tag{4}$$

Conditions $\dot{\xi}(1, t') = 1/M$ and $\dot{\xi}(2, t') = 0$ yield $1/M = A_1\omega_1 + A_2\omega_2$ and $0 = A_1\omega_1 - A_2\omega_2$, so that

$$A_1\omega_1 = A_2\omega_2 = 1/2M \tag{5}$$

Display 5.8.

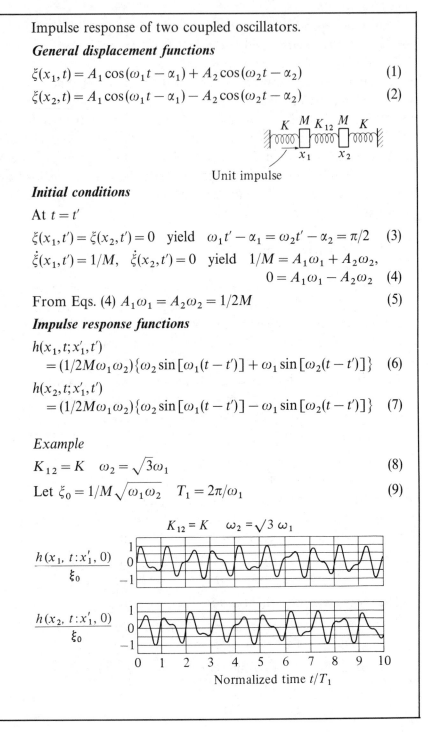

Impulse response of two coupled oscillators.

General displacement functions

$$\xi(x_1, t) = A_1 \cos(\omega_1 t - \alpha_1) + A_2 \cos(\omega_2 t - \alpha_2) \quad (1)$$

$$\xi(x_2, t) = A_1 \cos(\omega_1 t - \alpha_1) - A_2 \cos(\omega_2 t - \alpha_2) \quad (2)$$

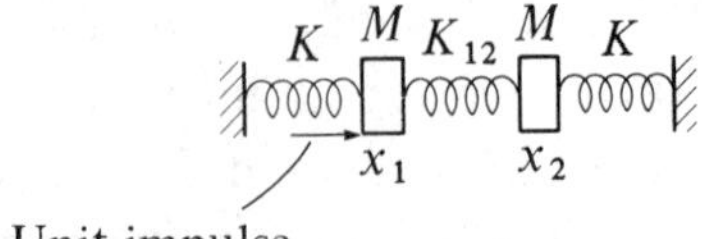

Initial conditions

At $t = t'$

$$\xi(x_1, t') = \xi(x_2, t') = 0 \quad \text{yield} \quad \omega_1 t' - \alpha_1 = \omega_2 t' - \alpha_2 = \pi/2 \quad (3)$$

$$\dot{\xi}(x_1, t') = 1/M, \quad \dot{\xi}(x_2, t') = 0 \quad \text{yield} \quad 1/M = A_1\omega_1 + A_2\omega_2, \quad 0 = A_1\omega_1 - A_2\omega_2 \quad (4)$$

From Eqs. (4) $A_1\omega_1 = A_2\omega_2 = 1/2M$ (5)

Impulse response functions

$$h(x_1, t; x_1', t') = (1/2M\omega_1\omega_2)\{\omega_2 \sin[\omega_1(t - t')] + \omega_1 \sin[\omega_2(t - t')]\} \quad (6)$$

$$h(x_2, t; x_1', t') = (1/2M\omega_1\omega_2)\{\omega_2 \sin[\omega_1(t - t')] - \omega_1 \sin[\omega_2(t - t')]\} \quad (7)$$

Example

$K_{12} = K \quad \omega_2 = \sqrt{3}\omega_1$ (8)

Let $\xi_0 = 1/M\sqrt{\omega_1\omega_2} \quad T_1 = 2\pi/\omega_1$ (9)

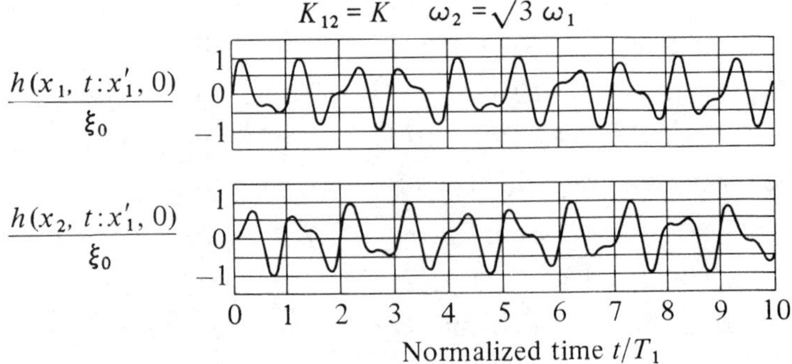

Inserting these expressions for the amplitudes in the general expressions for the displacement, we contain the **impulse response** functions

$$h(x_1, t; x'_1, t') = (1/2M\omega_1\omega_2)\{\omega_2 \sin[\omega_1(t - t')] + \omega_1 \sin[\omega_2(t - t')]\} \quad (6)$$

$$h(x_2, t; x'_1, t') = (1/2M\omega_1\omega_2)\{\omega_2 \sin[\omega_1(t - t')] - \omega_1 \sin[\omega_2(t - t')]\} \quad (7)$$

The arguments of these functions are of the form $(x, t; x', t')$ indicating that the displacement at x at the time t is produced by a unit impulse applied at x' at time t'.

For the special case when $K_{12} = K$, so that $\omega_2 = \sqrt{3}\omega_1$, we have plotted the impulse response functions as functions of t/T_1, where $T_1 = 2\pi/\omega_1$ is the period of the first normal mode. We note that both functions start out from zero displacement. The first has a slope corresponding to the velocity $1/M$ and the second with zero slope, corresponding to an initial velocity of zero.

Experimental modal analysis. The impulse response functions remain the same if the coordinates x_2 and x'_1 are interchanged. In other words, with the impulse applied at x_2, the displacement at x_1 will be the same as the displacement at x_2 when the impulse is applied at x_1.

This symmetry makes possible the following experimental technique for the determination of the normal modes of a system. A transducer, which produces an output signal proportional to the displacement, is placed at point x_1, and the output signal produced when the impulse is applied at the same point is stored in the memory of a computer. The impulse is then applied at x_2, and the corresponding signal output from the transducer at x_1 is stored also. By simple processing of these signals, the characteristics of the modes of the system can be determined.

In this simple example of two equal coupled oscillators and no damping, the signal 'processing' is quite simple. Addition of the displacement signals yields a harmonic function with the frequency of the first mode, and subtraction of the signals yields a harmonic signal with the frequency of the second mode. The amplitudes of the two signals are $1/M\omega_1$ and $1/M\omega_2$, respectively. This enables a determination not only of the mode frequencies but also of the mass M.

This technique can be used for a system of coupled oscillators with an arbitrary number of elements and even for continuous structures. The impulse is delivered by means of an instrumented hammer, which contains a piezoelectric element, the output of which determines the impact force as a function of time. Integration of this signal then yields the magnitude of the applied impulse.

The system is excited by impulses at different locations of the system, and the corresponding transducer signals obtained at a fixed location are stored. Since the impulses are known at each point, the transducer signal can be scaled, so that it is made to correspond to a unit impulse at each location.

The stored signals are then analyzed, and from such an analysis, the characteristics of the modes of the system are determined. The number of locations where impulses are delivered should be at least as large as the number of modes to be studied.

Response to arbitrary driving forces. After having obtained the impulse response of the system, we can determine the displacements produced by arbitrary driving forces applied to the two mass elements in complete analogy with the procedure for a single oscillator.

Thus, with reference to Display 5.9, with the driving force $F(x_1, t)$ applied at x_1, the displacements produced at x_1 and x_2 at time t are obtained as time integrals over the product of the impulse response function and the force, as indicated. The displacements produced by a force $F(x_2, t)$ at x_2 is obtained in analogous manner. If the forces act simultaneously, the resulting displacement is obtained by linear superposition.

As a specific example, it is of interest to calculate the frequency dependence of the displacement amplitudes produced by a harmonic force acting on element #1. This problem was considered in Displays 5.1–2, and we have now an opportunity to check the results obtained earlier in our study of the steady state displacement in forced harmonic motion.

The force is 'turned on' at $t=0$ and is $F = F(x_1)\cos(\omega t)$ for $t > 0$. The integral for the displacement at x_1 is then

$$\xi(x_1, t; x_1') = \int_0^t F(x_1', t')h(x_1, t; x_1', t')\,dt' A \int_0^t \{\omega_2 \sin[\omega_1(t-t')] + \omega_1 \sin[\omega_2(t-t')]\}\cos(\omega t')\,dt' \tag{8}$$

where $A = F(x_1)/2M\omega_1\omega_2$

Evaluation of the integral yields both the steady state motion and the transient. The latter contains the normal mode frequencies and can be omitted in this calculation, since we are interested only in the steady state motion. This motion has the same frequency as the driving force, and dominates after a certain time, since in reality the damping in the system will

Display 5.9.

Coupled oscillators with arbitrary driving forces.

Forces: $F(x_1,t)$ and $F(x_2,t)$ (1)

$$\xi(x_1,t)=\sum_i\left[\int_{-\infty}^{t} h(x_1,t;x_i',t')F(x_i',t')\,dt'\right]\quad i=1,2 \tag{2}$$

$$\xi(x_2,t)=\sum_i\left[\int h(x_2,t;x_i',t')F(x_i',t')\,dt'\right] \tag{3}$$

$h=$ impulse response function: See Eq. D.5.8.6–7.

Example

Harmonic force

$F(x_1,t)=F(x_1)\cos(\omega t)$ for $t>0$ $\quad F(x_1,t)=0$ for $t<0$ (4)

$F(x_2,t)=0$ (5)

$$\xi(x_1,t)=A\int_0^t\{\omega_2\sin[\omega_1(t-t')] + \omega_1\sin[\omega_2(t-t')]\}\cos(\omega t')\,dt' \tag{6}$$

$$\xi(x_2,t) = A\int\{\omega_2\sin[\omega_1(t-t')]-\omega_1\sin[\omega_2(t-t')]\}\cos(t')\,dt' \tag{7}$$

$$A=F(x_1)/2M\omega_1\omega_2 \tag{8}$$

$$\xi(x_1,t)=[F(x_1)/2M][1/(\omega_1^2-\omega^2)+1/(\omega_2^2-\omega^2)]\cos(\omega t) \tag{9}$$

$$\xi(x_2,t)=[F(x_1)/2M][1/(\omega_1^2-\omega^2)-1/(\omega_2^2-\omega^2)]\cos(\omega t) \tag{10}$$

$$\omega_1^2=K/M \quad \omega_2^2=(K+2K_{12})/M \tag{11}$$

make the transient decay. We leave it for one of the problems to show that the steady state displacements are those given in Eqs. D.5.9.(9)–(10).

5.5 Orthogonality. Configuration space

In our continued discussion let us denote the complex displacement amplitudes in the two normal modes by $\xi_1(1)$, $\xi_1(2)$ and $\xi_2(1)$, $\xi_2(2)$. As before, the arguments 1 and 2 stand for x_1 and x_2, the equilibrium locations of the two mass elements. As we have seen, $\xi_1(2)/\xi_1(1)=1$ and $\xi_2(2)/\xi_2(1)=-1$, and in this case of no damping the complex amplitude can be chosen to be real numbers.

Thus, if we regard the displacements as functions of the equilibrium position x, the modes can be described by 'wave' functions, the first with two equal 'spikes' at $x=x_1$ and $x=x_2$ and the second with spikes at the same points, equal in length but with different signs, as shown in Display 5.10.

The magnitudes of the amplitudes of the two modes may be different, of course, but regardless of these magnitudes, we note that the products of the function values at x_1 and x_2 add to zero, i.e.

$$\xi_1(1)\xi_2(1)+\xi_1(2)\xi_2(2)=0 \tag{1}$$

This sum of the products of the displacements of the two modes reminds us of the scalar product of two vectors with the components $\xi_1(1)$, $\xi_1(2)$ and $\xi_2(1)$, $\xi_2(2)$. When the scalar product is zero the vectors are orthogonal, and this geometrical property can be applied to the normal modes, if we refer them to a coordinate system in which the displacements $\xi(1)$ and $\xi(2)$ are used as the coordinates in a rectangular coordinate system. The space defined by these axes is called the **configuration space** of the system. Mode #1, with $\xi_1(1)=\xi_1(2)$ is then represented by a vector in the first quadrant making an angle of 45 degrees with the $\xi(1)$-axis and mode #2 is a vector in the third quadrant at an angle of 135 degrees with the $\xi(1)$-axis. These mode vectors are perpendicular to each other, hence the designation **orthogonal** modes.

The ratio between the time-dependent displacements of the two mass elements in the normal modes is the same as for the amplitudes, i.e. $\xi_1(1,t)=\xi_1(2,t)$ and $\xi_2(1,t)=-\xi_2(2,t)$. Thus, the point in configuration space $\xi_1(1,t)$, $\xi_1(2,t)$, representing the displacements at time t in the first normal mode, will move in harmonic motion with the frequency ω_1 along a straight line, which makes an angle of 45 degrees with the $\xi(1)$-axis. Similarly, the point $\xi_2(1,t)$, $\xi_2(2,t)$, representing the second normal mode, moves in harmonic motion of frequency ω_2 along a straight line which

Display 5.10.

Orthogonality and normalization of modes.

Normal modes

Reference Displays 5.5 and 5.6.

Denote (complex) amplitudes by $\xi(1)$, $\xi(2)$ (1)

MODE #1 $\omega_1^2 = a - c$ $\xi_1(2)/\xi_1(1) = -(\omega_1^2 - a)/c = 1$ (2)

MODE #2 $\omega_2^2 = a + c$ $\xi_2(2)/\xi_2(1) = -(\omega_2^2 - a)/c = -1$ (3)

MODE #1

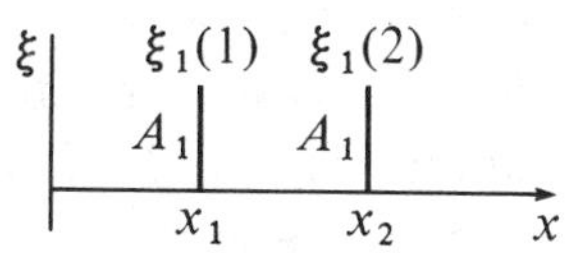

ξ $\xi_1(1)$ $\xi_1(2)$ A_1 A_1 x_1 x_2 x

MODE #2

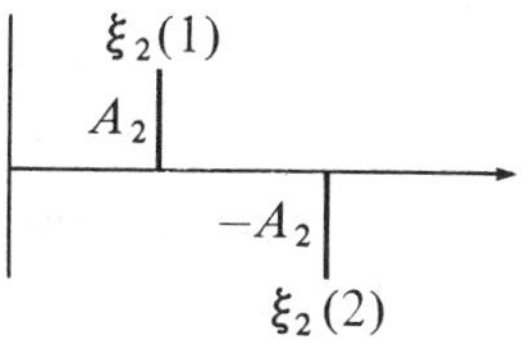

$\xi_2(1)$ A_2 $-A_2$ $\xi_2(2)$

Orthogonality

$\xi_1(1)\xi_2(1) + \xi_1(2)\xi_2(2) = 0$ (4)

$\xi_1 \cdot \xi_2 = 0$ (5)

Mode vector

$\xi_i = [\xi_i(1), \xi_i(2)]$ $i = 1, 2$ (6)

Normalization

$\xi_1 \cdot \xi_1 = \xi_1(1)\xi_1(1) + \xi_1(2)\xi_1(2) = 2A_1^2$ (7)

$\xi_1 \cdot \xi_1 = 1$ if $A_1 = 1/\sqrt{2}$ (8)

Mode unit vector

$\hat{\xi}_1 = [1/\sqrt{2}, 1/\sqrt{2}]$, $\hat{\xi}_2 = [1/\sqrt{2}, -1/\sqrt{2}]$ (9)

Normalized mode functions

$\xi_1(x_1, t) = \xi_1(x_2, t) = 1/\sqrt{2}\cos(\omega_1 t - \alpha_1)$ (10)

$\xi_2(x_1, t) = -\xi_2(x_2, t) = 1/\sqrt{2}\cos(\omega_2 t - \alpha_2)$ (11)

makes an angle of 135 degrees with the $\xi(1)$-axis, as shown in Display 5.11.

As we have seen, a general free motion can be expressed as the sum of the two normal modes, and the corresponding trajectory of the representative point $\xi(1,t)$, $\xi(2,t)$ in configuration space generally is considerably more complex than the normal mode lines.

As a specific example we reconsider the motion already analyzed in Display 5.6, where the displacements are plotted as functions of time. The ratio between the normal mode frequencies in this case is $\sqrt{3}$, and the motion is started by giving the system the initial displacements $\xi(1,0)=1$, $\xi(2,0)=0$.

The trajectory in configuration space is shown in Display 5.11. It starts out at $t=0$ from the point 1, 0 and continues as shown. In the first graph the trajectory extends over the first two periods (the period refers to the low frequency mode $T_1=2\pi/\omega_1$) and in the second, ten periods are included.

Since the ratio between the normal mode frequencies is an irrational number, $\sqrt{3}$, the trajectory will not be closed; it is displaced somewhat from one period to the next as it continues to cover the region within a square. Eventually, every point in the square will be encountered by the trajectory.

The fact that the normal mode trajectories are straight lines suggests the introduction of a new coordinate system with the axes along the normal mode lines. In the present example, these axes are rotated 45 degrees with respect to the original axes with the coordinates $\xi(1)$, $\xi(2)$. The new coordinates, denoted ξ_1 and ξ_2, called **normal coordinates**, are related to the old with the transformation

$$\xi(1)=(\xi_1-\xi_2)/\sqrt{2} \tag{2}$$

$$\xi(2)=(\xi_1+\xi_2)/\sqrt{2} \tag{3}$$

This transformation applies to the (complex) amplitudes as well as to the time dependent displacements, and if it is used in the equations of motion 5.1(3)–(4), with $F(1)=0$, the equations will be decoupled into the two harmonic-oscillator equations for the normal coordinates

$$\ddot{\xi}_1+\omega_1^2\xi_1=0 \tag{4}$$

$$\ddot{\xi}_2+\omega_2^2\xi_2=0 \tag{5}$$

The normal modes can be interpreted geometrically as vectors along the normal coordinates. It is often convenient to normalize these vectors to unit magnitude. In this case the components of the normal mode vectors will be $\hat{\xi}_1(1)=\hat{\xi}_1(2)=1/\sqrt{2}$ and $\hat{\xi}_2(1)=-\hat{\xi}_2(2)=1/\sqrt{2}$, the 'hats' signifying the normalization.

Display 5.11.

The motions in Display 5.6 described in terms of a trajectory in configuration space.

Reference Display 5.6

Initial displacements $\xi(1,0) = 1, \quad \xi(2,0) = 0$ (1)

Displacement functions

$\xi(1,t) = 0.5\cos(\omega_1 t) + 0.5\cos(\omega_2 t) \quad K_{12} = K$

$\omega_2 = \sqrt{3}\omega_1$ (2)

$\xi(2,t) = 0.5\cos(\omega_1 t) - 0.5\cos(\omega_2 t) \quad T_1 = 2\pi/\omega_1$ (3)

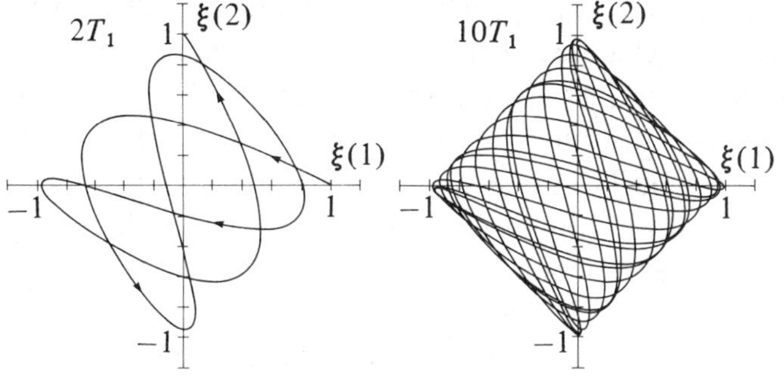

MODE 1 $\xi_1(1,t) = \xi_2(2,t) = 0.5\cos(\omega_1 t)$ (4)

MODE 2 $\xi_2(1,t) = -\xi_2(2,1) = 0.5\cos(\omega_2 t)$ (5)

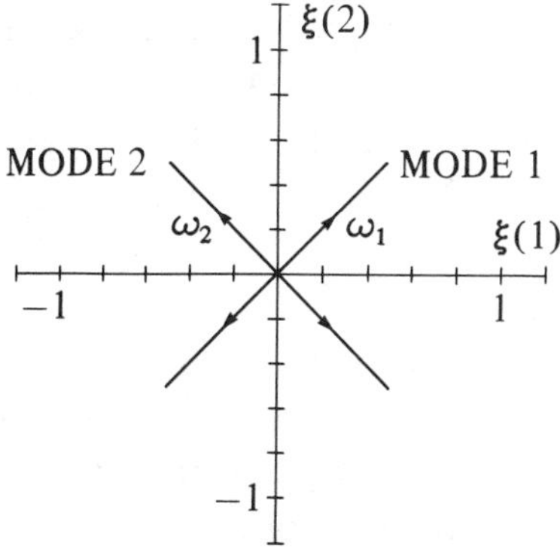

In complete analogy, an arbitrary displacement can be interpreted as a vector in configuration space, which can be decomposed into components along the direction of the normal mode vectors. Such a component is obtained simply as the scalar product of the displacement vector and the normal mode vector, as will be discussed later in connection with Fourier analysis.

Fourier analysis will be applied when we deal with the oscillations of continuous systems (strings, air columns in pipes, solid columns, electromagnetic transmission lines, etc.), in the decomposition of an oscillation into the normal modes of the system. A normal mode is then described by the x-dependence of the amplitude of oscillation at the normal mode frequency.

For example, the first normal mode of the transverse displacement of a string of length L clamped at the ends has the 'shape' given by $\sin(\pi x/L)$, as we shall find later. At a given time, all parts of the string move in the same direction in this mode, and the maximum amplitude occurs at the middle of the string. The same mode shape applies also to the longitudinal oscillations of an air column in a tube closed at both ends. This mode corresponds to the two equal spikes representing the 'wave function' of the two coupled oscillators in Display 5.10.

The second mode of the string is $\sin(2\pi x/L)$, in which the amplitude is zero at the center of the string and the two halves of the string move in opposite direction at the frequency of the second mode, which is twice that of the first mode. This mode corresponds to the two equal but opposite spikes of the second mode of the two coupled oscillators in Display 5.10 (in which case the ratio between the frequencies is not 2 but $\sqrt{3}$).

In complete analogy with the two coupled oscillators, the Fourier decomposition of a displacement into normal modes can be interpreted in terms of scalar multiplication of vectors in configuration space, as will be referred to later.

5.6 Unequal coupled oscillators

The general systematic theory of coupled oscillators is best developed in terms of the Lagrangian or Hamiltonian formulation of mechanics, which is beyond the scope of the present treatment. Therefore, we shall limit ourselves here merely to the case of two coupled oscillators, extending the analysis of the previous sections to include oscillators with different masses and spring constants.

Although the analysis is much the same as for two coupled identical oscillators, summarized in Display 5.5, we shall carry out the analysis in some detail since there are some aspects which deserve special mention.

Display 5.12.

Free motion of two unequal coupled oscillators.

$$M_1\xi(1,t) = -K_1\xi(1,t) - K_{12}[\xi(1,t) - \xi(2,t)] \tag{1}$$

$$M_2\xi(2,t) = -K_2\xi(2,t) - K_{12}[\xi(2,t) - \xi(1,t)] \tag{2}$$

Complex amplitude equations

Denote complex amplitudes by $\xi(1)$, $\xi(2)$

$$(-\omega^2 M_1 + K_1 + K_{12})\xi(1) - K_{12}\xi(2) = 0 \tag{3}$$

$$-K_{12}\xi(1) + (-\omega^2 M_2 + K_2 + K_{12})\xi(2) = 0 \tag{4}$$

Let $a_1 = (K_1 + K_{12})/M_1, \quad a_2 = (K_2 + K_{12})/M_2,$

$$b_1 = K_{12}/M_1, \quad b_2 = K_{12}/M_2 \tag{5}$$

$$(\omega^2 - a_1)\xi(1) + b_1\xi(2) = 0 \tag{6}$$

$$b_2\xi(1) + (\omega^2 - a_2)\xi(2) = 0 \tag{7}$$

Symmetrization: Let $c = K_{12}/\sqrt{M_1 M_2}$,

$\zeta(1) = \xi(1)\sqrt{M_1}$,

$$\zeta(2) = \xi(2)\sqrt{M_2} \tag{8}$$

$$(\omega^2 - a_1)\zeta(1) + c\zeta(2) = 0 \tag{9}$$

$$c\zeta(1) + (\omega^2 - a_2)\zeta(2) = 0 \tag{10}$$

Normal mode frequencies

$$(\omega^2 - a_1)(\omega^2 - a_2) - c^2 = 0 \tag{11}$$

$$\omega_\pm^2 = (a_1 + a_2)/2 \pm \sqrt{(1/4)(a_1 - a_2)^2 + c^2} \tag{12}$$

Let $\omega_1 = \omega_-, \quad \omega_2 = \omega_+$

Normal mode displacement functions

From Eqs. (9), (10)

$$\zeta_1(2)/\zeta_1(1) = -(\omega_1^2 - a_1)/c, \quad \zeta_2(2)/\zeta_2(1) = -(\omega_2^2 - a_1)/c \tag{13}$$

$$\zeta_1(1,t) = A_1\cos(\omega_1 t - \alpha_1) \tag{14}$$

$$\zeta_1(2,t) = A_1[-(\omega_1^2 - a_1)/c]\cos(\omega_1 t - \alpha_1) \tag{15}$$

$$\zeta_2(1,t) = A_2\cos(\omega_2 t - \alpha_2) \tag{16}$$

$$\zeta_2(2,t) = A_2[-(\omega_2^2 - a_1)/c]\cos(\omega_2 t - \alpha_2) \tag{17}$$

General solution $\quad \zeta(1,t) = \zeta_1(1,t) + \zeta_2(1,t) \qquad (18)$

$$\zeta(2,t) = \zeta_1(2,t) + \zeta_2(2,t) \tag{19}$$

With reference to Display 5.12, the masses and spring constants of the two oscillators are M_1, M_2, and K_1, K_2, respectively, and the coupling spring constant is K_{12}. We neglect friction forces, although they can be added later simply by assigning complex values to the spring constants. The differential equations for **free motion** of the system are then

$$M_1\ddot{\xi}(1,t) = -K_1\xi(1,t) - K_{12}[\xi(1,t) - \xi(2,t)] \tag{1}$$

$$M_2\ddot{\xi}(2,t) = -K_2\xi(2,t) - K_{12}[\xi(2,t) - \xi(2,t)] \tag{2}$$

With the complex displacement amplitudes denoted by $\xi(1)$ and $\xi(2)$, the complex amplitude equations are

$$(-\omega^2 M_1 + K_1 + K_{12})\xi(1) - K_{12}\xi(2) = 0 \tag{3}$$

$$-K_{12}\xi(1) + (-\omega^2 M_2 + K_2 + K_{12})\xi(2) = 0 \tag{4}$$

For convenience we introduce the parameters $a_1 = (K_1 + K_{12})/M_1$, $a_2 = (K_2 + K_{12})/M_2$, $b_1 = K_{12}/M_1$, and $b_2 = K_{12}/M_2$, in terms of which the complex amplitude equations become

$$\left.\begin{aligned}(\omega^2 - a_1)\xi(1) + b_1\xi(2) = 0\\ b_2\xi(1) + (\omega^2 - a_2)\xi(2) = 0\end{aligned}\right\} \tag{5}$$

The parameters b_1 and b_2 express the coupling between the oscillators, each being proportional to K_{12}. To simplify the equations further, we shall make the coupling terms 'symmetrical' by the introduction of the coupling constant

$$c = K_{12}/\sqrt{M_1 M_2} \tag{6}$$

in terms of which $b_1 = c\sqrt{M_2/M_1}$ and $b_2 = c\sqrt{M_1/M_2}$. As we note, the b-parameters differ from c by factors involving the mass ratio. These factors can be absorbed, however, by redefining the displacement variables as follows:

$$\zeta(1) = \xi(1)\sqrt{M_1}, \quad \zeta(2) = \xi(2)\sqrt{M_2} \tag{7}$$

making the complex amplitude equations symmetrical

$$\begin{aligned}(\omega^2 - a_1)\zeta(1) + c\zeta(2) = 0\\ c\zeta(1) + (\omega^2 - a_2)\zeta(2) = 0\end{aligned} \tag{8}$$

In order for these equations to have solutions for the displacement different from zero, the frequency must satisfy the following **frequency equation**

$$(\omega^2 - a_1)(\omega^2 - a_2) - c^2 = 0 \tag{9}$$

which follows by eliminating one of the displacement variables between the complex amplitude equations. The solutions to the frequency equation are

$$\omega_{\pm}^2 = (a_1 + a_2)/2 \pm \sqrt{(1/4)(a_1 - a_2)^2 + c^2} \tag{10}$$

If the oscillators are identical we have $a_1 = a_2$ and $\omega_{\pm}^2 = a \pm c$, as in Eq. D.5.5(6). To be consistent with the notation used earlier, we denote the normal mode frequencies by $\omega_1 = \omega_-$ and $\omega_2 = \omega_+$.

The frequency equation $(\omega^2 - a_1)(\omega^2 - a_2) - c^2 = 0$ can be written $(\omega^2 - \omega_1^2)(\omega^2 - \omega_2^2) = 0$, and it follows, that

$$\begin{aligned} &\omega_1^2\omega_2^2 = a_1 a_2 - c^2, \quad \omega_1^2 + \omega_2^2 = a_1 + a_2, \\ &\omega_1^2 - a_1 = -(\omega_2^2 - a_2) \end{aligned} \tag{11}$$

With the normal mode frequencies introduced into the complex amplitude equations we obtain the ratio between the displacement amplitudes of the normal modes

$$\begin{aligned} \zeta_1(2)/\underline{\zeta}_1(1) &= -(\omega_1^2 - a_1)/c = -c/(\omega_1^2 - a_2) \\ &= c/(\omega_2^2 - a_1) = \tan(\phi_1) \\ \zeta_2(2)/\underline{\zeta}_2(1) &= -(\omega_2^2 - a_1)/c = \tan(\phi_2) \end{aligned} \tag{12}$$

Each of these ratios has been written as the tangent of the angle between the corresponding normal mode line in configuration space and the $\zeta(1)$-axis. It follows that $\tan(\phi_1)\tan(\phi_2) = -1$, i.e. $\tan(\phi_1) = -\cot(\phi_2)$, which means that $\phi_1 - \phi_2 = \pi/2$. In other words, the normal mode vectors in configuration space are orthogonal.

It should be pointed out in this context that the main reason for introducing the new displacement variables ζ was to make the normal modes orthogonal. Had we used the original coordinates ξ, the product of the amplitude ratios would not have been -1, which is the condition for orthogonality.

Having determined the normal mode frequencies and the corresponding ratios between the displacement amplitudes in the normal modes, the general expressions for the time dependence of the normal mode displacements can be expressed as

$$\left.\begin{aligned} \zeta_1(1,t) &= A_1 \cos(\omega_1 t - \alpha_1) \\ \zeta_1(2,t) &= [-(\omega_1^2 - a_1)/c] A_1 \cos(\omega_1 t - \alpha_1) \\ \zeta_2(1,t) &= A_2 \cos(\omega_2 t - \alpha_2) \\ \zeta_2(2,t) &= [-(\omega_2^2 - a_1)/c] A_2 \cos(\omega_2 t - \alpha_2) \end{aligned}\right\} \tag{13}$$

where A_1, A_2, α_1, and α_2 are constants, which are determined by known values of displacement and velocity of the two oscillators at a given time. It should be remembered, though, that the original displacements are given by $\xi(1) = \zeta(1)/\sqrt{M_1}$ and $\xi(2) = \zeta(2)/\sqrt{M_2}$.

At any value of time t, the normal mode displacements $\zeta_1(1,t)$, $\zeta_1(2,t)$ and $\zeta_2(1,t)$, $\zeta_2(2,t)$, as we have seen, can be interpreted as the components of normal mode vectors in a configuration space, in which the axes are $\zeta(1)$ and $\zeta(2)$. Similarly, an arbitrary displacement $\zeta(1,t)$, $\zeta(2,t)$ can be treated as the components of a vector, and the decomposition of this displacement into components along the normal mode axes is completely analogous to the procedure discussed in the previous section.

Effect of damping. The results obtained for the free oscillations of the unequal coupled oscillators can be extended to include damping merely by letting the spring constants be complex, as was mentioned already in Chapter 3.

Thus, if a dashpot is placed in parallel with a spring, and if the displacements at the two ends of this combination are $\xi(1,t)$ and $\xi(2,t)$, the spring force (in the direction from 1 to 2) is $K[\xi(1,t)-\xi(2,t)]$ and the friction force $R[\dot{\xi}(1,t)-\dot{\xi}(2,t)]$ with the complex amplitudes $K[\xi(1)-\xi(2)]$ and $-i\omega R[\xi(1)-\xi(2)]$. The total complex force amplitude is then $(K-i\omega R)[\xi(1)-\xi(2)]=K'[\xi(1)-\xi(2)]$, where $K'=K-i\omega R$ is a complex spring constant.

With dashpot dampers connected to all springs in the system, we have to introduce complex spring constants $K_1'=K_1-i\omega R_1$, $K_2'=K_2-i\omega R_2$, and $K_{12}'=K_{12}-i\omega R_{12}$.

The corresponding parameters $a_1=(K_1+K_{12})/M_1$, $a_2=(K_2+K_{12})/M_2$, and $c=K_{12}/\sqrt{M_1M_2}$ also will be complex. The same holds true for the solutions to the frequency equation

$$(\omega^2-a_1)(\omega^2-a_2)-c^2=0 \tag{14}$$

It is important to realize that the parameters a_1, a_2, and c no longer are frequency independent, since each contains the factor $(-i\omega)$ in the complex spring constants. Consequently, the frequency equation will contain terms of first, second, third, and fourth power of frequency, and the general expression for the roots is somewhat involved.

For the purpose of illustration, we shall consider here only the case of equal oscillators, i.e. $a_1=a_2=a=(K_1'+K_{12}')/M$ and $c=K_{12}'/M$. The frequency equation

$$(\omega^2-a)^2-c^2=0 \tag{15}$$

then leads to the following relations for the two modes of motion:

Mode #1. $\omega^2-a=-c$ or $\omega^2-(a-c)=0$. With $a-c=K'/M=$

$(K - i\omega R)/M$ the corresponding frequency equation becomes

$$\omega^2 + i\omega R/M - \omega_0^2 = 0 \quad \text{where } \omega_0^2 = K/M \tag{16}$$

with the solutions

$$\omega_\pm = -i\gamma_1 \pm \omega_1', \quad \omega_1' = \sqrt{\omega_0^2 - \gamma_1^2}, \quad \gamma_1 = R/2M \tag{17}$$

The ratio between the complex displacement amplitudes in this first mode is

$$\xi_1(2)/\xi_1(1) = 1 \tag{18}$$

and the time dependence of the displacement is

$$\xi_1(1,t) = \xi_2(2,t) = A_1 e^{-\gamma_1 t} \cos(\omega_1' t - \alpha_1) \tag{19}$$

where A_1 and α_1 are to be determined from initial conditions.

Mode #2. $\omega^2 - a = c$ or $\omega^2 - (a + c) = 0$. With $a + c = (K' + 2K_{12}')/M$ the corresponding frequency equation becomes

$$\omega^2 + i\omega(R + 2R_{12})/M - \omega_2^2 = 0 \quad \text{where } \omega_2^2 = (K + 2K_{12})/M \tag{20}$$

with the solutions

$$\omega_\pm = -i\gamma_2 \pm \omega_2' \quad \omega_2' = \sqrt{\omega_2^2 - \gamma_2^2}, \quad \gamma_2 = (R + R_{12})/M \tag{21}$$

The ratio between the complex displacement amplitudes is

$$\xi_2(2)/\xi_2(1) = -1 \tag{22}$$

and the general time dependence is

$$\xi_2(1,t) = -\xi_2(2,t) = A_2 e^{-\gamma_2 t} \cos(\omega_2' t - \alpha_2) \tag{23}$$

In general motion of the system, the total displacement of each of the two mass elements is a superposition of the two modes, so that

$$\begin{aligned} \xi(1,t) &= A_1 e^{-\gamma_1 t} \cos(\omega_1 t - \alpha_1) + A_2 e^{-\gamma_2 t} \cos(\omega_2 t - \alpha_2) \\ \xi(2,t) &= A_1 e^{-\gamma_1 t} \cos(\omega_1 t - \alpha_1) - A_2 e^{-\gamma_2 t} \cos(\omega_2 t - \alpha_2) \end{aligned} \tag{24}$$

The four unknown constants A_1, A_2, α_1, and α_2 can be determined from known values of displacement and velocity of the two oscillators at a given time.

Not only are the frequencies of the two modes different but also the decay constants. The high frequency mode decays faster than the low frequency mode. Consequently, if we wait long enough after the impulse excitation of the system, the low frequency mode will be dominant.

This is illustrated by the example in Display 5.14, which is based on the analysis summarized in Display 5.13. A unit impulse is applied to mass

Display 5.13.

Free motion of damped coupled oscillators.

$$M\ddot{\xi}(1,t) = -K\xi(1,t) - K_{12}[\xi(1,t) - \xi(2,t)] - R\dot{\xi}(1,t)$$
$$- R_{12}[\dot{\xi}(1,t) - \dot{\xi}(2,t)] \quad (1)$$

$$M\ddot{\xi}(2,t) = -K\xi(2,t) - K_{12}[\xi(2,t) - \xi(1,t)] - R\dot{\xi}(2,t)$$
$$- R_{12}[\dot{\xi}(2,t) - \dot{\xi}(1,t)] \quad (2)$$

Complex amplitude equations

Denote complex amplitudes by $\xi(1), \xi(2)$ (3)

Let $a = [K + K_{12} - i\omega(R + R_{12})]/M, \quad c = (K_{12} - i\omega R_{12})/M$ (4)

$$(\omega^2 - a)\xi(1) + c\xi(2) = 0 \quad (5)$$
$$c\xi(1) + (\omega^2 - a)\xi(2) = 0 \quad (6)$$

K M K_{12} M K

R R_{12} R

Normal modes

Frequency equation

$$(\omega^2 - a)^2 - c^2 = 0 \quad \text{or } (\omega^2 - a) = \pm c \quad (7)$$

Mode #1

$$\omega^2 - a = -c \quad \text{or } \omega^2 - (a - c) = 0, \quad \omega^2 + i\omega R/M - \omega_1^2 = 0,$$
$$\omega_1^2 = K/M \quad (8)$$

Solution $\omega = i\gamma_1 \pm \omega'_1, \quad \gamma_1 = R/2M, \quad \omega'_1 = \sqrt{\omega_1^2 - \gamma_1^2}$ (9)

$$\xi_1(2)/\xi_1(1) = -(\omega_1^2 - a)/c = 1 \quad (10)$$

$$\xi_1(1,t) = \xi_1(2,t) = A_1 \exp(-\gamma_1 t)\cos(\omega'_1 t - \alpha_1) \quad (11)$$

Mode #2

$$\omega^2 - a = c \quad \text{or } \omega^2 - (a + c) = 0,$$
$$\omega^2 + i\omega(R + 2R_{12})/M - \omega_2^2 = 0, \quad \omega_2^2 = (K + 2K_{12})/M \quad (12)$$

Solution $\omega = -i\omega_2 \pm \omega'_2, \quad \gamma_2 = (R + 2R_{12})/2M,$

$$\omega'_2 = \sqrt{\omega_2^2 - \gamma_2^2} \quad (13)$$

$$\xi_2(2)/\xi_2(1) = -(\omega_2^2 - a)/c = -1 \quad (14)$$

$$\xi_2(1,t) = -\xi_2(2,t) = A_2 \exp(-\gamma_2 t)\cos(\omega'_2 t - \alpha_2) \quad (15)$$

General free motion

$$\xi(1,t) = \xi_1(1,t) + \xi_2(1,t) \quad (16)$$

$$\xi(2,t) = \xi_1(2,t) + \xi_2(2,t) \quad (17)$$

Display 5.14.

Impulse response function for damped coupled oscillators.

Example

With reference to the system of coupled oscillators in Display 5.13, determine the motion resulting from a unit impulse on element #1 at time $t=0$. Let $K_{12}=K$, $R_{12}=R$. For $D=R/\omega_1 M=0.03$, plot the displacements of the two mass elements as a function of time.

From Eqs. D.5.13.(16)(17), the general expressions for the two displacements are

$$\xi(1,t)=A_1 e^{-\gamma_1 t}\cos(\omega_1' t-\alpha_1)+A_2 e^{-\gamma_2 t}\cos(\omega_2' t-\alpha_2) \quad (1)$$

$$\xi(2,t)=A_2 e^{-\gamma_1 t}\cos(\omega_1' t-\alpha_1)-A_2 e^{-\gamma_2 t}\cos(\omega_2' t-\alpha_2) \quad (2)$$

The conditions $\xi(1,0)=\xi(2,0)=0$ yield $\alpha_1=\alpha_2=\pi/2$, and from the conditions $\dot{\xi}(1,0)=1/M$ and $\dot{\xi}(2,0)=0$ follows $A=1/2M\omega_1'$, $A_2=A_1\omega_1'/\omega_2'$. In terms of the characteristic displacement $\xi_0=1/M\sqrt{\omega_1\omega_2}$, the corresponding displacements can be written

$$\xi(1,t)=(\xi_2/2)[\sqrt{\omega_2'/\omega_1'}\,e^{-\gamma_1 t}\sin(\omega_1' t) + \sqrt{\omega_1'/\omega_2'}\,e^{-\gamma_2 t}\sin(\omega_2' t)] \quad (3)$$

$$\xi(2,t)=(\xi_0/2)[\sqrt{\omega_1'/\omega_2'}\,e^{-\gamma_1 t}\sin(\omega_1' t) - \sqrt{\omega_1'/\omega_2'}\,e^{-\gamma_2 t}\sin(\omega_2' t)] \quad (4)$$

Numerical calculations

Dimensionless time variable $=t/T_1$, where $T_1=2\pi/\omega_1$

$$\omega_1' t=\sqrt{\omega^2-\gamma_1^2}\,t=\omega_1 t\sqrt{1-(\gamma_1/\omega_1)^2}=\omega_1 t\sqrt{1-(D/2)^2}$$
$$\simeq \omega_1 t=2\pi(t/T_1), \quad \omega_2'/\omega_1'\simeq\omega_2/\omega_1=\sqrt{3}$$
$$\gamma_1 t=\gamma_1 T_1(t/T_1)=(\pi R/\omega_1 M)(t/T_1)=\pi D(t/T_1), \quad \gamma_2=3\gamma_1$$
$$D=0.03.$$

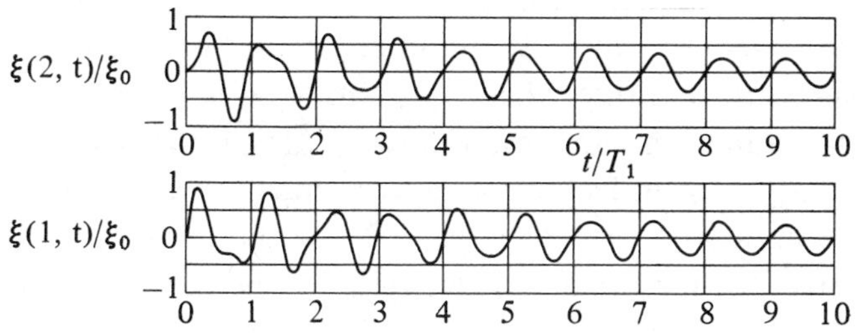

element #1 at time $t = 0$, and the subsequent displacements of the two elements are plotted as functions of time for the special case when all spring constants are the same and the damping parameter $D = R/2M$ has the value 0.03.

The decay constant for the anti-symmetrical mode will be three times larger than for the symmetrical mode, and after a few periods of oscillations, the symmetrical mode becomes dominant, so that the motion becomes approximately harmonic with a period of the low frequency mode. Note, that the displacements of the two particles are then approximately the same, indicating that the system essentially is in its symmetrical mode of motion.

Because of the difference in the damping parameters of the two modes, the forced harmonic motion of the system yields a larger resonance amplitude for the low frequency mode, as was demonstrated in Section 5.1.

Also in practice, the low frequency modes of a structure generally carry most of the oscillatory energy, and at resonance the amplitudes can be unacceptably large. A challenging problem for the engineer is to predict the response of a structure to various types of excitation and to incorporate vibration damping measures in the design so as to prevent large amplitude vibration of the structure.

Returning to the specific example in Display 5.14, the difference in the decay constants for the two modes readily can be understood without analysis. In the symmetrical mode there is no relative motion between the end points of the coupling spring and its associated dashpot. Consequently, the decay will be the same as for a single oscillator with the resistance coefficient R.

In the asymmetrical mode, on the other hand, the relative velocity of the end points of the coupling spring is twice the velocity of each of the mass elements. The coupling dashpot then will contribute a friction force, which is twice as large as from the other dashpots. Thus the motion of one of the mass elements is acted on by a total friction force, which is three times that in the symmetrical mode.

The situation is different if the damping is provided by direct friction on the mass elements rather than by dashpots (for example by means of aerodynamic dampers in the form of cardboards mounted on air suspended carts, frequently used in lecture demonstrations). In that case the friction force is determined by the absolute and not the relative velocity of the mass elements. For further details we refer to one of the problems.

Examples

E5.1 *Damped coupled oscillators.* The mass elements of two identical simple pendulums are connected with a dashpot. Determine the normal modes of the system.

Solution. This is a simplified version of the coupled oscillators considered in Display 5.13, and the characteristics of the normal modes can be determined without mathematical analysis.

As in the case of identical coupled oscillators without damping, discussed in Section 5.2, we have here a symmetrical and an anti-symmetrical mode. In the symmetrical mode, the pendulums have the same amplitude and direction of motion. Thus, there will be no relative motion between the piston and the cylinder in the dashpot and no friction damping. The frequency of the mode will be the same as for the free pendulum, $\omega_1 = \sqrt{g/L}$, and the displacements of the pendulums will be the same $\xi_1(x_1, t) = \xi_1(x_2, t)$, where $\xi_1(x_1, t) = A\cos(\omega_1 t - \alpha_1)$.

In the anti-symmetrical mode, the pendula move in opposite directions with the same amplitude. The relative velocity between the piston and the cylinder in the dashpot now will be twice the velocity of one pendulum. Consequently, the friction force on each of the mass elements will be $-2Ru$, and the decay constant of the motion will be $\gamma = 2R/2M = R/M$. Thus, the second mode of the system is characterized by $\xi_2(x_1, t) = -\xi_2(x_2, t)$, where $\xi_2 = B\exp(-\gamma t)\cos(\omega_2 t - \alpha_2)$ and $\omega_2 = \sqrt{\omega_1^2 - \gamma^2}$, with $\omega_1^2 = g/L$, $\gamma = R/M$.

Problems

5.1 *Forced harmonic oscillation.* (a) Show that the forced harmonic motion of the coupled oscillators in Fig. 1 yields the same displacements of the two mass elements as in the motion described in Display 5.1, if $|F(1)| = K\xi_0$.

(b) Determine the frequency dependence of the displacement amplitudes when the system is driven as indicated in Fig. 2.

(c) What are the equivalent electrical analogue circuits of the mechanical oscillators in (a) and (b)?

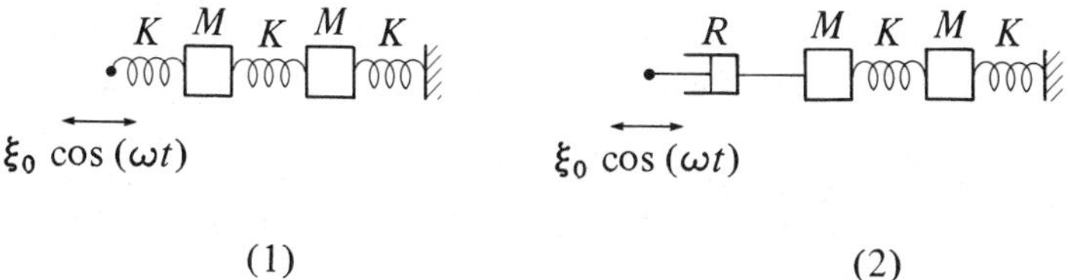

(1) (2)

5.2 *Forced harmonic motion and power transfer.* Two pendulums are coupled by means of a dashpot damper. One of the pendulums is driven in harmonic motion, as shown.

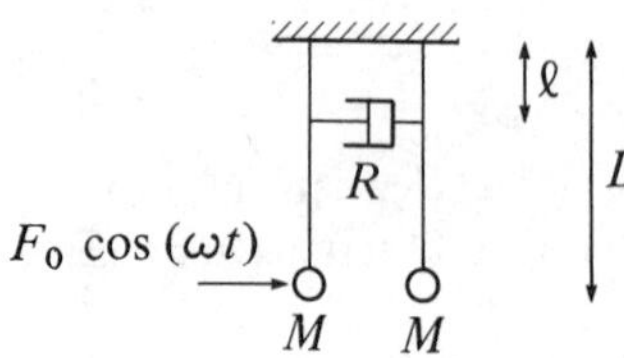

(a) Determine the frequency dependence of the displacement amplitudes of the pendulums, assuming that the angular deflections are small.
(b) Determine the average power transfer.

5.3 *Electric circuit.* An electric circuit is driven by a voltage source with harmonic time dependence as shown.

(a) With the complex amplitudes of voltage and input current being V and I, determine the input impedance $Z = V/I$.
(b) Determine the frequency dependence of the amplitudes of the charges on the two capacitors. What are these amplitudes at zero frequency? Compare these values with those obtained from electrostatics.
(c) What is the average power delivered by the voltage source? In particular, what is the power at the angular frequency $\omega_0 = 1/\sqrt{LC}$?

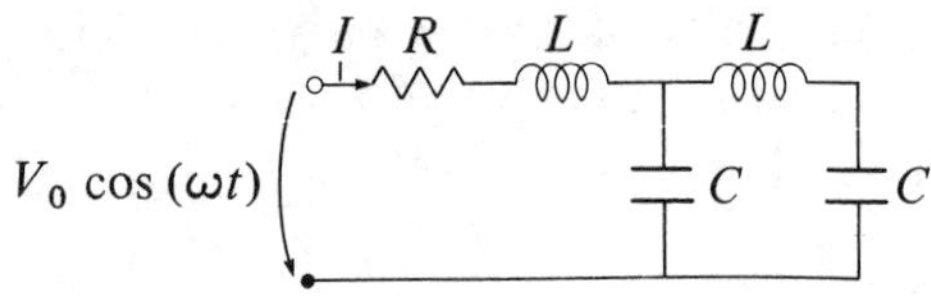

5.4 *Impulse response.* The left of the two coupled pendulums in Problem 5.2 receives an impulse J in the horizontal direction at $t = 0$. Determine the subsequent displacement functions for the two pendulums. What will be the motion for very large values of t?

5.5 *Weakly coupled oscillators.* Two simple pendulums of equal length $L = 1$ m are connected with a spring with a spring constant $K = 0.05\, Mg/L$. The pendulums are started by releasing one of them from a displaced position. The subsequent motion is characterized by an oscillatory energy exchange between the pendulums. What is the period of this transfer?

5.6 *Free damped oscillations.* Determine the frequencies and decay constants of the two modes of motion of the pendulums in Problem 5.2. What condition must be imposed on the values of R, M, L, and l in order for the motion to be oscillatory?

5.7 *Normal modes and coordinates.* Consider the free motion of the coupled oscillators in Display 5.12. Let $K_1 = 2K$, $K_2 = K_{12} = K$, $M_1 = M$, and $M_2 = 2M$.

(a) Determine the normal mode frequencies and the ratio between the displacements in the normal modes.

(b) Determine the normal coordinates and the corresponding unit vectors along the normal coordinate axes in configuration space.

5.8 *Free damped motion.* Let the mass elements of the coupled oscillators in Display 5.12 be air suspended carts of equal mass on a track. Each of the three springs has a spring constant K, and the resistance coefficient of each dashpot is R.

We now consider another set of coupled oscillators with the same masses and spring constants but with the dashpots replaced by aerodynamic dampers (cardboards) mounted on the carts. Determine the ratio between the decay constants of corresponding modes of free motion of the two systems.

5.9 *Normal modes.* Determine the normal modes of motion of the free coupled mass–spring elements shown in the figure. (Hint: The total momentum of the system is conserved).

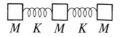

$M \quad K \quad M \quad K \quad M$

5.10 *Coupling of axial and torsional oscillations.* A dumbbell is attached to the end of a coil spring with its upper end clamped. Vertical oscillation of the system generates a torsional oscillation of the dumbbell about the axis of the spring, and, conversely, a torsional oscillation induces a vertical motion. This is a result of the elastic property of the coil spring that an extension ξ produces not only a vertical force $(-K_{11}\xi)$ but also a torque $(K_{21}\xi)$ on the dumbbell. Conversely, a torsional displacement of the dumbbell by an angle θ results not only in a torque $(-K_{22}\theta)$ but also an axial force $(K_{12}\theta)$ on the dumbbell. For small displacements, the total reaction force and torque resulting from a deformation of the spring are then given by the linear equations

$$F = -K_{11}\xi + K_{12}\theta$$
$$\tau = K_{21}\xi - K_{22}\theta$$

(a) Establish the equations for the coupled axial and torsional oscillations of the system and determine the normal modes and frequencies. Mass of dumbbell $= M$, moment of inertia $= I$.

(b) Discuss the case of weak coupling, i.e. with K_{12} and K_{21} much smaller than K_{11} and K_{22}. Determine the 'beat period', i.e. the period of the oscillatory exchange of the energy between the axial and torsional oscillations.

(c) How would you go about measuring the constants K_{ij}?

5.11 *Impulse response.* Derive Eqs. D.5.9(9)–(10) and compare the results with the steady state response functions obtained in Section 5.1.

PART 2

Waves

6

Fundamentals of waves

Extending the discussion of two coupled oscillators in the last section, we now consider the transmission of a disturbance on a long line of equally spaced identical mechanical or electrical elements and on the corresponding continuous transmission lines. This leads to the concept of a wave, and appropriate field variables and concepts are introduced for its description, including the complex wave amplitude function (in the case of harmonic time dependence). The relations between the field variables are expressed by field equations, from which a wave equation is obtained. These equations are found to have the same form for a variety of waves on both mechanical and electrical transmission lines.

6.1 What is a wave?

You have no doubt observed the activity on a line of cars stopping or starting at a traffic light. All cars in the line do not start at the same time as the light turns green; rather, the event of starting travels backwards through the line with a certain speed. This 'starting wave' is initiated by the first car in the line, and the speed of the wave depends on the reaction time of the drivers and the response characteristics of the cars, the inertial mass being one of the factors.

It is important to realize that no mass is transported by the wave; what travels is merely the act of starting. In fact, in this example, the transport of mass involved in the motion of the cars is in the opposite direction to the wave.

Similarly, when the end ($x = 0$) of a stretched rope is suddenly moved sideways (corresponding to the motion of the first car in the example above), the event of 'moving sideways' travels along the rope as a wave with a certain speed v, which depends on the tension in the rope and the mass.

The initial displacement will be repeated at a distance x from the end after a travel time x/v. Again, no mass is transported by the wave; in this case the mass elements move in the transverse direction.

Similarly, in a compressional wave in a fluid or solid, it is that state of being compressed that travels, and on an electric transmission line it is the electromagnetic field which is transmitted with a certain wave speed, which has little to do with the velocity of the electrons, which carry the current in the line. (For simple demonstrations, see Section 6.6)

6.2 Description of waves (kinematics)

In this section we introduce the variables and concepts used in the quantitative description of kinematical characteristics, common to all types of waves, and no attention will be paid to the particular 'forces' or the corresponding dynamical equations involved. The dynamical aspects will be considered in the next section.

Travelling waves. We use as a basis for our discussion the transverse motion of a string, rope, or coil spring under tension, as indicated in Display 6.1. This motion is simple to demonstrate and is familiar from everyday observations. A particularly effective and commonly used alternate model for wave demonstrations is a 'ladder' of closely spaced bars, mounted on a wire or metal strip, excited in torsional motion. The mathematical description of these motions is applicable also to other mechanical and electrical transmission lines, as will be discussed in detail later.

In the description of the motion of two coupled oscillators, we used the notations $\xi(x_1, t)$ and $\xi(x_2, t)$ for the displacements of the mass elements with the two equilibrium coordinates x_1 and x_2. In the case of a continuous transmission line (our string), we have 'oscillators' (mass elements) at **every** position x, and the displacement from equilibrium then will be expressed by $\xi(x, t)$, where x is now a **continuous** variable. Actually, to indicate that we deal with a transverse rather than a longitudinal displacement we shall use the notation $\eta(x, t)$.

Thus, the displacement $\eta(x, t)$ is a function of two independent continuous variables, x and t. The time dependence of the displacement at $x = 0$ (the left end of the string), is $\eta(0, t)$, and in the display, this displacement is in the form of a pulse of duration t_2, and it can be considered to be the source of the wave motion. For simplicity, we denote $\eta(0, t)$ by $D(t)$.

The observed displacement at location x is found to have the same time dependence except for a time delay, which is proportional to x. We express the delay as x/v, where v is the **wave speed**. Consequently, the mathematical

Display 6.1.

Transverse waves on a string.

Travelling wave in positive x-direction

Time dependence

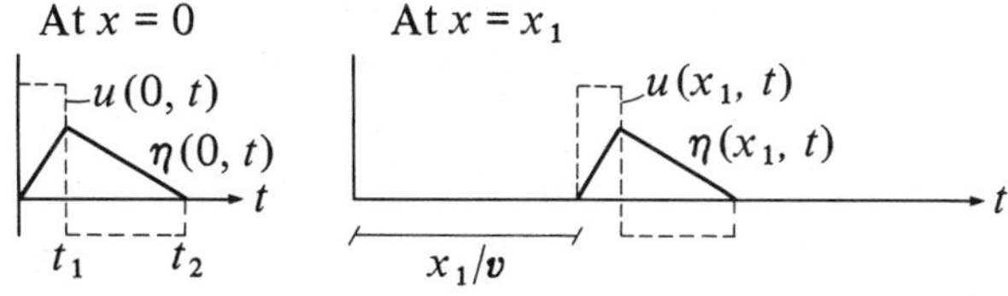

Space dependence

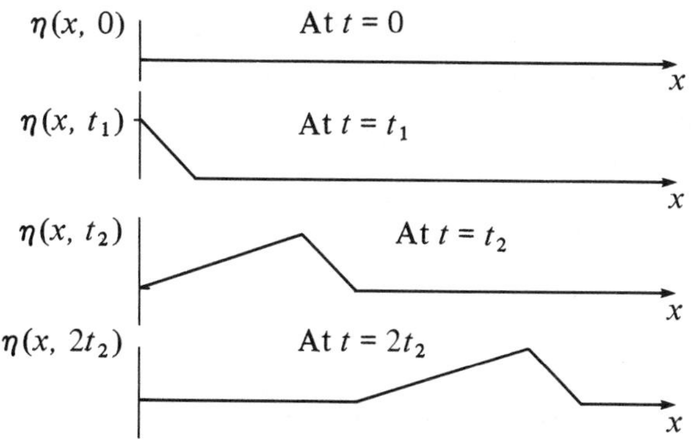

Wave functions	Displacement wave	$\eta(x, t) = D(t - x/v)$
	Velocity wave	$u(x, t) = \partial D(t - x/v)/\partial t$

Waves in positive and negative x-directions

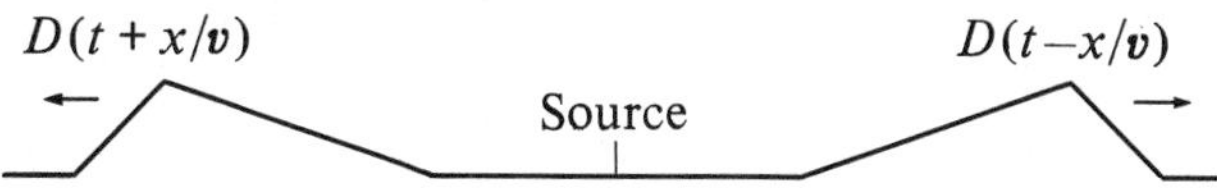

description of the displacement at x is

$$\eta(x, t) = \eta(0, t - x/v) \equiv D(t - x/v) \tag{1}$$

The fact that the wave shape does not change with location indicates that all parts of the pulse travel with the same speed. Thus, for a given time dependence of the displacement at $x = 0$, the shape of the wave, i.e. the x-dependence of the displacement at a given time, is obtained by moving the displacements at $x = 0$, as they occur, in the x-direction with the speed v, as indicated in the display. The displacement produced at $t = 0$ will form the 'wave front', and the functions of the displacement versus t and versus x essentially are mirror images of one another. Mathematically, this is expressed by the fact that the argument in the displacement function contains the variables x and t in the combination $t - x/v$, with different signs for t and x.

In addition to the displacement, there is a transverse velocity wave, which is the time derivative of the displacement. Furthermore, the transverse force applied at the end of the string is also transmitted as a wave. As we shall see shortly, this force is proportional to the slope of the string, i.e. to the space derivative of the displacement.

To indicate that the time derivative refers to a fixed position x, we use partial differentiation, so that the particle velocity of the string in the transverse direction at position x is

$$u(x, t) = \partial\eta(x, t)/\partial t \tag{2}$$

It should be emphasized that the wave we have considered so far, described by the function $D(t - x/v)$, travels in the **positive x-direction**. If t is increased by an amount Δt, the argument, and hence the displacement, remains the same if x is increased by the amount $\Delta x = v\Delta t$. In other words, if we 'sit' on a wave, we move forward with the velocity v in the positive x-direction

In a similar manner, a wave travelling in the **negative x-direction** is described by the wave function $D(t + x/v)$. If we drive a string at its midpoint ($x = 0$), two outgoing waves, $D(t - x/v)$ and $D(t + x/v)$ will be generated, as indicated schematically in the display.

In addition to the displacement wave $\eta(x, t)$ there is an associated velocity wave $u(x, t) = \partial\eta/\partial t$. Also the transverse force generating the wave at $x = 0$ is transmitted as a wave $F(x, t)$. The quantities $\eta(x, t)$, $u(x, t)$, and $F(x, t)$ will be called **field variables**, which are functions of the independent variables x, t.

A basic problem is to find the relationship between the field variables, and we shall see later, that these relations can be expressed in terms of two **field equations** containing the variables $u(x, t)$ and $F(x, t)$.

Analogous velocity and force variables are used in the description of other mechanical waves, and for electromagnetic waves, velocity and force are replaced by the analogous quantities current and voltage, the same analogy as was established in Chapter 3. We shall find that the two field equations will have the same form for all these waves, and by eliminating one of the field variables, we obtain a **wave equation** for each of the variables.

Harmonic time dependence. The important case of harmonic time dependence deserves special attention. Thus, if the displacement of the string at $x = 0$ is $\eta(0, t) = A_0 \cos(\omega t)$, the expression for the displacement at x, as we have seen, is obtained by replacing t by $t - x/v$, so that

$$\eta(x, t) = \eta(0, t - x/v) = A_0 \cos[\omega(t - x/v)] = A_0 \cos(\omega t - kx)$$
$$= A_0 \cos[2\pi(t/T - x/\lambda)] \qquad (3)$$

where we have introduced the propagation constant $k = \omega/v$, the period T, and the wavelength $\lambda = vT$.

With the period of the harmonic motion being T, the distance of wave travel in the time of one period will be vT. This distance, by definition, is the **wavelength** $\boldsymbol{\lambda = vT}$. It is the distance between two adjacent wave crests (or troughs) and is the period of the spatial wave pattern, as observed at a fixed time t.

The inverse of the time period T, i.e. the number of periods per unit time, is the **frequency** $\boldsymbol{\nu = 1/T}$. The inverse of the spatial period, the number of wavelengths per unit length, is usually called the **wave number**.

If we multiply the frequency by 2π we obtain the **angular frequency** $\boldsymbol{\omega = 2\pi\nu}$. The corresponding quantity for the spatial variation of the wave function is called the **propagation constant** $\boldsymbol{k = 2\pi/\lambda = \omega/v}$. We shall use this quantity rather than the wave number in expressing the spatial variation of a harmonic wave.

At each location x, the string performs harmonic motion, and a string element at x can be regarded as a harmonic oscillator with a certain amplitude and a certain phase angle. In forced harmonic motion, the frequency of oscillation is determined by the driving force, and with a single travelling wave carried by the string, the amplitude of oscillation will be the same at all locations. The **phase angle (phase lag)**, on the other hand, will be proportional to x, being $kx = 2\pi x/\lambda$ for a wave travelling in the positive x-direction.

In Display 6.2 is shown a sequence of 'snapshots' of a string driven at $x = 0$ by a displacement, which is started at $t = 0$ and is $\eta(0, t) = A_0 \cos(\omega t)$ for $t > 0$. The resulting displacement wave travels along the string with the speed v. For example, at $t = T/4$, the front of the wave has arrived at

Display 6.2.

Travelling harmonic wave.

Harmonic wave travelling in the positive x-direction

Displacement at $x = 0$

$$\eta(0, t) = A_0 \cos(\omega t) \tag{1}$$

Displacement at x

$$\begin{aligned} \eta(x, t) &= A_0 \cos[\omega(t - x/v)] = A_0 \cos(\omega t - kx) \\ &= A_0 \cos[2\pi(t/T - x/\lambda)] \end{aligned} \tag{2}$$

Propagation constant $\quad k = \omega/v = 2\pi/vT = 2\pi/\lambda$ (3)

Wavelength $\quad \lambda = vT = v/\nu \quad \nu = 1/T = \omega/2\pi =$ frequency
$T =$ period (4)

$$\eta(x, t) = A_0 \cos[2\pi(t/T - x/\lambda)]$$

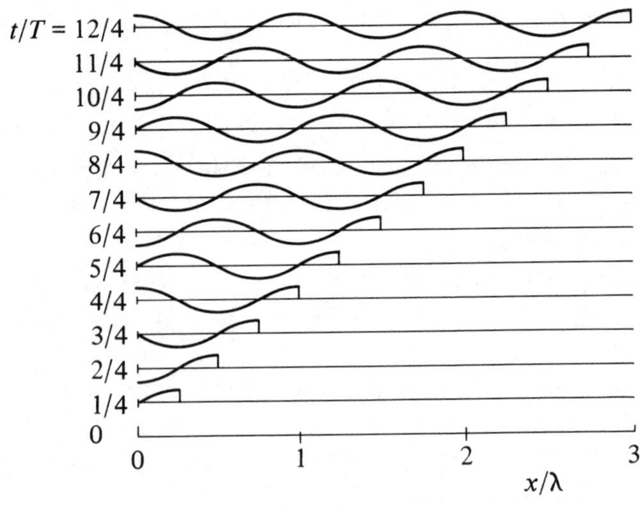

$x = vT/4 = \lambda/4$. The shape of the string at other times up to $t = 3T$ is shown at intervals of $T/4$.

It should be remarked that the sharp wave front shown in the display is an idealization. In reality, the front is rounded off somewhat (because of internal friction and stiffness). In a gas the width of the wave front in a shock wave is of the order of the mean-free-path in the gas.

Complex amplitude functions for travelling waves. A harmonic displacement wave travelling in the positive x-direction is

$$\eta_+(x,t) = A_0 \cos[\omega(t - x/v) - \alpha] = A_0 \cos[\omega t - kx - \alpha]$$

if the displacement at $x = 0$ is chosen to be $A_0 \cos(\omega t - \alpha)$, with the corresponding complex amplitude being $\eta_+(0) = A_0 \exp(\mathrm{i}\alpha)$. The complex amplitude at x is $\eta_+(x,k) = \eta_+(0,k)\exp(\mathrm{i}kx)$. Since $k = k(\omega)$, in this case $k = \omega/v$, we can express $\eta_+(x,k)$ also as $\eta_+(x,\omega)$.

Similarly, a harmonic wave travelling in the negative x-direction has the complex amplitude $\eta_-(0)\exp(-\mathrm{i}kx)$, where $\eta_-(0)$ is the complex amplitude at $x = 0$.

Since the complex amplitudes $\eta_\pm(x, k)$ are functions of x, we often call them **complex amplitude functions** to distinguish them from the constants $\eta_+(0)$ and $\eta_-(0)$, which will be called complex amplitude constants or simply complex wave amplitudes. The magnitudes of $\eta_+(0)$ and $\eta_-(0)$ will be called the wave amplitudes. Usually, there is little risk for misunderstanding if we express $\eta(x, k)$ simply as $\eta(x)$.

In the example in Display 6.2, the displacement at $x = 0$ is chosen to be $A_0 \cos(\omega t)$, i.e. with the phase angle α being zero. In that case the complex amplitude at $x = 0$, $\eta_+(0) = A_0$ is real.

The real space–time dependence of the displacement $\eta(x,t) = \mathrm{Re}\{\eta(x,k)\exp(-\mathrm{i}\omega t)\}$, where $\eta(x) = \eta(0)\exp(\mathrm{i}kx)$, is called the real wave function, the quantity $\eta(x,k)\exp(-\mathrm{i}\omega t)$, the complex wave function, and $\eta(x,k)$, the complex wave amplitude function.

It should be observed that the signs in the arguments of the exponentials of the complex amplitude functions:

$$\eta_+(x) = \eta_+(0)\exp(\mathrm{i}kx)$$
$$\eta_-(x) = \eta_-(0)\exp(-\mathrm{i}kx)$$

are the same as the directions of wave travel. This is a consequence of our choice of $\exp(-\mathrm{i}\omega t)$, rather than $\exp(\mathrm{i}\omega t)$, as the time factor in the complex wave function, as discussed in Chapter 2.

It should be noted also that the complex amplitude function $\eta_+(0)\exp(\mathrm{i}kx)$, representing a wave in the positive direction, moves in the complex plane along a circle of radius $|\eta_+(0)|$ in the counter-clockwise direction as x increases. Similarly, the complex amplitude function $\eta_-(0)\exp(-ikx)$, representing a wave travelling in the negative x-direction, moves along a circle of radius $|\eta_-(0)|$ in the clockwise direction as x increases. For $x=0$, the locations of the complex wave amplitude functions depend on the magnitude and phase angles of $\eta_+(0)$ and $\eta_-(0)$.

From the complex wave amplitude representation it follows immediately that the sum of two harmonic waves travelling in the same direction can be described as a single wave with a certain amplitude and phase angle. If the complex amplitudes of the two waves are $\eta_1(0)\exp(ikx)$ and $\eta_2(0)\exp(ikx)$ the sum will be $[\eta_1(0)+\eta_2(0)]\exp(ikx)=\eta(0)\exp(ikx)$, from which the amplitude and phase angle of $\eta(0)$ can be determined in the usual manner by the addition of complex numbers. For further details, including the combined energy of several travelling waves, we refer to the problem section.

Standing wave. In a similar manner we can determine the sum of two waves $A\exp(ikx)$ and $B\exp(-ikx)$ travelling in **opposite** directions. Of particular interest is the case when $A=B$. The sum then becomes $A[\exp(ikx)+\exp(-ikx)]=2A\cos(kx)$. This complex amplitude does not represent a travelling wave. First, the amplitude $2|A\cos(kx)|$ is now a function of x and not a constant as in a travelling wave, and furthermore, the phase angle is determined by the phase angle of A and the sign of $\cos(kx)$, and is not proportional to x, as in a travelling wave.

With $A=A_0\exp(i\alpha)$, the corresponding real wave function is

$$\eta(x,t)=2A_0\cos(kx)\cos(\omega t-\alpha), \tag{4}$$

which is called a **standing wave**. The amplitude varies from 0 to $2A_0$. The locations of zero amplitude are called **nodes**, and the distance between two adjacent nodes corresponds to an increase of kx by π, i.e. a distance of **half a wavelength**. The locations of the maximum amplitudes are called **antinodes**.

As mentioned, the phase angle is determined by α and the sign of $\cos(kx)$. In the region between two nodes the sign of $\cos(kx)$ and hence the phase angle is constant. Crossing a node produces a change of sign and a change of the phase angle by π. In other words, the string motions in the regions between adjacent node pairs are in opposite directions.

If the wave amplitude constants A and B are different, the sum $A\exp(ikx)+B\exp(-ikx)$ can be expressed as

$$\begin{aligned}&B[\exp(ikx)+\exp(-ikx)]+(A-B)\exp(ikx)\\&\quad=2B\cos(kx)+(A-B)\exp(ikx)\end{aligned}$$

Display 6.3.

Harmonic waves.

Travelling waves

Travel in positive x-direction

Real wave function $\quad \eta(x,t) = A_0 \cos(\omega t - kx - \alpha)$ (1)

Complex amplitude $\quad \eta(x,k) = A \exp(ikx)$
where $A = A_0 \exp(i\alpha)$ (2)

Wave amplitude constant $\quad A = A_0 \exp(i\alpha)$ (3)

Complex wave function $\quad A \exp(ikx - i\omega t)$ (4)

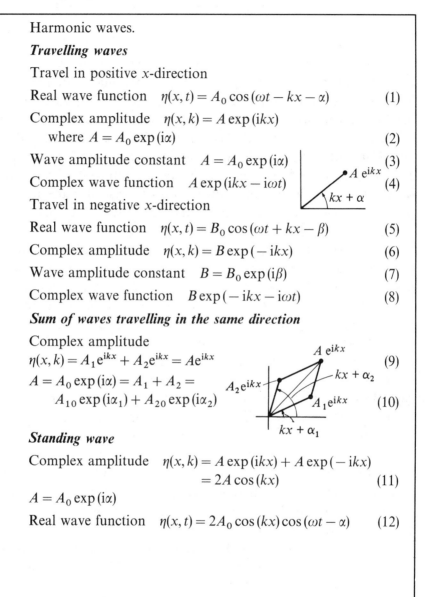

Travel in negative x-direction

Real wave function $\quad \eta(x,t) = B_0 \cos(\omega t + kx - \beta)$ (5)

Complex amplitude $\quad \eta(x,k) = B \exp(-ikx)$ (6)

Wave amplitude constant $\quad B = B_0 \exp(i\beta)$ (7)

Complex wave function $\quad B \exp(-ikx - i\omega t)$ (8)

Sum of waves travelling in the same direction

Complex amplitude

$\eta(x,k) = A_1 e^{ikx} + A_2 e^{ikx} = A e^{ikx}$ (9)

$A = A_0 \exp(i\alpha) = A_1 + A_2 =$
$A_{10} \exp(i\alpha_1) + A_{20} \exp(i\alpha_2)$ (10)

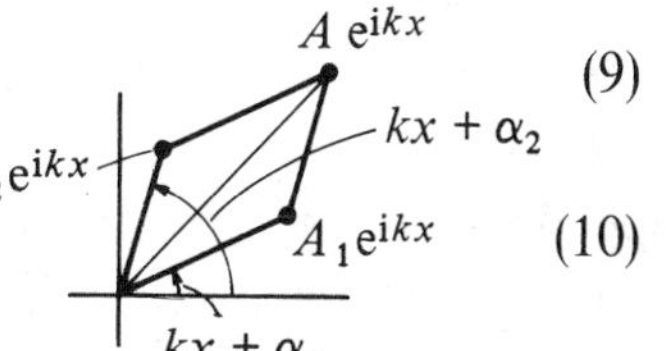

Standing wave

Complex amplitude $\quad \eta(x,k) = A \exp(ikx) + A \exp(-ikx)$
$= 2A \cos(kx)$ (11)

$A = A_0 \exp(i\alpha)$

Real wave function $\quad \eta(x,t) = 2A_0 \cos(kx) \cos(\omega t - \alpha)$ (12)

i.e. as the sum of a standing and a travelling wave. These observations concerning wave kinematics are summarized in Display 6.3

Wave diagrams. To further illustrate kinematical aspects of waves, we show in Display 6.4 the t- and x-dependence of the pressure in a wave pulse, such as might be generated by an explosion in a fluid contained in a pipe. The wave travels in the positive x-direction and the first figures refer to the time dependence of the pressure (in excess of the static pressure) at two different locations x_1 and x_2. The pulse is characterized by a sudden rise in pressure followed by a slower decay and a portion of 'under pressure'. The two pulses are delayed in time by the travel time $(x_2 - x_1)/v$, where v is the wave speed. (For simplicity, we assume that all parts of the wave pulse travel with the same speed, so that the shape of the pulse does not change with position).

The corresponding x-dependence of the pressure, i.e. the 'shape' of the pulse, at two different times is shown also, and we note that the t- and x-dependence of the pressure are essentially mirror images of one another.

It is instructive to show the progression of the wave pulse in a three-dimensional plot, as indicated. The axes are time t, location x, and the field variable, which in this case is the pressure p.

Imagine now that we have observers or pressure transducers placed at the locations $x_1, x_2, \ldots$, and that at each location the observed time dependence of the pressure is recorded. The pressure is displaced at each location as a function of time and this results in a succession of traces, which are displaced with respect to one another, as shown.

The arrival of the pressure front (or any other part of the wave) at a certain location is represented as a point in the t–x plane, and it is referred to as an 'event'. The events representing the arrival of the pressure front clearly lie on a straight line in the t–x plane. The line goes through the origin and is given by $t - x/v = 0$ or $t = x/v$.

The events corresponding to the trailing edge of the pulse are represented by the line $t - x/v = \tau$, where τ is the duration of the pulse. The lines are parallel with a slope $1/v$ with respect to the x-axis. Along these lines, or any other parallel line, the argument of the wave function and hence the value of the function remain constant. Such a line is called a **wave line**. Frequently, the t-axis is replaced by a vt-axis, in which case the slope of a wave line will be $+1$ for a wave travelling in the positive x-direction. Similarly, the wave line for a wave travelling in the negative x-direction will have a slope equal to -1.

Display 6.4.

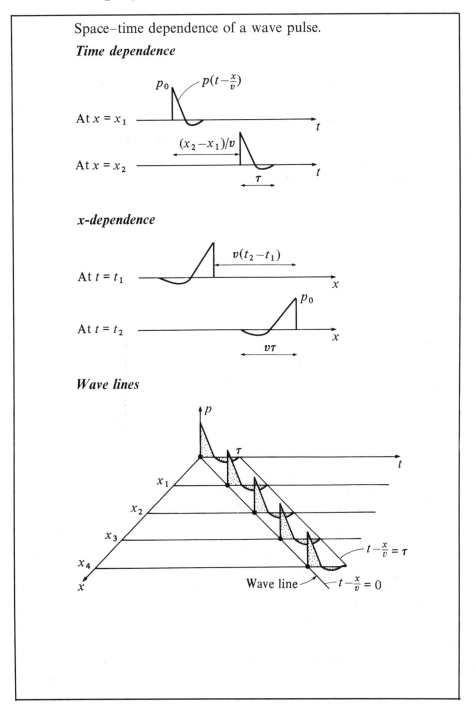

Space–time dependence of a wave pulse.

Time dependence

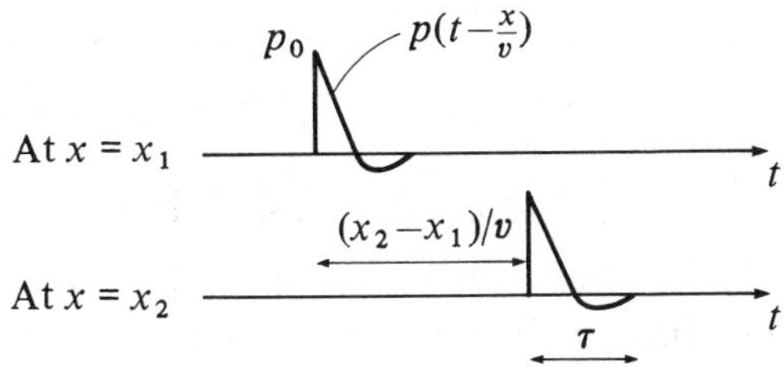

x-dependence

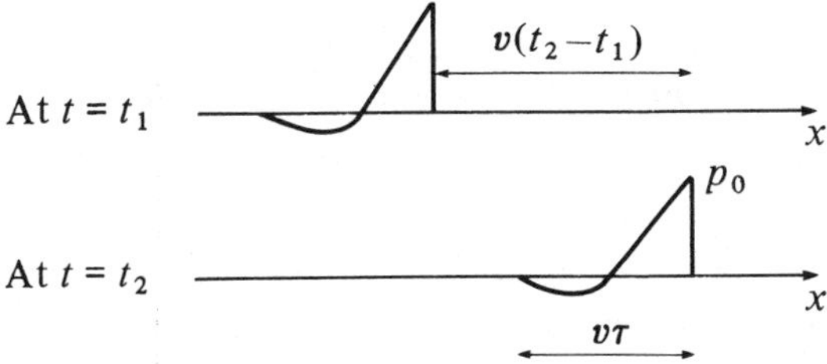

Wave lines

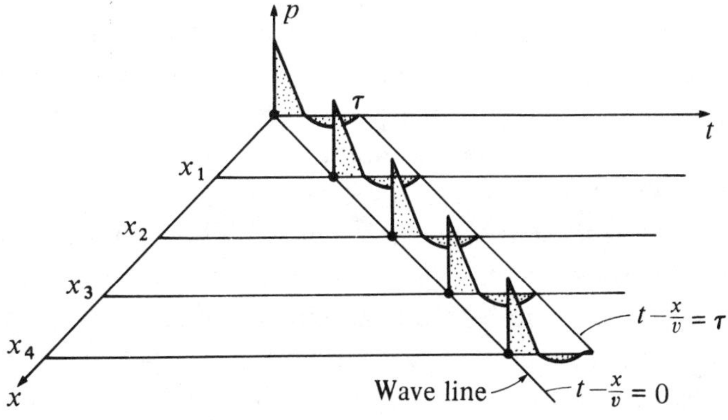

Wave lines can be useful in illustrating graphically certain features of wave propagation, as demonstrated by the simple examples in Display 6.5. In the first example are shown the wave lines, which correspond to waves travelling in the positive and negative x-directions. The slopes of the lines are $\pm 1/v$.

In the second example, a sound wave is incident on the boundary between two regions with different wave speeds, for example air and helium, as indicated. (The boundary can be considered to be a very thin sound transparent membrane). The wave speed in helium is about 3 times larger than in air, and the slope of the wave line of the transmitted wave, transmitted across the boundary, will be approximately 1/3 of the slope of the wave line of the incident wave in air.

In the third example are shown the wave lines of the waves emitted from a source moving in the positive x-direction. The velocity of the source is U and its trajectory is given by $t = x/U + \text{const.}$, as shown.

The source emits wave pulses at regular intervals, and the waves are transmitted both in the positive and the negative x-direction. The slopes of the wave lines are determined only by the wave speed v in the surrounding medium, and with the speed of the source smaller than the wave speed, the slope of each wave line is smaller than the slope of the source trajectory.

It is interesting to consider the time dependence of the wave trains recorded by observers at rest ahead of and behind the source. It is clear from the wave diagram that the number of wave lines (wave pulses) observed per second ahead of the source will be greater than behind it. Instead of representing wave pulses, the wave lines could equally well represent the crests of the waves emitted from a harmonic source in motion. In that case the observed pressure will be harmonic also with the observed frequency being higher ahead of the source than behind it. This shift in frequency is known as the **Doppler shift**, which is an important phenomenon throughout physics, as will be discussed in a subsequent chapter. It is used in a wide range of applications, both technical and scientific, for the measurement of the speed of moving objects, ranging from molecules to galactic objects.

In the case of a source of sound, such as an airplane, it is possible that the speed of the source can exceed the wave speed. In that case, the slope of the trajectory of the source will be smaller than the slope of the wave lines, and it follows that the wave lines for waves emitted in the positive and the negative x-direction will emerge on the same side of the trajectory line. The wave lines will then cross each other, which indicates interference between forward and backward running waves (see Problem 6.5).

Display 6.5.

Wave diagrams.

Single waves

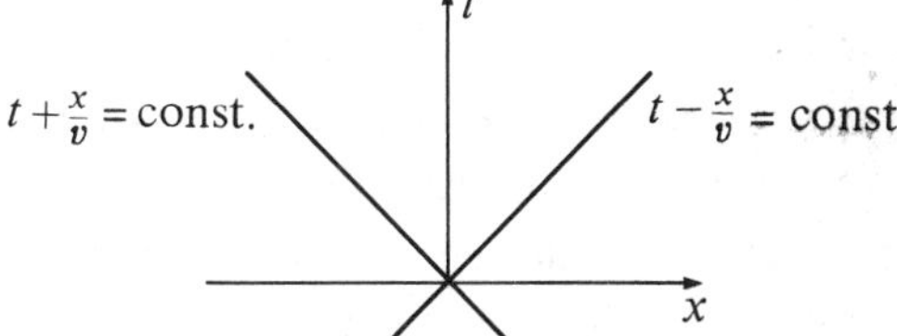

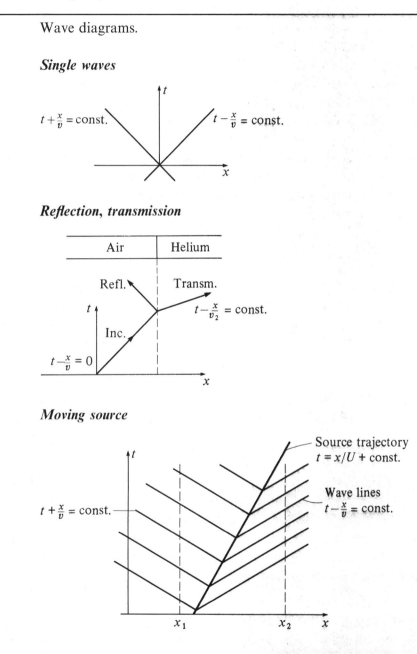

Reflection, transmission

Air
Helium
Refl.
Transm.
t
$t - \frac{x}{v_2} = \text{const.}$
Inc.
$t - \frac{x}{v} = 0$
x

Moving source

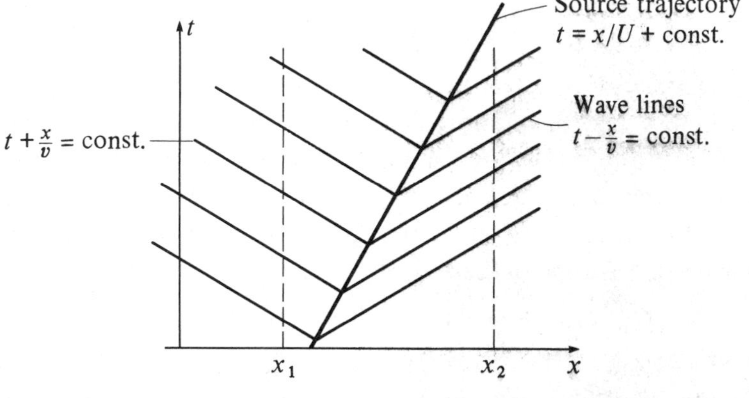

6.3 Dynamics. Field equations

In our study of the dynamics of waves, we shall start with a line of coupled oscillators, mechanical or electrical. In the case of mass–spring oscillators, the field variables involved in the equations of motion (Newton's law applied to each element) are the displacements and spring forces acting on the elements. Actually, we shall find it better to use velocity instead of displacement as a field variable, corresponding to the current on the electrical line. In terms of force and velocity as field variables, the equations will have the same form for all waves we are going to consider.

The equations thus obtained are valid for all values of the equilibrium distance d between the elements in the line, including the limit of d equal to zero, in which case the equations describe waves on a continuous transmission line.

Longitudinal waves. We shall analyze the motion of the mass–spring lattice shown in Display 6.6. In equilibrium the distances d between the mass elements are all the same and the location of the nth element is denoted by x_n. The displacement from equilibrium is $\xi(x_n, t)$.

The force acting on the left side of the nth element is $F(x_n, t)$ or $F(n, t)$ for short. It is counted positive in the positive x-direction. Similarly, the force on the left side of the $(n+1)$th element is $F(n+1, t)$. This force is a result of the compression of the spring between elements n and $n+1$. Consequently, the corresponding reaction force on the right hand side of the nth element will be $-F(n+1, t)$. The net force on the nth element is then $F(n, t) - F(n+1, t)$. Thus the equation of motion for the nth element is

$$M\dot{u}(n, t) = F(n, t) - F(n+1, t) \tag{1}$$

The force $F(n+1, t) = K[\xi(n, t) - \xi(n+1, t)]$ is the product of the spring constant K and the compression $\xi(n, t) - \xi(n+1, t)$ of the spring between elements n and $n+1$. This relation can be expressed in terms of the velocity rather than the displacement by differentiation with respect to time. Thus, if we introduce the compliance $C = 1/K$ of the spring, the equation of compression can be expressed as

$$C\dot{F}(n+1, t) = u(n, t) - u(n+1, t) \tag{2}$$

We now make the transition to a continuous transmission line, such as a spring or a fluid or solid column with a continuous mass distribution, by letting the 'lattice constant' d go to zero, keeping the average mass $\mu = M/d$ per unit length of the line constant.

The field variables u and F then can be regarded as continuous functions $u(x, t)$ and $F(x, t)$ of x and t. In the limit of $d = 0$ the ratio

Display 6.6.

Longitudinal wave motion.

Field variables

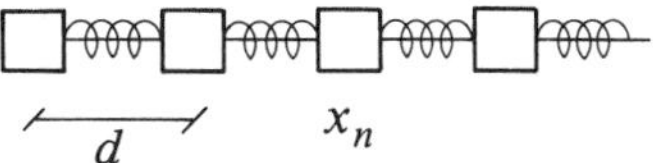

Equilibrium position $\quad x_n = nd$ (1)

Displacement $\quad \xi(x_n, t)$ denoted by $\xi(n)$ (2)

Spring force on the left side of element n $F(x_n, t) \equiv F(n)$ (3)

Spring force on the right side of element n $\quad -F(n+1)$ (4)

Velocity $\quad u(n) = \mathrm{d}\xi(n)/\mathrm{d}t$ (5)

Equations of motion

$-F(n+1)$
$F(n)$ $\quad F(n+1)$
$\xi(n)$ $\quad \xi(n+1)$

$M\dot{u}(n) = F(n) - F(n+1)$ (6)

Spring force relation $\quad F(n+1) = K[\xi(n) - \xi(n+1)]$ (7)

or

$(1/K)\dot{F}(n+1) = u(n) - u(n+1) \quad \dot{F} = \mathrm{d}F/\mathrm{d}t$ (8)

Transition to continuum. Field equations

$[\xi(n+1) - \xi(n)]/d = \partial\xi/\partial x \quad [F(n+1) - F(n)]/d = \partial F/\partial x$ (9)

Let $\mu = M/d$, average mass per unit length (10)

$\kappa = 1/Kd$, compliance per unit length (11)

$\mu \partial u/\partial t = -\partial F/\partial x$ (12)

$\kappa \partial F/\partial t = -\partial u/\partial x$ (13)

Wave equation

$\partial/\partial x$ of Eq. (12) and $\partial/\partial t$ of Eq. (13) yield

$\partial^2 F/\partial x^2 = (\mu\kappa)\partial^2 F/\partial t^2 = (1/v^2)\partial^2 F/\partial t^2$ (14)

$v^2 = 1/\mu\kappa$ (15)

Waves $F(t \pm x/v)$ satisfy Eq. (14) (16)

$v = 1/\sqrt{\mu\kappa}$ is the wave speed.

$[F(x_{n+1}, t) - F(x_n, t)]/d$ can then be expressed as $\partial F(x,t)/\partial x$, the partial derivative of F with respect to x (keeping t constant). Similarly, $[u(n+1, t) - u(n, t)]/d$ becomes $\partial u(x, t)/\partial x$.

Returning to the two equations above, we now replace the time derivatives $\dot{u}(x_n, t)$ and $F(x_{n+1}, t)$ by the partial derivatives $\partial u(x,t)/\partial t$ and $\partial F(x,t)/\partial t$, and by expressing the right hand sides of the equations in terms of the derivatives with respect to x, as shown above, we obtain the important field equations

$$\mu \partial u/\partial t = -\partial F/\partial x \qquad (3)$$

$$\kappa \partial F/\partial t = -\partial u/\partial x \qquad (4)$$

where $\mu (= M/d)$ is the mass per unit length and $\kappa (= C/d = 1/Kd)$ is the compliance per unit length. Usually, when we apply these equations to fluid and solid columns, the force refers to unit area. The mass per unit length then is the same as the mass density ρ, and the compliance per unit length becomes the compressibility of a fluid and the inverse of the Young's modulus E for a solid, as will be discussed later.

As we shall see shortly, these field equations apply also to transverse waves on a string with a tension S, if F is the force on the string in the transverse direction and $\kappa = 1/S$.

In a fluid, only longitudinal waves are generally involved, but in a solid transverse waves can occur also. A special form of transverse waves are torsional waves in an elastic column. The velocity u and force F then are replaced by the angular velocity and the torque, and μ and κ are replaced by the moment of inertia per unit length and the torsional compliance.

Surface waves on shallow water will be analyzed in a subsequent chapter, and we shall find that the field equations apply in their original form with F replaced by the pressure perturbation in the fluid produced by the variation in height of the water surface.

Similarly, for electromagnetic waves on a transmission line, the field equations are still valid with u and F replaced by the current and the voltage and μ and κ replaced by the inductance and capacitance per unit length, respectively.

If we express the current and voltage in terms of the corresponding magnetic and electric fields, the equations will be a special case of **Maxwell's equations**. The constants μ and κ then stand for the magnetic and electric permeabilities, respectively.

Wave equation. With two field variables F and u and two field equations, we can in principle solve for the field variables in terms of the independent variables x and t, and we shall give several examples of such solutions in subsequent chapters.

By combining the equations, we can eliminate one of the variables, say u, and obtain an equation for F alone. This is done by differentiating the first field equation with respect to x and the second with respect to t. We can then eliminate the term $\partial^2 u/\partial x \partial t = \partial^2 u/\partial t \partial x$ and obtain the **wave equation**

$$\partial^2 F/\partial x^2 = \mu\kappa \partial^2 F/\partial t^2 = (1/v^2)\partial^2 F/\partial t^2 \tag{5}$$

We have introduced $v = 1/\sqrt{\mu\kappa}$, which is the **wave speed**, as will be shown below.

In the study of the kinematics of wave motion in the previous section, we found that the general expressions for waves travelling in the positive and negative x-directions are $A(t - x/v)$ and $B(t + x/v)$, respectively. By differentiating twice with respect to x and t, we find that both functions satisfy the wave equation with the wave speed determined by $v = 1/\sqrt{\mu\kappa}$. (With $z = t \pm x/v$ we note that $\partial F/\partial t = \mathrm{d}F/\mathrm{d}z$ and $\partial F/\partial x = \pm(1/v)\mathrm{d}F/\mathrm{d}z$). Actually, the general solution to the wave equation is

$$F = A(t - x/v) + B(t + x/v) \tag{6}$$

where A and B represent waves travelling in the positive and negative x-directions. We shall discuss later how these functions can be determined from known source and boundary conditions.

Transverse waves. Returning to the mass–spring lattice, it is instructive to carry out a detailed analysis for transverse motion of the lattice, analogous to that for the longitudinal motion. In this case the lattice is assumed to have a constant tension S. The transverse displacement of the nth element is denoted by $\eta(x_n, t)$ or simply $\eta(n, t)$, as indicated in Display 6.7.

The transverse displacement does not produce a change in the length of the springs to first order in displacement, and in order to get a transverse force developed on an element, it is necessary to have the line 'pre-stressed'. A relative displacement of two adjacent elements then gives rise to a slope of the spring between the elements and a corresponding transverse component F of the tension force S. For small displacements the tension S can be assumed to be constant; the error is of second order in the displacement, as discussed further in one of the problems.

As mentioned, the transverse force acting on one side of a mass element in the lattice is the transverse component F of the tension force S, and this transverse component is proportional to the slope of the spring involved. To be consistent with the analysis of longitudinal motion, we denote by $F(x_n, t)$, or simply $F(n, t)$, the transverse force acting on the left side of the nth element in the positive y-direction. The corresponding force on the $(n + 1)$th element

Display 6.7.

Transverse wave motion on a string.

Field variables

Transverse displacement $\eta(x_n, t)$ denoted by $\eta(n)$ for short (1)

Transv. comp. of force on left side of element n $F(n)$ (2)

Transv. comp. of force on right side of element n $-F(n+1)$ (3)

Transv. velocity $u(n) = \mathrm{d}\eta(n)/\mathrm{d}t = \dot{\eta}(n)$ (4)

String tension $= S$

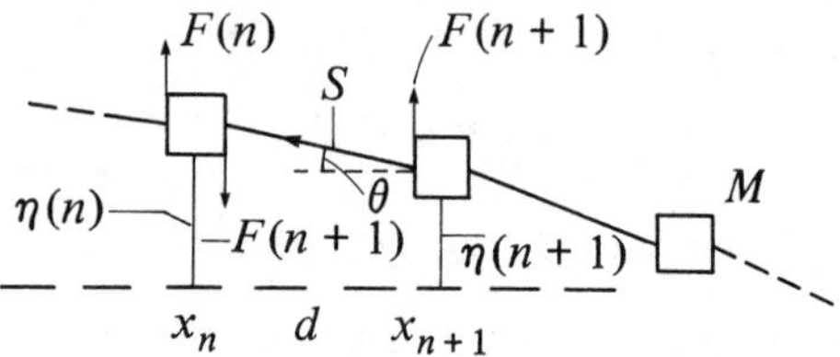

Equations of motion

$$M\dot{u}(n) = F(n) - F(n+1) \quad (5)$$

$$F(n+1) = S\sin\theta \simeq S\tan\theta = S[\eta(n) - \eta(n+1)]/d \quad (6)$$

or

$$(1/S)\dot{F}(n+1) = u(n) - u(n+1) \quad (7)$$

Transition to a continuous string. Field equations

Let $d \to 0$ as $\mu = M/d$ is kept constant

$$[\eta(n+1) - \eta(n)]/d = \partial\eta/\partial x \quad [F(n+1) - F(n)]/d = \partial F/\partial x \quad (8)$$

Let $\kappa = 1/S, \quad \mu = M/d$ (9)

$$\mu\partial u/\partial t = -\partial F/\partial x \quad (10)$$

$$\kappa\partial F/\partial t = -\partial u/\partial x \quad (11)$$

Wave equation

$$\partial^2 F/\partial x^2 = (1/v^2)\partial^2 F/\partial t^2 \quad (12)$$

$v = 1/\sqrt{\mu\kappa} = \sqrt{S/\mu}$ = transverse wave speed (13)

is $F(n+1,t)$. Since the springs are assumed to have negligible mass, the corresponding reaction force on the nth element is $-F(n+1,t)$, i.e. equal in magnitude but opposite in direction.

The net transverse force on the nth element is then $F(n,t)-F(n+1,t)$, which must equal $M\mathrm{d}u(n,t)/\mathrm{d}t$, where $u(n,t)$ is the transverse velocity of the nth element.

It remains to express F in terms of the tension S and the displacements. For small slopes of the spring between elements n and $n+1$, the force acting on the left hand side of element $n+1$ is $F(n+1,t)=S[\eta(n,t)-\eta(n+1,t)]/d$. As before, the velocity variable u rather than the displacement will be used, and the relation between F and u becomes $\dot{F}(n+1,t)=-(S/d)[u(n+1,t)-u(n,t)]$.

The transition to a continuous line is made in much the same way as for the longitudinal motion. Thus, we replace $[\eta(n+1,t)-\eta(n,t)]/d$ by $\partial\eta/\partial x$ and $[F(n+1,t)-F(n,t)]/d$ by $\partial F/\partial x$.

Then, with the average mass per unit length of the line being $\mu(=M/d)$ as before, the two field equations become

$$\mu\partial u/\partial t = -\partial F/\partial x \tag{7}$$

$$(1/S)\partial F/\partial t = -\partial u/\partial x \tag{8}$$

In other words, the equations are the same as for the longitudinal wave motion with $1/S$ taking the place of the compliance per unit length of the line.

These equations apply to a continuous spring or string under tension S. It should be emphasized that when dealing with problems of waves on a string it is recommended that the transverse force F and the velocity u are chosen as field variables. In terms of these variables the field equation pair is of the 'standard' form as presented here, and there is no need for a derivation of these equations in the study of specific problems.

Electromagnetic waves. For the sake of completeness, we derive the analogous equations for electromagnetic (EM) wave propagation on a line of inductances and capacitances and the corresponding field equations for a continuous line, such as a coaxial cable. The LC-line is illustrated in Display 6.8.

The field variables are now the electrical current $I(x_n, t)$ through the nth inductance and the voltage $V(x_n, t)$ over the nth capacitor. These variables will be denoted by $I(n,t)$ and $V(n,t)$ for short.

With the voltages over the capacitors n and $n+1$ being $V(n,t)$ and $V(n+1,t)$, the voltage across the nth inductance is $V(n,t)-V(n+1,t)$, and

Display 6.8.

Electromagnetic waves on a transmission line.

Field variables

Electric charge on capacitor n $Q(x_n, t)$ denoted by $Q(n)$ (1)

Electric current through the nth inductance $I(n)$. (2)

Voltage across the n the capacitor $V(n)$. (3)

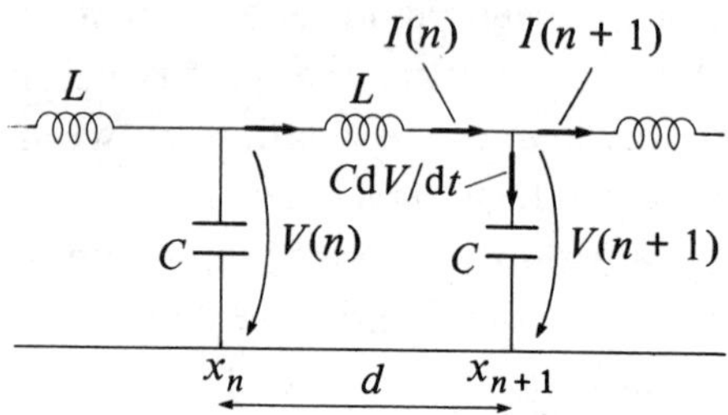

Equations

Voltage across nth inductance $V(n) - V(n+1)$

$LdI(n)/dt = V(n) - V(n+1)$ (4)

From $Q = CV$, after time differentiation, $CdV/dt = dQ/dt$ (5)

or

$CdV(n+1)/dt = I(n) - I(n+1)$ (6)

Transition to a continuous line. Field equations

Let $d \to 0$ keeping $L_0 = L/d$ constant. Let $C_0 = C/d$ (7)

Put $[V(n+1) - V(n)]/d = \partial V/\partial x$ and
$[I(n+1) - I(n)]/d = \partial I/\partial x$ (8)

Eqs. (4) and (6) reduce to the field equations

$L_0 \partial I/\partial t = -\partial V/\partial x$ (9)

$C_0 \partial V/\partial t = -\partial I/\partial x$ (10)

Wave equation

$\partial^2 V/\partial x^2 = (1/v^2)\partial^2 V/\partial t^2$ (11)

$v = 1/\sqrt{L_0 C_0}$ = wave speed

Newton's law in the mass–spring lattice is replaced by the induction law $L\dot{I}(n,t) = V(n,t) - V(n+1,t)$, where L is the inductance.

The relation between the force and the compression of the coupling spring in the line of mechanical oscillators is replaced by the relation between the charge Q on a capacitor and the voltage across it, $Q = CV$, where C is the capacitance. By taking the time derivative of this relation it can be replaced by $I = \mathrm{d}Q/\mathrm{d}t = C\mathrm{d}V/\mathrm{d}t$, where I is the current through the capacitor.

Applying these relations to capacitor $n+1$, we note that the current through this capacitor is the difference between the currents $I(n,t)$ and $I(n+1,t)$, so that $I(n,t) - I(n+1,t) = C\dot{V}(n+1,t)$.

We now go to the limit of a continuous line by letting the separation d between adjacent LC elements go to zero while keeping constant the average inductance $L_0 = L/d$ and capacitane $C_0 = C/d$ per unit length.

The differences $V(n+1,t) - V(n,t)$ and $I(n+1,t) - I(n,t)$ can then be replaced by $[\partial V(x,t)/\partial x]d$ and $[\partial I(x,t)/\partial x]d$. Then, in terms of the inductance L_0 and capacitance C_0 per unit length, the field equations for the continuous line become

$$L_0 \partial I/\partial t = -\partial V/\partial x \tag{9}$$

$$C_0 \partial V/\partial t = -\partial I/\partial x \tag{10}$$

The inductance and capacitance per unit length, correspond to the mass and compliance per unit length in a mechanical transmission line. For a coaxial cable and a parallel wire transmission line, the values of L_0 and C_0 readily can be expressed in terms of the geometrical parameters involved, as will be discussed in a subsequent chapter.

6.4 Harmonic waves and complex amplitude equations

In the important case of harmonic time dependence, we introduce the complex amplitudes of the field variables in the usual manner, as described in Section 6.2. We shall consider the 'standard' form of the field equations for $u(x,t)$ and $F(x,t)$

$$\mu \partial u/\partial t = -\partial F/\partial x \tag{1}$$

$$\kappa \partial F/\partial t = -\partial u/\partial x \tag{2}$$

Denoting the complex amplitudes of velocity and force by $u(x,k)$ and $F(x,k)$, or simply $u(x)$ and $F(x)$, the corresponding complex amplitude equations are

$$-\mathrm{i}\omega\mu u(x) = -\mathrm{d}F(x)/\mathrm{d}x \quad u(x) = (1/\mathrm{i}\omega\mu)\mathrm{d}F(x)/\mathrm{d}x \tag{3}$$

$$-\mathrm{i}\omega\kappa F(x) = -\mathrm{d}u(x)/\mathrm{d}x \quad F(x) = (1/\mathrm{i}\omega\kappa)\mathrm{d}u(x)/\mathrm{d}x \tag{4}$$

where we have made use of the fact that $\partial u(x, t)/\partial t$ has the complex amplitude $-\mathrm{i}\omega u(x)$ with a corresponding expression for the force.

The complex amplitudes are functions of x alone, and the derivative with respect to x can be expressed in terms of the ordinary (rather than the partial) derivative.

With the complex amplitude of $\partial^2 F/\partial t^2$ being $-\omega^2 F(x)$, the wave equation $\partial^2 F/\partial x^2 = (1/v^2)\partial^2 F/\partial t^2$ leads to the following equation for the complex amplitude $F(x)$,

$$\mathrm{d}^2 F/\mathrm{d}x^2 + k^2 F = 0 \tag{5}$$
$$k = \omega/v \tag{6}$$

which is an ordinary second-order differential equation. It has the same form as the harmonic oscillator equation (with a space derivative rather than a time derivative), and the general solution is

$$F(x, k) \equiv F(x) = A\mathrm{e}^{ikx} + B\mathrm{e}^{-ikx} \tag{7}$$

where A and B generally are complex amplitude factors.

As we recall from our study of the kinematics of waves, this solution represents the sum of a wave $A\exp(ikx)$ travelling in the positive and a wave $B\exp(-ikx)$ travelling in the negative x-direction.

Having obtained the solution for the complex force amplitude, we obtain the corresponding complex velocity amplitude from the first of the two field equations, as already indicated,

$$u(x, k) \equiv u(x) = (1/\mathrm{i}\omega\mu)\mathrm{d}F/\mathrm{d}x = (1/Z)[A\mathrm{e}^{ikx} - B\mathrm{e}^{-ikx}] \tag{8}$$
$$Z = \mu v = \sqrt{\mu/\kappa} \quad (k = \omega/v \quad v = 1/\sqrt{\mu\kappa}) \tag{9}$$

We have here introduced the **wave impedance** $\boldsymbol{Z = \mu v}$, which can be thought of as the ratio between the complex force amplitude and the complex velocity amplitude in a wave travelling in positive x-direction. This quantity is a real number and is sometimes called the wave resistance.

The wave amplitude constants A and B can be determined from the known complex values of F and u at a given value of x. For example, in terms of $F(0)$ and $u(0)$ at $x = 0$, we determine A and B from $F(0) = A + B$ and $u(0) = (1/Z)(A - B)$, so that $A = [F(0) + Zu(0)]/2$ and $B = [F(0) - Zu(0)]/2$.

Having obtained the complex amplitude functions $F(x) = F_0(x)\exp[\mathrm{i}\phi(x)]$ and $u(x) = u_0(x)\exp[\mathrm{i}\beta(x)]$, the real wave functions are

$$F(x, t) = F_0 \cos[\omega t - \phi(x)], \quad u(x, t) = u_0(x)\cos[\omega t - \beta(x)] \tag{10}$$

Several examples will be discussed in subsequent chapters.

Example. Forced motion of a real spring. As an example of the use of the general solutions for the complex amplitudes of force and velocity in a one-dimensional wave problem, we shall consider the forced motion of a spring, not the idealized kind without mass but a real spring with its own mass. It is clamped at one end and driven by the force $F = F_0 \cos(\omega t)$ at the other. We locate the clamped end at $x = 0$ and the driven end at $x = -L$, as shown in Display 6.9. The spring constant of the spring is K and its total mass is m. The corresponding compliance and mass per unit length are $\kappa = 1/KL$ and $\mu = m/L$.

The general expression for the complex amplitudes of force and velocity of the spring are Eqs. (7) and (8)

$$F(x) = A\mathrm{e}^{ikx} + B\mathrm{e}^{-ikx} \tag{11}$$

$$u(x) = (1/Z)(A\mathrm{e}^{ikx} - B\mathrm{e}^{-ikx}) \tag{12}$$

where the wave impedance $Z = \mu v = \sqrt{\mu/\kappa}$ and $k = \omega/v$.

The complex amplitude of the driving force at $x = -L$ is simply F_0 and the complex amplitude of the velocity at $x = 0$ must be zero. If we impose the 'boundary condition' for velocity on the general expression for $u(x)$ we obtain

$$u(0) = 0 = (1/Z)(A - B) \tag{13}$$

which yields $A = B$. This means that the corresponding expression for the complex force amplitude must be

$$F(x) = A\mathrm{e}^{ikx} + B\mathrm{e}^{-ikx} = 2A\cos(kx) \tag{14}$$

and imposing the condition $F(-L) = F_0$, we obtain

$$A = F_0/2\cos(kL) \tag{15}$$

Having determined the unknown constants A and B, we have formally solved the problem about the x-dependence of the complex force and velocity amplitudes,

$$F(x) = [F_0/\cos(kL)]\cos(kx) \tag{16}$$

$$u(x) = \mathrm{i}[F_0/Z\cos(kL)]\sin(kx) \tag{17}$$

where we have made use of $\exp(\mathrm{i}kx) - \exp(-\mathrm{i}kx) = 2\mathrm{i}\sin(kx)$.

It remains to discuss this solution. Of particular interest is the velocity (and the corresponding displacement) at the driven end. Thus, inserting $x = -L$ into $u(x)$, we obtain

$$u(-L) = -\mathrm{i}(F_0/Z)\tan(kL) \tag{18}$$

and with $u(x) = -\mathrm{i}\omega\xi(x)$ the complex displacement amplitude is

$$\xi(x) = (F_0/Z\omega)\tan(kL) \tag{19}$$

Display 6.9.

Harmonic waves and complex amplitude equations.

Field variables $u(x,t) \equiv u$ and $F(x,t) \equiv F$

Field equations

$$\mu \partial u/\partial t = -\partial F/\partial x \quad (1)$$

$$\kappa \partial F/\partial t = -\partial u/\partial x \quad (2)$$

Complex amplitude functions

$u(x,k) \equiv u(x)$ and $F(x,k) \equiv F(x)$

$$-\mathrm{i}\omega\mu u(x) = -\mathrm{d}F(x)/\mathrm{d}x \quad (3)$$

$$-\mathrm{i}\omega\kappa F(x) = -\mathrm{d}u(x)/\mathrm{d}x \quad (4)$$

Wave equation

$$\partial^2 F/\partial x^2 = (1/v)^2 \partial^2 F/\partial t^2 \quad (5)$$

Wave speed $v = 1/\sqrt{\mu\kappa}$ (6)

Complex amplitude equation

$$\mathrm{d}^2 F(x)/\mathrm{d}x^2 + k^2 F(x) = 0 \quad (7)$$

$$k = \omega/v \quad (8)$$

General solution

$$F(x) = A\mathrm{e}^{ikx} + B\mathrm{e}^{-ikx} \quad [= F_0(x)\mathrm{e}^{\mathrm{i}\phi(x)}],$$

$$F(x,t) = F_0 \cos(\omega t - \phi) \quad (9)$$

$$u(x) = (1/Z)(A\,\mathrm{e}^{ikx} - B\,\mathrm{e}^{-ikx}) \quad [= u_0(x)\,\mathrm{e}^{\mathrm{i}\beta(x)}],$$

$$u(x,t) = u_0 \cos(\omega t - \beta) \quad (10)$$

$$Z = \mu v = \sqrt{\mu/\chi} \quad (11)$$

Example Forced motion of spring

$u(0) = 0$ Yields $A = B$, (12)

so that

$$F(x) = 2A\cos(kx), \quad u(x) = \mathrm{i}(2A/Z)\sin(kx) \quad (13)$$

$$F(-L) = F_0 \quad \text{yields } 2A = F_0/\cos(kL) \quad (14)$$

$$F(x) = [F_0/\cos(kL)]\cos(kx), \quad (15)$$

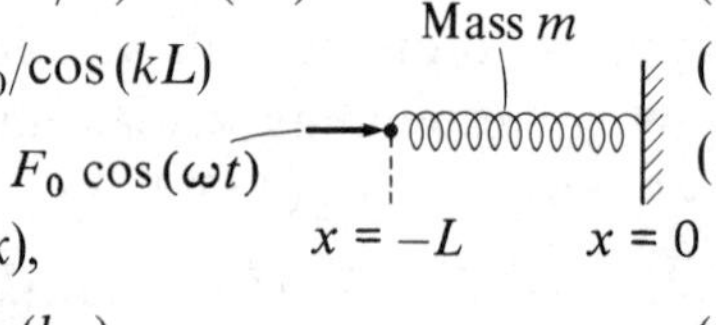

$$u(x) = \mathrm{i}[F_0/Z\cos(kL)]\sin(kx),$$

$$\xi(x) = -[F_0/Z\omega\cos(kL)]\sin(kx) \quad (16)$$

First resonance frequency

$$\omega_0 = (\pi/2)\sqrt{K/m} \quad (k_0 L = \pi/2) \quad (17)$$

In the limit of zero frequency, we can replace $\tan(kL)$ by kL, and with $k=\omega/v=\omega\sqrt{\mu\kappa}$, we obtain $\xi(x)\simeq F_0L/Zv=F_0L/\mu v^2=F_0L\kappa=F_0/K$, i.e. the 'static' deformation of the spring, where $\kappa L=1/K$.

As the frequency increases, the amplitude increases toward a resonance with infinite amplitude when $kL=\pi/2$. The corresponding resonance frequency is given by

$$\omega_0=(\pi/2)(v/L)=(\pi/2)\sqrt{1/\mu\kappa L^2}=(\pi/2)\sqrt{K/m}=\sqrt{K/M_e}$$
$$M_e=(2/\pi)^2m\simeq 0.405m \qquad (20)$$

In other words, the resonance frequency of the spring is the same as an idealized mass–spring oscillator with a spring constant K and a mass m, about 41% of m, attached to the end of the spring. The analogous problem with a mass M attached to the end of the spring will be analyzed in Chapter 8, where the resonance frequency for a 'real' oscillator will be derived.

6.5 Examples of waves

With proper choice of field variables and the parameters μ and κ, the pair of field equations $\mu\partial u/\partial t=-\partial F/\partial x$, $\kappa\partial F/\partial t=-\partial u/\partial x$ describe a variety of one dimensional waves, both mechanical and electromagnetic. In this section we shall list some examples and make some comments on the various waves. More detailed discussions will be given in subsequent chapters including studies of surface waves on liquids and solids.

The list is given in Display 6.10. The first example refers to longitudinal waves on an ordinary spring. Usually, in problems of oscillations, a spring is treated as a massless element with the sole function of providing a restoring force, and the masses involved in the system are lumped into separate elements. In reality, a spring has its own mass and should be treated as a continuous elastic column or transmission line. The dynamics of the spring then have to be analyzed with the aid of the field equations. If the total length L, mass M and spring constant K are given, the parameters μ and κ are $\mu=M/L$ and $\kappa=1/KL$. Note that L is the equilibrium length of the spring, which need not be the relaxed length.

The transverse wave on a spring or string under tension is considered next. The inverse of the tension S now corresponds to the compliance per unit length, and the wave speed is proportional to the square root of the tension. It should be emphasized that the field equations are valid only for small displacements so that the $\sin(\theta)$ can be replaced by $\tan(\theta)$, where θ is the inclination angle of the string.

The equations for the electric transmission line (coaxial cable, for

Display 6.10.

Examples of waves.

Force: $F(x, t)$. Velocity: $u(x, t)$.

Harmonic time dependence: $F(x, k) \equiv F(x)$, $u(x, k) \equiv u(x)$

Field equations $\mu \partial u/\partial t = -\partial F/\partial x$

$\kappa \partial F/\partial t = -\partial u/\partial x$

Harmonic: $u(x) = (1/\mathrm{i}\omega\mu)\,\mathrm{d}F(x)/\mathrm{d}x$ (1)

$F(x) = (1/\mathrm{i}\omega\kappa)\,\mathrm{d}u(x)/\mathrm{d}x$ (2)

Wave equation $\partial^2 F/\partial x^2 = (1/v^2)\partial^2 F/\partial t^2$

Harmonic: $\mathrm{d}^2 F/\mathrm{d}x^2 + (\omega/v)^2 F = 0$ (3)

Wave function $F = F_1\left(t - \dfrac{x}{v}\right) + F_2\left(t + \dfrac{x}{v}\right)$,

Harmonic: $F(x) = A\,\mathrm{e}^{ikx} + B\,\mathrm{e}^{-ikx}$ (4)

Wave speed and impedance $v = 1/\sqrt{\mu\kappa}$,

$Z = \mu v = \sqrt{\mu/\kappa}$ (5)

Longitudinal wave on a spring F = local spring force (6)

u = local velocity

Length L Total mass M — μ = mass per unit length $= M/L$

Spring constant K. — $\kappa = 1/KL$ = compliance per unit length

$v = 1/\sqrt{\mu\kappa} = L\sqrt{K/M}$

Transverse wave on a spring or string F = transverse component of S (7)

u = transverse velocity

Tension S — μ = mass per unit length

$\kappa \to 1/S$

$v = \sqrt{S/\mu}$, $Z = \mu v$

EM wave on a cable	$F \to V =$ voltage	(8)
	$u \to I =$ current	
	$\mu \to L_0 =$ inductance per unit length	
	$\kappa \to C_0 =$ capacitance per unit length	
	$v = 1/\sqrt{L_0 C_0}$, $Z = L_0 v$	
Sound wave in a fluid	$F \to p =$ sound pressure	
	$u =$ velocity	(9)
For gas: Static pressure $= P$	$\mu \to \rho =$ mass density	
Specific heat ratio $\gamma = c_p/c_v$	$\kappa =$ compressibility ($= 1/\gamma P$ for gas)	
	$v = 1/\sqrt{\kappa\rho}$ ($= \sqrt{\gamma P/\rho}$ for gas), $Z = \rho v$	
Longitudinal wave on a solid bar	$F \to -\sigma$ $\sigma =$ stress	(10)
	$u =$ velocity	
Young's modulus E.	$\mu \to \rho =$ mass density	
	$\kappa \to 1/E$	
	$v = \sqrt{E/\rho}$, $Z = \rho v$	
Torsional wave on a rod	$F \to \tau =$ torque	(11)
	$u \to w = \partial\theta/\partial t$	
	$\mu \to J =$ moment of inertia per unit length with respect to axis	
$J = \rho\int r^2\,\mathrm{d}A$ ($A =$ area)	$\kappa \to \rho/JG$	
	$v = \sqrt{G/\rho}$	

example) can be obtained directly from the electromechanical analogues discussed in Chapter 3. The inductance and capacitance per unit length then correspond to the mass and compliance per unit length of the mechanical transmission line.

For wave propagation in a liquid or gas column, the force variable F will be the sound pressure p multiplied by the area of the column. If the cross sectional area is unity, F and p will be the same, the mass per unit length will be the mass density, and the compliance per unit length will be the compressibility. For a gas, the compressibility is inversely proportional to the static pressure, as shown; it will be discussed further in a subsequent chapter. For a liquid, the compressibility is much smaller than for a gas, and the wave speed is greater.

Longitudinal stress waves in a solid bar are similar to sound waves in a fluid, but instead of the sound pressure the stress is used as a field variable. Like pressure, it is the force per unit area, but it is generally defined to have opposite sign. In other words, a compressional stress is negative.

The compliance per unit length is the inverse of the Young's modulus E, which can be thought of as the 'spring constant' of a bar of unit area and unit length.

The next example is the torsional wave on a rod or shaft. The torsion of the rod is expressed by the axial rate of change of the angular displacement of the rod about its axis, and the corresponding torque is proportional thereto. As referred to in the problem section, the constant of proportionality can be expressed in terms of the shear modulus G and the moment of inertia J per unit length. Then, if we introduce the local angular velocity of the rod (not to be confused with the angular frequency of harmonic time dependence), the field equations for torsional motion will be of the 'standard' form, with a corresponding wave equation. The moment of inertia J and the quantity ρ/JG take the place of μ and κ. The torsional wave speed will be $\sqrt{G/\rho}$, similar to the transverse wave speed on a string, with G taking the place of the string tension (see Chapter 20).

6.6 Simple demonstrations

Apart from such familiar examples as waves on water, ropes and strings, and the 'waves' of activity travelling through a line of falling dominoes or a line of cars stopping or starting at a traffic light, there are other demonstrations, which can be used not only for qualitative but also quantitative studies of some of the results we have obtained.

For example, wave motion on a mass–spring lattice can be studied by means of a set of spring coupled air suspended carts on a track, or by means of a torsion ladder. The wave speed can be measured as a function of the mass of the carts and the equilibrium distance between them, and it is instructive to compare the result obtained with the value predicted from the equivalent continuous transmission line. The calculation of the wave speed for the discrete lattice will be carried out in Chapter 8.

In another simple demonstration, the sound speed in a gas can be measured by sending short pulse trains through a tube (typically 4–5 m long and with a diameter of about 5 cm). A small loudspeaker at one end can be used as a sound source, and the sound pressure in the tube can be measured by a microphone. The microphone signal is displayed on one channel in a two channel oscilloscope with the other channel showing the input signal to the loudspeaker.

By moving the microphone along the tube, we observe how the time delay between the microphone signal and the input signal increases with the distance between the source and the microphone. The slope of the line representing time delay versus distance is the inverse of the wave speed.

Furthermore, we find that the shape of the wave trains (pulses) remains the same along the tube (as long as the carrier frequency of the wave train is sufficiently low, corresponding to a wavelength large compared to the tube diameter). This demonstrates that the pressure field indeed is of the form $p(t - x/v)$, where v is the wave speed.

During the course of the demonstration, with the tube tilted somewhat, it is simple to pass helium through the tube, injecting it at the lower end. As the air gradually is removed and replaced by helium, the time delay is seen to be reduced by more than a factor of two, indicating a corresponding increase in the sound speed. The result is found to be in good agreement with the predicted value in Display 6.10.

A similar experiment can be carried out with electric pulses on a long radio frequency cable, typically 100 m long (wound up on a spool). The voltage across the cable at the source end is displayed on an oscilloscope. The reflected pulse can easily be seen, and the time delay, corresponding to the round trip, can be measured and compared with the predicted wave speed given in Display 6.10. The values of the inductance and capacitance per unit length of the cable usually are supplied by the manufacturer; if not, they can be measured or calculated, as will be discussed later.

In this experiment, it is interesting to study the effect of different terminations on the reflected pulse. If the cable is open at the end, the reflected voltage pulse has the same sign as the incident, whereas for a short circuited termination, the polarity is reversed. It is noted also, that an appropriately chosen resistive termination will eliminate the reflected pulse altogether. The calculation of the reflection coefficient will be deferred to a subsequent chapter.

Also in the experiment with sound pulses, the reflections from the end of the tube are quite apparent. If the microphone is close to the end, the incident and reflected pulses will overlap, and, as the microphone is moved along the tube, destructive and constructive interferences are observed. These effects will be analyzed in the next chapter.

Linearization. It should be pointed out, that the field equations $\mu\partial u/\partial t = -\partial F/\partial x$ and $\kappa\partial F/\partial t = -\partial u/\partial x$ are valid only for small displacements in the mechanical transmission line. This was pointed out explicitly in the discussion of transverse wave motion.

This limitation applies also to longitudinal motion. For example, in the momentum equation $\mu\partial u/\partial t = -\partial F/\partial x$, it is assumed, that the mass per unit length is a constant. This is not quite correct, since the deformation produced by the wave will cause a small relative change of μ of the order of $\partial\xi/\partial x$, where ξ is the displacement. The correct form of the momentum equation $\partial(\mu u)/\partial t = -\partial F/\partial x$ then will contain an extra term with the product of $\partial\xi/\partial x$ and $\partial u/\partial t$, which is of second-order in the field variables. The omission of this second order term is justified for small signals. The resulting field equations then contain only linear terms and are said to be **linearized**.

6.7 Wave power

Single travelling harmonic wave. As a specific example, let us consider an electric transmission line, such as a coaxial cable which is placed along the x-axis. It is driven at $x = 0$ with the voltage $V(0, t) = V_0 \cos(\omega t)$.

The corresponding current in the cable is $I(0, t) = V(0, t)/Z = I_0 \cos(\omega t)$, where $I_0 = V_0/Z$. As explained in Chapter 6, $Z = L_0 v = \sqrt{L_0/C_0}$ is the **wave impedance**, where L_0 is the inductance per unit length, C_0 the capacitance per unit length, and $v = 1/\sqrt{L_0 C_0}$ is the wave speed.

The instantaneous power transferred to the cable from the voltage source is $P(t) = V(0, t)I(0, t) = V_0 I_0 \cos^2(\omega t)$. We are generally interested in the time average of the power over one period, which is

$$P = V_0 I_0 \langle\cos^2(\omega t)\rangle = \tfrac{1}{2}V_0 I_0 = \tfrac{1}{2}Z I_0^2 = \tfrac{1}{2}V_0^2/Z \qquad (1)$$

where the bracket signifies time averaging.

The power P is the average energy transferred to the line per second and it is carried by the wave. If there are no losses in the line, the same power P is found at every location x, since energy cannot pile up on the line in this steady state situation. Actually, this follows directly from the fact that at a position x the voltage is $V(x, t) = V_0 \cos(\omega t - kx)$ and the current is $I(x, t) = I_0 \cos(\omega t - kx)$. The line to the right of x can be considered to be driven by this voltage and corresponding current, and the time average power transferred will be the time average of $V(x, t)I(x, t)$, which will be $P = \frac{1}{2} V_0 I_0$, as before, independent of x.

Since we generally deal with the complex amplitude functions $V(x) = V_0 \exp(ikx)$ and $I(x) = I_0 \exp(ikx)$ rather than the real wave functions, it is convenient to express the time average power as

$$P = \tfrac{1}{2}\mathrm{Re}\{V(x)I^*(x)\} = \tfrac{1}{2} V_0 I_0 \tag{2}$$

with $I^* = I_0 \exp(-ikx)$

in complete analogy with the expression for the power transfer to an oscillator, dicussed in Chapter 3.

General wave field. In general, we have several waves present on the transmission line travelling in both the positive and the negative x-direction. As we have shown earlier, the sum of several harmonic waves with the same frequency travelling in the same direction can be replaced by a single travelling wave. Consequently, the general field on a transmission line can be considered to be the sum of one wave in the positive and one wave in the negative direction. If we denote the complex voltage amplitudes of these waves by $A \exp(ikx)$ and $B \exp(-ikx)$, the total complex voltage amplitude is

$$V(x) = A \exp(ikx) + B \exp(-ikx).$$

The corresponding expression for the complex current amplitude is obtained from the field equation $-i\omega L_0 I = -dV/dx$, and we obtain

$$I(x) = (1/Z)[A \exp(ikx) - B \exp(-ikx)]$$

where A and B generally are complex. The complex conjugate of the current is $I^* = (1/Z)[A^* \exp(-ikx) - B^* \exp(ikx)]$ and the time average power is

$$P = \tfrac{1}{2}\mathrm{Re}\{V(x)I^*(x)\} = (1/2Z)(|A|^2 - |B|^2) = P_+ - P_- \tag{3}$$

where we have introduced the squared amplitudes $|A|^2 = AA^*$ and $|B|^2 = BB^*$. The quantity $|A|^2/2Z = P_+$ we recognize as the power carried by the wave travelling in the positive direction and $|B|^2/2Z = P_-$ is the power carried by the wave travelling in the negative x-direction.

Wave energy density and flux. To investigate wave energy in more detail, we start from the two field equations

$$L_0 \partial I/\partial t = -\partial V/\partial x \tag{4}$$

$$C_0 \partial V/\partial t = -\partial I/\partial x \tag{5}$$

and derive from these an energy conservation relation, which will enable us to give a more detailed definition of the time dependent wave power and introduce a corresponding energy density.

Proceeding in much the same manner as in particle mechanics in the derivation of the 'work–energy' relation, we multiply Eq. (4) by I and Eq. (5) by V and add the equations to obtain

$$L_0 I\partial I/\partial t + C_0 V\partial V/\partial t = -I\partial V/\partial x - V\partial I/\partial x \tag{6}$$

or

$$\partial(L_0 I^2/2 + C_0 V^2/2)/\partial t = -\partial(VI)/\partial x \tag{7}$$

The term $L_0 I^2/2$ on the left hand side is the magnetic energy per unit length and the second term is the electrostatic energy $C_0 V^2/2$ per unit length. (The corresponding quantities on a mechanical transmission line will be the kinetic and potential energy per unit length, $\mu u^2/2$ and $\kappa F^2/2$, respectively.)

In other words, the left hand side of Eq. (7) can be interpreted as the time rate of change of the total energy W per unit length. Eq. (7) then states that the time rate of increases of $W(x, t)$ is balanced by a decrease in the instantaneous power $P(x, t) = V(x, t)I(x, t)$; i.e. what is lost in the energy flow $P(x, t)$ must show up as energy density (assuming no damping). Thus, on the basis of this energy balance equation we introduce the instantaneous **wave energy density** $\boldsymbol{W(x, t)}$ and **wave power** $\boldsymbol{P(x, t)}$ as

$$W(x, t) = L_0 I^2/2 + C_0 V^2/2 \tag{8}$$

$$P(x, t) = VI \quad [V = V(x, t) \quad I = I(x, t)] \tag{9}$$

which obey the conservation (continuity) equation

$$\partial W(x, t)/\partial t + \partial P(x, t)/\partial x = 0 \tag{10}$$

Since in this case we consider wave motion on a cable, the energy 'density' is strictly speaking the energy per unit length. When we apply the results obtained to a sound wave or electromagnetic wave in free space, the wave power refers to unit area, in which case it is called **energy flux (intensity)**, and the energy per unit length becomes the energy per unit volume.

We now apply this general result to the particular case of harmonic time dependence and average Eq. (10) over one period. In this case of steady state, the time average of $W(x, t)$ and of P will be independent of time, so that the first term in the equation vanishes. Denoting the time average power by $P(x)$, Eq. (10) states that $\partial P(x)/\partial x = 0$, i.e. $P(x) = P$ independent of x in accordance with the conclusions arrived at in the previous subsection.

With the complex amplitudes of voltage and current being $V(x) = V_0 \exp(ikx)$ and $I(x) = I_0 \exp(ikx)$, the time average of the power and the energy density for harmonic time dependence become

$$P = \tfrac{1}{2}\mathrm{Re}\{VI^*\} = \tfrac{1}{2}V_0 I_0 = \tfrac{1}{2}ZI_0^2 \quad (Z = L_0 v) \tag{11}$$

$$W = \tfrac{1}{2}\mathrm{Re}\{L_0 II^* + C_0 VV^*\} = \tfrac{1}{2}(L_0 I_0^2/2 + C_0 V_0^2/2) \tag{12}$$

For a single travelling wave in the positive x-direction we have $V_0 = L_0 v I_0$, so that $L_0 I_0^2 = C_0 V_0^2$ (recall $v^2 = 1/L_0 C_0$). We then obtain

$$P = vW \quad \text{(single harmonic wave)} \tag{13}$$

In other words, the energy per second, carried by the wave past a fixed point of observation, can be thought of as the energy density in the wave being convected at the wave speed.

6.8 Wave pulse

Many of the characteristics of wave motion can be understood from elementary considerations in terms of a wave pulse. To be specific, let us assume that, at $x = 0$, we generate a compression wave on a spring by moving the end forward with constant velocity u during the time Δt. Knowing from experiments that the motion at $x = 0$ is transmitted along the spring with a certain speed v, it follows that at the end of the interval Δt, the wave front will be at $x_2 = v\Delta t$. In the meantime, the driven end of the spring has moved forward to $x_1 = u\Delta t$. The length of the wave, in which the velocity is u throughout, is then $(v - u)\Delta t$, which we approximate by $v\Delta t$, assuming the 'material' velocity u to be much smaller than the wave speed v. The total momentum of the spring at time Δt is $\mu uv\Delta t$, where μ is the mass per unit length.

The impulse delivered by the force is $F\Delta t$, which must equal the momentum on the spring, i.e.

$$F = \mu vu \tag{1}$$

which is the familiar relation between force and velocity in a travelling wave.

We can express the force also in terms of the compression $u\Delta t$ of the spring. The portion of the spring which is compressed at time Δt is $(v - u)\Delta t \simeq v\Delta t$. Thus, if the compliance per unit length of the spring is κ, it follows that $u\Delta t = \kappa(v\Delta t)F$ or

$$F = u/\kappa v \tag{2}$$

Combining this equation with Eq. (1) we obtain

$$v = 1/\sqrt{\mu\kappa} \tag{3}$$

the well known expression for the wave speed.

The total energy in the wave pulse is the work done by the force, which is

Display 6.11.

Wave power and wave energy density

Harmonic time dependence

Complex force amplitude $F(x)$ (1)

Complex velocity amplitude $u(x)$ (2)

Wave power in the x-direction

$P = \frac{1}{2}\text{Re}\,\{F(x)u^*(x)\}$ (time average) (3)

Example

Single travelling wave $F(x) = F_0 e^{ikx}, \quad u(x) = u_0 e^{ikx}$ (4)

$P = \frac{1}{2}F_0 u_0 = \frac{1}{2}Zu_0^2 = \frac{1}{2}F_0^2/Z \quad Z = \mu v \quad v = 1/\sqrt{\kappa\mu}$ (5)

Example

Standing wave $F(x) = 2A\cos(kx)$ (6)

$u(x) = (l/\text{i}\omega\mu)\,\text{d}F/\text{d}x = \text{i}(2A/Z)\sin(kx)$ (7)

$P = \frac{1}{2}\text{Re}\,\{F(x)u^*(x)\} = 0$ (8)

Arbitrary time dependence

$W(x, t)/\partial t + \partial P(x, t)/\partial x = 0$ (9)

Energy density

$W(x, t) = \mu u^2/2 + \kappa F^2/2$ (10)

Power (Energy flow) $P = F(x, t)u(x, t)$ (11)

Example

Time average, single travelling harmonic wave:

$F(x) = F_0 \exp(\text{i}kx) \quad u(x) = u_0 \exp(\text{i}kx)$ (12)

$W = \frac{1}{2}(\frac{1}{2}\mu u_0^2 + \frac{1}{2}\kappa F_0^2), \quad P = \frac{1}{2}F_0 u_0 = \mu v u_0 \quad v = 1/\sqrt{\mu\kappa}$ (13)

$P = vW$ (14)

$Fu\Delta t = \mu u^2 v\Delta t$, which corresponds to a total energy per unit length

$$W = \mu u^2 \tag{4}$$

The kinetic energy per unit length is $\mu u^2/2$ which accounts for half of the total energy. The other half is the potential energy $\kappa F^2/2$. With $F = \mu vu$ and $v^2 = 1/\mu\kappa$, we see that the potential energy per unit length is equal to the kinetic energy per unit length.

Discussion. In equations 6.7(4)–(10), there is a subtle problem with the concepts of wave power and wave energy density when applied to mechanical waves. This stems from the fact that the field equations are 'linearized', i.e. terms of second and higher order are neglected. As far as the resulting calculated values of the field variables are concerned, this omission is justified for small displacements, as was mentioned in the discussion of transverse wave motion of a string and in the analysis of the pulse wave above.

The trouble is that wave power and wave energy density are products of the field variables and are of second order, i.e. of the same order as the terms omitted from the original equations. One might wonder, therefore, whether the quantities P and W in the 'energy equation' 6.7(10), when applied to mechanical waves, represent true power and energy density, or if they should be regarded as interesting (and useful) properties of the energy-like products of the solutions to the linearized field equations. A more detailed study shows that W and P indeed are valid energy quantities as long as the medium carrying the wave has no mean motion with respect to the source. Reference to this question will be made in a subsequent chapter.

In Display 6.11 we have summarized the material in this section using as field variables the 'generalized' force and velocity.

6.9 Waves in two and three dimensions

By analogy with our approach to one-dimensional waves in Section 6.3, we use mass–spring lattices in the derivation of the field- and wave equations in two and three dimensions. Thus, for the study of waves and oscillations of a membrane, it is instructive to start from set a finite mass elements in the form of a two-dimensional mass–spring lattice in Display 6.12. The elements are all the same of mass m and they are connected by identical springs with spring constant K, lined up in the x- and y-directions. In equilibrium, the lattice is stretched so that there is a bias force S (tension) in the springs, and the distance between neighboring elements is d. For

Display 6.12.

Wave equation for a rectangular membrane.

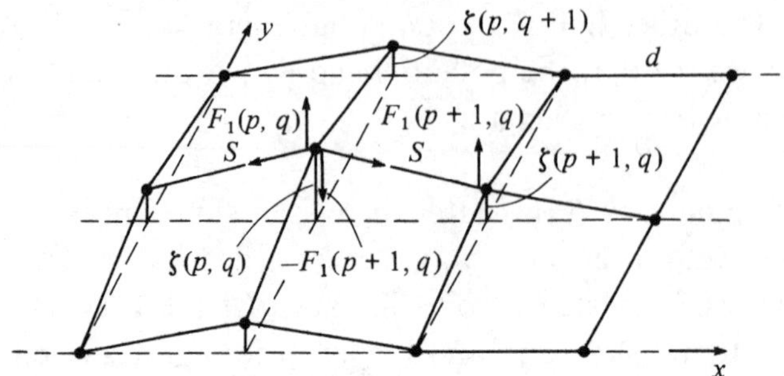

Equilibrium coordinates of element p, q, x_p, y_q (1)

Displacement in z-direction, $\zeta(x_p, y_q)$ (2)

Force on m by 'x-springs'

$$F_1(p, q) - F_1(p + 1, q) \simeq -(\partial F_1/\partial x)d \tag{3}$$

Force on m by 'y-springs'

$$F_2(p, q) - F_2(p, q + 1) \simeq -(\partial F_2/\partial y)d \tag{4}$$

Bias tension in springs S (5)

$$\begin{aligned} F_1(p + 1, q) &= S\sin(\theta_1) \simeq S\tan(\theta) \\ &= S[\zeta(p, q) - \zeta(p + 1, q)]/d \\ &\simeq -S\partial\zeta/\partial x \end{aligned} \tag{6}$$

Similar expression obtains for $F_2(p, q + 1)$

Equation of motion for m:

$$\mathrm{d}^2\zeta/\partial t^2 = F_1(p, q) - F_1(p + 1, q) + F_2(p, q) - F_2(p, q + 1) \tag{7}$$

Continuum equation for membrane $\mu = m/d^2 \quad \sigma = S/d$

$$\partial^2\zeta/\partial t^2 = (\sigma/\mu)(\partial^2\zeta/\partial x^2 + \partial^2\zeta/\partial y^2) \tag{8}$$

Harmonic time dependence $v^2 = \sigma/\mu$

$$v^2\zeta + (\omega/v)^2\zeta = 0 \tag{9}$$

Normal modes $\zeta_{mn} = A_{mn}\sin(k_m x)\sin(k_n y)$,

$$\omega_{mn} = [(m\pi/a)^2 + (n\pi/b)^2]^{\frac{1}{2}} \tag{10}$$

simplicity, we shall assume that the tension is the same in the two directions.

The lattice is located in the xy-plane, and the equilibrium position of a mass element is given by $x_p = pd$, $y_q = qd$, $z = 0$, where p and q are integers. A transverse displacement of the element in the z-direction (i.e. out of the equilibrium plane) is denoted by $\zeta(p, q)$. The element is acted on by four springs which provide restoring forces on the element. Following the sign convention used in the discussion of the transverse motion of a one-dimensional lattice, as summarized in Display 6.7, the force in the positive z-direction from the 'x-spring' on the left side of the element p, q is denoted by $F_1(p, q)$ and the force on the corresponding side from the 'y-spring' is $F_2(p, q)$. The forces on the adjacent elements at $p + 1, q$ and $p, q + 1$ are $F_1(p + 1, q)$ and $F_2(p, q + 1)$ and the corresponding reaction forces in the z-direction on element p, q are then $-F(p + 1, q)$ and $-F(p, q + 1)$.

The equation of motion for element p, q then becomes

$$m\,\mathrm{d}^2\zeta/\mathrm{d}t^2 = F_1(p, q) - F_1(p + 1, q) + F_2(p, q) - F_2(p, q + 1) \qquad (1)$$

Having assumed that the tensions in the springs are the same and equal to S, the force $F_1(p + 1, q)$ in the z-direction from the x-spring on the left side of element $p + 1, q$ can be expressed as

$$F_1(p + 1, q) = S\sin(\theta_1) \simeq S\tan(\theta_1) = S[\zeta(p, q) - \zeta(p + 1, q)]/d \quad (2)$$

where θ_1 is the inclination angle of the spring between element $p, q + 1$ and p, q, as shown in the display. (Compare the derivation for the one-dimensional lattice in Display 6.7). Similarly, the force contribution on element $p, q + 1$ from the y-spring is

$$F_2(p, q + 1) \simeq S[\zeta(p, q) - \zeta(p, q + 1)]/d \qquad (3)$$

As already indicated, the corresponding reaction forces on element p, q are the same but with opposite sign. Eqs. (1)–(3) can be regarded as the field equations for the lattice.

Following the development in Display 6.7 for the one-dimensional lattice, we proceed directly to the limiting case of a continuous membrane by letting d go to zero as we keep the average mass per unit area $\mu = m/d^2$ and the average tension $\sigma = S/d$ over a cell constant. The coordinates x_p, y_q are now treated as continuous variables x, y, and, from Eqs. (1)–(3) and with $S = \sigma d$, the differences on the right hand sides of the equations (1)–(3) can be expressed in terms of the partial derivatives $[-\partial F_1/\partial x - \partial F_2/\partial y]d$, $-\sigma\partial\zeta/\partial x$, and $-\sigma\partial\zeta/\partial y$, respectively.

Under these conditions, making use of Eqs. (2) and (3), we note that Eq. (1) results in the wave equation for the displacement ζ

$$\partial^2\zeta/\partial t^2 = v^2(\partial^2\zeta/\partial x^2 + \partial^2\zeta/\partial y^2) \qquad (4)$$

$$v^2 = \sigma/\mu$$

This wave equation is the basis for the mathematical description of wave motion and oscillations of a membrane with a uniform tension σ and a mass per unit area equal to μ.

For harmonic time dependence, with $\zeta(x,y,t) = \mathrm{Re}\{\zeta(x,y,\omega)\exp(-\mathrm{i}\omega t)\}$, the wave equation for the complex amplitude function $\zeta(x,y,\omega)$ takes the form

$$\partial^2\zeta/\partial x^2 + \partial^2\zeta/\partial y^2 + (\omega/v)^2\zeta = 0 \tag{5}$$

often referred to as the Helmholtz equation.

A plane harmonic wave travelling in a direction specified by the unit vector $\hat{r}$ and the corresponding coordinate r will have a complex amplitude of the form $\exp(\mathrm{i}kr)$, where $k = \omega/v$. If the unit vector $\hat{r}$ makes an angle θ with the x-axis, it follows that $r = x\cos(\theta) + y\sin(\theta)$ so that the wave function becomes $\exp[(\mathrm{i}kx\cos(\theta) + \mathrm{i}ky\sin(\theta)]$. If we define a propagation vector $\vec{k}$ with the components $k_x = k\cos(\theta)$ and $k_y = k\sin(\theta)$, we can express the wave function as $\exp(\mathrm{i}\vec{k}\cdot\vec{r})$.

Three-dimensional waves. In analogous manner, we can obtain the field equations for compressional waves in a fluid by starting from a rectangular three-dimensional mass–spring lattice. In this case there is no need for a bias tension in the springs. The transition to a continuum is made in such a way that the average mass density $m/d^3 = \rho$ and the average spring force per unit area, $F/d^2 = p = \Delta P$ (corresponding to the membrane tension σ), are kept constant. The compressibility of the medium is $\kappa = -(1/V)\Delta V/\Delta P$, and the relative volume change can be expressed as $\partial\xi/\partial x + \partial\eta/\partial y + \partial\zeta/\partial z$, where ξ, η, and ζ are the components of the displacement of a mass element. By accounting for the equation of motion of the mass element m, we arrive at the wave equation for the pressure

$$(1/v^2)\partial^2 p/\partial t^2 = \partial^2 p/\partial x^2 + \partial^2 p/\partial y^2 + \partial^2 p/\partial z^2 \tag{6}$$

$$v = (1/\kappa\rho)^{\frac{1}{2}}$$

where κ is the compressibility and ρ the density of the continuum. The details of the derivation will be given in Chapter 10 in a somewhat different manner together with a discussion of spherical waves, and the normal modes in three dimensions will be considered in Chapter 15.

Examples

E6.1 *Maximum wave speed on a string.* Consider a string of length L clamped at both ends. To make the normal mode frequencies of the string as high as possible, the transverse wave speed and hence the tension S in the string should be made as large as possible. The maximum possible tension corresponds to a stress in the string material equal to the tensile strength of the material (stress = force per unit area, see Chapter 20). For steel the tensile strength is $\sigma' = 3.2 \times 10^{10}$ and for aluminum 2.8×10^9 dyne/cm^2. The mass density of steel is 7.8 and for aluminum 2.7 g/cm^3.

If A is the cross sectional area of the string the stress is $\sigma = S/A$, and if the mass density is ρ, the mass per unit length of the string is $\mu = A\rho$. The wave speed then can be expressed as

$$v = \sqrt{S/\mu} = \sqrt{\sigma/\rho} \tag{1}$$

The maximum wave speed is obtained when the stress equals the tensile strength, independent of the diameter of the string. Using the values for the tensile strength given above, we obtain for the maximum wave speed the value $v' = 6.4 \times 10^4$ cm/s for steel and for aluminum 3.2×10^4 cm/s. In other words, the maximum wave speed for a steel string is larger than the speed of sound in air, but for an aluminum string it is somewhat smaller.

E6.2 *Radiation damping.* A mass M is attached to the end of a long coil spring. It is given an initial velocity $u(0)$ at $t = 0$. Determine the subsequent motion.

Solution. We assume the spring to be so long that we can neglect reflections from the end. Then, as M moves, a wave is produced on the spring which carries energy from M.

To obtain the corresponding reaction force on M, we note that the ratio between the force and the particle velocity on the spring is the wave impedance $Z = F/\dot{u} = \mu v = \sqrt{\mu/\kappa}$, where μ is the mass per unit length of the spring, v the wave speed, and κ the compliance per unit length.

The reaction force on M will be $-Zu$ and the equation of motion of M becomes $M\mathrm{d}u/\mathrm{d}t = -Zu$ with the solution

$$u(t) = u(0)\exp[-(Z/M)t] \tag{1}$$

The velocity decays exponentially. It is interesting to note that the total displacement of M will be $\int u(t)\,\mathrm{d}t = u(0)M/Z$, where the range of integration is from zero to infinity.

The damping of the mass element in this case is due to the wave energy generated by M, and it is usually referred to as radiation damping. For electrical and acoustical oscillators, the generation of EM and acoustic waves results in a similar damping of the oscillators involved, as will be discussed in Chapters 9 and 17. In forced motion of an oscillator, the radiation damping shows up in the width of the frequency response curve for the amplitude.

E6.3 *Superposition of waves.* Two harmonic voltage waves $V_1(x, t) = 2\cos(\omega t - kx)$ and $V_2(x, t) = 4\cos(\omega t - kx - \pi/3)$ V are travelling simultaneously on a coaxial cable with a wave impedance $Z = 50\,\Omega$.

(a) What are the complex amplitude (functions) of these waves?
(b) The sum of the waves can be expressed as a single wave $C\cos(\omega t - kx - \gamma)$.

What are the numerical values of the amplitude C and the phase angle γ?
(c) When each wave is transmitted separately on the cable, what is the time average power carried by each wave?
(d) What is the power carried by the total wave field when the waves are transmitted simultaneously?

Solution (a) It follows from the definition of the complex amplitude that $V_1(x) = 2\exp(ikx)$ and $V_2(x) = 4\exp[i(kx + \pi/3)] = [4\exp(i\pi/3)\cdot\exp(ikx)]$.
(b) The complex amplitude of the sum of the waves is

$$2\exp(ikx) + [4\exp(i\pi/3)]\exp(ikx) = 2[1 + 2\exp(i\pi/3)]\exp(ikx)$$
$$= 2[2 + i\sqrt{3}]\exp(ikx) = 2\sqrt{7}\exp(i\gamma)\exp(ikx)$$
$$\gamma = \arctan(\sqrt{3}/2)$$

Thus the amplitude is $C = 2\sqrt{7}$ and $\gamma = \arctan(\sqrt{3}/2) = 40.9$ degrees.
(c) The time average of the power transmitted by a wave with a voltage amplitude V_0 is $\frac{1}{2}V_0^2/Z$. The power carried by each wave separately is $\frac{1}{2}4/50 = 0{\cdot}04$ watt and $\frac{1}{2}16/50 = 0.16$ watt.
(d) The power carried by the total wave field is $\frac{1}{2}[2\sqrt{7}]^2/50 = 0.28$ watt.

It is instructive to calculate the amplitude of the resulting wave from the fact that $V = V_1 + V_2$ and $|V|^2 = (V_1 + V_2)(V_1^* + V_2^*) = V_1V_1^* + V_2V_2^* + V_1V_2^* + V_2V_1^* = 4 + 16 + 8[\exp(-i\pi/3) + \exp(i\pi/3)] = 20 + 16\cos(\pi/3) = 28$.

Problems

6.1 *Wave kinematics.* The transverse wave speed on a stretched string is 10 m/sec. The transverse displacement at $x = 0$ is $\eta(0, t) = 0.1(t^2 - t^3)$m, when $0 < t < 1.0$ sec and zero for all other times. (t is measured in seconds).

(a) Plot the transverse displacement as a function of time at $x = 0$.
(b) Plot the displacement as a function of x at $t = 1.0$ sec.
(c) What is the mathematical expression for the displacement as a function of time at $x = 10$ m? What are the displacements at this point at $t = 1, 1, 5$, and 3 sec?
(d) What is transverse velocity of the string at $x = 10$ m and $t = 1.5$ sec?
(e) What is the slope of the string at $x = 10$ m and $t = 1.5$ sec?

6.2 *Harmonic wave.* The end of string is driven at $x = 0$ with the transverse displacement $\eta(0, t) = \eta_0\cos(\omega t)$ with a frequency $\nu = 10$ Hz and an amplitude 0.2 m. The wave speed on the string is 10 m/sec and the mass per unit length 0.001 kg/m.

(a) What is the displacement as a function of time at $x = 1$ m?
(b) Sketch the shape of the string at $t = 0.5$ sec. What is the wavelength?
(c) What are the transverse velocity and acceleration as functions of time at $x = 1$ m?
(d) What is the phase difference between the harmonic oscillations of the string at $x = 0$ and $x = 0.2$ m?

6.3 *Sum of travelling waves.* Consider two harmonic waves $A\cos(\omega t - kx)$ and $2A\cos(\omega t - kx - \pi/4)$. Prove that the sum is a travelling wave $B\cos(\omega t - kx - \alpha)$ and determine B and α. Carry out the calculation first algebraically and then with the aid of complex amplitudes.

6.4 *Wave diagram.* A pressure pulse of duration 5 milliseconds is generated at $t = 0$ at the left end of the tube shown in the figure. The tube contains air and helium, as shown, and the gases are separated by a thin membrane, which can be considered to be transparent to the wave.

(a) Accounting for the reflections at the boundary between the gases and at end walls, make a wave diagram, which extends over the first 50 milliseconds.
(b) Indicate in the $t - x$ plane the regions where you expect interference to occur between reflected and incident waves.

Air	Helium
4 m	12 m

6.5 *Moving source. Wave diagram and Doppler shift.* A source emits short pulses at a rate of ν per second as it moves through the air in the x-direction. The duration τ of each pulse is short compared to $T = 1/\nu$.

Indicate in a wave diagram the wave lines representing the waves emitted in the positive and negative x-directions. From the diagram determine geometrically the time between two successive pulses as recorded by observers on the x-axis ahead of and behind the source. Determine also the corresponding number of pulses per second (Doppler shift) and express the result in terms of ν, the velocity U of the source, and the sound speed v when

(a) $U < v$. Subsonic motion of the source.
(b) $U > v$. Supersonic motion of the source.
(c) Is it possible that wave interference will occur in (a) or (b) between waves emitted in the two directions? If so indicate in the t–x plane the regions where such interference takes place.

6.6 *Longitudinal wave on a spring.* A 1 m long spring of a certain material has a spring constant $K = 100$ N/m and a mass 0.25 kg. The end $x = 0$ of a long spring of this material is driven by a longitudinal displacement $\xi(0, t) = 0.02\,[(t/T) - (t/T)^2]$ during the time between $t = 0$ and $t = T = 0.02$ sec. The displacement is zero at all other times.

(a) What are the particle velocity of the spring and the driving force as functions of time at $x = 0$?
(b) During what time interval is the wave pulse passing the point $x = 10$ m?
(c) What region of the spring is occupied by the wave pulse at $t = 2$ sec.?

6.7 *Longitudinal and transverse waves.* The relaxed length of a coil spring is l, the mass M, and the spring constant K. The spring is stretched to a length L and kept at this length.

(a) What is the ratio between the transverse and longitudinal wave speed on the spring? Can the transverse wave speed be larger than the longitudinal?
(b) What is the time of travel of a longitudinal wave pulse from on end of the spring to the other, and how does it depend on the length L?

6.8 *Longitudinal and torsional waves.* The elastic modulus (Young's modulus) of steel is $E = 1.7 \times 10^{11}\,\mathrm{N/m^2}$, and the shear modulus is $G = 7.6 \times 10^{10}\,\mathrm{N/m^2}$. The mass density is $\rho = 7.8\,\mathrm{g/cm^3}$.

(a) What is the ratio between the longitudinal and the torsional wave speed on a steel shaft?
(b) What is the stress amplitude in a harmonic longitudinal wave in the shaft when the velocity amplitude is 0.1% of the longitudinal wave speed?
(c) Suppose the shaft diameter is $D = 2.5\,\mathrm{cm}$. What is the torque amplitude required to produce a torsion wave with an angular displacement amplitude of 0.001 radians in a harmonic travelling wave?

6.9 *Molecular model of wave propagation.* As a naive model of wave motion in a gas, one might consider the transmission of an impulse along a line of air suspended carts, which represent the molecules. The carts, each with a mass M, are initially at rest with a distance d between adjacent carts. An impulse J is delivered to the first cart in the line.

(a) Describe the subsequent motion and determine the wave speed of the motion along the line.
(b) In the motion in (a), the wave speed depends on the strength of the impulse. For sound, on the other hand, the wave speed is independent of the strength (at least at sufficiently weak signals). What is the basic deficiency of the model? Under what conditions might the model be reasonable?

6.10 *Sound speed and thermal motion.* The thermal kinetic energy of a molecule for each translation degree of freedom is $mv_x^2/2 = k_B T/2$, where k_B is the Boltzmann constant and T the absolute temperature. What is the relation between the sound speed and the thermal speed v_x?

6.11 *Nonlinearity.* In the derivation of the field equations for transverse motion of a spring or string (Display 6.7), the relation between the force and the displacement was based on the assumption of small amplitudes, with the total length and the tension of the spring assumed to be independent of the displacement. Furthermore, the slope of the spring was assumed small so that $\sin(\theta)$ could be replaced by $\tan(\theta)$.

Rederive the relation between the force and the displacement without making these assumptions and show that the resulting field equations no longer are linear. What are some of the consequences of the nonlinearity? For example, can a longitudinal and a transverse wave be transmitted along the spring without interaction?

6.12 *Wave energy on a string.* (a) Calculate the time average power in watts carried by the wave described in Problem 6.2.

(b) What is the average wave energy per unit length of the string?
(c) What happens to the power and the wave energy density if the tension of the string is doubled?

6.13 *Energy flux.* A longitudinal harmonic wave with a frequency of 10^4 Hz is generated in a steel bar by a piezo-electric crystal mounted at the end of the bar. What should be the displacement amplitude of the bar in order that the energy flux (intensity) of the wave be 10 watt/cm^2? Density $\rho = 7.8$ g/cm^3. Wave speed $v = 5300$ m/sec.

6.14 *Transducer.* A transducer, regarded here as a vibrating piston, is applied first to an air column in a long tube and then to the end of a long steel bar with the same area as the tube. What is the ratio between the energy fluxes in the waves generated by the piston in the two cases if:

(a) the driving force amplitude on the piston is the same in the two cases, and
(b) if the displacement amplitude is the same in the two cases? The mechanical impedance of the piston is Z_m. Discuss the conditions for maximum power transfer from the piston.

6.15 *Radiation load on an oscillator.* A mass–spring oscillator ($M = 2$kg, $K = 32$ N/m) is connected to a long string (tension $\tau = 100$ N, mass per unit length $\mu = 0.25$ kg/m). The mass is sliding on a horizontal frictionless guide bar, as shown.

(a) What is the nature of the effect of the string on the oscillator, is it equivalent to a mass- , stiffness- , or resistive-load?
(b) What is the Q-value of the oscillator, accounting for the effect of the string?
(c) The oscillator is started from an initial displacement $A = 5$ cm. Indicate the shape and length of the wave on the string at the time when the amplitude of the oscillator has decreased to the value $1/e$ of the initial displacement.

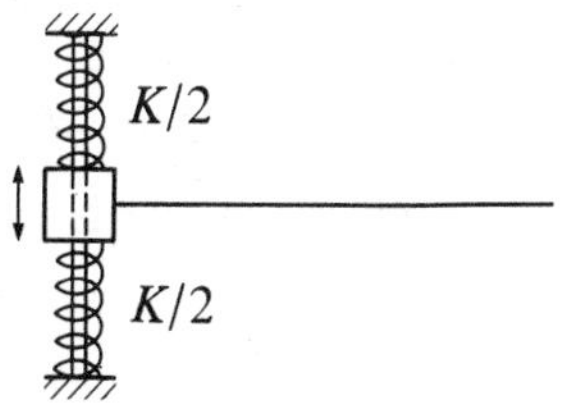

6.16 *Power carried by superimposed waves.* Consider two 'force' waves on a transmission line, $F_1(x, t) = A\cos(\omega t - kx - \alpha)$ and $F_2(x, t) = B\cos(\omega t - kx - \beta)$. When alone, the average power transmitted by the first wave is P_1 and by the second P_2.

(a) What is the average power when both waves are present simultaneously? Express the dependence of the power on the phase difference $(\alpha - \beta)$.
(b) If the phase difference is varied, what are the resulting maximum and minimum values of the resulting power?
(c) If N waves of equal amplitude are present, what then is the maximum value of the total power in terms of the power of a single wave?

6.17 *Wave pulse.* The voltage applied at $t = 0$ to the beginning of a cable increases linearly from zero to 1000 volts in 0.1 microseconds and then falls suddenly to zero.

(a) Sketch the time dependence of the corresponding current pulse at the beginning of the cable ($x = 0$).
(b) Sketch the x-dependence of the voltage pulse $V(x, t)$ on the cable at $t = 0.2\,\mu$sec.
(c) What is the peak power in watts delivered by the source?
(d) What is the total energy in joules carried by the pulse? Cable inductance: 2.4×10^{-7} H/m, capacitance: 9×10^{-11} F/m.

6.18 *Maximum wave speed on a membrane.* Estimate the maximum transverse wave speed on a steel membrane. Assume the membrane stretched uniformly to a tension which corresponds to the tensile strength (force per unit cross sectional area) of steel, 100 000 lb per sq. in. The mass density of steel is 7.8 g/cm^3.

6.19 *Inhomogeneous transmission line. Whirling string.* The basic field equations 6.3(3) and 6.3(4) are valid also in the case when the transmission line is inhomogeneous, so that the mass (inductance) μ and the compliance (capacitance) per unit length κ are functions of x.

(a) Show that the wave equation 6.3(5) for the force (voltage) is now replaced by $\partial^2 F/\partial t^2 = (1/\kappa)\partial(1/\mu \partial F/\partial x)/\partial x$. Similarly, show that the 'wave equation' for the velocity (current) variable is $\partial^2 u/\partial t^2 = (1/\mu)\partial(\kappa^{-1}\partial u/\partial x)\partial x$.
(b) We recall that the field equations for the transverse velocity of a string are obtained if κ is replaced by the tension S in the string. Consider now a uniform string which is whirling in a horizontal plane with one end ($x = 0$) fixed on the axis of rotation and the other end free ($x = L$). The angular whirling velocity Ω is constant. In the frame of reference of the whirling string, derive the wave equation for the velocity field of small transverse oscillations. The tension in the string is now due to the centrifugal force on the string and it will vary with position. Start by determining the x-dependence of the tension S. (The problem of a vertical chain hanging from one end and oscillating in a vertical plane is analogous. See Problem 8.14).

7

Wave reflection, transmission, and absorption

In the previous chapter we dealt with waves on a homogeneous transmission line with little attention to what happens at the junction between two transmission lines or other boundaries. These aspects will be treated quantitatively in this chapter with derivations of the reflection and transmission coefficients at a junction and the energy absorption coefficient at a termination.

7.1 Wave reflection at a junction

Frequently transmission lines of different characteristics are connected. A fluid column may be bounded by solid columns, two electrical cables of different characteristics may be connected. It is of interest to study what happens when an incident wave encounters such a junction.

It is instructive to carry out some exploratory experiments with the aid of the 'wave ladder' for torsional waves, mentioned in the last chapter. Such a ladder consists of a long metal band upon which closely spaced rods are mounted. The wave speed is determined by the torsion constant of the band and the moment of inertia of the rods. The wave speed of the torsion wave is comparatively low, and the wave can be readily observed. Similar experiments can be carried out on other transmission lines but the wave speeds generally are so high, however, that the waves cannot be observed without the aid of various transducers and related instrumentation.

Two ladders with rods of different lengths will have different wave speeds and wave impedances, and to study wave reflection and transmission at a junction, we join the ladders and generate a torsion wave on one of them. As the wave reaches the junction, one can clearly see the formation of a reflected and a transmitted wave.

A similar result is obtained for a transverse wave on a string or spring ('slinky') or a longitudinal wave on a spring. In our analysis we consider longitudinal waves on the 'standard' transmission line of a spring.

At a junction between two different springs, both the force variable F and the velocity variable u must be continuous. (If the force were not continuous, there would be a net force on the junction, and since it is massless, infinite acceleration would result. The displacement and the velocity must be continuous, since the springs are connected at the junction).

As indicated in Display 7.1, we have placed the junction at $x=0$, and the wave is incident from the left. The parameters and field variables of the two transmission lines are designated by the subscripts 1 and 2. The two wave speeds are v_1 and v_2 and the wave impedances Z_1 and Z_2.

We shall consider harmonic time dependence with the frequency ω. The corresponding two propagation constants on lines 1 and 2 are $k_1=\omega/v_1$ and $k_2=\omega/v_2$.

The wave field on the left side of the junction is the sum of an incident and a reflected wave, and we express the complex force field as

$$F_1(x)=A\exp(\mathrm{i}k_1x)+B\exp(-\mathrm{i}k_1x) \tag{1}$$

On the right hand side, the wave field consists of a single wave travelling in the positive x-direction with the force field

$$F_2(x)=C\exp(\mathrm{i}k_2x) \tag{2}$$

The corresponding velocity fields are obtained from the field equation $-\mathrm{i}\omega\mu u=-\mathrm{d}F/\mathrm{d}x$ with the results

$$u_1(x)=(1/Z_1)[A\exp(\mathrm{i}k_1x)-B\exp(-\mathrm{i}k_1x)] \tag{3}$$

$$u_2(x)=(1/Z_2)C\exp(\mathrm{i}k_2x) \tag{4}$$

We now apply the conditions of continuity of force and velocity at the junction, $F_1(0)=F_2(0)$ and $u_1(0)=u_2(0)$, and obtain from Eqs. (1)–(4)

$$A+B=C \tag{5}$$

$$(1/Z_1)(A-B)=(1/Z_2)C \tag{6}$$

From these equations follow the expressions for the **reflection coefficient R_F** and the **transmission coefficient T_F** for the force wave

$$R_F=B/A=(Z_2-Z_1)/(Z_1+Z_2) \tag{7}$$

$$T_F=C/A=2Z_2/(Z_1+Z_2) \tag{8}$$

The corresponding coefficients for the velocity wave follow from Eqs. (3)–(4),

$$R_u=-B/A=-R_F=(Z_1-Z_2)/(Z_1+Z_2) \tag{9}$$

$$T_u=(C/Z_2)/(A/Z_1)=2Z_1/(Z_1+Z_2) \tag{10}$$

Display 7.1.

Wave reflection and transmission at a junction between two transmission lines.

Reflection and transmission coefficients

Wave impedances $Z_1 = \mu_1 v_1 = \sqrt{\mu_1/\kappa_1}$

$Z_2 = \mu_2 v_2 = \sqrt{\mu_2/\kappa_2}$

For force $R_F = B/A = (Z_2 - Z_1)/(Z_1 + Z_2)$ (1)

$T_F = C/A = 2Z_2/(Z_1 + Z_2)$ (2)

For velocity $R_u = -R_F$ (3)

$T_u = 2Z_1/(Z_1 + Z_2)$ (4)

$A\,e^{ik_1x}$ $\longrightarrow$ Ce^{ik_2x} $\longrightarrow$

$\longleftarrow$ Be^{-ik_1x} $x = 0$

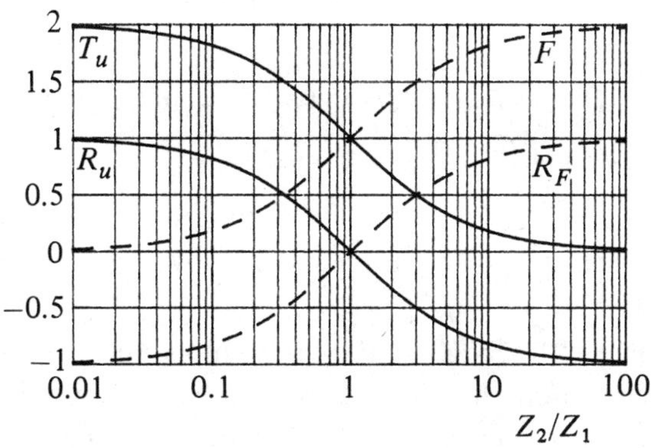

The reflection and transmission coefficients are functions of the ratio Z_2/Z_1, which are plotted in the display.

In terms of the reflection coefficients R_F, R_u, the total force amplitude at the junction is $F(0) = A(1 + R_F)$ and the velocity amplitude is $u(0) = (A/Z_1)(1 + R_u)$. These values equal the amplitudes of the corresponding transmitted waves.

As Z_2/Z_1 goes to infinity, the force reflection coefficient goes to unity, and the total force at the junction will be twice the amplitude of the incident wave. The reflection coefficient for the velocity, on the other hand, goes to -1, and the corresponding total velocity amplitude at the junction will approach zero.

Since the field to the right of the junction is a single travelling wave, the amplitude will be constant and equal to the amplitude at the junction. To the left of the junction, on the other hand, the field is a superposition of an incident and a reflected wave. The corresponding amplitude distributions for force and velocity are the magnitudes of the complex amplitudes in Eqs. (1) and (3). In terms of the reflection coefficient, the complex amplitude of the force field becomes

$$\begin{aligned} F_1(x) &= A\mathrm{e}^{\mathrm{i}k_1x}(1 + R_F\mathrm{e}^{-\mathrm{i}2k_1x}) \\ &= A\mathrm{e}^{\mathrm{i}k_1x}[1 + R_F\cos(2k_1x) - \mathrm{i}R_F\sin(2k_1x)] \end{aligned} \tag{11}$$

with the amplitude

$$\begin{aligned} |F_1(x)| &= |A|\{[1 + R_F\cos(2k_1x)]^2 + [R_F\sin(2k_1x)]^2\}^{\frac{1}{2}} \\ &= |A|\{1 + R_F^2 + 2R_F\cos(2k_1x)\}^{\frac{1}{2}} \end{aligned} \tag{12}$$

The expression for the velocity amplitude is obtained by replacing R_F by R_u. The resulting amplitude distributions are shown in Display 7.2 for some different values of ratio Z_2/Z_1.

In the first graph, with $Z_2/Z_1 = 100$, the impedance of the second transmission line is so high that it acts essentially as a rigid wall. This means that the velocity amplitude at the junction will be zero and the corresponding reflection coefficient will be close to -1. The force reflection coefficient, on the other hand, will be $+1$.

With the amplitudes of the wave in the incident and reflected waves considered to be the same, the sum will be a standing wave with the complex force amplitude

$$F_1(x) = A[\exp(\mathrm{i}k_1x) + \exp(-\mathrm{i}k_1x)] = 2A\cos(k_1x)$$

with the magnitude

$$|F_1(x)| = 2|A\cos(k_1x)|$$

Display 7.2.

Amplitude distribution of force and velocity in the springs in Display 7.1. Dashed line: Velocity. Solid line: Force.

$$F_1(x) = Ae^{ik_1x} + Be^{-ik_1x} = Ae^{ik_1x}(1 + R_F e^{-i2k_1x})$$

$$F_2(x) = Ce^{ik_2x} \tag{1}$$

$$|F(x)| = \{|A|[1 + R_F\cos(2k_1x)]^2 + [R_F\sin(2k_1x)]^2\}^{1/2} \tag{2}$$

$$|F(x)| = |C| \tag{3}$$

$$|u_1(x)| = \{(|A|/Z_1)[(1 - R_F\cos(2k_1x)]^2 + [R_F\sin(2k_1x)]^2\}^{1/2} \tag{4}$$

$$|u_2(x)| = |C|/Z_2 \tag{5}$$

$$R_F = (Z_2 - Z_1)/(Z_1 + Z_2) \quad Z = \mu v = \sqrt{\mu/\kappa} \quad k = \omega/v \tag{6}$$

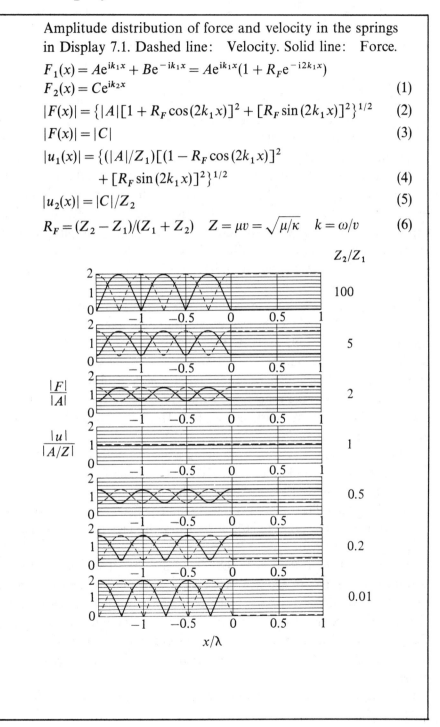

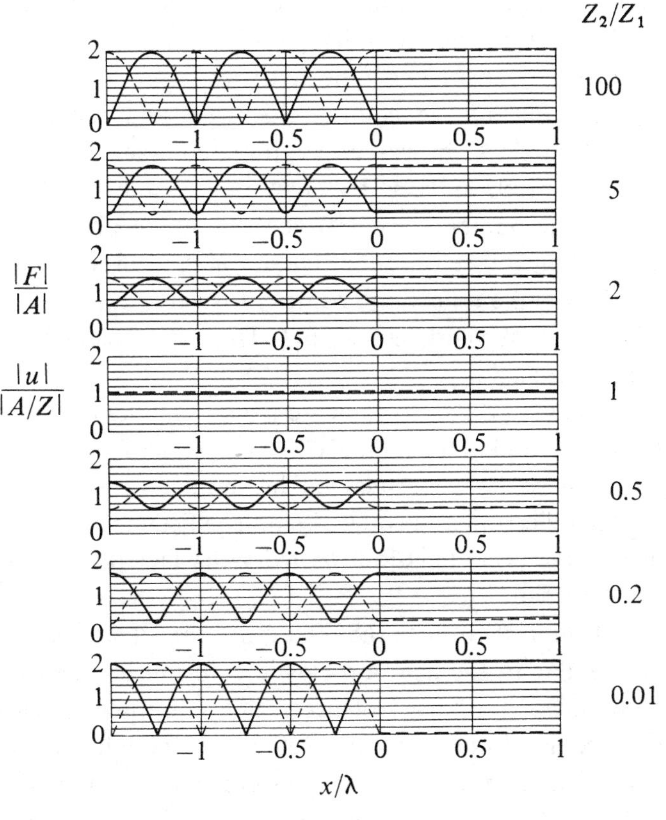

The maxima of this function are given by $\cos(k_1x) = \pm 1$, i.e. $k_1x = n\pi$, where $n = 0, 1, 2, \ldots$ With $k_1 = 2\pi/\lambda_1$, the corresponding locations are $x = n\lambda_1/2$ where λ_1 is the wavelength on the first transmission line.

The related standing wave for the velocity field is obtained in a similar manner or from the field equation $-\mathrm{i}\omega\mu u(x) = -\mathrm{d}F/\mathrm{d}x$, which yields $u(x) = (A/Z_1)\sin(k_1x)$.

With a rigid termination, the force amplitude transmitted across the junction will be twice the incident, but since the velocity is zero, no power will be transmitted; it is all reflected.

As the ratio Z_2/Z_1 is reduced, both a force and a velocity wave are transmitted across the junction. The amplitude of the reflected wave no longer will be the same as the incident. The maximum and minimum values of the force amplitude will be $A(1 + R_F)$ and $A(1 - R_F)$ with similar expressions for the maximum and minimum velocity amplitudes. These values follow without calculations if we view the x-dependence in the complex plane of the two complex amplitudes $A\exp(\mathrm{i}k_1x)$ and $AR_F \exp(-\mathrm{i}k_1x)$ of the incident and reflected waves. As x increases, the first moves along a circle in the counter-clockwise and the second in the clockwise direction. At some value of x these complex numbers will 'line up' and the magnitude is then the sum of the individual magnitudes. Similarly, when the numbers are 180 degrees out of phase, the resulting magnitude is the difference of the individual magnitudes.

When $Z_2/Z_1 = 1$, the reflection coefficient is zero, and the transmission coefficient for both force and velocity is 1. The two transmission lines then are said to be 'matched'.

A further decrease of Z_2/Z_1 to values less than 1 makes the force reflection coefficient negative, and in the limit $Z_2/Z_1 = 0$, we get $R_F = -1$ and $R_u = 1$.

Our discussion so far has involved harmonic waves, but the results obtained for the reflection and transmission coefficients apply also to wave pulses of arbitrary form. This follows from the fact that Z_1 and Z_2 are frequency independent.

It may be of interest to note that the reflection of waves at a junction is somewhat similar to the head-on elastic collision between two masses. One of the particles, the target of mass M_2, is initially at rest, and the other, the projectile of mass M_1 and velocity U, collides with the target. Conservation of momentum and energy yields the velocities U' and U'' of the two bodies after the collision in terms of the initial velocity U of the projectile.

Defining the reflection and transmission coefficients as $R_U = U'/U$ and

$T_U = U''/U$, we obtain

$$R_U = (M_1 - M_2)/(M_1 + M_2)$$
$$T_U = 2M_1/(M_1 + M_2)$$

These expressions have the same form as the wave reflection and transmission coefficients with the masses playing the roles of the wave impedances.

If the two masses are the same, the reflection coefficient becomes zero, and the transmission coefficient 1. The incident momentum is then transferred completely to the target, and the projectile comes to rest after the collision.

7.2 Energy absorption at a termination

A problem of considerable importance concerns the transfer of wave energy to a termination of a transmission line. For an electric transmission line, the termination is simply an electric circuit, a combination of resistances, inductances, and capacitances. In the case of a mechanical transmission line, the termination can be a similar combination of mechanical elements, such as a damped oscillator circuit. For transmission lines terminated by 'antennas', the termination is determined by the radiation impedance of the antenna, and similar radiation impedances apply to radiators of mechanical waves, such as organ pipes.

In Display 7.3 we have treated the transmission line as an electrical cable, terminated by the impedance Z. Examples of analogous terminations for mechanical transmission lines with longitudinal and transverse waves are shown also.

It is convenient to express the termination impedance Z in terms of the characteristic impedance $Z_0 = L_0 v = \sqrt{L_0/C_0}$ of the transmission line. The corresponding normalized impedance is

$$\zeta = Z/Z_0 = \theta + i\chi \tag{1}$$

with the real and imaginary parts θ and χ, which generally are known functions of frequency.

The voltage on the transmission line is the sum of an incident and a reflected wave, with the complex amplitude

$$V(x) = A\exp(ikx) + B\exp(-ikx) = A[\exp(ikx) + R_V\exp(-ikx)] \tag{2}$$

where $k = \omega/v$, v = wave speed $= 1/\sqrt{L_0C_0}$, and $R_V = B/A$ the voltage reflection coefficient.

The corresponding complex current amplitude is obtained from the field

Display 7.3.

Reflection and absorption coefficients at a termination of a transmission line.

Complex amplitudes

Voltage $V(x) = Ae^{ikx} + B^{-ikx}$ (1)

Current $I(x) = (A/Z)(Ae^{ikx} - Be^{-ikx})$ (2)

Termination impedance Z, Normalized $\zeta = Z/Z_0$

$$Z_0 = L_0 v = \sqrt{L_0/C_0} \quad (3)$$

Voltage reflection coeff. $R_V = (\zeta - 1)/(\zeta + 1)$ (4)

Absorption coefficient

$$\alpha = 1 - |R_V|^2 = 1 - R_V R_V^* = 4\theta/[(1+\theta)^2 + \chi^2], \quad \zeta = \theta + i\chi \quad (5)$$

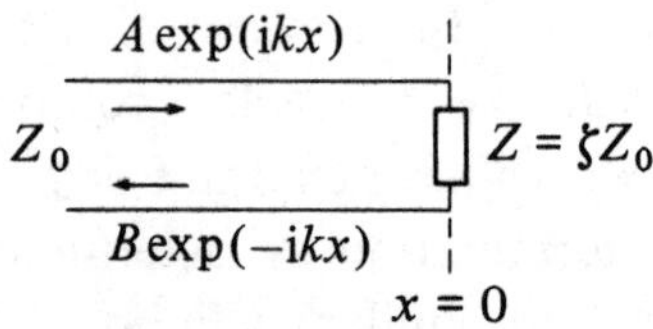

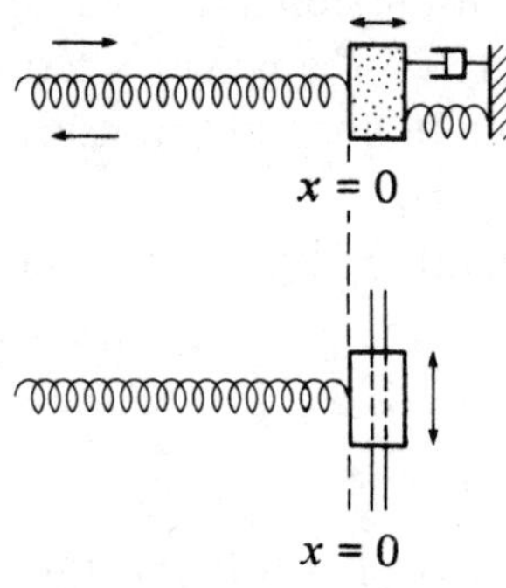

equation $-i\omega L_0 I = -dV/dx$,

$$I(x) = (A/Z_0)[\exp(ikx) - R_V \exp(-ikx)] \quad (3)$$

with $Z_0 = L_0 v = \sqrt{L_0/C_0}$ = wave impedance.

Reflection coefficient. The termination is placed at $x = 0$, and the total voltage and current amplitude at that location must be such that their ratio equals the known termination impedance, i.e. $Z = V(0)/I(0) = Z_0(1 + R_V)/(1 - R_V)$ from which follows

$$R_V = (Z - Z_0)/(Z + Z_0) = (\zeta - 1)/(\zeta + 1) \quad (4)$$

The corresponding reflection coefficient for the current wave follows from Eq. (3) and is $R_I = -R_V$. The reflection coefficients generally are complex numbers, with an amplitude and phase angle which can be expressed in terms of the (known) values of the magnitude and phase angle of the termination impedance.

With R being the reflection coefficient for a field variable, the amplitude of the total field at the termination, at $x = 0$, will be $(1 + R)$ times the amplitude of the incident wave.

If the termination impedance is infinite, which corresponds to an open electric line, the current at the termination will be zero and $R_I = -1$. The corresponding voltage reflection coefficient is $R_V = 1$. In the other limit of zero impedance, the electrical line is short circuited at the end, and the voltage is zero and $R_V = -1$.

For a mechanical transmission line, an infinite termination impedance corresponds to a rigid termination (a spring attached to a wall), and zero impedance corresponds to a free end.

Absorption coefficient. If the total complex amplitudes of voltage and current are $V(x)$ and $I(x)$, the time average power transmitted in the positive x-direction is $P = \frac{1}{2}\mathrm{Re}\{VI^*\}$. If we apply this result to the wave field on the transmission line with $V(x) = A\exp(ikx) + B\exp(-ikx)$ and $I = (1/Z_0)[A\exp(ikx) - B\exp(-ikx)]$, we obtain the power

$$P = (1/2Z_0)(|A|^2 - |B|^2) \quad (5)$$

where we have used $AA^* = |A|^2$ and $BB^* = |B|^2$. The first term is the power carried by the incident wave and the second the reflected power.

The net energy flow P must be absorbed by the termination, and the absorption coefficient is defined as the ratio between the absorbed power P and the incident power $P_i = (1/2Z_0)|A|^2$, $\alpha = P/P_i = (|A|^2 - |B|^2)/|A|^2 = 1 - |R_V|^2 = 1 - R_V R_V^*$.

The squared magnitude of the reflection coefficient can be calculated from Eq. (4), and with $Z/Z_0 = \zeta = \theta + i\chi$, we obtain $R_V R_V^* = [(\theta - 1)^2 + \chi^2]/[(\theta + 1)^2 + \chi^2]$, so that

$$\alpha = 1 - R_V R_V^* = 4\theta/[(1 + \theta)^2 + \chi^2] \tag{6}$$

The termination impedance generally is frequency dependent. In the case of a resonator circuit, the reactance χ will be zero at resonance, so that $\alpha = 4\theta/(1 + \theta)^2$. Thus, if the termination is purely resistive and equal to the wave impedance of the transmission line, i.e. if $\zeta = \theta = 1$, we get $\alpha = 1$. All the incident energy is then absorbed by the termination which is said to be 'matched' to the transmission line.

Such **impedance matching**, $Z = Z_0$, is always desirable, of course, when energy transfer is involved on electrical as well as mechanical transmission lines. It is an essential aspect in the design of systems ranging from audio circuits to the devices for the extraction of wave energy from the sea.

If the termination impedance is mismatched, it is possible to improve the power transfer by means of a 'transformer', which serves to reduce the discontinuity in impedance between the transmission line and the termination.

Again, the problem is much like that encountered in the energy transfer in an elastic head-on collision between a moving projectile and a stationary target. As was mentioned earlier, all the incident energy is transferred to the target only if its mass M_2 is the same as the mass M_1 of the projectile. If this condition is not met, the energy transfer can be improved by the insertion of a third particle between the projectile and the target, as indicated in Display 7.4. The energy transfer will be a maximum if the mass M' of the inserted particle is the geometric mean $M' = \sqrt{M_1 M_2}$ of the masses of the colliding particles. Further improvement in the energy transfer can be achieved by the use of several bodies arranged along a line with the masses varying monotonically between M_1 and M_2.

In analogous manner it is possible to increase the power transfer to the termination of a transmission line by the insertion of a transmission line element with an appropriately chosen wave impedance Z'. It can be shown that it is possible to absorb all the incident wave power if the length of the transformer line is an odd number of quarter wavelengths and if the wave impedance of the transformer line is

$$Z' = \sqrt{Z_1 Z_2} \tag{7}$$

i.e. the geometric mean of the wave impedance of the primary transmission line and the termination impedance. In a similar manner the coating of a lens in optics serves to reduce the reflection of light at the lens surface.

Display 7.4.

Termination impedance Z_2 (1)

Transmission line impedances:

Primary line Z_1 (2)

Transformer line Z' (3)

If transformer line length is an odd number of quarter wavelengths, complete absorption of incident wave power by termination if

$$Z' = \sqrt{Z_1 Z_2} \tag{4}$$

M_1 M' M_2

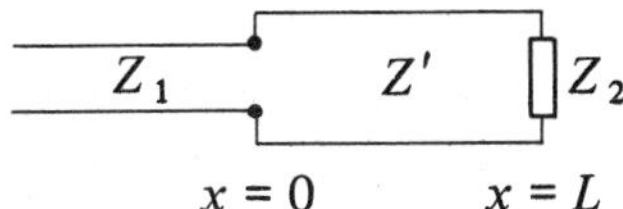

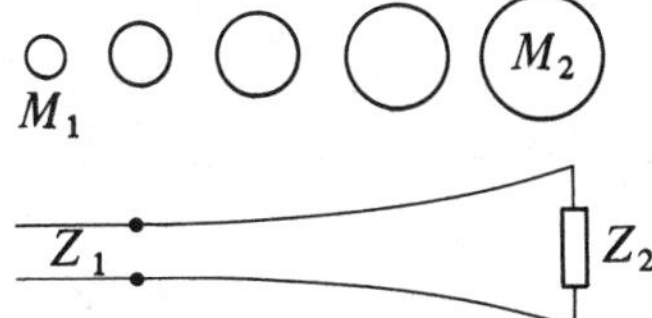

This complete energy transfer occurs only at discrete frequencies, however. In order to obtain good energy transfer over a wide range of wavelengths, and not only when L is an odd number of quarter wavelengths, several transmission line elements can be used in series with monotonically varying values of the line impedance. Frequently, a tapered line is used with a continuously varying characteristic impedance, as indicated schematically in the display. Acoustic and microwave horns are of this kind, providing an improved energy transfer between the source and the radiation field.

7.3 Wave reflection and elastic collisions

At the end of Section 7.1 it was mentioned in passing, that the expressions for the reflection and transmission coefficients at a junction are similar in form to the corresponding coefficients for a head-on elastic collision between two particles. This analogy will now be explained, by considering the reflection of a wave pulse.

As in Section 6.8, a longitudinal wave pulse is generated by a force $F(t)$, which is constant during the interval Δt and zero at all other times. It is applied to the first line, on which the wave speed is v_1 and the mass per unit length is μ_1. The corresponding quantities on the second line are v_2 and μ_2.

As explained in Section 6.8, the momentum transferred to the spring by the force spring is $F\Delta t = \mu_1 u_i v_1 \Delta t$, where u_i is the material velocity of the incident pulse, which is constant during the interval Δt and zero at all other times. The corresponding energy in the pulse is $\mu_1 u_i^2 v_1 \Delta t$.

As the pulse arrives at the junction, a reflected and a transmitted pulse are produced, each with duration Δt. The material velocities in these pulses in the positive x-direction are denoted by u_r and u_t, respectively. The values of the momentum in these pulses are $\mu_1 u_r v_1 \Delta t$ and $\mu_2 u_t v_2 \Delta t$ with corresponding expressions for the energies.

Introducing the wave impedance $Z_1 = \mu_1 v_1$ and $Z_2 = \mu_2 v_2$, we can express the equations for momentum and energy conservation as

$$Z_1 u_i = Z_1 u_r + Z_2 u_t \tag{1}$$

$$Z_1 u_i^2 = Z_1 u_r^2 + Z_2 u_t^2 \tag{2}$$

From the first of these equations follows $Z_1(u_i - u_r) = Z_2 u_t$ and from the second $Z_1(u_i^2 - u_r^2) = Z_2 u_t^2 = Z_1(u_i - u_r)(u_i + u_r)$, i.e. $u_i + u_r = u_t$.

From these relations we can determine the velocity reflection and transmission coefficients

$$R_u = u_r/u_i = (Z_1 - Z_2)/(Z_1 + Z_2) \tag{3}$$

$$T_u = u_t/u_i = 2Z_1/(Z_1 + Z_2) \tag{4}$$

in agreement with the results obtained in Section 7.1.

In an elastic head-on collision between a projectile of mass M_1 and velocity u_i and a stationary target of mass M_2, the equations for conservation of momentum and energy will have the same form as Eqs. (1) and (2) with Z replaced by M and with u_r and u_t being the velocities of the projectile and the target, respectively, after the collision.

Examples

E7.1 *Reflection and transmission coefficients.* Two transmission lines with the wave impedances Z_1 and Z_2 are joined at $x = 0$. A resistance R is connected across the junction between the conductors of the transmission lines. A travelling wave is incident from the left and is partially reflected from the junction. Determine the reflection and transmission coefficients for the voltage and current waves.

Solution. A voltage wave with the complex amplitude $V_1 \exp(ikx)$ travels toward the junction on the first transmission line and the corresponding current wave is $(V_1/Z_1)\exp(ikx)$.

This first transmission line can be considered to be terminated at $x = 0$ by an impedance Z which is the parallel combination of R and the wave impedance Z_2 of the transmission line to the right of the junction. This termination impedance is $Z = RZ_2/(R + Z_2)$.

According to the general formula for the reflection coefficient R_V for the voltage wave in Eq. 7.2.(4), we obtain

$$R_V = (Z - Z_1)/(Z + Z_1) = [R(Z_2 - Z_1) - Z_1 Z_2]/[R(Z_1 + Z_2) + Z_1 Z_2] \tag{1}$$

and the current reflection coefficient is $R_I = -R_V$.

With the complex amplitude of the incident voltage wave being $V_1 \exp(ikx)$, the total voltage across the junction will be the sum of V_1 and the reflected voltage $R_V V_1$ at $x = 0$. The total voltage across the junction $V_2 = (1 + R_V)V_1$ is transmitted to the second line to the right of the junction.

The transmitted voltage wave is then $V_2 \exp(ikx)$ and the corresponding transmission coefficient becomes

$$T_V = V_2/V_1 = 2RZ_2/[R(Z_1 + Z_2) + Z_1 Z_2] \tag{2}$$

The transmission coefficient for the current becomes

$$T_I = (V_2/Z_2)/(V_1/Z_1) = (Z_1/Z_2)T_V \tag{3}$$

As a check, we note that for an infinite value of R (which is obtained if the resistance is removed), the voltage reflection coefficient becomes $R_V = (Z_2 - Z_1)/(Z_2 + Z_1)$, as it should, and for $R = 0$ we get $R_V = -1$.

If the wave impedance Z_2 is larger than Z_1, it follows from Eq. (1) that the reflected wave will be eliminated if the resistance is chosen to be $R = Z_1 Z_2/(Z_2 - Z_1)$. If Z_2 is smaller than Z_1, the reflection can be eliminated by means of a resistance in series with the two lines rather than in parallel. We leave the detailed analysis of this case for the reader.

Problems

7.1 *Reflection of sound.* A tube is filled with air and helium, and we assume that there is a well defined boundary between the two gases, provided by a thin (sound transparent) membrane perpendicular to the tube axis. A harmonic sound wave is incident on this boundary from the air. Determine the pressure reflection and transmission coefficients at the boundary. If the sound pressure of the incident wave is A, what then is the sound pressure amplitude at the boundary? Densities: Air, 1.293 kg/m^3, He, 0.170 kg/m^3. Sound speeds: Air, 340 m/sec, He, 998 m/sec.

7.2 *Reflection of longitudinal and transverse waves.* Two long coil springs A and B of equal length and mass but with different spring constants K_A and K_B are connected and stretched. It is found that the length of A is doubled and the length of B tripled. A wave is incident on the junction from A. What are the force reflection and transmission coefficients if the wave is (a) transverse and (b) longitudinal?

7.3 *Wave reflection, solids.* One end of a copper bar is joined to the end of an aluminum bar. The cross sections of the bars are the same. A wave pulse with a total energy of 10 joules in the copper bar is incident on the junction.

(a) How much energy is transmitted into the aluminum bar?
(b) If instead the wave is incident on the junction from the aluminum bar, what energy will be transmitted into the copper bar?
(c) Which of the two cases will yield the larger velocity amplitude at the junction?
$Cu{:}\rho = 8900,\ E = 1.26 \times 10^{11}$; Al:$\rho = 270\,\mathrm{kg/m^2}\ \ Y = 7.2 \times 10^{10}\,\mathrm{N/m^2}$

7.4 *Power transmission.* In a particular case of wave reflection at the junction of two transmission lines, it is desired that 20% of the incident wave energy is to be transmitted across the junction. If the wave impedance of the transmission line of the incident wave is Z_1, how should the impedance of the second line be chosen? Is there more than one possible choice?

7.5 *Transmission of sound from air to water.* What fraction of the power of a sound wave in air is transmitted into water at normal incidence?
(For data see Display 6.12)

7.6 *Absorption by a termination.* A coil spring of mass M is driven at one end in transverse motion, and the other end is attached to a ring also with a mass M, which can slide on a horizontal bar, normal to the direction of the spring. We assume that the friction force on the ring from the bar has the magnitude Ru, where u is the velocity of the ring. The spring has a spring constant K and is stretched to a length which is much longer than the relaxed length.

(a) Show that the wave impedance of the spring is approximately $M\omega_0$, where $\omega_0 = \sqrt{K/M}$.

(b) Show that the absorption coefficient of the termination can be expressed as $\alpha = 4D/[(1+D)^2 + (\omega/\omega_0)^2]$, where $D = R/\omega_0 M$.

7.7 *Reflection of a wave pulse.* Two strings, with the masses per unit length $\mu_1 = 0.1$ kg/m and $\mu_2 = 0.4$ kg/m, are connected and kept under a tension $\tau = 10$ N. At $t = 0$ a transverse displacement of the end of the light string is started with a constant velocity $u(0)$ immediately followed by a displacement back to the origin with a velocity $-u(0)/2$. The duration of this 'triangular' pulse is 0.3 sec.

(a) Sketch the time dependence of the transverse displacement at $x = 0$.

(b) How long must be the minimum length of the light string in order that the entire wave pulse can be carried by the string without interference from the reflected pulse?

(c) Sketch the x-dependence of the pulse in (b) at $t = 0.3$ sec.

(d) Sketch the reflected and transmitted pulses after the process of reflection is completed.

(e) Suppose we denote the time when the reflection starts by T. Sketch the x-dependence of the displacement of the two strings at the times T, $T + 0.1$, $T + 0.2$, and $T + 0.3$ sec.

7.8 *Transmission line 'resonances'.* A long transmission line has a wave impedance Z_1. A transmission line element of length L and wave impedance $Z = 10Z_1$ is inserted in the line as shown.

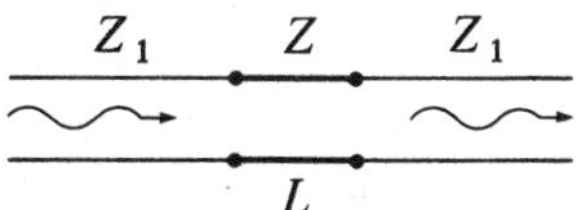

A wave is incident from the left on the element. Calculate the frequency dependence of the fraction of the power transmitted across the line element. Identify the frequencies of maximum and minimum power transmission (resonances and anti-resonances).

7.9 *Coaxial cable.* In a lecture demonstration, a voltage pulse was applied at the beginning ($x = 0$) of a coaxial cable of length $L = 428$ ft (rolled up on a spool). The voltage at $x = 0$ was displayed on an oscilloscope.

When the cable was open at $x = L$, a reflected pulse appeared at $x = 0$ after a time of 1.2 microseconds. (The pulse had the same sign as the applied pulse. When the cable was short-circuited at $x = L$, the reflected pulse had the opposite sign, all in accordance with the values of 1 and -1 for the voltage reflection coefficient). When the cable was terminated with a resistance of 50 ohms, no reflected pulse occurred.

From this information, determine the inductance and capacitance per unit length of the cable.

7.10 *Standing wave on a Lecher line.* In a lecture demonstration, a standing wave was generated on a transmission line (Lecher line) consisting of two parallel wires, separated a distance of 19.4 cm. The frequency of the generator (applied to one end of the line) was 75 MHz. The standing wave on the line was observed by means of a light bulb connected to the wires and moved along the line. The distance between two adjacent electric field nodes (light bulb dark) was found to be 198 cm.

(a) Determine the speed of the EM wave on the transmission line.
(b) Determine the location of the voltage node closest to the end of the line when the line is open and when it is shortcircuited.
(c) From the inductance and capacitance of such a line, calculate the wave speed and compare with the measured value. The diameter of each wire was 1.5 mm.

7.11 *Reflection.* Suppose that the light bulb used in the demonstration in Problem 7.11 is located at $x = 0$ and that the transmission line extends in both directions from the bulb. A voltage wave $V = V_0 \cos(\omega t - kx)$ (complex amplitude $V(x) = V_0 \exp(ikx)$) is incident on the bulb. Determine the voltage and current waves reflected from the bulb and the corresponding transmitted waves in terms of the bulb resistance R.

7.12 *Measurement of absorption coefficient.* Consider an electrical (Lecher line) or acoustical uniform transmission line (duct) which is terminated at the end by an element with a certain impedance and corresponding absorption coefficient. A source at the other end of the line produces a plane wave field on the line which can be regarded as the superposition of an incident and a reflected wave. The wave amplitude along the line is determined by a detector and the ratio between the maximum and minimum values is determined. If this ratio is denoted by n, show that the absorption coefficient of the termination is $\alpha = 4n/(n+1)^2$.

8

Resonances and normal modes

In the analysis of coupled oscillators in Chapter 5, we started with the forced harmonic motion, which led to the identification of resonances and corresponding normal modes, which were analyzed further in the study of the free motion of the system. It was shown, for example, how an arbitrary motion could be expressed as a superposition of normal modes.

We now proceed in the same manner in the study of continuous systems, and we shall start with the forced harmonic motion of a spring clamped at one end. As a related example, we re-examine the mass–spring oscillator, now accounting for the mass of the spring. The resonances and modes of motion are identified and used in the decomposition of an arbitrary displacement of a string in free motion. Similarly, the oscillation produced by an arbitrary continuous harmonic force distribution is analyzed, the force function being decomposed into normal mode functions.

8.1 Forced harmonic waves

To illustrate the application of the general complex amplitudes of force and velocity in one-dimensional wave motion

$$F(x) = A\exp(ikx) + B\exp(-ikx)$$
$$u(x) = (1/Z)[A\exp(ikx) - B\exp(-ikx)]$$

we considered as an example, in Section 6.4, the forced harmonic motion of a real spring clamped at one end. We shall start by re-examining this problem in more detail, comparing the solutions obtained when the spring is driven in harmonic motion, first with a source of constant (frequency independent) velocity amplitude and then with a source of constant force amplitude.

The analysis applies equally well to the transverse motion of a string clamped at one end, an air column in a pipe closed at one, or an open electrical cable.

Constant velocity amplitude source. As before, the spring considered has the length L, spring constant K, and mass m. It is clamped at one end, $x = 0$, and driven at the other, $x = -L$ with a velocity $u(-L, t) = u_0 \cos(\omega t)$, the corresponding complex amplitude being simply u_0. The related complex displacement amplitude is $\xi_0 = u_0/(-\mathrm{i}\omega)$.

From the boundary condition $u(0) = 0$, it follows from the general expressions for the complex amplitude of force and velocity

$$F(x) = A \exp(\mathrm{i}kx) + B \exp(\mathrm{i}kx) \quad k = \omega/v \quad v = 1/\sqrt{\mu\kappa} \tag{1}$$

$$u(x) = (1/Z)[A \exp(\mathrm{i}kx) - B \exp(-\mathrm{i}kx)] \quad \mu = m/L$$

$$\kappa = 1/KL \quad Z = \mu v \tag{2}$$

that $A = B$ and that $F(x) = 2A \cos(kx)$ and $Zu(x) = \mathrm{i}2A \sin(kx)$. The complex velocity amplitude at $x = -L$ is $u(-L) = u_0$, and from this condition follows $-\mathrm{i}2A \sin(kL) = Zu_0$. This yields $2A = \mathrm{i}Zu_0/\sin(kL)$, so that

$$F(x) = \mathrm{i}[Zu_0/\sin(kL)] \cos(kx) \tag{3}$$

$$u(x) = -[u_0/\sin(kL)] \sin(kx) \tag{4}$$

where $u_0 = -\mathrm{i}\omega\xi_0$.

The complex driving force amplitude is $F(-L) = \mathrm{i}Zu_0 \cot(kL)$. At low frequencies, such that $kL = \omega L/v$ is much less than unity, we have $\cot(kL) \simeq 1/kL$, so that $F(-L) \simeq \mathrm{i}Zu_0/kL = K\xi_0$, where we have used $Z = \mu v$, $v^2 = l/\mu\kappa$, and $1/\kappa L = K$. In other words, as the frequency goes to zero, the force amplitude goes to the same value as the force required for a static displacement of the spring, as it should.

At a position x different from the end points 0 and $-L$, the velocity and displacement amplitude goes to infinity at a frequency such that $kL = \omega L/v = n\pi$, where n is an integer. Thus, these resonance frequencies and corresponding wavelengths are given by

$$\omega_n = n\pi v/L \quad \nu_n = nv/2L \quad (n = 1, 2, \ldots) \tag{5}$$

$$\lambda_n = v\nu_n = 2L/n \tag{6}$$

At the nth resonance, the length of the spring is equal to n half wavelengths.

It is important to realize that in this case of no damping, the amplitude will be infinite for any finite displacement amplitude of the driver. This suggests, that the spring can oscillate with finite amplitude at a resonance frequency without a driver, i.e. with the spring clamped at both at $x = 0$ and $x = -L$. The corresponding shape of this motion is expressed by the complex displacement amplitude

$$\xi_n(x) = A_n \sin(k_n x) = A_n \sin(n\pi x/L) \tag{7}$$

$$k_nL = n\pi \quad \nu_n = nv/2L \tag{8}$$

As we shall discuss in more detail later in this chapter, this is the nth mode of longitudinal oscillations of a spring clamped at both ends. It describes also the modes of transverse oscillations of a string, the acoustic oscillations of a gas column in a tube with closed ends, and the electromagnetic oscillations of a transmission line with both ends open.

Constant force amplitude source. The spring is now driven by force $F(-L, t) = F_0 \cos(\omega t)$ with a constant amplitude. It was considered in Section 6.4 and summarized in Display 6.9. Starting from the general expressions for the complex amplitudes $F(x)$ and $u(x)$ in Eqs. (1)–(2) and applying the condition $u(0) = 0$ and $F(-L) = F_0$, we obtained

$$F(x) = [F_0/\cos(kL)] \cos(kx) \tag{9}$$

$$u(x) = \mathrm{i}[F_0/Z \cos(kL)] \sin(kx)$$

$$\xi(x) = -[F_0/Z\omega \cos(kL)] \sin(kx) \tag{10}$$

The resonance frequencies in this case are given by $kL = \omega L/v = (2n-1)\pi/2$, where n is an integer, i.e.

$$\omega_n = (2n-1)\pi v/2L \quad \nu_n = (2n-1)v/4L \tag{11}$$

$$\lambda_n = 4L/(2n-1) \quad (n = 1, 2, \ldots) \tag{12}$$

In this case resonance occurs when the length of the spring is an odd number of quarter wavelengths.

For a finite driving force, the amplitude goes to infinity at the resonances and a finite motion can occur (in the absence of damping) only if the driving force is zero, i.e. if the end of the spring is free. The corresponding shape of the spring at the resonance frequencies is given by

$$u_n(x) = B_n \sin[(2n-1)\pi x/2L] \tag{13}$$

which represents the nth mode of free motion of a spring clamped at one end ($x = 0$) and free at the other ($x = -L$). The same wave function applies to the nth normal mode of sound in a fluid column in a tube closed at one end and open at the other and to EM waves on a cable open at one end and shortcircuited at the other.

The condition of a string under tension with one end clamped and the other free can be approximately achieved if the tension is provided by means of a second string which is much lighter than and attached to the 'free' end of the first string.

For a summary of this section we refer to Display 8.1.

Display 8.1.

Forced harmonic oscillations of transmission lines 'clamped' at one end.

Source with constant velocity amplitude

Spring driven by $u(-L, t) = u_0 \cos(\omega t)$ (1)

Complex amplitude functions:

$$u(x) = -[u_0/\sin(kL)]\sin(kx) \quad (2)$$

$$F(x) = \mathrm{i}[Zu_0/\sin(kL)]\cos(kx) \quad (3)$$

Resonances:

$$\sin(k_nL) = 0, \quad k_nL = n\pi \quad (4)$$

$$\nu_n = \omega_n/2\pi = nv/2L, \quad \lambda = 2L/n \quad (n = 1, 2, \ldots) \quad (5)$$

$$(v = 1/\sqrt{\mu\kappa} = L\sqrt{K/m}, \quad m = \mu L, \quad \kappa = 1/KL, \quad Z = \mu v) \quad (6)$$

Source with constant force amplitude

Spring driven by $F(-L, t) = F_0 \cos(\omega t)$ (7)

Complex amplitude functions:

$$F(x) = [F_0/\cos(kL)]\cos(kx) \quad (8)$$

$$u(x) = \mathrm{i}[F_0/Z\cos(kL)]\sin(kx) \quad (9)$$

Resonances:

$$\cos(k_nL) = 0, \quad k_nL = (2n-1)\pi/2 \quad (10)$$

$$\nu_n = \omega_n/2\pi = (2n-1)v/4L, \quad \lambda_n = 4L/(2n-1) \quad (n = 1, 2, \ldots) \quad (11)$$

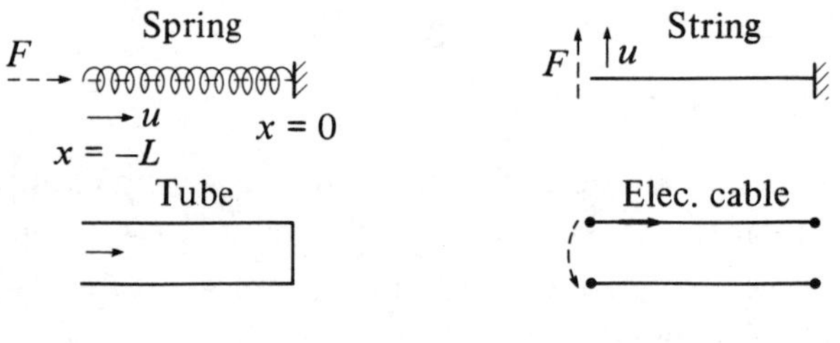

Spring

F

u

$x = 0$

$x = -L$

String

F u

Tube

Elec. cable

8.2 Forced harmonic motion of the real mass–spring oscillator

The analysis of the frequency response and impulse response of a mass–spring oscillator in Chapters 3 and 4 referred to an idealized oscillator in which the mass of the spring was neglected. As a result, the oscillator had only one degree of freedom and had only one resonance frequency.

In the real oscillator, we account for the mass of the spring, which is then treated as a continuous transmission line. A mass M is attached to the end of the spring, which is clamped at the other end. As before, the clamped end is at $x=0$ and the other end, carrying M, is at $x=-L$. The spring has a spring constant K and a mass m, the compliance and mass per unit length being $\kappa=1/KL$ and $\mu=m/L$.

As in the previous section, the general expression for the complex amplitudes of the force and velocity on the spring are

$$F(x)=A\exp(ikx)+B\exp(-ikx) \quad (1)$$

$$u(x)=(1/Z)[A\exp(ikx)-B\exp(-ikx)] \quad (2)$$

where $Z=\mu v$, the wave impedance of the spring. With the condition $u(0)=0$ we get $A=B$, as before, and

$$F(x)=2A\cos(kx) \quad (3)$$

$$u(x)=\mathrm{i}(2A/Z)\sin(kx) \quad (4)$$

The complex amplitude of the external force is simply F_0, and the complex amplitude of the net force on M (at $x=-L$) is then $F_0-F(-L)$. The complex amplitude version of Newton's law applied to M is then

$$-\mathrm{i}\omega u(-L)M=F_0-F(-L) \quad (5)$$

With $u(-L)=-\mathrm{i}(2A/Z)\sin(kL)$ and $F(-L)=2A\cos(kL)$, as given by Eqs. (3)–(4), we obtain $2A=F_0/[\cos(kL)-(\omega M/Z)\sin(kL)]$ from Eq. (5).

We write $\omega M/Z=kvM/Z=(kL)M/m$, where $m=\mu L$ is the spring mass, so that $2A=F_0/[\cos(kL)-(M/m)(kL)\sin(kL)]$. By insertion into Eq. (4), the complex velocity amplitude $u(-L)$ of M and the corresponding displacement amplitude $\xi(-L)=u(-L)/(-\mathrm{i}\omega)$ can be expressed in terms of kL. For example, for the complex displacement amplitude we get $\xi(-L)=\xi'/[kL\cot(kL)-(M/m)(kL)^2]$, where $\xi'=F_0/K$.

The resonance frequency of the idealized oscillator is $\omega_0=\sqrt{K/M}$, and it is instructive to use this frequency as reference in expressing the frequency response of the real oscillator. We note that $(kL)^2M/m=\omega^2L^2\mu\kappa M/m=\omega^2M/K=\omega^2/\omega_0^2=\Omega^2$, where $\Omega=\omega/\omega_0$ is the normalized frequency, so that $kL=\sqrt{m/M}\;\Omega=\delta\Omega$, where $\delta=\sqrt{m/M}$. In terms of these quantities we obtain for the frequency dependence of the complex displacement

Display 8.2.

Frequency response of a real mass–spring oscillator driven by the force $F_0 \cos(\omega t)$, $m =$ spring mass, $M =$ load mass, $\xi' = F_0/K$, $\Omega = \omega/\omega_0$, $\delta = m/M$

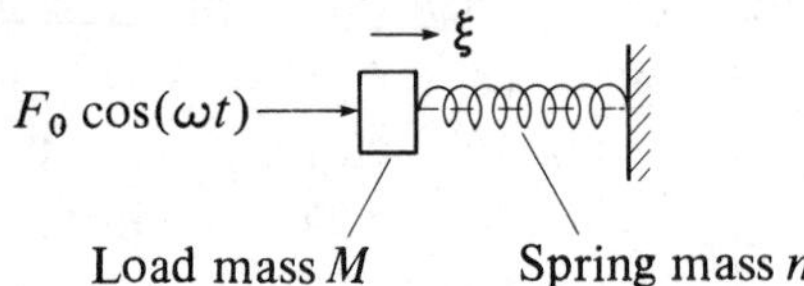

$\xi/\xi' = 1/[\delta\Omega \cot(\delta\Omega) - \Omega^2]$
($\xi =$ compl. displ. ampl. of M) (1)

$\xi' = F_0 K, \quad \delta = \sqrt{m/M}, \quad \Omega = \omega/\omega_0, \quad \omega_0 = \sqrt{K/M}$ (2)

If $\delta \ll 1, \quad \xi/\xi' \simeq 1/(1 - \Omega^2)$ (3)

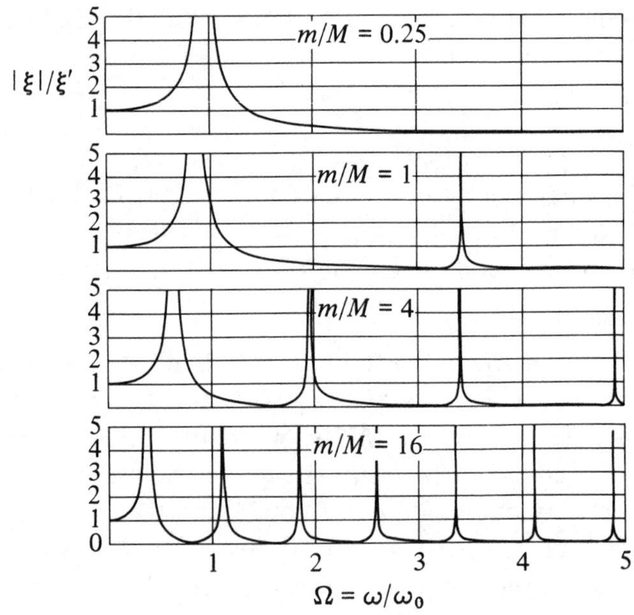

amplitude

$$\xi(-L) = \xi'/[\delta\Omega \cot(\delta\Omega) - \Omega^2]$$

$$\xi' = F_0/K, \quad \delta = \sqrt{m/M}, \quad \Omega = \omega/\omega_0, \quad \omega_0 = \sqrt{K/M} \qquad (6)$$

Since we have not accounted for damping, this complex amplitude is either positive or negative. A positive value indicates that the displacement is in phase with the 'driving force' and if it is negative, the phase lag is 180 degrees.

It should be noted that for small values of m/M, the equation reduces to the well known frequency response function $\xi'/(1-\Omega^2)$ for displacement of the idealized mass–spring oscillator (since $\cot(\delta\Omega) \simeq 1/\delta\Omega$ for $\delta \ll 1$). Unlike the idealized oscillator, however, the real oscillator has more than one resonance since the spring, treated as a continuous transmission line, has an infinite number of degrees of freedom.

In Display 8.2 we have plotted the frequency dependence of the displacement amplitude $|\xi|$ of M in terms of the 'static' displacement $\xi' = F_0/K$. In the first curve the ratio m/M between the mass m of the spring and the mass M of the 'load' at the end of the spring is 0.25; for this value of m/M the response curve is not much different from that of the idealized oscillator. In fact, in the frequency range shown, only one resonance occurs and the frequency is only slightly below the value $\omega_0 = \sqrt{K/M}$ of the idealized oscillator.

As the spring mass is increased, additional resonances appear. For example, if the spring mass is equal to the load mass, the first additional resonance frequency is about $3.4\omega_0$. The bandwidths of the higher resonances are considerably smaller than for the fundamental resonance and decrease with increasing frequency.

8.3 Free oscillations and normal modes

In forced harmonic motion of a continuous 'transmission line', as shown in the last section, resonances represent important responses of the system, in which the length of the line is related to the wavelength in a particular way. In the absence of damping, the amplitude is infinite at each resonance, and this indicates that at these discrete frequencies, the system can oscillate by itself in free motion.

The corresponding oscillations are the **normal modes** of the system, and the characteristics of each mode, such as x-dependence of the amplitude, can be derived from the response to the forced harmonic motion. It is useful, however, to study the normal modes separately by an analysis of the free motion of the system.

As in the case of coupled oscillators, a normal mode is characterized by the fact that all elements of the system oscillate in harmonic motion with the same frequency. The amplitude of the various elements is described by a complex amplitude function (wave function). For the two equal coupled oscillators in Chapter 4, the amplitude functions for the displacement were given by $\xi_n(x)$, with $n=1$ and 2 for the first and second modes. These functions were zero except at the locations x_1 and x_2 of the mass elements. For the first mode, we had $\xi_1(x_1)=\xi_1(x_2)$ and for the second, $\xi_2(x_1)=-\xi_2(x_2)$. The time dependence of each of the modes can be expressed as

$$\xi_n(x,t)=\mathrm{Re}\{\xi_n(x)\exp(-\mathrm{i}\omega_n t)\}=\xi_{n0}(x)\cos(\omega_n t-\alpha_n) \qquad (1)$$

This expression is the same as for a single oscillator except that we have now two locations $x=x_1$ and $x=x_2$ where the function is different from zero (the locations of the mass elements) and there are two values of n, i.e. two modes of motion. The amplitude factor $u_{n0}(x)$ and the phase angle α_n in each mode are determined from the initial conditions of displacement and velocity, in the same manner as for the single oscillator, as explained in Chapter 4. If the system is started from rest, the phase angle α_n will be zero, so that the time dependence of each mode is expressed by the time function $\cos(\omega_n t)$. If, on the other hand, the system is started from zero displacement with an initial velocity, the time function is given by $\sin(\omega_n t)$.

The description of the normal modes for a continuous system is much the same, except that the complex amplitude function, the wave function, is now a continuous function of x; there is an 'oscillator' at every location x. The wave function describes the shape of the continuous system in the normal mode. It depends on the boundary conditions of the system, and some examples are given below.

It should be pointed out, that the description of normal modes given above applies to all types of wave motion, mechanical as well as electrical, in accordance with the analogies discussed in previous chapters.

Spring clamped at both ends. We start with the example of a spring (or string) clamped at $x=0$ and $x=L$. Since we are now interested in normal modes, the time dependence is harmonic, and in that case the general solutions to the field equations are expressed by the complex amplitudes for the force and velocity variables by Eqs. 8.1.(1)–(2), i.e.

$$F(x)=A\exp(ikx)+B\exp(-ikx)$$
$$u(x)=(1/Z)[A\exp(ikx)-B\exp(-ikx)]$$

which were the basis for the study also of the forced harmonic motion in the last section.

From the boundary condition $u(0)=0$ it follows that the complex velocity amplitude must be of the form of Eq. 8.1.(4), i.e. $u(x)=U\sin(kx)$, where U is a constant. Since we assume U to be different from zero, the boundary condition $u(L)=0$ at the other end of the spring can be satisfied only for discrete values k_n of k given by $k_nL=n\pi$, where n is an integer. With $k_n=\omega_n/v=2\pi/\lambda_n$, the corresponding frequencies and wavelengths are

$$k_nL=n\pi \quad (n=1,2,3\ldots) \tag{1}$$

$$\nu_n=\omega_n/2\pi=nv/2L \tag{2}$$

$$\lambda_n=2L/n \tag{3}$$

The related complex amplitude functions are expressed as

$$\xi_n(x)=D_n\sin(k_nx) \quad D_n=D_{n0}\exp(\mathrm{i}\alpha_n) \tag{4}$$

$$u_n(x)=U_n\sin(k_nx) \tag{5}$$

$$F_n(x)=F_n\cos(k_nx) \tag{6}$$

where $U_n=-\mathrm{i}\omega_nD_n$ and $F_n=-\mathrm{i}ZU_n=-\omega_nZD_n$, which follow from Eqs. 8.1.(3)–(4) or directly from the complex amplitude field equation $-\mathrm{i}\omega\mu u(x)=-\mathrm{d}F(x)/\mathrm{d}x$.

The corresponding expression for the time dependence of the real displacement is

$$\xi_n(x,t)=\mathrm{Re}\{\xi_n(x)\exp(-\mathrm{i}\omega_nt)\}=D_{n0}\sin(k_nx)\cos(\omega_nt-\alpha_n) \tag{7}$$

$$D_n=D_{n0}\exp(\mathrm{i}\alpha_n) \tag{8}$$

with similar expressions for $u_n(x,t)$ and $F_n(x,t)$.

As an example let us assume that the spring is started from rest at $t=0$ for a displacement $\xi(x,0)=A\sin(\pi x/L)$. Since the initial velocity is zero, the phase angle α_1 must be zero, and the time dependence of the subsequent displacement becomes $\xi_1(x,t)=A\sin(\pi x/L)\cos(\omega_1t)$. If instead, the spring had been started with a velocity $u(x,0)=U\sin(\pi x/L)$ from zero displacement, the time dependence of the displacement becomes

$$\xi(x,t)=(U/\omega_1)\sin(\pi x/L)\sin(\omega_1t),\ \text{for}\ \alpha_1=\pi/2.$$

Both ends free. If both ends of the spring in the previous example are free rather than clamped, the complex force amplitude must be zero at the ends. The wave function for force will have the same form as the wave function for velocity in the previous case, and the normal mode frequencies will be the same as before.

An open ended (organ) pipe and a transmission line shorted at both ends belong to the type of transmission lines characterized by the spring free at both ends. A string cannot be kept under tension with the ends free,

Display 8.3.

Normal modes of a spring (string)

Normal modes

$$u_n(x,t) = \text{Re}\{u_n(x)\exp(-\text{i}\omega_n t)\} = u_{n0}(x)\cos(\omega_n t - \beta_n) \qquad (1)$$

$$u_n(x) = u_{n0}(x)\exp(\text{i}\beta_n) = \text{complex amplitude function} \qquad (2)$$

Examples of complex amplitude functions (wave functions)

Clamped–clamped spring $[u_n(0) = u_n(L) = 0]$

$$u_n(x) = U_n \sin(k_n x) \quad k_n = n\pi/L \quad \nu_n = nv/2L \quad \lambda_n = 2L/n \qquad (3)$$

$$F_n(x) = F_n \cos(k_n x) \quad U_n = U_{n0}\exp(\text{i}\beta_n) \quad F_n = -\text{i}ZU_n \qquad (4)$$

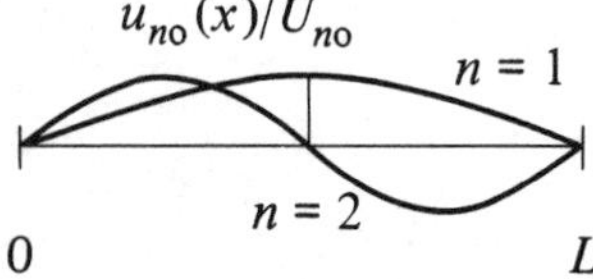

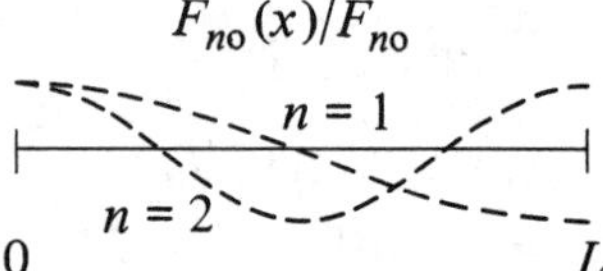

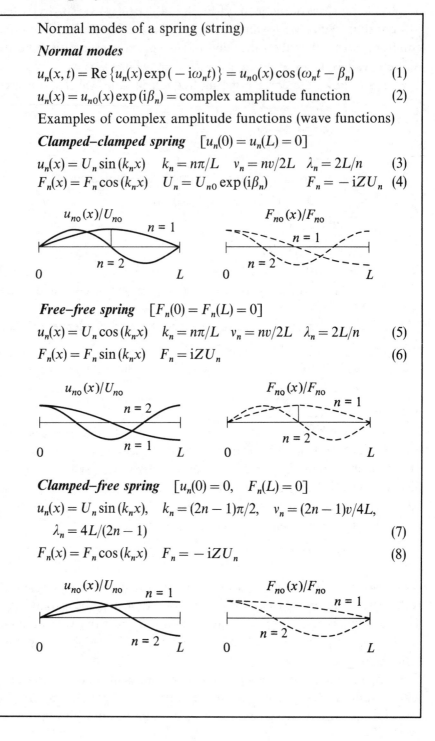

Free–free spring $[F_n(0) = F_n(L) = 0]$

$$u_n(x) = U_n \cos(k_n x) \quad k_n = n\pi/L \quad \nu_n = nv/2L \quad \lambda_n = 2L/n \qquad (5)$$

$$F_n(x) = F_n \sin(k_n x) \quad F_n = \text{i}ZU_n \qquad (6)$$

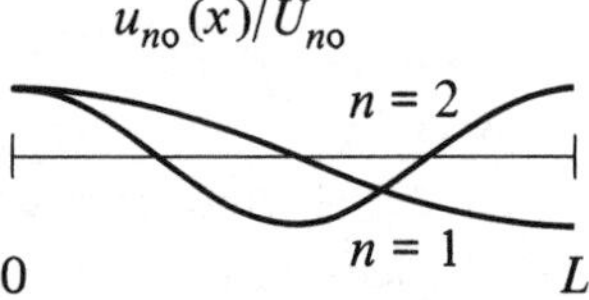

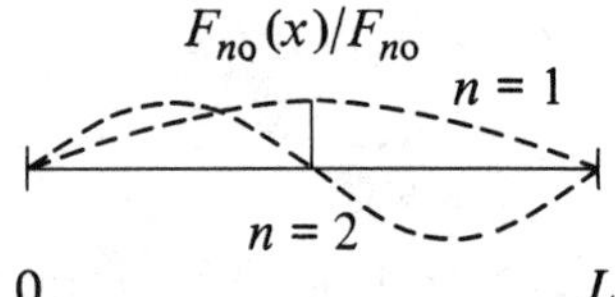

Clamped–free spring $[u_n(0) = 0, \quad F_n(L) = 0]$

$$u_n(x) = U_n \sin(k_n x), \quad k_n = (2n-1)\pi/2, \quad \nu_n = (2n-1)v/4L,$$
$$\lambda_n = 4L/(2n-1) \qquad (7)$$

$$F_n(x) = F_n \cos(k_n x) \quad F_n = -\text{i}ZU_n \qquad (8)$$

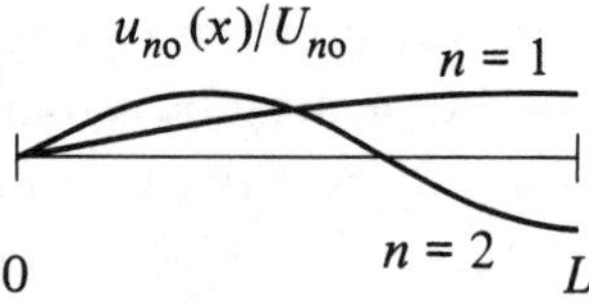

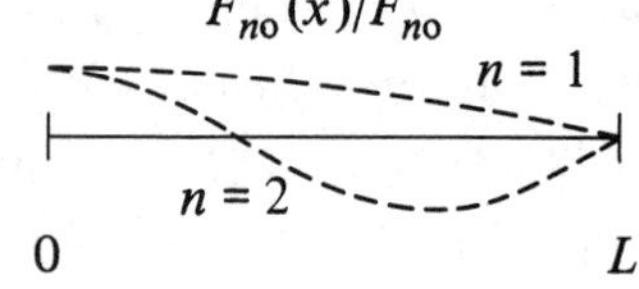

but this condition can be approximated if the ends are connected to long (lossy) strings under tension, which have much smaller mass per unit length than the string under consideration.

One end clamped, the other free. The spring under consideration is clamped at $x = 0$ and the free end is at $x = L$. As before, the general expressions for the complex amplitudes of force and velocity are

$$F(x) = A\exp(ikx) + B\exp(-ikx)$$
$$u(x) = (1/Z)[A\exp(ikx) - B\exp(-ikx)]$$

From the condition $u(0) = 0$ follows $A = B$, and since $F(L) = 0$, we get $2A\cos(kL) = 0$. With A different from zero, this condition can be fulfilled only for discrete values k_n of k given by $k_nL = (2n - 1)\pi/2$, where n is an integer. This requirement identifies the normal mode frequencies $\omega_n = vk_n$ and the wavelengths λ_n of the system,

$$\nu_n = \omega_n/2\pi = (2n - 1)v/4L \tag{9}$$
$$\lambda_n = 2\pi/k_n = 4L/(2n - 1) \tag{10}$$

The complex amplitude functions for the normal modes are

$$u_n(x) = U_n\sin[(2n - 1)\pi x/2L] \tag{11}$$
$$F_n(x) = F_n\cos[(2n - 1)\pi x/2L] \tag{12}$$

where $F_n = -i\omega_n ZU_n$ and $Z = \mu v$.

At the fundamental frequency, corresponding to $n = 1$, one quarter wavelength equals the length of the spring. The normal mode frequencies are the same as the resonance frequencies obtained when the spring was driven by a harmonic force with constant amplitude. For comparison we note that the normal mode frequencies for the clamped–clamped spring are the same as the resonance frequencies of the spring clamped at one end and driven at the other with a harmonic displacement of constant amplitude.

The normal mode frequencies and wave functions apply to the oscillations of the air column in a pipe closed at one end and open at the other and to EM waves on a cable, shortcircuited at one end and open at the other.

For a summary we refer to Display 8.3.

The real mass–spring oscillator. We now add a mass M to the end of the spring at $x = L$ in the previous example. As before, the clamped end is at $x = 0$, and since the velocity must be zero at this end, the expressions for the complex amplitudes of force and velocity must be $F(x) = 2A\cos(kx)$ and $u(x) = i(2A/Z)\sin(kx)$, as in the previous example.

The boundary condition at $x = L$ is now expressed by the equation of

Display 8.4.

Normal mode frequencies of a real mass-spring oscillator.

Frequency equation

$(kL)\tan(kL) = m/M, \quad m = \mu L, \quad k = \omega/v$

$$v = 1/\sqrt{\mu\kappa} = L\sqrt{K/m} \tag{1}$$

$$m/M \ll 1 \quad (kL)^2 \simeq m/M \quad \omega^2 = v^2/mML^2 = K/M \quad \text{(lowest mode)} \tag{2}$$

$$M = \infty \quad \tan(kL) = 0 \quad k_nL = n\pi \tag{3}$$

$$M = 0 \quad \tan(kL) = \infty \quad k_nL = (2n-1)\pi/2 \tag{4}$$

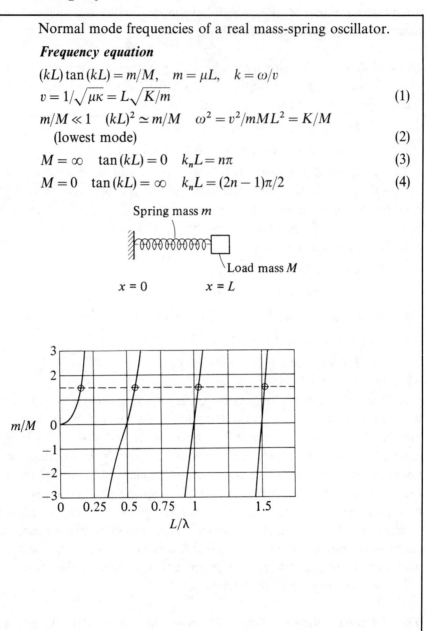

Spring mass m

Load mass M

$x = 0$ $\qquad$ $x = L$

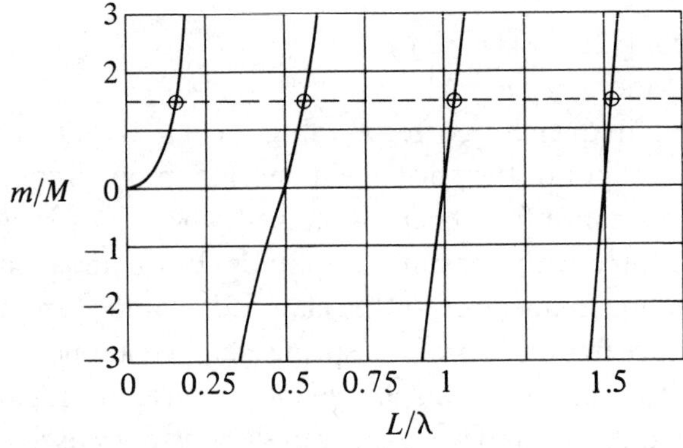

motion for M, which in complex amplitude form is $F(L) = -\mathrm{i}\omega M u(L)$. With $u(L) = \mathrm{i}(2A/Z)\sin(kL)$ and $F(L) = 2A\cos(kL)$ this condition becomes

$$\begin{aligned} 2A\cos(kL) &= -\mathrm{i}\omega M(\mathrm{i}2A/Z)\sin(kL) \\ &= 2A(\omega M/Z)\sin(kL) = 2A(M/m)(kL)\sin(kL) \end{aligned} \qquad (13)$$

where we have used $Z = \mu v$, $m = \mu L$, and $k = \omega/v$.

The equation for the determination of the normal mode frequencies then takes the form

$$(kL)\tan(kL) = m/M \qquad (14)$$

This equation can be obtained also from the analysis of the forced harmonic motion in Section 8.2 by putting the denominator equal to zero in the frequency response equation 8.2(6). In that equation we worked with the normalized frequency $\Omega = \omega/\sqrt{K/M}$, which is related to kL through the relation $kL = \sqrt{m/M}\,\Omega$. At present it is more instructive, however, to use the quantity kL or the corresponding quantity L/λ as the frequency variable.

If the spring mass m is much smaller than the attached mass M, $m/M \ll 1$, kL will be small also, and $\tan(kL)$ can be approximated by kL. The frequency equation (14) then reduces to $(kL)^2 = m/M$ with the corresponding frequency given by $\omega^2 = v^2 m/ML^2 = K/M$, as it should.

If M is infinite, the attached mass acts like a rigid termination so that the spring in effect will be clamped also at $x = L$, the frequency equation yields $\tan(kL) = 0$. The corresponding mode frequencies are given by $kL = n\pi$, i.e. the same as obtained for the clamped–clamped spring. In the other limit, with $M = 0$, the frequency equation becomes $\tan(kL) = \infty$, with the frequencies given by $kL = (2n-1)\pi/2$, the same as obtained earlier for the clamped–free spring.

For an arbitrary value of m/M, the frequency equation has to be solved numerically or graphically. In Display 8.4, the solutions are obtained from the intersection of the function $(kL)\tan(kL)$ and the constant m/M plotted versus L/λ. In particular, we have illustrated the case when $m/M = 1.5$. We note that with increasing frequency, the normal mode frequencies approach the values for the clamped–clamped spring, $\lambda_n \simeq 2L/n$. This is to be expected, since the velocity of M decreases with increasing frequency because of inertia.

8.4 Normal mode expansion. Fourier series

In the last section we found that the time dependence of a field variable in a normal mode is of the form

$$\xi_n(x,t) = \xi_{n0}(x)\cos(\omega_n t - \alpha_n) = \mathrm{Re}\{\xi_n(x)\exp(-\mathrm{i}\omega_n t)\}$$

It is harmonic with all elements in the system oscillating with the frequency of the mode and with amplitudes determined by the amplitude function (wave function) $\xi_{n0}(x)$. The general motion of the system is a superposition of the normal modes of oscillation, and we are now interested in the decomposition of an arbitrary displacement into these modes.

Two equal coupled oscillators. It is instructive to review the decomposition problem of the coupled oscillators discussed in Chapter 5.

As we recall, the complex amplitude function for the two modes of the system, $\xi_n(x)$, is zero everywhere except at the locations x_1 and x_2 of the two mass elements. For the first mode we have $\xi_1(x_1)=\xi_1(x_2)=A_1$ and $\xi_2(x_1)=A_2$, $\xi_2(x_2)=-A_2$, where A_1 and A_2 are the amplitudes of the two mass elements, the 'spikes' in the wave functions at x_1 and x_2.

The amplitude functions have the important property of being orthogonal, which means that $\xi_1(x_1)\xi_2(x_1)+\xi_1(x_2)\xi_2(x_2)=0$. As indicated in Chapter 5, the sum of these products can be interpreted as the scalar product of two vectors with the components $\xi_1(x_1)$, $\xi_1(x_2)$ and $\xi_2(x_1)$, $\xi_2(x_2)$. Since the scalar product is zero the vectors are said to be orthogonal.

The scalar product can be expressed as $\sum \xi_1(x_i)\xi_2(x_i)$ where x_i represent the locations where the functions are different from zero, i.e. the equilibrium positions x_1 and x_2 of the mass elements of the two oscillators.

The orthogonality condition then takes the form

$$\sum_i \xi_1(x_i)\xi_2(x_i)=0 \quad i=1,2 \tag{1}$$

The only contributions to the sum in this example of two coupled oscillators come from x_1 and x_2, the contribution being $\xi_1(x_1)\xi_2(x_1)=A_1A_2$ and $\xi_1(x_2)\xi_2(x_2)=-A_1A_2$, adding up to zero.

Similarly, we get

$$\sum_i \xi_1^2(x_i)=2A_1^2 \quad \text{and} \quad \sum_i \xi_2^2(x_i)=2A_2^2 \tag{2}$$

As far as the 'shape' of a normal mode amplitude function is concerned, the magnitude of the amplitude A_1 or A_2 is not important, and it is convenient to normalize the functions by choosing $A_1=A_2=1/\sqrt{2}$ so that the sums in Eq. (2) become unity. We shall denote the normalized functions by $\Psi_1(x)$ and $\Psi_2(x)$. They have the properties

$$\sum \Psi_1(x_i)\Psi_2(x_i)=0 \tag{3}$$

$$\sum \Psi_1^2(x_i)=\sum \Psi_2^2(x_i)=1 \quad (\text{Sum over } i=1,2) \tag{4}$$

The decomposition of an arbitrary displacement of the two oscillators in terms of the normal mode functions was discussed from different points of view in Chapter 5. We now choose the procedure, which readily can be generalized to an arbitrary number of coupled oscillators and to the continuous transmission lines considered in this section. Thus, the decomposition of an arbitrary displacement $\xi(x)$ into the normal modes is expressed as

$$\xi(x_i) = a_1\Psi_1(x_i) + a_2\Psi_2(x_i) = \sum_n a_n\Psi_n(x_i) \quad i = 1, 2 \tag{5}$$

where the problem is to determine the expansion coefficients a_1 and a_2.

We recall that the projection of a vector **A** on a unit vector **U** is the scalar product $\mathbf{A}\cdot\mathbf{U} = A_1U_1 + A_2U_2$. The coefficients a_1 and a_2 are obtained in similar manner. Thus, to obtain the projection a_1 of the 'displacement vector' ξ, with the components $\xi(x_i)$, on the 'unit vector' Ψ_1, with the components $\Psi_1(x_i)$, we multiply both sides of Eq. (5) by $\Psi_1(x_i)$ and sum over i to obtain what corresponds to the sum of products of the vector components in a scalar product. Since the wave functions Ψ_n are orthogonal and normalized, only a_1 will remain on the right hand side of Eq. (6). Similarly, by multiplication by $\Psi_2(x_i)$ and summation only a_2 will remain, so that

$$a_1 = \sum \xi(x_i)\Psi_1(x_i) \tag{6}$$

$$a_2 = \sum \xi(x_i)\Psi_2(x_i) \quad \text{(Sum over } i = 1, 2) \tag{7}$$

As an example consider the case in Chapter 5, in which the oscillators were started from the initial displacements $\xi(x_1) = 1$, $\xi(x_2) = 0$. With $\Psi_1(x_1) = \Psi_1(x_2) = 1/\sqrt{2}$ and $\Psi_2(x_1) = -\Psi_2(x_2) = 1/\sqrt{2}$ we obtain $a_1 = a_2 = 1/\sqrt{2}$. The subsequent motion of the system is then

$$\begin{aligned}\xi(x_1, t) &= a_1\Psi_1(x_1)\cos(\omega_1 t) + a_2\Psi_2(x_1)\cos(\omega_2 t)\\ &= 1/2[\cos(\omega_1 t) + \cos(\omega_2 t)]\\ \xi(x_2, t) &= a_1\Psi_1(x_2)\cos(\omega_1 t) + a_2\Psi_2(x_2)\cos(\omega_2 t)\\ &= 1/2[\cos(\omega_1 t) - \cos(\omega_2 t)]\end{aligned}$$

which is consistent with the result obtained in Chapter 5.

Spring clamped at both ends. We found in the last section that the amplitude function for the nth mode of a spring or string clamped at both ends is $\xi_n(x) = A\sin(k_n x)$, where $k_n = n\pi/L$. The first mode ($n = 1$) corresponds to the symmetrical mode of the two coupled oscillators, in which the elements of the systems are in phase, i.e. move in the same direction for all values of t. Similarly, the second mode of the string ($n = 2$) corresponds to the asymmetrical mode of the two oscillators, in which the two halves of the system move in opposition.

The modes of the spring are orthogonal just as for the two coupled oscillators. For the latter the orthogonality condition was expressed in terms of the scalar product type sum involving the components of the wave functions evaluated at the locations of the mass elements. For a continuous system there is an infinite number of mass elements, and the scalar product is now expressed as an integral of the product of the wave functions.

Let us determine the integral of the product $\xi_n(x)\xi_m(x)$, where n and m are two different mode numbers. With the limits of integration 0 and L, we have

$$\int \sin(m\pi x/L)\sin(n\pi x/L)\,\mathrm{d}x$$

$$= \tfrac{1}{2}\int [\cos((n-m)\pi x/L) - \cos((n+m)\pi x/L)]\,\mathrm{d}x = 0$$

where we have used $\sin(a)\sin(b) = \frac{1}{2}[\cos(ab) - \cos(a+b)]$.

If $n = m$ then it follows that $\int \sin^2(n\pi x/L)\,\mathrm{d}x = L/2$, which means that the normalized nth wave function is

$$\Psi_n(x) = \sqrt{2/L}\,\sin(n\pi x/L) \tag{8}$$

Consequently, these functions satisfy the orthogonality and normalization conditions

$$\int \Psi_m(x)\Psi_n(x)\,\mathrm{d}x = 0 \quad m \neq n \tag{9}$$

$$\int \Psi_n^2(x)\,\mathrm{d}x = 1 \quad n = m \tag{10}$$

where the limits of integration are 0 and L.

The decomposition of an arbitrary displacement $\xi(x)$ of the spring or string in terms of the normal modes is completely analogous to the decomposition for the two coupled oscillators. Thus, we have

$$\xi(x) = \sum a_n \Psi_n(x) \tag{11}$$

To determine the coefficients a_n, we proceed as for the coupled oscillators. Consequently to find the coefficient a_m, we multiply both sides of the equation by $\Psi_m(x)$ and integrate over the range 0 to L. Since the normal mode functions are orthogonal, only one term on the right hand side will be different from zero, namely the term $n = m$. Actually, since the functions are also normalized, the integral of the squared wave function is 1, so that the right hand side is reduced to a_m. The left hand side is $\int \xi(x)\Psi_m(x)\,\mathrm{d}x$, which

Display 8.5.

Normal mode functions for coupled oscillators and for a string clamped at both ends.

Two equal coupled oscillators

Normal mode amplitude functions (normalized)

$$\Psi_1(x_1) = \Psi_2(x_2) = 1/\sqrt{2} \tag{1}$$

$$\Psi_2(x_1) = -\Psi_2(x_2) = 1/\sqrt{2} \tag{2}$$

Orthogonality

$$\Psi_1(x_1)\Psi_2(x_1) + \Psi_1(x_2)\Psi_2(x_2) = 0 \tag{3}$$

$$\sum \Psi_1(x_i)\Psi_2(x_i) = 0 \quad i = 1, 2 \tag{4}$$

Normalizations

$$\sum \Psi_1^2(x_i) = \sum \Psi_1^2(x_i) = 1 \tag{5}$$

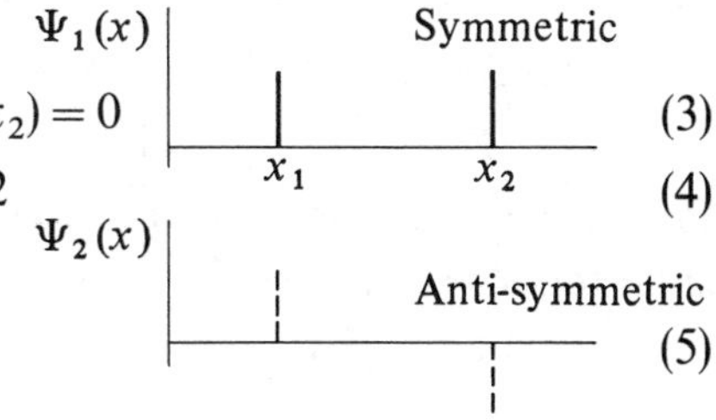

Normal mode expansion

$$\xi(x_i) = a_1\Psi_1(x_i) + a_2\Psi_2(x_i) = \sum_n a_n\Psi_n(x_i) \tag{6}$$

$$a_n = \sum_i \xi(x_i)\Psi_n(x_i) \quad n = 1, 2 \tag{7}$$

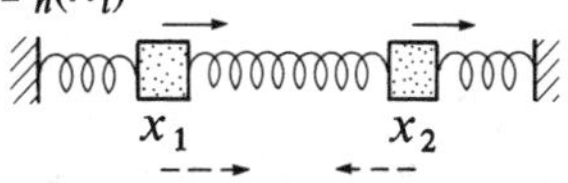

String clamped at both ends

Normal mode amplitude functions (normalized)

$$\Psi_n(x) = \sqrt{2/L}\sin(n\pi x/L) \tag{8}$$

Orthogonality

$$\int \Psi_m(x)\Psi_n(x)\,\mathrm{d}x = 0 \quad m \neq n \tag{9}$$

Normalization condition

$$\int \Psi_n^2(x)\,\mathrm{d}x = 1 \tag{10}$$

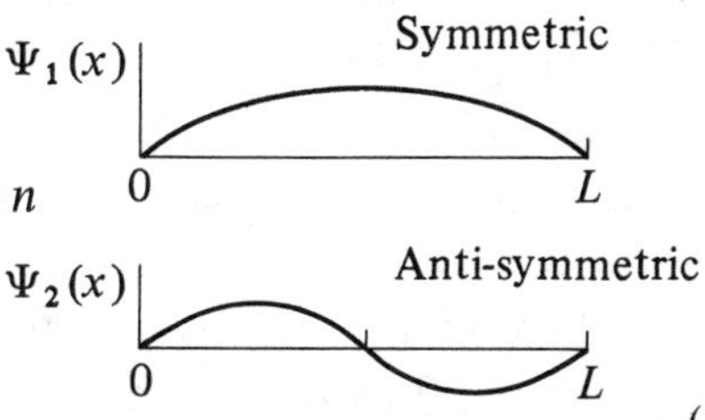

Normal mode expansion

$$\eta(x) = \sum a_n\Psi_n(x) \tag{11}$$

$$a_n = \int \eta(x)\Psi_n(x)\,\mathrm{d}x \tag{12}$$

means that

$$a_m = \int \xi(x)\Psi_m(x)\,dx = \sqrt{2/L}\int \xi(x)\sin(m\pi x/L)\,dx \qquad (12)$$

where the limits of integration are 0 and L.

For a summary we refer to Display 8.5.

Example. As an illustration of the decomposition of a displacement into normal modes, we shall consider a string, which at $t = 0$ is started from the displacement shown in Display 8.6 and described by the function

$$\eta(x) = 2Ax/L \qquad 0 < x < L/2 \qquad (13)$$

$$\eta(x) = 2A[1-(x/L)] \quad L/2 < x < L \qquad (14)$$

The maximum displacement is A at $x = L/2$. The expansion into the normal modes is expressed as

$$\eta(x) = \sum a_n\Psi_n(x) \qquad (15)$$

$$a_n = \int_0^L \eta(x)\Psi_n(x)\,dx \quad \text{where} \quad \Psi_n(x) = \sqrt{2/L}\sin(n\pi x/L) \qquad (16)$$

Although the normalized wave functions are convenient in general discussions, they offer no advantage in numerical calculations. Rather, they are somewhat inconvenient to use, since the normalization factor $\sqrt{2/L}$ has to be carried along. Nevertheless we use these functions in this example but suggest that the reader repeat the analysis using the expansion $\eta(x) = \sum A_n \sin(n\pi x/L)$.

The displacement $\eta(x)$ is symmetrical with respect to the center of the string, and only the coefficients which refer to the wave functions with the same symmetry will be different from zero, i.e. the functions with odd values of n. For these values the integral from 0 to L can be replaced by twice the integral from 0 to $L/2$, so that

$$a_n = 2\sqrt{2/L}(2A/L)\int_0^{L/2} x\sin(n\pi x/L) \quad n = 1, 3, 5, \ldots \qquad (17)$$

Through 'integration by parts' we obtain

$$a_n = A(8/\pi^2)(1/n^2)(-1)^{(n-1)/2} \quad n = 1, 3, 5\ldots \qquad (18)$$

so that Eq. (15) becomes

$$\eta(x) = A(8/\pi^2)[\sin(\pi x/L) - (1/9)\sin(3\pi x/L) + (1/25)\sin(5\pi x/L) - \cdots] \qquad (19)$$

(Consequence At $x = L/2$, $\eta(L/2) = A$, i.e.
$\pi^2 = 8[1 + (1/9) + (1/25) + \cdots]$)

Display 8.6.

Example of normal mode expansion of the displacement of a string.

Displacement function

$$\eta(x) = 2Ax/L \qquad 0 < x < L/2$$
$$\eta(x) = 2A[1-(x/L)] \quad L/2 < x < L \tag{1}$$

Normal mode expansion (Fourier series):

$$\eta(x) = \sum a_n \Psi_n(x) \quad \Psi_n(x) = \sqrt{2/L}\sin(n\pi x/L) \tag{2}$$

$$a_n = \int_0^L \eta(x)\,\Psi_n(x)\,\mathrm{d}x = (8/\pi^2)(1/n^2)(-1)^{(n-1)/2} \quad n = 1, 3, 5\ldots \tag{3}$$

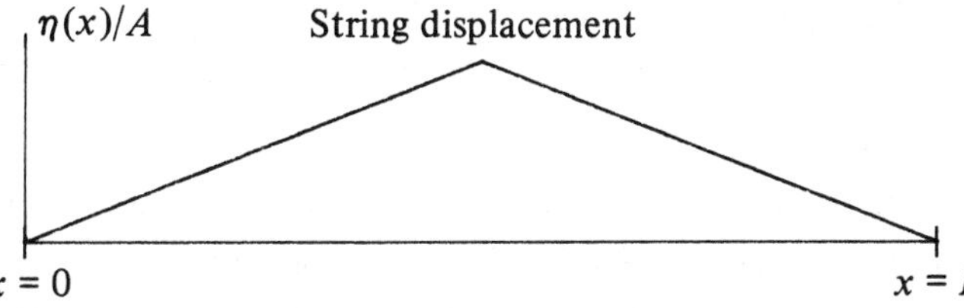

Sum of two terms in the normal mode expansion

Sum of three terms

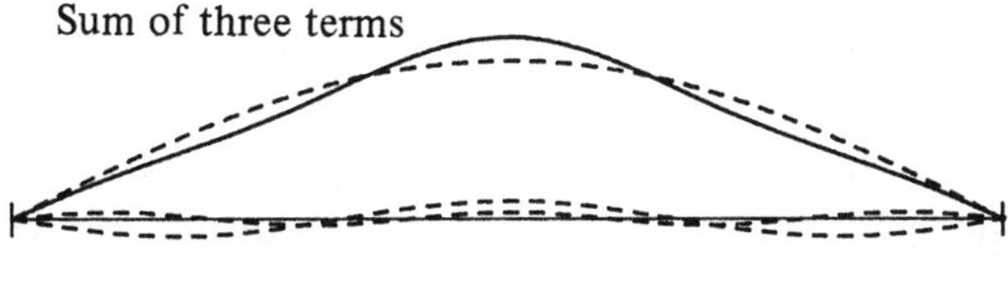

Sum of ten terms

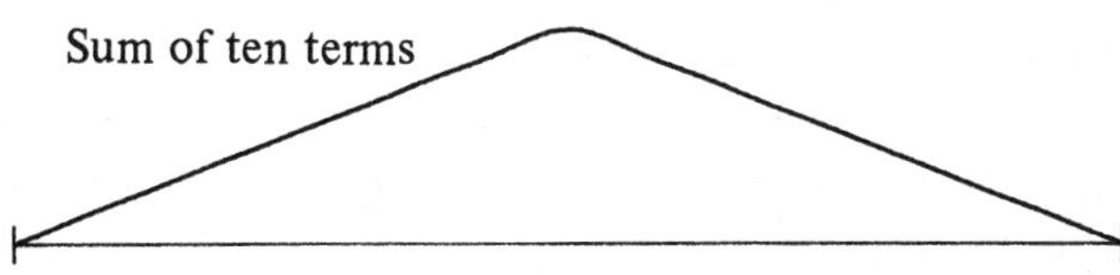

The result of some computations are shown in the display. The first graph is the actual original displacement of the string. The second and third graphs show the sums of the first two and three terms, respectively, in the normal mode expansion in Eq. (19). Finally, in the fourth graph is shown the sum of the first ten terms. It is apparent that the actual displacement is approached as the number of terms is increased. As in all problems of this kind, rapid changes in the function or its derivatives require a large number of terms in the expansion to be reproduced closely by the sum.

Each individual mode of the string will have harmonic time dependence with a frequency equal to the mode frequency $\nu_n = nv/2L$, where v is the wave speed. The time dependence of the total displacement is the sum of the contributions from the various modes

$$\eta(x, t) = \sum a_n \sin(n\pi x/L)\cos(n2\pi t/T) \tag{20}$$

$$a_n = A(8/\pi^2)(1/n^2)(-1)^{(n-1)/2} \quad n = 1, 3, 5, \ldots$$

$$T = 2L/v$$

In Display 8.7 we have shown computed 'snapshots' of the string at time intervals of $T/10$, where T is the fundamental period. It is quite interesting to find that the motion starts in the center portion of the string, and the motion spreads towards the ends. As a result, a flat portion of the string is produced, and the length of this portion increases with time.

The motion is illustrated further in the next graph, which shows the time dependence over one period of different points on the string. It should be noted that the midpoint of the string moves back and forth with constant speed. For comparison, we have shown also the time dependence of the midpoint in the fundamental mode of motion.

This result can be obtained without any mathematical analysis by a simple superposition of waves, as will be demonstrated below.

Superposition of waves. When we release a long string from an initial displacement $D(x)$ over a finite portion of a string, this displacement gives rise to one wave $\frac{1}{2}D(x - vt)$ travelling in the positive x-direction and one wave $\frac{1}{2}D(x + vt)$ travelling in the negative x-direction, as discussed in Chapter 6. At $t = 0$ the waves add up to the initial displacement $D(x)$. If there are boundaries present, these waves will be reflected and will interfere with each other as they 'collide'. The total displacement is obtained through linear superposition of the two waves.

We now apply this procedure to the string problem in Display 8.7. As the string is released from the initial triangular displacement $\eta(x)$, with the

Display 8.7.

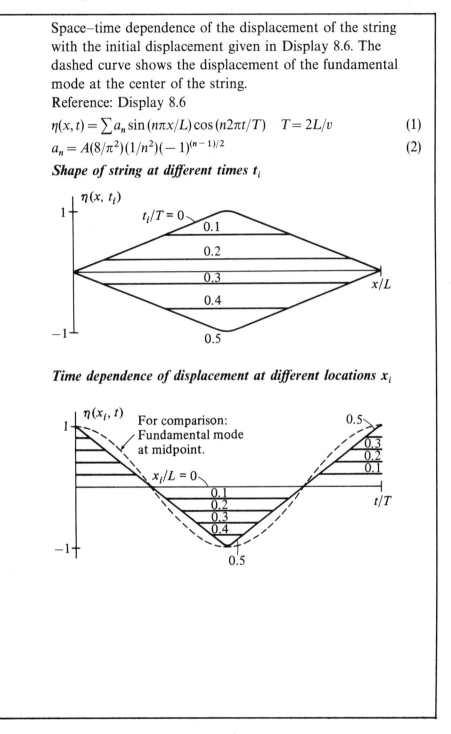

Space–time dependence of the displacement of the string with the initial displacement given in Display 8.6. The dashed curve shows the displacement of the fundamental mode at the center of the string.
Reference: Display 8.6

$$\eta(x,t) = \sum a_n \sin(n\pi x/L)\cos(n2\pi t/T) \quad T = 2L/v \tag{1}$$

$$a_n = A(8/\pi^2)(1/n^2)(-1)^{(n-1)/2} \tag{2}$$

Shape of string at different times t_i

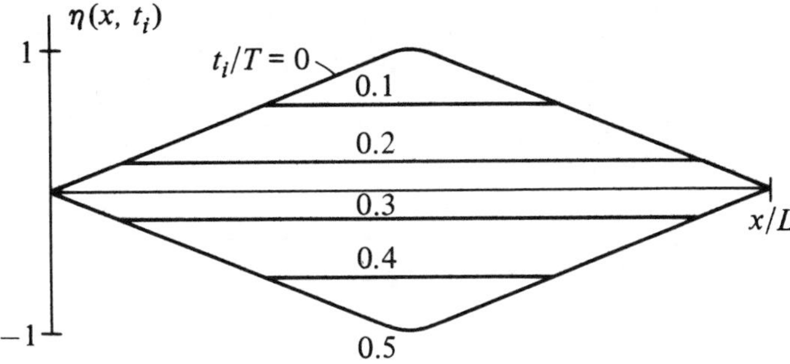

Time dependence of displacement at different locations x_i

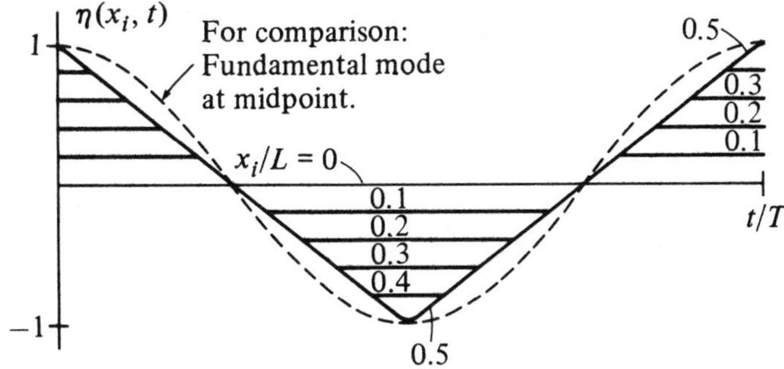

Display 8.8.

String oscillation in Display 8.7 explained as a superposition of elementary travelling waves A and B. A_r, B_r = reflected waves.

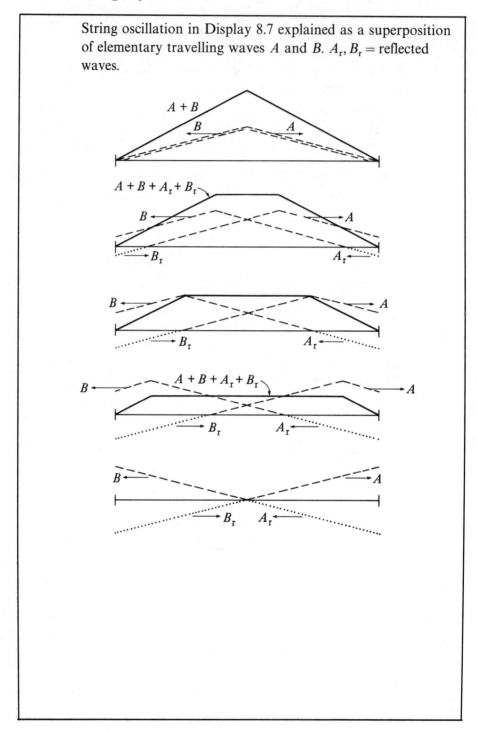

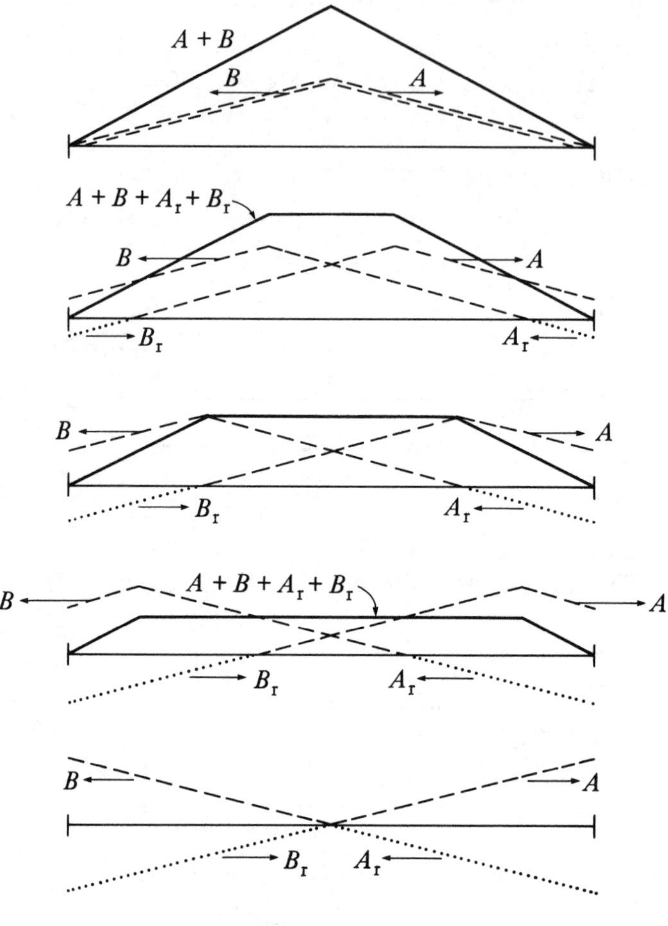

value A at the center of the string, a wave $\frac{1}{2}\eta(x - vt)$ will travel to the right, and since the leading end of it is already in contact with the boundary $x = L$ at $t = 0$, a reflected wave occurs from the start. Similarly, the wave $\frac{1}{2}\eta(x + vt)$ travelling to the left will be reflected at the left boundary, at $x = 0$. The waves will be reflected repeatedly as they travel back and forth between the boundaries. The motion will be periodic, however, and we need to consider only the first reflection to understand the motion.

The situation is illustrated in Display 8.8, where we have shown the superposition of the incident and reflected waves. We find that the time dependence of the resulting total displacement indeed will be of the form obtained in Display 8.7, where we obtained the total displacement through a 'brute force' addition of normal modes.

This procedure of superposition of incident and reflected waves can be used for any arbitrary initial displacement $\eta(x)$, not necessarily symmetrical, as in the present case.

8.5 Continuous harmonic force distribution

Another important aspect of wave motion involves the harmonic motion produced by a continuous source distribution. As a specific example, we shall study the forced motion of a string clamped at both ends. It can be demonstrated as illustrated schematically in Display 8.9. A metal string, clamped at both ends, passes between the pole pieces of a magnet, and a transverse force on the string is obtained through the interaction of an electric current through the string and the magnetic field.

The string runs in the x-direction with the ends at $x = 0$ and $x = L$. The magnetic field $B(x)$ is in the y-direction, and the force per unit length $f(x, t) = I(t)B(x)$ is then in the z-direction. The current $I(t)$ through the string is produced by an oscillator and power amplifier, with the terminals applied to the ends of the string. The x-dependence of the magnetic field strength is assumed known.

The presence of an external force distribution results in a source term in the first field equation (Newton's law), which now reads

$$\mu \partial u/\partial t = -\partial F/\partial x + f(x, t) \tag{1}$$

As before, F is the transverse component of the tension force (F is counted positive in the positive z-direction, and is transmitted from the left to the right across an imagined cut in the string) and μ is the mass per unit length of the string.

The second field equation

$$\kappa \partial F/\partial t = -\partial u/\partial x \quad (\kappa = 1/S) \tag{2}$$

Display 8.9.

Forced harmonic motion of string clamped at both ends.

Complex amplitude functions

Driving force per unit length $f(x) = f_n \sin(k_n x)$ (1)

Displacement:

$$\eta(x) = \sum [A_n/(1 - \Omega_n^2)] \sin(k_n x) \quad k_n = n\pi/L \quad (2)$$

$$A_n = (1/\pi n)^2 (f_n L/S)L, \quad \Omega_n = \omega/\omega_n, \quad \omega_n = 2\pi(v/2L) \quad (3)$$

Example

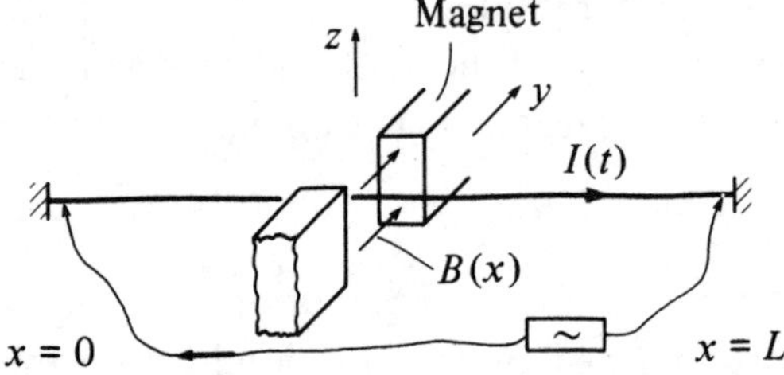

$$f(x) = f_0 = BI \quad x' < x < x' + \Delta x' \quad (4)$$

$$f(x) = 0 \quad (5)$$

elsewhere

$$A_n = (8/\pi^2)(1/n^2) A \sin(k_n x') \quad A = f_0 \Delta x'/4S \quad (6)$$

is unchanged; it basically describes the characteristics of the string, the compliance per unit length being represented by $1/S$, where S is the string tension (see Chapter 6).

The wave equation for the velocity variable is obtained in the usual manner by differentiating Eq. (1) with respect to time and Eq. (2) with respect to x. Elimination of F then gives

$$\partial^2 u/\partial x^2 - (1/v)^2 \partial^2 u/\partial t^2 = -\kappa \partial f/\partial t \qquad (3)$$
$$v^2 = 1/\mu\kappa$$

where the right hand side usually is referred to as the source term.

Since we are now interested in harmonic time dependence, we introduce the complex amplitudes $u(x)$ and $f(x)$ to obtain

$$\mathrm{d}^2 u(x)/\mathrm{d}x^2 + k^2 u(x) = \mathrm{i}\omega\kappa f(x) \quad k = \omega/v \qquad (4)$$

where ω is the angular frequency of the driving force.

If we expand the complex amplitude functions $u(x)$ and $f(x)$ in terms of the normal mode functions of the string, Eq. (4) is reduced to a harmonic oscillator equation for each mode, as follows. With

$$u(x) = \sum u_n \sin(k_n x) \qquad (5)$$
$$f(x) = \sum f_n \sin(k_n x) \quad \text{where } k_n = n\pi/L = \omega_n/v \qquad (6)$$

we obtain $\mathrm{d}^2 u/\mathrm{d}x^2 = -\sum k_n^2 u_n \sin(k_n x)$, and Eq. (4) reduces to

$$\sum (k^2 - k_n^2) u_n \sin(k_n x) = \mathrm{i}\omega\kappa \sum f_n \sin(k_n x)$$

which is satisfied for all x only if the amplitude factor u_n of the nth mode is

$$u_n = \mathrm{i}\omega\kappa f_n/(k^2 - k_n) = -\mathrm{i}\omega\kappa f_n/k_n^2(1 - \Omega_n^2) \qquad (7)$$
$$\Omega_n = \omega/\omega_n \quad \omega_n = 2\pi(nv/2L)$$

The corresponding complex displacement amplitude, $\xi_n = u_n/-\mathrm{i}\omega$, is

$$\xi_n = A_n/(1 - \Omega_n^2) \quad A_n = \kappa f_n/k_n^2 \qquad (8)$$

which has the same form as the familiar frequency response for the **single harmonic oscillator**, where $\boldsymbol{\Omega_n = \omega/\omega_n}$ is the normalized frequency in the nth mode.

By analogy with the single oscillator, we have introduced the 'static' displacement amplitude A_n of the nth mode,

$$A_n = \kappa f_n/k_n^2 = (1/\pi n)^2 (f_n L/S) L \qquad (9)$$

which is the maximum displacement of the string in the limit of zero frequency when it is driven with a harmonic force with the amplitude distribution of the nth mode, i.e. $f(x) = f_n \sin(n\pi x/L)$. It should be noted that this static displacement is inversely proportional to n^2.

This analysis illustrates the considerable importance of the single harmonic oscillator. By decomposing a displacement of a continuous system into its normal modes, the response of the system to external force can be described in terms of harmonic oscillator response functions, one for each mode.

As a specific example, let us assume that the magnetic field is constant and equal to B over a small region between x' and $x'+\Delta x'$ and zero elsewhere. The force per unit length of the string is then constant in this region, $f(x)=f_0$, and zero elsewhere.

The coefficients f_n in the normal mode expansion $f(x)=\sum f_n \sin(n\pi x/L)$ are then given by Eq. (6), i.e.

$$f_n=(2/L)\int f(x)\sin(n\pi x/L)\mathrm{d}x \simeq (2/L)(f_0\Delta x')\sin(n\pi x'/L) \qquad (10)$$

since $\sin(n\pi x/L)$ can be considered to have the constant value $\sin(n\pi x'/L)$ in the small region in the vicinity of x', the location of the magnet. (It should be remarked that, consistent with Eqs. (5) and (6), we have not used the normalized wave functions in the expansion).

Using Eq. (10) for f_n, we obtain

$$\xi(x)=\sum \xi_n \sin(nx/L)=\sum[A_n/(1-\Omega_n^2)]\sin(n\pi x/L) \qquad (11)$$

$$A_n=(1/n\pi)^2(f_nL/S)L=(8/\pi^2)(1/n^2)A\sin(n\pi x'/L) \qquad (12)$$

where $A=f_0\Delta x'/4S$ is the displacement of the center of the string, $x=L/2$, at zero frequency, $\Omega_n=0$, when the magnet is located at the center, i.e. $x=x'=L/2$. Quantity $f_0\Delta x'$ is the complex amplitude of the net force on the string. Statically, it is balanced by the two transverse components of the tension S (one on each side of the force), each of magnitude $[A/(L/2)]S$, where $A/(L/2)$ is the slope of the string.

It follows from Eq. (11) that the nth mode of the string cannot be excited by the driving force at a location x' such that $n\pi x'/L=m\pi$, where m and n are integers, i.e. at $x'=(m/n)L$. This is the location of a displacement node of the nth mode. For example, the second mode ($n=2$) cannot be excited when $x'=L/2$.

8.6 Normal modes in two and three dimensions

Rectangular membrane. As shown in Section 8.3, the normal mode function of a string of length L clamped at $x=0$ and $x=L$, was obtained by the superposition of waves travelling in opposite directions with their amplitudes and phases chosen so as to satisfy the boundary conditions of zero displacement at the ends. The resulting normal mode (standing wave) function for the nth mode was of the form $\sin(k_nx)=\sin(n\pi x/L)$, and the corresponding normal mode frequency was $\omega_n = vk_n = vn\pi/L$, where $n = 1, 2, 3, \ldots$ and v the wave speed.

The wave equation for a membrane was derived in Section 6.9 and, for harmonic time dependence of the displacement $\zeta(x, y, t) = \text{Re}\{\zeta(x, y, \omega) \times \exp(-\text{i}\omega t)\}$, the complex amplitude $\zeta(x, y, \omega)$ is satisfied by Eq. 6.9(5), i.e.

$$\partial^2\zeta/\partial x^2 + \partial^2\zeta/\partial y^2 + (\omega/v)^2\zeta = 0 \tag{1}$$

For a rectangular membrane clamped along the edges $x = 0$, $x = a$, and $y = 0$, $y = b$, the complex amplitude function $\zeta(x, y, \omega)$ which satisfies this equation is

$$\zeta_{mn} = A_{mn} \sin(k_m x) \sin(k_n y) \tag{2}$$

where $v = (\sigma/\mu)^{\frac{1}{2}}$ is the wave speed, σ = tension, μ = mass per unit area, and

$$k_m = m\pi/a, \quad k_n = n\pi/b, \quad m, n = 1, 2, 3, \ldots \tag{3}$$

These are the normal mode functions of the membrane, analogous to the normal mode function $\sin(n\pi x/L)$ for the clamped string, referred to above. The corresponding normal mode frequencies are obtained by inserting the wave function 2 into Eq. (2),

$$\omega_{mn} = v(k_m^2 + k_n^2)^{\frac{1}{2}} = v[(m\pi/a)^2 + (n\pi/b)^2]^{\frac{1}{2}} \tag{4}$$

The mnth mode is characterized by a propagation vector with a direction given by $\tan(\theta) = k_n/k_m = na/mb$, where θ is the angle between the propagation vector and the x-axis. A plane wave travelling in this particular direction and opposite direction will 'come back on itself' after reflections from the boundaries of the membrane. Furthermore, the mnth mode will have $m - 1$ nodal lines perpendicular to the x-axis and $n - 1$ nodes perpendicular to the y-axis. The normal mode can be regarded as a standing wave, and crossing a nodal line results in a change of sign of the function.

Rectangular cavity. The normal modes of a fluid in a rectangular cavity with rigid walls is of particular interest and will be discussed separately in Chapter 15. For harmonic time dependence, with $p(x, y, z, t) = \text{Re}\{p(x, y, z, \omega) \exp(-\text{i}\omega t)\}$, the three-dimensional wave equation 6.9.(6) reduces to

$$\partial^2 p/\partial x^2 + \partial^2 p/\partial y^2 + \partial^2 p/\partial z^2 + (\omega/v)^2 p = 0 \tag{5}$$
$$v = (1/\kappa\rho)^{\frac{1}{2}}$$

for the complex pressure amplitude $p = p(x, y, z, \omega)$, where κ is the compressibility and ρ the mass density of the fluid. For rigid cavity walls (where the sound pressure amplitude must be a maximum and the corresponding normal fluid velocity zero, as explained in Chapter 15), the normal mode wave functions which satisfy Eq. (5) are

$$\zeta_{lmn} = A_{lmn} \sin(k_l x) \sin(k_m y) \sin(k_n z) \tag{6}$$
$$k_l = l\pi/a, \quad k_m = m\pi/b, \quad k_n = n\pi/c, \quad l, m, n = 1, 2, 3 \ldots$$

with the walls of the cavity defined by $x=0$, $x=a$, $y=0$, $y=b$, $z=0$, $z=c$.

8.7 Density of normal modes

The number of normal modes with frequencies below a certain value or the number of modes in a given frequency range plays an important role in many areas of physics and engineering ranging from the theory of specific heat to the acoustics of concert halls and characteristics of lasers.

As an example of a one-dimensional system we consider a string of length L, clamped at $x=0$ and $x=L$. As was shown in Section 8.3, the wavelength of the nth normal mode of the string is given by $\lambda_n = 2\pi/k_n$, where $k_n = n\pi/L$ and the corresponding frequency is $\nu_n = \omega_n/2\pi = vk_n$, where v is the wave speed. It follows that the number of modes with k_n less than k is $N(k) = (L/\pi)k$, and, with $k = 2\pi\nu/v$, the corresponding number of modes with a frequency ν_n less than ν is $N(\nu) = 2L\nu/v$.

We define the density of modes (or 'states') in 'frequency space' as $n(\nu) = \mathrm{d}N(\nu)/\mathrm{d}\nu$ and obtain

$$n(\nu) = \mathrm{d}N(\nu)/\mathrm{d}\nu = (L/v)2 \quad \text{(one-dimensional)} \tag{1}$$

The corresponding density of modes in 'k-space' is $n(k) = \mathrm{d}N(k)/\mathrm{d}k = L/\pi$.

In two dimensions, we consider as an example the modes of a rectangular membrane, as discussed in the previous section. It was shown that the characteristic propagation vector $\mathbf{k}_{mn}$ of the mn mode has the components k_m, k_n and each mode can be described by a point in two-dimensional k-space, in which the axes are k_m and k_n. The spacing between adjacent k-values on the two axes are π/a and π/b, respectively, and the average 'area' in k-space occupied by one mode is π^2/ab. It follows then, that for sufficiently large value of k, the number of normal modes with k_{mn} less than k can be expressed as $N(k) = (\pi k^2/4)/(\pi^2/ab)$, where $\pi k^2/4$ is the area in k-space enclosed by the circle $k =$ constant in the quadrant between the positive axes k_m and k_n.

The corresponding number of modes with a frequency less than ν is then obtained by introducing $k = 2\pi\nu/v$ in $N(k)$ so that $N(\nu) = (A/v^2)\pi\nu^2$ and the density of modes in frequency space is then

$$n(\nu) = \mathrm{d}N(\nu)/\mathrm{d}\nu = (A/v^2)2\pi\nu \quad \text{(two-dimensional)} \tag{2}$$
$$A = ab = \text{area of membrane.}$$

The density of modes in k-space is $(1/2\pi)Ak$.

In a completely analogous manner we obtain the density of modes in the cavity in the previous section,

$$n(\nu) = (V/v^3)4\pi\nu^2 \tag{3}$$

$$V = abc = \text{cavity volume}$$

which will be discussed in detail in Chapter 15 (Display 15.7).

It should be noted that the density of modes increases with the size of the system, the length L of the string, the area A of the membrane, and the volume V of the cavity. In some engineering applications, this size effect can be of considerable importance in regard to the risk of exciting resonances and generating instabilities of oscillation of structures, as discussed in Chapter 21.

Examples

E8.1 *The Aeolian tone.* The wake behind a cylinder in fluid flow is known to be periodic under certain conditions, consisting of vortices which are shed periodically from one side to the other of the cylinder (Karman vortices, see Chapter 21). The frequency of oscillation is known from experiments to be $\nu = 0.2U/d$, where U is the flow velocity and d the diameter of the cylinder. As a result of the vortex shedding, a periodic reaction force is produced on the cylinder in the direction normal to the flow.

Consider the case when the cylinder is a string, clamped at both ends. For what tension in the string will the fundamental mode frequency equal the vortex shedding frequency?

With L being the length of the string, the fundamental mode frequency is

$$\nu_1 = v/2L$$

where $v = \sqrt{S/\mu}$ is the wave speed. Introducing the density ρ of the string material, the mass per unit length of the string can be expressed as $\mu = (\pi d^2/4)\rho$ so that $\nu_1 = (1/Ld)\sqrt{S/\pi\rho}$. Thus, the condition for resonance excitation of the string by the flow, $\nu = \nu_1$, yields

$$S = (0.04\pi)\rho U^2 L^2$$

which is independent of the string diameter.

This is the mechanism of tone generation in the Aeolian harp, and the phenomenon is often referred to as the 'Aeolian tone'.

Flow induced oscillations of this kind can become a serious problem in many branches of engineering, where the oscillations can lead to structural failure. Examples of such flow induced oscillations are found in tube heat exchangers, chimneys, bridges, etc. A closely related phenomenon was responsible for the Tacoma bridge disaster.

In tubes, chimneys, and similar structures, the bending stiffness takes the place of the string tension, as will be discussed in Chapter 21.

A familiar example of flow excitation of bending motion is the wind driven oscillation of a car radio antenna, which in a certain speed range of the car can result in large amplitudes.

Problems

8.1 *Organ pipe modes.* A coil spring of mass $m = 10$ g and length 0.5 m is free at both ends. What should be the spring constant K in N/m in order that the normal mode frequencies of the spring be the same as those of the acoustic modes in an open ended pipe of length $L = 1$ m? The speed of sound in air is 340 m/sec.

8.2 *String modes.* A string with tension $S = 40$ N, total mass $M = 0.1$ kg and length $L = 1$ m, is clamped at one end ($x = 0$) and attached to a ring at the other end ($x = L$). The ring is free to slide on a guide bar which is perpendicular to the string. Neglect friction and the mass of the ring.

(a) What is the wave speed on the string?
(b) What is the frequency of the fundamental mode?
(c) What is the x-dependence of the displacement and the velocity of the string in the fundamental mode?

8.3 *String in a magnetic field.* In a lecture demonstration, a horizontal string, clamped at one end and running over a pulley at the other end and supporting a weight of mass $M = 7$ kg, was driven in harmonic motion through the interaction of an alternating current through the wire and a magnetic field provided by a permanent magnet. The length of the wire was $L = 2.34$ m and the mass per unit length $\mu = 0.0039$ kg/m (steel, diameter 0.08 cm).

Calculate the frequency of the current which will excite at resonance the second normal mode of the string. (The measured frequency in the demonstration was 58 Hz $\pm$ 5%).

Where should the magnet be located to make the amplitude of the second mode as large as possible?

8.4 *Spring oscillator.* A 0.5 m long spring of mass $m = 0.5$ kg with one end free and the other clamped, oscillates in its fundamental longitudinal mode on a frictionless table. The frequency is 10 Hz and the amplitude of the end of the spring is 2 cm.

(a) What is the amplitude of the force on the support at the clamped end of the spring?
(b) What is the total energy of oscillation?

8.5 *Resonance of a pile.* A pile consists of a 10 m long wooden uniform cylinder. It is driven at one end by a harmonic force with an amplitude which is independent of frequency. For what frequency will the displacement amplitude at the other end of the pile be a maximum?

Young's modulus of wood, $E = 10^{11}$ dyne/cm^2. Density, $\rho = 0.74$ g/cm^3.

If instead, the pile is driven in such a way that the displacement amplitude of the driven end is independent of frequency, what then is the optimum driving frequency?

8.6 *Mode expansion and free motion of a string.* A string, clamped at the ends at $x = 0$ and $x = L$, is given an initial displacement A, which is the same over the entire string (the slopes of the strings at $x = 0$ and $x = L$ are theoretically infinite).

(a) Decompose this 'square wave' displacement into the normal mode amplitude functions of the string.
(b) What is the general expression for the time dependence of the displacement after the release of the string at $t = 0$?
(c) With reference to Section 8.4, construct graphically the shape of the string at the times $t = T/16$, $T/8$, and $T/4$ by superposition of travelling waves which are reflected from the boundaries (see the example in Display 8.8).

8.7 *Forced harmonic motion of a string.* With reference to Display 8.9, let the magnetic field extend over the entire length of the string. The electric current through the string is harmonic.

(a) For a constant value of the current amplitude, what is the frequency dependence of the displacement amplitude at the midpoint of the string?
(b) What is the frequency dependence of the transverse force amplitude on the string support at $x = 0$?

8.8 *Strings with magnetic interaction.* Two identical parallel stretched strings of mass M and length L and separation d are clamped at both ends. Each carries an electric harmonic current $I_0 \cos(\omega t)$.

(a) At what string tension is the fundamental mode excited through the interaction of the currents?
(b) Suppose that a DC-current is superimposed on the oscillatory current. What, then, is the required tension?
(c) Determine the static displacement of each string in (a). (You may like to make use of the result in Problem 8.7).

8.9 *Real mass–spring oscillator.* In a mass–spring oscillator, the mass of the spring is $m = 1.5\,\text{kg}$ and the mass of the 'load' is $M = 1\,\text{kg}$. The spring constant is $K = 150\,\text{N/m}$.

(a) Determine the frequencies of the first 4 modes of the oscillator.
(b) If the spring mass is neglected in calculating the frequency, of the fundamental mode, what will be the percentage error? (The graphical solution for the frequencies in Display 8.4 may be used).

8.10 *Forced harmonic motion.* The mass–spring oscillator in Problem 8.9 is driven by a harmonic force $F_0 \cos(\omega t)$ applied to the load mass M. In addition, there is a friction force proportional to the velocity of M.

(a) What is the complex amplitude equation for the displacement of the oscillator?

(b) Make a qualitative graph of the frequency dependence of the displacement amplitude.

8.11 *Membrane modes.* A membrane has a uniform tension σ (force per unit length), which corresponds to the tensile stress T (force per unit area) in the membrane material. The mass density of the material is ρ.

(a) What is the corresponding transverse wave speed in the membrane?
(b) If the membrane is clamped along the edges at $x = 0$, $x = a$, and $y = 0$, $y = b$, what is the corresponding frequency of the (12) mode of oscillation if $a = 10$ cm and $b = 20$ cm. The stress is $10^7\ \mathrm{N/m^2}$ and the density $7800\ \mathrm{kg/m^3}$.

8.12 *Soap film.* The surface tension of a soap film is 70 dyne/cm. (The corresponding membrane tension is twice the surface tension, see Section 18.2). If the film has a thickness of 1 micron (10^{-6} m) and is held in a square metal frame of length 10 cm, what is the frequency of the fundamental mode of oscillation of the film?

8.13 *Density of modes.* A rectangular room has the dimensions $a = 6$ m, $b = 8$ m, and $c = 3$ m. Estimate the number of acoustic modes in the room in the frequency range between 1000 and 1200 Hz. The speed of sound is 340 m/sec.

8.14 *Oscillating chain; x-dependent tension.* A chain is hanging from its upper fixed end under the influence of gravity. The connections between the links in the chain are friction free. Under these conditions, the chain can be treated approximately as a string as long as the wavelength on the chain is much larger than the distance between the links. The length of the chain is L, the mass M, and $\mu = M/L$.

(a) Consider small oscillations in a vertical plane with the displacement $\eta(x, t) = \mathrm{Re}\{U(x)\exp(-i\omega t)\}$. Show that the equation for the complex amplitude is $(SU_x)_x + \varepsilon\omega^2 U = 0$, where S is the tension and the subscript x signifies differentiation with respect to x. With the origin of x placed at the free end of the chain (and measured positive upwards), show that this equation reduces to

$$U_{xx} + (1/x)U_x + (\omega^2/g)U/x = 0 \qquad (1)$$

(b) By making the coordinate transformation $\eta = (x/L)^{\frac{1}{2}}$ show that this equation is transformed to

$$U_{\eta\eta} + (1/\eta)U_\eta + \beta^2 U = 0 \qquad (2)$$

where $\beta = 2\omega/\omega'$ and $\omega' = (g/L)^{\frac{1}{2}}$.

Discussion

In the absene of the second term, the solution to Eq. (2) is simply a combination of the harmonic functions $\cos(\beta\eta)$ and $\sin(\beta\eta)$. With the second term present, the corresponding solutions are known as Bessel's functions of the first and second kind. The function which is relevant in the present case, finite at $x = 0$, corresponds to $\cos(\beta\eta)$ and is denoted by $J_0(\beta\eta)$. Like $\cos(\beta\eta)$, this function starts out from a maximum value of unity at $\eta = 0$, but, unlike $\cos(\beta\eta)$, decays in an oscillatory manner to zero as η goes to infinity. The zero crossings of the function are known to occur at values of the argument equal to 2.405, 5.52, 8.654, 11.792,.... By imposing the boundary condition that these zeroes must fall at $x = L$, i.e. $\eta = 1$, we obtain the frequencies of the various modes of motion of the string, starting with the first mode for which $2\omega/\omega' = 2.405$ or $\omega_1 \approx 1.21\sqrt{g/L}$. This fundamental mode frequency should be compared with the frequency of oscillation of a rigid pendulum $\sqrt{3g/2} \simeq 1.23\sqrt{g/L}$.

The frequencies of the higher modes of motion are obtained in an analogous manner, and it should be noted that they are not multiples of the fundamental mode frequency. As the mode number increases, however, the separation of arguments of adjacent zeroes approaches η, just as for the harmonic functions.

Orthogonality. The modal functions associated with these different frequencies, U_n are orthogonal. This can be shown in the conventional manner by starting from the wave equation $(SU_x)_x = -\varepsilon\omega_1^2 U$. Considering two different modes with the frequencies ω_1 and ω_2 with the corresponding wave functions U_1 and U_2, we obtain

$$(SU_{1x})_x = -\varepsilon\omega_1^2 U_1$$
$$(SU_{2x})_x = -\varepsilon\omega_2^2 U_2$$

Multiplying these equation by U_2 and U_1, respectively, and subtracting one from the other yields $\varepsilon(\omega_1^2 - \omega_2^2)\int U_1 U_2 \mathrm{d}x = \int[U_1(SU_{2x})_x - U_2(SU_{1x})]\,\mathrm{d}x$. After integration by parts and realizing that $U = 0$ at $x = L$ and $S = 0$ (and U_x) at $x = 0$, we find that the right hand side vanishes so that $\int U_1 U_2 \mathrm{d}x = 0$, which is the condition for orthogonality of the two modes.

9

Electromagnetic waves

The analysis in Chapter 6 of EM wave propagation on a transmission line, such as a coaxial cable, is reexamined and interpreted in terms of Maxwell's equations. In the process, an effort is made to illucidate some aspects of wave emission from a dipole and an accelerated charge. Other characteristics of EM waves, such as dispersion and polarization, are also analyzed and illustrated by examples, including a mechanical analogue of polarization and a polarizer.

9.1 Review of units

A variety of systems of units are still in use, and when we wish to get a numerical answer to a problem, we have to keep track of various conversion factors to enable us to express numerical values in one and the same system of units. So far, this has not been much of a problem, because we have dealt with a relatively small number of physical quantities.

We shall now include electromagnetic variables to a greater extent than before, and the number of physical quantities involved is increased. The question of units then needs further attention, and we shall pause to review this question.

CGS system. In the CGS system of units, all physical quantities are expressed in terms of length, mass, and time, and the letters CGS signify the units Centimeter, Gram and Second.

The unit of **force**, **1 dyne**, is the force which produces an acceleration of 1 cm/sec^2 of a mass of 1 gram, and the **unit of energy, dyne-cm (erg)**, is the work done by 1 dyne acting over a distance of 1 cm.

The units of electrical quantities are defined in terms of the electrostatic or the magnetic forces. Thus, we have two options leading to what is generally referred to as the electrostatic and the electromagnetic CGS systems.

In the **electrostatic CGS system** of units (ese), the definition of the charge unit is based on Coulomb's law in the form $F = q_1 q_2/r^2$. With the physical dimensions of length, mass, and time denoted by M, L, and T, the physical dimension $[q_e]$ of charge (e signifies 'electrostatic') is then $[F^{\frac{1}{2}}]\mathrm{L} = \mathrm{M}^a\mathrm{L}^b\mathrm{T}^c$ where $a = 1/2$, $b = 3/2$, and $c = -1/2$. The corresponding **unit of charge, 1 ese**, is defined as the charge which repels an equal charge by the force of 1 dyne when the charge separation r is 1 cm.

The unit of **electric current** follows then from the fact that current is the charge transport per second, and similarly, the unit of **electric field** E_e follows from the definition of electric field, $F = E_e q_e$. Similarly, the **magnetic induction** B_e is defined from the relation $F = I_e B_e$ as the force per unit current I_e.

In the **electromagnetic CGS system** of units (eme), the electrical quantities are expressed in terms of mechanical, through the force between two current elements, $\mathrm{d}F = (I_1\,\mathrm{d}s_1)(I_2\,\mathrm{d}s_2)/r^2$, where we have assumed the elements to be parallel. Quantity I is the current and $\mathrm{d}s$ the length of the current element. The unit of current is the current which repels an equal current by the force of 1 dyne when $\mathrm{d}s_1 = \mathrm{d}s_2 = r = 1$ cm. The dimension of current will be $[I_m] = [F^{\frac{1}{2}}]$.

The unit of charge is the charge transported by unit current in one second, and the dimension is $[q_m] = [F^{\frac{1}{2}}]/T$. It follows that

$$[q_e]/[q_m] = L/T$$

has the dimension of velocity. Experimentally, the ratio q_e/q_m is found to be 3×10^{10} cm/sec and it is known to be the speed of light. This means that if one and the same charge is expressed in esu and emu, the number of esus will be 3×10^{10} times larger than the number of emu s. The electromagnetic unit of charge, therefore, must be 3×10^{10} larger than the electrostatic unit, 1 eme $= 3 \times 10^{10}$ esu.

The units of the electric and magnetic fields are obtained, as before, from the relations $F = q_m E_m$ and $F = I_m B_m$. The unit of the magnetic field thus defined is 1 Gauss. ($B = 1$ Gauss when $F = 1$ dyne and $I = 1$ eme).

The '**Gaussian**' version of the CGS system is a mixture of the electrostatic and electromagnetic systems, in which the magnetic quantities are expressed in emu and the others in esu. The force on a charged particle moving in an electric and magnetic field is given by $\mathbf{F} = q_e(\mathbf{E} + \mathbf{u} \times \mathbf{B}_e) = q_m(\mathbf{E}_m + \mathbf{u} \times \mathbf{B}_m)$. If we used mixed units this force can be expressed as

$$\mathbf{F} = q_e[\mathbf{E}_e + (\mathbf{u}/c) \times \mathbf{B}_m] \qquad (1)$$

where we have introduced the speed of light, $c = q_e/q_m$.

It is important to note that in the Gaussian system of units the dimensions of magnetic induction and the electric field are the same, $[B_m] = [E_e]$. In physics, particularly in relativity, this is conceptually appealing, since a B-field in one frame of reference will generate an E-field in another frame of reference in relative motion. Actually, E and B can be regarded as components of one and the same field vector.

As we recall, the source of the E-field is the total electrical charge, including free and induced, and the source of B is the total current, the sum of the 'free' (conduction and convection) currents and the internal (molecular) currents. The induced charges and internal currents usually can be considered to be proportional to E and B, respectively. The constants of proportionality are material constants, which generally have to be determined experimentally. They are contained in the permittivity and permeability, mentioned below.

The fields produced by the free charges and the free currents are the displacement D and the magnetic field H, respectively. In vacuum, where there are no induced charges and currents, we have $E = D$ and $B = H$ in the Gaussian system. For a point charge q we have $D = q/r^2$ and for a current element $dH = I\,ds/r^2$ at a location on the normal to the current element. In matter, $E = (1/\chi_e)D$ and $B = \chi_m H$, where χ_e is the electrical permittivity and χ_m the magnetic permeability.

The MKSA system. In the MKSA system, or MKS system, as it is sometimes called, the letters stand for **M**eter, **K**ilogram, **S**econd, and **A**mpere, as the units of length, mass, time, and electric current. Unlike the CGS system, the MKSA system includes as basic a fourth physical quantity, namely electric current (or equivalently, electric charge) with corresponding unit Ampere (Coulomb).

In Display 9.1, some of the definitions, dimensions, and units of some of the most commonly used physical quantities are summarized.

As in the Gaussian or magnetic CGS system, the basis for the definition of the unit of current is the force between two equal parallel current elements, i.e. $dF = \text{const.}\cdot(I_1\,ds_1)(I_2\,ds_2)/r^2$. The constant of proportionality, denoted by $\mu_0/4\pi$, is now chosen to be 10^{-7}, and with this choice, the unit of electric current, the Ampere, can be considered to be current $I_1 = I_2$, which gives rise to an interaction force of 10^{-7} N/m, when the current elements are 1 meter apart. The constant will have the dimension $[F]/[I^2]$. In principle, the value of the constant is arbitrary. It has been chosen here so that the unit of current will be consistent with earlier definitions of the Ampere, from charge deposition in electrolysis, for example.

Display 9.1.

Quantities and units in MKSA and International (SI) Systems.

Quantity	*Unit*	
Length, L	Meter, m	$= 100\,\text{cm}$
Mass, M	Kilogram, kg	$= 10^3\,\text{g}$
Time, T	Second, s	
Force, f	kgm/s^2, Newton, N	$= 10^5$ dyne
Energy, W	Nm, Joule = Wattsec	$= 10^7$ erg
Power, P	Nm/s = Watt	$= 10^7$ erg/s
E1. current, j	Ampere, A	$= 3 \times 10^9$ esu
E1. charge, q	As = Coulomb, Q	$= 3 \times 10^9$ esu
E1. field, E	N/Q = V/m	$= (10^{-5}/3)$ esu
Voltage, V	joule/Q = Volt	$= (1/300)$ esu
Magn. ind. B	Vs/m^2 = Weber/m^2	$= 10^4$ Gauss
Magn. flux, Φ	Vs = Weber	$= 10^8$ Gauss cm^2
Resistance, R	V/A = Ohm	$= (10^{-11}/9)$ esu
Capacitance, C	Q/V = Farad	$= 9 \times 10^{11}$ ese
Inductance, L	Weber/A = Henry, H	$= (10^{-11}/9)$ esu

El. Displacement D Q/m^2

Magn. field, H A/m

$\mu_0 = 4\pi \times 10^{-7}$ H/m

$4\pi\varepsilon_0 \approx 10^{-9}/9$F/m

From the unit of current, the **unit of electric charge** is defined as one Ampere-second, called one **Coulomb**.

The definitions of the electric and magnetic fields can be considered to be based on the expression for the force on a charge moving in an electromagnetic field, i.e. $\mathbf{F} = q(\mathbf{E} + \mathbf{u} \times \mathbf{B})$, which has the same form in the pure CGS systems. (In the mixed CGS system u is replaced by u/c). The fields thus defined together with the related quantities of resistance, capacitance, and inductance are summarized in the display.

It should be realized, that after the definition of current and electric charge, the constant of proportionality in Coulomb's law no longer can be chosen arbitrarily. Rather, it is fixed by the predetermined units of force, electric charge, and length and has to be obtained from direct measurements of the force produced by given charges or from the value of the speed of light through $c = \sqrt{1/\varepsilon_0\mu_0}$.

With the Coulomb law expressed as $F = (1/4\pi\varepsilon_0)q_1q_2/r^2$, the value of the constant $4\pi\varepsilon_0$ is approximately $(10^{-9}/9)$, and the unit is Farad/m. For comparison, we recall the expression for the force between the two current elements used in the definition of the unit of current, $\mathrm{d}F = (\mu_0/4\pi)(I_1\,\mathrm{d}s_1)(I_2\,\mathrm{d}s_2)/r^2$. The constants ε_0 and μ_0 are called the permittivity and permeability of vacuum, respectively, with $E = D/\varepsilon_0$ and $B = \mu_0 H$ in vacuum.

In the presence of matter, the displacement D and the magnetic field H will be unchanged, but the fields E and B will be altered by replacing ε_0 and μ_0 by $\varepsilon = \chi_\varepsilon\varepsilon_0$ and $\mu = \chi_m\mu_0$.

As we shall see, the permittivity ε_0 and the permeability μ_0 are related to the electromagnetic wave speed, $c = 1/\sqrt{\mu_0\varepsilon_0}$. Thus, by comparison with wave propagation in an elastic continuum, we note that the permeability corresponds to the density (or mass per unit length) and the permittivity to the compressibility (or compliance per unit length).

The International System of Units, SI (Système International) is the same as the MKSA system in regard to the four basic units of length, mass, time, and electric current and the units of the quantities derived from them. The essential difference is that in SI there is a special unit, candela, for the luminosity of a light source.

9.2 EM waves on a cable reconsidered

Our discussion of EM wave propagation on a coaxial cable in Chapter 6 was based on the field equations of a periodic lattice of inductances and capacitances (and the corresponding mass–spring lattice). We found that the field equations for voltage $V(x, t)$ and the current $I(x, t)$ on the cable are

$$L_0 \partial I/\partial t = -\partial V/\partial x \quad (1)$$

$$C_0 \partial V/\partial t = -\partial I/\partial x \quad (2)$$

where L_0 is the inductance and C_0 the capacitance per unit length of the cable.

The current in the cable produces a magnetic field in the annular region between the inner and outer conductor of the cable. The magnetic field is in the circumferential direction. It is proportional to the current I and varies inversely as the distance r from the axis of the cable, as obtained from Ampere's law. The magnetic field outside the cable is zero (why?).

The electric field is radial and is proportional to the voltage V, and, like the magnetic field, varies inversely as the distance r from the axis.

The electric and magnetic field energy per unit length readily can be calculated. By definition of the inductance L_0 per unit length, the magnetic field energy in the annulus is equal to $\frac{1}{2}L_0 I^2$ and, similarly, the electric field energy is $\frac{1}{2}C_0 V^2$. From the calculated field energies, L_0 and C_0 can be found. We assume that the annular region is filled with air so that the permeability and permittivity can be set equal to μ_0 and ε_0.

In particular, we consider the case when the separation d between the inner and outer conductors is small compared to the radius b of the inner conductor. In that case the field between the conductors will be approximately the same as between two parallel plates with a separation d, as indicated in Display 9.2.

The electric field between the plates is then simply $E = V/d$, and the magnetic field between two plane parallel conductors is $H = I_d$, where I_d is the current per unit length of the conductor (perpendicular to the current). This result follows directly from Ampere's law. (The contribution from each conductor is $I_d/2$; the contributions add between the conductors and cancel each other outside). The perimeter of one of the conductors in the coaxial cable is $l \simeq 2\pi a \simeq 2\pi b$, and the corresponding current per unit length of the perimeter is $I_d = I/l$, where I is the total current.

The electric field energy per unit length along the axis of the cable is $\frac{1}{2}\varepsilon_0 E^2 (ld) = \frac{1}{2}\varepsilon_0 (V/d)^2 (ld)$. It can be expressed also as $\frac{1}{2}C_0 V^2$, and it follows that the capacitance per unit length of the cable in the parallel plate approximation is $C_0 = \varepsilon_0/d$.

Similarly, the magnetic field energy per unit length along the axis is

Display 9.2.

Field equations for E and H, coaxial cable.

$$L_0 \partial I/\partial t = -\partial V/\partial x \tag{1}$$

$$C_0 \partial V/\partial t = -\partial I/\partial x \tag{2}$$

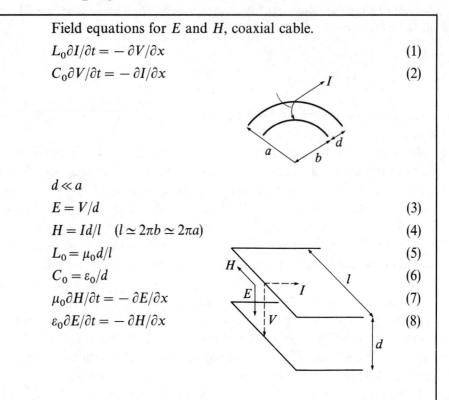

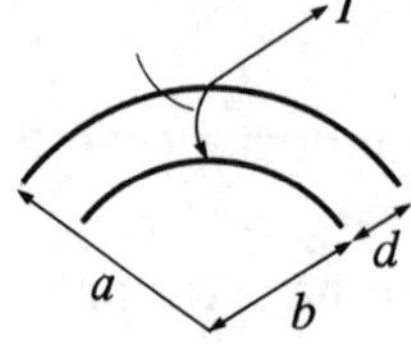

$d \ll a$

$$E = V/d \tag{3}$$

$$H = Id/l \quad (l \simeq 2\pi b \simeq 2\pi a) \tag{4}$$

$$L_0 = \mu_0 d/l \tag{5}$$

$$C_0 = \varepsilon_0/d \tag{6}$$

$$\mu_0 \partial H/\partial t = -\partial E/\partial x \tag{7}$$

$$\varepsilon_0 \partial E/\partial t = -\partial H/\partial x \tag{8}$$

$\frac{1}{2}\mu_0 H^2(ld) = \frac{1}{2}\mu_0 (I/l)^2(ld)$. It can be expressed also as $\frac{1}{2}L_0 I^2$, and it follows that the inductance per unit length of the cable is $L_0 = \mu_0 d/l$.

We now introduce these values, $L_0 = \mu_0 d/l$ and $C_0 = \varepsilon_0/d$, together with $V = Ed$ and $I = H$, into the field equations (1) and (2) to obtain

$$\mu_0 \partial H/\partial t = -\partial E/\partial x \tag{3}$$

$$\varepsilon_0 \partial E/\partial t = -\partial H/\partial x \tag{4}$$

These equations for E and H have the same form as the 'standard' field equations for one-dimensional waves with E being analogous to the force variable F and H to the velocity variable u.

Eliminating the magnetic field H by differentiating Eq. (3) with respect to x and Eq. (4) with respect to t, we obtain the wave equation

$$\partial^2 E/\partial x^2 = (1/c^2)\partial^2 E/\partial t^2 \tag{5}$$

$$c = 1/\sqrt{\mu_0 \varepsilon_0} \tag{6}$$

The general solution to this equation, as we know from Chapter 6, is the sum of one wave travelling in the positive x-direction and one wave in the negative x-direction, each having the speed $c = 1/\sqrt{\mu_0 \varepsilon_0}$. Inserting the numerical values for μ_0 and ε_0 we obtain $c = 3 \times 10^{10}$ m/sec, the speed of light.

In the presence of matter, with a dielectric material between the conductors, for example, we have to replace μ_0 and ε_0 by $\mu = \chi_m \mu_0$ and $\varepsilon = \chi_e \varepsilon_0$, where χ_m and χ_e are the relative permeability and permittivity of the dielectric material. The wave speed is then reduced to $v = 1/\sqrt{\mu_0 \varepsilon_0}$. Typically, in a cable this wave speed is about 60% of the speed of light.

9.3 Maxwell's equations

In order to demonstrate that the field equations represent the Maxwell equations, we express E and H as components of field vectors. Thus, we introduce a right handed rectangular coordinate system x, y, z as indicated in Display 9.3, in terms of which we obtain $E = -E_z$ and $H = H_y$. The field equations 9.2.3 and 9.2.4 then can be expressed as

$$\partial E_z/\partial x = \partial B_y/\partial t \tag{1}$$

$$\partial H_y/\partial x = \partial D_z/\partial t \tag{2}$$

where we have used $B = \mu_0 H$ and $D = \varepsilon_0 E$.

In our example we have $[\text{curl}(E)]_y = \partial E_x/\partial z - \partial E_z/\partial x = -\partial E_z/\partial x$ and $[\text{curl}(H)]_z = \partial H_y/\partial x - \partial H_x/\partial y$, and each of Eqs. (1) and (2) can be considered

Display 9.3.

Coaxial cable and Maxwell's equations.

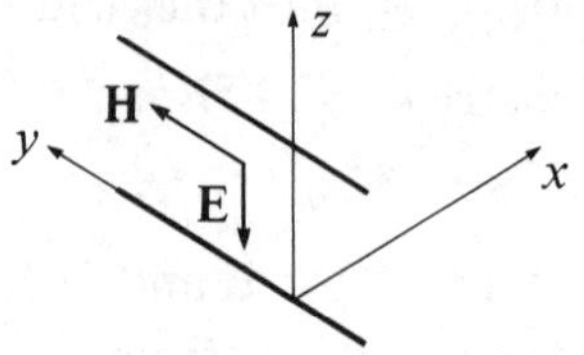

$$H_y = H \quad (1)$$

$$E_z = -E \quad (2)$$

$$\partial E_z/\partial x = \mu_0 \partial H_y/\partial t \quad (3)$$

$$\partial H_y/\partial x = \varepsilon_0 \partial E_z/\partial t \quad (4)$$

$$[\mathrm{curl}\,(E)]_y = \partial E_x/\partial z - \partial E_z/\partial x = -\partial E_z/\partial x \quad (5)$$

$$[\mathrm{curl}\,(H)]_z = \partial H_y/\partial x - \partial H_x/\partial y = \partial H_y/\partial x \quad (6)$$

***Maxwell's equations* (*vacuum, no current or charge density*)**

$$\mathrm{curl}\,(\mathbf{E}) = -\partial \mathbf{B}/\partial t \quad \text{(Eq. (3))} \quad (7)$$

$$\mathrm{curl}\,(\mathbf{H}) = \partial \mathbf{D}/\partial t \quad \text{(Eq. (4))} \quad (8)$$

$$\mathrm{div}\,(\mathbf{B}) = 0, \quad (9)$$

$$\mathrm{div}\,(\mathbf{D}) = 0 \quad (10)$$

Wave equation:

$$\nabla^2 \mathbf{E} = (1/c^2)\partial^2 \mathbf{E}/\partial t^2 \quad (11)$$

to be a component of

$$\text{curl}(\mathbf{E}) = -\partial\mathbf{B}/\partial t \tag{3}$$

$$\text{curl}(\mathbf{H}) = \partial\mathbf{D}/\partial t \tag{4}$$

which are the two Maxwell equations, which express Faraday's induction law and (a generalization) of Ampere's law (when there is no volume distribution of currents), respectively.

The equations are supplemented by the conditions

$$\text{div}(\mathbf{B}) = 0 \tag{5}$$

$$\text{div}(\mathbf{D}) = 0 \tag{6}$$

the first expressing the fact that there are no free magnetic poles, and the second indicating that in this case there is no free charge density present. If a free charge density ρ_c does exist, Eq. (6) is replaced by $\text{div}(\mathbf{D}) = \rho_c$.

Differentiating Eq. (4) with respect to t and taking the curl of Eq. (3) we can eliminate the magnetic field to obtain

$$\begin{aligned}\varepsilon_0\partial^2\mathbf{E}/\partial t^2 &= \text{curl}(\partial\mathbf{H}/\partial t) = -(1/\mu_0)\,\text{curl}[\text{curl}(\mathbf{E})]\\ &= (1/\mu_0)[\nabla^2\mathbf{E} - \text{grad}(\text{div}(\mathbf{E}))].\end{aligned}$$

With $\text{div}(\mathbf{E}) = 0$, this results in the wave equation for the E-field

$$\nabla^2\mathbf{E} = (1/c^2)\partial^2\mathbf{E}/\partial t^2 \quad c^2 = 1/\mu_0\varepsilon_0 \tag{7}$$

where $\nabla^2\mathbf{E} = (\partial^2/\partial x^2 + \partial^2/\partial y^2 + \partial^2/\partial z^2)\mathbf{E}$.

9.4 Plane EM waves

The equations 9.3.1 and 9.3.2 hold at any point in the space between the conductors. There is nothing, in principle, that prevents us from making the separation d as large as we wish, and we expect the equations to apply also to the electromagnetic field in free space. This is indeed the case, as we shall discuss further in the next section.

Wave impedance. To study some of the characteristics of the electromagnetic waves we limit ourselves here to the plane wave, described by Eqs. 9.3(1) and 9.3(2), i.e.

$$\mu_0\partial H_y/\partial t = \partial E_z/\partial x \tag{1}$$

$$\varepsilon_0\partial E_z/\partial t = \partial H_y/\partial x \tag{2}$$

and the corresponding wave equation

$$\partial^2 E_z/\partial x^2 = (1/c^2)\partial^2 E_z/\partial t^2 \tag{3}$$

The general solution for a wave travelling in the positive x-direction is $E_z = F(t - x/c)$. We note that $\partial E_z/\partial x = -(1/c)\partial E_z/\partial t$ and insertion of this

Display 9.4.

Plane electromagnetic wave. Wave impedance.

Plane EM wave

$E_z = F(t - x/c)$ (1)

$H_y = -E_z/Z$ (2)

Wave impedance

$Z = \mu_0 c = \sqrt{\mu_0/\varepsilon_0} \simeq 120\pi$ Ohm (3)

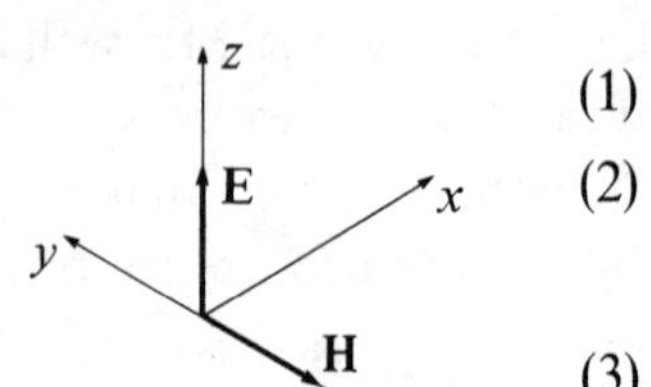

Display 9.5.

Wave energy and wave momentum.

Wave energy flux. The Poynting vector

Average power input to cable $= VI$ (1)

$V = Ed$ (2)

$I = (Hl)$ (3)

$\langle VI \rangle = (dl)\langle EH \rangle$ (4)

$S = \langle EH \rangle$ (energy flux) (5)

General energy equation:

$\partial W(x,t)/\partial t + \partial S(x,t)/\partial x = 0$ (6)

$W(x,t) = \mu_0 H^2/2 + \varepsilon_0 E^2/2$ (energy density) (7)

$S(x,t) = EH$ (energy flux) (8)

Stationary field: $\partial W/\partial t = 0 \quad \partial \langle S(x,t) \rangle/\partial x = \partial S/\partial x = 0$ (9)

Poynting vector

$\mathbf{S} = \mathbf{E} \times \mathbf{H}$ (10)

Wave momentum flux

$\mathbf{G} = \mathbf{S}/c$ (11)

Harmonic time dependence:

$E = E_0 \cos(\omega t - kx), \quad H = H_0 \cos(\omega t - kx)$

$S = \langle S(x,t) \rangle = \frac{1}{2}E_0 H_0 = \frac{1}{2}ZH_0^2 = \frac{1}{2}E_0^2/Z \quad Z = \mu_0 c$ (12)

expression into Eq. (1) yields

$$H_y = -(1/\mu_0 c)E_z = -(1/Z)E_z \tag{4}$$

$$Z = \mu_0 c = \sqrt{\mu_0/\varepsilon_0} \simeq 120\pi \text{ Ohm} \tag{5}$$

The ratio between the magnitudes of the electric and magnetic fields is the **wave impedance** $Z = \mu_0 c$, as expected, since E and H correspond to the force and velocity variables, respectively. The relationship between the directions of the E- and H-fields in the plane wave is shown in Display 9.4.

Energy flux. Poynting vector. With reference to our discussion in the last section of the wave on a cable or between two plane parallel plates, the power transferred from the source to the cable is $P(t) = V(t)I(t)$. The time average power $P = \langle P(t) \rangle$ for harmonic time dependence must be carried by the wave. We can express this power in terms of the electric and magnetic fields by inserting $V = Ed$ and $I = lH$, where l is the perimeter of the cable introduced earlier, and we obtain $P = (ld)\langle EH \rangle$, where the bracket indicates time average. This means, that the average power in the wave per unit area, the **energy flux**, is $\langle S \rangle = \langle EH \rangle = Z\langle H^2 \rangle = \langle E^2 \rangle / Z$, where Z is the wave impedance given in Eq. (5).

The energy balance problem is best studied by making use of the field equations (1) and (2), in complete analogy with the procedure carried out in Chapter 6. Thus, we multiply Eq. (1) by H_y and Eq. (2) by E_z and add the equations to obtain

$$\partial W/\partial t + \partial S/\partial x = 0 \tag{6}$$

$$W(x,t) = \tfrac{1}{2}\mu_0 H_y^2 + \tfrac{1}{2}\varepsilon_0 E_z^2 \tag{7}$$

$$S(x,t) = E_z H_y \tag{8}$$

The first term is the time rate of change of the sum of the electric and magnetic energy density. An increase in this total energy density $W(x,t)$ in the field is balanced by a corresponding decrease in the quantity $S(x,t) = E_z H_y$, which is the energy flux in the field. This relation is valid at every time t.

For a 'stationary' field, such as for harmonic time dependence, the time average value of $\partial W/\partial t$ is zero, which means that the time average value of S is independent of x.

The energy flux is in the positive x-direction, and this direction can be incorporated in the energy flux by the introduction of the vector

$$\mathbf{S} = \mathbf{E} \times \mathbf{H} \tag{9}$$

being the vector product of $\mathbf{E}$ and $\mathbf{H}$. The energy flux vector $\mathbf{S}$ is usually referred to as **the Poynting vector.**

The Poynting vector for the energy flux is obtained directly if the energy balance equation is derived from the Maxwell equations (9.3.(3)–(4)) by scalar multiplication of these equations by E and H, respectively. We leave the details for one of the problems.

Wave momentum. Waves carry not only energy but also momentum. In determining the momentum flux in an electromagnetic wave, we treat the wave as a stream of photons. A single photon has the energy $h\nu$, where h is Planck's constant and ν the frequency. The momentum of a photon is $h/\lambda = h\nu/c$, where λ is the wavelength. The number flux of photons required to produce the energy flux S is then $S/(h\nu)$, and the corresponding momentum flux is then

$$G = S/c \tag{10}$$

The corresponding **radiation pressure** on a perfect absorber will be G and on a perfect reflector $2G$.

For a summary we refer to Display 9.5.

9.5 Generation of EM waves

In the previous sections we discussed some properties of electromagnetic waves without much regard to how they were generated. In this case of wave propagation on a cable or between two parallel plates, the wave mode we have considered had a radial electric field and a circumferential magnetic field. The corresponding field vectors were perpendicular to each other and normal to the direction of propagation.

In order for a wave to be generated in the cable, it is necessary that a radial electric and/or a circumferential magnetic field be produced by the source. This is achieved simply by connecting the output terminals of the generator with wires to the inner and outer conductors of the cable, as indicated in (*a*) in Display 9.6. The current distribution in the conductors at the contact points will be complex, but after only a short distance it has become uniform and axial, leading to a circumferential magnetic field. (This asssumes, that the wave-length is long compared to the separation of the conductors and to the circumference of the cable, as will be explained in a subsequent chapter on wave guides).

But the wave can be generated even without electrical contact between the source and the cable conductors. The essential requirement for wave generation is that an appropriate field be produced, which is compatible with the field of the electromagnetic wave mode in the cable. For example, in case (*b*) in the display, an electric dipole antenna is used to generate a radial

Display 9.6.

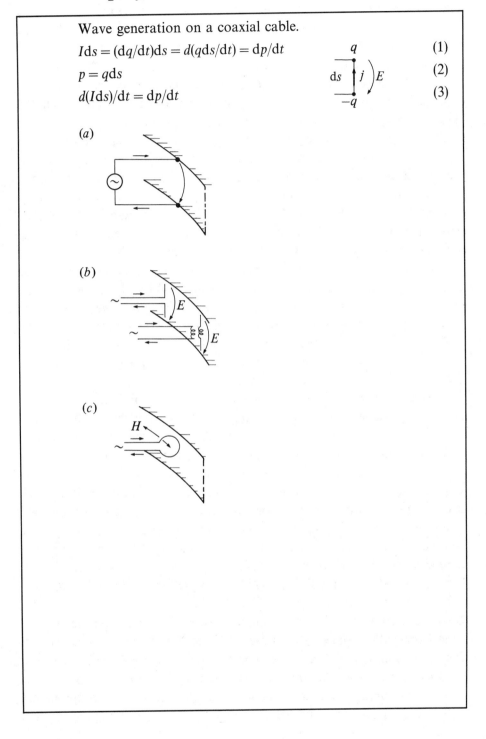

Wave generation on a coaxial cable.

$$Ids = (dq/dt)ds = d(qds/dt) = dp/dt \tag{1}$$

$$p = qds \tag{2}$$

$$d(Ids)/dt = dp/dt \tag{3}$$

(a)

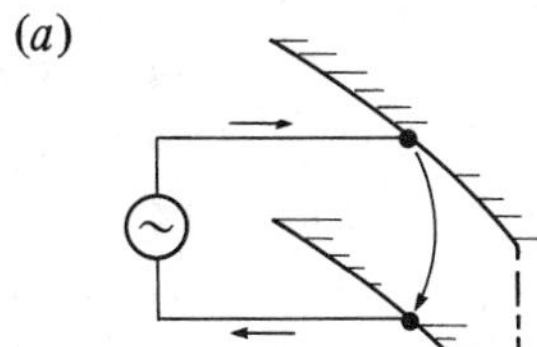

(b)

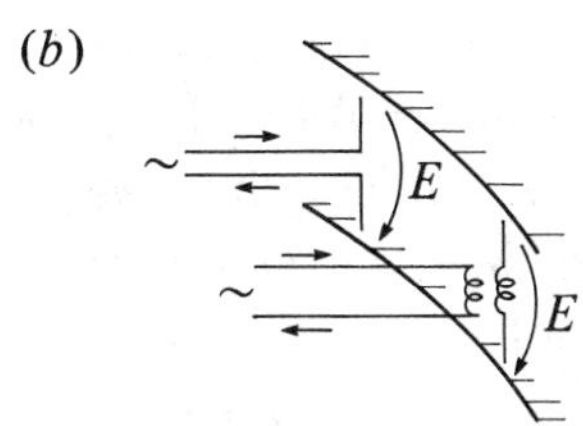

(c)

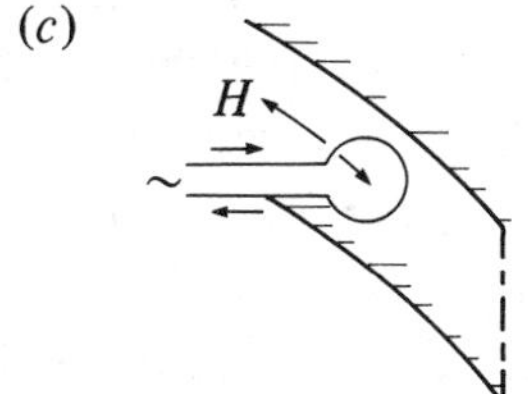

electric field, which couples to the electromagnetic wave in the cable. Again, the field close to the antenna is complicated, but it becomes uniform within a short distance from the antenna if the wavelength is large compared to the transverse cable dimensions. The antenna consists simply of a radial current element which is connected directly or via a transformer to the generator.

Instead of an electric dipole, a magnetic dipole antenna can be used, as indicated in (*c*). In principle, it consists of a circular loop with its axis in the direction of the magnetic field in the electromagnetic wave.

The electric dipole antenna can be regarded as a current element, i.e. a straight conductor of length ds carrying a time dependent current I. Conservation of electric charge requires that the charges at the end of the element will be equal and opposite in sign, $\pm q$. The charge q varies with time in accordance with $dq/dt = I$. In other words, the antenna acts like a dipole with a time dependent dipole moment $p = q ds$. The wire element is on the average electrically neutral; when the conduction electrons flow toward one end, making it negative, the stationary positive background charge will be uncovered at the other end, making it positive. With the velocity of the electrons being u and their charge density per unit length ρ', we have $I = \rho' u$.

A DC current cannot be carried by the electric dipole antenna, and the EM radiation field is expected to be proportional to the rate of change of I and hence to the second time derivative of the dipole moment and to the acceleration of the electrons since $I = \rho' u$.

The electromagnetic field between the two parallel conductors considered here is a plane wave. The field from a dipole source obviously is more complex. Nevertheless, it will couple to the plane wave because it does contain a uniform part, which is represented by the first term in the Fourier expansion of the field in terms of the transverse coordinates.

The other terms in the Fourier expansion will generate fields, which decrease rapidly with distance from the antenna at the frequencies we are interested in here, at which the wavelength is large compared to the transverse dimensions of the transmission line. Only in the microwave regime, in which the wavelength is sufficiently small, several wave modes can be transmitted along the line, as will be discussed in a subsequent chapter.

Dipole radiation. If the dipole antenna is placed between two parallel plate conductors rather than at the entrance of a coaxial cable, it will generate a wave of cylindrical symmetry, as indicated in figure (*a*) in Display 9.7. At a distance r from the antenna the area of a cylindrical 'control surface' surrounding the antenna is $(2\pi r)d$, where d is the separation of the plates.

If we consider a stationary field, in which S is the time average energy flux

Display 9.7.

Radiation from an electric dipole antenna.

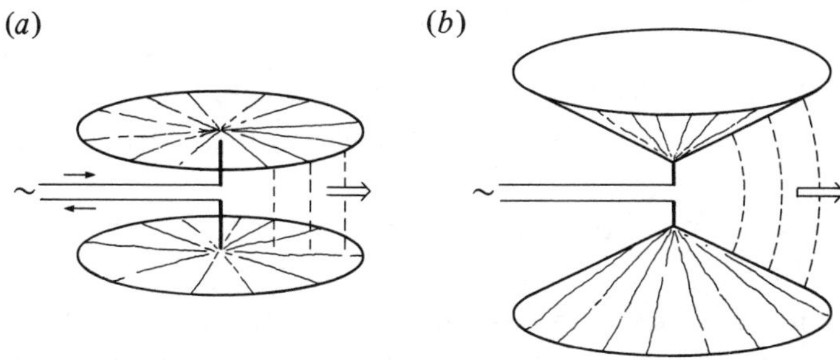

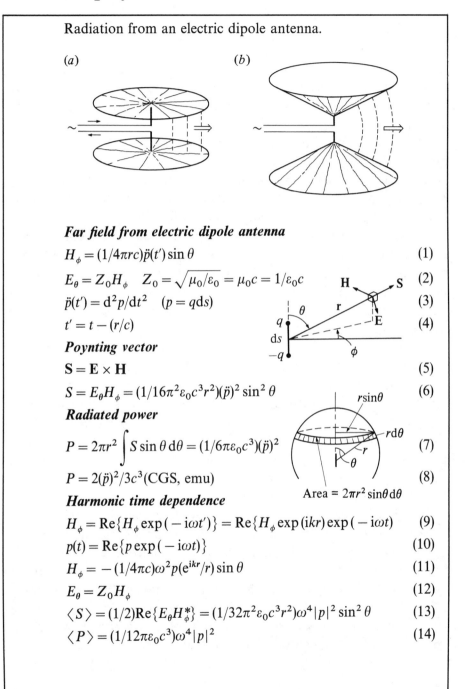

Far field from electric dipole antenna

$$H_\phi = (1/4\pi rc)\ddot{p}(t')\sin\theta \quad (1)$$

$$E_\theta = Z_0 H_\phi \quad Z_0 = \sqrt{\mu_0/\varepsilon_0} = \mu_0 c = 1/\varepsilon_0 c \quad (2)$$

$$\ddot{p}(t') = \mathrm{d}^2 p/\mathrm{d}t^2 \quad (p = q\mathrm{d}s) \quad (3)$$

$$t' = t - (r/c) \quad (4)$$

Poynting vector

$$\mathbf{S} = \mathbf{E} \times \mathbf{H} \quad (5)$$

$$S = E_\theta H_\phi = (1/16\pi^2\varepsilon_0 c^3 r^2)(\ddot{p})^2 \sin^2\theta \quad (6)$$

Radiated power

$$P = 2\pi r^2 \int S \sin\theta \,\mathrm{d}\theta = (1/6\pi\varepsilon_0 c^3)(\ddot{p})^2 \quad (7)$$

$$P = 2(\ddot{p})^2/3c^3 (\text{CGS, emu}) \quad (8)$$

Harmonic time dependence

$$H_\phi = \mathrm{Re}\{H_\phi \exp(-\mathrm{i}\omega t')\} = \mathrm{Re}\{H_\phi \exp(\mathrm{i}kr)\exp(-\mathrm{i}\omega t) \quad (9)$$

$$p(t) = \mathrm{Re}\{p \exp(-\mathrm{i}\omega t)\} \quad (10)$$

$$H_\phi = -(1/4\pi c)\omega^2 p(\mathrm{e}^{\mathrm{i}kr}/r)\sin\theta \quad (11)$$

$$E_\theta = Z_0 H_\phi \quad (12)$$

$$\langle S \rangle = (1/2)\mathrm{Re}\{E_\theta H_\phi^*\} = (1/32\pi^2\varepsilon_0 c^3 r^2)\omega^4 |p|^2 \sin^2\theta \quad (13)$$

$$\langle P \rangle = (1/12\pi\varepsilon_0 c^3)\omega^4 |p|^2 \quad (14)$$

and P is the time average radiated power, conservation of energy requires that $S(2\pi rd) = P =$ constant at every value of r.

This means that the energy flux S decreases as $1/r$. Since S is proportional to the square of the electric field (or magnetic field), it follows that the electric field strength will decrease as $1/\sqrt{r}$. Actually, sufficiently far from the antenna the wave locally is approximately plane, so that $S = Z\langle H^2\rangle = \langle E^2\rangle/Z$, where $Z = \mu_0 c$. In this '**far-field**' regime, we then have $\langle H^2\rangle = P/Z(2\pi rd)$.

We now deform the parallel plates into a cylindrical 'horn', as shown in figure (*b*). Far away from the dipole, the area of a cylindrical control surface surrounding the source at a distance r from the source is proportional to r^2. Consequently, the average energy flux will decrease as $1/r^2$, and the corresponding electric and magnetic fields will decrease as $1/r$.

The electric field vector must be perpendicular to the conducting surfaces, and the electrical field lines will have the form indicated schematically in the figure. In the far field, the magnetic field vector is perpendicular to the electric field vector, as in a plane wave. Thus, with the electric field in the positive θ-direction, as indicated, the magnetic field is in the positive ϕ-direction, where θ is the polar angle and ϕ the azimuth angle.

We now go to the limit of a horn with an angle of 180 degrees. This leaves us with a dipole in free field. At $\theta = 0$, i.e. on the line going through the dipole, the electric field must be zero (otherwise we would have a discontinuity in the direction of the field). Actually, guided by the behaviour of the static dipole field, we expect the electric field to be proportional to $\sin(\theta)$, i.e. proportional to the component of the dipole moment that can be 'seen' by the observer, i.e. proportional to the projection of the dipole moment normal to the radius vector to the observer. The maximum field is then obtained in the direction $\theta = 90$ degrees, normal to the dipole.

Summarizing the previous discussion, we note that the radiated field is expected to be proportional to the second derivative of the dipole moment with respect to time. Furthermore, the angular dependence is $\sin(\theta)$ and in the far field the r-dependence is given by $1/r$. With cylindrical symmetry, the field should be independent of the aximuthal angle ϕ. In other words, we expect the magnetic field to be of the form $H = \text{const}\cdot(\ddot{p}/r)\sin(\theta)$ and the electric field to be $E = ZH$. A detailed study of the problem shows that the constant of proportionality is $1/4\pi c$, so that

$$H_\phi(r, t) = (1/4\pi rc)\ddot{p}(t')\sin(\theta) \quad (p = q\,\mathrm{d}s) \tag{1}$$

$$E_\theta(r, t) = ZH_\phi \tag{2}$$

$$t' = t - r/c \quad Z = \mu_0 c.$$

It should be noted that if the time dependence of the dipole moment is $p(t)$, the time dependence of the fields a distance r away is expressed by the retarded time $t' = t - r/c$, corresponding to a wave travel time r/c.

To determine the total radiated power we have to integrate the energy flux S over a spherical control surface surrounding the source. The differential surface element on the sphere is $r\sin(\theta)\,\mathrm{d}\theta r\,\mathrm{d}\phi$, and since the field is independent of ϕ we can immediately integrate over this angle to make the resulting surface element $2\pi r^2 \sin(\theta)\,\mathrm{d}\theta$. The expression for the radiated power is then

$$P = 2\pi r^2 \int_0^{\pi} S\sin(\theta)\,\mathrm{d}\theta = (1/6\pi\varepsilon_0 c^3)\ddot{p}^2 = (1/4\pi\varepsilon_0)2\ddot{p}^2/3c^3 \tag{3}$$

In the important case of harmonic time dependence, the expressions for the complex amplitudes of the fields are (from Eqs. 1 and 2)

$$H_\phi = -(1/4\pi c)\omega^2 p(\mathrm{e}^{\mathrm{i}kr}/r)\sin(\theta) \tag{4}$$

$$E_\theta = ZH \tag{5}$$

The corresponding expression for the time average of the energy flux and of the radiated power are

$$S = \tfrac{1}{2}\mathrm{Re}\{E_\theta H_\phi^*\} = (1/32\pi^2\varepsilon_0 c^3 r^2)\omega^4 |p|^2 \sin^2(\theta) \tag{6}$$

$$P = (1/12\pi\varepsilon_0 c^3)\omega^4 |p|^2. \tag{7}$$

For a given magnitude of the dipole moment, the radiated power is proportional to the **fourth power** of the frequency. This is essential in the explanation of the blue of the sky and red of the sunset, as will be discussed in Chapter 17. The dipole moments involved in that case are induced by the incident light and the dipole radiation represents the scattered light.

Radiation from an accelerated charge. In the previous discussion the dipole moment was produced by a time dependent current in an antenna. A time dependent dipole moment can be produced also by a moving single charge. The dipole moment p of the charge with respect to the origin is the product of the charge and the distance to the origin, and the rate of change of the dipole moment will be $\dot{p} = qu$, were u is the velocity of the particle.

Eqs. (1) and (2) indicate that the radiated field is proportional to $\ddot{p}$, and we assume this to be true also when the dipole moment is that of a single charge. With $\dot{p} = qu$ it follows that the field radiated from a moving charge is proportional to the acceleration of the particle. Thus, if we denote the acceleration by $a(t)$, we obtain $\ddot{p} = qa$, and the expressions for the magnetic and electric fields take the form

$$H_\phi(r,t) = (1/4\pi rc)qa(t')\sin(\theta) \tag{8}$$

Display 9.8.

Radiation from an accelerated charge.

Acceleration of charge $= a(t)$

$$H_\phi(r,t) = (1/4\pi rc)qa(t')\sin(\theta) \tag{1}$$

$$E_\theta(r,t) = ZH_\phi \tag{2}$$

$$t' = t - r/c. \quad Z = \mu_0 c \tag{3}$$

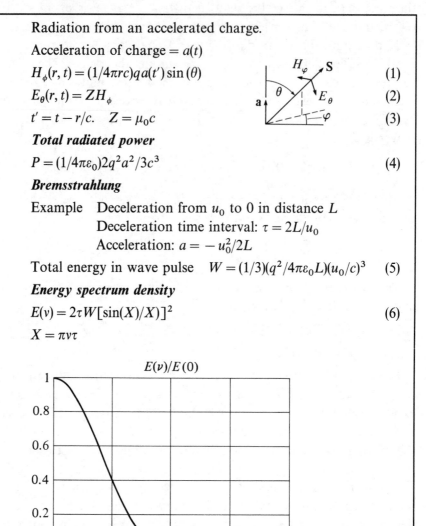

Total radiated power

$$P = (1/4\pi\varepsilon_0)2q^2a^2/3c^3 \tag{4}$$

Bremsstrahlung

Example Deceleration from u_0 to 0 in distance L

Deceleration time interval: $\tau = 2L/u_0$

Acceleration: $a = -u_0^2/2L$

Total energy in wave pulse $W = (1/3)(q^2/4\pi\varepsilon_0 L)(u_0/c)^3$ (5)

Energy spectrum density

$$E(\nu) = 2\tau W[\sin(X)/X]^2 \tag{6}$$

$X = \pi\nu\tau$

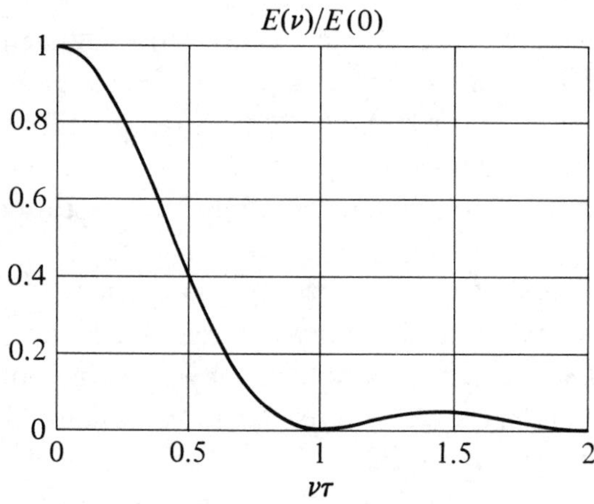

$$E_\theta(r, t) = ZH_\phi \tag{9}$$
$$t' = t - r/c, \quad Z = \mu_0 c$$

where the angles θ and ϕ are defined as before.

The expression for the instantaneous radiated power in Eq. (3) now takes the form

$$P = (1/4\pi\varepsilon_0)2q^2a^2/3c^3 \tag{10}$$

and if the displacement of the particle is harmonic, $\eta = \eta_0 \cos(\omega t)$, the time average of the power will be that given in Eq. (7) with $|p|$ replaced by $q\eta_0$.

Bremsstrahlung. When a moving charged particle encounters an obstacle or target so that it will be decelerated, it will emit a 'pulse' of electromagnetic radiation with a time dependence which is the same as that of the acceleration. (German verb 'bremsen' means 'to brake' and 'Strahlung' means 'radiation', thus, 'Bremsstrahlung' can be translated as 'braking radiation').

The energy spectrum of the radiation will be continuous, as will be discussed in terms of an example below, with a frequency range which often extends into the X-ray region. In the X-ray generators used in medical diagnostics, the radiation is produced by an electron beam impinging upon a metal target (tungsten). Electronic excitation in the target gives rise to spectral lines, characteristic of the target, which are superimposed on the continuous Bremsstrahlung spectrum.

As a quantitative illustration, we consider as a simple model a linear deceleration of a velocity of the charged particle from u_0 to zero. The deceleration takes place over a short distance L so that the corresponding acceleration is $a = -u_0^2/2L$ in the short time interval from 0 to $\tau = 2L/u_0$.

The power emitted during this time interval is obtained from Eq. (10), and multiplying by τ, we get the total energy emitted

$$W = (1/3)(q^2/4\pi\varepsilon_0 L)(u_0/c)^3 \tag{11}$$

The corresponding energy spectrum can be calculated from the known time dependence of the radiation, which is the same as that of a square pulse of duration τ.

The Fourier amplitude of such a pulse was calculated in Section 3.7 (Display 3.13), and the corresponding spectrum density $E(\nu)$ can be calculated from Eq. 3.7(21). We leave it as a problem to show that in this case we obtain

$$E(\nu) = 2\tau W[\sin(X)/X]^2 \tag{12}$$
$$X = \pi\nu\tau$$

where W is the total radiated energy, as given in Eq. (11).

The frequency dependence of the energy spectrum is expressed by the factor $(\sin(X)/X)^2$, which starts from unity at $\nu = 0$ and decreases to zero at $\nu_1 = 1/\tau$ and at multiples thereof. Between these zeroes are maxima, which decrease with increasing frequency. Further details of the frequency dependence can be deduced from the result shown in Display 3.13.

It follows that the bulk of the radiated energy falls in the frequency range below $\nu_1 = 1/\tau = u_0/2L$. We note that with $L = 1$ Å and $u_0 = 0.01c$, we obtain $\nu_1 = 5 \times 10^{15}$ Hz.

9.6 Polarization

In a longitudinal wave the field vector, such as displacement or the corresponding velocity, is aligned with the direction of propagation. In a transverse wave, on the other hand, the field vector is normal to the propagation vector, and these two vectors define a plane, the **plane of polarization**.

Waves on strings and electromagnetic waves are examples of transverse waves which we have discussed so far, and other examples will be dealt with later (hydromagnetic waves in Chapter 18 and elastic waves in Chapter 20). In each case we can define a plane of polarization of the wave.

In ordinary light, the direction of the electric field vector varies randomly in space and time in planes perpendicular to the direction of propagation; all directions are equally probable. This is an expression of the random orientation of the atomic dipole radiators, which are the sources of the light.

For purpose of illustration, let us consider a wave on a string which runs in the z-direction. The end of the string is driven by a harmonic force in the x-direction. The displacement of the string will be in the same plane at every location and will remain in this plane of polarization. The wave is then said to be **linearly polarized**.

If the end of the string is driven by one force in the x-direction and one in the y-direction, the displacement wave will still be linearly polarized if the two forces are in phase. The inclination of the plane of polarization is determined by the ratio between the amplitudes of the two forces.

On the other hand, if the two forces are not in phase, the motion no longer will be linearly polarized. In the special case when the amplitudes of the forces are the same but their phases differ by $\pi/2$, the motion of the string will be circular. This will be the case if $F_x = F_0 \cos(\omega t)$ and $F_y = F_0 \sin(\omega t)$, with F_y lagging behind F_x by $\pi/2$.

The displacement vector of the string at each location z now will turn about the z-axis with the angular velocity ω in the positive (clock-wise) direction. The wave is said to be circularly polarized with positive or 'right handed' direction of rotation.

If the roles of the forces are interchanged with the x-component lagging behind the y-component by 90 degrees, the direction of rotation will be reversed.

With a 90 degree phase lag and different amplitudes of the two force components, each mass element of the string will move along an ellipse with the angular velocity ω. The wave is said to be **elliptically polarized**. Actually, the motion of the string is similar to that of the two-dimensional oscillator discussed in Display 1.9. The basic features of polarization, which we have illustrated in terms of the displacement of a string, apply equally well to the electromagnetic wave, with the electric field vector taking the place of the displacement vector of the string.

Linear polarization. To study the problem of polarization in more detail, we consider a wave travelling in the z-direction with the field vector along the x-axis. This vector can be the electric field in a plane EM wave or the displacement of a string, for example.

Considering here a harmonic wave, we express the real field variable as $X(z, t) = X_0 \cos(\omega t - kz - \alpha)$. The corresponding field vector is

$$\mathbf{X}(z, t) = \hat{\mathbf{x}} X_0 \cos(\omega t - kz - \alpha) \tag{1}$$

where $\hat{\mathbf{x}}$ is the unit vector along the x-axis.

Similarly, for a wave with the field vector in the y-direction we have

$$\mathbf{Y}(z, t) = \hat{\mathbf{y}} Y_0 \cos(\omega t - kz - \beta) \tag{2}$$

where $\hat{\mathbf{y}}$ is the unit vector along the y-axis.

The complex amplitude functions for the waves in Eqs. (1)–(2) become, with $X = X_0 \exp(\mathrm{i}\alpha)$ and $Y = Y_0 \exp(\mathrm{i}\beta)$,

$$\mathbf{X}(z) = \hat{\mathbf{x}} X \exp(\mathrm{i}kz) \tag{3}$$

$$\mathbf{Y}(z) = \hat{\mathbf{y}} Y \exp(\mathrm{i}kz) \tag{4}$$

These waves are both linearly polarized, and if X and Y are real (or have the same phase angle), the two waves are in phase. In that case superposition of the waves will be a linearly polarized wave also. The corresponding vector field becomes

$$\mathbf{L}(z) = \hat{\mathbf{l}} L \exp(\mathrm{i}kz) \quad (X, Y \text{ real}) \tag{5}$$

$$L = \sqrt{X^2 + Y^2}, \quad \hat{\mathbf{l}} = \hat{\mathbf{x}} \cos(\phi) + \hat{\mathbf{y}} \sin(\phi), \quad \tan(\phi) = Y/X$$

where $\hat{\mathbf{l}}$ is the unit vector in the direction of the field vector and L the magnitude of the vector.

Circular polarization. As already indicated, the superposition of two waves, linearly polarized in the x- and y-directions, respectively, results in a

Display 9.9.

Complex amplitude description of polarized waves.

Linearly polarized travelling wave

Complex amplitude of field vector:

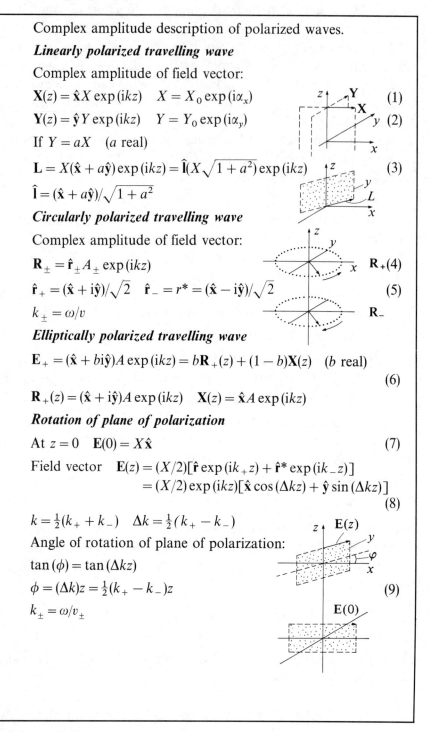

$$\mathbf{X}(z) = \hat{\mathbf{x}} X \exp(\mathrm{i}kz) \quad X = X_0 \exp(\mathrm{i}\alpha_x) \tag{1}$$

$$\mathbf{Y}(z) = \hat{\mathbf{y}} Y \exp(\mathrm{i}kz) \quad Y = Y_0 \exp(\mathrm{i}\alpha_y) \tag{2}$$

If $Y = aX$ (a real)

$$\mathbf{L} = X(\hat{\mathbf{x}} + a\hat{\mathbf{y}}) \exp(\mathrm{i}kz) = \hat{\mathbf{l}}(X\sqrt{1 + a^2}) \exp(\mathrm{i}kz) \tag{3}$$

$$\hat{\mathbf{l}} = (\hat{\mathbf{x}} + a\hat{\mathbf{y}})/\sqrt{1 + a^2}$$

Circularly polarized travelling wave

Complex amplitude of field vector:

$$\mathbf{R}_\pm = \hat{\mathbf{r}}_\pm A_\pm \exp(\mathrm{i}kz) \qquad \mathbf{R}_+ \tag{4}$$

$$\hat{\mathbf{r}}_+ = (\hat{\mathbf{x}} + \mathrm{i}\hat{\mathbf{y}})/\sqrt{2} \quad \hat{\mathbf{r}}_- = r^* = (\hat{\mathbf{x}} - \mathrm{i}\hat{\mathbf{y}})/\sqrt{2} \tag{5}$$

$$k_\pm = \omega/v \qquad \mathbf{R}_-$$

Elliptically polarized travelling wave

$$\mathbf{E}_+ = (\hat{\mathbf{x}} + b\mathrm{i}\hat{\mathbf{y}})A \exp(\mathrm{i}kz) = b\mathbf{R}_+(z) + (1 - b)\mathbf{X}(z) \quad (b \text{ real}) \tag{6}$$

$$\mathbf{R}_+(z) = (\hat{\mathbf{x}} + \mathrm{i}\hat{\mathbf{y}})A \exp(\mathrm{i}kz) \quad \mathbf{X}(z) = \hat{\mathbf{x}} A \exp(\mathrm{i}kz)$$

Rotation of plane of polarization

$$\text{At } z = 0 \quad \mathbf{E}(0) = X\hat{\mathbf{x}} \tag{7}$$

$$\begin{aligned}\text{Field vector} \quad \mathbf{E}(z) &= (X/2)[\hat{\mathbf{r}} \exp(\mathrm{i}k_+ z) + \hat{\mathbf{r}}^* \exp(\mathrm{i}k_- z)] \\ &= (X/2) \exp(\mathrm{i}kz)[\hat{\mathbf{x}} \cos(\Delta kz) + \hat{\mathbf{y}} \sin(\Delta kz)]\end{aligned} \tag{8}$$

$$k = \tfrac{1}{2}(k_+ + k_-) \quad \Delta k = \tfrac{1}{2}(k_+ - k_-)$$

Angle of rotation of plane of polarization:

$$\tan(\phi) = \tan(\Delta kz)$$

$$\phi = (\Delta k)z = \tfrac{1}{2}(k_+ - k_-)z \tag{9}$$

$$k_\pm = \omega/v_\pm$$

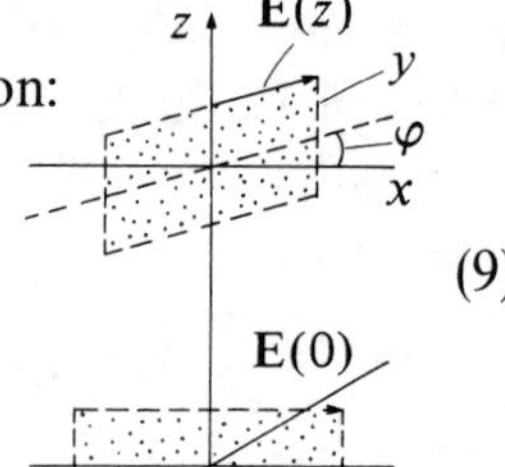

circularly polarized wave if there is a phase difference of $\pi/2$ between the two waves.

If we let the first wave be $\mathbf{X}(z,t) = \hat{\mathbf{x}}X_0 \cos(\omega t - kz)$ and wish the second wave to lag behind the first by the angle $\pi/2$, we have to choose $\mathbf{Y}(z,t) = \hat{\mathbf{y}}Y_0 \sin(\omega t - kz)$. Furthermore, with the amplitudes of the waves the same, $X_0 = Y_0 = A$, the sum $\mathbf{R}_+ = \mathbf{X} + \mathbf{Y}$ becomes

$$\mathbf{R}_+(z,t) = A[\hat{\mathbf{x}} \cos(\omega t - kz) + \hat{\mathbf{y}} \sin(\omega t - kz)] \tag{6}$$

Changing the sign of the Y-wave is equivalent to making it run ahead of the X-wave by the phase angle $\pi/2$, and the resulting wave

$$\mathbf{R}_-(z,t) = A[\hat{\mathbf{x}} \cos(\omega t - kz) - \hat{\mathbf{y}} \sin(\omega t - kz)] \tag{7}$$

represents a circularly polarized wave with the field vector rotating around the z-axis in the negative direction with the angular velocity ω (left handed polarization).

The complex amplitude vectors of these circularly polarized waves can be expressed in a simple and interesting manner.

We start with the complex amplitudes for the linearly polarized waves in Eqs. (3) and (4) and put $X = Y = A$. In the complex amplitude description, a phase lag of $\pi/2$ is produced by a factor i (see Chapter 2, E2.1) and it follows that the complex amplitude of $\mathbf{R}_+(z,t)$ is $\mathbf{R}_+(z) = \mathbf{X}(z) + \mathrm{i}\mathbf{Y}(z) = A(\hat{\mathbf{x}} + \mathrm{i}\hat{\mathbf{y}}) \exp(ikz)$. This expression can be written in a compact manner if we introduce the complex (rotation) unit vector $\hat{\mathbf{r}} = (1/\sqrt{2})(\hat{\mathbf{x}} + \mathrm{i}\hat{\mathbf{y}})$, and we obtain from Eq. (6)

$$\mathbf{R}_+(z) = (A\sqrt{2})\hat{\mathbf{r}} \exp(ikz) \tag{8}$$

$$\hat{\mathbf{r}} = (\hat{\mathbf{x}} + \mathrm{i}\hat{\mathbf{y}})/\sqrt{2}$$

Similarly, with $r^* = (\hat{\mathbf{x}} - \mathrm{i}\hat{\mathbf{y}})/\sqrt{2}$, the complex amplitude of the wave $\mathbf{R}_-(z,t)$ is

$$\mathbf{R}_-(z) = (A\sqrt{2})\hat{\mathbf{r}}^* \exp(ikz) \quad \hat{\mathbf{r}}^* = (\hat{\mathbf{x}} - \mathrm{i}\hat{\mathbf{y}})/\sqrt{2} \tag{9}$$

These expressions show that the magnitude of the field vector is $A\sqrt{2}$, and the direction of rotation is indicated by the unit rotation vector $\hat{\mathbf{r}}$ (positive direction) and its complex conjugate $\hat{\mathbf{r}}^*$ (negative rotation).

It should be noted also that the squared magnitude of the field vector can be expressed formally as the scalar product of $\mathbf{R}$ and the complex conjugate $\mathbf{R}^*$, i.e. $|\mathbf{R}|^2 = \mathbf{R}\cdot\mathbf{R}^* = 2A^2$.

As we have seen, a circularly polarized wave can be obtained by superimposing two linearly polarized waves. Similarly, it is possible to express a linearly polarized wave as the superposition of two circularly polarized waves of equal amplitude and rotating in opposite directions.

Thus, it follows from Eqs. (3), (4) and (8), (9) that

$$\mathbf{X}(z) = A\hat{\mathbf{x}} \exp(ikz) = [\mathbf{R}_+(z) + \mathbf{R}_-(z)]/2 \tag{10}$$

$$\mathbf{Y}(z) = Ai\hat{\mathbf{y}} \exp(ikz) = [\mathbf{R}_+(z) - \mathbf{R}_-(z)]/2 \tag{11}$$

Elliptical polarization. An elliptically polarized wave results from the superposition of two linearly polarized waves of different amplitudes, 90 degrees out of phase, and with the E-vectors perpendicular to each other. The complex amplitude is then of the form $\mathbf{E} = X\hat{\mathbf{x}} \pm iY\hat{\mathbf{y}}$. The special case $X = Y$ yields circular polarization.

Rotation of the plane of polarization. Propagation of an EM wave in a material in a magnetic field results in the rotation of the plane of polarization (Faraday effect), as will be illustrated for a plasma shortly. In some materials, notably sugar solutions and quartz, the plane of polarization will be rotated even in the absence of a magnetic field. There are two forms of sugar and quartz which produce rotation in opposite directions.

The fact that the direction of rotation of the electric field vector in an anisotropic material influences the wave speed is a basic reason for the rotation of the plane of polarization. Therefore, to calculate the angle of rotation of a linearly polarized wave transmitted through the material, we decompose this wave into two circularly polarized waves, as shown in Eq. (10).

With the two wave speeds denoted by v_+ and v_- and the corresponding propagation constant $k = \omega/v$ by k_+ and k_-, the complex amplitudes of the circularly polarized waves are $\mathbf{R}_+(z) = \hat{\mathbf{r}}A \exp(ik_+z)$ and $\mathbf{R}_-(z) = \hat{\mathbf{r}}^* B \exp(k_-z)$. The constants A and B are chosen so that the total field vector at $z = 0$ is in the x-direction and equal to $X\hat{x}$ with no component in the y-direction. The corresponding constants are then $A = B = X/2$.

Thus, at $z = 0$ we start with a linearly polarized wave with the field vector in the x-direction, and we wish to determine the angle of rotation of the plane of polarization at location z.

The complex amplitude of the total field vector is

$$\begin{aligned}\mathbf{E}(z) &= (X/2)[\hat{\mathbf{r}} \exp(ik_+z) + \hat{\mathbf{r}}^* \exp(ik_-z)] \\ &= (X/2)\exp(ikz)[\hat{\mathbf{r}} \exp(i\Delta kz) + \hat{\mathbf{r}}^* \exp(-i\Delta kz)] \\ &= (X/2)\exp(ikz)[\hat{\mathbf{x}} \cos(\Delta kz) + \hat{\mathbf{y}} \sin(\Delta kz)] \qquad (12) \\ k &= \tfrac{1}{2}(k_+ + k_-) \quad \Delta k = \tfrac{1}{2}(k_+ - k_-)\end{aligned}$$

It follows that after having started with a field at $z = 0$ given by $\mathbf{E}(0) = \hat{\mathbf{x}}X$, we find that the field at z contains also a component in the y-direction. This means that the plane of polarization has been rotated (see Eq. (5)), with the angle of rotation ϕ determined by the coefficients of $\hat{\mathbf{x}}$ and $\hat{\mathbf{y}}$, i.e. $\tan(\phi) = \tan(\Delta kz)$ or

$$\phi = (\Delta k)z = \tfrac{1}{2}(k_+ - k_-)z \tag{13}$$

$$k_+ = \omega/v_+, \quad k_- = \omega/v_-$$

If $k_+ > k_-$, i.e. if $v_+ < v_-$, the angle of rotation will be positive and if $v_+ > v_-$, the angle will be negative. The angle of rotation is related to the properties of the material involved, and measurement of ϕ can be used for diagnostic purposes. For example, in a magnetized plasma, the angle ϕ will be proportional to the magnetic field and the electron density, and the latter quantity is frequently determined from measurement of ϕ.

Polarizer. In Chapter 12 we shall discuss reflection of electromagnetic waves at a plane boundary between two materials and it will be shown that the reflection and transmission coefficients depend on the plane of polarization of the E-vector in the incident wave. At a certain angle of incidence, the Brewster angle, the reflected wave will be linearly polarized regardless of the polarization of the incident wave. This is one example of a polarizer.

In other polarizers the phenomena of double refraction, notably in quartz and calcite crystals, can be used to produce polarized light over a rather wide range of frequencies, as discussed in details in most books on optics. In Polaroid film, invented by Land in 1932, one component of linear polarization is completely absorbed, whereas the other is transmitted with little loss. This effect is limited to a smaller frequency range than the polarizers utilizing the double refracting crystals.

There exist also 'polarizers' for mechanical waves, and as a simple example we consider a wave on a string which passes through a thin slot in a plate perpendicular to the string.

Let us consider first a linearly polarized wave, with the displacement vector making an angle ϕ with the x-axis. As we have seen, such a wave can be considered to be a superposition of two waves with the displacements in the x- and y-direction respectively. The complex amplitude of the wave can be written

$$\boldsymbol{\eta}(z) = \hat{\mathbf{l}}L\exp(ikx) = L[\hat{\mathbf{x}}\cos(\phi) + \hat{\mathbf{y}}\sin(\phi)]\exp(ikz) \tag{14}$$

where L is the resulting displacement amplitude. The component in the x-direction is $X = L\cos(\phi)$, and if we neglect the friction in the slot, this component is transmitted through the slot without reflection.

The y-component, on the other hand, will be reflected, since at the slot there can be no displacement in the y-direction, corresponding to a displacement (velocity) reflection coefficient of minus one. Thus, if the slot is located at $z = 0$, the expression for the complex amplitude of the reflected displacement wave is

$$\boldsymbol{\eta}_2(z) = -[L\sin(\phi)]\hat{\mathbf{y}}\exp(\mathrm{i}kz) \tag{15}$$

The total field to the left of the slot is then

$$\begin{aligned}\boldsymbol{\eta}(z) &= \boldsymbol{\eta}_1(z) + \boldsymbol{\eta}_2(z)\\ &= L[\hat{\mathbf{x}}\cos(\phi) + \hat{\mathbf{y}}\sin(\phi)]\exp(\mathrm{i}kz)\\ &\quad - \hat{\mathbf{y}}L\sin(\phi)\exp(-\mathrm{i}kz)\\ &= \hat{\mathbf{x}}X\exp(\mathrm{i}kz) + \hat{\mathbf{y}}Y2\mathrm{i}\sin(kz) \qquad (16)\\ X &= L\cos(\phi), \quad Y = L\sin(\phi)\end{aligned}$$

The first term represents a travelling wave and the second term a standing wave. It should be noted in this context, that unless waves are polarized in the same direction, they cannot interfere to cancel each other and produce nodes of zero displacement.

It follows from Eq. (15), that the real displacement function is

$$\begin{aligned}\boldsymbol{\eta}(z,t) &= \mathrm{Re}\{\boldsymbol{\eta}(z)\exp(-\mathrm{i}\omega t)\}\\ &= \hat{\mathbf{x}}X\cos(\omega t - kz) + \hat{\mathbf{y}}2Y\sin(kz)\sin(\omega t)\end{aligned} \tag{17}$$

The wave transmitted through the slot polarizer is

$$\boldsymbol{\eta}_3(z) = \hat{\mathbf{x}}X\exp(\mathrm{i}kz) \quad X = L\cos(\phi) \tag{18}$$

with the real displacement $\boldsymbol{\eta}_3(z,t) = \hat{\mathbf{x}}X\cos(\omega t - kz)$.

From the complex displacement amplitude we obtain the complex velocity amplitude $\mathbf{u}(z) = -\mathrm{i}\omega\boldsymbol{\eta}(z)$. The complex force amplitude then follows from the field equation

$$-\mathrm{i}\omega\kappa\mathbf{F}(z) = -\mathrm{d}\mathbf{u}/\mathrm{d}z \tag{19}$$

The wave power is the scalar product of the real functions $\mathbf{F}(z,t)$ and $\mathbf{u}(z,t)$, $P = \mathbf{F}(z,t)\cdot\mathbf{u}(z,t) = F_x(z,t)u_x(z,t) + F_y(z,t)u_y(z,t)$. The time average of $F_x u_x$ can be expressed in terms of the complex amplitudes $F_x(z)$ and $u_x(z)$ in the usual manner, $\frac{1}{2}\mathrm{Re}\{F(z)u^*(z)\}$, with a similar expression for the y-components. It follows then that the time average of the total power can be written

$$P = \tfrac{1}{2}\mathrm{Re}\{\mathbf{F}(z)\cdot\mathbf{u}^*(z)\} \tag{20}$$

where scalar multiplication is implied.

Example. Circular to linear polarization. As another example we consider again the string and the slot polarizer with a frictionless slot in the x-direction, as before. This time, however, the incident wave is circularly

Display 9.10.

Circularly polarized wave on a string transmitted through a slot-polarizer resulting in a linearly polarized transmitted wave.

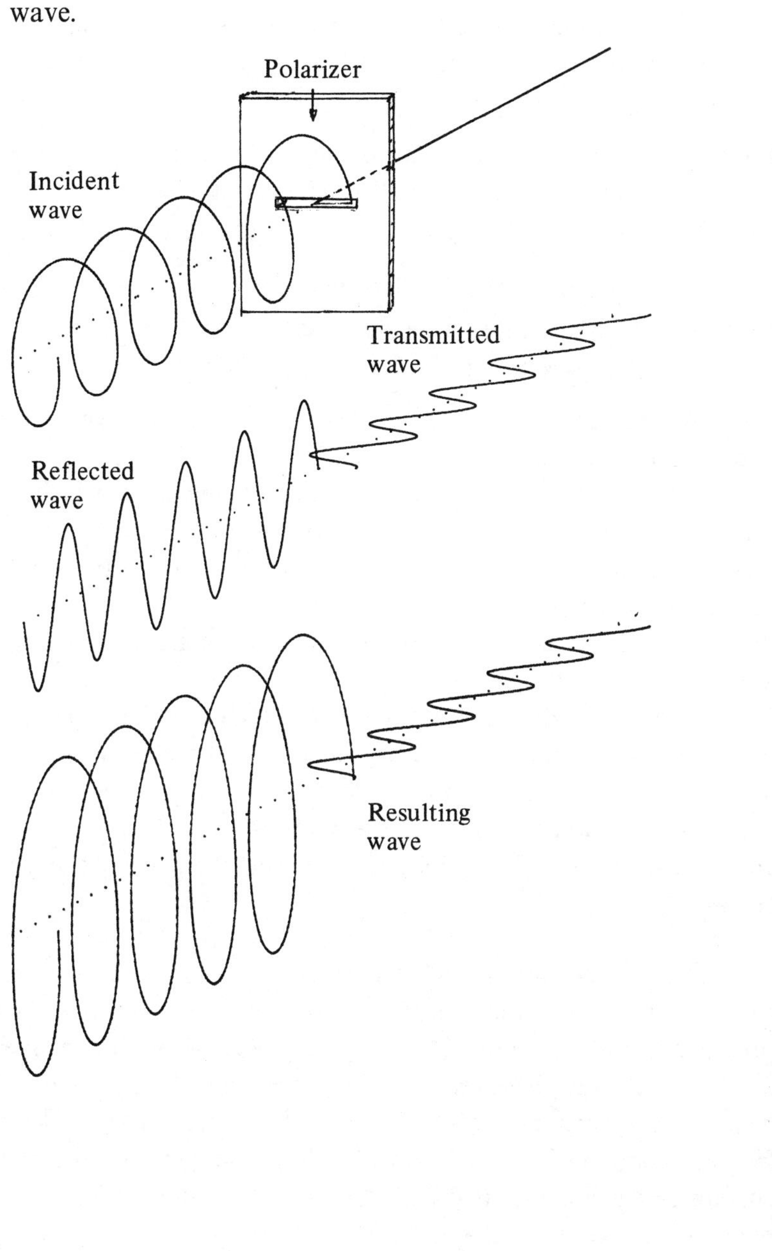

polarized. The complex amplitude of the incident wave is

$$\boldsymbol{\eta}_1(z) = A(\hat{\mathbf{x}} + \mathrm{i}\hat{\mathbf{y}}) \exp(\mathrm{i}kz) \tag{21}$$

With the slot oriented in the x-direction, the x-component of the displacement wave will be transmitted through the slot without reflection. For the y-component the reflection coefficient will be -1, however, and the complex amplitude of the reflected wave is

$$\boldsymbol{\eta}_2(z) = -\mathrm{i}\hat{\mathbf{y}}A \exp(-\mathrm{i}kz) \tag{22}$$

where, as before, we have assumed the slot to be located at $z = 0$.

The total displacement vector to the left of the slot is

$$\begin{aligned}\boldsymbol{\eta}(z) &= (\hat{\mathbf{x}} + \mathrm{i}\hat{\mathbf{y}})A \exp(\mathrm{i}kz) - \mathrm{i}\hat{\mathbf{y}}A \exp(-\mathrm{i}kz) \\ &= \hat{\mathbf{x}}A \exp(\mathrm{i}kz) - \hat{\mathbf{y}}2A \sin(kz)\end{aligned} \tag{23}$$

with the corresponding real displacement

$$\boldsymbol{\eta}(z, t) = \hat{\mathbf{x}}A \cos(\omega t - kz) - \hat{\mathbf{y}}2A \sin(kz) \cos(\omega t) \tag{24}$$

and the real displacements for the incident and reflected waves are obtained in analogous manner.

In Display 9.10 we have plotted the spatial dependence of the location of the tip of the displacement vector of the string. The top curve shows the incident circularly polarized wave and in the second figure the reflected and the transmitted waves are shown. These waves are both linearly polarized, in the x- and y-directions, respectively. The total displacement field is shown in the last figure.

The analysis readily can be extended to the case when there is friction in the slot so that some of the wave energy is absorbed in much the same manner as in a Polaroid film. We leave this as a problem for the reader.

It should be pointed out, that the analysis for an EM wave is completely analogous with the electric field vector taking the place of the displacement vector.

9.7 Dispersion of EM waves

Our discussion of EM waves so far has dealt mainly with propagation in vacuum. The wave speed is then independent of the frequency, and the phase velocity is the same as the group velocity. In matter, however, the wave interacts with charged particles, notably electrons, both free and bound. We shall consider first the case of free electrons, as encountered in an ionized gas (plasma) or in a conductor, including the effect of an external constant magnetic field. This is followed by an analysis of the case of elastically bound electrons, illustrating the basic features of resonance dispersion and the related absorption.

Free electrons. In an ionized gas, called a **plasma**, electrons are dissociated from their ions and move about in thermal motion. An EM wave in the plasma will produce forces on the charged particles which will affect wave propagation. The electrons, with much smaller mass than the ions, generally dominate the resulting interaction, and it is often a good aproximation to neglect the motion of the ions. If this is done, the electric current in the plasma is contributed only by the motion of the electrons, and if the velocity is **u**, the corresponding current density is

$$\mathbf{I} = -Ne\mathbf{u} \tag{1}$$

where N is the number density of the electrons and $(-e)$ the charge of an electron.

The Maxwell equations then take the form

$$\mathrm{curl}\,(\mathbf{H}) = \varepsilon_0 \partial\mathbf{E}/\partial t - Ne\mathbf{u} \tag{2}$$

$$\mathrm{curl}\,(\mathbf{E}) = -\mu_0 \partial\mathbf{H}/\partial t \tag{3}$$

where **E** and **H** are the electric and magnetic fields.

The electric force on an electron is $-e\mathbf{E}$ and the magnetic force is $-e\mathbf{u} \times \mathbf{B}$. We assume that there is no static external field in the plasma so that both u and H are the time dependent field variables. The magnetic force then becomes of second order, and if we consider relatively weak fields, it can be neglected in comparison to the electric force. Under such conditions, the equation of motion for an electron becomes

$$m\mathrm{d}\mathbf{u}/\mathrm{d}t = -e\mathbf{E} \tag{4}$$

We shall consider a plane harmonic wave travelling in the z-direction with the E field in the x-direction. The H field is then in the y-direction, so that the complex amplitudes of the field vectors are of the form

$$\mathbf{E}(x, y, z) = \hat{\mathbf{x}}E(x, y)\exp(\mathrm{i}kz) \tag{5}$$

$$\mathbf{H}(x, y, z) = \hat{\mathbf{y}}H(x, y)\exp(\mathrm{i}kz) \tag{6}$$

where $\hat{\mathbf{x}}$ and $\hat{\mathbf{y}}$ are the unit vectors in the x- and y-directions. To simplify the writing somewhat we shall use the short notations E and H for $E(x, y)$ and $H(x, y)$.

It follows that $\mathrm{curl}\,(\mathbf{H}) = \hat{\mathbf{x}}(-\mathrm{i}k)H$ and $\mathrm{curl}\,(\mathbf{E}) = \hat{\mathbf{y}}\cdot(\mathrm{i}k)E$, and the complex amplitudes of Eqs. (2)–(4) become

$$-\mathrm{i}kH = -\mathrm{i}\omega\varepsilon_0 E - Neu \tag{7}$$

$$\mathrm{i}kE = \mathrm{i}\omega\mu_0 H \tag{8}$$

$$-\mathrm{i}\omega mu = -eE \tag{9}$$

Expressing u and H in terms of E from Eqs. (8) and (9), we obtain from

Eq. (7) the dispersion relation

$$k^2 = (\omega^2 - \omega_p^2)/c^2 \tag{10}$$

$$\omega_p^2 = Ne^2/\varepsilon_0 m \quad c^2 = 1/\mu_0\varepsilon_0$$

where $\omega_p/2\pi$ is a characteristic frequency, known as the **electron plasma frequency**, which will be discussed further in Chapter 19.

It is important to note that the propagation constant k is real only if the frequency exceeds the plasma frequency. The inertia of the electrons then limits the velocity and the current density, so that the displacement current $\varepsilon_0 \partial E/\partial t$ is dominant.

At lower frequencies, on the other hand, the conduction current will take over so that the plasma will respond in much the same way as a conductor. The wave amplitude then decays exponentially with z as $\exp(-k_i z)$, where

$$k_i = (\omega/c)\sqrt{(\omega_p/\omega)^2 - 1} \tag{11}$$

which reduces to $k_i \simeq \omega_p/c$ for $\omega \ll \omega_p$. In the high frequency regime, $\omega \gg \omega_p$, we get $k \simeq \omega/c$ as in vacuum. The inertia of the electrons then prevents them from moving.

The dispersion relation can be written also as

$$\omega^2 = \omega_p^2 + c^2k^2 \tag{12}$$

and the expression for the phase velocity becomes

$$v_p = \omega/k = c/[1 - (\omega_p/\omega)^2]^{\frac{1}{2}} \tag{13}$$

This velocity is larger than the speed of light c, but this is not in contradiction with relativity, since, as we shall see in Chapter 14, the energy in a dispersive wave travels with another velocity, the **group velocity**, which is given by

$$v_g = \mathrm{d}\omega/\mathrm{d}k = c[1 - (\omega_p/\omega)^2]^{\frac{1}{2}} \tag{14}$$

which is smaller than c; in fact, $v_p v_g = c^2$.

Wave dispersion in a plasma is of importance not only in laboratory plasmas but also in nature. For example, radio waves will penetrate the ionosphere only if the frequency exceeds the plasma frequency of the free electrons. At lower frequencies, there will be strong reflection from the ionosphere. The free electrons in a conductor behave much like a plasma with a plasma frequency which is larger than that of visible light and the opaqueness of the conductor to visible light is intimately related to the amplitude decay below the plasma frequency.

Since the plasma is isotropic, the direction of propagation will not affect the dispersion relation, and our choice of the particular direction for the E field in our derivation is not essential. In fact, we can obtain the wave equation and the corresponding dispersion relation directly from Eqs. (2)–

(4) by taking the curl of Eq. (2) and the time derivative of Eq. (1). We can then eliminate the magnetic field and obtain a wave equation for the electric field after having expressed u in terms of E from Eq. (4). This leads directly to the dispersion relation (10). We leave the details of the derivation as an exercise.

Magnetized plasma. Rotation of the plane of polarization. The isotropy of the plasma will be eliminated if a static magnetic field is applied. The field now defines a direction in the plasma and the wave transmission characteristics depend on the direction of propagation with respect to the magnetic field. Due to the magnetic force on an electron the velocity u generally will not be in the direction of E, and the general motion will be considerably more complicated than in the previous case with no magnetic field.

We shall consider here only the case when the direction of propagation of the wave is in the same direction as the magnetic field, and we choose this direction to be the z-direction. Thus, with the static magnetic field written as $\mathbf{H} = \hat{\mathbf{z}}H_0$, the total magnetic field becomes $\mathbf{H}' = \hat{\mathbf{z}}H_0 + \mathbf{H}(x, y, t)$ and the magnetic force on an electron $-e\mathbf{u} \times \mathbf{B}'$, where $\mathbf{B}' = \mu_0 \mathbf{H}'$. Again neglecting the nonlinear term $-e\mathbf{u} \times \mathbf{B}$, we obtain the linearized equations

$$\operatorname{curl}(\mathbf{H}) = \varepsilon_0 \partial \mathbf{E}/\partial t - Ne\mathbf{u} \tag{15}$$

$$\operatorname{curl}(\mathbf{E}) = -\mu_0 \partial \mathbf{H}/\partial t \tag{16}$$

$$m\mathrm{d}\mathbf{u}/\mathrm{d}t = -e[\mathbf{E} + (\mathbf{u} \times \hat{\mathbf{z}})B_0] \quad B_0 = \mu_0 H_0 \tag{17}$$

The velocity u then is confined to the xy-plane, although the direction will not remain fixed, and the same applies to $\mathbf{E}$ and $\mathbf{H}$. Consequently, the complex amplitude of the E-vector will be of the form

$$\mathbf{E}(x, y, z) = \mathbf{E}(x, y) \exp(\mathrm{i}kz) \tag{18}$$

with similar expressions for $\mathbf{H}(x, y, z)$ and $\mathbf{u}(x, y, z)$. As before, we shall use the short notations $\mathbf{E}, \mathbf{H}$ and $\mathbf{u}$ for $\mathbf{E}(x, y)$, $\mathbf{H}(x, y)$, and $\mathbf{u}(x, y)$.

With the field vectors all confined to the xy plane, and introducing the propagation vector $\mathbf{k} = \hat{\mathbf{z}}k$, we obtain $\operatorname{curl}(\mathbf{E}) = \mathrm{i}\mathbf{k} \times \mathbf{E}$ and $\operatorname{curl}(\mathbf{H}) = \mathrm{i}\mathbf{k} \times \mathbf{H}$, so that the complex amplitudes of Eqs. (15)–(17) can be written

$$\mathrm{i}\mathbf{k} \times \mathbf{H} = -\mathrm{i}\omega\varepsilon_0 \mathbf{E} - Ne\mathbf{u} \tag{19}$$

$$\mathrm{i}\mathbf{k} \times \mathbf{E} = \mathrm{i}\omega\mu_0 \mathbf{H} \tag{20}$$

$$-\mathrm{i}\omega m\mathbf{u} = -e[\mathbf{E} + (\mathbf{u} \times \hat{\mathbf{z}})\mu_0 H_0] \tag{21}$$

From these equations we can determine the relationship between the field variables $\mathbf{E}, \mathbf{H}$, and $\mathbf{u}$ and the corresponding dispersion relation $k = k(\omega)$. The calculation can be carried out in many different ways. One possibility is to eliminate $\mathbf{H}$ between Eqs. (19) and (20) by multiplying Eq. (20) vectorially by $\mathbf{k}$ and making use of the relation $\mathbf{k} \times (\mathbf{k} \times \mathbf{E}) = -k^2\mathbf{E}$. Insertion of the resulting expression for $\mathbf{k} \times \mathbf{H}$ into Eq. (19) then yields a relation between $\mathbf{E}$ and $\mathbf{u}$ which reduces Eq. (21) to the form

$$\mathbf{u} = \gamma(\hat{\mathbf{z}} \times \mathbf{u}) \tag{22}$$

where $\gamma = \mathrm{i}(\omega_c/\omega)/[1 + (\omega_p^2/2)/(K^2 - 1)]$ and $K = k/(\omega/c)$. We have here introduced the plasma frequency and the cyclotron frequency given by $\omega_p^2 = Ne^2/\varepsilon_0 m$ and $\omega_c = eB_0/m$, where $B_0 = \mu_0 H_0$.

If we express the velocity in terms of its components, we have $\mathbf{u} = \hat{\mathbf{x}}u_x + \hat{\mathbf{y}}u_y$ and $\hat{\mathbf{z}} \times \mathbf{u} = -\hat{\mathbf{x}}u_y + \hat{\mathbf{y}}u_x$, and the component form of Eq. (22) becomes

$$u_x = -\gamma u_y \tag{23}$$

$$u_y = \gamma u_x \tag{24}$$

from which follows $\gamma^2 = -1$ or $\gamma = \pm\mathrm{i}$.

In other words there are two possible solutions for the complex velocity amplitude,

$$\mathbf{u}_+ = u_x(\hat{\mathbf{x}} + \mathrm{i}\hat{\mathbf{y}}), \quad \mathbf{u}_- = u_x(\hat{\mathbf{x}} - \mathrm{i}\hat{\mathbf{y}}) \tag{25}$$

From the discussion of circularly polarized waves in Section 9.6 we recognize these solutions as representing waves of circular polarization with positive and negative rotations about the z-axis, respectively.

After having obtained $\gamma = \pm\mathrm{i}$, we can determine the corresponding dispersion relations for the two characteristic modes of wave propagation, and we find

$$k_\pm = (\omega/c)[1 - a/(1 \pm b)]^{\frac{1}{2}} \tag{26}$$

$$a = \omega_p^2/\omega^2, \quad b = \omega_c/\omega$$

If the static magnetic field is zero, the result reduces to the dispersion relation obtained earlier, in which case we dealt with a linearly polarized wave. With a magnetic field present, a linearly polarized wave cannot be maintained in a fixed plane since the direction of the field vectors change as the wave propagates.

In the discussion of polarization in the previous section, we expressed a linearly polarized wave as a superposition of two circularly polarized waves (with opposite directions of rotation), and if two circularly polarized waves travelled with different speeds, we showed that this resulted in a rotation of the plane of polarization.

Actually we found, that if the propagation constants of the two circularly polarized waves are k_+ and k_-, the angle of rotation of the plane of polarization in a distance z is

$$\phi = \tfrac{1}{2}(k_+ - k_-)z \tag{27}$$

We have now an opportunity to apply this result to the magnetized plasma, using the expressions for the propagation constants in Eq. (26). For a given value of z and magnetic field and a corresponding cyclotron frequency, a measurement of the angle of rotation makes it possible to determine the plasma frequency and hence the electron density.

If the magnetic field is small, so that $\omega_c/\omega \ll 1$, we can expand the square root in Eq. (26) so that the expression for the angle of rotation can be written

$$\phi = \tfrac{1}{2}(k_+ - k_-)z \simeq \tfrac{1}{2}\frac{\omega_c z}{c}\frac{1}{\sqrt{1-a}} \tag{28}$$

$$a = \omega_p^2/\omega^2$$

In Display 9.11 we have illustrated the rotation of the plane of polarization of a linearly polarized wave which enters a magnetized plasma. As it enters the plasma the wave does not remain in the same plane of polarization as a result of the magnetic interaction. As we have seen, the wave within the plasma can be described as the superposition of two circularly polarized waves, which travel with different wave speeds. In the example given, the difference is such that the plane of polarization is turned by an angle of π in a distance of 16 wavelengths of the incident wave. In this case the incident wave is polarized along the x-axis and the same holds true for the transmitted wave.

Bound electrons. An atomic electron is often modelled classically as elastically bound with a spring constant K and a corresponding resonance frequency $\omega_0 = \sqrt{K/m}$, where m is the mass of the electron.

As in a standard mass–spring oscillator, we introduce also a friction damping, and the interaction of the bound electron with a linearly polarized wave then results in the complex velocity amplitude of the electron,

$$\begin{aligned}\mathbf{u} &= -e\mathbf{E}/(R - \mathrm{i}\omega m + \mathrm{i}K/\omega)\\ &= (-e\mathbf{E}/-\mathrm{i}\omega m)/(1 - 1/\Omega^2 + \mathrm{i}D/\Omega) = -e\mathbf{E}/(-\mathrm{i}\omega m')\\ m' &= m(1 - 1/\Omega^2 + \mathrm{i}D/\Omega), \quad \Omega = \omega/\omega_0, \quad D = R/\omega_0 m\end{aligned} \tag{29}$$

where $\mathbf{E}$ is the complex amplitude of the electric field.

We now consider a material with N such oscillators per unit volume and wish to determine the dispersion of an EM wave in this material. The field equations will be the same as in the case of free electrons except that the mass m is replaced by m', as defined in Eq. (29). Accordingly, the dispersion relation is obtained from the dispersion relation (10) for the free electron case by replacing $\omega_p^2 = Ne^2/\varepsilon_0 m$ by $\omega_p'^2 = Ne^2/\varepsilon_0 m' = \omega_p^2\Omega^2/(\Omega^2 - 1 + \mathrm{i}D\Omega)$, so

Display 9.11.

Rotation of the plane of polarization in a magnetized plasma.

$$\mathrm{curl}\,(\mathbf{H}) = \varepsilon_0 \partial \mathbf{E}/\partial t - Ne\mathbf{u} \quad (1)$$

$$\mathrm{curl}\,(\mathbf{E}) = -\mu_0 \partial \mathbf{H}/\partial t \quad (2)$$

$$m\mathrm{d}\mathbf{u}/\mathrm{d}t = -e[\mathbf{E} + (\mathbf{u} \times \hat{\mathbf{z}})\mathbf{B}_0] \quad B_0 = \mu_0 H_0 \quad (3)$$

Complex amplitude equations

$$\mathrm{i}\mathbf{k} \times \mathbf{H} = -\mathrm{i}\omega\varepsilon_0 \mathbf{E} - Ne\mathbf{u} \quad (4)$$

$$\mathrm{i}\mathbf{k} \times \mathbf{E} = \mathrm{i}\omega\mu_0 \mathbf{H} \quad (5)$$

$$-\mathrm{i}\omega m\mathbf{u} = -e[\mathbf{E} + (\mathbf{u} \times \hat{\mathbf{z}})B_0] \quad (6)$$

$$\mathbf{u} = \gamma(\hat{\mathbf{z}} \times \mathbf{u}), \quad \gamma = \mathrm{i}(\omega_c/\omega)/(1 + \omega_p^2/c^2K^2), \quad K^2 = k^2 - (\omega/c)^2 \quad (7)$$

$$\gamma^2 = -1, \quad \gamma = \pm \mathrm{i}, \quad \pm \omega_c/\omega = 1 + \omega_p^2/c^2K_\pm^2 \quad (8)$$

Dispersion relation

$$k_\pm = (\omega/c)[1 - a/(1 \pm b)]^{1/2} \quad a = \omega_p^2/\omega^2 \quad b = \omega_c/\omega \quad (9)$$

$\omega_c/\omega \ll 1$

$$k_\pm \simeq (\omega/c)[1 - a \pm ab]^{\frac{1}{2}} \simeq (\omega/c)\sqrt{1-a}\,[1 \pm \tfrac{1}{2}ab/(1-a)] \quad (10)$$

Rotation of plane of polarization

$$\phi = \tfrac{1}{2}(k_+ - k_-)z \simeq \tfrac{1}{2}(2\pi z/\lambda)ab/\sqrt{1-a}, \quad \omega_c/\omega \ll 1 \quad (11)$$

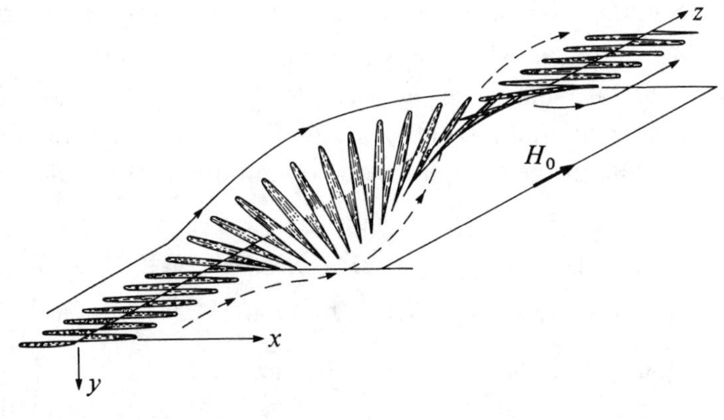

Display 9.12.

Dispersion of EM waves propagating through a material with a resonance (absorption line) at the frequency $\omega_0/2\pi$.

$\mathrm{curl}(\mathbf{H}) = \varepsilon_0 \partial\mathbf{E}/\partial t - Ne\mathbf{u}$ (1)

$\mathrm{curl}(\mathbf{E}) = -\mu_0 \partial\mathbf{H}/\partial t$ (2)

$m\mathrm{d}\mathbf{u}/\mathrm{d}t + R\mathbf{u} + K\boldsymbol{\xi} = -e\mathbf{E}$ (3)

Complex amplitude equations

$\mathbf{H}(x, y, z) = \mathbf{H}(x, y)\exp(\mathrm{i}kz)$ (similar for $\mathbf{E}$ and $\mathbf{u}$) (4)

Let $\mathbf{H}(x, y) = \mathbf{H}$. (Similar for $\mathbf{E}$ and $\mathbf{u}$) (5)

$\mathrm{i}\mathbf{k} \times \mathbf{H} = -\mathrm{i}\omega\varepsilon_0\mathbf{E} - Ne\mathbf{u}$ (6)

$\mathrm{i}\mathbf{k} \times \mathbf{E} = \mathrm{i}\omega\mu_0\mathbf{H}$ (7)

$-m\mathrm{i}\omega\mathbf{u} + R\mathbf{u} + \mathrm{i}(K/\omega)\mathbf{u} = -e\mathbf{E}$ (8)

Dispersion relation

$k = (\omega/c)[1 - a_0/(1 - \Omega^2 - \mathrm{i}D\Omega)]^{\frac{1}{2}} = (\omega/c)(K_\mathrm{r} + \mathrm{i}K_\mathrm{i})$ (9)

$\Omega = \omega/\omega_0, \quad K = k/(\omega/c)$

$a_0 = \omega_\mathrm{p}^2/\omega_0^2, \quad D = R/\omega_0 m, \quad \omega_\mathrm{p}^2 = Ne^2/\varepsilon_0 m, \quad \omega_0^2 = K/m$ (10)

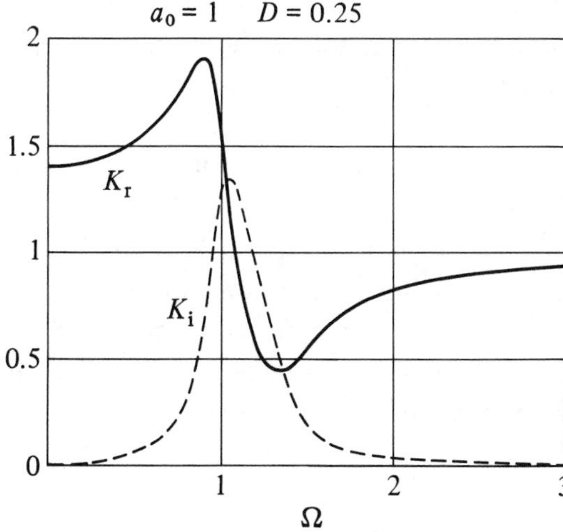

that

$$\begin{aligned} k/(\omega/c) &= \sqrt{1-(\omega_p'/\omega)^2} = [1 + a_0/(1-\Omega^2 - \mathrm{i}D\Omega)]^{\frac{1}{2}} \\ &= \sqrt{A + \mathrm{i}B} = K_r + \mathrm{i}K_i \end{aligned} \tag{30}$$

$$A = 1 + a_0(1-\Omega^2)/N, \quad B = a_0 D\Omega/N,$$
$$N = (1-\Omega^2)^2 + (D\Omega)^2, \quad a_0 = \omega_p^2/\omega_0^2$$

The normalized propagation constant K_r represents the index of refraction. It approaches unity as the frequency goes to infinity, as for a plasma, and approaches $\sqrt{1+a_0}$ as the frequency goes to zero.

In a solid the quantity $a_0 = \omega_p^2/\omega_0^2$ generally is larger than 1, in a gas it is much less than 1. As an illustration, let the frequency ω_0 correspond to the ionization potential of hydrogen. Then, introducing the Bohr radius r_0 and the interparticle distance l given by $l^3 = 1/N$, we find $a_0 = 16\pi(r_0/l)^3$.

In Display 9.12 we have shown the frequency dependence of the real and imaginary parts of the propagation constant for the special case when $a_0 = \omega_p^2/\omega_0^2 = 1$ and $D = R/\omega_0 m = 0.25$. It illustrates the basic features of **resonance dispersion**. At low frequencies the real part of the normalized propagation constant K_r, which is the same as the index of refraction, starts out being essentially independent of frequency. As the resonance frequency is approached, K_r increases, which means that the phase velocity decreases. In the region of resonance, there is a sudden drop in K_r to a region with an index of refraction less than 1 and a corresponding phase velocity larger than c. In this region the response of the electron is mass controlled and yields a response similar to that of the free electron plasma.

Close to resonance, the imaginary part of the propagation constant goes through a maximum, which is approximately equal to the value at resonance, $K_i = \sqrt{a_0/D}$. In a solid or liquid, with a_0 of the order of magnitude unity, the corresponding attenuation will be quite large and the material becomes essentially opaque even for very small thicknesses, of the order of a few wavelengths. A material has a distribution of resonances, and it can be opaque over an extended range of frequencies.

Examples

E9.1 *Rotation of plane of polarization.* With reference to the figure in Display 9.11, what can you say about the difference between the speeds of propagation of circularly polarized waves in the medium through which the waves are transmitted?

Solution. The figure refers to a wave travelling through a plasma, in which there is a uniform magnetic field in a finite region, as shown. The wave travels in the direction of the magnetic field and is linearly polarized as it enters the magnetic field with the E-field in the x-direction. The wave exists from the magnetized region with the plane of polarization turned 180 degrees.

By inspection, we note that the length z of the magnetized region is 16 wavelengths. According to Eq. 9.6.(13),, the general expression for the angle of rotation of the plane of polarization is

$$\phi = \tfrac{1}{2}(k_+ - k_-)z \tag{1}$$

where k_+ and k_- are the propagation constants for circularly polarized waves turning about the z-axis in the positive and negative directions, respectively.

With the corresponding wave speeds denoted by v_+ and v_-, we have, with $k_+ = \omega/v_+$ and $h_- = \omega/v_-$.

$$\phi = \tfrac{1}{2}\omega(v_- - v_+)/v_+v_- = \tfrac{1}{2}[(v_- - v_+)/v]2\pi z/\lambda \tag{2}$$

where we have put $v_+ \simeq v_- = v$ in the denominator and $\omega/v = 2\pi/\lambda$.

With $\phi = \pi$ and $z/\lambda = 16$, we obtain

$$(v_- - v_+)/v = 1/16 \tag{3}$$

In other words, the difference between the wave speeds of circularly polarized waves turning in the negative and positive directions is about 6.3%.

Problems

9.1 *Plane EM wave.* The intensity of a plane harmonic EM wave is $I = 1\,\mathrm{W/cm^2}$. Determine

(a) the amplitudes of the electric and the magnetic field,
(b) the radiation pressure on a perfect absorber,
(c) the radiation pressure on a perfect reflector.

9.2 *Cylindrical wave.* With reference to Display 9.7(a), the power radiated by an EM dipole antenna between two plane parallel conducting planes $P = 1\,\mathrm{kW}$. The separation between the planes is $d = 0.5\,\mathrm{m}$ and the wavelength is much larger than d.

(a) What is the intensity of the EM wave at a distance of $R = 10\,\mathrm{m}$ from the source?
(b) What are the corresponding values of E and H?

9.3 *Dipole radiation.* A harmonic dipole at $x = y = z = 0$ points in the z-direction. It radiates a power $P = 1\,\mathrm{kW}$ into free space. What is the intensity and the corresponding values of E and H in the EM wave at the locations (a) $x = 10\,\mathrm{m}$, $y = z = 0$, (b) $x = y = z = 10\,\mathrm{m}$, and (c) $x = y = 0$, $z = 10\,\mathrm{m}$?

9.4 *Radiation from accelerated electron.* An electron, travelling in the x-direction with a velocity $0.1\,c$, is accelerated to a speed of $0.2\,c$ in a time $\tau = 1\,\mu\mathrm{sec}$. What is the total EM energy radiated, assuming the acceleration to be constant, if the final velocity is

(a) in the positive x-direction?
(b) in the negative x-direction?
(c) in the positive y-direction?

9.5 *EM wave in a plasma.* A fully ionized plasma has an electron density of $N = 10^{16}$ electrons/cm^3.

(a) A plane EM wave is found to travel through the plasma without attenuation with a phase velocity which is twice the group velocity. What is the frequency of the wave?
(b) What happens if the frequency is reduced by a factor of 4?

9.6 *Polarization.* Two harmonic dipoles A and B of equal strength are located at the origin $x = y = z = 0$. One is oriented along the z-axis and the other along the y-axis. Describe the polarization of the resulting wave in the far field at a location on the x-axis if

(a) A and B are in phase,
(b) B lags behind A by the phase angle $\pi/2$.
(c) Answer the questions in (a) and (b) if the point of observation is at $x = z = L$, $y = 0$.

9.7 *Rotation of the plane of polarization.* A plane EM wave travels along the constant magnetic field in a homogeneous plasma. After a travel distance of L, where the plane of polarization of the wave has been rotated an angle $\phi = \pi/6$, determine the magnetic field. Assume $\omega \gg \omega_p$.

9.8 *Electromagnetic interaction.* A linearly polarized EM wave, travelling in the z-direction, has the electric field in the x-direction, $E_x = E_0 \cos(\omega t - kz)$, where $k = \omega/c$.

(a) Determine the corresponding H-field.
(b) The wave interacts with a charged particle (charge q, mass m) with the equilibrium position at $x = y = z = 0$. Determine the particle displacement produced by the wave. Account for the effect of the E-field only.
(c) To get an idea of the effect of the magnetic field, use the displacement obtained in (b) and calculate the magnetic force and the corresponding 'secondary' displacement of the particle for small values of t.

9.9 *Polarizer.* As an illustration of a polarizer we consider a slot in a plate which is used as a polarizer for a wave on a string which runs through the slot. The string is horizontal and perpendicular to the plate. The slot is inclined at an angle ϕ, as shown. The displacement of the incident wave is in the horizontal plane with the displacement $A \cos(\omega t - kx)$. Neglect friction in the slot. The tension in the string is S and the mass per unit length is μ.

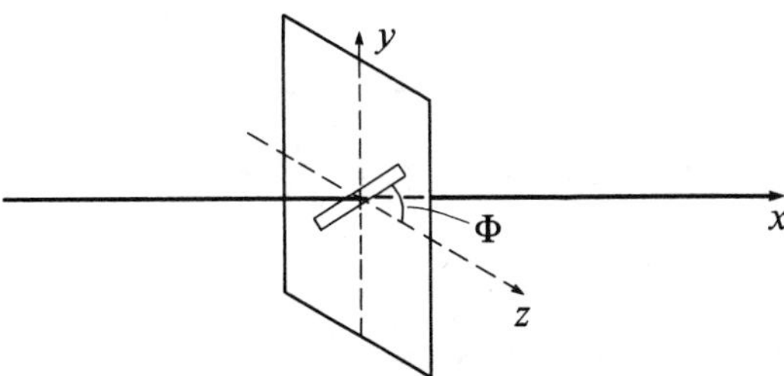

(a) What is the amplitude of the transmitted wave?

(b) What fraction of the incident wave energy is transmitted?

(c) Determine the total displacement function in front of the plate and describe the state of polarization of this wave.

(d) Repeat the analysis for the case when there is a friction force Ru on the string in the slot, where u is the velocity of the string.

(e) Repeat the analysis for the case when the incident wave is circularly polarized.

10

Acoustic waves in fluids

In Chapter 6 we summarized the characteristics of waves on several types of continuous 'transmission lines'. All these waves could be described by the same field equations and wave equation, which were obtained from the analysis of wave motion on a lattice of masses and springs (or inductances and capacitances) by letting the lattice spacing go to zero. One-dimensional waves in fluids and solids were included amongst these waves.

In this chapter we shall put the theory of acoustic waves in a fluid on firmer ground by starting from the equations of fluid motion. This will enable us to extend the previous analysis to waves in more than one dimension, and to determine the conditions, under which the linear wave theory is valid.

10.1 Description of fluid motion

From the standpoint of particle mechanics, the microscopic motion of a fluid (liquid, gas, plasma) is a problem of an enormous number of particles (molecules), and as such it is not tractable for analysis from first principles, unless some simplifying assumptions are made. In the macroscopic theory of fluid motion, the first step toward such a simplification involves a decomposition of the fluid motion into an internal (molecular) or 'thermal' motion, and a macroscopic motion.

The details of the internal motion and intermolecular forces are avoided by describing the relevant effects in terms of appropriate statistical averages, which are expressed in terms of **thermodynamic state variables**, pressure P, temperature T, and density ρ. These variables are related through an **equation of state** $P = (\rho, T)$, which generally has to be determined experimentally. This means that only **two** of the thermodynamic variables are

independent. For an ideal gas, the equation of state is known to be

$$P = r\rho T \tag{1}$$

r = gas constant per unit mass = R/M
R = gas constant per mole = 8.3 joule/K = 8.3×10^7 erg/K
M = molar mass ($\simeq$ 29 g for air)

One molar mass contains $N_0 = 6.02 \times 10^{23}$ molecules. If the mass of one molecule is m, we have $M = N_0 m$, and the number of molecules per unit volume is $N = \rho/m$. In terms of these quantities, the equation of state for the ideal gas can be written as

$$P = Nk_B T \tag{2}$$

k_B = Boltzmann constant = $R/N_0 = 1.38 \times 10^{-23}$ joule/K
$N_0 = 6.02 \times 10^{23}$ (Avogadro's number)
$N = \rho/m$ = number of molecules per unit volume

We have here introduced the important Boltzmann constant k_B, which can be considered to be the 'gas constant' per particle.

The macroscopic motion can be described in two different ways. The first is a direct adaptation of the method used in particle mechanics. Thus, the fluid is regarded as a collection of macroscopic elements, the positions of which are described as a function of time. This method is called the **Lagrangian description** of fluid motion.

In the second method, the **Eulerian description**, the velocity **U** of the fluid is used as a field variable and is regarded as a function of the fixed coordinates x, y, z and of time t. The velocity at time t concerns the fluid element which happens to be at x, y, z at time t. At a later time another particle has moved into position and its velocity is recorded. (In the Lagrangian description the time dependence refers to one and the same fluid element).

From the known Eulerian time dependence of the velocity throughout the fluid, the trajectory of a particular fluid element can be determined. The Eulerian description is the most common, and we shall use it here.

The field variables are regarded as continuous functions of the space coordinates x, y, z and time t. In three dimensions, the velocity has three components, which, together with the two independent thermodynamic variables (say, P and ρ), make a total of five field variables, which we wish to determine as functions of the independent coordinates x, y, z, and t. To make this possible, we need five independent equations. These equations can be obtained from the conservation laws for mass (one equation), momentum (three equations), and energy (one equation). If motion only in one dimension is involved, the number of required equations is reduced to three.

10.2 Fluid equations

We shall restrict the discussion to motion in one dimension, but the generalization to three dimensions is straight forward, as will be demonstrated in Section 10.3, at the end of the chapter.

Conservation of mass. In expressing the conservation laws, we consider a 'control volume' in space between the planes at x and $x + \mathrm{d}x$, as shown in Display 10.1. It has the form of a parallelepiped with the surface normal to the x-axis having unit area.

The mass of fluid per unit time entering the volume through the surface at x at time t is

$$M(x,t) = \rho(x,t)U(x,t) \tag{1}$$

which is called the **mass flux**. The corresponding rate of mass leaving the volume through the surface at $x + \mathrm{d}x$ at the same time t is $M(x + \mathrm{d}x, t)$. The difference between $M(x,t)$ and $M(x + \mathrm{d}x, t)$ is the net mass influx to the volume, and it can be expressed as $M(x,t) - M(x + \mathrm{d}x, t) = -(\partial M/\partial x)\,\mathrm{d}x$ for sufficiently small $\mathrm{d}x$.

The total mass inside the control volume at time t is $\rho(x,t)\,\mathrm{d}x$ (or $\frac{1}{2}(\rho(x,t) + \rho(x + \mathrm{d}x, t))$, which goes to $\rho(x,t)\,\mathrm{d}x$ as $\mathrm{d}x$ goes to zero. As a result of the net influx of mass into the volume the mass inside has to increase, and the time rate of change is equal to the influx,

$$\partial\rho/\partial t + \partial M/\partial x = 0 \tag{2}$$

It should be pointed out that the thermal motion does not contribute to the mass flux. Eq. (2) is sometimes referred to as the continuity equation, in which a time rate of change of a density is balanced by a spatial rate of a flux.

Conservation of momentum. The conservation of momentum equation can be expressed in a similar manner in terms of a momentum density and a momentum flux. An important difference is that although the thermal motion does not contribute to the mass flux, it does contribute to the momentum flux, which, by definition, is the pressure P in the fluid. (The momentum density ρU is the same as the mass flux, and there is no contribution to it by the thermal motion).

As indicated above, the rate of mass transported through the surface at x is $M = \rho u$. The momentum of this mass is MU, which is transferred to

Display 10.1.

Fluid equations

Field variables

Velocity, $U(x,t)$ (1)

Density, $\rho(x,t)$ (2)

Pressure, $P(x,t)$ (3)

Temperature, $T(x,t)$ (4)

Equation of state: $P = P(\rho, T)$ (5)

Conservation of mass

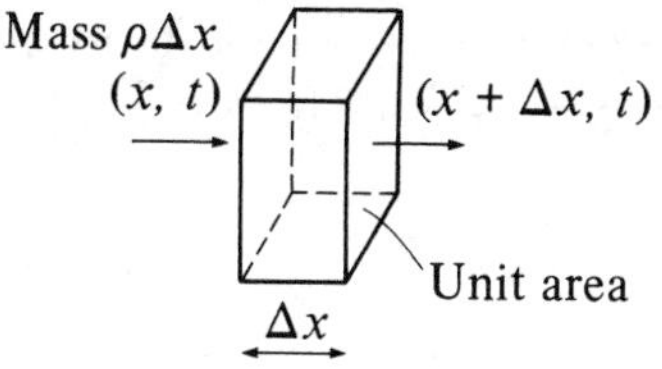

Mass density ρ (6)

Mass flux $M = \rho U$ (7)

$\partial\rho/\partial t + \partial M/\partial x = 0$ (8)

Conservation of momentum

Momentum density $M = \rho U$ (9)

Momentum flux $G = P + MU = P + \rho U^2$ (10)

$\partial M/\partial t + \partial G/\partial x = 0$ (11)

Conservation of energy

Energy density W (12)

Energy flux S (13)

$\partial W/\partial t + \partial S/\partial x = 0$ (14)

Isentropic flow (No viscosity and heat conduction)

$P = P(\rho)$ (15)

Fluid equations become

$\partial\rho/\partial t = -\partial M/\partial x \quad M = \rho U$ (16)

$\partial M/\partial t = -\partial G/\partial x \quad G = P + \rho U^2$ (17)

$P = P(\rho)$ (18)

the control volume from the left. In addition, the momentum transfer due to the thermal motion is expressed by the pressure $P(x, t)$, which is a result of the collisions between the particles on the left with the particles inside the control volume. Thus, the total rate of momentum transferred to the particles in the control volume through the surface at x is

$$G = P + MU = P + \rho u^2 \tag{3}$$

which is called the **momentum flux.**

Conservation of energy. The energy density $W = \rho U^2/2 + \rho E$ in a fluid is the sum of the kinetic energy $\rho u^2/2$ of the (macroscopic) motion and the internal energy ρE of the fluid. In a gas, the internal energy is dominated by the kinetic energy of the thermal motion of the molecules, the potential energy between the molecules being small (and equal to zero, by definition, in an ideal gas).

If we introduce an energy flux S, we can express conservation of energy as

$$\partial W/\partial t + \partial S/\partial x = 0 \tag{4}$$

by analogy with the conservation of mass, the energy density taking the place of the mass density and the energy flux replacing the mass flux. In order to determine S, we have to establish how the internal energy varies during the motion.

The internal energy of an element can be increased in two ways, by heat flow into the element from the surrounding fluid (or from external sources) and by the work of compression by the pressure of the surrounding fluid. This energy balance and the experimental fact that the internal energy is a function of the thermodynamic state variables (such as temperature and density), is called **the first law of thermodynamics**. By combining it with the momentum and mass conservation equations, one can express the energy balance by Eq. (4), in which W and S can be expressed in terms of the various field variables.

Combining the conservation equations with the equation of state, $P = p(\rho, T)$, we have four equations for the four variables U, ρ, P, and T. Although, in principle, this should make it possible to solve the equations for the four variables, the fact that the equations are nonlinear (because of the terms ρU and ρU^2) makes the problem difficult. In fact, a general solution is not known and only a few exact solutions to specific problems exist.

With viscosity and heat conduction included as well as external forces, the equations in their three-dimensional form, should be capable of describing the many complex motions of fluids, from the motion of the atmosphere to flow in capillaries. Many professional life times have been spent in the study of these equations, and research is still ongoing with considerable efforts devoted to computer simulation of fluid flow and approximate and numerical methods of solving the equations in particular situations.

The reason for our present interest in the equations is that we wish to use them as a basis for a study of the acoustic waves in a fluid. We consider the special case of an ideal gas without viscosity and heat conduction. In that case, we can combine the first law of thermodynamics and the equation of state to eliminate temperature variable T and express the pressure as function of density alone $P = P(\rho)$. Under these conditions the fluid equations reduce to

$$\partial\rho/\partial t = -\partial M/\partial x \quad (M = \rho U) \tag{5}$$

$$\partial M/\partial t = -\partial G/\partial x \quad (G = P + \rho U^2) \tag{6}$$

$$P = P(\rho) \tag{7}$$

for the three variables U, P, and ρ.

10.3 Acoustic field equations

After this brief review of the equations of fluid flow, we are prepared to discuss the equations for acoustic waves in a fluid.

The unperturbed field variables are assumed to be time independent and are given the subscript 0. The acoustic perturbations of density, velocity, mass flux, and pressure are denoted by $\hat{\rho}, u, m$, and p. We shall assume that the density perturbation is small, so that $\hat{\rho} \ll \rho_0$. Under these conditions, we obtain the corresponding pressure perturbation by expanding $P(\rho + \hat{\rho})$ in a Taylor series

$$P(\rho_0 + \hat{\rho}) \approx P_0 + (\mathrm{d}P/\mathrm{d}\rho)\hat{\rho} + (\tfrac{1}{2})(\mathrm{d}^2P/\mathrm{d}\rho^2)\hat{\rho}^2 + \cdots$$

$$\simeq P_0 + (\mathrm{d}P/\mathrm{d}\rho)\hat{\rho} \tag{1}$$

$$P_0 = P(\rho_0)$$

$$p = (\mathrm{d}P/\mathrm{d}\rho)\hat{\rho} \tag{2}$$

The compressibility, which corresponds to the compliance per unit length of a spring, is defined as

$$\kappa = (1/\rho_0)\,\mathrm{d}\rho/\mathrm{d}P \tag{3}$$

and if we use this quantity in Eq. (2), we obtain

$$p = (1/\kappa\rho_0)\hat{\rho} \tag{4}$$

For an ideal gas the compressibility is known to be inversely proportional to the static pressure P_0, where P_0 is of the order of $\rho_0 v_t^2$, where v_t is the thermal speed of the molecules.

Display 10.2.

Acoustic field equations and wave equation.

Acoustic field variables

Velocity $u(x, t)$ (1)

Density $\hat{\rho} = \rho - \rho_0$ (2)

Pressure $p = P - P_0$ (3)

Isentropic conditions

Compressibility $\kappa = (1/\rho_0)\,\mathrm{d}P/\mathrm{d}\rho$ (4)

$P(\rho) = P(\rho_0 + \hat{\rho}) \simeq P(\rho_0) + (\mathrm{d}P/\mathrm{d}\rho)\hat{\rho} = P_0 + (\rho_0\kappa)\hat{\rho}$ (5)

$p = P - P = (1/\rho_0\kappa)\hat{\rho}$ (6)

Acoustic field equations

One dimension

$\kappa\partial p/\partial t = -\partial u/\partial x$ (from $\partial\hat{\rho}/\partial t + \rho_0\partial u/\partial x = 0$) (7)

$\rho_0\partial u/\partial t = -\partial p/\partial x$ (8)

$\partial^2 p/\partial x^2 = (1/v^2)\partial^2 p/\partial t^2$ (9)

$v^2 = 1/\rho_0\kappa$ (10)

Three dimensions

$\kappa\partial p/\partial t = -\operatorname{div}(\mathbf{u})$ (11)

$\rho_0\partial\mathbf{u}/\partial t = -\operatorname{grad}(p)$ (12)

$\nabla^2 p = (1/v^2)\partial^2 p/\partial t^2$ (13)

Harmonic time dependence

$\nabla^2 p(x) + k^2 p(x) = 0$ (14)

$k = \omega/v$ (15)

Energy density and flux

$W = \frac{1}{2}\rho_0 u^2 + \frac{1}{2}\kappa p^2$ (16)

$S = p\mathbf{u}$ (17)

The perturbation in the mass flux is $m_1 = (\rho_0 + \hat{\rho})(U_0 + u) - \rho_0 U_0$. We assume that the unperturbed fluid to be at rest, $U_0 = 0$, so that $m = \rho_0 u + \hat{\rho} u \simeq \rho_0 u$, where we have made use of the assumption that $\hat{\rho} \ll \rho_0$. Similarly, the perturbation in the total momentum flux $G = \rho U^2 + P$ is $G_1 \simeq p$, since, as will be clear later, $\rho_0 u^2$ is negligible compared to p if u is much smaller than the wave speed.

Under these conditions we have $\partial\hat{\rho}/\partial t = (\kappa\rho_0)\partial p/\partial t$, and the fluid equations (5)–(7) reduce to

$$\rho_0 \partial u/\partial t = -\partial p/\partial x \tag{5}$$

$$\kappa \partial p/\partial t = -\partial u/\partial x \tag{6}$$

which are the familiar field equations for the acoustic field, which we arrived at in Chapter 6 from studies of the continuous transmission line. Unlike the complete fluid equations, the acoustic equations are linear in the field variables and the process of going from the fluid equations to the acoustic equations by omitting nonlinear terms is usually referred to as linearization.

The analysis readily can be generalized to include the case when the unperturbed fluid is in motion, i.e. when U_0 is not zero, which is referred to in one of the problems.

In the case of three-dimensional motion, the fluid and acoustic equations have to be modified to account for the y- and z-components of the velocity.

In the mass conservation equation $\partial\rho/\partial t = -\partial(\rho U/\partial x)$, the net mass flux entering the control volume now contains a contribution from each of the velocity components, and the term $\partial(\rho U)/\partial x$ has to be replaced by $\partial(\rho U_x) + \partial(\rho U_y)/\partial y + \partial(\rho U_z)/\partial z = \operatorname{div}(\rho \mathbf{U})$. The corresponding linearized version $\partial\hat{\rho}/\partial t = -\rho_0 \operatorname{div}(\mathbf{u})$, with $\hat{\rho} = \kappa\rho_0 p$, is

$$\kappa \partial p/\partial t = -\operatorname{div}(\mathbf{u}) \tag{7}$$

Similarly, the linearized momentum equation $\rho_0 \partial u/\partial t = -\partial p/\partial x$ has to be supplemented with analogous equations for the y- and z-components of the velocity. Since the components of $\operatorname{grad}(p)$ are $\partial p/\partial x, \partial p/\partial y, \partial p/\partial z$, it follows that the three momentum equations for the velocity components can be combined into the vector equation

$$\rho_0 \partial \mathbf{u}/\partial t = -\operatorname{grad}(p) \tag{8}$$

By taking the time derivative of the first of these equations and divergence of the second, we can eliminate velocity to obtain the three-dimensional wave equation

$$\nabla^2 p = (1/v^2)\partial^2 p/\partial t^2 \tag{9}$$

where we have used $\operatorname{div}[\operatorname{grad}(p)] = \nabla^2 p = \partial^2 p/\partial x^2 + \partial^2 p/\partial y^2 + \partial^2 p/\partial z^2$.

Quantity v is the speed of sound, as obtained in Chapter 6. For a gas the compressibility is $\kappa = 1/\gamma P_0$, where, for a diatomic gas, $\gamma = 1.4$. Accounting for the equation of state $P_0 = r\rho_0 T_0$, we can express the sound speed as

$$v = \sqrt{\gamma P_0/\rho_0} = \sqrt{\gamma r T_0} = \sqrt{\gamma R T_0/M} = \sqrt{\gamma k_B T_0/m} \qquad (10)$$

where, as before, R is the gas constant per mole, M the molar mass, k_B the Boltzmann constant, and m the mass of a molecule.

It is important to note that for a given gas the sound speed depends only on the temperature T_0. Since T_0 is proportional to the average kinetic energy $mv_t^2/2$ of the molecules, the sound speed will be of the same order of magnitude as the thermal molecular speed.

With $R = 8.3 \times 10^7$ erg/K we obtain for the sound speed of air ($M = 29$ g and $\gamma = 1.4$) the value 330.7 m/sec at a temperature of 273 K (0 °C).

Wave energy density and energy flux. By analogy with the discussion of wave energy density and energy flux for a one-dimensional wave in Chapter 6, we can derive an energy conservation relation from the field equations (7) and (8). Thus, with scalar multiplication of Eq. (7) by **u** and multiplication of Eq. (8) by p, we obtain by adding the equations

$$\partial W/\partial t + \operatorname{div}(\mathbf{S}) = 0 \qquad (11)$$

where the wave energy density W and the wave energy flux S are

$$W = \tfrac{1}{2}\rho_0 u^2 + \tfrac{1}{2}\kappa p^2 \qquad (12)$$

$$\mathbf{S} = p\mathbf{u} \qquad (13)$$

We have here used $p\operatorname{div}(\mathbf{u}) + \mathbf{u}\cdot\operatorname{grad}(p) = \operatorname{div}(p\mathbf{u})$ and put $\mathbf{u}\cdot\mathbf{u} = \mathbf{u}^2$. The wave energy flux S is sometimes called the acoustic intensity.

In the special case of a one-dimensional wave (plane wave) the ratio between pressure and velocity is the wave impedance $Z = \rho_0 v$ so that $S = Zu^2 = p^2/Z$. For air at 0 °C and atmospheric pressure, $\rho_0 = 0.00129$ g/cm^3 and $v = 330.7$ m/sec, and we obtain the wave impedance $Z = \rho_0 v \simeq$ 430 MKS.

Discussion. As indicated at the end of Section 6.8, our field equations for mechanical waves were linearized, i.e. terms of second and higher order in the field variables were omitted. Wave energy density and energy flux, being products of solutions to the linearized equations, are themselves of second order, and the question arises whether they are 'true' energy quantities, correct to second order. In other words, if second order quantities had been included in the original equations, would the expressions for W and S in Eqs. (12) and (13) still satisfy the conservation equation (11)? It can be shown, that this is indeed true as long as the mean velocity U_0 of the fluid is zero. If U_0 is not zero, the expressions for W and S have to be modified. For further discussion of this and related matters, including a mass flow paradox, we refer to Problems 10.8 and 10.9.

Returning to Eq. (12) for the energy density, we note that term $\rho_0 u^2/2$ is the kinetic energy and $\kappa p^2/2$ the potential energy per unit volume. The latter is analogous to the energy of compression of a spring. For an ideal gas, the acoustic potential energy density corresponds to a change in the thermal kinetic energy of the molecules.

The acoustic energy flux (intensity) is a vector with the same direction as the oscillatory fluid velocity.

In the case of harmonic time dependence we are generally interested in the time averages $\langle W \rangle$ and $\langle \mathbf{S} \rangle$. By taking the time average of the energy equation (11) over one period, we note that $\langle W \rangle$ is independent of time, and Eq. (11) reduces to div $\langle \mathbf{S} \rangle = 0$. This means that there is no acoustic energy being created or lost in the bulk of the fluid (we have assumed no losses and no source distribution outside the primary source).

An equivalent statement is that the integral of the normal component of $\langle S \rangle$ over a closed surface which does not contain any source will be zero, i.e. the average power entering one part of the surface must by balanced by the power out through another part of the surface. If the surface encloses a source, the integral will be constant, independent of the location and equal to the power emitted by the source.

For a plane wave we have $p = Zu$, and the kinetic and potential energy densities are the same. In that case we have

$$S = Zu^2 = \rho_0 u^2 v = Wv \quad \text{(plane wave)} \tag{14}$$

with S being in the direction of propagation.

10.4 Spherical waves

The spherically symmetrical acoustic wave and its source play the same basic role as does the electric monopole in the study of the electrostatic field. As we recall, the field from an arbitrary charge distribution can be expressed in terms of a superposition of monopole fields, and a similar result applies to harmonic waves, as will be discussed later in this chapter.

Much of what we have learned about plane waves can be applied to spherical waves. One reason is that with complete spherical symmetry, the field variables depend only on one spatial coordinate, the radial distance r from the origin. Furthermore, at large distances from the origin, the curvature of the spherical wave front is small, and the wave can then be approximated locally by a plane wave.

Display 10.3.

Spherical waves

$$\nabla^2 p = (1/r^2)(\partial/\partial r)(r^2 \partial p/\partial r) = (1/v^2)\partial^2 p/\partial t^2 \tag{1}$$

$$p(r, t) = (1/r)\Psi(r, t) \tag{2}$$

$$\nabla^2 \Psi = (1/v^2)\partial^2 \Psi/\partial t^2 \tag{3}$$

Harmonic time dependence:

$$\nabla^2 p(r) + k^2 p(r) = 0 \tag{4}$$

$$p = (A/r)\exp(\mathrm{i}kr) + (B/r)\exp(-\mathrm{i}kr) \tag{5}$$

$$u(r) = (1/\mathrm{i}\omega\rho_0)\mathrm{d}p(r)/\mathrm{d}r \tag{6}$$

Outgoing wave:

$$p(r) = A\exp(\mathrm{i}kr) \tag{7}$$

$$u(r) = (A/\rho_0 v)(1/r)[1 + \mathrm{i}(1/kr)]\exp(\mathrm{i}kr) \tag{8}$$

In the case of spherical symmetry we have

$$\nabla^2 p = (1/r^2)\partial/\partial r(r^2 \partial p/\partial r)$$

as we recall by evaluating div [grad(p)].

(By definition, div (**A**) is the limit value of the ratio between the surface integral of the normal component of the vector A and the volume of a volume element. We select a volume element between the spherical surfaces at r and $r + \mathrm{d}r$, occupying the solid angle $\Delta\Omega$. The only contributions to the surface integral of **A** = grad (p) come from the spherical surfaces at r and $r + \mathrm{d}r$. The positive contribution is $[(\partial p/\partial r)r^2\Delta\Omega]$ evaluated at $r + \mathrm{d}r$, and the negative contribution (the 'inflow') is the same quantity evaluated at r. The net outflow is the difference between these contributions, $[(\partial/\partial r)r^2(\partial p/\partial r)]\Delta\Omega\,\mathrm{d}r$. Division by the volume element $r^2\Delta\Omega\,\mathrm{d}r$, yields the expression for div (grad (p)) $= \nabla^2 p$, as given above).

With the expression for $\nabla^2 p$ derived above, the wave equation for p becomes

$$(1/r^2)\partial/\partial r(r^2\partial p/\partial r) = (1/v^2)\partial^2 p/\partial t^2 \tag{1}$$

In this form the equation does not resemble the wave equation we are familiar with. It can be transformed into the familiar form, however, by a change of variables. The source of the radially symmetrical field is located at the origin $r = 0$. Conservation of acoustic power then requires that the energy flux must decrease as $1/r^2$ and the acoustic pressure as $1/r$. It is natural, therefore, to introduce a new field variable defined by

$$p = (1/r)\Psi(r, t) \tag{2}$$

It follows then from Eq. (1) that

$$\partial^2\Psi/\partial r^2 = (1/v^2)\partial^2\Psi/\partial t^2 \tag{3}$$

For harmonic time dependence the corresponding complex amplitude equation is

$$\mathrm{d}^2\Psi(r)/\mathrm{d}r^2 + k^2\Psi(r) = 0 \quad k = \omega/v \tag{4}$$

This is the familiar equation for one-dimensional wave motion with the general solution $\Psi(r) = A\exp(\mathrm{i}kr) + B\exp(-\mathrm{i}kr)$. The corresponding general solution for the complex pressure amplitude is

$$p(r) = (A/r)\exp(\mathrm{i}kr) + (B/r)\exp(-\mathrm{i}kr) \tag{5}$$

The general pressure field consists of two spherical waves, one outgoing and one incoming, with amplitudes proportional to $1/r$.

The corresponding velocity field is obtained from the momentum equation $\rho\partial u/\partial t = -\partial p/\partial r$ with the complex amplitude equation

$$-\mathrm{i}\omega\rho u(r) = -\mathrm{d}p(r)/\mathrm{d}r \tag{6}$$

As an example, let us consider only an outgoing wave so that $p = (A/r)\exp(ikr)$. From Eq. (6) we obtain the complex amplitude for the velocity $u(r) = (A/-i\omega\rho)[(1/r^2) - (ik/r)]\exp(ikr)$ or

$$u(r) = (A/\rho v)(1/r)[1 + i(1/kr)]\exp(ikr) \quad \text{(outgoing wave)} \tag{7}$$

For small values of r, in the **near field**, in which $kr \ll 1$, the velocity varies as $1/r^2$ and lags behind the pressure by an angle of $\pi/2$. For large values of r, in the **far field**, in which $kr \gg 1$, we have $p/u = \rho v$ as in a plane wave with the pressure and velocity being in phase.

10.5 The acoustic spectrum

Oscillations in matter cover a range of frequencies from zero to a frequency, which is determined by the distance between the molecules. As we shall see in a subsequent chapter, the upper frequency limit for a propagating acoustic wave in a solid corresponds to a wavelength equal to twice the intermolecular spacing, which typically is 10^{-8} cm. With a sound speed in the solid of the order of 10^5 cm/sec, the upper frequency limit will be approximately 10^{13} Hz.

In a gas, the lower wavelength limit of a propagating wave is of the order of the mean free path, which, in air at 1 atmosphere, is approximtely 10^{-5} cm. With a sound speed of 340 m/sec, the corresponding upper frequency limit will be approximately 10^9 Hz.

The portion of the acoustic frequency spectrum which can be heard by the average human is 20–20 000 Hz which covers approximately 10 octaves. The sensitivity of the ear is frequency dependent with the highest sensitivity occurring in the vicinity of 4000 Hz.

Oscillations and waves at frequencies below 20 Hz and above 20 000 Hz usually are referred to as the **infrasound** and **ultrasound**, respectively. Infrasonic waves are generated by several natural phenomena, earthquakes, wind, ocean waves, etc., but they can also be man-made by means of various devices and machine components oscillating at low frequencies. Among other sources can be mentioned airplanes flying through turbulent air.

Waves and oscillations at ultrasonic frequencies can be generated in nature by the pressure fluctuations in turbulent flow. Under controlled conditions, ultrasonic waves are usually produced by piezoelectric or magnetostriction transducers at frequencies up to approximately 1000 MHz.

Ultrasonic waves have many applications. In medicine they can be used as a diagnostic tool to supplement X-rays, and they are used extensively in industry for nondestructive testing of materials. It is also possible to perform surgery by means of focussed ultrasound and an ultrasonic microscope has been developed.

Ultrasonically driven 'chip hammers' can be used as drills in dentistry, for example, and in piezoelectric semiconductors, both in bulk and on the surface, the waves can be amplified by means of a superimposed electric field. Attenuation of the waves then can be eliminated and this has led to applications in the computer field, for example.

High-intensity sound is known to produce emulsification and agglomeration and it affects many processes, particularly in the chemical industry, and a rapidly growing engineering field (usually referred to as 'sonics') has emerged.

The spectrum of thermal fluctuations in matter covers the acoustic wave regime with wavelengths of the order of magnitude of the wavelength of light, i.e. about 10^{-5} cm. The waves travel in all directions in the bulk of the material and the corresponding density fluctuations can be used for scattering of laser light. From studies of the scattered light, the compressibility of the material as well as viscosity coefficients can be determined. This method, Brillouin scattering, will be discussed in a subsequent chapter.

10.6 Measurements

As has been shown in the previous section, an acoustic field in a fluid (gas or liquid) is characterized by a time dependent perturbation in density and a corresponding pressure $p(t)$. The sound pressure is usually only a very small fraction of the static pressure. In ordinary speech, for example, the sound pressure at the listener's ear is of the order of one millionth of the static pressure.

The sound pressure is measured by means of transducers, which convert the pressure fluctuations to an electrical signal. In a **dynamic** microphone, a coil in a magnetic field is connected to the moving element, and this results in an induced voltage, which is proportional to the velocity of the coil.

In a **crystal** microphone, a piezo-electric element is strained by the incident sound pressure, and a potential difference is produced between the terminals of the element.

In the **condenser** microphone, the sound pressure acts on a moveable membrane, which constitutes one of the 'plates' in a capacitor. The spacing between the capacitor plates, and hence the capacity, is then forced to deviate from the static value and a time dependent current is produced through the capacitor circuit. The charge on the plates is maintained by means of a 'polarizing' DC voltage, typically of the order of 200 Volt.

By proper design, the electrical signals from the microphones can be made proportional to the sound pressure over a wide range of frequencies and sound pressures.

In most sound measurement equipment, the signal from the microphone is processed so that the meter reading is made proportional to the root mean square (rms) value $\sqrt{\langle p^2(t)\rangle}$, where the bracket signifies time average.

The human ear is a remarkable 'transducer', making it possible to hear sound over a wide range of frequencies and sound pressures. The frequency range is approximately 20–20 000 Hz. The **threshold of hearing** for the average human at a frequency of 1000 Hz corresponds to an rms value of 0.0002 dyne/cm^2 = 0.00002 N/m^2, which is approximately 2×10^{-10} atm. (For harmonic time dependence the rms value is $1/\sqrt{2}$ times the amplitude). The ear is able to handle sound pressure amplitudes up to almost a million times the threshold value.

It is customary to use the threshold value as a reference, and express the mean square value of a sound pressure $p(t)$ as the ratio $\langle p^2(t)\rangle/p_0^2$. If this ratio is set equal to 10^{B}, the exponent $\mathrm{B} = \log(\langle p^2(t)\rangle/p_0^2)$ is said to express the sound pressure level in Bels. The number of deciBels is 10 times this value and the **sound pressure level in dB** is defined as

$$\mathrm{dB} = 10 \cdot \log[\langle p^2(t)\rangle/p_0^2] = 20 \cdot \log(p_{\mathrm{rms}}/p_0) \tag{1}$$

where $p_{\mathrm{rms}} = \sqrt{\langle p^2(t)\rangle}$

It is clear that the threshold of hearing corresponds to a sound pressure level of zero dB. The level of normal speech at the listener's ear is about 60 dB and the level of pain is about 130 dB. A harmonic wave with an amplitude equal to the atmospheric pressure would have the rms value $10^6/\sqrt{2}$ dyne/cm and a sound pressure level of $20 \cdot \log(10^6/\sqrt{2}p_0) = 191$ dB.

Spectrum density. A harmonic sound pressure $p(t)$ is uniquely determined by the amplitude, frequency, and the phase angle. Often $p(t)$ is not harmonic but an irregular or 'random' function of time, the noise from a jet, for example. In such a case the time dependence cannot be described in functional form. Rather, the signal is described in terms of its statistical properties. One such property is the mean square value $p(t)$ and the corresponding rms value of a sufficiently long sample of the signal. If the rms value is independent of when the signal sample is taken, the function $p(t)$ is said to be **stationary**.

The mean square value is proportional to the energy density in the sound field. The signal can be Fourier decomposed, and the total energy density can be considered built up from contributions distributed over the entire frequency spectrum. The contribution from a frequency band $\mathrm{d}\nu$ is $E(\nu)\mathrm{d}\nu$,

Display 10.4.

Examples of noise spectra measured with filters with different bandwidth. (From L.D. Mitchell and G.A. Lynch, (*Machine Design*, May 1, 1969).)

Spectrum density $E(\nu)$

$$\langle p^2(t) \rangle = \int_0^\infty E(\nu)\,\mathrm{d}\nu \tag{1}$$

Band level

$$\text{Band level, dB} = 10 \cdot \log\left[\int_{\nu_1}^{\nu_2} E(\nu)\,\mathrm{d}\nu / p_0^2\right]$$

$$(p_0 = 0.0002\,\text{dyne/cm}^2) \tag{2}$$

$$\nu_2 = 2^{1/n}\nu_1 \tag{3}$$

$n = 1$ Octave band

$n = 3$ One third octave band

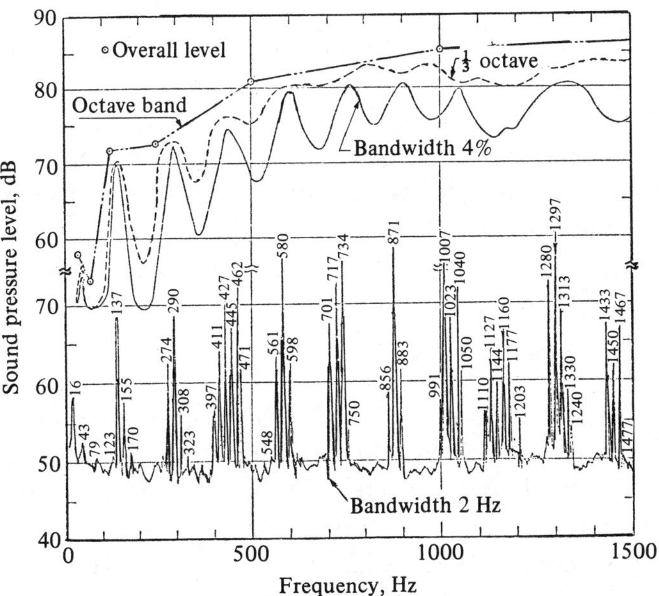

which defines $E(\nu)$, the **spectrum density** of the pressure field, and we have

$$\langle p^2(t) \rangle = \int_0^\infty E(\nu)\,\mathrm{d}\nu \tag{2}$$

The spectrum density can be thought of as the contribution to the mean square pressure per cycle in the spectrum.

Some analyzers, with constant percentage bandwidth, display the value of the integral of $E(\nu)$ over the frequency band $\nu_2 - \nu_1$, where $\nu_2 = 2^{1/n}\nu_1$. The frequency bandwidth is then $1/n$th of an octave, and the bandwidth increases with frequency; the percentage bandwidth is constant.

In a **white noise** spectrum, the energy density $E(\nu)$ is constant over a frequency range, which covers the audible range. For **'pink' noise**, the spectrum density is inversely proportional to frequency, so that the energy spectrum as measured by an analyzer with constant percentage bandwidth becomes constant.

The choice of filter bandwidth in the measurement of a spectrum depends on the degree of resolution or detail required in the analysis of the signal. As an illustration is shown in Display 10.4 the spectra of the noise produced by two cylinders in rolling contact as measured by different analyzers.

The lower curve is obtained with a narrow band analyzer with a constant bandwidth of 2 Hz. The spectrum thus obtained contains considerable details, and it is possible to identify most of the peaks as combination tones involving sums and differences of multiples of the cylinder frequencies.

The upper curves are the spectra obtained with constant percentage bandwidth analyzers with bandwidths of 4%, 1/3 octave, and 1 octave, respectively. In these spectra, the fine structure of the frequency dependence is lost, particularly at higher frequencies (since the bandwidth increases in proportion to the frequency).

***Frequency response of the ear.* dB*A*.** Instruments are available which measure a frequency weighted value of the mean square sound pressure, as expressed by $\int W(\nu)E(\nu)\,\mathrm{d}\nu$, where $W(\nu)$ is the weighting function. Of particular importance is the function shown in Display 10.5. It is called the *A*-weighting function and is designed to account for the frequency dependence of the 'sensitivity' of the average ear. The corresponding level is called the *A*-weighted sound pressure level or simply the sound level, and is expressed in dB*A*

$$\mathrm{dB}A = 10\cdot\log\left[\int A(\nu)E(\nu)\,\mathrm{d}\nu/p_0^2\right]$$

where, as before, p_0 is the rms threshold value 0.0002 dyne/cm^2.

Display 10.5.

Frequency weighting function used in the measurement of sound level.

Sound level

Sound level, $\mathrm{dB}A = 10\cdot\log\left[\int A(\nu)E(\nu)\,\mathrm{d}\nu/p_0^2\right]$ (1)

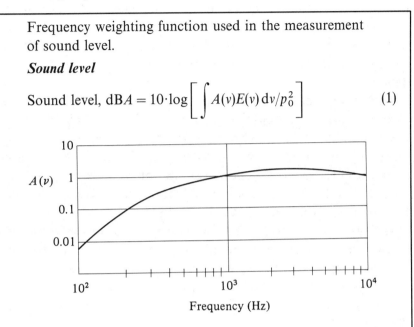

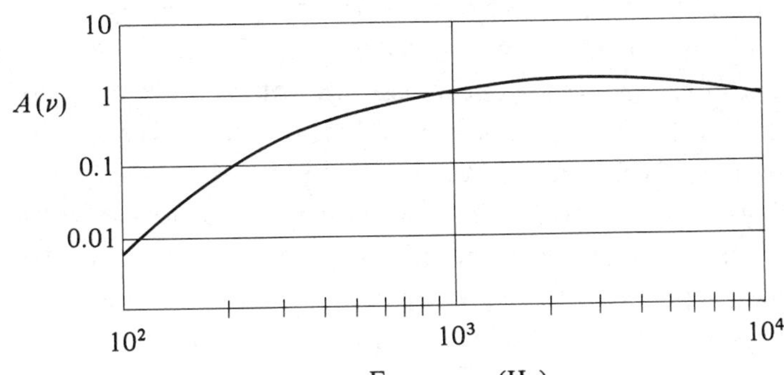

This quantity has been found to correlate rather well with the physiological and psychological effects of noise on humans, and is now used as the physical measure of noise in most industrial and environmental noise rules and legislation. For example, to prevent noise induced hearing damage, the average level of the daily exposure on the work place should not exceed 90 dBA.

The various characteristics of hearing are only partly understood. The relation between the subjective quantities loudness and pitch and the sound pressure and frequency have been established empirically. The A-weighting function is approximately the inverse of the frequency dependence of the loudness, at least at low sound pressures. For example, a sound pressure at 1000 Hz has the same loudness as a 200 Hz tone with 3.2 times the sound pressure (i.e. 10 dB higher level). (See Display 10.5).

With experimentally determined frequency dependence of the ear drum, and with the drum placed at the end of a 2.5 cm long ear canal, the calculated frequency dependence of the velocity amplitude of the ear drum turns out to be approximately the same as the frequency dependence of the sensitivity weighting function $A(v)$ in Display 10.5 which is related to loudness. (Problem 10.7).

Problems

10.1 *Threshold of hearing.* The average threshold of hearing of the human ear for a pure tone at a frequency of 1000 Hz corresponds to an rms pressure amplitude of 0.0002 dyne/cm^2.

(a) What is the corresponding threshold intensity in watt/cm^2?

(b) It is known that at the threshold intensity a tone burst with a duration of about one cycle (at 1000 Hz) can be detected by the average ear.

Assume a value for the cross sectional area of the ear and estimate the corresponding energy intercepted by the ear. Compare it to the average thermal translation energy $3k_BT/2$ of an air molecule.

($k_B = 1.38 \times 10^{-23}$ joule/K. T = 300 K).

10.2 *Spherical wave.* A monopole source has an acoustic power output of 1 kW (typical power for a jet engine in a commercial airplane). Determine the sound pressure at a distance of 100 m from the source. What is the corresponding sound pressure level in dB?

10.3 *Conservation of wave energy.* In an isothermal atmosphere, the density varies with height y above ground level as $\rho = \rho_0 \exp(-mgy/k_BT)$, where $m = M/N$, M = molar mass of air, N = Avogadro's number, g = acceleration of gravity, k_B = Boltzmann's constant, and T = absolute temperature.

A plane sound wave travels in the vertical direction in this atmosphere. How do the amplitudes of pressure and air velocity vary with y? (Hint: Neglect reflection and make use of conservation of acoustic energy).

10.4 *The speed of sound.* A 30 m long steam pipe in a power plant connects two large plenum chambers and is similar acoustically to an organ pipe open at both ends. Determine the frequencies of the first three axial acoustic modes in the pipe if the temperature of the steam is 1000 K.

10.5 *Sound intensity and pressure.* A 1 MHz ultrasonic transducer in water generates a sound beam which is focussedby means of a lens. The cross sectional area of the beam is 4 cm^2 and the sound intensity is 1 W/cm^2. Estimate the sound intensity and sound pressure at the focus, assuming that the beam area at the focus is λ^2, where λ is the wavelength.

Having obtained the sound pressure, discuss the validity of your estimate. The static pressure in the water is 1 atm., and the sound speed is 1482 m/sec.

10.6 *Spectrum density.* The spectrum density in 'pink noise' is proportional to the inverse of the frequency. Show that the rms value of the sound pressure in a frequency band equal to one octave is independent of the center frequency of the octave.

10.7 *Frequency response of ear drum.* The length of the human ear canal (terminated by the ear drum) is $L \approx 2.5$ cm. The ear is exposed to a sound pressure $p(t) = A\cos(2\pi\nu t)$ with an amplitude independent of the frequency ν. Treat the ear canal as a straight uniform tube which is terminated by a rigid wall. What then is the frequency dependence of the sound pressure amplitude at the termination? If, instead, the impedance of the ear drum is z (rather than ∞), what is then the frequency dependence of the velocity amplitude of the ear drum?

10.8 *Mass flux paradox.* The space-time dependence in the density wave in a harmonic sound wave is given by $\hat{\rho} = \hat{\rho}_0 \cos(\omega t - kx)$.

(a) What is the corresponding first order velocity wave $u(x, t)$?
(b) From $\hat{\rho}(x, t)$ and $u(x, t)$, show that the time average of the mass flux in the sound wave is $\hat{\rho}_0 u_0/2$, where u_0 is the velocity amplitude.
(c) Explain qualitatively and quantitatively this paradox.

10.9 *Acoustic wave energy density and flux in a moving fluid.* A plane sound wave with the fluid velocity $\mathbf{u}(x, t)$ travels in a fluid which has a constant mean velocity $(\mathbf{U}_0)$. Derive an energy conservation equation, analogous to Eq. 10.3.11.

11

Wave interference and diffraction

Our discussion of waves so far has dealt mainly with the field from a single source. We now turn to the study of the superposition of waves from several sources and the resulting interference patterns. The character of these patterns depends intimately on the correlation of the waves, as will be illustrated by simple examples.

The results will be applied to the diffraction of waves by one or more slits in a screen, with particular emphasis on the important problem of diffraction limited image resolution.

11.1 Interference of two plane waves

Display 11.1 refers to sound waves, which are generated in a tube by two identical loudspeakers located at $x = -L/2$ and $x = L/2$. The speakers are driven by harmonic input voltages from one and the same generator so that the sound from each individual source at the location of the source will have the same amplitude and the same phase. The tube is long enough or provided by absorbers so that reflections from the ends can be neglected. If the diameter of the tube is small compared to the wavelength, the waves can be considered to be plane except in a small near field region close to the speakers. This near field will be ignored in the present discussion. (Further discussion of this question will be given in Chapter 15 on wave guides and cavities).

In the regions to the left and to the right of the two sources, the waves from the individual sources travel in the same direction; between the sources, on the other hand, the directions are opposite. In this latter region, we express the complex pressure amplitudes of the waves as $A \exp[ik(x + L/2)]$ and $A \exp[-ik(x - L/2)]$, each of the waves having the complex amplitude A at the location of its source, $x = -L/2$ and $x = L/2$, respectively. The sum

Display 11.1.

Interference of plane harmonic waves.

Complex pressure amplitude

$-L/2 < x < L/2 \quad p(x) = 2A\exp(ikL/2)\cos(kx)$ (1)

$x > L/2 \quad p(x) = 2A\cos(kl/2)\exp(ikx)$ (2)

$x < -L/2 \quad p(x) = 2A\cos(kL/2)\exp(-ikx)$ (3)

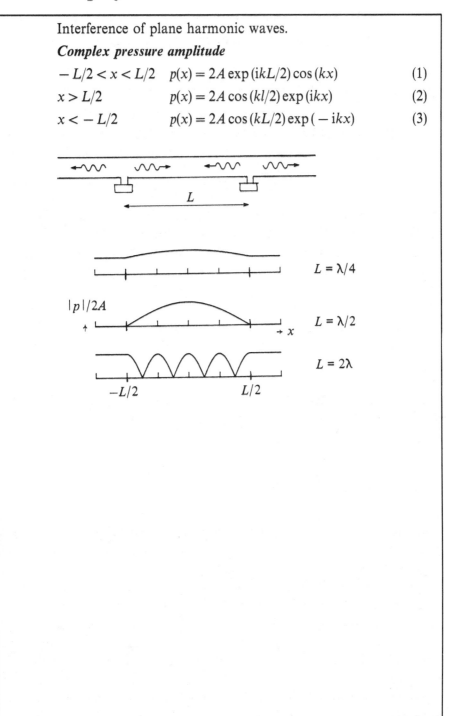

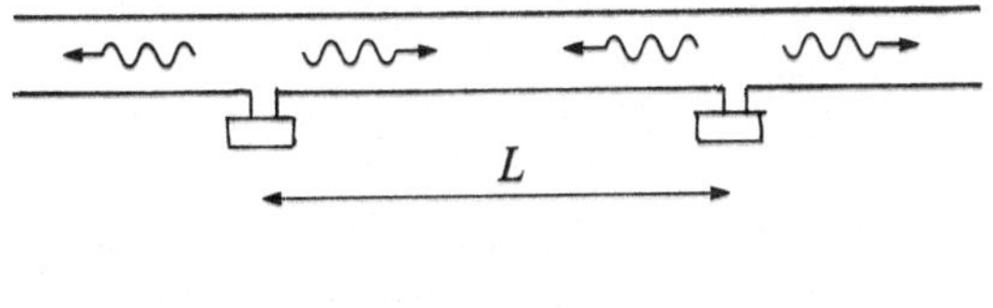

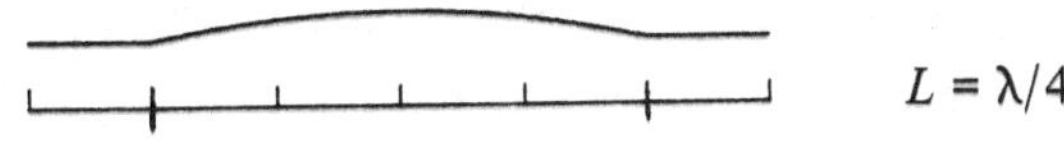

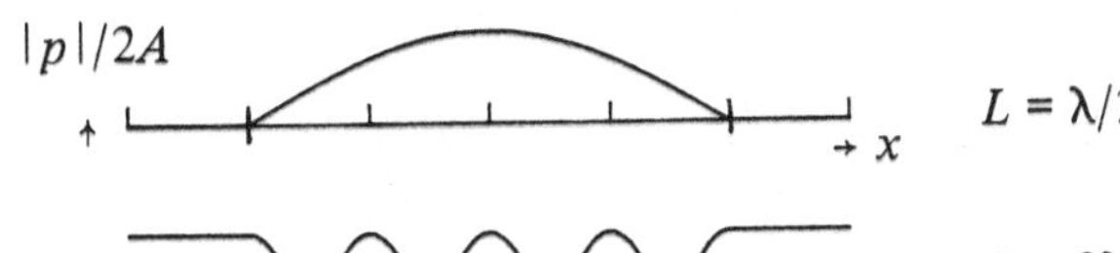

of these complex pressure amplitudes is then

$$p(x) = 2A \exp(ikL/2) \cos(kx) \quad (-L/2 < x < L/2) \tag{1}$$

which is a standing wave. The amplitude (i.e. the magnitude of the complex amplitude) varies from 0 to $2A$; it is 0 in the pressure nodes and $2A$ in the antinodes. The antinodes are located at $x = n\lambda/2$, corresponding to $kx = n\pi$, where $n = 0, 1, 2, \ldots$, and the nodes are at $x = (2n+1)\lambda/4$. By varying the phase difference between the two input voltages to the speakers, this pressure pattern can be moved along the x-axis, as discussed in one of the problems.

In the region to the right of the two sources, the total complex pressure amplitude is $A \exp[ik(x + L/2)] + A \exp[ik(x - L/2)]$ or

$$p(x) = 2A \cos(kL/2) \exp(ikx) \quad (x > L/2) \tag{2}$$

It is a wave travelling to the right with an amplitude which depends on the separation of the sources. If the separation is such that $kL/2 = (2n+1)\pi/2$, i.e. $L = (2n+1)\lambda/2$, the waves from the two sources are displaced an odd number of half wavelengths with respect to each other and hence cancel each other. The same applies to the region to the left of the sources. The possibility of cancelling sound by sound in this manner has received attention lately as a means of reducing noise, particularly at low frequencies.

If, on the other hand, the separation of the sources is an integer number of full wavelengths, so that $kL/2 = n\pi$, the two wave contributions interfere constructively, and the amplitude of the travelling wave will be $2A$. These characteristics are illustrated in the display, where we have plotted the x-dependence of the pressure amplitude in the two regions for the source separations $L = \lambda/4$, $\lambda/2$, and 2λ.

In determining the phase angle of the complex amplitude, the sign of the cosine functions must be accounted for, a minus sign corresponding to a phase angle of π.

Correlation function. In practice we often have to deal with noise fields in which the time dependence of the field variables is not harmonic but irregular so that it cannot be expressed by a known function of time. Rather, the function is described in terms of its statistical characteristics, of which the mean square value is one, as mentioned in the last chapter.

The problem of wave interference then has to be re-examined. Thus, in the present example, we express the time dependence of the wave emitted from the two sources by $A(t)$ and $B(t)$. In other words, the pressure at $x = -L/2$ from the source at that location will be $2A(t)$, so that the waves emitted in the positive and negative x-directions will be $A[t \pm (x + L/2)/v]$. Similarly, the waves from the source at $x = L/2$ will be $B[t \pm (x - L/2)/v]$.

We now wish to study the interference of these waves in the region between the sources at the location x. To simplify the expressions somewhat, we introduce $t' = t - (L/2 - x)/v$ and the difference in travel times from the two sources to x, $\tau = 2x/v$. The two waves then can be written $A(t' - \tau)$ and $B(t')$, and the mean square pressure at x becomes

$$\langle p^2(t') \rangle = \langle A^2(t' - \tau) \rangle + \langle B^2(t') \rangle + 2\langle A(t' - \tau)B(t') \rangle \tag{3}$$

where the brackets indicate time average over t, the averages being functions of x.

For a statistically stationary function, the mean square value is independent of time so that $\langle A^2(t' - \tau) \rangle = \langle A^2(t') \rangle$. Furthermore, by introducing the **cross correlation function**

$$\Psi_{12}(\tau) = \langle A(t' - \tau)B(t') \rangle \tag{4}$$

we can express the mean square pressure as

$$\langle p^2(t') \rangle = \langle A^2(t') \rangle + \langle B^2(t') \rangle + 2\Psi_{12}(\tau) \tag{5}$$

If the sources are uncorrelated, the cross correlation function is zero at all locations.

If the sources are driven by one and the same input signal, obtained from a common oscillator–amplifier unit, we get $A(t) = B(t)$. Eq. (4) then defines the **auto-correlation function**

$$\Psi(\tau) = \langle A(t')A(t' - \tau) \rangle \tag{6}$$

If the delay τ is zero, the auto-correlation function equals the mean square value of $A(t')$. If $A(t')$ is periodic, the auto-correlation function will be periodic also. For a noise-like function, however, $\Psi(\tau)$ generally goes to zero as τ goes to infinity. For a completely random function, $\Psi(\tau)$ is zero for all values of τ different from zero. This is an idealization, however, and in reality, there is no discontinuity at the origin. Often the correlation function varies with τ as a decaying oscillation. In such a case, the value of τ, at which $\Psi(\tau)$ first becomes zero generally is called the correlation time of the function.

In the present case, the time delay $\tau = 2x/v$ is converted into a distance x, and $\tau = 0$ corresponds to $x = 0$, i.e. the midpoint between the sources. At this location, the travel times from the sources are the same, and with the speakers driven in phase, the pressure contributions from the two sources will be the same, so that the resulting mean square value will be $4\langle A^2(t) \rangle$. The x-dependence of $\langle p^2(t') \rangle - 2\langle A^2(t') \rangle$ in the region between the sources, $-L/2 < x < L/2$, has the shape of the auto-correlation function, as indicated schematically in Display 11.2.

Display 11.2.

Correlation function

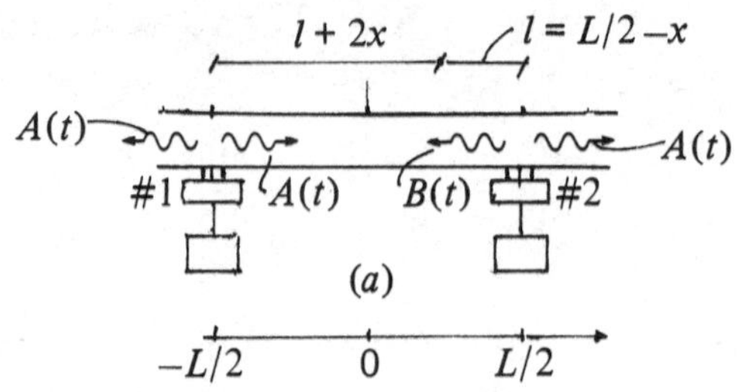

Let $t' = t - (L/2 - x)/v$ (1)

Let $\tau = (L/2 + x)/v - (L/2 - x)/v = 2x/v$ (2)

$\langle p^2(t')\rangle = \langle A^2(t' - \tau)\rangle + \langle B^2(t')\rangle + 2\langle A(t' - \tau)B(t')\rangle$ (3)

Stationary process $\langle A^2(t' - \tau)\rangle = \langle A^2(t')\rangle$ (4)

Cross-correlation function $\Psi_{12}(\tau) = \langle A(t' - \tau)B(t')\rangle$ (5)

$\langle p^2(t')\rangle = \langle A^2(t')\rangle + \langle B^2(t')\rangle + 2\Psi_{12}(\tau)$ (6)

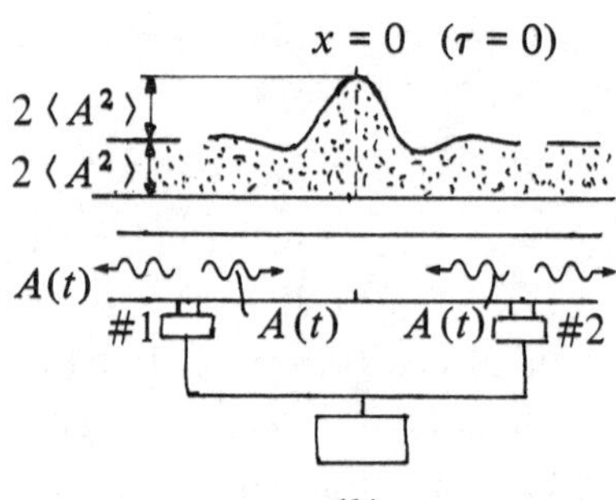

Auto-correlation function

$A(t) = B(t)$ (7)

$\Psi(\tau) = \langle A(t')A(t' - \tau)\rangle$ (8)

$\langle [A(t') + A(t' - \tau)]^2\rangle = 2\langle A^2(t')\rangle + 2\Psi(\tau)$ (9)

Example Harmonic time dependence

$A(t') = A\cos(\omega t')$

$\Psi(\tau) = A^2\langle \cos(\omega t')\cos[\omega(t' - \tau)]\rangle = (A^2/2)\cos(\omega\tau)$ (10)

$\langle p^2\rangle = A^2[1 + \cos(\omega\tau)] = 4\langle A^2(t')\rangle\cos^2(\omega\tau/2)$ (11)

$\sqrt{\langle p^2\rangle} = 2\sqrt{\langle A^2(t')\rangle}\cos(kx)$ (12)

11.2 Interference of waves from an array of sources

We now turn to the important problem of the interference of waves from an array of sources in free space. An array can be one- two-, or three-dimensional; the basic characteristics of the wave field are essentially the same for all and will be illustrated here by an analysis of the one-dimensional array.

We shall consider point sources of sound, since the exact expression for the spherical wave field from a point source is known from our discussion in Chapter 10. As far as the angular distribution of the intensity in the far field is concerned, however, the results obtained for sound are applicable also to the field from electric dipole sources in a plane perpendicular to the dipoles.

Two sources. For simplicity we start with two sources located at $y=-c$ and $y=c=d/2$, where d is the separation of the sources, as indicated in Display 11.3.

According to Eq. 10.4(5), the complex pressure amplitude of an outgoing spherical wave from a point source is $p(r)=(A/r)\exp(ikr)$ where r is the distance from the sources to the point of observation and $k=\omega/v=2\pi/\lambda$. The total complex amplitude from the two sources in the xy-plane through the sources is then

$$\begin{aligned} p &= (A/r_1)\exp(ikr_1)+(A/r_2)\exp(ikr_2) \\ &= (A/r_1)\exp(ikr_1)\{1+(r_1/r_2)\exp[ik(r_2-r_1)]\} \qquad (1) \\ r_1^2 &= x^2+(y-c)^2, \quad r_2^2=x^2+(y+c)^2, \quad c=d/2 \end{aligned}$$

where r_1 and r_2 are the distances from the sources to the point of observation, as shown.

According to the discussion of spherical waves in Chapter 10 (Eq. 10.4(7)), the velocity is simply the pressure divided by the wave impedance Z, just as in a plane wave, if the distances between the sources and the point of observation are much larger than the wavelength. In that case, $|p|^2/2Z$ is the energy flux in the field, and with $r_1 \simeq r_2 \gg d$ we obtain

$$\begin{aligned} S &= (|p|^2/2Z) \\ &= (1/2Z)|A/r_1|^2 2\{1+\cos[k(r_2-r_1)]\} \quad (r_1, r_2 \gg d, \lambda) \qquad (2) \end{aligned}$$

In this regime, the far field or **Fraunhofer regime**, for a fixed value of $r_1=r$, the energy flux depends only on the travel path difference r_2-r_1, which is approximately $d\sin(\theta)$, where the angle θ is defined in the display. If $r_2-r_1=n\lambda$, where λ is the wavelength and n an integer, we have $\cos[k(r_2-r_1)]=1$ and the energy flux is a maximum, equal to 4 times the energy flux produced by one source alone.

Display 11.3.

Interference of the fields from two point sources. The hyperbolas are curves of constant phase difference between the two fields.

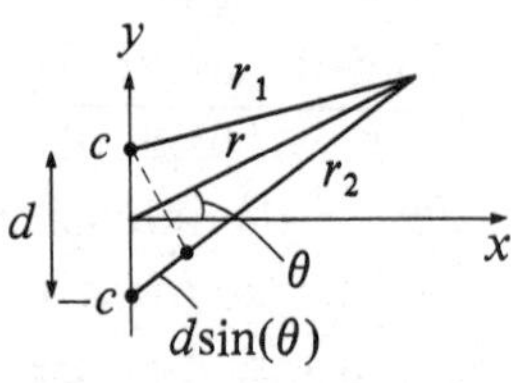

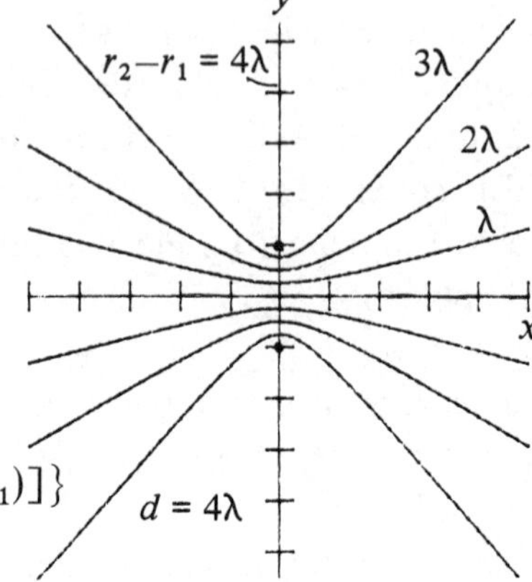

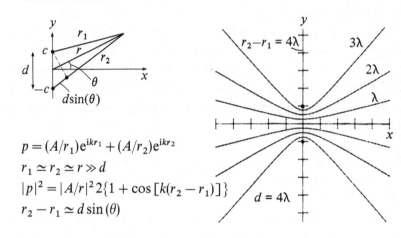

$$p = (A/r_1)e^{ikr_1} + (A/r_2)e^{ikr_2}$$

$$r_1 \simeq r_2 \simeq r \gg d$$

$$|p|^2 = |A/r|^2 2\{1 + \cos[k(r_2 - r_1)]\}$$

$$r_2 - r_1 \simeq d\sin(\theta)$$

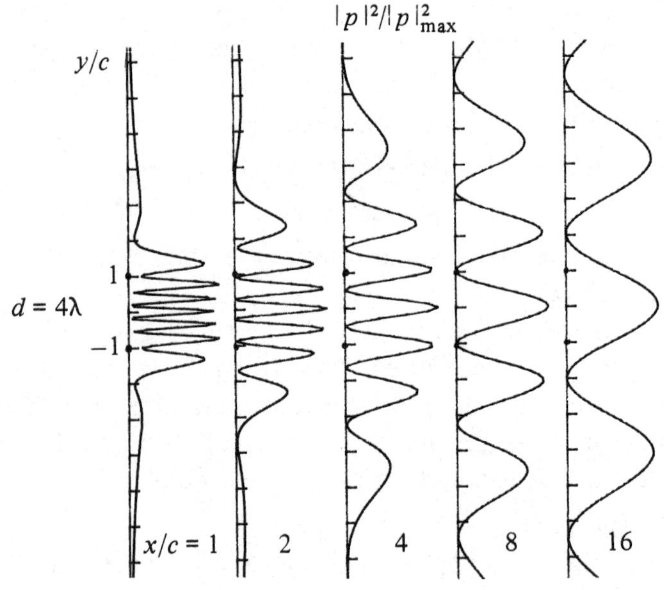

It should be emphasized that although Eq. (2) was obtained from Eq. (1) by using the approximation $r_2/r_1 = 1$ for the amplitude ratio in the far field, the difference $r_2 - r_1$ in the phase factor $\exp[ik(r_2 - r_1)]$ is essential, since a variation of $(r_2 - r_1)$ by a wavelength changes this factor from 1 to -1 and the total energy flux from 0 to 2. In fact, the interference pattern in the far field is due solely to the variation of the phase difference between the two waves at the point of observation.

Closer to the source, in the **Fresnel zone**, the difference in amplitude of the interfering wave must be accounted for and the interference pattern becomes more complex. The interfering waves no longer can be considered to be parallel, defined by the same angle θ, as in the Fraunhofer regime, but reach the observation point from different directions.

The points of observation for which $r_2 - r_1$, and hence the phase difference, is a constant lie on a hyperbola with the foci at the locations $y = \pm c$ of the sources. With the travel path difference $r_2 - r_1$ denoted by $2b$, the 'standard' form of the equation for the hyperbola is $y^2/b^2 - x^2/a^2 = 1$, where $a^2 = c^2 - b^2 = c^2[1 - (b/c)^2]$ and $d = 2c$. The hyperbolas along which the two waves are in phase are then obtained with $2b = n$. These hyperbolas are shown in the display for the special case $d = 2c = 4\lambda$.

For this value of the source separation d, we have shown the squared amplitude in the xy-plane as a function of y for some different distances x from the sources. For each x, the quantity plotted is $|p|^2/|p|^2_{max}$, where $|p|_{max}$ refers to the particular value of x involved. Close to the sources, the field in the region between the sources is a superposition of two spherical waves, which travel in approximately opposite directions. The resulting total amplitude distribution is then not much different from the standing wave obtained in the analogous one-dimensional problem in Display 11.1. Further out, in the Fraunhofer regime, the interference pattern is determined by $1 + \cos[kd\sin(\theta)]$, where $\tan(\theta) = y/x$. In this regime, the hyperbolas can be approximated by their asymptotic straight lines $y = \pm(b/a)x$.

The angular distribution function $1 + \cos[kd\sin(\theta)]$ can be written as $2\cos^2(\delta)$, where $\delta = (kd/2)\sin(\theta)$. The maxima in the far field occur in the directions of **constructive** interference where the path difference is an integer number of wavelengths. In the minima the two field contributions interfere **destructively**, corresponding to a path difference of an odd number of half wavelengths. The distribution will be considered again as a special case in the analysis of the field from a linear array of N sources.

Array of N sources. The analysis above now will be extended to a linear array of N sources, as summarized in Display 11.4. We shall focus attention

Display 11.4.

Interference field from an array of N harmonic sources.

Complex amplitude functions

$\Psi_i = (A/r_i)\exp(ikr_i)$ (1)

$\Psi = \sum \Psi_i \simeq (A/r)[\exp(ikr_1) + \exp(ikr_2) + \cdots \exp(ikr_N)]$ (2)

Let $\Delta = r_{1+1} - r_i = d\sin(\theta)$ and $s = \exp(ik\Delta)$ (3)

$\Psi \simeq (A/r)\exp(ikr_1)[1 + \exp(ik\Delta) + \exp(i2k\Delta)$
$+ \cdots \exp(ik(N-1)\Delta)]$ (4)

$\Psi \simeq (A/r)\exp(ikr_1)\sum s^n \quad n = 0$ to $N-1$ (5)

Recall sum of geometric series:

$\sum s^n = (1 - s^N)/(1 - s)$
$= [1 - \exp(ikN\Delta)]/[1 - \exp(ik\Delta)]$
$= \exp[ik(N-1)\Delta/2][\sin(kN\Delta/2)/\sin(k\Delta/2)]$ (6)

Intensity distribution

Let $\delta = k\Delta/2 = (kd/2)\sin(\theta) = (\pi d/\lambda)\sin(\theta)$ (7)

$S(\theta) = C[\sin(N\delta)/\sin(\delta)]^2 \quad C = \text{constant}$ (8)

$S(0) = C\cdot N^2$ (9)

$S(\theta) = S(0)[\sin(N\delta)/N\sin(\delta)]^2$ (10)

Maxima

$S(\theta) = S(0)$ for $\delta = n\pi$, i.e. (11)

$d\sin(\theta) = n\lambda$ (12)

Special case. $N = 2$

$\sin(2\theta) = 2\sin(\theta)\cos(\theta)$ (13)

$S(\theta) = S(0)\cos^2(\theta)$ Compare Eq. 11,2(2)

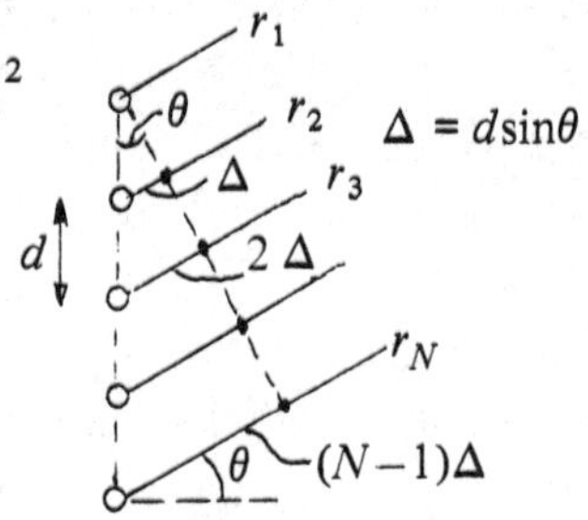

on the far field, so that the distance between the array and the receiver is large compared to the wavelength and the length of the array. The path lengths are labelled with the indices $i = 1$ to N. The complex amplitude from the ith source will be denoted by Ψ_i, where Ψ represents the pressure in an acoustic field and the electric field in the EM case.

The path length difference for adjacent sources can be considered to be constant and equal to $\Delta = d\sin(\theta)$, as before, and the total complex amplitude at the receiver becomes

$$\begin{aligned}\Psi = \sum \Psi_i &= (A/r)\,[\exp(\mathrm{i}kr_1) + \exp(\mathrm{i}kr_2) + \cdots \exp(\mathrm{i}kr_N)]\\ &= (A/r)\exp(\mathrm{i}kr_1)[1 + \exp(\mathrm{i}k\Delta) + \exp(\mathrm{i}2k\Delta)\\ &\quad + \cdots \exp(\mathrm{i}k(N-1)\Delta)]\end{aligned}$$

which is a geometric series with the ratio between adjacent terms being $s = \exp(\mathrm{i}k\Delta)$.

The sum of the geometric series is $(1 - s^N)/(1 - s)$ so that $\Psi = [A\exp((\mathrm{i}kr_1)/r)][1 - \exp(\mathrm{i}kN\Delta/2)]/[1 - \exp(\mathrm{i}k\Delta/2)]$.

In the numerator we take out the factor $\exp(\mathrm{i}kN\Delta/2)$ and in the denominator the factor $\exp(\mathrm{i}k\Delta/2)$. Combining these factors with $\exp(\mathrm{i}kr_1)$ we obtain $\exp(\mathrm{i}kr)$, where r is the average distance between the array and the receiver. The remaining factor in the sum can then be expressed as $\sin(Nk\Delta/2)/\sin(k\Delta/2)$, and the total complex amplitude as

$$\Psi = [A\exp(\mathrm{i}kr)/r][\sin(N\delta)/\sin(\delta)] \tag{3}$$

$$\delta = k\Delta/2 = (kd/2)\sin(\theta) = (\pi d/\lambda)\sin(\theta) \tag{4}$$

With the complex amplitude Ψ representing pressure or electric field, the time average energy flux will be $S = |\Psi|^2/2Z$, where Z is the wave impedance. The maximum value of S (obtained when $\delta = 0, \pi, 2\pi, \ldots,$) is $S(0) = N^2|A|^2/2r^2$, i.e. N^2 times larger than the energy flux produced by a single source.

Then, by expressing $|A|^2/2r^2$ as $S(0)/N^2$, the angular distribution of the intensity can be expressed as

$$S(\theta) = S(0)[\sin(N\delta)/N\sin(\delta)]^2 \tag{5}$$

where δ is defined in Eq. (4)

Again the maxima are obtained in the directions where the path difference between adjacent sources is an integer number of wavelengths, which results in constructive interference of the individual waves.

Examples of the angular dependence of the energy flux are described by the 'radiation patterns' in Display 11.5. A maximum is always obtained for $\delta = 0$, i.e. for $\theta = 0, \pi$. In order to get additional maxima, it is necessary that the source separation d be at least one wavelength.

Display 11.5.

Examples of radiation patterns from source arrays

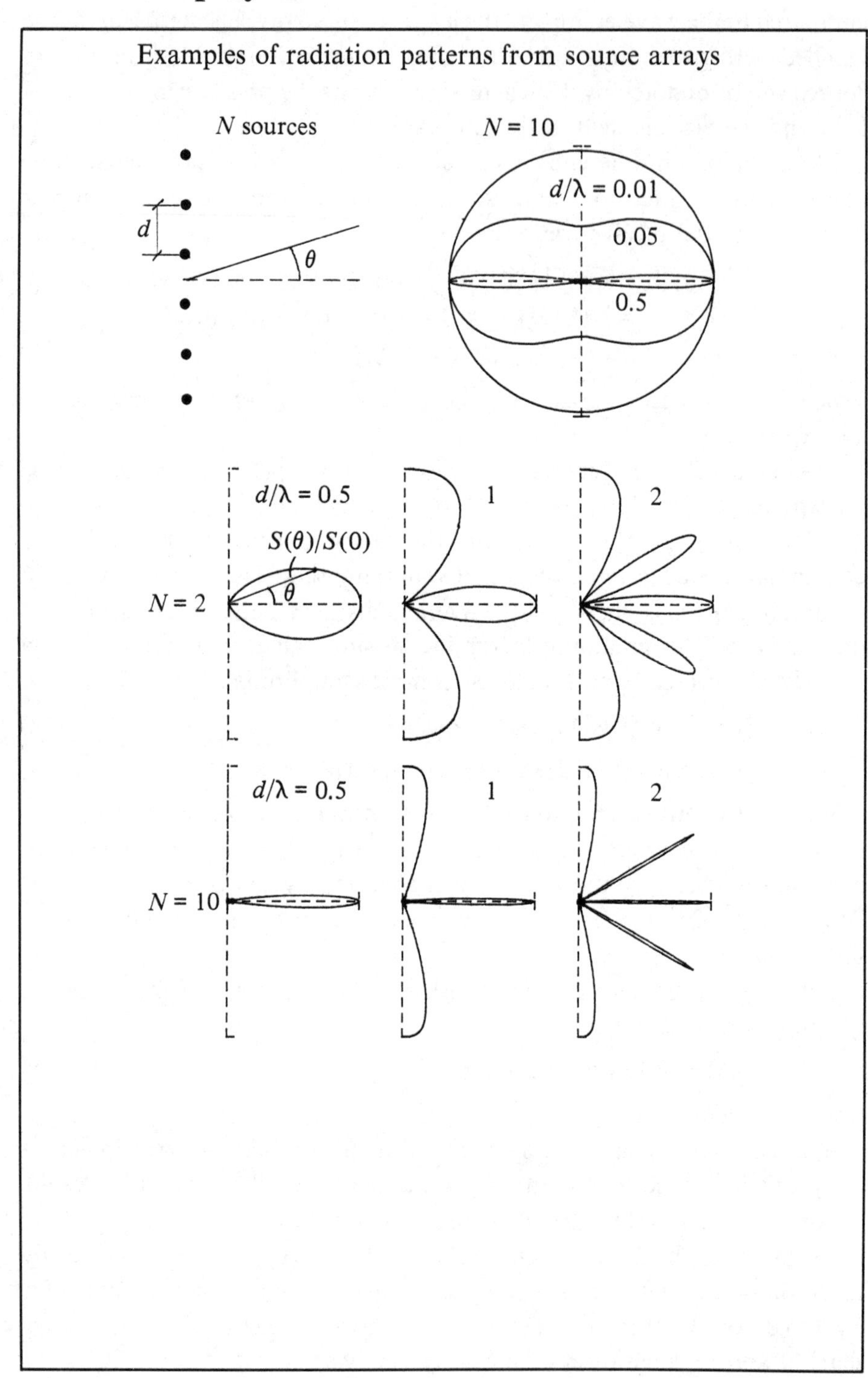

The first set of patterns in the display refers to source separations less than half a wavelength. When the separation is quite small compared to a wavelength, the radiation pattern has almost radial symmetry, as for a single source. As d increases, the intensity in the 90 degree direction decreases and reaches a minimum (destructive interference) for $d = 0.5\lambda$. It is important to note that the sharpness of the radiation pattern increases rapidly with increasing source separation; with $d = 0.5\lambda$ the 'beam width' with $N = 10$ is about 23 degrees. In general, the beam width is defined as the angular separation of the first zeroes in the radition pattern, which are given by $N\delta = \pm\pi$, i.e.

$$\sin(\theta) = \pm\lambda/Nd \quad \text{(zeroes of central beam)} \tag{6}$$

Thus, for large values of N, the beam width is approximately $2\theta = 2\lambda/Nd$ radians.

If the source separation is increased to one wavelength, an additional lobe in the radiation pattern is obtained in the direction along the array, as shown in the next two sets of patterns. The first of these corresponds to $N = 2$ and the second to $N = 10$. The number of lobes in the patterns is determined only by d/λ.

For a given d, the effect of increasing N (and hence the total length of the array) is to increase the sharpness of the lobes. It should be noted that in the last two sets of radiation patterns, we have considered only the angular range from -90 to 90 degrees. With $d = m\lambda$, the total number of lobes in the range from 0 to 360 degrees is $4m$.

Phased array. In the discussion so far, it was assumed that the source in the array had the same phase angle, and phase difference at the receiver was then due solely to the difference in the paths of wave travel.

Then, if the source separation in the array is less than one wavelength, the maximum energy flux is in the 'forward' direction $\theta = 0$ (with its mirror image in the backwards direction). The width of the beam will be approximately $2\lambda/Nd$ radians, and the maximum energy flux will be N^2 times the energy flux for one source.

It is of considerable practical interest to realize that the direction of the beam can be changed by the introduction of a phase difference between adjacent sources. In Display 11.6, we have a constant phase lag ϕ in the direction of increasing N.

The calculation of the resulting complex amplitude at the receiver is completely analogous to the calculation summarized in Display 11.5. The phase difference resulting from the difference in path length is $k\Delta = kd\sin(\theta)$

Display 11.6.

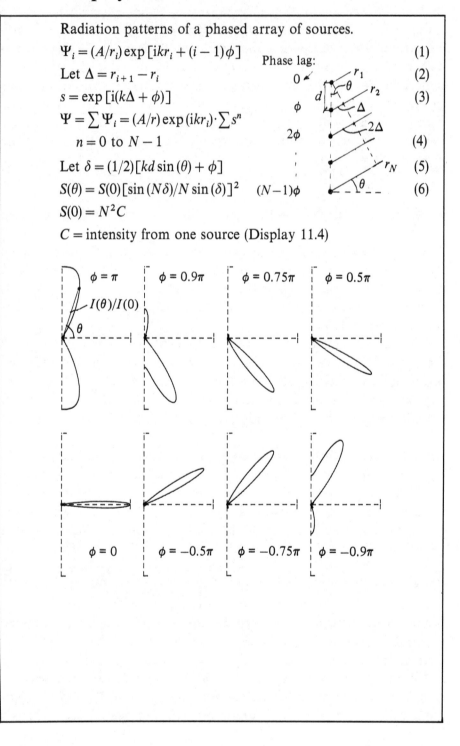

Radiation patterns of a phased array of sources.

$\Psi_i = (A/r_i)\exp[ikr_i + (i-1)\phi]$ (1)

Let $\Delta = r_{i+1} - r_i$ (2)

$s = \exp[i(k\Delta + \phi)]$ (3)

$\Psi = \sum \Psi_i = (A/r)\exp(ikr_i)\cdot\sum s^n$

$n = 0$ to $N-1$ (4)

Let $\delta = (1/2)[kd\sin(\theta) + \phi]$ (5)

$S(\theta) = S(0)[\sin(N\delta)/N\sin(\delta)]^2$ (6)

$S(0) = N^2C$

C = intensity from one source (Display 11.4)

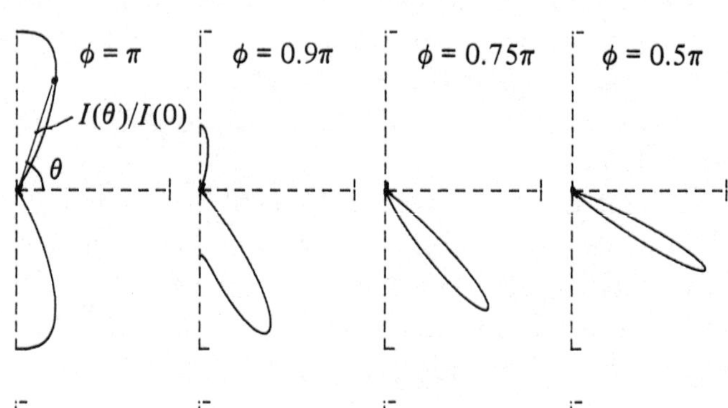

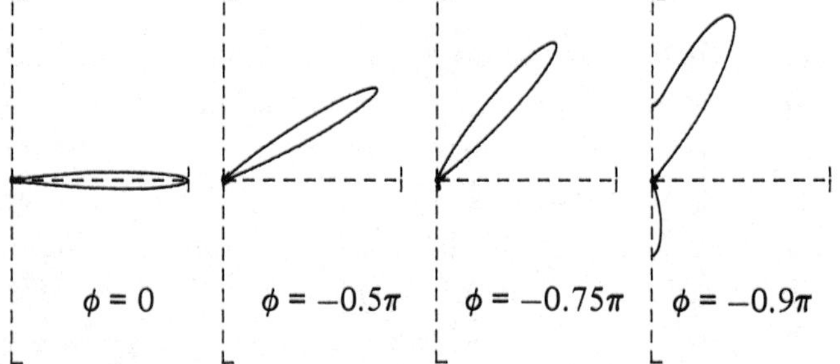

as before. To this phase difference we now have to add ϕ, and the angular distribution of the energy flux is obtained by replacing δ in Eqs. (4) and (5) by $\delta' = \delta + \phi/2$, i.e. $\delta' = (1/2)[kd \sin(\theta) + \phi]$.

As a specific numerical example, we have computed the angular dependence of the energy flux from an array with $N = 10$ and $d/\lambda = 0.5$. Varying the phase lag ϕ from $-\pi$ to $+\pi$ we can change the beam direction over an angle of 180 degrees. It is interesting to note that the beam width gradually decreases as the beam is turned toward the forward direction, corresponding to $\theta = 0$.

With a two-dimensional array of sources arranged in rows and columns, and a variable phase difference between the sources in both the rows and the columns, the beam can be moved in any direction in space. This leads to many important practical applications, such as target tracking, navigation, radiation diagnostics etc.

11.3 Continuous source distribution

From the result of the angular distribution of the energy flux from an array of N equal sources in the last section, we can determine the radiation pattern from a uniform line source by letting the number N go to infinity and the source separation d go to zero in the array.

Thus, with the length of the line source being b, we have $b = Nd$ in this limit and, correspondingly, the denominator in the angular distribution function $[\sin(N\delta)/N \sin(\delta)]$ for the N-source energy flux goes to $N\delta = (kb/2) \sin(\theta) \equiv \beta$, and the function becomes $[\sin(\beta)/\beta]$. Thus, the energy flux from the uniform line source in a plane containing the source becomes

$$S(\theta) = S(0)[\sin(\beta)/\beta]^2 \tag{1}$$

$$\beta = (kb/2)\sin(\theta) = (\pi b/\lambda)\sin(\theta)$$

The maximum $S(0)$ of the energy flux is obtained when $\beta = 0$, i.e. in the direction $\theta = 0$ (and π). Unlike the radiation pattern of the N source array, the continuous source pattern has only this single main lobe. The energy fluxes in the secondary, or 'sides lobes', are quite weak, by comparison. In fact, in the radiation pattern shown in Display 11.7 they are not clearly resolved.

If the line source distribution is nonuniform, we cannot obtain the result from the N-source array. Rather, we introduce a line 'source strength' $A(h)$ per unit length so that the elementary complex field amplitude from a length $\mathrm{d}h$ of the source is $\mathrm{d}\Psi = [A(h)\,\mathrm{d}h/r] \exp[ikr(h)]$, where $r(h) = r_0 - h \sin(\theta)$ is the distance from the element $\mathrm{d}h$ to the receiver, as indicated in Display 11.8.

Display 11.7.

Radiation pattern of a uniform continuous source distribution

N line sources (Ref. Display 11.4)

$$S(\theta) = S(0)[\sin(N\delta)/N\sin\delta]^2 \quad (1)$$

$$\delta = (kd/2)\sin\theta, \quad k = 2\pi/\lambda \quad (2)$$

Uniform continuous source

$$d \to 0, \quad N \to \infty, \quad Nd \to b \quad (3)$$

$$N\sin\delta \to N\delta = (kNd/2)\sin\theta = \beta \quad (4)$$

$$\beta = (kb/2)\sin\theta \quad (5)$$

$$S(\theta) = S(0)[\sin\beta/\beta]^2 \quad (6)$$

Beam width

$$S_{max} = (0) \quad (7)$$

$$S_{min} = 0 \quad \text{at} \quad \beta = \pm n\pi \quad (8)$$

$$\text{First zeroes} \quad \beta = \pm\pi, \quad \theta = \pm\theta_1 \quad (9)$$

$$\sin\theta_1 = \lambda/b \quad (10)$$

$$\theta_1 \simeq \lambda/b \quad \text{for} \quad b \gg \lambda \quad (11)$$

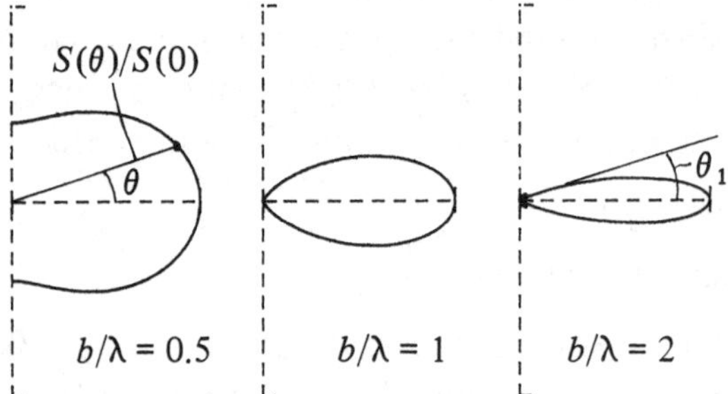

Display 11.8.

Radiation from a nonuniform continuous source distribution

Nonuniform continuous source distribution

Width of source b

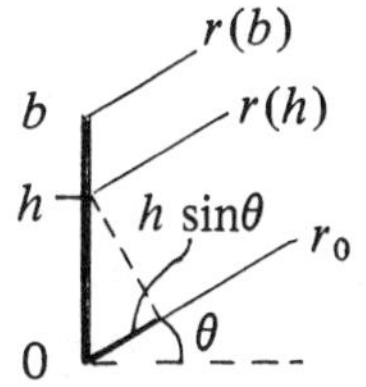

Source strength per unit width $A(h)$ (1)

$r(h) = r_0 - h\sin(\theta)$ (2)

$\Delta\Psi = [A(h)\,dh]\exp[ikr(h)]/r(h)$ (3)

$$\Psi = [\exp(ikr_0)/r]\int_0^b A(h)\exp[-ikh\sin(\theta)]\,dh \quad (4)$$

r = average distance (see Eq. (9))

Example. Uniform source

$A(h) = A = \text{constant}$ (5)

$$\Psi = [\exp(ikr_0)/r]\int \exp[-ikh\sin(\theta)]\,dh \quad (6)$$

$$= [A\exp(ikr_0)/r][\exp(-ikb\sin(\theta)) - 1)]/[-ik\sin(\theta)] \quad (7)$$

$$= [A\exp(ikr)/r]\,2\sin[\tfrac{1}{2}kb\sin(\theta)]/ik\sin(\theta) \quad (8)$$

where $r = r_0 - (b/2)\sin(\theta)$ (9)

Let $\beta = (kb/2)\sin(\theta)$ (10)

$\Psi = [A\exp(ikr)/r][\sin(\beta)/\beta]$ (11)

$S(\theta) = S(0)\,[\sin(\beta)/\beta]^2$ (12)

Linear phase lag

Phase lag at h Kh, $K = \omega/V$ (13)

Source function $A(h) = A\exp(iKh)$ (14)

Let $\beta' = \beta - Kb/2 = (b/2)[k\sin(\theta) - K]$ (15)

$S(\theta) = S_{max}[\sin(\beta')/\beta']$ (16)

$S = S_{max}$ for $\beta' = 0$, i.e. at $\theta = \theta'$, (17)

where

$(kb/2)\sin(\theta') = Kb/2$, $\sin(\theta') = K/k = v/V$ (18)

The total complex field amplitude is then obtained as the integral over the source

$$\Psi = (1/r)\exp(ikr_0)\int_0^b A(h)\exp[-ikh\sin(\theta)]\,dh \tag{2}$$

In the special case of a constant source strength $A(h) = A$, we recover the result $S(\theta) = S(0)[\sin(\beta)/\beta]$ for the energy flux of the uniform source in Eq. (1). Similarly, if we introduce a phase lag in the source function proportional to h, so that $A(h) = A\exp(iKh)$, the angular distribution of the energy flux is obtained by replacing β by $\beta' = \beta - Kb/2 = [k\sin(\theta) - K]b/2$. For a positive value of K, this corresponds to a shift in the radiation pattern in the positive direction. Actually, since the maximum occurs when $\beta' = 0$, it follows that the center of the main lobe is at the angle given by

$$\sin(\theta') = K/k. \tag{3}$$

A phase lag proportional to h is characteristic of a 'source wave' which travels in the positive h-direction. If the speed of this wave is V, we have $K = \omega/V$, and it follows that the expression for the location of the main lobe can be written $\sin(\theta') = v/V$. It is important to note, that there will be a lobe only if $V > v$. This result is intimately related to the refraction of a wave at a boundary, as we shall see in a subsequent chapter.

Near field and far field. The analysis above referred to the far field (Fraunhofer) regime. Here the point of observation is so far from the source that the waves emitted from different parts of the source can be considered to be parallel, and in the amplitude factors one and the same value for the distance r was used. These assumptions simplify the mathematical analysis considerably, and for a uniform source distribution it was possible to express the resulting field amplitude in terms of elementary functions in closed form.

In the near field these assumptions are no longer valid; the total field depends not only on the phase relationship between the elementary waves but also on the differences in the amplitudes.

In Display 11.9, we have shown the result of a numerical calculation of the squared sound pressure amplitude produced by a uniform line source located between $y = -b/2$ and $+b/2$. The field is presented in terms of the y-dependence of $|p|^2/|p|^2_{max}$ at different distances x from the source in the xy-plane, which contains the source. The value of $|p|_{max}$ refers to the particular x-value involved in each case.

The length of the source is $b = 10\lambda$, and the distances x are given by $x' = x/(b/2) = 1$, 2, 4, 8, 16, and 32. At $x' = 32$ (i.e. $x = 16b$) we are at the

Display 11.9.

Near field and far field from a uniform line source.

Length of line source $b = 2L$ (1)

$$p(x, y) = \int_{-L}^{+L} A(e^{ikr}/r)\,dh = \int_{-1}^{+1} A(e^{ikr'}/r')\,dh' \quad (2)$$

$$h' = h/L, \quad x' = x/L, \quad y' = y/L \quad (3)$$

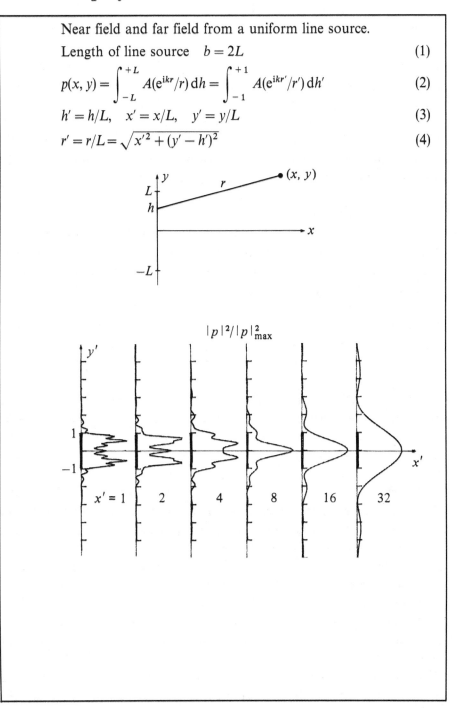

$$r' = r/L = \sqrt{x'^2 + (y' - h')^2} \quad (4)$$

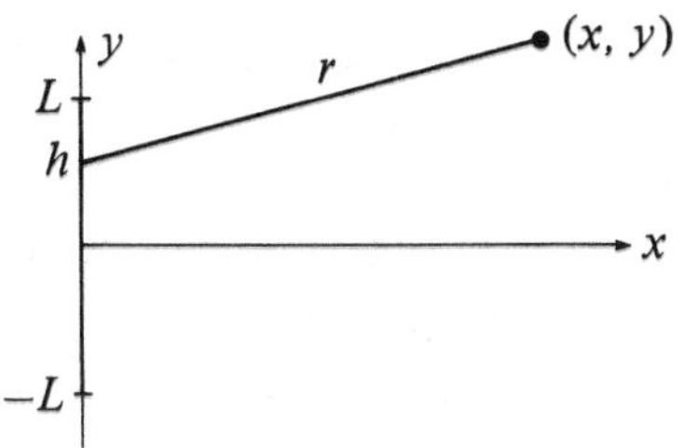

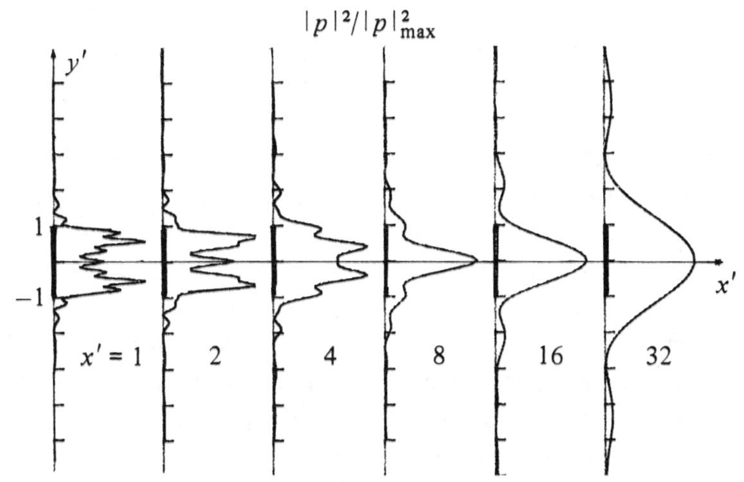

beginning of the Fraunhofer regime, in which the field will be independent of x when plotted versus y/x.

With x' between 0 and approximately 16 (i.e. $x = 8b$), the field falls off sharply for $|y| > b/2$, i.e. the field is like a 'beam'. The beam is not uniform but has a structure, which is a result of interference of waves from different parts of the source. Further away from the source, the field spreads out so that it extends beyond the source dimension in the y-direction. The width of the beam in this Fraunhofer regime is approximately $x(\lambda/b)$, and if this is to exceed the width b of the source, we must have $x > b^2/\lambda$, which can be used to define approximately the boundary between the Fresnel and the Fraunhofer regimes. In this case, with $b/\lambda = 10$, we obtain $b^2/\lambda = 10b$.

11.4 **Diffraction**

As mentioned in Chapter 4, a harmonic function of time is an idealization. Similarly, a plane wave in free space is an idealization, since it extends over all space and thus contains an infinite amount of energy. It is characterized by a single direction of propagation and, if it is harmonic, by a single wavelength.

Nevertheless, the plane wave is of major importance in the study of waves, since an arbitrary wave field can be decomposed into plane waves in much the same manner as an arbitrary function of time can be decomposed into harmonic components.

A sufficiently long train of a $\cos(\omega t)$-function can be used as an approximation of a harmonic function, and, similarly, a plane wave can be approximated by a wave of finite extent, so that it is possible to carry out experiments which can be interpreted in terms of the characteristics of a single plane wave.

Of particular interest in this section is the study of the transmission of a plane wave through slits or apertures in a plane screen. A rigorous analysis of this problem is beyond the scope of this chapter. Fortunately, it is possible to understand important features of the transmitted field without a rigorous analysis, by using the results obtained in the study of radiation from extended sources in the previous sections.

Qualitatively, we know from everyday observations with sound waves and waves on water, that the waves emerging from the slits generally do not form a sharp beam but are 'bent around the corners' and thus spread out in all directions.

In determining analytically the corresponding angular distribution from our previous study of radiation, it is assumed that the wave field to the right of the screen, the diffraction field, is the same as the field produced by a continuous distribution of 'secondary' sources in the slits. This procedure can be considered to be the analytical counterpart of **Huygens**' principle, according to which the wave front of the total field emerging from the slits is constructed geometrically as the envelope of the spherical wavelets emitted from the secondary sources in the slit.

Although this procedure will make it possible to determine the angular distribution of energy flux in the diffraction field, it will not yield the magnitude of the wave amplitude in terms of the amplitude of the primary wave incident on the screen. In fact, it does not even deal with the question of whether the source strength in the slits should be proportional to the force or the velocity variable in the field (corresponding to pressure and velocity in a sound wave and electric and magnetic field in an EM wave) or a combination of them.

The calculation of the radiation field from the source distributions in the previous sections was limited to the far field, in which the distance to the receiver is so large that the waves emitted from the various points in the slit region can be considered to be parallel. The variation of the field amplitude is due to the interference between the various waves and this, in turn, depends on the path differences of wave travel, as before.

This far-field interference pattern is usually called the **Fraunhofer diffraction field**. It can be observed on a screen far away from the slit screen or in the focal plane of a lens placed on the right hand side of the slit. It should be noted that the lens does not yield an image of the slit but rather an 'image' of the parallel waves travelling in the different directions. These waves are brought together in the focal plane of the lens and the interference between waves can then be observed on a screen in the focal plane.

The interference between the elementary waves is not limited to the Fraunhofer or far field regime, but it occurs in the entire region. In calculating the amplitude distribution in the wave field, the direction of the elementary waves no longer can be considered to be parallel, and the integrals involved usually cannot be expressed by elementary functions. This region is usually referred to as the near field or **Fresnel regime**.

For light waves, Fresnel diffraction patterns can be demonstrated experimentally as follows. A light source, formed by illuminating a narrow slit by an intense light beam, for example, is placed in the object plane of a lens, so that a sharp image of the source slit is formed in the image plane of the lens. Note, that the light rays are not parallel on either side of the lens. A second screen with a slit is now inserted between the source screen and the lens, so that the diverging field reaching the lens from the source is intercepted and reduced in size in the lateral direction by the slit. The image of the source no longer will be sharp but bordered by a diffraction pattern in the form of bright and dark bands.

In both the Fraunhofer and Fresnel regimes, the interference pattern is a result of the constructive and destructive interference which ocurs as a result of the travel path differences between the interfering elementary waves. In the Fraunhofer regime it is a matter of path differences between waves travelling in the **same** direction. In the Fresnel regime, on the other hand, the interfering waves travel in **different** directions. In fact, in a plane very close to the slit plane, in the region of a slit, the waves from different parts of the slit travel in opposite directions, and the interference is like that of a standing wave. As the distance from the slit increases this 'standing' wave pattern becomes less pronounced, the intensity variation becomes more uniform until we arrive in the Fraunhofer regime, where we get a pronounced pattern resulting from the interference of elementary waves travelling in the same direction.

Diffraction occurs not only as a result of wave interaction with slits but with all inhomogeneities and objects which cause a change in the direction of the flow of wave energy. The phenomenon is also called scattering. The two terms are essentially synonymous. Scattering often refers to wave interaction with (randomly distributed) small inhomogeneities or particles; Rayleigh scattering of light being a typical example. Diffraction generally involves single objects or symmetrical arrays; classical examples are diffraction by the edge of a screen, X-ray diffraction by crystals, and the diffraction by single and multiple slits in a screen.

Scattering will be discussed in connection with a study of wave generation in Chapter 17, where scattering of both acoustic and electromagnetic waves (including Rayleigh scattering) will be discussed. We shall then be able to determine not only the angular distribution of the amplitude of the scattered field but also the absolute value of the amplitude.

N parallel narrow slits. The slits are contained in a plane opaque screen and are assumed to be identical, parallel, and equidistant. A plane wave is incident on the screen from the left in the direction normal to the screen. The screen prevents the wave from continuing as a plane wave behind the screen, since only the small portion of the wave front which falls on the slits will contribute to the field behind the screen.

To determine this diffracted field, we treat the slits as sources. Narrow slits with a width much smaller than a wavelength simulate line sources, and with the incident harmonic wave being normal to the screen, the line sources will have the same amplitude and phase. The result obtained in Eq. 11.2(5) can

Display 11.10.

Diffraction by N narrow slits.

N narrow slits

$S(\theta) = S(0)[\sin(N\delta)/N \sin\delta]^2$ (1)

$\delta = (kd/2)\sin\theta = (\pi d/\lambda)\sin\theta$ (2)

$\theta = \arctan(y/L)$ (3)

Maxima

$S(\theta) = S(0)$ (4)

for

$\delta = n\pi$ (5)

$\sin\theta = 2n\pi/kd = n\lambda/d$ (6)

$\theta \simeq y/L = n\lambda/d$ if $d \gg \lambda$ (7)

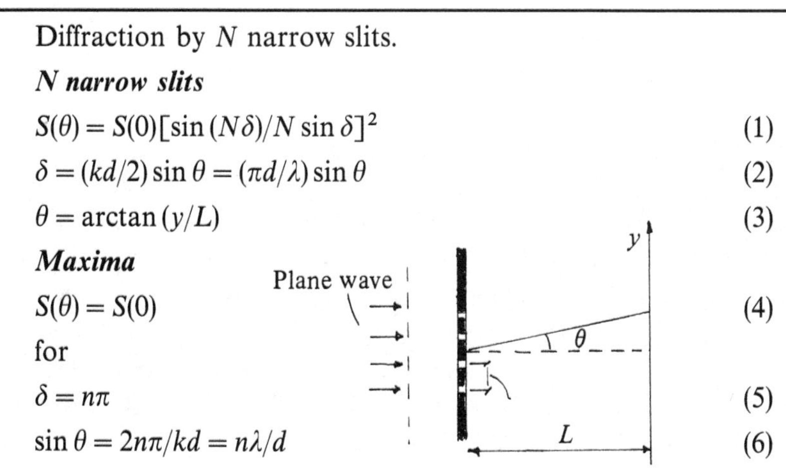

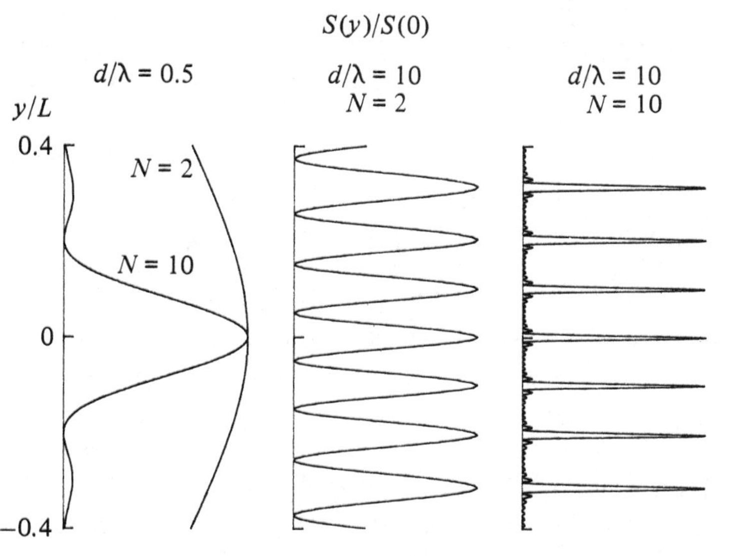

then be used for the angular distribution of the energy flux in the far zone of the diffracted field.

Instead of expressing the directivity pattern as a function of the angle θ, we are now interested in the energy flux distribution on a plane at a distance L from the slits. The coordinate along the plane is denoted by y, $\tan(\theta) = y/L$. We are usually interested only in the central portion of the diffraction pattern so that $\theta \simeq y/L$. In the case of diffraction of light the pattern can be recorded on a photographic plate or by means of a photomultiplier. In both cases the pattern thus obtained is proportional to the energy flux.

In Display 11.10 we have plotted the intensity distribution thus obtained in the range $y/L = -0.4$ to 0.4 for $N = 2$ and 10 and $d/\lambda = 0.5$, and 10. When the ratio d/λ between the slit separation and the wavelength is less than 1, there is only one major lobe in the diffraction field, which occurs in the forward direction $\theta = 0$. With d/λ larger than 1, there will be an additional major lobe for every location y, at which the difference in path length to two adjacent slits is an integer number of wavelengths, i.e.

$$\theta \simeq y/L = n\lambda/d \quad (d \gg \lambda) \tag{1}$$

For a given d/λ, the essential effect of increasing the number of slits is to make the major lobes sharper.

In addition to the main lobes in the diffraction pattern, there are secondary lobes with greatly reduced amplitudes, as can be seen in the diffraction pattern for $d/\lambda = 10$, $N = 10$.

Single 'wide' slit. The slit width in the previous section was assumed small compared to the wavelength. We now consider a single slit of arbitrary width b in an opaque screen and a plane harmonic wave incident on the screen from the left. With the direction of the harmonic wave being normal to the screen, the field across the slit will be uniform both in amplitude and phase (to a first approximation). We are interested in the angular distribution of the energy flux on the right hand side of the screen as observed in the far field on a plane parallel with the slit screen, as indicated in Display 11.11.

Treating the slit as a uniform source distribution, we obtain the same angular distribution as for the continuous line source in Eq. 11.3.1, and the corresponding y-dependence of the energy flux on the plane is obtained by using $\tan(\theta) = y/L$, or, for the central portion of the field, $\theta \simeq y/L$.

Unlike the diffraction pattern from N narrow slits, the present pattern has only one major lobe, and with the incident wave being normal to the slit screen, the center of this lobe is at $\theta = 0$. As shown in the display, we have computed the diffraction pattern for some different values of the slit width ranging from 0.5 to 10 wavelengths.

Display 11.11.

Diffraction by a single slit.

Single slit

Ref. Displays 11.7 and 11.8.

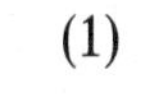

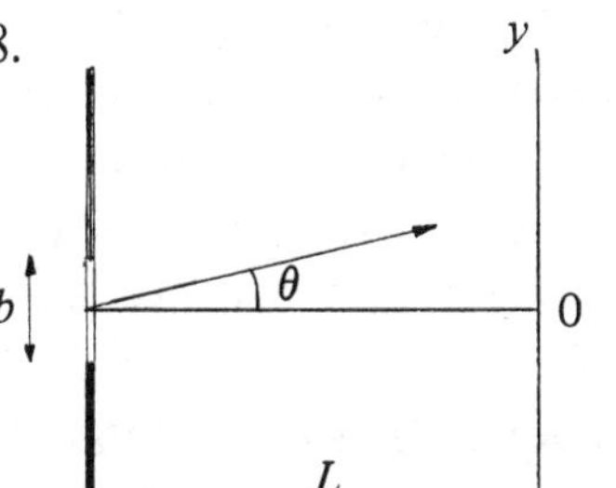

$S(\theta)/S(0) = (\sin\beta/\beta)^2$ (1)

$\beta = (kb/2)\sin\theta$ (2)

$\theta = \arctan(y/L)$ (3)

Minima

$\beta = n\pi, \quad n = 1, 2, \ldots$ (4)

$\sin\theta = 2n\pi/bk = n\lambda/b$ (5)

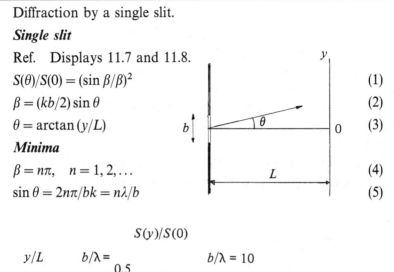

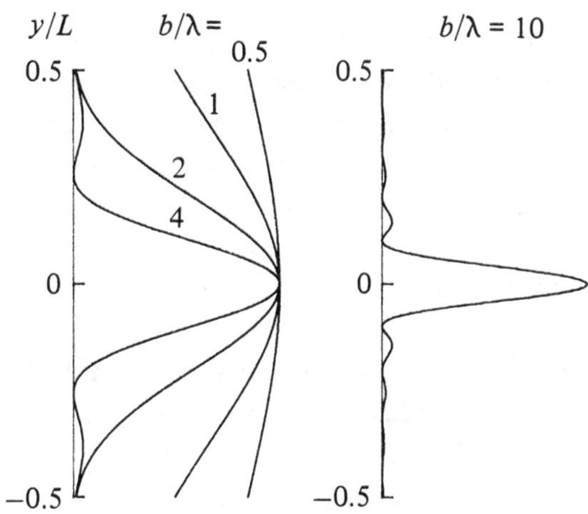

If b is considerably smaller than the wavelength, the radiation is almost the same in all directions. As the slit width increases, the width of the lobe decreases, and there will be directions in which destructive interference results in zero energy flux. In order to have at least one such minimum, the slit width must be larger than one wavelength.

To understand the occurrence of a minimum, we can divide the slit into two halves. Then if the total width is larger than one wavelength, we can find pairs of elementary strips in these halves, for which the path difference to the receiver in a certain direction is half a wavelength, leading to destructive in interference.

With $\beta = (kb/2)\sin(\theta) = (\pi b/\lambda)\sin(\theta)$, the angular distribution of the energy flux, $S(\theta) = S(0)[\sin(\beta)/\beta]^2$, will have minima for

$$\beta = n\pi \tag{2}$$

or

$$\sin(\theta) = n\lambda/b \quad \theta \simeq n\lambda/b \quad (n = 1, 2, \ldots.)$$

The width of the main lobe is defined by the angle between the minima at $\theta \simeq -\lambda/b$ and $\beta \simeq \lambda/b$, so that the **beam width** $= \mathbf{2\lambda/b}$. It is important to realize that the beam width **increases** as the slit width **decreases.**

The diffraction pattern by a single slit can be observed in simple experiments using the edges of razor blades to form a slit or simply a slit cut in a piece of paper. If a monochromatic light source is viewed through the slit, the central lobe and the 'bands' of minimum intensity can be observed.

N wide slits. With reference to the analysis of the radiation field from a nonuniform source distribution in Section 11.3, we can express the angular distribution of the energy flux in the far field from a set of N wide slits. The sum of the complex amplitude contributions from the slits will be a geometric series, as for N narrow slits.

The difference is now that each term in the series is an integral over a wide slit. This integral is the same in each term and can be represented as a common factor. The remaining geometric series is the same as for the narrow slits, and it follows that the resulting angular distribution of the energy flux is the product of the distribution functions $[\sin(N\delta)/N\sin(\delta)]^2$ and $[\sin(\beta)/\beta]^2$ for the N narrow slits and for the single wide slit, i.e.

$$S(\theta) = S[\sin(N\delta)/N\sin(\delta)]^2[\sin(\beta)/\beta)]^2 \tag{3}$$

$$\delta = (kd/2)\sin(\theta)$$

$$\beta = (kb/2)\sin(\theta)$$

Examples of computed diffraction patterns with $N = 10$, $d/\lambda = 10$, and $b/\lambda = 0.5$, 1, and 2, are shown in Display 11.12. The resulting patterns should

Display 11.12.

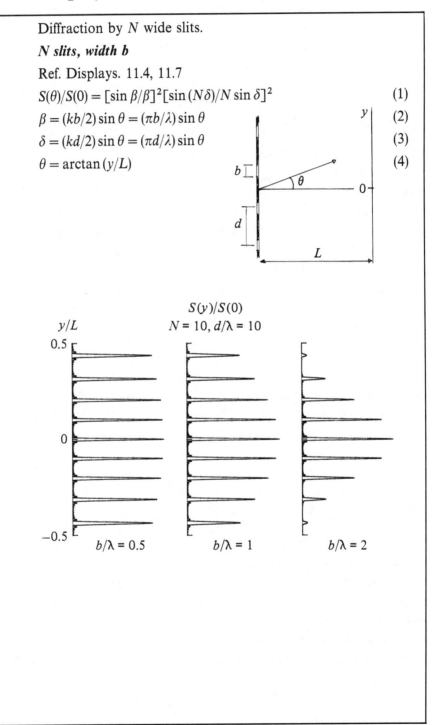

Diffraction by N wide slits.

N slits, width b

Ref. Displays. 11.4, 11.7

$$S(\theta)/S(0) = [\sin\beta/\beta]^2[\sin(N\delta)/N\sin\delta]^2 \quad (1)$$

$$\beta = (kb/2)\sin\theta = (\pi b/\lambda)\sin\theta \quad (2)$$

$$\delta = (kd/2)\sin\theta = (\pi d/\lambda)\sin\theta \quad (3)$$

$$\theta = \arctan(y/L) \quad (4)$$

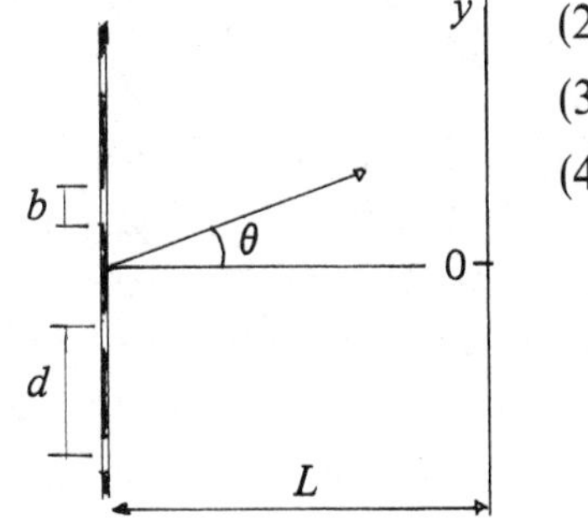

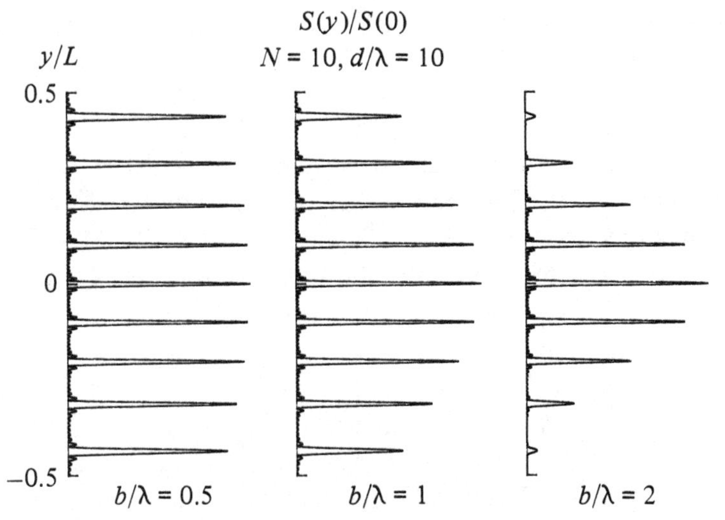

be compared with those obtained from the corresponding values of the parameters in Displays 11.10 and 11.11. It should be noted that the diffraction pattern for the single wide slit forms the 'envelope' of the amplitude distribution of the peaks in the diffraction pattern.

In a more systematic mathematical analysis of diffraction it becomes apparent that the diffraction pattern can be expressed in terms of the Fourier transform of the source distribution, in which case the result can be obtained in a more elegant manner.

11.5 Diffraction limited resolution

Returning to the diffraction pattern for a single wide slit, we now treat the case, when the direction of propagation of the incident wave makes an angle ϕ with the normal (x-axis) to the slit screen, as indicated in Display 11.13. An incident wave front then reaches the lower end of the slit first, at $y = 0$, and to reach a point y in the slit, the wave has to travel a distance $y \sin(\phi)$; the corresponding phase lag is $(2\pi/\lambda)y \sin(\phi) = ky \sin(\phi)$. Thus, although the amplitude distribution across the slit will be constant, the phase lag will increase with y as $ky \sin(\phi)$.

With reference to the analysis of the radiation field from a uniform source distribution in Display 11.8, we note the angular distribution of the resulting energy flux is obtained by replacing $\beta = (kb/2) \sin(\theta)$ by $\beta' = (kb/2)[\sin(\theta) - \sin(\phi)]$, which for small angles corresponds to $(kb/2)(\theta - \phi)$. This means, that the center of the main lobe in the diffraction pattern is shifted an angle $\theta = \phi$, as expected.

We now apply this result to two waves with the angles of incidence $-\phi$ and ϕ, as indicated. The waves may represent the light from two distant sources (stars, for example), and the fields are assumed to be uncorrelated. The total energy flux in the diffraction field is the sum of the energy fluxes of the individual sources. If a lens is placed in the slit (or aperture), images of the source (stars) will be formed in the focal plane of the lens. The images will not be 'points', however, but will exhibit the diffraction pattern of the slit, since the wave fronts are not infinitely extended but limited in width by the slit.

Of particular interest here is the question of how large the angular separation $\psi = 2\phi$ of the sources must be in order for the images to be resolved in the resulting diffraction pattern.

If the angular separation is zero ($\psi = 0$), the diffraction patterns of the two beams will fall on top of each other, and the shape will be the same as for a single waves. As the angular separation increases, the individual patterns will be separated. But the total energy flux does not show two maxima, until the angular separation exceeds a certain value.

Display 11.13.

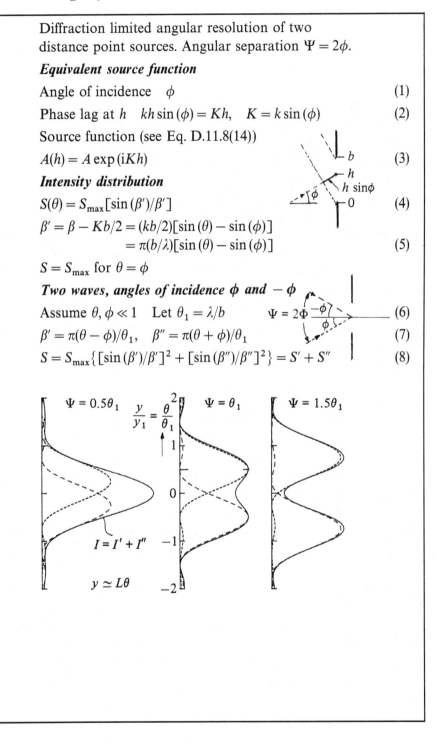

Diffraction limited angular resolution of two distance point sources. Angular separation $\Psi = 2\phi$.

Equivalent source function

Angle of incidence $\quad \phi$ (1)

Phase lag at $h \quad kh\sin(\phi) = Kh, \quad K = k\sin(\phi)$ (2)

Source function (see Eq. D.11.8(14))

$A(h) = A\exp(\mathrm{i}Kh)$ (3)

Intensity distribution

$S(\theta) = S_{\max}[\sin(\beta')/\beta']$ (4)

$$\beta' = \beta - Kb/2 = (kb/2)[\sin(\theta) - \sin(\phi)] = \pi(b/\lambda)[\sin(\theta) - \sin(\phi)] \quad (5)$$

$S = S_{\max}$ for $\theta = \phi$

Two waves, angles of incidence ϕ and $-\phi$

Assume $\theta, \phi \ll 1$ Let $\theta_1 = \lambda/b$ (6)

$\beta' = \pi(\theta - \phi)/\theta_1, \quad \beta'' = \pi(\theta + \phi)/\theta_1$ (7)

$S = S_{\max}\{[\sin(\beta')/\beta']^2 + [\sin(\beta'')/\beta'']^2\} = S' + S''$ (8)

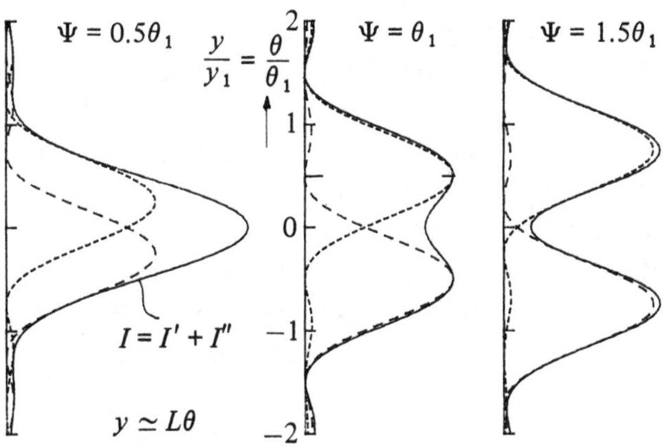

To illustrate this quantitatively, we have plotted the individual diffraction patterns together with the total intensity distribution for three different values of the angular separation ϕ of the waves. The separation has been expressed in terms of the 'half width' θ_1 of an individual diffraction pattern, i.e. the angular separation of the maximum and the first minimum in the pattern.

The resulting diffraction patterns have been plotted for $\psi = 0.5\theta_1, \theta_1$, and $1.5\theta_1$. For $\psi = 0.5\theta_1$, it has only one maximum, and the two sources clearly are not resolved. For $\psi = 1.5\theta_1$, on the other hand, we get two distinct maxima in the pattern, and the images of the two sources are well resolved.

In the intermediate case, $\psi = \theta_1$, the maximum in one diffraction pattern falls on the first minimum of the other. The resulting pattern has two distinct maxima with a separation equal to the angular separation of the sources. By definition, the angle $\psi = \theta_1$ is the smallest angular separation of the sources which is considered resolvable; the minimum angular source separation being equal to the half width of the diffraction pattern. This is usually called the **Rayleigh criterion** for resolution.

It should be emphasized, that the resolution limit $\psi = \theta_1 \simeq \lambda/b$ is valid for a slit. For a **circular aperture** the diffraction pattern will have circular symmetry, and it can be shown that the first circle of zero energy flux corresponds to the angular half width $\boldsymbol{\theta_1 = 1.22\lambda/D}$ of the diffraction pattern, where D is the diameter of the aperture.

The diffraction limited image resolution, which we have arrived at, does not depend on the distance L from the slit to the image plane. It should be realized, however, that it was implied in our analysis that L was large enough to be in the Fraunhofer region. As the image plane is moved closer to the slit, the character of the diffraction pattern changes in much the same manner as illustrated in Display 11.9.

In the Fraunhofer region (far-field), the width of the main lobe is $2\theta_1 L$. As L decreases toward zero, inside the Fresnel region (near field), the image size will not go to zero, however, but will approach the width of the slit (see Display 11.9).

As an example, consider a typical lecture demonstration of single slit diffraction pattern when the slit width is 0.05 cm and the distance to the image plane is 5 m. With a wavelength of 5×10^{-5} cm, the width of the main lobe in the far field will be 2×10^{-3} radians and the corresponding linear dimension 1 cm, i.e. 20 times the slit width. The near field is reached when the predicted image size equals the slit-width and the corresponding value for L is 25 cm.

Such a small value of L is encountered in a **pinhole camera**, which is simply

Display 11.14.

Image resolution. Pinhole camera.

Fraunhofer region (far-field)

Slit $\phi_1 > \lambda/D$ (1)

Circular aperture $\psi_1 > 1.22\lambda/D$ (2)

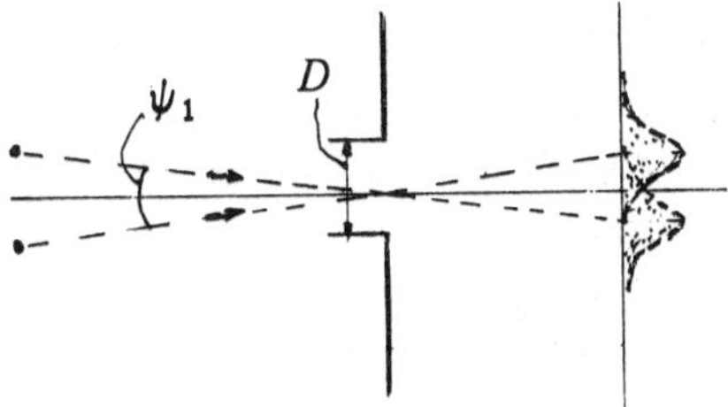

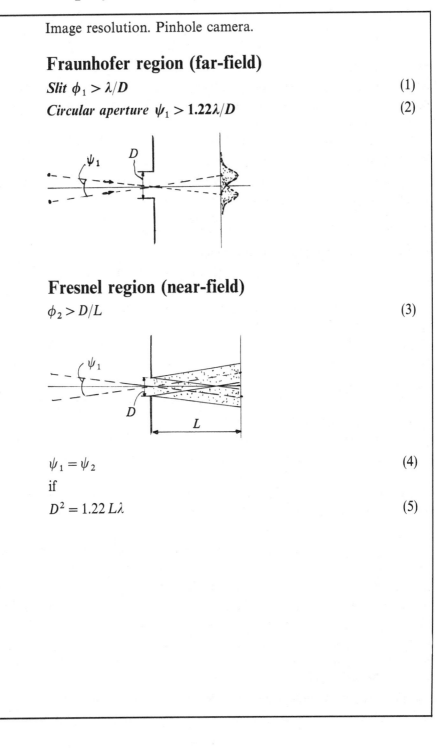

Fresnel region (near-field)

$\phi_2 > D/L$ (3)

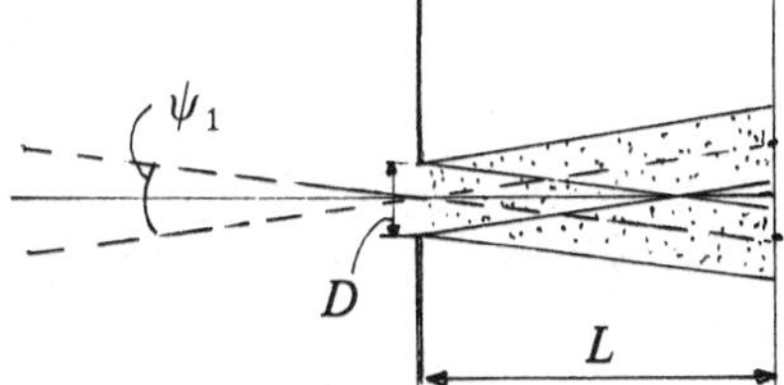

$\psi_1 = \psi_2$ (4)

if

$D^2 = 1.22\, L\lambda$ (5)

a rectangular box with a small aperture (pinhole) in one of its walls. If the film is so close to the pinhole that it is in the near field of diffraction, the image on the film is simply the projection of the aperture on the film, i.e. a circular disc with the diameter D of the aperture. Two incident plane waves (from two distant point sources) with a certain angular separation will then produce two circular spots on the field, each with a diameter D. These images can be resolved if the distance between their centers is larger than D. Thus, the angular resolution in the near field is D/L.

In order to improve this resolution, we can decrease D and/or increase L. But as L increases the far field diffraction region is eventually reached, in which the angular resolution will be $1.22\lambda/D$. The best resolution with the pinhole camera is then obtained when the near field resolution equals the far field resolution, i.e. when $D/L = 1.22\lambda/D$, i.e. $D^2 = 1.22\lambda L$.

11.6 Coherence and incoherence

In our discussion of wave interference, the sources of the fields were assumed to be harmonic with the same frequency and with a constant (time independent) phase difference. Such sources and fields are said to be **coherent**.

By contrast, consider a field produced by a set of wave trains of finite length of the same frequency but with phase angles which fall at random between 0 and 2π. An example is the superposition of tones of finite duration, sounded at random by the individual members of an ensemble of wind and/or string instruments.

Ordinary light is another example. It is produced as a result of randomly occurring spontaneous electronic transitions in the atoms, each transition resulting in the emission of a wave train of light. Typically, the duration of these wave trains is of the order of $\tau \simeq 10^{-8}$ sec, and the corresponding length of each train will be $c\tau \simeq 3$ m. Such waves and their sources are said to be **incoherent**.

The electromagnetic 'noise' produced by electrons in random motion in an electron gas or the acoustic noise generated by the random fluid motion in a jet are other examples of incoherent radiation. The time dependence of these functions cannot be expressed in a deterministic manner in functional form, as already mentioned in Chapter 10.

If incoherent sources were used in a linear array, there would be no distinct interference pattern of the kind obtained in the previous section. Rather, the time average of the energy flux simply will be the sum of the energy fluxes of the individual sources. In other words, if the field contributions from the individual sources are Ψ_i, the time average of the square of the total field is simply $\sum\langle\Psi_i^2\rangle$, the cross terms $2\sum\sum\langle\Psi_i\Psi_j\rangle$ adding up to zero because of the random phases involved. With reference to the discussion in Section 11.1, this result follows if the cross correlation function of the fields from different sources is zero.

It is still possible, though, to produce an interference pattern by two or more slits illuminated by an incoherent light source, provided that the signals arriving at the observation point from different slits are displaced in time by an amount less than the correlation time in the light. As indicated above, this corresponds to a total path difference less than approximately 3 m for ordinary light, a condition which is usually fulfilled. Under these conditions, the fields in the various slits, and the corresponding equivalent sources, are not uncorrelated, since they will be exposed by wave fronts belonging to the same wave trains.

If the light contains a band of frequencies, the diffraction pattern will be 'blurred', since the angular positions of the maxima and minima in the pattern depend on the wavelength. Even in a 'monochromatic' beam of ordinary light, there is a certain frequency band width (line width), and the diffraction pattern will not be as well defined as when a (single mode) laser light is used, for which the line width is considerably smaller.

The quality of a diffraction pattern is often expressed in terms of the 'visibility', a quantity, which is determined by the ratio of the maximum and minimum intensities in the pattern.

Apart from the frequency bandwidth, the finite size of a source can also contribute to a reduction in the visibility. The extended source can be regarded as a collection of point sources, and the waves from these sources will have slightly different angles of incidence on the slit screen, and their diffraction patterns will be displaced on the image plane. The total diffraction pattern thus consists of a large number of patterns, which are slightly displaced with respect to each other, and, as a result, the minima in the pattern will not have zero intensity.

Examples

E11.1 *Application of single slit diffraction.* In the discussion in Section 11.4, the minima in the diffraction pattern from a single slit were found to occur at angles of diffraction given by

$$\sin(\theta) = n\lambda/b \qquad (1)$$

where b is the slit width and λ the wavelength. If the distance to the screen upon which the diffraction pattern is observed is L and the coordinate along the screen perpendicular to the slit is y, we have $\sin(\theta) \simeq y/L$ for small angles, and it follows that the nth minimum occurs at the y-coordinate

$$y_n = n\lambda L/b \tag{2}$$

The separation between two adjacent minima is $(L/b)\lambda$, which increases with decreasing slit width b, as illustrated in Fig. 1.

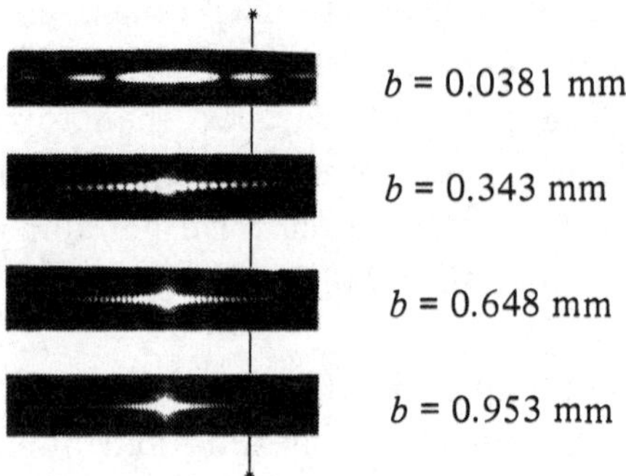

Fig. 1. Single slit diffraction pattern for different slit widths. Distance to screen, $L = 1.5$ m. $\lambda = 6380$ Å. Courtesy of Diffracto Ltd., Troy, MI, U.S.A.

Measurement of separation. It follows from Eq. (2) that for a given wavelength λ, measurement of L and the distance y_n to the nth minimum, makes it possible to determine the width of the slit

$$b = n(L/y_n)\lambda \tag{3}$$

In the lecture demonstration referred to in Problem 11.1, the screen was at a distance $L = 900$ cm and the distance from the center ($y = 0$) to the 10th minimum (fringe) was 50 cm. With a wavelength of the laser light, $\lambda = 6380 \times 10^{-8}$ cm, the slit width in that particular case was $b = n(L/y_n)\lambda = 0.011$ cm.

Thus, using the single diffraction apparatus as a 'proximity sensor', the measured separation of 50 cm for a slit width of 0.011 cm, corresponds to a 'magnification' of about 5000.

The variation of the separation of the edges of two bodies readily can be displaced through the diffraction of a laser beam which has been 'fanned' by means of a lens to cover the entire length of the 'slit' between the bodies. An example is given in Fig. 2.

Measurement of displacement. For a given location y in the image plane of the diffraction pattern, the slit width is proportional to the number of fringes between $y = 0$ and y_n. Therefore, at a fixed position y, a variation in the slit width results in a change in this fringe number and by counting the number of fringes which move by the fixed position, the corresponding change in the slit width b can be determined.

Thus, if the fringe number is n before and n' after the displacement of one (or both) of the slit edges, we obtain $b' - b = (n' - n)(L/y)\lambda$,

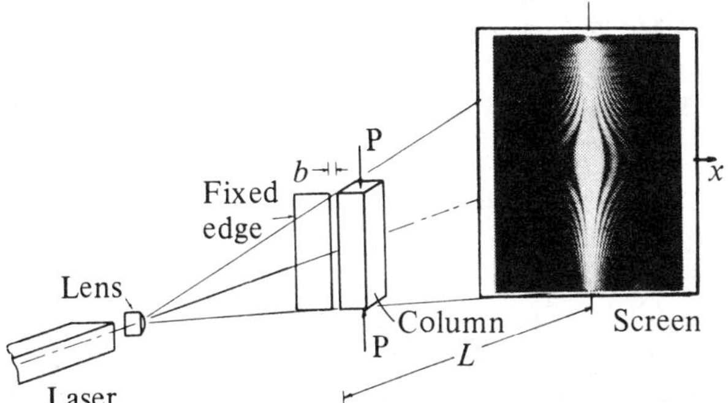

Fig. 2. Diffraction pattern of the 'slit' between the edges of two bodies showing the variation in the slit width. Courtesy of Diffracto Ltd., Troy, MI, U.S.A.

or

$$(b' - b)/b = (n' - n)/n \tag{4}$$

In other words, the relative change $\Delta b/b$ in the slit width is equal to the relative change $\Delta n/n$ in the fringe number.

For example, in the lecture demonstration referred to in Problem 11.1, a change of one fringe at the location y_{10} corresponds to a change in the slit width of $(b/10) = 0.0011$ cm $\simeq 10$ microns. In Fig. 2, the deflection of the center of the beam which forms one edge in the 'slit' is 0.05 mm.

Consequently, single slit diffraction can be used as a 'strain gauge' to measure small deformation and any physical quantity which can be related thereto. A sensitivity of about 1 micron per fringe can be achieved.

Measurement of vibration amplitude. If one of the edges in the slit is oscillating so that the slit width varies with time, the fringe pattern will shrink and expand in the same manner, and the output from a photo detector at a fixed position will yield an output proportional to the intensity of the fringes as they move by the detector.

The intensity decreases with increasing n, and the output signal will vary in amplitude in accordance with the variation in fringe number. If the time dependence is periodic, we obtain a periodic output signal with an amplitude from which the amplitude of vibration can be determined.

This procedure can be used in a variety of vibration measurements and for absolute calibration of vibration transducers over a wide frequency range from zero to hundreds of kilocycles.

E11.2 *Waves as 'probes.'* In the application of diffraction for measurement of separation in E11.1, the light wave can be considered to be a 'probe' or length scale. It can be used in many other applications, and an example is discussed in Chapter 18, where laser light is used to measure the wavelength of the thermal 'roughness' of a liquid surface.

It should be realized that such measurements are possible only when the slit width or the surface wavelength is larger than the wavelength of the incident light. To make possible measurement of smaller dimensions by wave probes, radiation of shorter wavelengths is required.

This is realized in X-ray diffraction of the interatomic distances in matter in bulk; with gamma rays, the length scale is brought down even further.

Bragg reflection of X-rays. W.L. Bragg (1912) expressed the condition for constructive interference in X-ray diffraction from a crystal in a simple manner by regarding the atomic planes in a crystal as reflecting surfaces, as illustrated schematically in the figure. In this figure the dashed lines represent surfaces of equal phase and the solid lines are the rays, i.e. the normals to the phase surfaces.

In this model constructive interference occurs when the travel distance between the waves reflected from adjacent planes is an integer number of wavelengths. (Strictly, the wavelength corresponds to the wave speed in the crystal, but generally, the index of refraction is close to unity, so that the wavelength can be taken to be the free space wavelength). With the distance between adjacent planes denoted by *d*, the condition for constructive interference, the **Bragg condition**, is then expressed by

$$2d \sin(\theta) = n\lambda \qquad (1)$$

where θ is the angle between the wave normal and the atomic planes, as shown, and n is an integer.

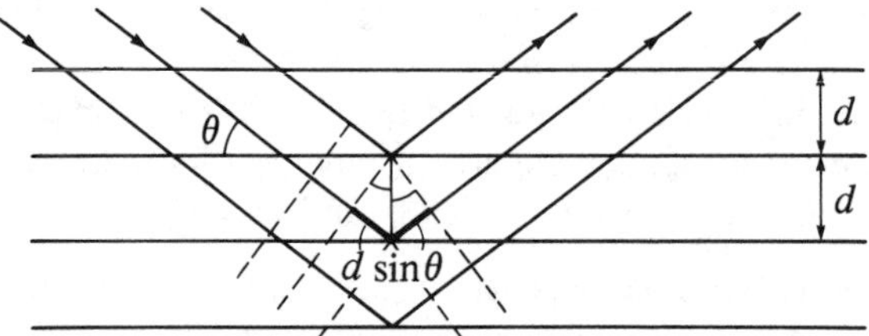

Brillouin scattering. Thermal fluctuations in a solid can be interpreted as a superposition of acoustic waves in the solid travelling in 'all possible' directions. The wave fronts are represented by planes of compression and rarefaction with corresponding periodic variation of the density. The period is equal to the wavelength of sound in the crystal. This wavelength is such that 'Bragg reflection' from these planes can be obtained with light waves.

Since the planes are travelling with the sound speed, the light reflected from the planes will be Doppler shifted, and by measuring this shift, the sound speed can be determined. Furthermore, measurement of the angle at which constructive scattering occurs determines the wavelength of the sound. In this form, the Bragg reflection usually is referred to as Brillouin scattering. By means of it, light can be used as a 'probe' for the measurement of the sound speed and the corresponding compressibility as a function of frequency in the megacycle range.

Radar and Sonar. Electromagnetic waves at larger wavelengths (in the cm-range) are used also as probes in the detection of objects such as airplanes (RADAR) and in the 'mapping' of the electron distribution in the ionosphere. In analogous manner, acoustic waves are used both in the air and in the ocean in various applications of 'sound ranging' (SONAR), and the use of acoustic waves in seismology in well known in probing the interior of the Earth. At shorter wavelengths, sound is often used as a diagnostic tool in medicine and in non-destructive testing of materials. Actually, ultrasonic microscopes have been developed.

Matter waves as probes. As discussed in Chapter 16, the wave nature of material particles, such as electrons and neutrons, makes possible the use of particles as 'wave probes' through the interaction of particles with other particles or matter in bulk (the electron microscope is a well known example).

Problems

11.1 *Diffraction by slits.* In a lecture demonstration, the light from a He–Ne laser, wavelength $\lambda = 6380 \times 10^{-8}$ cm, was used in diffraction experiments with single and multiple slits and gratings. The diffraction pattern was observed on a screen a distance $L = 900$ cm from the slit or grating plane.

(a) The pattern obtained from a single slit is shown schematically in Fig. 1. From it determine the slit width.

0 5 10 15 cm

(b) The pattern obtained with a grating is shown schematically in Fig. 2. From it determine the separation between the 'slits' in the grating.

11.2 *Wave interference (antenna).* An 'antenna' consists of $N = 10$ line sources placed along a straight line with the distance between adjacent sources $d = 10\,\lambda$, where λ is the wavelength of the emitted wave.

(a) How many intensity maxima are there in the radiation field around the antenna (from 0 to 360 degrees)?
(b) If N is increased to 20, what is the change in the number of intensity maxima?

11.3 *Image resolution.* Estimate the minimum separation of two points on the Moon such that they can be resolved by a telescope with an aperture diameter $D = 5$ m. The distance to the Moon is 3.84×10^8 m and the wavelength of the light can be taken to be $\lambda = 5500$ Angstrom ($1\,\text{Å} = 10^{-8}$ cm).

11.4 *Continuous source distribution.* Consider an acoustic line source of length H

along the z-axis, between $z = -H/2$ and $z = H/2$, with a continuous monopole source distribution. Derive expressions for the sound pressure field in the xy-plane

(a) when the source distribution has the same phase along the entire length and
(b) when the phase is constant only within small patches of length ΔH but with the phases in the various patches uncorrelated (randomly distributed in the range between 0 and π).

11.5 *Resolution* (*antenna array*). A radio astronomical interferometer consists of an array of 32 antennas, seven meters apart and placed along a straight line. What is the angular resolution of the array for the 21 cm line of radiation? (The angular resolution is defined as the half-width of the main lobe of the array, i.e. the angular separation of the maximum and the adjacent minimum in the directivity pattern of the array).

12

Refraction and reflection

A wave incident on a plane boundary between two regions produces a disturbance which travels along the boundary with a speed which depends on the angle of incidence of the wave. The boundary then acts like a travelling source distribution, and the field transmitted into the second region has the same directional characteristics as obtained from the phased array in the last section. If the wave speeds are different in the two regions the direction of propagation of the transmitted wave will be different from that of the incident, and this effect is called refraction. It is a wave kinematical effect, and the result obtained applies to all kinds of waves.

To determine the amplitudes of the refracted and reflected waves, the dynamical equations have to be considered, and the quantitative results depend on the particular wave involved. For EM waves, the result depends also on the polarization of the incident wave.

12.1 Trace velocity and Snell's law

Refraction of light is well known from everyday observations, and it can be demonstrated quantitatively in simple experiments with laser beams, as illustrated in Display 12.1. The scattering of light out of the beam is sufficient to make it visible, and the path of the beam can easily be observed.

In the first example the laser beam enters a container of water from above. In the process, the beam is bent toward the normal to the water surface. This phenomenon is an example of **refraction** of waves.

In the second demonstration, the beam first goes through the glass wall of the container, being refracted at both interfaces and then is incident on the water surface from below. As the beam leaves the water, it is bent away from the normal. If the angle of incidence is increased, a critical value is

Display 12.1.

Refraction. Trace velocity and Snell's law.

Simple demonstrations

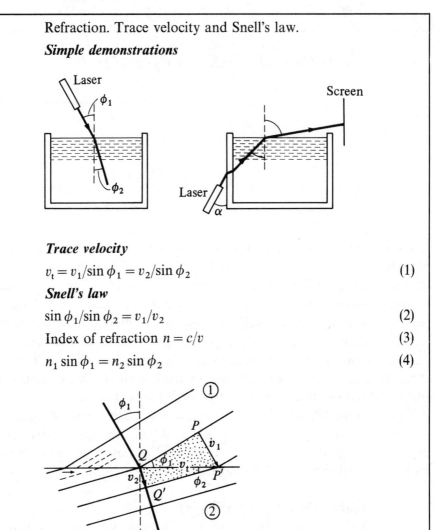

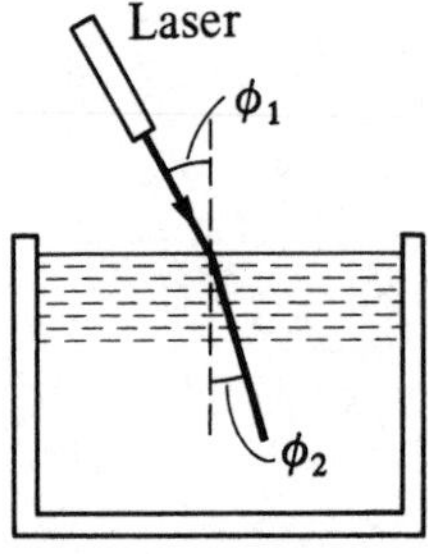

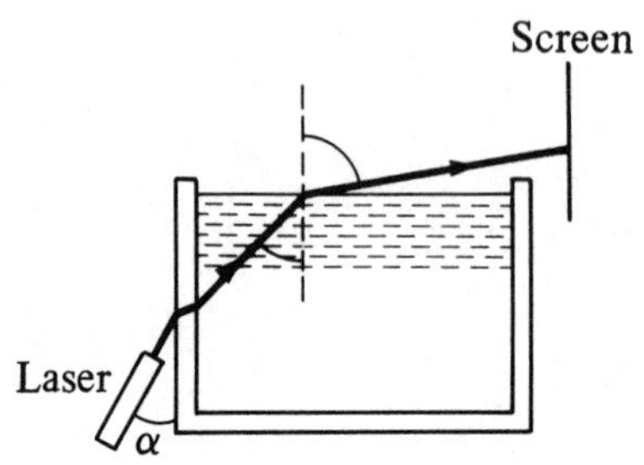

Trace velocity

$$v_t = v_1/\sin\phi_1 = v_2/\sin\phi_2 \quad (1)$$

Snell's law

$$\sin\phi_1/\sin\phi_2 = v_1/v_2 \quad (2)$$

Index of refraction $n = c/v$ (3)

$$n_1 \sin\phi_1 = n_2 \sin\phi_2 \quad (4)$$

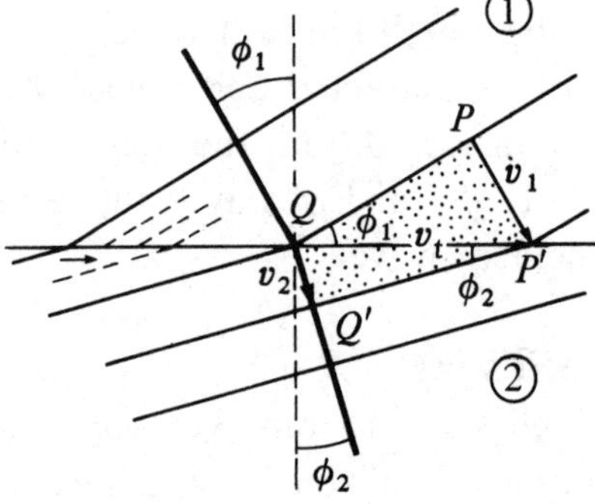

reached, at which the light leaves parallel with the water surface. Above this critical angle, no light is transmitted but is totally reflected from the water surface.

To analyze the refraction phenomenon, we refer to the second half of Display 12.1. Here we have shown the wave fronts, i.e. the surfaces of constant phase, of a plane wave, which is incident on a plane boundary between two regions, in which the wave speeds are v_1 and v_2, respectively. The normal to the constant phase surfaces indicates the direction of propagation, and it makes an angle ϕ_1 with the normal to the boundary. This angle is called the angle of incidence.

Let us follow the progress of an incident wave front. It travels foward with the wave speed (phase velocity) v_1 and intersects the boundary at the point Q. As the wave front moves forward from point P to P', the point Q moves along the boundary to P'. The velocity of Q along the boundary will be called the **trace velocity** $\mathbf{v}_t$, and it follows from the diagram that $v_t = v_1/\sin(\phi_1)$.

The **refracted** wave in the second region can be considered to be generated by the incident wave which produces a disturbance which travels along the boundary with the trace velocity. Like the incident wave, the refracted wave will be a plane wave, and the direction of propagation of this wave makes an angle ϕ_2 with the normal to the boundary. The wave fronts of the incident and refracted waves must be connected at the boundary, and it follows that the trace velocity for the two waves must be the same so that

$$v_t = v_1/\sin(\phi_1) = v_2/\sin(\phi_2) \tag{1}$$

This 'conservation' of the trace velocity leads to the relation between the angles of incidence and refraction $\sin(\phi_2) = (v_2/v_1)\sin(\phi_1)$ which is usually called **Snell's law**. It should be noted that if many parallel boundaries are involved between regions of different wave speeds, the trace velocity will be the same throughout. Therefore, if v_2 and ϕ_2 refer to the final region, the relation between the final angle of refraction and the original angle of incidence is given by Eq. (1), regardless of the number and properties of the layers involved.

The speed of light v in a material such as glass is typically 0.6–0.7 of the speed c in vacuum, and it generally depends on frequency, which then is true also for the **index of refraction**, which is defined as $n = c/v$. Therefore, if the incident light is 'white', the different frequencies and corresponding colors will be refracted at different angles; i.e. the light is broken up into its color spectrum. The frequency dependence of the phase velocity is known as **dispersion**, which will be discussed in some detail in a subsequent chapter.

12.2 Acoustic reflection and transmission coefficients

The discussion of refraction in the previous section was purely kinematical; the dynamical equations of motion did not enter into the discussion, and Snell's law is valid regardless of the particular wave or medium involved.

The determination of the amplitudes of the reflected and refracted waves, however, is a dynamical problem, and requires knowledge of the field equations for the waves involved. We shall illustrate the calculation of these amplitudes for acoustic and electromagnetic waves reflected from a plane boundary between two uniform regions.

We start with the acoustic wave. The boundary is placed in the yz-plane, and the direction of propagation is in the xy-plane, as indicated in Display 12.2. The time dependence is harmonic and the angle of incidence is ϕ_1. The direction of propagation as well as the wavelength can be expressed in terms of a propagation vector $\mathbf{k}_1$, with the magnitude $k_1 = 2\pi/\lambda_1$ and the direction specified by the angle ϕ_1.

If we designate by s the coordinate along the direction of propagation, the expression for the complex amplitude of the incident wave can be written $p_i = A\exp(ik_1 s)$. The distance s is measured from the origin in the xy-plane to the 'wave front', i.e. the surface of constant phase corresponding to $k_1 s$ is perpendicular to $\mathbf{k}_1$, as indicated in the display.

Let x, y be the coordinates of the vector r from the origin to a point of the wave front with the constant phase angle $k_1 s$. The distance s is then the projection of $\mathbf{r}$ on the direction of $\mathbf{k}_1$ and can be expressed as the scalar product of $\mathbf{r}$ and the unit vector $\mathbf{k}_1/k_1$, i.e. $s = \mathbf{r}\cdot(\mathbf{k}_1/k_1)$. In other words, the complex amplitude of the incident wave can be expressed as

$$p_i = A\exp(ik_1 s) = A\exp(i\mathbf{k}_1\cdot\mathbf{r}) = A\exp(ik_{1_x}x + ik_{1_y}y) \tag{1}$$

$$k_{1_x} = k_1\cos(\phi_1) \quad k_{1_y} = k_1\sin(\phi_1) \quad k_1 = \omega/v_1 = 2\pi/\lambda_1$$

The propagation vector for the reflected wave has the components $-k_1\cos(\phi_1)$ and $k_2\sin(\phi_2)$, and the complex amplitude function is

$$p_r = B\exp(-k_{1_x}x + ik_{1_y}y) \tag{2}$$

where k_{1_x} and k_{1_y} are given in Eq. (1).

In the same manner we find the transmitted (refracted) complex amplitude function to be

$$p_t = C\exp(ik_{2_x}x + ik_{2_y}y) \tag{3}$$

$$k_{2_x} = k_2\cos(\phi_2) \quad k_{2_y} = k_2\sin(\phi_2) \quad k_2 = \omega/v_2$$

In region 1 the total complex amplitude is the sum of the incident and

Display 12.2.

Reflection of sound at a plane boundary.

Constant phase surface

$$ks = \mathbf{k}\cdot\mathbf{r} = k_x x + k_y y \quad (k = 2\pi/\lambda = \omega/v) \qquad (1)$$

$$k_x = k\cos\phi \quad k_y = k\sin\phi \qquad (2)$$

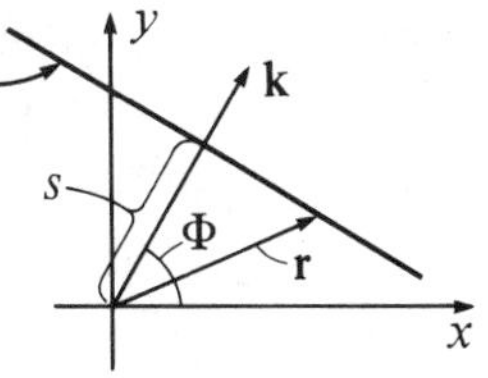

Pressure field

$$p_{\mathrm{i}} = A\exp(\mathrm{i}k_{1_x}x + \mathrm{i}k_{1_y}y) \qquad (3)$$

$$p_{\mathrm{r}} = B\exp(-\mathrm{i}k_{1_x}x + \mathrm{i}k_{1_y}y) \qquad (4)$$

$$p_{\mathrm{t}} = C\exp(\mathrm{i}k_{2_x}x + \mathrm{i}k_{2_y}y) \qquad (5)$$

$$k_{1_x} = (\omega/v_1)\cos\phi_1 \qquad k_{2_x} = (\omega/v_2)\cos\phi_2 \qquad (6)$$

Velocity field

$$\rho\partial u_x/\partial t = -\partial p/\partial x \qquad (7)$$

$$-\mathrm{i}\omega\rho u_x = -\partial p/\partial x \qquad (8)$$

$$u_x = (1/\mathrm{i}\omega\rho)\partial p/\partial x \qquad (9)$$

$$u_{\mathrm{i}_x} = (k_{1_x}/\omega\rho_1)p_{\mathrm{i}} = (p_{\mathrm{i}}/Z_1)\cos\phi_1 \qquad (10)$$

$$u_{\mathrm{r}_x} = -(k_{1_x}/\omega\rho_1)p_{\mathrm{r}} = -(p_{\mathrm{r}}/Z_1)\cos\phi_1 \qquad (11)$$

$$u_{\mathrm{t}_x} = (k_{2_x}/\omega\rho_2)p_{\mathrm{t}} = (p_{\mathrm{t}}/Z_2)\cos\phi_2 \qquad (12)$$

$$Z = \rho v \qquad (13)$$

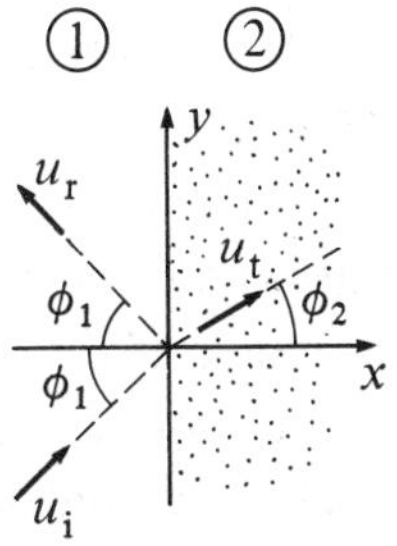

Boundary conditions

Trace velocity matching:

$$(\omega/v_1)\sin\phi_1 = (\omega/v_2)\sin\phi_2 \quad \text{or} \quad k_{1_y} = k_{2_y} \qquad (14)$$

Continuity of p and u_x at $x = 0$:

$$p_{\mathrm{i}} + p_{\mathrm{r}} = p_{\mathrm{t}} \quad u_{\mathrm{i}_x} + u_{\mathrm{r}_x} = u_{\mathrm{t}_x} \quad \text{at } x = 0. \qquad (15)$$

$$A + B = C \quad [(A - B)/Z_1]\cos\phi_1 = (C/Z_2)\cos\phi_2 \qquad (16)$$

Reflection and transmission coefficients for pressure

$$R = B/A = (Z_2\cos\phi_1 - Z_1\cos\phi_2)/(Z_2\cos\phi_1 + Z_1\cos\phi_2) \qquad (17)$$

$$T = C/A = 1 + R = 2Z_2\cos\phi_1/(Z_2\cos\phi_1 + Z_1\cos\phi_2) \qquad (18)$$

$$\sin\phi_2 = (v_2/v_1)\sin\phi_1 \qquad (19)$$

the reflected waves and in region 2 it is simply the refracted wave,

$$p_1 = p_i + p_r = [A\exp(ik_{1_x}x) + B\exp(-ik_{1_x}x)]\exp(ik_{1_y}y) \tag{4}$$

$$p_2 = p_t = Ce^{ik_{2x}x}e^{ik_{2y}y} \tag{5}$$

The corresponding velocity fields are obtained from the equation of motion $\rho\partial u_x/\partial t = -\partial p/\partial x$ and the corresponding complex amplitude equation $-i\omega\rho u_x = -dp/dx$ with similar expressions for the y-component. By introducing the wave impedances $Z = \omega\rho/k = \rho v$ in the two regions, we can express the x-components of the velocity fields as

$$u_{1_x} = (1/Z_1)\cos(\phi_1)[A\exp(ik_{1_x}x) - B\exp(-ik_{1_x}x)]\exp(ik_{1_y}y) \tag{6}$$

$$u_{2_x} = (1/Z_2)\cos(\phi_2)[C\exp(ik_{2_x}x)]\exp(ik_{2_y}y) \tag{7}$$

Reflection and transmission coefficients. At the boundary, at $x = 0$, the pressure amplitudes in the two regions must be the same. A pressure difference across the boundary would lead to an infinite acceleration of the (massless) boundary, which is unacceptable. In addition, the normal velocity (or displacement) must be the same on both sides at the boundary since we assume that there is no penetration of one medium into the other. Since we have neglected viscosity in the present case, it is possible that 'slippage' may occur at the boundary in the tangential direction.

If these **boundary conditions** are to be satisfied for all values of y, it is necessary that the factors containing the y-dependence must be the same on both sides. This requires that

$$k_{1_y}\sin(\phi_1) = k_{2_y}\sin(\phi_2) \tag{8}$$

or

$$v_1/\sin(\phi_1) = v_2/\sin(\phi_2) \tag{9}$$

which is nothing but Snell's law of refraction.

From Eqs. (4), (5) and (6), (7), the continuity of pressure and normal velocity at $x = 0$ lead to the conditions

$$A + B = C$$

$$[(A - B)/Z_1]\cos(\phi_1) = (C/Z_2)\cos(\phi_2)$$

From these relations we obtain the reflection and transmission coefficients for pressure

$$R = B/A = \frac{Z_2\cos(\phi_1) - Z_1\cos(\phi_2)}{Z_2\cos(\phi_1) + Z_1\cos(\phi_2)} \tag{10}$$

$$T = C/A = 2Z_2\cos(\phi_1)/[Z_2\cos(\phi_1) + Z_1\cos(\phi_2)] \tag{11}$$

$$\sin(\phi_2) = (v_2/v_1)\sin(\phi_1) \tag{12}$$

Whereas Snell's law involves only the ratio between the wave speeds in the two regions, the reflection and transmission coefficients depend on the wave impedances. If the wave impedance of the second region is much larger than that of the first, the pressure reflection coefficient will be approximately equal to 1. For example, this is true for a sound wave incident from air on a water surface. If the sound is incident from the water side, the reflection coefficient becomes close to -1. In either case, the transmission coefficient is nearly zero. It should be noted also, that as the angle of incidence approaches 90 degrees, the pressure reflection coefficient approaches -1 for any finite value of Z_2/Z_1.

12.3 EM reflection and transmission coefficients

The analysis of the reflection and refraction of an electromagnetic wave is analogous to that in the previous section, but since this wave is transverse, we must consider the role of the polarization of the wave. We shall start with the electric field vector in the z-direction, parallel with the reflecting boundary, as shown in the first figure in Display 12.3. The corresponding direction of the magnetic field follows from the field equations or simply from the Poynting vector $\mathbf{S} = \mathbf{E} \times \mathbf{H}$, which is in the direction of propagation.

Under these conditions the electric field components of the incident, reflected, and transmitted fields are

$$E_{i_z} = A \exp(ik_{1_x}x + ik_{1_y}y)$$
$$E_{r_z} = B \exp(-ik_{1_x}x + ik_{1_y}y)$$
$$E_{t_z} = C \exp(ik_{2_x}x + ik_{2_y}y) \quad k_1 = \omega/c_1 \quad k_2 = \omega/c_2 \tag{1}$$

As discussed in Chapter 9, the ratio between the magnitudes of the electric and magnetic fields in a plane wave is simply the wave impedance $Z = \sqrt{\mu/\varepsilon}$, and the direction of the magnetic field is perpendicular to the electric field and the direction of propagation, as already indicated. Thus, with reference to the display, we find the components of the magnetic fields of the incident, reflected, and transmitted waves to be

$$H_{i_y} = -(E_{i_z}/Z_1)\cos(\phi_1)$$
$$H_{r_y} = (E_{r_z}/Z_1)\cos(\phi_1)$$
$$H_{t_y} = -(E_{t_z}/Z_2)\cos(\phi_2) \tag{2}$$

The relations between the amplitudes A, B, and C are determined by the **boundary conditions** that the tangential components of the electric and magnetic fields must be continuous across the boundary. In regard to the E-field, we note that a line integral along paths which run parallel and

Display 12.3.

Reflection of an EM wave at a plane boundary.

***E* in *z*-direction**

$E_{i_z} = A\exp(ik_{1_x}x + ik_{1_y}y)$ (1)

$E_{r_z} = B\exp(-ik_{1_x}x + ik_{1_y}y)$ (2)

$E_{t_z} = C\exp(ik_{2_x}x + ik_{2_y}y)$ (3)

$H_{i_y} = -(E_{i_z}/Z_1)\cos\phi_1$ (4)

$H_{r_y} = (E_{r_z}/Z_1)\cos\phi_1$ (5)

$H_{t_y} = -(E_{t_z}/Z_2)\cos\phi_2$ (6)

$k_x = (\omega/c)\cos\phi \quad k_y = (\omega/c)\sin\phi$ (7)

$Z = \sqrt{\mu/\varepsilon} = \mu c$ (8)

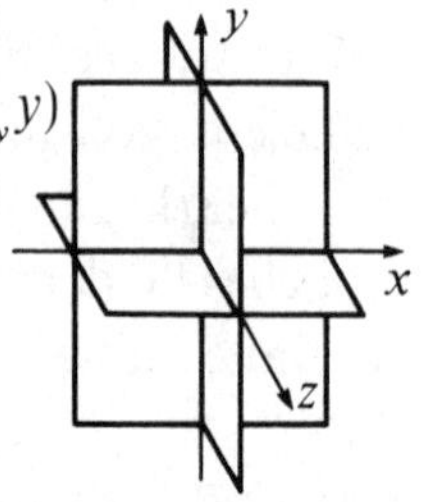

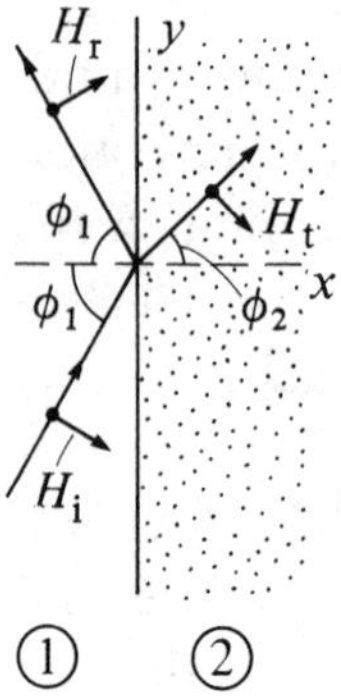

Boundary conditions: E_z and H_y continuous at the boundary $x = 0$.

$A + B = C$ (9)

$[(A - B)/Z_1]\cos\phi_1 = (C/Z_2)\cos\phi_2$ (10)

Reflection and transmission coefficients for E:

$R = B/A = (Z_2\cos\phi_1 - Z_1\cos\phi_2)/(Z_2\cos\phi_1 + Z_1\cos\phi_2)$ (11)

$T = C/A = 2Z_2\cos\phi_1/(Z_2\cos\phi_1 + Z_1\cos\phi_2)$ (12)

$\sin\phi_1 = (c_1/c_2)\sin\phi_2$ (13)

***H* in *z*-direction**

$E_{i_y} = E_i\cos\phi_1$ (14)

$E_{r_y} = -E_r\cos\phi_1$ (15)

$E_{t_y} = E_t\cos\phi_2$ (16)

$H_{i_z} = E_i/Z_1$ (17)

$H_{r_z} = E_r/Z_1$ (18)

$H_{t_z} = E_t/Z_2$ (19)

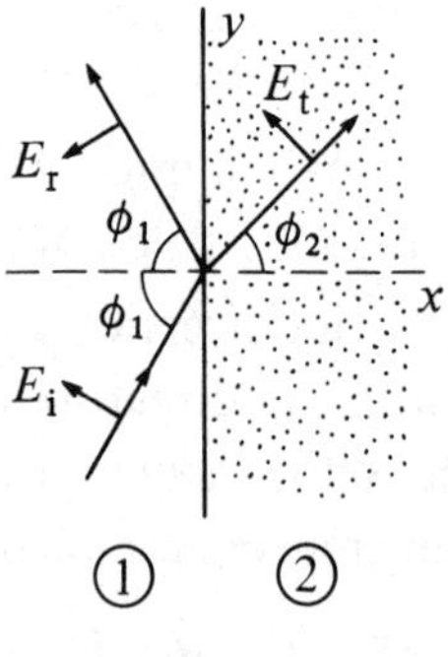

$R' = B'/A' = (Z_1\cos\phi_1 - Z_2\cos\phi_2)/(Z_1\cos\phi_1 + Z_2\cos\phi_2)$ (20)

$T' = C'/A' = 2Z_1\cos\phi_1/(Z_1\cos\phi_1 + Z_2\cos\phi_2)$ (21)

Brewster angle: $R' = 0$ if $\tan\phi_1 = n = c_1/c_2$; then $\phi_1 + \phi_2 = \pi/2$ (22)

arbitrarily close to the boundary must equal the time rate of change of the magnetic flux going through the area enclosed by the path. Since this area can be made arbitrarily small and since the magnetic field is finite, the line integral must be zero, and this requires that the tangential components of the E-field be the same along the two sides of the boundary. A similar argument applies to the magnetic field, where the flux now refers to flux of the displacement current.

Applying these conditions to the z-component of the E-field and the y-component of the H-field, we obtain from Eqs. (1) and (2)

$$A + B = C$$
$$[(A - B)/Z_1]\cos(\phi_1) = (C/Z_2)\cos(\phi_2) \tag{3}$$

We note that these relations are exactly the same as for the acoustic case, and the reflection and transmission coefficients for the electric field are

$$R = B/A = [Z_2\cos(\phi_1) - Z_1\cos(\phi_2)]/[Z_2\cos(\phi_1) + Z_1\cos(\phi_2)] \tag{4}$$

$$T = C/A = 2Z_2\cos(\phi_1)/[Z_2\cos(\phi_1) + Z_1\cos(\phi_2)] \tag{5}$$

In the second part of the display is shown schematically the conditions when the magnetic rather than the electric field of the incident wave is in the z-direction, parallel with the boundary. The electric field vectors of the incident, reflected, and transmitted waves are then directed as shown.

The calculation of the reflection and transmission coefficients by applying the boundary conditions is similar to the derivation of the results in Eqs. (4) and (5), and we leave the details for one of the problems. The resulting reflection and transmission coefficients for the electric field are

$$R' = A'/B' = [Z_1\cos(\phi_1) - Z_2\cos(\phi_2)]/[Z_1\cos(\phi_1) + Z_2\cos(\phi_2)] \tag{6}$$

$$T' = C'/A' = 2Z_1\cos(\phi_1)/[Z_1\cos(\phi_1) + Z_2\cos(\phi_2)] \tag{7}$$

To distinguish this case from the previous, we have denoted the amplitude factors of the incident, reflected, and transmitted electric fields by A', B', and C', and the corresponding reflection and transmission coefficients by R' and T'.

The reflection and transmission coefficients for the magnetic field are defined in similar manner, and we leave the calculations for one of the problems.

The Brewster angle. According to Eq. (6), the reflection coefficient R' has the interesting property that it becomes zero at a certain angle of incidence given by $Z_1\cos(\phi_1) - Z_2\cos(\phi_2) = 0$. Inserting $Z_2/Z_1 \simeq c_2/c_1 = \sin(\phi_2)/\sin(\phi_1)$, we obtain $\sin(2\phi_1) = \sin(2\phi_2)$. It is satisfied for $\phi_1 = \phi_2$, in which case the two regions are the same, and this is of no particular interest. The other solution is $2\phi_2 = \pi - 2\phi_1$ or $\phi_1 + \phi_2 = \pi/2$, so that $\sin(\phi_2) = \cos(\phi_1)$. From Snell's law we then obtain

$$\tan(\phi_1) = c_1/c_2 = n \tag{8}$$

The corresponding angle is called the Brewster angle. In introducing the index of refraction n, we have assumed that the wave speed c_1 is equal to the speed of light in vacuum.

The difference in the reflection characteristics for incident waves with different planes of polarization can be demonstrated by letting an incident light beam pass through a polaroid filter before it is reflected from a glass plate. With the glass plate turned about a vertical axis and with the incident electric field in the horizontal plane, the reflected beam is seen to vanish at the Brewster angle. On the other hand, with the electric field in the vertical plane, the angular dependence of the intensity of the reflected light is weak and no angle of zero intensity is found.

The fact that the reflection coefficient is zero at the Brewster angle can be understood physically from what we learned in Chapter 9 about the radiation from an oscillating charge. The electrons in the region of refraction are driven in harmonic motion by the refracted electric field in the direction of the transmitted E-field and hence perpendicular to the direction of propagation. The reflected wave can be considered to be a result of the radiation (coherent scattering) from these electrons, acting as dipole radiators.

As we found in Chapter 9, the radiation field from an oscillating charge is zero in the direction of motion of the charge, which is perpendicular to the direction of propagation of the refracted wave. The reflected wave travels in this direction of zero radiation if $\phi_1 + \phi_2 = \pi/2$, and this, as we have seen corresponds to the Brewster angle of incidence given by Eq. (8).

12.4 Total reflection

From Snell's law, the angle of refraction is obtained from

$$\sin(\phi_2) = (v_2/v_1)\sin(\phi_1) = v_2/v_t \qquad v_t = v_1/\sin(\phi_1) \tag{1}$$

If the wave is refracted into a medium with a larger wave speed, $v_2 > v_1$, as is the case when light emerges from glass or water into air, the right hand side of Eq. (1) will be equal to 1 when the angle of incidence is given by $\sin(\phi_1) = v_1/v_2$. The angle of refraction is then 90 degrees. According to the analysis in the last section, the reflection coefficient then becomes 1, and all the incident wave energy is reflected from the boundary. This is referred to as total reflection and the corresponding angle of incidence is often called the **critical angle**.

For an angle of incidence larger than the critical, the right hand side of Eq. (1) becomes larger than 1, and there will be no real value of the angle of refraction that satisfies this equation. Actually, $\cos(\phi_2) = \sqrt{1 - \sin^2(\phi_1)}$ will be purely imaginary,

$$\cos(\phi_2) = \mathrm{i}\sqrt{(v_2/v_1)^2 \sin^2(\phi_1) - 1} = \mathrm{i}\sqrt{(v_2/v_t)^2 - 1} \qquad (2)$$
$$v_t = v_1/\sin(\phi_1)$$

According to Eqs. 12.2(3)–(5) and 12.3(1), the refracted field is of the form $C\exp[\mathrm{i}k_2 x\cos(\phi_2)]\exp[\mathrm{i}k_2 y\sin(\phi_2)]$. The second exponential factor can be written $\exp[\mathrm{i}(\omega/v_t)y]$ since, according to Snell's law, $\sin(\phi_2) = (v_2/v_1)\sin(\phi_1) = v_2/v_t$.

This means that when total reflection occurs, the field amplitude Ψ in the second region decays exponentially away from the boundary, and the surfaces of constant phase are perpendicular to the boundary, travelling with the trace velocity in the y-direction

$$\Psi = C\exp(\mathrm{i}k_{2_x}x)\exp(\mathrm{i}k_{2_y}y) = C\exp(-\alpha x)\exp[\mathrm{i}(\omega/v_t)y] \qquad (3)$$
$$\alpha = (\omega/v_2)\sqrt{(v_2/v_t)^2 - 1}$$

where we have used Eq. (2) in expressing $k_{2x} = (\omega/v_2)\cos(\phi_2)$.

Thus, total reflection does not mean that the field in the second region is zero; there is some penetration of the wave through the boundary, which results in a field which decays exponentially away from the boundary. The power carried by this field is zero, however, and we leave it for one of the problems to prove this in a formal manner. It is clear, that during the transient period in the approach to steady state, the wave energy in this exponential layer is built up and some power must cross the boundary. We shall not pursue a detailed analysis of this problem here.

Total reflection of light can be produced in a prism or some other configuration with boundaries which are not parallel. As we have mentioned, the exponentially decaying field in total reflection does not carry any power. If the exponential field is created in a narrow air gap between two glass surfaces, however, there will be power penetration through the gap into the second glass plate. This **barrier penetration** will be discussed further in connection with similar transmission characteristics in wave propagation in wave guides and matter wave propagation through potential barriers.

12.5 Refraction in a moving medium

In a moving fluid, the phase velocity of a sound wave with respect to a fixed coordinate system is the sum of the local sound speed in the fluid and the component of the fluid velocity in the direction of propagation of the wave. A spatial variation of the fluid velocity produces a corresponding variation of the phase velocity, and refraction can result.

An electromagnetic wave can be refracted in similar manner in a moving fluid. In that case, however, the resulting phase velocity is obtained through the relativistic addition of the local speed of light and the component of the velocity of the material in the direction of propagation.

We shall analyze here the idealized case of two fluids moving with different speeds in the same direction and separated by a plane boundary, as indicated in Display 12.4. The fluid velocities in the two regions are U_1 and U_2, and we have assumed that the sound speeds in the two fluids are different, v_1 and v_2. A sound wave is obliquely incident on the plane boundary between the fluids, and we wish to determine the angle of refraction. It should be noted that we now specify the direction of propagation in terms of the angle between the wave normal and the boundary rather than the normal to the boundary. In a moving medium this choice has an advantage as will be clear as we proceed.

As indicated above, the phase velocity is the sum of the sound speed and the component of the fluid velocity along the direction of propagation

$$v_p = v + U\cos(\psi) \tag{1}$$

which applies to both fluids involved.

As in Section 12.1 we use the continuity of the trace velocity across the boundary to determine the relationship between the angles of incidence and refraction. As before, the trace velocity is the velocity of the intersection point Q of a wave front with the boundary,

$$v_t = v_p/\cos(\psi) = U + v/\cos(\psi) \tag{2}$$

Using this expression and the continuity of the trace velocity, we obtain the relation

$$v_t = U_1 + v_1/\cos(\psi_1) = U_2 + v_2/\cos(\psi_2) \tag{3}$$

$$\cos(\psi_2) = v_2/v_t' \tag{4}$$

$$v_t' = v_1/\cos(\psi_1) + U_1 - U_2 = v_1/\cos(\psi_1) + U' \tag{5}$$

where $U' = U_1 - U_2$ is the fluid velocity and v_t' the trace velocity relative to the second fluid. Eq. (3) can be regarded as a generalized Snell's law, which, when written in the form of Eq. (4), has the same form as Eq. 12.4(1).

Because of the relative velocity of the two fluids, the refraction will not be

Display 12.4.

Refraction of sound at the boundary between two moving fluids.

Phase velocity $v_p = v + U\cos(\psi)$ (1)

Trace velocity $v_t = v_p/\cos(\psi) = U + v/\cos(\psi)$ (2)

Continuity of trace velocity

$v_t = U_1 + v_1/\cos(\psi_1) = U_2 + v_2/\cos(\psi_2)$ (3)

Let $v_t' = v_t - U_2 = v_1/\cos(\psi_1) + U'$ (4)

$U' = U_1 - U_2$ (5)

$\cos(\psi_2) = v_2/v_t'$ (6)

Total reflection

$\psi_2 = 0, \pi$ (7)

$\cos(\psi_1) = v_1/(U' \pm v_2)$ (8)

Shadow zones

$\psi_1 = 0, \pi$ (9)

$\cos(\psi_2) = v_2/v_t' = v_2/(U' \pm v_1)$ (10)

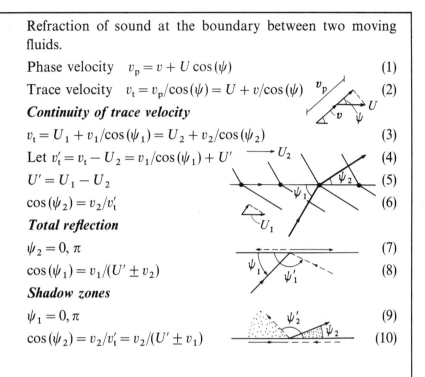

symmetrical with respect to the normal to the boundary. In the range 0–90 for the angle ψ, the contribution $v_1/\cos(\psi_1)$ to the trace velocity v_t' is in the same direction as $U_1 - U_2$, whereas in the range 90–180 degrees, it is in the opposite direction. This sign difference is automatically accounted for by the function $\cos(\psi_1)$, which is asymmetrical with respect to the normal to the boundary. Had we used the angle of incidence ϕ_1 with the normal to the boundary and the corresponding angle function $\sin(\phi_1)$, it would have been necessary to introduce a negative and positive direction for the angle to account for the asymmetry.

As an illustration, let us consider total reflection. The angle of refraction then is either 0 or π, and in this case of fluid motion, we have to consider these cases separately. From Eq. (4), the corresponding critical angles of incidence are then determined by $v_t' = \pm v_2$, and Eq. (5) then yields

$$\cos(\psi_1) = v_1/(U' \pm v_2) \quad [\psi_2 = 0, \pi, \quad U' = U_1 - U_2]. \tag{6}$$

If there is no relative fluid motion so that $U' = 0$, this relation gives two angles ψ_c and $\pi - \psi_c$, which are symmetrical with respect to the normal. When relative motion is present, however, the two critical angles will not have this symmetry.

As a numerical example, let $v_2/v_1 = 1.5$ and $U'/v_1 = 0.3$. The critical angle of incidence which corresponds to $\psi_2 = 0$ is then given by $\cos(\psi_1) = 1/(0.3 + 1.5)$ or $\psi_1 = 56.3°$, and with $\psi_2 = 180°$, we get $\cos(\psi_1) = 1/(0.3–1.5)$ or $\psi_1 = 146.4°$. The corresponding angles with respect to the normal to the boundary are 33.7° and 56.4°. For angles of incidence outside the range defined by these angles, total reflection will occur.

Conversely, waves incident along the boundary under the angles $\psi_1 = 0°$ and 180° will be refracted according to Eqs. (4) and (5) so that $\cos(\psi_2) = v_2/(U' \pm v_1)$. As an example, let $v_2/v_1 = 0.7$ and $U'/v_1 = -0.2$: we obtain $\psi_2 = 29°$ and 125.7°, respectively. The region outside this angular range cannot be reached by a refracted ray, regardless of the angle of incidence. Such regions are usually termed **shadow zones**.

12.6 Inhomogeneous medium

In the previous section we considered the refraction of a wave at the boundary between two homogeneous regions. In each of these regions, the surfaces of constant phase were parallel with each other and the corresponding 'ray', i.e. the normal to these surfaces, was a straight line, which suffered a sudden change in direction at the boundary.

With a continuous rather than a sudden change in the wave speed and the corresponding index of refraction, there will be a gradual change in direction of the wave, as indicated qualitatively in Display 12.5.

Display 12.5.

Refraction in a vertically stratified medium

Wave speed gradient

$v = v(z)$ (1)

Ray curvature independent of the direction of wave. (Isotropy). (*a*)

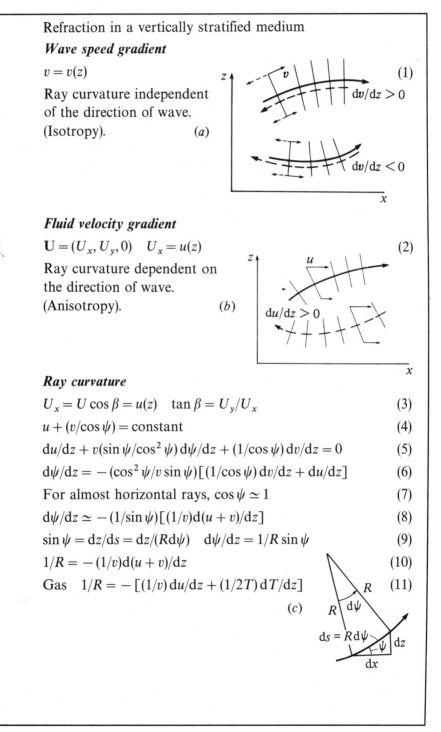

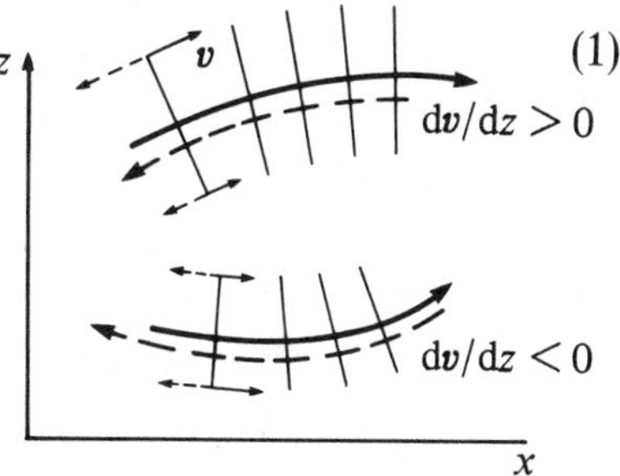

Fluid velocity gradient

$\mathbf{U} = (U_x, U_y, 0) \quad U_x = u(z)$ (2)

Ray curvature dependent on the direction of wave. (Anisotropy). (*b*)

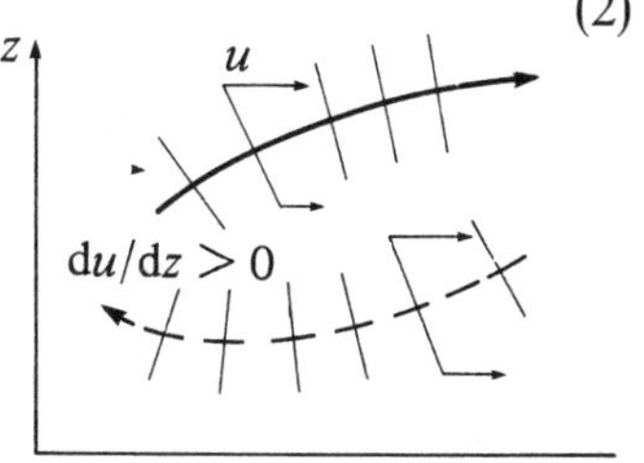

Ray curvature

$U_x = U\cos\beta = u(z) \quad \tan\beta = U_y/U_x$ (3)

$u + (v/\cos\psi) = \text{constant}$ (4)

$\mathrm{d}u/\mathrm{d}z + v(\sin\psi/\cos^2\psi)\,\mathrm{d}\psi/\mathrm{d}z + (1/\cos\psi)\,\mathrm{d}v/\mathrm{d}z = 0$ (5)

$\mathrm{d}\psi/\mathrm{d}z = -(\cos^2\psi/v\sin\psi)[(1/\cos\psi)\,\mathrm{d}v/\mathrm{d}z + \mathrm{d}u/\mathrm{d}z]$ (6)

For almost horizontal rays, $\cos\psi \simeq 1$ (7)

$\mathrm{d}\psi/\mathrm{d}z \simeq -(1/\sin\psi)[(1/v)\mathrm{d}(u+v)/\mathrm{d}z]$ (8)

$\sin\psi = \mathrm{d}z/\mathrm{d}s = \mathrm{d}z/(R\,\mathrm{d}\psi) \quad \mathrm{d}\psi/\mathrm{d}z = 1/R\sin\psi$ (9)

$1/R = -(1/v)\mathrm{d}(u+v)/\mathrm{d}z$ (10)

Gas $\quad 1/R = -[(1/v)\,\mathrm{d}u/\mathrm{d}z + (1/2T)\,\mathrm{d}T/\mathrm{d}z]$ (11)

(*c*)

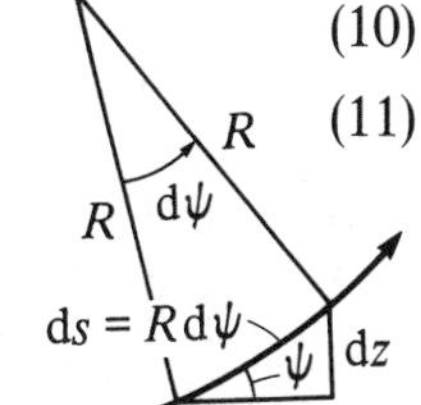

Thus, in Fig. (*a*) is shown the behaviour of the wave fronts in a stationary medium in which the wave speed varies with the height z. If the wave speed increases with z, $dv/dz > 0$, the point P on a wave front will travel faster than the point Q, and as a result the wave front and the corresponding ray will turn downwards. The same holds true for a wave travelling in the opposite direction, as indicated. Similarly, if the wave speed decreases with height, $dv/dz < 0$, the ray will turn upwards.

In Fig. (*b*), the medium (fluid) is moving in the horizontal direction with a velocity U, parallel with the xy-plane. The ray shown lies in the xz-plane, and it will be affected by the x-component $U_x = u(z)$ of the fluid velocity. The corresponding trace velocity of the wave is constant, independent of z,

$$v_t = u + v/\cos(\psi) = \text{constant} \tag{1}$$

where $v(z)$ is the wave speed and $\psi(z)$ is the angle of the ray with the x-axis, as shown.

Differentiation with respect to z yields

$$du/dz + v[\sin(\psi)/\cos^2(\psi)]\, d\psi/dz + [1/\cos(\psi)]\, dv/dz = 0$$

or

$$d\psi/dz = -[\cos^2(\psi)/v\sin(\psi)][du/dz + (dv/dz)/\cos(\psi)] \tag{2}$$

This can be regarded as the differential equation for a ray, where $\tan(\psi)$ is the slope dy/dx of the ray.

For almost horizontal rays we have $\cos(\psi) \simeq 1$, and the differential equation for ψ reduces to

$$d\psi/dz = -[1/\sin(\psi)][(1/v)d(u+v)/dz] \tag{3}$$

It follows from Fig. (*c*), that the angle ψ can be related to the radius of curvature R of a ray as $\sin(\psi) = dz/ds = ds/(R\,d\psi)$ or

$$d\psi/dz = 1/R\sin(\psi) \tag{4}$$

Accounting for Eq. (3), we can express the radius of curvature in terms of the gradient of the total phase velocity,

$$1/R = -(1/v)d(u+v)/dz \tag{5}$$

In the particular case of a sound wave in an ideal gas, the wave speed is proportional to the square root of the absolute temperature, as was explained in Chapter 10, and Eq. (5) can be written

$$1/R = -[(1/v)\,du/dz + (1/2T)\,dT/dz] \tag{6}$$

A positive value of the curvature corresponds to a ray which turns upwards; the slope of the ray then increases with x.

As an important application, we consider sound propagation in the atmosphere, in which the temperature and the wind speed vary with height z.

Display 12.6.

Refraction of sound in the atmosphere.

$$\mathrm{d}(u+v)/\mathrm{d}z > 0 \quad (v = \text{sound speed}, \ u = U_x) \tag{1}$$

$$\mathrm{d}(u+v)/\mathrm{d}z < 0 \tag{2}$$

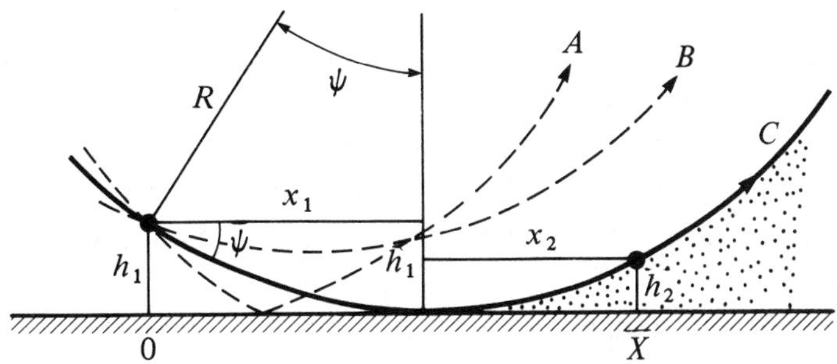

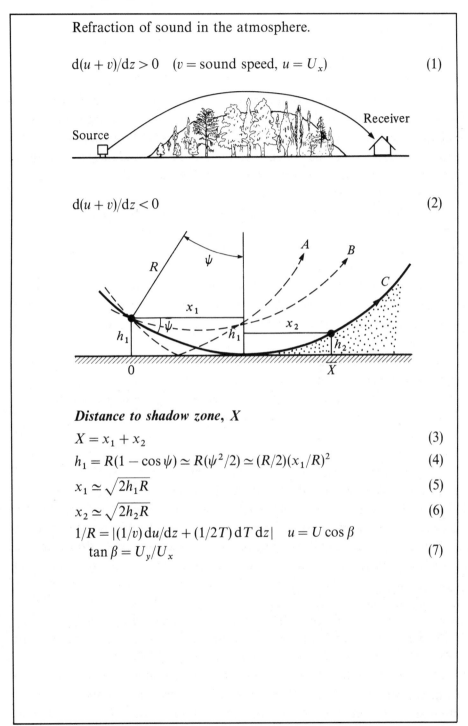

Distance to shadow zone, X

$$X = x_1 + x_2 \tag{3}$$

$$h_1 = R(1 - \cos\psi) \simeq R(\psi^2/2) \simeq (R/2)(x_1/R)^2 \tag{4}$$

$$x_1 \simeq \sqrt{2h_1 R} \tag{5}$$

$$x_2 \simeq \sqrt{2h_2 R} \tag{6}$$

$$1/R = |(1/v)\,\mathrm{d}u/\mathrm{d}z + (1/2T)\,\mathrm{d}T\,\mathrm{d}z| \quad u = U\cos\beta$$
$$\tan\beta = U_y/U_x \tag{7}$$

In the top figure in Display 12.6, the wave speed increases with height, and the sound is refracted downwards; it can reach an observer, who normally would be shielded from the sound source.

If no wind is present such refraction occurs when the temperature increases with height, a condition referred to as temperature inversion in meteorology.

Wind with a velocity component in the direction of sound propagation will produce a similar effect provided that the wind speed increases with height, which is the normal behaviour (the wind speed must be zero at the ground surface). It should be realized that a uniform wind does not produce refraction; it is the variation of the wind speed with height which is responsible for the ray curvature.

If the total wave speed decreases with height, the rays will be bent upwards, as indicated in the second figure in the display. A negative temperature gradient dT/dz, usually called 'lapse rate' in meteorology, will give rise to such a refraction.

This type of refraction is caused also by wind if it has a velocity component in the direction from the observer to the source, i.e. opposite the direction of sound propagation. Again an increase of the wind speed with height is assumed.

It should be noted that the refraction caused by a temperature gradient is independent of direction of sound propagation. For wind, on the other hand, the medium becomes anisotropic, as far as wave propagation is concerned. For example, if the wind causes a downward refraction when the sound goes from source to receiver, it will produce refraction upwards, when the direction of propagation is reversed; reciprocity does not apply in this case.

For a certain angle of emission, a sound ray, denoted by C in the figure, will be horizontal as it makes contact with the ground surface. The region beyond the contact point between the ground surface and C becomes a geometrical 'shadow', into which no ray (such as A or B) can penetrate.

This type of shadow formation plays an important role in problems of sound propagation in the atmosphere. It is of interest to obtain an expression for the distance to the shadow zone from the source. The source S, an airplane, for example, and the observer O are placed at different heights h_1 and h_2, and the distance to be determined is denoted by $X = x_1 + x_2$, as indicated. Quantities x_1 and x_2 are the distances between the contact point of the C-ray and the source and observer, respectively.

If the wind and temperature gradient are constant, independent of z, the radius of curvature of the rays in Eq. (6) will be constant, and the rays will be circular. For a small angle of emission, $\psi \simeq x_1/R$, and for a circular ray,

$h_1 = R[1 - \cos(\psi)] \simeq R\psi^2/2 = (R/2)(x_1/R)^2$. A similar expression is obtained for h_1, with x_1 replaced by x_2. Consequently,

$$x_1 \simeq \sqrt{2h_1 R} \quad (7)$$

$$x_2 \simeq \sqrt{2h_2 R} \quad (8)$$

$$X = x_1 + x_2 \quad (9)$$

$$1/R = (1/v)\,\mathrm{d}u/\mathrm{d}z + (1/2T)\,\mathrm{d}T/\mathrm{d}z \quad (10)$$

It should be recalled that $u = U_x$ is the component of the flow velocity in the x-direction. With the angle between $\mathbf{U}$ and the x-axis equal to β, we have $u = U\cos(\beta)$. If $\beta = 0$, the flow is in the same direction as the sound, and if $\mathrm{d}u/\mathrm{d}z$ and $\mathrm{d}T/\mathrm{d}z$ have the same sign, the refractive effects of wind and temperature cooperate. On the other hand, if $\beta = 180°$, these effects oppose each other and may cancel each other in certain directions. For numerical examples we refer to the problem section.

Examples

E12.1 *Refraction by a prism.* A light beam is refracted by a glass prism as shown schematically in the figure, and we wish to determine the resulting angle of deflection of the beam by the prism, particularly for small angles of incidence.

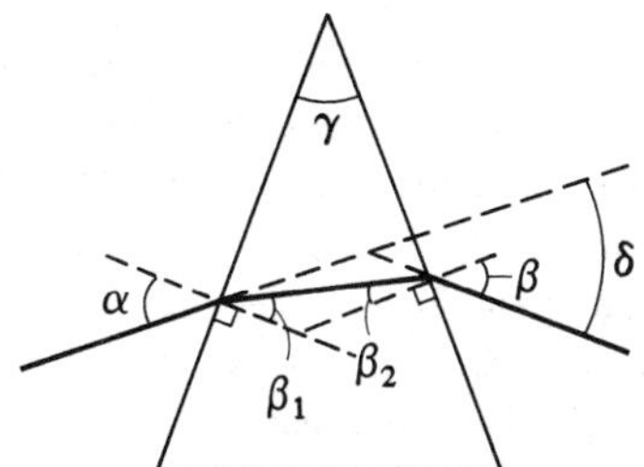

The angle of incidence of the beam entering the prism is α and the corresponding angle of refraction is β_1. Similarly, at the exit surface, the angle of incidence is denoted by β_2 and the angle of refraction of the ray leaving the prism is β. The angle of deflection at the first surface is $\alpha - \beta_1$ and at the second surface it is $\beta - \beta_2$, making the total angle of deflection equal to $\delta = \alpha + \beta - (\beta_1 + \beta_2)$. Furthermore, the prism angle γ is the same as the angle between the normals to the prism surfaces and we find $\gamma = \beta_1 + \beta_2$. Consequently, the angular deflection is

$$\delta = \alpha + \beta - \gamma \quad (1)$$

Application of Snell's law at the two surfaces yields $\sin(\alpha) = n\sin(\beta_1)$ and $\sin(\beta) = n\sin(\beta_2)$. For small angles these relations reduce to $\alpha \simeq n\beta_1$ and $\beta \simeq n\beta_2$ and the corresponding expression for the total angular deflection in Eq. (1)

becomes

$$\delta = (n - 1)\gamma \tag{2}$$

In other words, the deflection is directly proportional to the prism angle for small angles of incidence. The factor $n - 1$ expresses the relative change in the wave speed suffered by the wave as it enters the prism. For glass the index of refraction is typically $n = 1.5$, and for this value the angle of deflection will be equal to half the prism angle.

E12.2 *Optical lens.* The optical lens in Fig. 1 consists of an element with rotational symmetry about an axis and with spherical surfaces. The radii of curvature of these surfaces are denoted by R_1 and R_2. A point source of light is located on the axis a distance a from the front surface of the lens. Light is emitted in all directions from this source, and we shall be interested in a ray which enters the lens close to the axis, so that the angle of incidence will be small. We assume also that the thickness of the lens is small compared to the radii. After refraction by the lens the ray meets the axis at a distance from the lens, which we denote by b.

Fig. 1

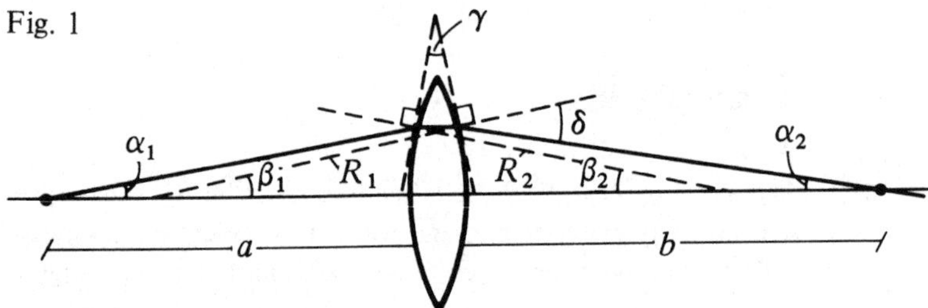

The lens can be regarded locally as a prism with an angle determined by the slope of the spherical surfaces at the location in question. On the assumptions made above, the ray will enter and leave the lens at approximately the same distance r from the axis, and the total deflection of the ray should be the same as that obtained in E12.1 for the 'local prism', i.e. $\delta = (n - 1)\gamma$, where γ is the prism angle. The local normals to the lens surfaces intercept the axis under the angles β_1 and β_2, and, as for the prism, $\gamma = \beta_1 + \beta_2$. Consequently, the angular deflection can be expressed as

$$\delta \simeq (n - 1)(\beta_1 + \beta_2) \tag{1}$$

The angles of the incident and refracted rays with the lens axes are denoted by α_1 and α_2 in Fig. 2, and in terms of these angles the total deflection is

$$\delta \simeq \alpha_1 + \alpha_2 \tag{2}$$

Since $r = a\tan(\alpha_1) = b\tan(\alpha_2)$ the small angle approximation yields $\alpha_1 \simeq r/a$ and $\alpha_2 \simeq r/b$. Similarly we obtain $\beta_1 \simeq r/R_1$ and $\beta_2 \simeq r/R_2$. Inserting these relations into Eqs. (1) and (2) and equating the two expressions for δ thus obtained, we obtain

$$1/a + 1/b = 1/f = (n - 1)(1/R_1 + 1/R_2)$$
$$f = R_1R_2/(n - 1)(R_1 + R_2) \tag{3}$$

We note that if the point source is placed at infinity, we get $b = f$; the image of the point source then will fall at the distance f from the lens. This distance is a property of the lens and is called the focal length; the corresponding point of intersection with the axis is the focal point of the lens. In other words, an incident ray which is parallel with the axis will be refracted into the focal point of the lens.

The image of a point source located off the axis is obtained as indicated in Fig. 2 from the intersection of rays A and B. The first of these is parallel with the axis and will be refracted through the focal point of the lens. The second ray goes through the center of the lens, where the lens surfaces are parallel (prism angle zero), and will not be deflected. The distance of the intersection point of these rays to the lens is denoted by b. It follows from the figure that $f/r_1 = (b-f)/r_2$ and, since $r_1/r_2 = a/b$, the relation between a and b is again given by Eq. (3).

Fig. 2

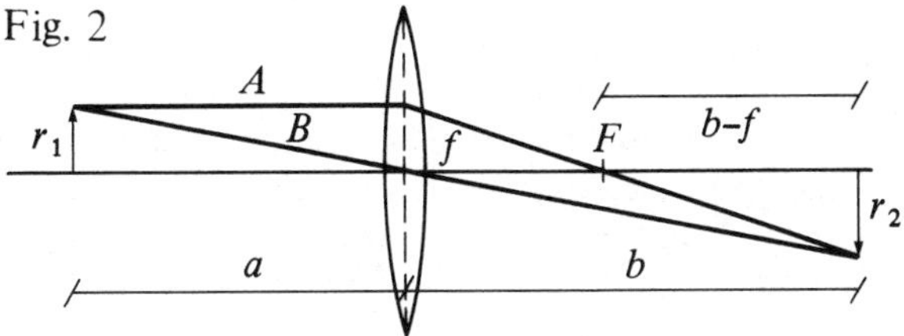

This means that every point in a plane perpendicular to the axis at a distance a from the lens will be mapped onto a plane a distance b from the lens. This plane is called the image plane. Note that the image of an object will be inverted.

E.12.3 *Acoustic lens.*

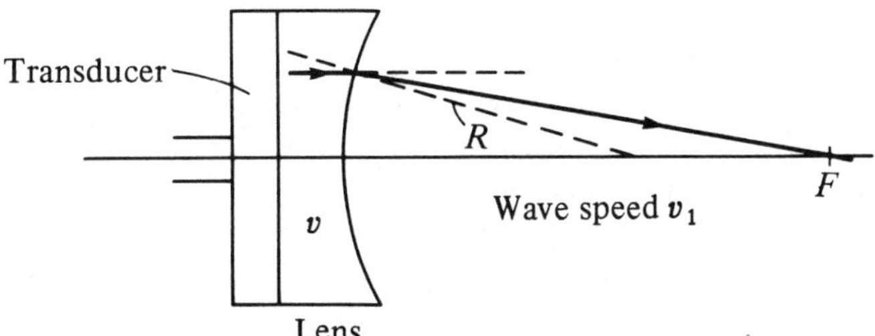

In many applications of ultrasonic waves in liquids, the waves are focussed by means of lenses. A lens usually is made of plastic with the sound speed v in the lens being larger than the wave speed v_1 in the surrounding liquid. Under such conditions, a focussing lens must have concave rather than convex surfaces; a common arrangement is shown schematically in the figure. If the radius of curvature of the lens in this case is R, the focal length of the lens will be $f = R/(1-n)$, where the index of refraction $n = v_1/v$ is now less than 1.

Problems

12.1 *Refraction of light.* A coin at the bottom of a water tank is viewed by an observer above the water surface. Discuss the apparent location of the coin as a function of the location of the observer. Let the coin be at $x = y = z = 0$ and the water surface at $z = 1$ m. The observer is in a horizontal plane 1 m above the water surface at various distances r from the z-axis. The index of refraction of water is $n = 1.33$.

12.2 *Total reflection.* With reference to Display 12.1, determine the range of the angle α for which total reflection occurs at the water surface. The index of refraction for water is $n = 1.33$.

12.3 *Refraction and reflection of sound.* A plane sound wave is incident on the plane boundary between air and helium (separated by means of a thin membrane with negligible mass and tension).

(a) Determine the critical angle of incidence, ϕ_c, above which total reflection occurs.
(b) What is the angle of refraction when the angle of incidence is $\phi_c/2$? Determine also the pressure reflection and transmission coefficients.
(c) Determine the pressure field in He when the angle of incidence is $2\phi_c$.
(d) Sketch the pressure reflection and transmission coefficients as functions of the angle of incidence.

12.4 *Electromagnetic wave reflection and refraction.* The yz-plane represents the boundary between air ($x < 0$) and a region ($x > 0$) of glass with the index of refraction $n = 1.5$. A linearly polarized monochromatic plane light wave is incident on the boundary at an angle of incidence of 60 degrees. The direction of propagation is contained in the xy-plane. Calculate the reflection and transmission coefficients for the E-field, if the E-vector is

(a) in the xz-plane, and (b) in the yx-plane.
(c) In each of these two cases, determine the fraction of the incident power per unit area of boundary which is transmitted into the glass. Under what condition will this fraction be 100%?
(d) Discuss the nature of the wave field in the air ($x < 0$) and obtain the ratio between the maximum and minimum electric field amplitudes in this region.

12.5 *Effect of motion on refraction.* A sound source is located on the axis of an air jet which moves with velocity U with respect to the surrounding ambient air. The temperature in the jet is 2000 Kelvin. Consider a sound wave (ray) emitted by the source in a direction which makes an angle α with the jet axis. It will be refracted at the jet boundary as it emerges from the jet. In what direction will it emerge, if U is equal to the speed of sound within the jet and $\alpha = -45, 90, 45$, and 10 degrees?

12.6 *Relativistic motion.* With reference to Display 12.4, derive Snell's law of refraction of an EM wave which is incident on the boundary of a medium (index of refraction n) which moves with the velocity U parallel with the boundary. What happens when U approaches the speed of light c?

12.7 *Inhomogeneous medium.* A sound source is located at a height $z = 1$ m above ground. A listener is located upwind from the source at a height 2 m above ground. Beyond what distance between the source and the listener will the listener be in a 'shadow' zone, if the average vertical wind gradient in the first few meters above ground is $dU/dz = -3\,\text{sec}^{-1}$?

12.8 *Acoustic lens.* Derive the expression for the focal length of the ultrasonic lens discussed in E12.3.

12.9 *Focal length in terms of lens diameter and thickness.* With reference to the discussion in E12.2, show that the expression in Eq. (3) can be written $f = (n-1)^{-1}D^2/8d$, where D is the diameter of the lens and d the thickness.

13

The Doppler effect

So far in our discussion of wave generation and propagation, we have tacitly assumed, that there is no relative motion between the source and the receiver, and the observed frequency has always been the same as the emitted.

When relative motion between source and receiver does exist, the observed and emitted frequencies are generally different, however. This is known as the Doppler effect. It is geometric (kinematic) in nature, and one might think that it should be the same for all types of waves. As we shall see, however, the effect is somewhat different for mechanical and electromagnetic waves, and this is related to the fact that for an EM wave in vacuum, the wave speed is independent of the motion of the frame of reference. A sound wave basically is a result of intermolecular collisions, and the wave speed depends on the mean motion of the medium and hence upon the motion of the frame of reference.

The Doppler effect is the basis for numerous important applications in physics and engineering, and we shall describe briefly some of them.

13.1 Mechanical waves

We shall consider first the Doppler effect for mechanical waves, such as sound waves in a gas or surface waves on a liquid, for which the wave phase velocity is the sum of the local wave velocity and the component of the medium velocity in the direction of wave travel. By local wave velocity we mean the velocity of the wave with respect to the medium.

For the purpose of illustrating the basic ideas involved, we start with a medium which is at rest with respect to our frame of reference and with source and receiver located on the x-axis. These restrictions will be relaxed later.

Moving source, stationary observer. The source under consideration emits a harmonic wave with a frequency ν in the source bound reference frame ('rest frame' of the source). The source moves with constant velocity u in the positive x-direction, and an observer O is located on the x-axis. We let the source be at $x=0$ at $t=0$ so that the source position is $x=ut$.

As indicated in Display 13.1, the wave front emitted at $t=0$ will be at $x=vT$ after one period, where v is the speed of the wave with respect to the medium. At this time the location of the source is $x=uT$, at which time it starts to emit the next period of the wave. The distance between the first and the second wave fronts is then $(v-u)T$, which, by definition, is the wavelength.

The wave fronts shown in the display are emitted in the direction of motion of the source, and we note, that the wavelength is reduced as a result of the motion of the source. In the opposite direction, the wavelength is increased as a result of the motion.

The wave fronts move forward with the wave speed v, and the time difference T' between the passage of the successive wave fronts at the stationary observer O will be $T'=(v-u)T/v$, and the corresponding Doppler shifted frequency is $1/T'$, i.e.

$$\nu' = \nu/(1-m) \qquad (1)$$

$$m = u/v$$

where m is the Mach number of the source. As the source velocity approaches the wave speed, the observed frequency goes to infinity. The reason is, of course, that the source is always catching up with the wave fronts, so that the distance between them will be zero.

It should be emphasized that for waves travelling in the negative x-direction, the component of the fluid velocity along the direction of propagation is $-u$, and the wavelength is increased and the Doppler shifted frequency is decreased.

Stationary source, moving observer. In this case the observer moves in the positive x-direction with a velocity u_o. With the source at rest, the distance between two wave fronts emitted in a time interval T will be vT. These two wave fronts move forward with the velocity $(v-u_o)$ relative to the observer. Thus, the time required for the two wave fronts to pass the observer will be $T'=vT/(v-u_o)$. The corresponding observed frequency is

$$\nu' = 1/T' = \nu(1-m_0) \qquad m_0 = u_0/v \qquad (2)$$

If the observer moves with the wave speed in the positive x-direction, the

Display 13.1.

Doppler effect for mechanical waves.

Moving source, stationary observer

$\lambda' = (v - u)T = vT(1 - m) = \lambda(1 - m)$ (1)

$m = u/v$ (2)

Observed period $T' = \lambda'/v = (\lambda/v)(1 - m) = T(1 - m)$ (3)

Observed frequency $\nu' = 1/T' = \nu/(1 - m)$ (4)

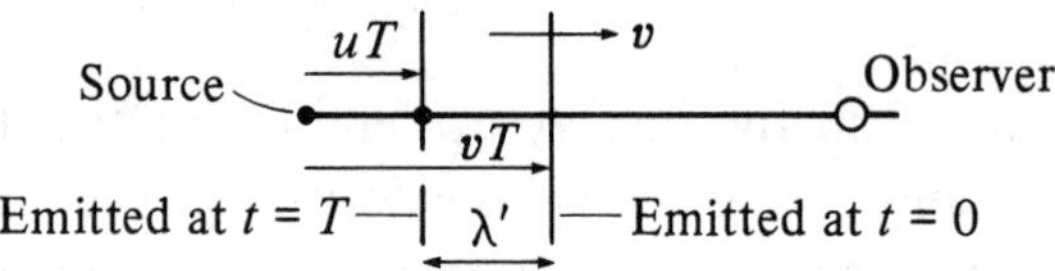

Stationary source, moving observer

Observed period $T' = \lambda/(v - u_o) = T/(1 - m_o)$ (5)

$m_o = u_o/v$ (6)

Observed frequency $\nu' = 1/T' = \nu(1 - m_o)$ (7)

v v
O u_o
Emitted at $t = T$ $\lambda = vT$ Emitted at $t = 0$

Source and observer moving

Observed period $T' = T(v - u)/(v - u_o)$ (8)

Observed frequency $\nu' = \nu(1 - m_o)/(1 - m)$ (9)

$m = u/v \quad m_o = u_o/v$ (10)

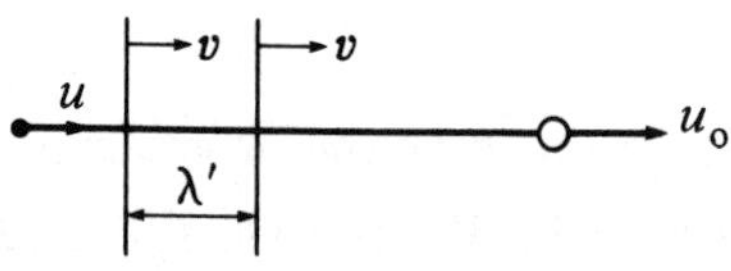

observed frequency will be zero; motion toward the source with the wave speed makes the observed frequency 2ν.

Both source and observer moving. As before, we denote by u and u_0 the velocities of the source and receiver, respectively. The distance between two wave fronts, emitted a time T apart, will be $(v-u)T$. The two wave fronts travel with the speed $(v-u_0)$ with respect to the observer, and the time required for the wave fronts to pass O will be $T' = (v-u)T/(v-u_0)$ and the corresponding observed frequency is

$$\nu' = \nu(1-m_0)/(1-m) \tag{3}$$

$$m = u/v, \quad m_0 = u_0/v$$

For small values of m and m_0 we obtain $\nu' \simeq \nu(1-m_0+m) = [1+(m-m_0)]\nu$, which depends only on the relative velocity of the source and receiver. For electromagnetic waves in vacuum, we shall find that under all conditions, the Doppler shift depends only on the relative speed between the source and the receiver. It is a consequence of the fact, that in the EM case, the wave speed, the speed of light, is the same in all frames of reference.

Angular dependence of the Doppler shifted frequency. In the previous examples, the motions of source and observer were along the x-axis, and the wave reaching the observer was emitted either in the direction of motion of the source or in the opposite direction.

In general, the observer is not located on the path of motion of the source, and as the source moves by, the direction of the wave emitted toward the observer varies continuously, and there is a corresponding variation in the observed frequency. As an example, we consider a source moving with constant velocity u along the x-axis and an observer O at rest at the position x, y, as indicated in Display 13.2.

At time t the location of the source is $x_s = ut$. The wave reaching the observer at this time t was emitted at an earlier time t_e from the emission point $x_e = ut_e$. The distance from this point to the observer O is denoted by R, and the time of flight from x_e to O is $t - t_e = R/v$, where v is the wave speed. During this time the source moves from x_e to x_s, and we can express $t - t_e$ also as $t - t_e = (x_s - x_e)/u$.

The distance R is determined from the relation $R^2 = y^2 + (x - x_e)^2$, in which $x - x_e = x - x_s + (x_s - x_e) = x - ut + u(R/v)$. This yields the equation $R^2 = y^2 + [x - ut + u(R/v)]^2$ for R with the solution

Display 13.2.

Doppler effect. Source moving passed a stationary observer.

Calculation of R

$v =$ wave speed $\quad u =$ source velocity

Note $\quad R/v = (x_s - x_e)/u = t - t_e \qquad (1)$

$$R^2 = h^2 + (x - x_e)^2 = h^2 + [x - x_s + (x_s - x_e)]^2$$
$$= h^2 + [x - ut + u(t - t_e)]^2 = h^2 + [x - ut + mR]^2 \qquad (2)$$

$m = u/v \quad$ Let $X = x - ut \qquad (3)$

Eq. (2): $\quad R^2(1 - m^2) - 2mXR = h^2 + X^2 \qquad (4)$

Solution $\quad R = [mX/(1 - m^2)] \pm [(mX/(1 - m^2)^2$
$$+ (h^2 + X^2)/(1 - m^2)]^{\frac{1}{2}} \qquad (5)$$

Only + sign gives a positive R. Thus,

$$R = (mX + R_1)/(1 - m^2) \qquad (6)$$

$$R_1 = [X^2 + (1 - m^2)h^2]^{\frac{1}{2}} \qquad (7)$$

Doppler shifted frequency

$$\nu' = \nu/(1 - m\cos\phi) \qquad (8)$$

$$R\cos\phi = x - ut_e = x - ut + u(t - t_e) = X + mR \qquad (9)$$

Insert into Eq. (6)

$$(1 - m^2)R = mR(\cos\phi - m) + R_1 \qquad (10)$$

$$R_1 = R(1 - m\cos\phi) \qquad (11)$$

Eq. (8): $\quad \nu' = R/R_1 \qquad (12)$

$$\nu' = \nu[1 + mX/\sqrt{X^2 + (1 - m^2)h^2}]/(1 - m^2) \qquad (13)$$

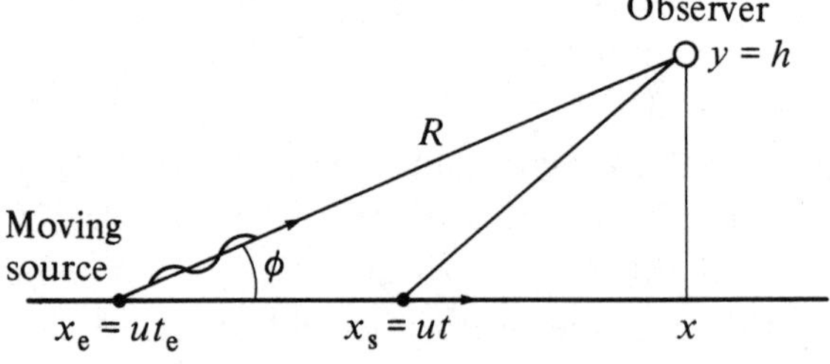

$$R = [m(x - ut) + R_1]/(1 - m^2)$$
$$R_1 = [(x - ut)^2 + (1 - m^2)y^2]^{\frac{1}{2}} \qquad (4)$$
$$m = u/v$$

With the angle between the line x_e–O and the x-axis denoted by ϕ, the component of the source velocity in the direction of wave propagation toward the observer is $u\cos(\phi)$, and the Doppler shifted observed frequency will be

$$\nu' = \nu/[1 - m\cos(\phi)] \qquad (5)$$
$$m = u/v$$

With $R\cos(\phi) = x - x_e = x - ut_e$, this frequency can be expressed in terms of the observer coordinates and time. Furthermore, since $t - t_e = R/v$, we can write $x - ut_e = R\cos(\phi) = x - ut + u(t - t_e) = x - ut + mR$ which yields $x - ut = R(\cos(\phi) - m)$. Insertion in the expression for R in Eq. 4 results in $1 - m\cos(\phi) = R/R_1$, and Eq. 5 then gives

$$\nu' = \nu/[1 - m\cos(\phi)] = (R/R_1)\nu \qquad (6)$$
R, R_1: See Eq. (4).

As an example, we have computed the Doppler shifted frequency at the observer location $x = 0, y = h$ as a function of time for some different velocities of a source which moves along the x-axis. The results are shown in Display 13.3.

For large negative values of the source location x_s, the component of the source velocity in the direction of the observer is approximately u, and the corresponding Doppler shifted frequency is $\nu/(1 - m)$. Similarly, after the source has passed the observer, the frequency approaches the value $\nu/(1 + m)$ asymptotically.

The Doppler shift is zero for $\phi = \pi/2$. When the source crosses the y-axis, the observed frequency is still upshifted, and the reason is, of course, that the sound was emitted at an earlier time, corresponding to an emission angle ϕ less than $\pi/2$; at the time of emission, the component of the source velocity in the direction of the observer was not zero. Actually, an upshifted frequency is observed even after the source has crossed the y-axis and it decreases to the source frequency when the emission point falls on the y-axis.

In regard to the graph in the display, it should be pointed out that the normalized time coordinate can be interpreted also as the normalized position coordinate x_s/h of the source at the time of observation. Other aspects of this example are left for the problem section.

Display 13.3.

Doppler shifted frequency from moving source observed at O as a function of time or source location ($x_s = ut$).

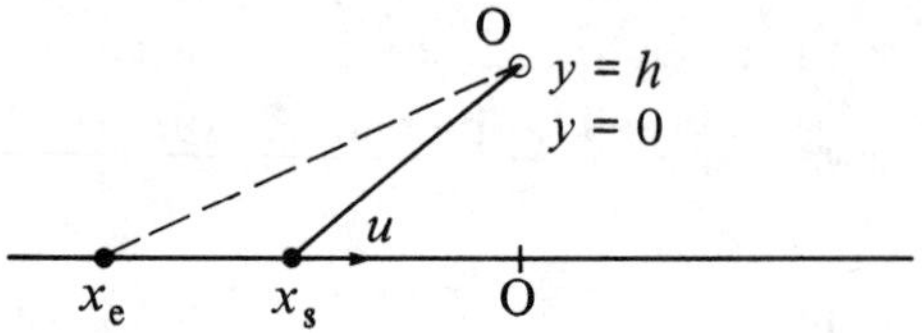

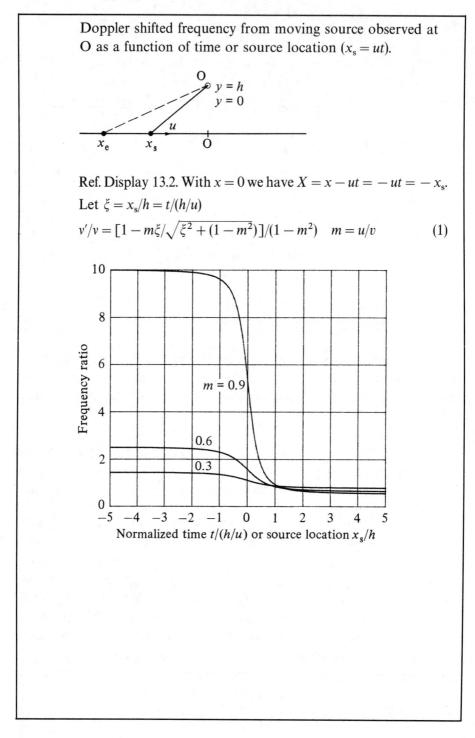

Ref. Display 13.2. With $x = 0$ we have $X = x - ut = -ut = -x_s$.

Let $\xi = x_s/h = t/(h/u)$

$$v'/v = [1 - m\xi/\sqrt{\xi^2 + (1 - m^2)}]/(1 - m^2) \quad m = u/v \tag{1}$$

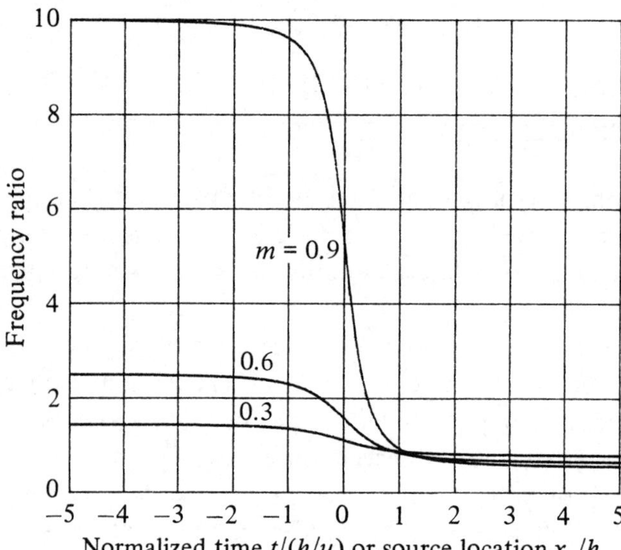

Moving medium. We now assume that the medium (fluid) carrying the wave is moving with a velocity U in the positive x-direction. The velocity of a wave front (phase velocity) then will be $v + U$ in the positive x-direction and $v - U$ in the negative x-direction. These two velocities can be expressed as $v_p = v + U\cos(\phi)$, where ϕ is the angle between the medium velocity U and the direction of wave propagation.

Using this phase velocity v_p instead of v in Eq. (3), we obtain

$$\begin{aligned} \nu' &= \nu[1-(u_0/v_p)]/[1-(u/v_p)] \\ &= \nu[1-(m_0 - M\cos(\phi))]/[1-(m - M\cos(\phi))] \qquad (7) \\ m_0 &= u_o/v, \quad m = u/v, \quad M = U/v \end{aligned}$$

When both the source and the receiver are at rest, $m = m_0 = 0$, there is no Doppler shift produced by the motion of the medium, although this motion does affect the wavelength. This can be understood as follows. The separation of two wave fronts, emitted a time T apart, will be $\lambda' = (v + U)T = \lambda(1 + M)$, which is the wavelength of the sound in the downstream direction from the source. The wavelength is increased by the factor $1 + M$. The wave speed is increased by the same factor, however, and the time it takes for the wavelength to pass the observer will be equal to the original period T, so that the observed frequency will be unshifted. Thus, if there is no relative motion between the source and the receiver, the motion of the medium *per se* will not produce a Doppler shift.

The result obtained readily can be generalized to more general motions of source, observer, and the medium, described by the vector velocities $\mathbf{u}, \mathbf{u}_0$, and $\mathbf{U}$.

The simplest way of deriving the expression for the corresponding Doppler shifted frequency is to consider the motion from a frame of reference in which the medium is at rest. The velocities of the source and the observer with respect to the medium are then $(\mathbf{u} - \mathbf{U})$ and $(\mathbf{u}_0 - \mathbf{U})$. The Doppler shift depends on the velocity components in the direction of wave travel, and we shall denote by $\hat{\mathbf{k}}$ the unit vector for the direction of propagation of the wave to the observer from the emission point of the source. The corresponding relevant velocity components are then $(\mathbf{u} - \mathbf{U})\cdot\hat{\mathbf{k}}$ and $(\mathbf{u}_0 - \mathbf{U})\cdot\hat{\mathbf{k}}$. We can then apply Eq. (3) to obtain

$$\begin{aligned} \nu' &= \nu[1-(\mathbf{m}_0 - \mathbf{M})\cdot\hat{\mathbf{k}}]/[1-(\mathbf{m} - \mathbf{M})\cdot\hat{\mathbf{k}}] \qquad (8) \\ \mathbf{m} &= \mathbf{u}/v, \quad \mathbf{m}_0 = \mathbf{u}_0/v \quad \mathbf{M} = \mathbf{U}/v \end{aligned}$$

In the special case when the motions of the source and the observer are on the x-axis, we have $\mathbf{m}\cdot\hat{\mathbf{k}} = m$ and $\mathbf{m}_0\cdot\hat{\mathbf{k}} = m_0$, and Eq. (8) reduces to Eq. (7).

Display 13.4

Acoustic Doppler effect in a moving fluid.

Doppler shift once again

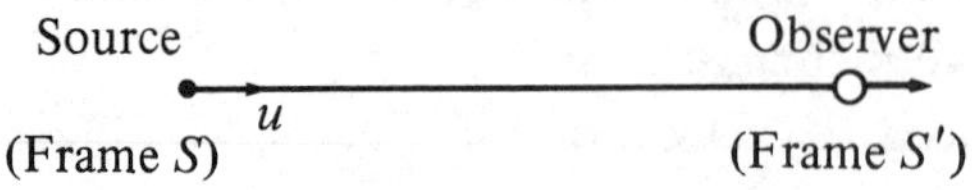

Source moving, velocity u. Observer stationary.
In source bound frame of reference
Fluid velocity $= -u$. Observer velocity $= -u$.

Phase velocity $\quad v_p = v - u, \quad k = \omega/v_p$ (1)

Galilean transformation $\quad x = x' - ut$ (2)

Wave function:

In S $\quad \exp(ikx - i\omega t)$ (3)

In S' $\quad \exp(ikx' - i\omega t - ikut) = \exp(ikx' - i\omega' t)$ (4)

$\omega' = \omega + ku = \omega[1 + (u/v_p)] = \omega v/(v - u) = \omega/(1 - m)$

$\quad m = u/v$ (5)

Generalization. Moving medium

Source velocity $= \mathbf{u}$ (6)

Observer velocity $= \mathbf{u}_0$ (7)

Medium (fluid) velocity $= \mathbf{U}$ (8)

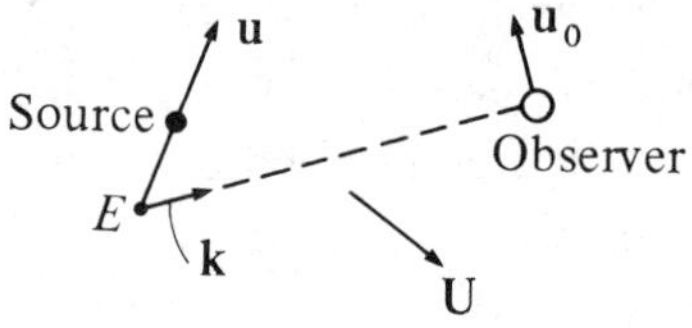

In source bound frame S:

Fluid velocity $= (\mathbf{U} - \mathbf{u})$ (9)

Observer velocity $= \mathbf{u}_0 - \mathbf{u}$ (10)

Phase velocity, $v_p = v + (\mathbf{U} - \mathbf{u})\cdot\hat{\mathbf{k}}$ (11)

Galilean transformation $\quad \mathbf{r} = \mathbf{r}' + (\mathbf{u}_0 - \mathbf{u})t$ (12)

$\mathbf{k} = (\omega/v_p)\hat{\mathbf{k}}$ (13)

$\exp(i\mathbf{k}\cdot\mathbf{r} - i\omega t) = \exp(i\mathbf{k}\cdot\mathbf{r}' - i\omega' t)$ (14)

$\omega' = \omega - \mathbf{k}\cdot(\mathbf{u}_0 - \mathbf{u}) = \omega[1 - (\mathbf{u}_0 - \mathbf{u})\cdot\hat{\mathbf{k}}/v_p]$

$\quad = \omega[1 - (\mathbf{m}_0 - \mathbf{M})\cdot\hat{\mathbf{k}}]/[1 - (\mathbf{m} - \mathbf{M})\cdot\hat{\mathbf{k}}]$ (15)

$\mathbf{M} = \mathbf{U}/v, \quad \mathbf{m} = \mathbf{u}/v, \quad \mathbf{m}_0 = \mathbf{u}_0/v$ (16)

Display 13.5.

Doppler shift. Subsonic and supersonic source motion.

From Eq. 13.1(4)

With $X = x - x_s = x - ut$

$$R = (mX \pm R_1)/(1 - m^2) \tag{1}$$

$$R_1 = [X^2 + (1 - m^2)h^2]^{\frac{1}{2}} \tag{2}$$

Subsonic motion, $m = u/v < 1$

Only one acceptable (positive) solution for R, corresponding to the plus sign in Eq. (1)

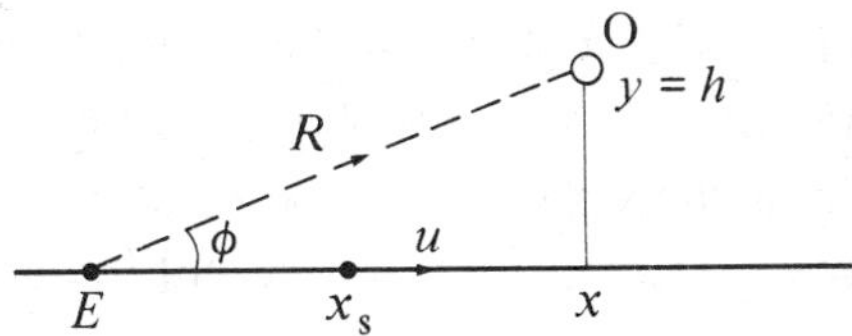

Supersonic motion, $m > 1$

Two solutions for R R' and R''

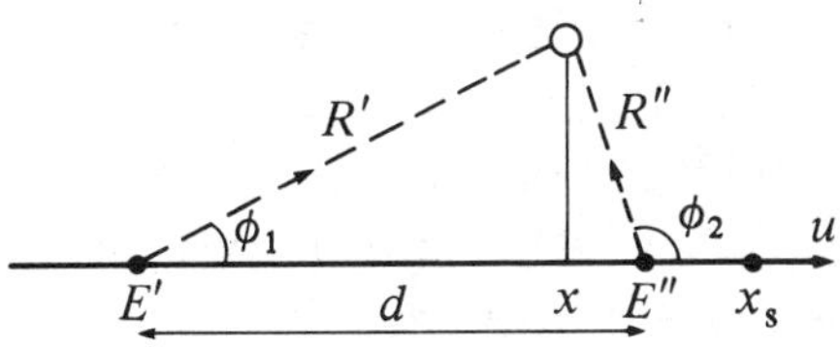

Note $R'/v = d/u + R''/v$ (3)

Doppler shifted frequency

$$\nu' = \nu/(1 - m\cos\phi) \tag{4}$$

Supersonic motion of the source. As we have seen, the wave front reaching the observer O at time t was emitted a time R/v earlier from the emission point E of the source, which is at a distance R (not known *a priori*) from the observer.

We have also seen, that the distance R is obtained from a second order equation. With the source moving at a subsonic speed, only one solution to this equation had physical meaning, as given by Eq. (4). It corresponds to a single emission point E.

For supersonic motion of the source, i.e. with $m > 1$, both roots R' and R'' to the R-equation are acceptable, and they correspond to two different emission points E' and E'', illustrated in Display 13.5.

With the velocity u of the source exceeding the wave speed v, it is possible to make the two travel times R'/v and $R''/v + d/u$ the same, where d is the distance between the two emission points E' and E''. Thus, signals emitted from E' and E'' will arrive at 0 at the same time. The total wave field at 0 then will be superposition of two contributions of different frequencies. We leave further details for the reader.

13.2 Electromagnetic waves

For a mechanical wave, such as a sound wave in a gas or surface wave on liquid, the Doppler shift resulting from the motion of the source is not the same as that caused by the motion of the observer, even if the relative velocity of the source and the observer is the same in the two cases. The reason can be traced to the fact that the phase velocity of the wave depends on the motion of the frame of reference.

For an EM wave in vacuum, on the other hand, the wave speed c (the speed of light) is independent of the velocity of the frame of reference. This experimental fact is contained in the kinematics of special relativity, as expressed by the Lorentz–Einstein (L–E) transformation, which we shall review briefly.

Galilean and Lorentz–Einstein transformations. We consider a stationary frame of reference S' and a frame S which moves with a velocity V in the x'-direction. The axes of the two frames coincide at $t = 0$ so that the location of the origin of S will be $x' = Vt$, as indicated in Display 13.6.

In the Galilean transformation, the time intervals measured by clocks in the two systems are assumed to be the same, independent of V, and the relationship between the coordinates in S and S' are $x' = x + Vt$, $y' = y$, $z' = z$, and $t' = t$. In order to be able easily to compare this transformation with the L–E transformation, we introduce $\beta = V/c$ and express the Galilean

Display 13.6.

Coordinate transformations. Relativistic addition of velocities

Galilean transformation

Stationary frame, $S': x', y', z'; t$

Moving frame, $S: x, y, z; t$

$$x' = x + Vt \quad y' = y \quad z' = z \tag{1}$$

$$t' = t \tag{2}$$

$$\text{Let } \beta = V/c \tag{3}$$

$$x' = x + \beta ct \quad y' = y \quad z' = z \tag{4}$$

$$ct' = ct \tag{5}$$

$$u' = \mathrm{d}x'/\mathrm{d}t = u + V \tag{6}$$

Lorentz–Einstein transformation

$$x' = \gamma(x + \beta ct) \quad y' = y \quad z' = z \tag{7}$$

$$ct' = \gamma(ct + \beta x) \tag{8}$$

$$\gamma = 1/\sqrt{1 - \beta^2} \tag{9}$$

Relativistic addition of velocities

$$\mathrm{d}x' = \gamma(\mathrm{d}x + \beta c\,\mathrm{d}t) \quad \mathrm{d}y' = \mathrm{d}y \quad \mathrm{d}z' = \mathrm{d}z \tag{10}$$

$$\mathrm{d}t' = \gamma(\mathrm{d}t + \beta\,\mathrm{d}x/c) \tag{11}$$

$$u'_x = \mathrm{d}x'/\mathrm{d}t' = (u_x + V)/[1 + (\beta u_x/c)] \quad u_x = \mathrm{d}x/\mathrm{d}t \tag{12}$$

$$u'_y = \mathrm{d}y'/\mathrm{d}t' = u_y/\gamma[1 + (\beta u_x/c)] \quad u_y = \mathrm{d}y/\mathrm{d}t \tag{13}$$

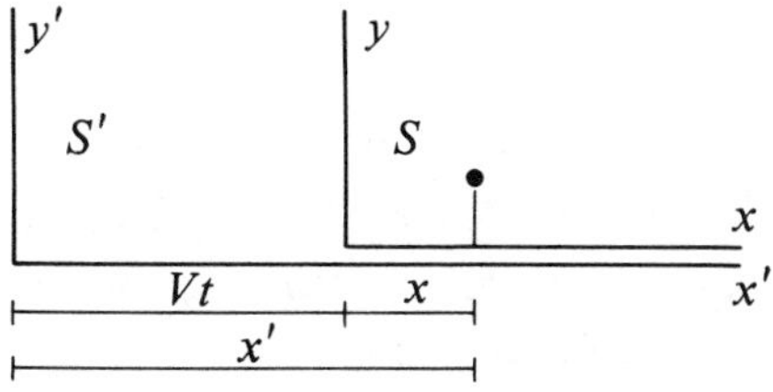

transformation as

$$x' = x + \beta ct \tag{1}$$
$$ct' = ct$$

with $y' = y$ and $z' = z$.

According to this transformation, an object which has a velocity u in the x-direction in S will have the velocity

$$u' = u + V \tag{2}$$

with respect to S'. If we apply this relation to the light wave emitted from a source fixed in S, the speed of this wave in the x-direction in S' is $c + U$, where c is the speed of light in S. This is not consistent with experiments, however, according to which the speed of light is the same in all frames of reference in relative motion (Michelson–Morley experiment).

The Lorentz–Einstein transformation is designed to make the speed of light the same in S and S', independent of the relative velocity of S and S'. In this transformation, the space and time coordinates, or rather x and ct, are treated in analogous manner, and the transformation is made symmetrical and linear in the variables x and ct. Thus, the Galilean transformation in Eq. (1) is replaced by

$$x' = \gamma(x + \beta ct)$$
$$ct' = \gamma(ct + \beta x)$$
$$\gamma = 1/\sqrt{1 - \beta^2} \quad \beta = V/c \tag{3}$$

with $y' = y$, $z' = z$, where the constant γ is chosen to make the speed of light the same in S and S'. This transformation reduces to the Galilean transformation for small values of β.

Relativistic addition of velocities. The motion of a particle is described by $x(t)$, $y(t)$, and $z(t)$ with respect to S and by $x'(t')$, $y'(t')$, and $z'(t')$ with respect to S'. The corresponding velocities in the x-direction are $u_x = \mathrm{d}x/\mathrm{d}t$ and $u'_x = \mathrm{d}x'/\mathrm{d}t'$, with similar expressions for the other components.

From the L–E transformation in Eq. (3), we obtain

$$\mathrm{d}x' = \gamma(\mathrm{d}x + \beta c\,\mathrm{d}t)$$
$$c\,\mathrm{d}t' = \gamma(c\,\mathrm{d}t + \beta\,\mathrm{d}x)$$

from which follows

$$\mathrm{d}x'/\mathrm{d}t' = (\mathrm{d}x + \beta c\,\mathrm{d}t)/(c\,\mathrm{d}t + \beta\,\mathrm{d}x)$$

or

$$u'_x = (u_x + U)/(1 + \beta u_x/c) \tag{4}$$

In similar manner we obtain

$$u'_y = u_y/\gamma(1 + \beta u_x/c)$$
$$u'_z = u_z/\gamma(1 + \beta u_x/c) \tag{5}$$

With $u_x = c$ we obtain $u'_x = c$, as required. This invariance of the speed of light is contained in the property of the L–E transformation, $(\mathrm{d}s')^2 - (c\,\mathrm{d}t')^2 = (\mathrm{d}s)^2 - (c\,\mathrm{d}t)^2$, with $(\mathrm{d}s)^2 = (\mathrm{d}x)^2 + (\mathrm{d}y)^2 + \mathrm{d}z^2$.

Relativistic transformation of directions of EM waves. We establish also the relation between the directions of propagation of an EM wave in the two frames. With reference to Display 13.7, the direction is specified by the angle ϕ between the propagation vector $\mathbf{k}$ and the x-axis. The wave speed is the same in both systems, i.e. the speed of light c. The trace velocity of a wave front along the x-axis in the S-system is $u_x = c/\cos(\phi)$ and $u'_x = c/\cos(\phi')$ in the S'-system. We have already established the relation between the velocities u_x and u'_x in Eq. (3), and by using it in the expressions for the trace velocities, we obtain

$$\cos(\phi') = [\cos(\phi) + \beta]/[1 + \beta\cos(\phi)] \tag{6}$$

This relation can be obtained also in the following manner. Treating the electromagnetic wave as a stream of photons in the direction of propagation, we treat them as particles with the x-components of velocity $u_x = c\cdot\cos(\phi)$ and $u'_x = c\cdot\cos(\phi')$ in S and S'. Insertion of these expressions into Eq. (5) leads to the relation between ϕ and ϕ' in Eq. (6).

This approach, although valid in this case, cannot be used for mechanical waves, because the relative motion of the two frames of reference necessarily makes the medium (fluid) move with respect to at least one of the frames, and the phase velocity then depends on the direction of wave travel.

EM Doppler effect. In deriving the EM Doppler effect, we start from the complex wave function $\exp(\mathrm{i}\mathbf{k}\cdot\mathbf{r} - \mathrm{i}\omega t) = \exp[\mathrm{i}\Psi(x, y, t)]$, which describes the space–time dependence of the wave function through the phase function

$$\Psi(x, y, t) = \mathbf{k}\cdot\mathbf{r} - \omega t = kx\cos(\phi) + ky\sin(\phi) - \omega t \tag{7}$$

We now use the L–E transformation and obtain the description of the same wave in the S' system with the phase function

$$\Psi(x', y', t') = \Psi(x', y') - [(\omega/c)\gamma\beta ct'\cos(\phi) - \gamma\omega t'] \tag{8}$$

where we have expressed explicitly only the portion of the function which contains the time t'. It remains to express ϕ in terms of ϕ' through the angle transformation in Eq. (6), and we obtain (see Display 13.7)

$$\omega' = \omega\sqrt{1 - \beta^2}/[1 - \beta\cos(\phi')] \tag{9}$$

Display 13.7.

Electromagnetic Doppler effect.

Transformation of direction of EM wave propagation

$u_x = c/\cos\phi$ (trace velocity of wave) (1)

$u'_x = c/\cos\phi'$ (2)

$\cos\phi'/\cos\phi = u_x/u'_x = u_x[1 + (\beta u_x/c)]/(u_x + U)$ (3)

$\cos\phi' = (\cos\phi + \beta)/(1 + \beta\cos\phi) \quad \beta = U/c$ (4)

EM Doppler effect

$\exp(\mathrm{i}\mathbf{k}\cdot\mathbf{r} - \mathrm{i}\omega t) = \exp[\mathrm{i}\Psi(x, y, t)]$ (5)

$\Psi(x, y, t) = (\omega/c)(x\cos\phi + y\sin\phi) - \omega t$ (6)

$x = \gamma(x' - \beta ct') \quad y = y'$ (7)

$ct = \gamma(ct' - \beta x')$ (8)

$\cos\phi = (\cos\phi' - \beta)/(1 - \beta\cos\phi')$ (9)

Time dependent part of $\Psi(x, y, t)$:

$-(\omega/c)\gamma\beta ct'(\cos\phi' - \beta)/(1 - \beta\cos\phi') - \gamma\omega t' = -\omega' t'$ (10)

$\omega' = \omega\sqrt{1 - \beta^2}/(1 - \beta\cos\phi')$ (11)

$\phi' = 0 \quad \omega' = \omega\sqrt{(1 + \beta)/(1 - \beta)} \quad \omega' \simeq \omega(1 + \beta) \quad \text{for } \beta \ll 1$ (12)

$\phi' = \pi \quad \omega' = \omega\sqrt{(1 - \beta)/(1 + \beta)} \quad \omega' \simeq \omega(1 - \beta) \quad \text{for } \beta \ll 1$ (13)

$\phi' = \pi/2 \quad \omega' = \omega\sqrt{1 - \beta^2}$ (14)

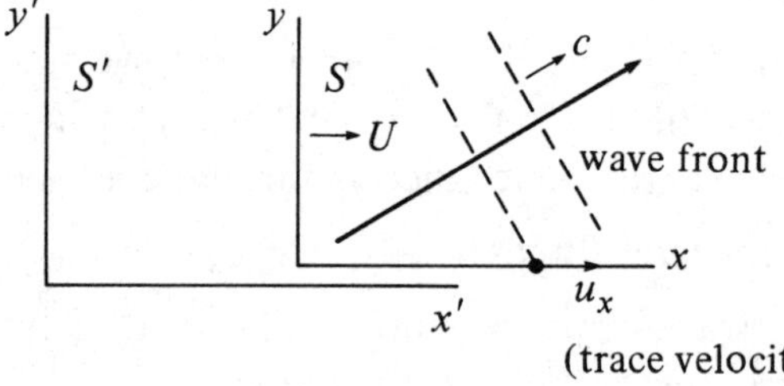

For small values of β, this expression reduces to the result for the mechanical Doppler effect.

The important new feature is the frequency shift in the direction $\phi' = \pi/2$ perpendicular to the velocity of the source at the point of emission. Unlike the case of a mechanical wave, we now get a Doppler shift different from zero with the Doppler shifted frequency being $\omega' = \omega\sqrt{1-\beta^2}$. It is due to the relativistic time dilatation, according to which the time scale is 'stretched' in the moving frame (clocks are running slower). Thus, the number of seconds between two events (the passage of two adjacent wave fronts), as measured on a clock in the moving frame S, will be smaller and the corresponding frequency ω larger than in the stationary frame S'.

For radiation in the forwards and backwards directions, corresponding to $\phi' = 0$ and ϕ, the observed frequency is upshifted and downshifted, respectively,

$$\begin{aligned}
\phi' &= 0: \quad \omega' = \omega\sqrt{(1+\beta)/(1-\beta)} \quad \omega' \simeq \omega(1+\beta) \quad \text{for } \beta \ll 1 \\
\phi' &= \pi: \quad \omega' = \omega\sqrt{(1-\beta)/(1+\beta)} \quad \omega' \simeq \omega(1-\beta) \quad \text{for } \beta \ll 1 \\
\phi' &= \pi/2: \quad \omega' = \omega\sqrt{1-\beta^2}
\end{aligned} \tag{10}$$

Problems

13.1 *Hearing Doppler shift.* In the frequency range between 600 and 4000 Hz, the smallest pitch change which can be resolved by a normal human ear corresponds to a relative frequency change of approximately $\Delta\omega/\omega \simeq 0.003$.

A sound source, which emits a tone of 1000 Hz, moves along a straight line with constant speed. What is the lowest speed of the source which produces a pitch change which can be heard by a stationary observer as the source moves by?

13.2 *Tone from airplane.* A propeller airplane emitting a tone of frequency ν flies at a constant speed U in the x-direction at a constant height $y = H$. As the plane crosses the y-axis, an observer at $x = 0$, $y = 0$ receives the Doppler shifted frequency 2ν and at a time τ later the unshifted frequency ν. From these data determine:

(a) the Mach number $M = U/v$ of the airplane, where v is the sound speed, and
(b) the height H.

13.3 *Thermal line broadening.* Discuss how the temperature of a gas can be determined from the observed broadening of a spectral line emitted from the gas.

13.4 *Red shift and Hubble's law.* Consider a galaxy at a distance r from us. Suppose this galaxy emits a spectral line of wavelength λ_0 and that we detect this line at a (larger) wavelength λ. The quantity

$$z = (\lambda - \lambda_0)/\lambda_0 = \Delta\lambda/\lambda_0$$

is called the redshift.

Hubble discovered from analysis of experimental data that the redshift is proportional to the distance r to the galaxy,

$$(\Delta\lambda/\lambda_0) = (H/c) = r/R \quad (R = c/H)$$

where H is Hubble's constant.

Interpret the redshift as a Doppler shift and obtain an expression for the velocity U of the galaxy in terms of r/R. Use the relativistically correct expression for the Doppler effect

$$\nu = \nu_0[(1-\beta)/(1+\beta)]^{\frac{1}{2}} \quad \beta = U/c,$$

where ν is the observed frequency. For what value of U/c is $r = R$?

14

Periodic structures. Dispersion

The wave velocities encountered so far have been frequency independent and determined solely by mass density and compressibility, or quantities equivalent thereto. In this chapter we discuss waves for which the wave speed does depend on frequency; in particular, we consider wave transmission on a mass–spring lattice (or its electrical analogue). The concepts of phase velocity and group velocity are introduced and the relationship between wave energy flow, wave energy dnsity and group velocity is discussed.

Finally, the effect of friction damping on wave transmision is considered briefly with particular attention to continuous transmission lines.

14.1 Wave transmission on a mass–spring lattice

In deriving the field equations for waves on a continuous transmission line in Chapter 6, we started with the equations of motion for a one-dimensional mass–spring lattice and then went directly to the continuum limit of these equations.

We now return to the equations 6.3(1)–(2), which are

$$M\partial u(n,t)/\partial t = F(n,t) - F(n+1,t) \tag{1}$$

$$C\partial F(n+1,t)/\partial t = u(n,t) - u(n+1,t) \tag{2}$$

where n is short for the equilibrium coordinate x_n of element #n in the lattice, as indicated in Display 14.1.

Quantity $u(n,t)$ is the velocity of element #n, $F(n,t)$ is the force acting on the left side of this element, $C = 1/K$ is the compliance of the springs between the mass elements, and M is the mass of each element.

Instead of going to the continuum limit, expressing the differences $u(n+1) - u(n)$ and $F(n+1) - F(n)$ as spatial derivatives, as in Chapter 6, we leave the

Display 14.1.

Dispersion relation for longitudinal waves on a periodic lattice.

Field equations

Let $F(x) = F \quad F(x+l) = F_1$ (1)

$u(x) = u \quad u(x+l) = u_1$ (2)

$-i\omega M u = F - F_1$ (3)

$-i\omega(1/K)F_1 = u - u_1$ (4)

Wave in positive x-direction

$F(x) = F(0)\exp(ikx)$ (5)

$u(x) = u(0)\exp(ikx) \quad x = nl$ (6)

$F_1 = F(x+l) = F(x)\exp(ikl)$ (7)

$u_1 = u(x+l) = u(x)\exp(ikl)$ (8)

Dispersion relation

From Eqs. (3)–(4), (7)–(8):

$(1 - e^{ikl})F - i\omega M u = 0$ (9)

$-i\omega(1/K)F + (1 - e^{-ikl})u = 0$ (10)

$(1 - e^{ikl})(1 - e^{-ikl}) = (\omega/\omega_0)^2$ (11)

$\omega_0^2 = K/M$ (12)

$1 - \cos(kl) = \omega^2/2\omega_0^2$ (13)

$\omega/\omega_0 = \pm 2\sin(kl/2)$ (14)

$$\omega/k = v\frac{\sin(kl/2)}{kl/2} \quad (15)$$

$v = \omega_0 l = l\sqrt{K/M}$ (16)

$kl \ll 1 \quad \omega \simeq vk$ (17)

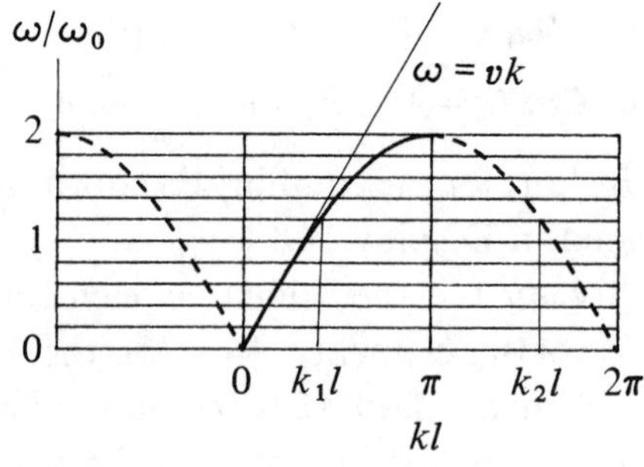

equations (1)–(2) in their original form and study the conditions under which a harmonic wave can be transmitted on the lattice.

Since we shall deal with harmonic time dependence, we rewrite Eqs. (1) and (2) in terms of the complex amplitudes and obtain

$$-\mathrm{i}\omega M u(n) = -F(n+1) \tag{3}$$

$$-\mathrm{i}\omega C F(n+1) = u(n) - u(n+1) \tag{4}$$

For a wave travelling in the positive x-direction the x-dependence of the complex amplitudes will be of the form

$$F(x) = F(0)\exp(\mathrm{i}kx) \tag{5}$$

$$u(x) = u(0)\exp(\mathrm{i}kx) \quad x = nl \tag{6}$$

The position coordinate x refers to the discrete values $x = nl$, where l is the equilibrium distance between adjacent elements.

We note that with the complex amplitude at $x = nl$ being $u(n) = u(0)\exp(\mathrm{i}kx)$, the complex amplitude $u(n+1)$ at $x = (n+1)l$ is obtained simply by multiplying $u(n)$ by $\exp(\mathrm{i}kl)$. Thus, if we introduce $F(n+1) = F(n)\exp(\mathrm{i}kl)$ and $u(n+1) = u(n)\exp(\mathrm{i}kl)$ into Eqs. (5) and (6) we obtain

$$(1 - \mathrm{e}^{\mathrm{i}kl})F - \mathrm{i}\omega M u = 0 \tag{7}$$

$$-\mathrm{i}\omega C F + (1 - \mathrm{e}^{-\mathrm{i}kl})u = 0 \tag{8}$$

In order for these equations to have solutions for $F = F(n)$ and $u = u(n)$ different from zero, it is necessary that the determinant of the coefficients in the equations be zero, i.e.

$$(1 - \mathrm{e}^{\mathrm{i}kl})(1 - \mathrm{e}^{-\mathrm{i}kl}) = (\omega/\omega_0)^2 \tag{9}$$

which follows by eliminating one of the variables in Eqs. (7)–(8).

This condition must be fulfilled to make it possible for a harmonic wave to be propagated along the lattice; it is called the **dispersion relation**, and it can be written as

$$\left.\begin{aligned} &1 - \cos(kl) = (\omega^2/2\omega_0^2) \\ \text{or}\quad & \\ &\omega/\omega_0 = \pm 2\sin(kl/2) \\ &\omega_0^2 = K/M \end{aligned}\right\} \tag{10}$$

where we have used the identity $1 - \cos(kl) = 2\sin^2(kl/2)$. Unlike the relation $\omega = vk$, familiar from waves on continuous transmission lines in previous chapters, the dispersion relation (10) is nonlinear. It does reduce to $\omega = vk$ in the limit of $kl = 2\pi l/\lambda = 0$, which corresponds to the continuum.

The plus and minus signs in Eq. (8) account for the fact that the dispersion relation includes waves travelling in both the positive and negative x-directions.

The dispersion relaton, in the dimensionless form ω/ω_0 vs. kl, is plotted in the display. For each value of ω we get several possible values of k. But these different values of k lead to the same waves as the k-values in the main branch of the dispersion relation corresponding to $-\pi < kl < \pi$.

Consider, for example, the two different values k_1l and k_2l, shown in the figure. We note that $k_2l = 2\pi - k_1l$, and it follows that $\exp(\mathrm{i}k_2x) = \exp(\mathrm{i}2\pi)\exp(-\mathrm{i}k_1x) = \exp(-\mathrm{i}k_1x)$. Thus, the only difference between the resulting wave functions is that they represent waves travelling in opposite directions. Nothing new is added, since these two waves are accounted for by the k-values in the main branch of the dispersion relation.

Cut-off frequency and exponential decay. So far it has been tacitly assumed that the propagation constant k is real, so that the magnitude of $\exp(\mathrm{i}kx)$ and the corresponding wave amplitude is constant, independent of x.

It follows then from the dispersion relation (10) and the graph in the display, that only frequencies below a maximum value of $2\omega_0$ can be transmitted as a travelling wave with constant amplitude. What will happen if the frequency of the driving force is above this 'cut-off' frequency $2\omega_0$?

Clearly, the time dependence of the motion of the driven element as well as the rest of the lattice will be harmonic. But since the amplitude along the lattice cannot be constant it must decrease with x, and this suggests that we might be able to describe the x-dependence of the amplitude in terms of a wave function $\exp(\mathrm{i}kx)$ with a complex value of k, $k = k_r + \mathrm{i}k_i$. We then get $\exp(\mathrm{i}kx) = \exp(-k_ix)\exp(\mathrm{i}k_rx)$, i.e. an exponential decay of the amplitude.

This turns out to be true. The dispersion relation (10), regarded as an equation for k in terms of ω, is satisfied with a complex k-value, $k = k_r + \mathrm{i}k_i$, and we obtain

$$\begin{aligned}\cos(k_rl + \mathrm{i}k_il) &= \cos(k_rl)\cos(\mathrm{i}k_il) - \mathrm{i}\sin(k_rl)\sin(\mathrm{i}k_il)\\ &= 1 - 2(\omega/2\omega_0)^2 \end{aligned} \tag{11}$$

The function $\cos(\mathrm{i}k_il)$ is real ($\cos(x)$ is an even function of x) and from its power series expansion we find that

$$\cos(\mathrm{i}k_il) = \cosh(k_il)$$

Thus, the left hand side of Eq. (11) is a complex number, and the right hand side is real. The quantities k_r and k_i are then obtained from the two equations for the real and imaginary parts of the equation.

We consider separately the two cases when the driving frequency is below and above the cut-off value $2\omega_0$, respectively. For $\omega < 2\omega_0$, the magnitude of the right hand side is always less than unity, and we can satisfy the

Display 14.2.

Complex propagation constant $k = k_r + ik_i$ for waves on a periodic mass–spring (or electrical LC) lattice.

Dispersion relation and complex propagation constant

$$\cos(kl) = 1 - (1/2)(\omega/\omega_0)^2 \quad (1)$$

$$k = k_r + ik_i \quad (2)$$

$$\cos(k_r l + ik_i l) = \cos(k_r l)\cosh(k_i l) - i\sin(k_r l)\sinh(k_i l)$$
$$= 1 - (1/2)(\omega/\omega_0)^2 \quad (3)$$

$\omega < 2\omega_0$:

$$k_i l = 0 \quad \cos(k_r l) = 1 - (1/2)(\omega/\omega_0)^2 \quad (4)$$

$\omega > 2\omega_0$:

$$k_r l = \pi \quad \cosh(k_i l) = (1/2)(\omega/\omega_0)^2 - 1 \quad (5)$$

Complex kl-plane

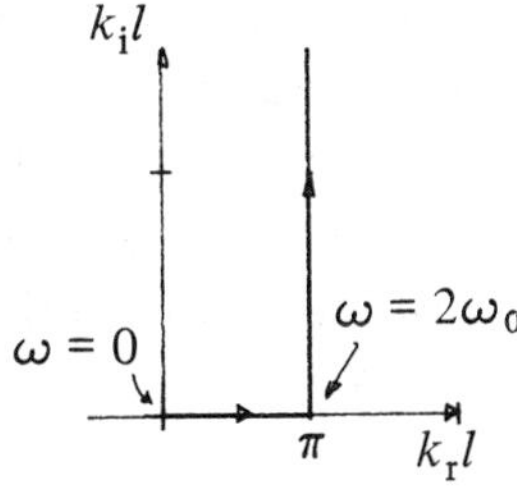

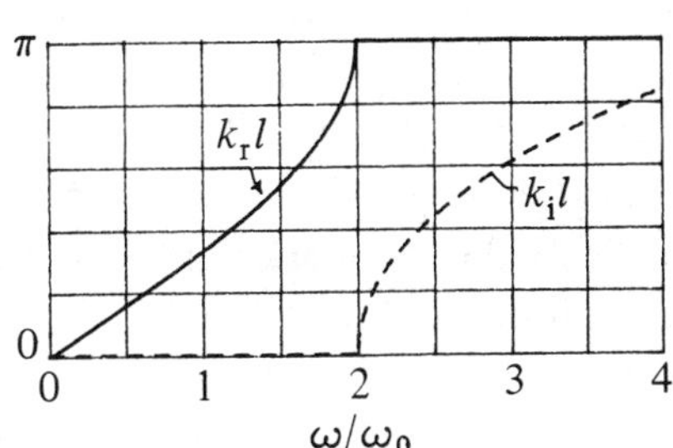

Amplitude ratio

$\omega < 2\omega_0$:

$u(x + l)/u(x) = \exp(ik_r l)$

$\omega > 2\omega_0$:

$u(x + l)/u(x) = -\exp(-k_i l)$

equation by making $k_i = 0$ and

$$\cos(k_r l) = 1 - 2(\omega/2\omega_0)^2 \quad k_i = 0 \quad (\omega < 2\omega_0)$$
$$u_0^2 = K/M \tag{12}$$

Above the cut-off frequency, the magnitude of the right hand side is always larger than unity, and the equation is satisfied by making $k_r l = \pi$ and

$$\cosh(k_i l) = 2(\omega/2\omega_0)^2 - 1 \quad k_r l = \pi \quad (\omega > 2\omega_0) \tag{13}$$

The choice of $k_r l = \pi$ rather than any other multiple of π is made so that the two values of $k_r l$ obtained from Eqs. (12) and (13) will be the same at the cut-off frequency.

The result expressed by Eqs. (12) and (13) can be considered to be the general form of the dispersion relation for the lattice, covering the entire range of frequencies. As the frequency increases, the location of the complex propagation constant k in the complex plane follows a path as shown. Below the cut-off frequency it runs along the real axis and then turns in the direction of the positive imaginary axis at the cut-off frequency. The frequency dependence of the real and imaginary parts of k have also been plotted in Display 14.2.

Above the cut-off frequency the amplitude decays exponentially with $u(n+1)/u(n) = -\exp(-k_i l)$, the minus sign indicating countermotion and arising from $\exp(ik_r l)$ with $k_r l = \pi$. No power is then transmitted along the lattice and it behaves like a low-pass filter. The results apply equally well to an electrical *LC* network with *L* and *C* being analogous to the mass *M* and the compliance *C*.

14.2 Wave packet and group velocity

The space–time dependence of a harmonic wave travelling in the positive x-direction is determined by the phase function $\Phi(x, t) = (kx - \omega t)$. This function increases with x and decreases with t and remains constant if $k\,dx = \omega\,dt$, i.e. for $dx/dt = \omega/k$. This means that in a frame of reference moving with the velocity ω/k, the phase function is constant. This is the wave speed we have dealt with in previous chapters, and it is called the **phase velocity**

$$v_p = \omega/k \tag{1}$$

For a linear dispersion relation, $\omega = vk$, the phase velocity, $v_p = v$, is the slope of the line $\omega = vk$ in the ω–k plane.

It should be realized that for a linear dispersion relation, v_p is independent of frequency, and if a wave is a superposition of a group of harmonic waves with different frequencies, they all travel together with the same speed.

Display 14.3.

Phase velocity and group velocity, periodic structure.

Phase velocity

Complex wave function $A\exp(\mathrm{i}kx - \mathrm{i}\omega t)$ (1)

Phase $\Phi(x,t) = kx - \omega(k)t = k[x - (\omega/k)t] = k[x - v_\mathrm{p}t]$ (2)

Phase velocity

$v_\mathrm{p} = \omega(k)/k$ (3)

Group velocity

Two waves $A\exp(\mathrm{i}\Phi_1)$ and $A\exp(\mathrm{i}\Phi_2)$ (4)

Let $\Phi_1 = \Phi - \mathrm{d}\Phi/2 \quad \Phi_2 = \Phi + \mathrm{d}\Phi/2$ (5)

$A\mathrm{e}^{\mathrm{i}\Phi_1} + A\mathrm{e}^{\mathrm{i}\Phi_2} = A\mathrm{e}^{\mathrm{i}\Phi}(\mathrm{e}^{-\mathrm{i}\mathrm{d}\Phi/2} + \mathrm{e}^{\mathrm{i}\mathrm{d}\Phi/2}) = 2A\cos(\mathrm{d}\Phi/2)\mathrm{e}^{\mathrm{i}\Phi}$ (6)

$\mathrm{d}\Phi = x\,\mathrm{d}k - t\,\mathrm{d}\omega = \mathrm{d}k[x - (\mathrm{d}\omega/\mathrm{d}k)t] = \mathrm{d}k[x - v_\mathrm{g}t]$ (7)

$v_\mathrm{g} = \mathrm{d}\omega/\mathrm{d}k$ (8)

$A\mathrm{e}^{\mathrm{i}\Phi_1} + A\mathrm{e}^{\mathrm{i}\Phi_2} = 2A\cos[(\mathrm{d}k/2)(x - v_\mathrm{g}t)]\cdot\mathrm{e}^{\mathrm{i}k(x - v_\mathrm{p}t)}$ (9)

Example, lattice

$\omega(k) = 2\omega_0\sin(kl/2)$ (10)

$v_\mathrm{p} = \omega/k = v\dfrac{\sin(kl/2)}{kl/2}$ (11)

$v_\mathrm{g} = \mathrm{d}\omega/\mathrm{d}k = v\cos(kl/2)$ (12)

$v_\mathrm{g} = v\sqrt{1 - (\omega/2\omega_0)^2}$ (13)

$v = \omega_0 l$ (14)

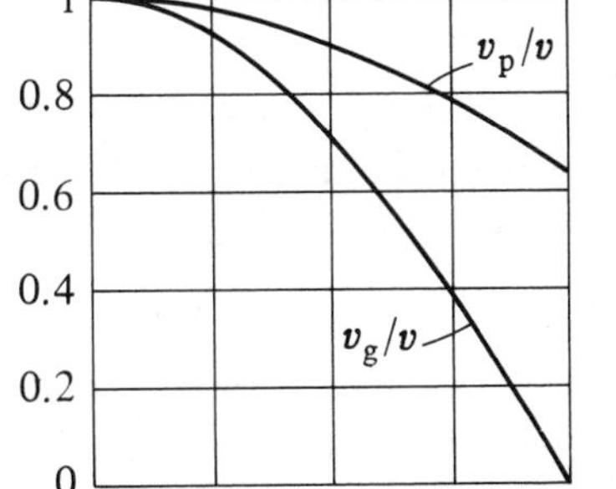

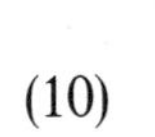

For a nonlinear dispersion relation, $\omega = \omega(k)$, on the other hand, the phase velocity is frequency dependent. For example, from the dispersion relation $\omega/\omega_0 = \pm \sin(kl/2)$ for the lattice (see Eq. 1(10)), we obtain

$$\begin{aligned} v_p &= \omega/k = v[\sin(kl/2)/(kl/2)] \\ v &= l\sqrt{K/M} = 1/\sqrt{\mu\kappa}, \\ \mu &= M/l \quad \kappa = 1/Kl \end{aligned} \tag{2}$$

which approaches the continuum value v as kl goes to zero and decreases to the value $2v/\pi$, as kl goes to π, i.e. $l = \frac{1}{2}\lambda$.

If the wave consists of a group of individual harmonic waves of different frequencies and hence speeds, the resulting wave shape will change with position, and the motion of the wave group cannot be described in terms of a single wave velocity.

A special case of particular interest involves a wave group which extends over a frequency range which is small compared to the center frequency of the group. As we shall see, the resulting field can then be interpreted as a wave $A(x, t)\cos(\omega t - kx)$ with a phase velocity determined by the average frequency of the group and an amplitude $A(x, t)$ which varies with x and t in such a way that it remains constant when viewed from a coordinate system which moves with the velocity

$$v_g = d\omega/dk \tag{3}$$

evaluated at the center frequency of the group; it is called the **group velocity.**

It follows from the dispersion relation Eqn. 1.(10) that the group velocity for such a **wave packet** on a lattice is

$$v_g = d\omega/dk = v\cos(kl/2) \tag{4}$$

In Display 14.3, we have plotted v_g and v_p as functions of kl. It is significant to note, that for $kl = \pi$, the group velocity is zero although the phase velocity is not.

As an example, we consider first the superposition of two harmonic waves with the same amplitudes A but with slightly different frequencies $\omega - d\omega/2$ and $\omega + d\omega/2$. The corresponding phase functions are $\Phi_1 = \Phi - d\Phi/2$ and $\Phi_2 = \Phi + d\Phi/2$, where $\Phi = k(\omega)x - \omega t$ and $d\Phi = x\,dk - t\,d\omega = [x - (d\omega/dk)t]dk = (x - v_g t)dk$. The sum of the two complex wave functions is then

$$\begin{aligned} Ae^{i\Phi_1} + Ae^{i\Phi_2} &= 2A\cos[(dk/2)(x - v_g t)]e^{ik(x-vt)} = A(x, t)e^{ik(x-vt)} \\ A(x, t) &= 2A\cos[(dk/2k)k(x - v_g t)] \end{aligned} \tag{5}$$

with the corresponding real wave function being $A(x, t)\cos(kx - \omega t)$.

For a given t and for small values of dk/k, the amplitude factor $A(x, t)$

Display 14.4.

Superposition of two travelling waves with slightly different frequencies. Propagation of 'beats'.

Example

Ref. Eq. 14.2(5)

$$F = A\exp(\mathrm{i}\Phi_1) + A\exp(\mathrm{i}\Phi_2) = 2A\cos[(\mathrm{d}k/2)(x - v_g t)]\mathrm{e}^{\mathrm{i}k(x - v_p t)} \quad (1)$$

$$F(x,t) = 2A\cos[(\mathrm{d}k/2k)k(x - v_g t)]\cos[k(x - v_p t)] \quad (2)$$

$$F(x,t)/2A = \cos[\alpha 2\pi(x' - \beta t')]\cos[2\pi(x' - t')] \quad (3)$$

$$x' = x/\lambda \quad t' = t/T \quad \alpha = \mathrm{d}k/2k \quad \beta = v_g/v_p \quad (4)$$

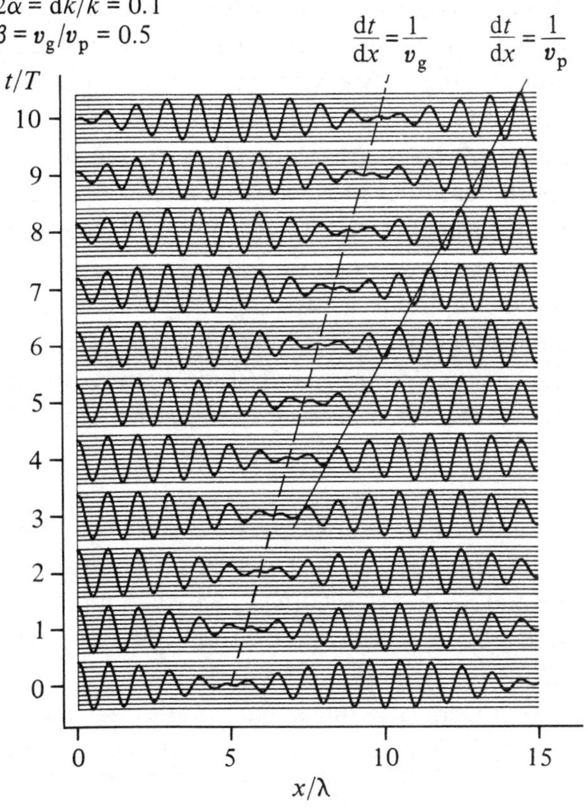

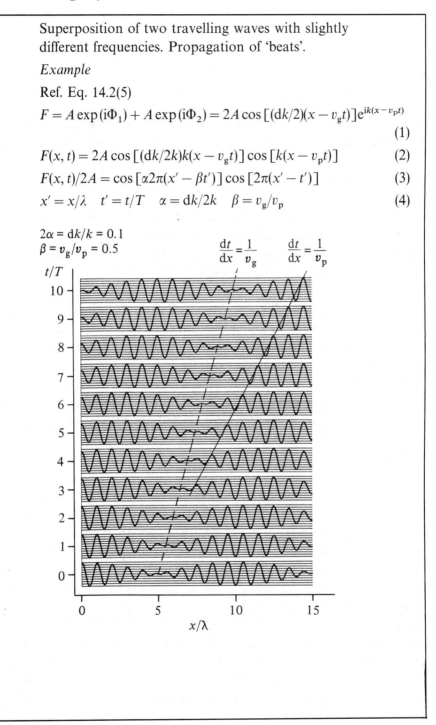

varies slowly with x; the distance D between adjacent maxima (or zeroes) of the magnitude $|A(x, t)|$ is given by $D\,dk/2 = \pi$ or $D = (k/dk)\lambda$.

In the numerical example shown in Display 14.4 we have used $dk/k = 0.1$ so that the period of $|A(x, t)|$ is 10. Furthermore, we have chosen $v_g/v_p = 0.5$.

Within the 'envelope' of the amplitude function $|A(x, t)|$, the x-dependence of the wave function is determined by the factor $\cos(kx - \omega t)$, which has the period T and wavelength λ. In one period T, this wave pattern moves forward a distance λ and the wave speed is $v = \lambda/T$, the phase velocity. The corresponding wave line, with the slope $1/v_p$, is shown as a solid line in the figure.

The envelope of the amplitude function, on the other hand, moves forward with the group velocity v_g, since the space–time dependence of the amplitude function $A(x, t)$ is expressed by the factor $(t - x/v_g)$. The corresponding wave line, with the slope $1/v_g$, is shown dashed in the figure.

It is instructive to reconsider the x-dependence of the amplitude function by viewing the motion of the complex amplitudes $A\exp[i(k - dk/2)x]$ and $A\exp[i(k + dk/2)x]$ in the complex plane as x increases. We let A be real so that both functions are on the real axis at $x = 0$, and their sum is $2A$.

As x increases, both quantities move along a circle of radius A. They do not stay together, however, but the angle between them, $x\,dk$, increases with x. When $x\,dk = \pi$, the two complex numbers are in opposite directions so that the total wave amplitude is zero. A further increase of $x\,dk$ by π, brings the two numbers together again, and the process is repeated, yielding a periodic variation in the resulting amplitude between 0 and $2A$. The distance D between maxima (or zeroes) is given by $D\,dk = 2\pi$, or $D = (k/dk)\lambda$, as before.

Wave packet. Next, instead of two waves, we consider a large number of waves, each with an amplitude A, distributed uniformly over the small frequency range $\Delta\omega$ and the corresponding range Δk. Again, we assume that all the waves are in phase at $x = 0$, so that the total amplitude at this location becomes NA, where N is the number of waves.

As x increases, the various complex amplitudes spread apart in the complex plane so that at location x they are evenly distributed in the complex plane over the angular range from 0 to $x\Delta k$. As a result, the total amplitude will be smaller than NA, and it goes to zero when $x\Delta k = 2\pi$, and at any other location such that $x\Delta k$ is a multiple of 2π. The complex amplitudes are then distributed uniformly over the range 2π. Between these locations, the angular distribution of the complex amplitudes is not quite uniform, and the amplitude, although small, is not zero.

As x goes to infinity, the distribution tends to be uniform, and the amplitude goes to zero. At a given time, the spatial amplitude distribution is in the form of a wave packet with the bulk of the energy concentrated in a range Δx such that $\Delta x \Delta k \simeq 2\pi$.

To put these observations into mathematical form, we express the wave packet $F(x, t)$ at a given position x as a Fourier integral over frequency

$$F(x,t) = \int_{-\infty}^{\infty} A(\omega) \exp\,[\mathrm{i}(kx - \omega t)]\,\mathrm{d}\omega \tag{6}$$

With $\mathrm{d}\omega = (\mathrm{d}\omega/\mathrm{d}k)\,\mathrm{d}k = v_g\,\mathrm{d}k$, this integral can be expressed also as an integral over k

$$\begin{aligned} F(x,t) &= \int_{-\infty}^{\infty} F(k) \exp\,[\mathrm{i}(kx - \omega t)]\,\mathrm{d}k \\ &= \int_{0}^{\infty} F(k) \exp\,[\mathrm{i}(kx - \omega t)]\,\mathrm{d}k + \text{C.C.} \end{aligned} \tag{7}$$

where C.C. stands for the complex conjugate.

The Fourier amplitude $F(k)$ is now assumed to be zero except in a small frequency range $\omega_0 - (\Delta\omega/2) < \omega < \omega_0 + (\Delta\omega/2)$ about the center frequency ω_0, where $\Delta\omega/\omega_0 \ll 1$. The corresponding range in k has the width Δk about the value $k_0 = k(\omega_0)$. In this range we let $F(k)$ be a real constant, $F(k) = F(k_0)$.

It is convenient to introduce the new variable $k' = k - k_0$, and in the small k-range considered we can then put $\omega(k) = \omega(k_0) + (\mathrm{d}\omega/\mathrm{d}k)k' = \omega_0 + v_g k'$.

The integral in Eq. (7) then reduces to

$$F(x,t) = F(k_0) \exp\,[\mathrm{i}(k_0 x - \omega_0 t)] \int_{-\Delta k/2}^{\Delta k/2} \exp\,[\mathrm{i}(x - v_g t)k']\,\mathrm{d}k' \tag{8}$$
$$+\,\text{C.C.}$$

which can be expressed as

$$F(x,t) = A(x,t) \cos\,(k_0 x - \omega_0 t) \tag{9}$$

$$A(x,t) = [F(k_0)\Delta k] \sin\,(X)/X \tag{10}$$

$$X = (x - v_g t)(\Delta k/2) \tag{11}$$

The amplitude function $A(x, t)$ has the maximum value $F(k_0)\Delta k$ at $x = v_g t$, and this maximum travels forward with the group velocity $v_g = \mathrm{d}\omega/\mathrm{d}k$, evaluated at $\omega = \omega_0$. At a given time the width of the amplitude function is $\Delta x \simeq 2\pi/\Delta k = (k/\Delta k)\lambda$, as explained above, and with $\Delta k/k_0 \ll 1$, it follows that the wave packet is many wavelengths wide.

Within the wave packet the variation of the wave function is then determined mainly by the factor $\cos\,(k_0 x - \omega_0 t)$, representing a wave which

Display 14.5.

Wave packets $\Delta k/k = 0.1$

(*a*) $v_g/v_p = 1/4$ (*b*) $v_g/v_p = 4$ (*c*) $v_g/v_p = 4$

$$F(x, t) = \int_0^\infty F(k) \exp[i(kx - \omega t)] \, dk + \text{C.C.} \tag{1}$$

$$F(k) = F(k_0) \quad \text{for } k_0 - (\Delta k/2) < k < k_0 + (\Delta k/2) \quad \Delta k/k_0 \ll 1 \tag{2}$$

$F(k) = 0$ for other values of k

$$F(x, t) = A(x, t) \cos(k_0 x - \omega_0 t) \quad \omega_0 = \omega(k_0) \tag{3}$$

$$A(x, t) = \sin(X)/X \quad X = (x - v_g t)(\Delta k/2) \quad v_g = d\omega/dk \tag{4}$$

$$X = 2\pi[x/\lambda_0 - (v_g/v_p)t/T_0]\Delta k/2k_0$$

$$\omega_0 = 2\pi/T_0 \quad k_0 = 2\pi/\lambda_0$$

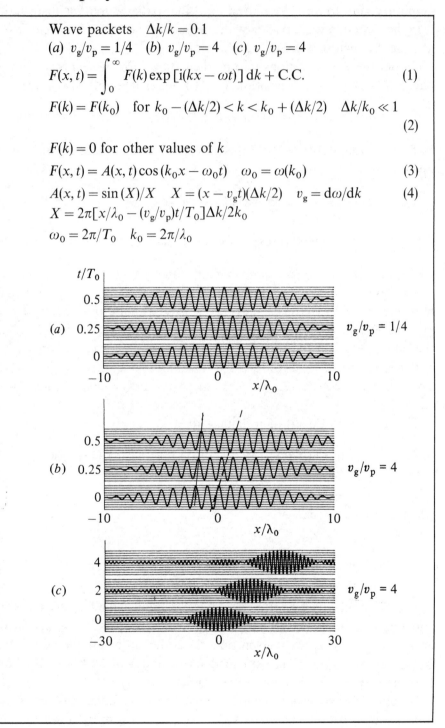

moves forward with the phase velocity $v_p = \omega_0/k_0$. In other words, there will be a wave within the wave packet which moves relative to the amplitude function with the velocity $v_p - v_g$. In the case of a mass–spring lattice, the group velocity v_g is always smaller than the phase velocity v_p, as illustrated in Display 14.3. In some cases, as for surface waves on a liquid, to be discussed in Chapter 18, the group velocity can be larger than the phase velocity.

The motion of a wave packet is illustrated in Display 14.5, in which case $\Delta k/k_0 = 0.1$ and $v_g/v_p = 1/4$ and 4.

14.3 Wave impedance and energy flow

The wave impedance, as we recall, is defined as the ratio between the complex force amplitude and the complex velocity amplitude in a single wave travelling in the positive x-direction. This ratio is obtained from either of the field equations 1(7)–(8), $Z = F/u = (-i\omega M)/[1 - \exp(ikl)] = (-i\omega M)\exp(-ikl/2)2i\sin(kl/2)$.

Making use of Eq. 2(2) for the phase velocity.

$$\omega/k = v\sin(kl/2)/(kl/2)$$

we can write the wave impedance simply as

$$Z = F/u = (\mu v)\exp(-ikl/2) \quad \mu = M/l \tag{1}$$

where $k = k_r + ik_i$ is given in Eqs. 14.1(12)–(13) and Display 14.2.

At low frequencies, $kl \ll 1$, the impedance is approximately equal to μv, the familiar value for a continuous transmission line. At frequencies below cut-off, the propagation constant k is real, and the impedance $(\mu v)\exp(-ikl/2)$ moves along a circle of radius μv in the complex plane as the frequency increases, ending up at $-i\mu v$ on the negative imaginary axis at the cut-off, as shown in Display 14.6. Beyond this frequency, the impedance follows the negative imaginary axis toward infinity.

The power carried by the wave is the product of the force and the velocity, and, as was shown in Chapter 3, the time average value in a harmonic wave can be written

$$P = \tfrac{1}{2}\mathrm{Re}\{Fu^*\} = \tfrac{1}{2}|u|^2\,\mathrm{Re}\{Z\} = \tfrac{1}{2}|F|^2\,\mathrm{Re}\{1/Z\} \tag{2}$$

Below the cut-off frequency, according to Eq. (1), the real part of the impedance is $(\mu v)\cos(k_r l/2) = (\mu v)[1 - \sin^2(k_r l/2)]^{\frac{1}{2}}$, and making use of Eq. 1(10), we obtain,

$$P = \tfrac{1}{2}\mathrm{Re}\{Fu^*\} = \tfrac{1}{2}\mu v|u|^2[1 - (\omega/2\omega_0)]^{\frac{1}{2}} \quad \omega < 2\omega_0 \tag{3}$$

Above the cut-off frequency $2\omega_0$, the real part of the impedance and the power is zero. In this frequency range, with the impedance being purely

Display 14.6.

Wave impedance and wave power. Periodic structure.

Wave impedance

$$Z = F/u = (-i\omega M)/(1 - e^{ikl}) = (-i\omega M)e^{-ikl/2}/(2i)\sin(kl/2) \quad (1)$$

$$\omega/k = v\frac{\sin(kl/2)}{kl/2} \text{ (Eq. 14.8(2)) With } \mu = M/l \quad v = \omega_0 l$$

$$k = k_r + ik_i \quad (2)$$

$$Z = \mu v \exp(-ikl/2) = \mu v \exp(k_i l/2)\exp(-ik_r l/2) \quad (3)$$

$$\omega \ll 2\omega_0 \quad (k_i = 0, \quad k_r l \ll 1): \quad Z \simeq \mu v \quad (4)$$

$$\omega = 2\omega_0 \quad (k_i = 0, \quad k_r l = \pi): \quad Z = -i\mu v \quad (5)$$

$$\omega \gg 2\omega_0 \quad (k_i l \simeq \log(\omega/\omega_0)^2, \quad k_r = 0): \quad Z \simeq -i\mu v \exp(k_i l/2) \quad (6)$$

Z-plane
$Z = Z_r + iZ_i$

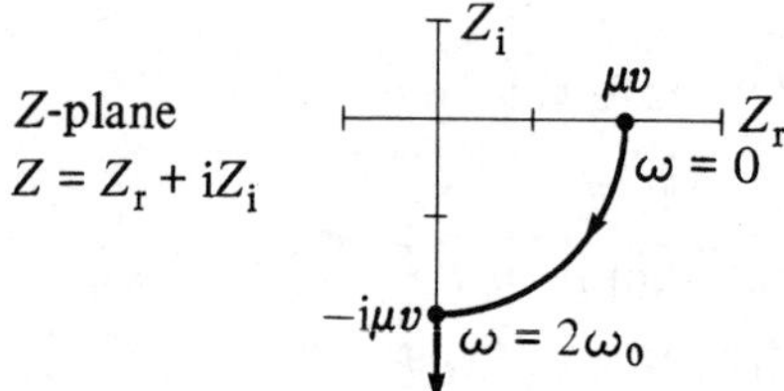

Wave power

$$P = (1/2)\operatorname{Re}\{Fu^*\} = (1/2)|u|^2 \operatorname{Re}\{Z\} = (1/2)|F|^2 \operatorname{Re}\{1/Z\} \quad (7)$$

$$\omega < 2\omega_0 \quad P = (1/2)|u|^2 \mu v \cos(k_r l/2)$$
$$= (1/2)|\underline{u}|^2 \mu v \sqrt{1 - (\omega/2\omega_0)^2} \quad (8)$$

$$\omega \geqslant 2\omega_0 \quad P = 0 \quad (9)$$

Wave power and group velocity

$$\langle E_k \rangle = (\mu/2)\langle u^2 \rangle = (\mu/4)|u|^2 \quad (10)$$

$$\langle E_p \rangle = (K/2l)\langle [\xi(x) - \xi(x+l)]^2 \rangle$$
$$= (K/4l)|\xi|^2 |1 - e^{ikl}|^2 \quad (11)$$
$$= (K/4l)|\xi|^2 [2\sin(kl/2)]^2$$
$$= (K/4l)|\xi|^2 (\omega/\omega_0)^2 = (\mu/4)|u|^2 \quad (12)$$

$$\langle E \rangle = (\mu/2)|u|^2 \quad (13)$$

From Eq. (8) and $v_g = v\cos(k_r l/2)$

$$P = \langle E \rangle v_g \quad (14)$$

imaginary, the force and the velocity are 90 degrees out of phase and the time average of the product will be zero.

In order to examine this question, we start by determining the wave energy density, i.e. the average energy per unit length of the lattice. The time average of the kinetic energy is

$$E_k = (\mu/2)\langle u^2 \rangle = (\mu/4)|u|^2$$

and for the potential energy we get

$$E_p = (K/2l)[\xi(x) - \xi(x+l)]^2 = (K/4l)|\xi|^2|1 - e^{ik}|^2$$
$$= (K/4l)|\xi|^2[2\sin(kl/2)]^2 = (K/4l)|\xi|^2(\omega/\omega_0)^2 = (\mu/4)|u|^2$$

so that the total energy density is $E = (\mu/2)|u|^2$, as expected.

According to Eq. 2.(4), the group velocity is $v_g = v\cos(k_r l/2)$, and by comparing with Eqs. (2) and (3), we find that the power can be expresed as the product of the average energy density and the group velocity.

The periodic lattice we have considered is an example of a **dispersive medium**, which, by definition, is one in which the dispersion relation is nonlinear and the corresponding wave speed dependent on frequency. Wave dispersion can occur also in continuous media, and examples will be given in subsequent chapters. Actually, the continuous transmission lines, considered in previous chapters, become dispersive if damping is included, as will be demonstrated in the next section.

14.4 Wave attenuation

The effects of damping on the motion of oscillators were discussed in Chapters 3–5, but little attention has been paid to damping of waves, which will be considered in this section. We use as a model for this analysis longitudinal wave motion on a mass–spring lattice.

We include two types of friction forces. The first is due to contact friction and is assumed to be $-ru$, proportional to the velocity u of the element.

The second friction force can be expressed in terms of a dashpot damper which is in parallel with each spring and is proportional to the difference between the velocities of adjacent mass elements. Thus, the dashpot between elements n and $n+1$ yields the friction force $-R[u(n+1) - u(n)]$ on element $n+1$, and the spring force is $K[\xi(n) - \xi(n+1)]$, where ξ is the displacement. The total force on the left hand side of element $n+1$ is then

$$F(n+1) = K[\xi(n) - \xi(n+1)] + R[u(n) - u(n+1)]$$

There is a corresponding force $F(n)$ acting on the left hand side of element n. The force acting on the right side is the reaction force $-F(n+1)$ from the spring and dashpot between the elements n and $n+1$. The equation of motion of element n is then

$$M\,du(n)/dt = -ru(n) + F(n) - F(n+1) \tag{1}$$

The expression for $F(n+1)$ given above can be expressed as

$$dF(n+1)/dt = K[u(n) - u(n+1)] + R\,d[u(n) - u(n+1)]/dt \tag{2}$$

The corresponding complex amplitude equations follow if we replace d/dt by $-i\omega$. We note, that the first friction force $-ru(n)$ can be incorporated in the inertial force term by replacing M by $M' = M(1+i\beta)$, where $\beta = r/\omega M$. Similarly, by replacing K by $K' = K(1-i\delta)$, where $\delta = \omega R/K$, the field equations (1) and (2) can be written in the 'standard' form of Eqs. 1(3)–(4)

$$-i\omega M' = F(n) - F(n+1) \tag{3}$$

$$-i\omega C' F(n+1) = u(n) - u(n+1) \tag{4}$$

$$M' = M(1+i\beta), \quad \beta = r/\omega M \tag{5}$$

$$C' = 1/K' \quad K' = K(1-i\delta) \quad \delta = \omega R/K \tag{6}$$

In the continuum limit these equations become

$$-i\omega\mu' u = -\partial F/\partial x \tag{7}$$

$$-i\omega\kappa' F = -\partial u/\partial x \tag{8}$$

$$\mu' = M'/l \quad \kappa' = 1/K'l \tag{9}$$

By considering a wave travelling in the positive x-direction, with the complex amplitudes $F(x) = F(0)\exp(ikx)$ and $u(x) = u(0)\exp(ikx)$, we obtain from Eqs. (7) and (8) the wave impedance Z and the dispersion relation $k = k(\omega)$

$$Z = F/u = k/\omega\kappa' = \omega\mu'/k \tag{10}$$

$$k = \omega\sqrt{\mu'\kappa'} = (\omega/v)\sqrt{(1+i\beta)/(1-i\delta)} = k_r + ik_i, \quad v = 1/\sqrt{\mu\kappa} \tag{11}$$

The propagation constant k is now complex, and the x-dependence of the complex amplitude will be of the form

$$u(x) = u(0)e^{-k_i x}e^{ik_r x} \tag{12}$$

The imaginary part of the propagation constant leads to an exponential decay of the amplitude. This decrease is referred to as wave attenuation, and k_i is called the attenuation constant.

If the friction parameters β and δ are small, we can neglect the product of these quantities and expand the radical to first order in these parameters to obtain

$$k \simeq (\omega/v)[1 + i(\beta+\delta)/2] = k_r + ik_i \quad \beta, \delta \ll 1 \tag{13}$$

Display 14.7.

Wave attenuation.

Field equations

$M\mathrm{d}u/\mathrm{d}t = F - F_1 - ru$ (1)

$(1/K)\mathrm{d}F_1/\mathrm{d}t = u - u_1 + (R/K)\,\mathrm{d}(u - u_1)/\mathrm{d}t$ (2)

Complex amplitude equations:

$-\mathrm{i}\omega Mu = F - F_1 - ru$ (3)

$-\mathrm{i}\omega(1/K)F_1 = u - u_1 - \mathrm{i}(R\omega/K)(u - u_1)$ (4)

Let $M' = M(1 + \mathrm{i}\beta) \quad \beta = r/\omega M$ (5)

$K' = K(1 - \mathrm{i}\delta) \quad \delta = \omega R/K$ (6)

$-\mathrm{i}\omega M'u = F - F_1$ (7)

$-\mathrm{i}\omega(1/K')F_1 = u - u_1$ (8)

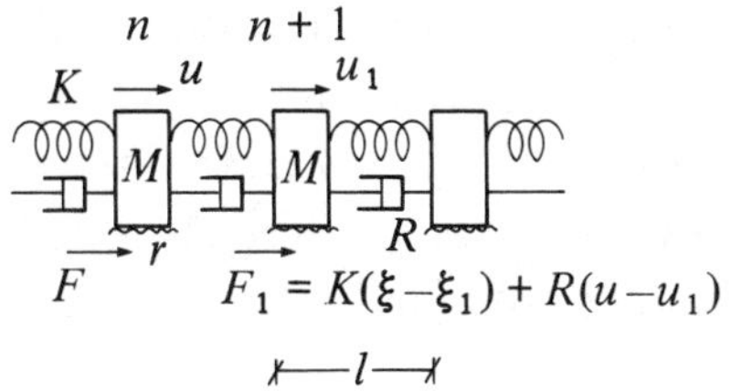

Continuum limit

$F - F_1 \simeq -(\partial F/\partial x)l; \quad u - u_1 \simeq -(\partial u/\partial x)l$ (9)

Let $\mu' = M/l = \mu(1 + \mathrm{i}\beta) \quad \beta = r/\omega M = r/\omega l\mu \quad \mu = M/l$ (10)

$\kappa' = 1/K'l = 1/(1 - \mathrm{i}\delta) \quad \kappa = 1/Kl$ (11)

$-\mathrm{i}\omega\mu' u = -\,\partial F/\partial x$ (12)

$-\mathrm{i}\omega\kappa' F = -\,\partial u/\partial x$ (13)

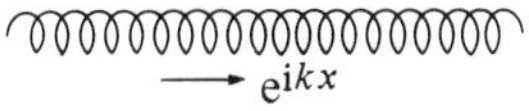

Dispersion relation

With $F(x) = F(0)\mathrm{e}^{\mathrm{i}kx} \quad u(x) = u(0)\mathrm{e}^{\mathrm{i}kx} \quad \partial/\partial x \to \mathrm{i}k$ (14)

From Eqs. (12), (13):

$Z = F/u = k/\omega\kappa' = \omega\mu'/k$ (15)

$k^2 = \omega^2\mu'\kappa'$ (16)

$k = \pm(\omega/v)\sqrt{(1 + \mathrm{i}\beta)/(1 - \mathrm{i}\delta)} = k_\mathrm{r} + \mathrm{i}k_\mathrm{i} \quad v = 1/\sqrt{\mu\kappa}$ (17)

$u(x) = u(0)\mathrm{e}^{-k_i x}\,\mathrm{e}^{\mathrm{i}k_r x}$ (18)

$\beta \ll 1 \quad \delta \ll 1$

$k \simeq (\omega/v)[1 + \mathrm{i}(\beta + \delta)/2] = k_\mathrm{r} + \mathrm{i}k_\mathrm{i}$ (19)

In this limit, the two friction effects are additive, and we note that in this approximation (i.e. to first order in β and δ) the real part of the propagation constant, and hence the phase velocity, is not affected by friction.

This analysis can be applied also to electromagnetic waves on a cable, if we let r be the series resistance per unit length of the cable and R the 'leakage' resistance per unit length across the cable in parallel with the capacitance per unit length. As before, μ and κ then stand for the inductance and capacitance per unit length, respectively.

For an elastic continuum, such as a solid rod, the dashpot resistance accounts for the internal damping in the solid, and is often expressed in terms of a complex Young's modulus. The contact friction, expressed by the coefficient r in the analysis above, is only present if the solid is in contact with an external medium and is generally negligible.

In the case of wave propagation in a fluid embedded in a porous material, we have both internal and contact friction. As in the case of a solid, the internal friction is accounted for in terms of a complex modulus of elasticity or complex compressibility. The contact friction force is proportional to the velocity of the fluid relative to the porous structure.

14.5 Normal modes of a one-dimensional lattice

In the discussion of coupled oscillators in Chapter 5, we dealt mainly with two degrees of freedom and the normal mode frequencies were obtained as solutions of a second-order equation for the squared frequency. For N coupled oscillators, this second-order equation is replaced by an equation of order N, and generally only a numerical solution can be obtained.

On the other hand, if the symmetry is high, as in the case of a line of identical oscillators, certain relations between the motions of the oscillators can be established *a priori* from symmetry, and the order of the frequency equation can be reduced.

For the infinitely long lattice, discussed in the previous sections, such a symmetry made it possible to satisfy the equations of motion by a single harmonic wave, the ratio between the complex amplitudes of force and velocity being the same for each element, independent of location.

The results thus obtained for the travelling wave will be applied now in the analysis of the motion of N identical mass elements coupled by springs, as shown in Display 14.8. As before, the equilibrium separation between adjacent elements is l, and the first element is at $x = l$. The springs at the ends of the lattice are held fixed at $x = 0$ and $x = (N + 1)l$, as shown.

For a single wave travelling on the lattice in the positive x-direction, the

Display 14.8.

Normal modes of a one-dimensional lattice of N elements.

Complex displacement amplitude

Single wave $\quad \xi(x_j) = B\exp(\mathrm{i}kx_j) \quad x_j = jl$ (1)

Dispersion relation $\quad \omega = 2\omega_0 \sin(kl/2)$ (2)

Total displacement:

$$\xi(x_j) = B\exp(\mathrm{i}kx_j) - B\exp(-\mathrm{i}kx_j) = A\sin(kx_j) \quad A = 2\mathrm{i}B \tag{3}$$

Boundary conditions

$x_j = 0 \qquad \xi = 0$ (4)

$x_j = (N+1)l \quad \xi = 0$ (5)

From Eq. (3):

$k(N+1)l = n\pi$ (6)

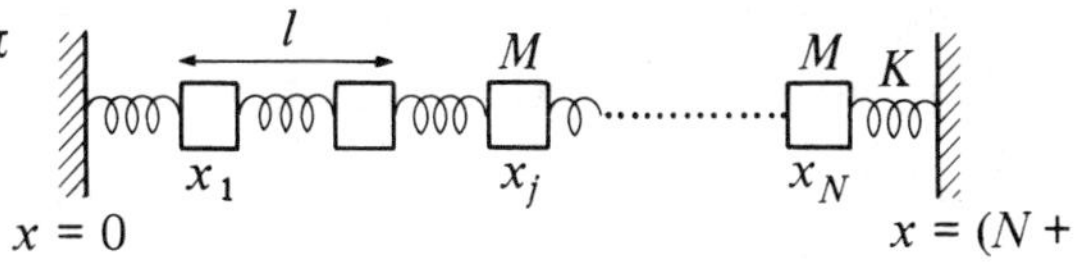

l M M K x_1 x_j x_N $x = 0$ $x = (N+1)l$

nth mode frequency

$k_n = n\pi/(N+1)l$ (7)

$\xi_n(x_j) = A_n \sin(k_n x_j) \quad A_n = A_n \exp(\mathrm{i}\alpha_n)$ (8)

$\omega_n = 2\omega_0 \sin[n\pi/2(N+1)]$ (9)

Real wave function of nth mode

$$\xi_n(x_j, t) = \mathrm{Re}\{\xi_n(x_j)\mathrm{e}^{-\mathrm{i}\omega_n t}\} = A_n \sin(k_n x_j)\cos(\omega_n t - \alpha_n) \tag{10}$$

Example, $N = 2$

Mode # 1 $\quad n = 1$

$$\xi_1(x_1) = A\sin(\pi/3) = A\sqrt{3/2} \quad \xi_1(x_2) = A\sin(2\pi/3) = \xi_1(x_1) \tag{11}$$

$\omega_1 = 2\omega_0 \sin(\pi/6) = \omega_0 = \sqrt{K/M}$ (12)

Mode # 2 $\quad n = 2$

$\xi_2(x_1) = A\sin(2\pi/3) \quad \xi_2(x_2) = A\sin(4\pi/3) = -\xi_2(x_1)$ (13)

$\omega_2 = 2\omega_0 \sin(2\pi/6) = \sqrt{3}\,\omega_0$ (14)

complex displacement amplitude of the jth element is

$$\xi_+(x_j) = B_+ \exp(ikx_j) \quad x_j = jl \tag{1}$$

provided that the propagation constant k is consistent with the dispersion relation for the lattice (see Eq. 1(10))

$$\omega = 2\omega_0 \sin(kl/2) \quad \omega_0 = \sqrt{K/M} \tag{2}$$

This single wave, although satisfying the field equations for the lattice, does not make the displacement vanish at the ends $x = 0$ and $x = (N+1)l$, as required, and to satisfy also these conditions, we have to include a wave $B_- \exp(-ikx_j)$, with $B_- = -B_+ = B$, travelling in the negative x-direction. The total complex displacement amplitude is then

$$\xi(x_j) = B\exp(ikx_j) - B\exp(-ikx_j) = A\sin(kx_j) \tag{3}$$

where $A = 2iB$.

With our choice $B_- = -B_+$, the boundary condition at $x = 0$ is automatically satisfied, and to make the displacement (velocity) zero also at the other end, $x = (N+1)l$, the propagation constant k must fulfill the condition

$$k = k_n = n\pi/(N+1)l \quad n = 1, 2, \ldots N \tag{4}$$

Each value of n corresponds to a particular mode of motion of the lattice. For the nth mode the x-dependence of the complex amplitude is

$$\xi_n(x_j) = A_n \sin(k_n x_j) \tag{5}$$

$$A_n = A_{n0} \exp(i\alpha_n)$$

and the frequency is

$$\omega_n = 2\omega_0 \sin[n\pi/2(N+1)] \tag{6}$$

Although n is an integer, there are only N different modes of motion, corresponding to $n = 1, 2, \ldots N$. If $n = N+1$, the displacement is identically zero for every mass element in the lattice, i.e. for all values of j. With $n = N+2$, we have $jn\pi/(N+1) = j[\pi + \pi/(N+1)]$, which yields the same displacement as for $n = 1$, and $n = N+1+s$ produces the same mode as $n = s$.

If we introduce the magnitude A_{n0} and the phase angle α_n of the complex amplitude constant in Eq. (5), we can express the time dependence of the real displacement in the nth mode as

$$\xi_n(x_j, t) = A_{n0} \sin(k_n x_j) \cos(\omega_n t - \alpha_n) \tag{7}$$

As a check, we compare this result with that obtained in the special case of two coupled oscillators, summarized in Display 5.1. Thus, with $N = 2$ we obtain for the two modes, corresponding to $n = 1$ and $n = 2$,

Display 14.9.

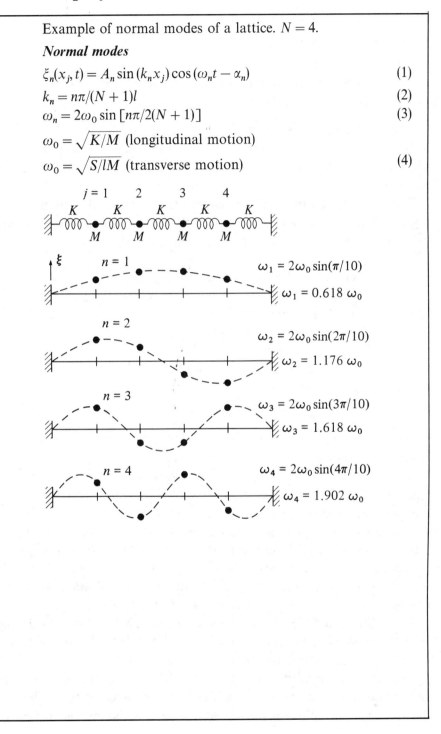

Example of normal modes of a lattice. $N = 4$.

Normal modes

$$\xi_n(x_j, t) = A_n \sin(k_n x_j) \cos(\omega_n t - \alpha_n) \quad (1)$$

$$k_n = n\pi/(N+1)l \quad (2)$$

$$\omega_n = 2\omega_0 \sin[n\pi/2(N+1)] \quad (3)$$

$$\omega_0 = \sqrt{K/M} \text{ (longitudinal motion)}$$

$$\omega_0 = \sqrt{S/lM} \text{ (transverse motion)} \quad (4)$$

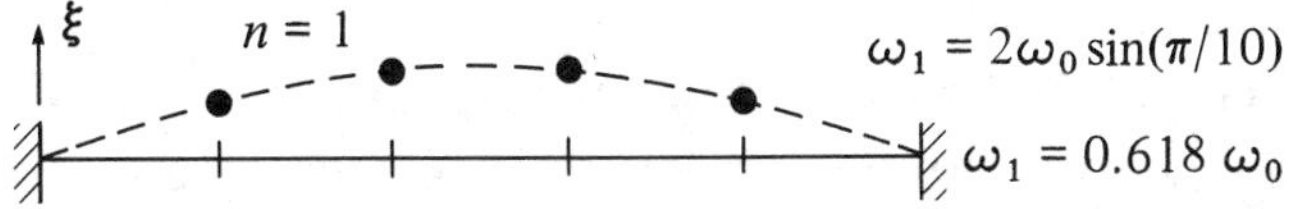

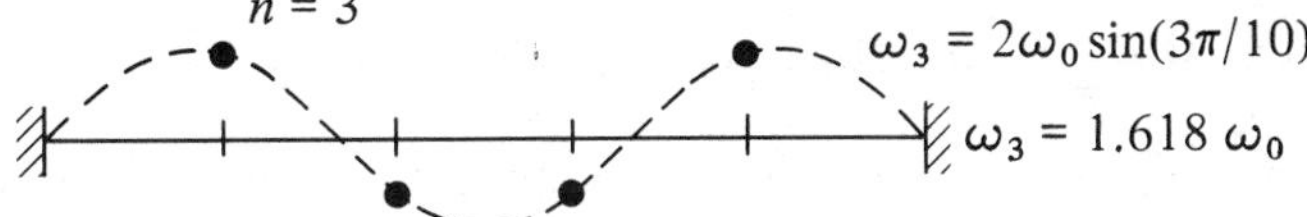

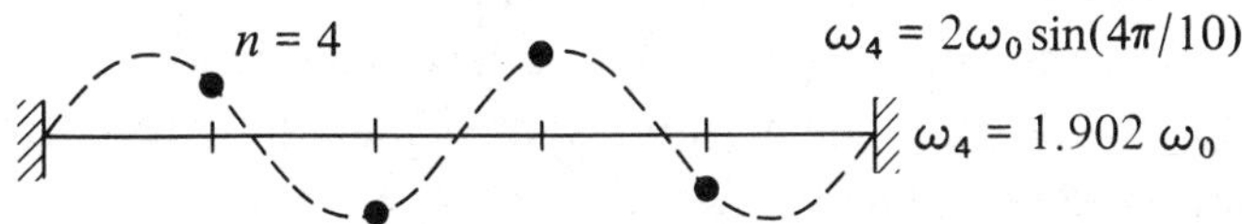

Mode # 1:

$$\xi_1(x_1) = A\sin(\pi/3) = A\sqrt{3}/2 \quad \xi_1(x_2) = A\sin(2\pi/3) = \xi_1(x_1) \tag{8}$$

$$\omega_1 = 2\omega_0 \sin(\pi/6)$$

Mode # 2:

$$\xi_2(x_1) = A\sin(2\pi/3) \quad \xi_2(x_2) = A\sin(4\pi/3) = -\xi_2(x_1) \tag{9}$$

$$\omega_2 = 2\omega_0 \sin(2\pi/6) = \sqrt{3}\omega_0$$

which is in agreement with the results in Display 5.1.

As another illustration, we have shown in Display 14.9 the normal mode displacements and the corresponding frequencies of a lattice with four elements.

Although the results obtained referred to longitudinal motion, they are applicable also to transverse motion if the spring constant is replaced by S/l, where S is the tension of the springs in the lattice.

14.6 General motion and impulse response of a periodic lattice

The general motion of a lattice of N coupled oscillators can be constructed as a linear combination of normal modes, and the displacement and velocity of element j can be written as

$$\xi(x_j, t) = \sum_1^N A_n \sin(k_n x_j) \sin(\omega_n t - \beta_n) \tag{1}$$

$$u(x_j, t) = \sum_1^N (A_n\omega_n) \sin(k_n x_j) \cos(\omega_n t - \beta_n) \tag{2}$$

where we have expressed the time dependence by $\sin(\omega_n t - \beta_n)$ rather than $\cos(\omega_n t - \alpha_n)$, but this change merely involves replacing α_n by $\beta_n - \pi/2$. (The real amplitude constant of the nth mode has been denoted by A_n rather than by A_{n0} to simplify the writing somewhat, and the corresponding complex amplitude is then $A_n \exp[\mathrm{i}(\beta_n - \pi/2)]$.)

The general motion contains $2N$ constants, N amplitudes and N phase angles. They can be determined from $2N$ initial conditions, specifying the displacements and velocities of the N elements at $t = 0$. Since the motion must be uniquely determined by these conditions, it follows that the linear combination of the modes is indeed the general solution.

As an illustration of the determination of the $2N$ constants, we consider here the important motion produced by a unit impulse on one of the elements, corresponding to $j = s$.

The velocities of all the elements immediately after the impulse are all zero except for element $j = s$, which is given the velocity $1/M$. The initial displacements of all the elements are zero, and if the impulse is delivered at $t = 0$, the initial conditions can be expressed as

$$\xi(x_j, 0) = 0 \quad \text{for all } j \tag{3}$$

$$u(x_j, 0) = 0 \quad \text{for } j \neq s \tag{4}$$

$$u(x_s, 0) = 1/M \tag{5}$$

With Eqs. (1) and (2), these conditions lead to the following equations for the determination of the constants A_n and β_n

$$0 = \sum_1^N A_n \sin(k_n x_j) \sin(\beta_n) \quad \text{yields} \quad \beta_n = 0 \tag{6}$$

$$u(x_j, 0) = \sum_1^N \omega_n A_n \sin(k_n x_j) \tag{7}$$

We are interested in solutions in which not all the amplitudes are zero, and it follows then from Eq. (6) that all the phase angles β_n must be zero, as indicated. This result has been used in Eq. (7) for the velocity.

To determine the amplitudes A_n, we multiply both sides of Eq. (7) by $\sin(k_m x_j)$ and sum over j from 1 to N. Since $u(x_j, 0) = 0$ except for $j = s$, the only contributing term on the left hand side will be $u(x_s, 0) \sin(k_m x_s) = (1/M) \sin(k_m x_s)$.

On the right side, the sum over j from 1 to N, $\sum_1^N \sin(k_n x_j) \sin(k_m x_j) = (1/2) \sum_1^N [\cos((k_n - k_m)x_j) - \cos((k_n + k_m)x_j)]$ becomes zero if n is different from m and equals $N/2$ if $n = m$. This result follows from the fact that each term in the sum, corresponding to a certain value of j, is always cancelled by another term (corresponding to another value of j). This cancellation can easily be seen in terms of the corresponding complex amplitudes, which are distributed uniformly on the unit circle in the complex plane, making their sum zero if $n \neq m$.

With $n = m$, the first cosine term on the right hand side of Eq. (7) becomes 1, and the sum over j from 1 to N yields the result N. Thus, to recapitulate, after multiplication of Eq. (7) by $\sin(k_m x_j)$ and summation over j and n, the right hand side becomes $NA_m/2$, and the resulting expression for A_m becomes

$$A_m = (2/N\omega_m) \sum_1^N u(x_j, 0) \sin(k_m x_j) = (2/N)(1/\omega_m M) \sin(k_m x_s) \tag{8}$$

Having obtained the amplitude A_m and phase angles β_m, we obtain from Eq. (1)

$$\xi(x_j, t) = G(x_j, t; x_s) = \sum_1^N (2/N)(1/\omega_m M) \sin(k_m x_s) \sin(k_m x_s) \sin(\omega_m t)$$

$$\omega_m = 2\omega_0 \sin(k_m l/2) \tag{9}$$

$$k_m = m\pi/(N+1)l \quad \omega_0 = \sqrt{K/M}$$

If, instead, the impulse is delivered at t', the corresponding response function is obtained by replacing $\sin(\omega_n t)$ by $\sin[\omega_m(t-t')]$.

It should be noted that the impulse response function is symmetrical in x_s and x_j. In other words, the displacement of the element j produced by an impulse on element s is the same as the displacement of element s produced by the same impulse applied to element j.

It follows from the impulse response function, sometimes called the Green function, that an impulse excites all the modes of the system, and that the amplitude of the mth mode is proportional to $\sin(k_m x_s)$. If $k_m x_s$ is a multiple of π, the amplitude of the mth mode will be zero; the impulse on x_s then does not 'couple' to the mth mode.

Since the frequencies of the modes are different, the time dependence of the total displacement of an element will appear quite complex, just as in the example with the two coupled oscillators in Display 5.6.

From the impulse response function, we can determine the displacement produced by an arbitrary force in complete analogy with the procedure discussed in Chapter 4 for the single oscillator. In particular, it is interesting to consider the displacement resulting from a harmonic driving force. We find that the frequency response of the displacement of the driven element will be the sum of N terms, each being similar to the response of a single oscillator with the resonance frequency equal to the frequency of the particular mode involved. At a driving frequency close to one of the mode frequencies, the displacement will be dominated by the contribution from this mode, and the system responds like a single oscillator.

Problems

14.1 *Wave dispersion on a mass–spring lattice.* In a mass–spring lattice of identical elements, the spring constant $K = 10^5$ N/m and the mass $M = 0.1$ kg. The separation between the elements is $l = 0.1$ m.

(a) What is the cut-off frequency of the lattice for longitudinal waves?
(b) What is the phase velocity at zero frequency and at the cut-off frequency?
(c) At what frequency is the phase velocity equal to twice the group velocity?
(d) The lattice is driven at a frequency equal to twice the cut-off frequency. What then is the ratio between the complex displacement amplitudes of the mass elements at $x = 0$ and at $x = 4l$?

14.2 *Wave impedance.* Prove that the wave impedance of a mass–spring lattice approaches $-i\omega M$ as the frequency goes to infinity.

14.3 *Power transmission.* The first element, at $x = 0$, of the lattice in Problem 14.1 is driven in harmonic motion with the displacement $\xi(0, t) = \xi_0 \cos(\omega t)$, where $\xi_0 = 0.1l$ and $\omega = \omega_0 = \sqrt{K/M}$.

(a) What is the wave impedance of the lattice?
(b) Determine the amplitude and phase angle of the driving force.
(c) What is the power transmitted along the lattice?
(d) What are the phase velocity and the group velocity?
(e) What is the average wave energy per unit length of the lattice?
(f) If, instead, the driving frequency had been $\omega = 4\omega_0$, what then would have been the results in (a), (b), and (c)?

14.4 *Surface waves.* The dispersion relation for the waves on the surface of a liquid with a depth much greater than the wavelength is given by $\omega^2 = gk + \sigma k^3/\rho$, as will be shown in a subsequent chapter (g = acceleration of gravity, σ = surface tension, ρ = mass density, and $k = 2\pi/\lambda$).

(a) Determine the relation between the phase velocity and the group velocity for long waves (gravity waves), for which the effect of surface tension can be neglected.
(b) Do the same for short waves (ripples or capillary waves), for which the effect of gravity can be neglected.
(c) The phase velocity cannot be smaller than a certain minimum value. Determine this value and the corresponding wavelength for surface waves on water ($\sigma = 72$ CGS units, $g = 981\ \text{cm/sec}^2$, $\rho = 1\ \text{g/cm}^3$).

14.5 *Bending waves.* As will be shown in a subsequent chapter, the dispersion relation for bending waves on a thin plate can be written in the form $\omega/k = (v/\sqrt{12})kh$, where $v^2 = E/\rho(1 - \sigma^2)$, E = Young's modulus, σ = Poisson ratio, ρ = mass density, h = plate thickness, and $k = 2\pi/\lambda$.

(a) Determine the relation between the phase velocity and the group velocity.
(b) Express the phase velocity in terms of the frequency of the wave.
(c) Consider a plate with $h = 1$ cm, $E = 1.7 \times 10^{12}\ \text{dyne/cm}^2$, $\rho = 7.8$ g/cm, and $\sigma = 0.2$. At what frequency is the phase velocity equal to the sound speed in the surrounding air (340 m/sec)?

14.6 *String embedded in an elastic material.* A string with tension S and a mass μ per unit length, is embedded in an elastic material, which produces a transverse restoring force $\gamma\eta$ per unit length of the string proportional to its transverse displacement η. Determine the dispersion relation for the string and the phase and group velocities.

14.7 *Wave attenuation.* A spring carries a longitudinal wave, which is attenuated as a result of the damping mechanisms discussed in Section 14.4.

(a) Starting from Eq. 14.4(11), prove Eq. 14.4(13).

(b) Under the conditions of validity of Eq. 14.4(13), discuss the frequency dependence of the attenuation per wavelength, which is determined by $k_i\lambda$, where $\lambda = 2\pi/k_r$. At what frequency is this attenuation a minimum, and determine this minimum value?

14.8 *Mass–spring lattice with springs of non-negligible mass.* Re-examine the problem of wave propagation on a mass–spring lattice, accounting for the mass m of each spring.

14.9 *Normal modes of a mass–spring lattice.* A mass–spring lattice has 5 mass elements which are coupled with springs, as shown. The ends of the lattice are held fixed. The mass of each element is $M = 0.1$ kg and the spring constant of each spring is $K = 10$ N/m.

Determine the angular frequencies of the normal modes of longitudinal oscillations of the lattice and the displacements of the mass elements in each mode.

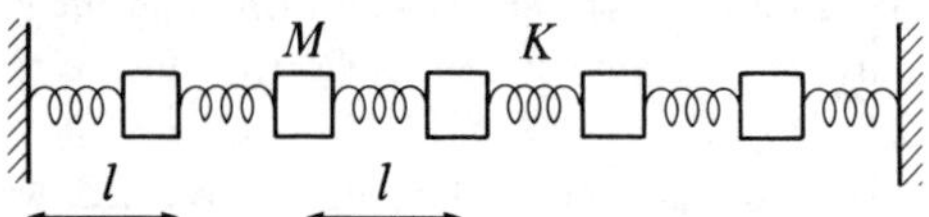

14.10 *Impulse response of a lattice.* The mass–spring lattice in Problem 14.9 is stretched to a tension $S = 1$ N. The distance between the mass elements in the stretched state is $l = 0.4$ m. An impulse $I = 0.1$ N sec is applied in the transverse direction on the center element at $t = 0$.

(a) What normal modes will be excited by the impulse?
(b) For each of the modes, identify the mass element with the largest displacement amplitude and determine this amplitude.

14.11 *Transverse motion of mass loaded string.* A string, clamped at both ends, carries three equal masses evenly spaced along the string. The tension in the string is S, the mass of each element is M, and the total length of the string is L.

At $t = 0$, the system is released from a displaced position, with the center mass element a distance D from its equilibrium position, as shown.

(a) Determine the subsequent motion.
(b) Compare the normal mode frequencies with those of a uniform string with the same tension, length and total mass.

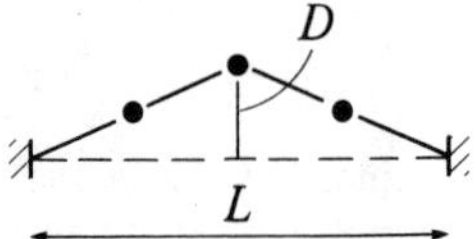

15

Wave guides and cavities

In our discussion of plane waves of sound on a fluid column in previous chapters, it was generally assumed that the wave was travelling along the axis (x-axis) of a pipe with rigid walls. We shall now study wave propagation in a pipe (wave guide) under more general conditions, involving a superposition of waves reflected back and forth between the walls in the guide. Both acoustic and electromagnetic waves will be considered. Of particular interest is the x-dependence of such a field and questions concerning phase and group velocity and energy flow.

If a wave guide is closed at both ends we obtain a cavity, in which the waves are reflected from all the walls in the cavity. If the walls are perfect reflectors, there will be no net energy flow in any direction in steady state, and the wave field in the cavity can be decomposed into 'standing' waves or normal modes, each with a frequency response like that of a harmonic oscillator.

15.1 **Propagating and decaying waves**

In the previous chapter we found that a wave could be transmitted along a lattice without attenuation at frequencies below a certain 'cut-off' frequency. Above this frequency, the wave amplitude decayed exponentially, and no energy could be transmitted in steady state, the force being 90 degrees out of phase with the velocity.

In a wave guide, there is a similar phenomenon of wave decay, but in this case the decay occurs at frequencies below a certain cut-off frequency. This decay, as we shall see, is intimately related to the phenomenon of total reflection at a boundary, and it is instructive to review this since it provides physical insight into the transmission characteristics of a wave guide.

Display 15.1.

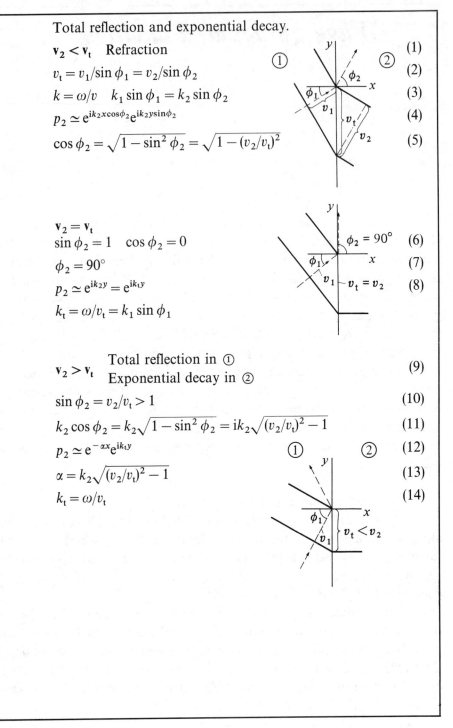

Total reflection and exponential decay.

$\mathbf{v}_2 < \mathbf{v}_t$ Refraction (1)

$$v_t = v_1/\sin\phi_1 = v_2/\sin\phi_2 \qquad (2)$$

$$k = \omega/v \quad k_1 \sin\phi_1 = k_2 \sin\phi_2 \qquad (3)$$

$$p_2 \simeq e^{ik_2 x\cos\phi_2} e^{ik_2 y\sin\phi_2} \qquad (4)$$

$$\cos\phi_2 = \sqrt{1-\sin^2\phi_2} = \sqrt{1-(v_2/v_t)^2} \qquad (5)$$

$\mathbf{v}_2 = \mathbf{v}_t$

$$\sin\phi_2 = 1 \quad \cos\phi_2 = 0 \qquad (6)$$

$$\phi_2 = 90° \qquad (7)$$

$$p_2 \simeq e^{ik_2 y} = e^{ik_t y} \qquad (8)$$

$$k_t = \omega/v_t = k_1 \sin\phi_1$$

$\mathbf{v}_2 > \mathbf{v}_t$ Total reflection in ① Exponential decay in ② (9)

$$\sin\phi_2 = v_2/v_t > 1 \qquad (10)$$

$$k_2\cos\phi_2 = k_2\sqrt{1-\sin^2\phi_2} = ik_2\sqrt{(v_2/v_t)^2-1} \qquad (11)$$

$$p_2 \simeq e^{-\alpha x} e^{ik_t y} \qquad (12)$$

$$\alpha = k_2\sqrt{(v_2/v_t)^2-1} \qquad (13)$$

$$k_t = \omega/v_t \qquad (14)$$

Thus, in Display 15.1, we reconsider the reflection and refraction of an incident wave at the boundary between two regions with different wave speeds v_1 and v_2. The wave, here considered to be a sound wave, is incident in region 1 with the angle of incidence ϕ_1. The trace velocity, as we recall from Chapter 12, is the velocity of the wave 'ripple' along the boundary, created by the incident wave, and the requirement

$$v_t = v_1/\sin(\phi_1) = v_2/\sin(\phi_2)$$

that the trace velocities in the two regions be the same is the law of refraction, which we can express also as

$$k_1 \sin(\phi_1) = k_2 \sin(\phi_2) \tag{1}$$

where $k = \omega/v$ is the propagation constant.

In region 1 the trace velocity $v/\sin(\phi_1)$ is always larger than the wave speed v_1. In region 2, on the other hand, there is no such condition; the wave speed v_2 may be larger or smaller than trace velocity v_t. In normal refraction, we have $v_2 < v_t$, which is the case illustrated in the first figure in the display. We have chosen $v_2 > v_1$, so that the angle of refraction is larger than the angle of incidence.

The x- and y-components of the propagation vector in region 2 are $k_{2x} = k_2 \cos(\phi_2)$ and $k_{2y} = k_2 \sin(\phi_2)$ and the x, y-dependence $\exp(ik_{2x}x)$. $\exp(ik_{2y}y)$ of the complex amplitude of the refracted field can be written in the form

$$p_2 \simeq \exp[ik_2 x \cos(\phi_2)] \exp[ik_2 y \sin(\phi_2)] \tag{2}$$

$$\cos(\phi_2) = [1 - \sin^2(\phi_2)]^{\frac{1}{2}} = [1 - (v_2/v_t)^2]^{\frac{1}{2}} \tag{3}$$

If $v_2 < v_t$, there is always a real value of the angle ϕ_2, and in the limiting case $v_2 = v_t$, we have $\phi_2 = \pi/2$, and the direction of the refracted wave is along the boundary.

If $v_2 > v_t$, $\cos(\phi_2)$ becomes imaginary, and, according to Eq. (2), the wave amplitude of the refracted wave will decay exponentially with the penetration depth x. The surfaces of constant amplitude are then parallel with the boundary, and the surfaces of constant phase are perpendicular thereto. Under these conditions, no steady power is carried by the refracted wave, and the incident wave is totally reflected.

If two waves are incident under the angles ϕ_1 and $-\phi_1$, as shown in Display 15.2, the wave ripple created along the boundary will be a standing wave with nodes and antinodes. In the display the amplitude distribution shown refers to the axial fluid velocity. This distribution can be regarded as the source of the wave transmitted into the second region.

The transmitted wave produced by the two incident waves, or, equi-

Display 15.2.

Wave field between two rigid planes as the superposition of two refracted plane waves.

Frequency ν

Wave speed v

Trace velocity v_t

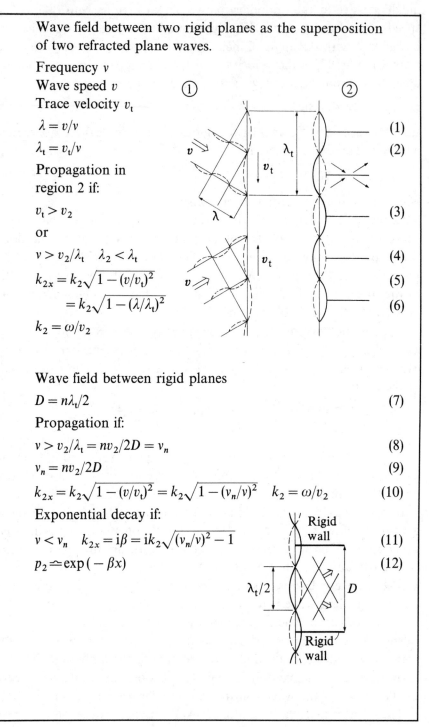

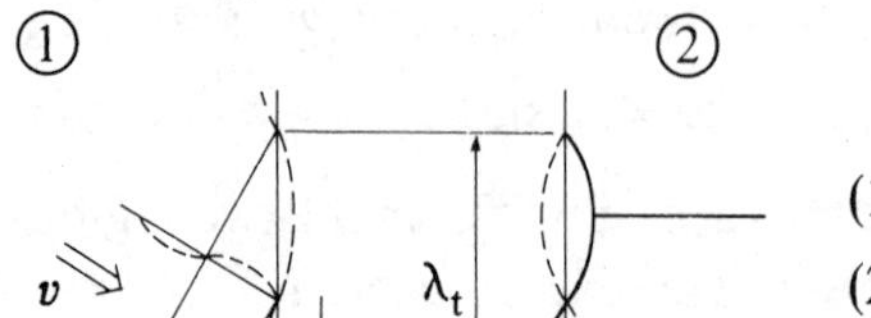

$\lambda = v/\nu$ (1)

$\lambda_t = v_t/\nu$ (2)

Propagation in region 2 if:

$v_t > v_2$ (3)

or

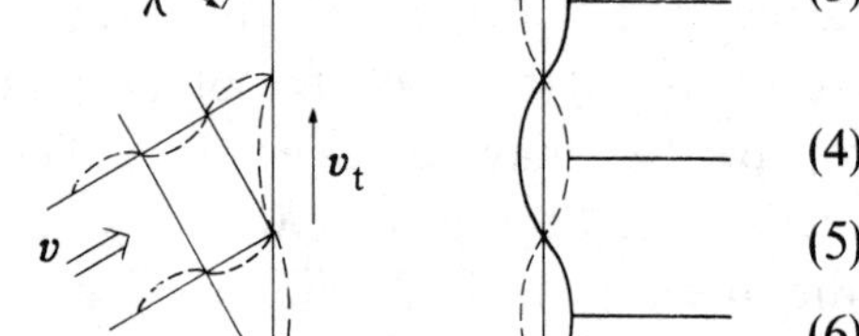

$\nu > v_2/\lambda_t \quad \lambda_2 < \lambda_t$ (4)

$k_{2x} = k_2\sqrt{1-(v/v_t)^2}$ (5)

$\qquad = k_2\sqrt{1-(\lambda/\lambda_t)^2}$ (6)

$k_2 = \omega/v_2$

Wave field between rigid planes

$D = n\lambda_t/2$ (7)

Propagation if:

$\nu > v_2/\lambda_t = nv_2/2D = \nu_n$ (8)

$\nu_n = nv_2/2D$ (9)

$k_{2x} = k_2\sqrt{1-(v/v_t)^2} = k_2\sqrt{1-(\nu_n/\nu)^2} \quad k_2 = \omega/v_2$ (10)

Exponential decay if:

$\nu < \nu_n \quad k_{2x} = i\beta = ik_2\sqrt{(\nu_n/\nu)^2 - 1}$ (11)

$p_2 \simeq \exp(-\beta x)$ (12)

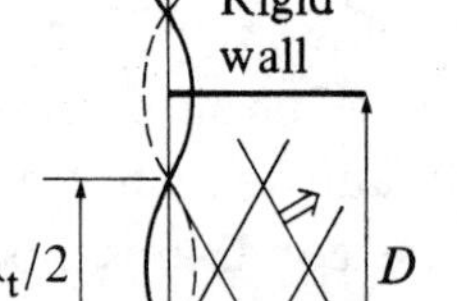

valently, by the boundary ripple, will be a standing wave also with respect to the dependence on the y-coordinate. The horizontal lines in the figure indicate nodal planes, at which the velocity component in the y-direction is zero.

Then, insertion of rigid boundaries in these nodal planes will not alter the wave field. The region between two adjacent planes a distance D apart, as shown in the second figure in the display, can be regarded as a two-dimensional duct, which is driven at one end by a source with an axial velocity amplitude distribution represented by the standing wave ripple along the boundary.

The distance between nodal planes is an integer number of half wavelengths of the boundary ripple,

$$D = n\lambda_t/2 = nv_t/2\nu \tag{4}$$

where ν is the frequency.

As we have seen, the condition for a propagated as opposed to a decaying refracted field is $v_2 < v_t = \lambda_t \nu$. Accounting for Eq. (4), this condition can be written

$$\nu > v_2/\lambda_t = nv_2/2D = \nu_n \tag{5}$$

where we have introduced the 'cut-off' frequency ν_n of the nth mode in the duct.

This mode is characterized by n nodal pressure planes in the duct. The duct walls are always nodal planes for the velocity field and anti-nodes for the pressure. For $n = 1$, the nodal pressure plane is along the middle of the duct, parallel with the boundaries. It should be noted, that for $n = 0$ we obtain the plane wave considered in previous chapters, in which the y-component of the velocity is zero and the pressure is uniform across the duct.

If the frequency exceeds the cut-off frequency of the nth mode, it follows from Eq. (3) that $k_{2_x} = k_2 \cos(\phi_2)$ can be written

$$k_{2x} = k_2[1 - (v_2/v_t)^2]^{\frac{1}{2}} = k_2[1 - (\nu_n/\nu)^2]^{\frac{1}{2}} \tag{6}$$

$$k_2 = \omega/v_2$$

will be real, and the magnitude of the wave amplitude will be independent of x. At frequencies below the cut-off frequency, on the other hand, the propagation constant becomes imaginary, and the amplitude will decay exponentially with x,

$$p_2(x) = p_2(0)\exp(-\beta_n x) \tag{7}$$

$$\beta_n = k_2[(\nu_n/\nu)^2 - 1]^{\frac{1}{2}}$$

The fundamental mode, $n = 0$, corresponds to an infinite value of the trace velocity and the ripple wavelength, and this means that the velocity amplitude at the beginning of the duct is uniform. In terms of the refraction analogy, this mode corresponds to the refracted wave at normal incidence for which there is no cut-off frequency in the acoustic case.

The decay of the wave in a wave guide can be viewed also as a result of destructive interference between waves emitted from the positive and negative portions of the source (ripple) at the beginning of the guide. If the width D is less than half a wavelength, constructive interference cannot occur at any location in the duct, and this results in exponential decay.

There is also a connection between the field in a duct and the field from an array of sources, discussed in Chapter 11. If we place a line source midway between the duct walls in our two dimensional duct, parallel with the walls, the wave field produced in the duct is the same as the field from the source and its array of image sources obtained through multiple reflections of the source in the duct walls. The condition for obtaining a propagating mode in the duct is then analogous to the condition for the formation of a lobe in the radiation field of the array.

15.2 Acoustic waves in a rectangular wave guide

After the introductory observations in the last section, we now proceed to a more conventional treatment of wave propagation in wave guides, and start with sound waves in a rectangular duct with rigid walls. The duct is along the x-axis, and the duct walls are at $y=0, y=a$ and $z=0$, $z=b$, as indicated in Display 15.3.

As discussed in Chapter 10, the sound pressure $p(x, y, z, t)$ satisfies the three dimensional wave equation

$$\nabla^2 p = \partial^2 p/\partial x^2 + \partial^2 p/\partial y^2 + \partial^2 p/\partial z^2 = (1/v^2)\partial^2 p/\partial t^2 \tag{1}$$

and with harmonic time dependence, the corresponding (Helmholtz) equation for the complex amplitude $p(x, y, z, \omega)$ is

$$\nabla^2 p + k^2 p = 0 \quad \text{or} \quad \partial^2 p/\partial x^2 + \partial^2 p/\partial y^2 + \partial^2 p/\partial z^2 + k^2 p = 0 \tag{2}$$
$$k = \omega/v$$

From the solution to this equation, the corresponding complex amplitude $u(x, y, z, \omega)$ of the velocity field follows from the momentum equation $\rho\partial\mathbf{u}/\partial t = -\nabla p$, which, for the complex amplitudes becomes

$$-\rho\omega\mathbf{u} = -\nabla p \tag{3}$$
$$-i\rho\omega u_x = -\partial p/\partial x, \quad \text{etc.}$$

There is an infinite set of solutions to the wave equation, but we are interested only in those which have zero velocity normal to the duct walls.

Display 15.3.

Sound propagation in a rectangular wave guide.

$$\nabla^2 p = \partial^2 p/\partial x^2 + \partial^2 p/\partial y^2 + \partial^2 p/\partial z^2 = (1/v^2)\partial^2 p/\partial t^2 \quad (1)$$

$$p(x, y, z, t) = \mathrm{Re}\,\{p(x, y, z)\cdot \mathrm{e}^{-\mathrm{i}\omega t}\} \quad (2)$$

$$\nabla^2 p + (\omega/v)^2 p = 0 \quad (3)$$

$$\rho \partial \mathbf{u}/\partial t = -\nabla p \quad (4)$$

$$-\mathrm{i}\omega\rho u_x = -\partial p/\partial x, \text{ etc.} \quad (5)$$

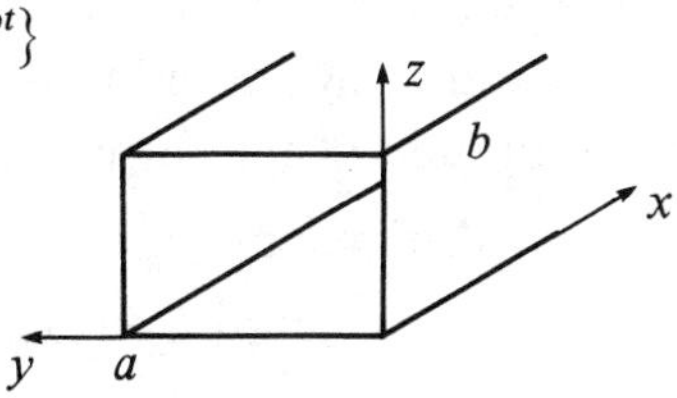

Boundary conditions:

$$\partial p/\partial y = 0 \quad \text{for} \quad y = 0 \quad \text{and} \quad y = a \quad (6)$$

$$\partial p/\partial z = 0 \quad \text{for} \quad z = 0 \quad \text{and} \quad z = b \quad (7)$$

$$p(x, y, z) = X(x)Y(y)Z(z) \quad (8)$$

$$X''/X + Y''/Y + Z''/Z = -(\omega/v)^2 \quad (X'' = \mathrm{d}^2X/\mathrm{d}x^2, \text{ etc.}) \quad (9)$$

$$X''/X = -k_x^2 \quad Y''/Y = -k_y^2 \quad Z''/Z = -k_z^2 \quad (10)$$

$$k_x^2 + k_y^2 + k_z^2 = (\omega/v)^2 \quad (11)$$

$$P_{mn} = A_{mn} \cos(k_y y)\cos(k_z z)\mathrm{e}^{\mathrm{i}k_x x} \quad (12)$$

$$k_y a = m\pi \quad k_y = m\pi/a \quad k_z b = n\pi \quad k_z = n\pi/b \quad (13)$$

$$k_x = k_{mn} = \sqrt{(\omega/v)^2 - k_m^2 - k_n^2} = (\omega/v)\sqrt{1 - (\omega_{mn}/\omega)^2} \quad (14)$$

$$\omega_{mn} = v\sqrt{k_m^2 + k_n^2} = v\sqrt{(m\pi/a)^2 + (n\pi/b)^2} \quad (15)$$

General wave field:

$$p = \sum\sum p_{mn} \quad \text{Sum over } m, n \quad (16)$$

From the set of solutions thus obtained, it remains to select the one which also satisfies the conditions at the source (say at $x=0$) and at the end of the duct, which may be $x=\infty$.

As indicated in the previous section, we expect the wave field in the duct to be composed of waves reflected back and forth from the duct walls. In Chapters 10 and 12 we noted that a plane wave in three dimensions has a complex amplitude of the form

$$\begin{aligned}\exp(i\mathbf{k}\cdot\mathbf{r}) &= \exp(ik_x x + ik_y y + ik_z z)\\ &= \exp(ik_x x)\exp(ik_y y)\exp(ik_z z),\end{aligned}$$

and the wave field in the duct is expected to be a superposition of such waves, but the combination, of course, must be such that the various boundary conditions are satisfied.

For a wave travelling in the positive x-direction, the x-dependence of the complex amplitude must be of the form $\exp(ik_x x)$. In the y- and z-directions, on the other hand, we must have waves travelling in both the negative and positive directions, producing standing waves as represented by the factors $\cos(k_y y)$ and $\cos(k_z z)$.

Since we have placed the duct walls at $y=0$ and $z=0$ these factors (without any phase angles) satisfy the requirement of zero normal velocity at the walls at $y=0$ and $z=0$, since these velocity components are proportional to $\partial p/\partial y$ and $\partial p/\partial z$, i.e. proportional to $\sin(k_y y)$ and $\sin(k_z z)$.

In order for the normal velocity to be zero also at the walls at $y=a$ and $z=b$ we must have

$$k_y a = m\pi \quad \text{and} \quad k_z b = n\pi \tag{4}$$

where m and n are integers. Each combination m, n specifies a pressure amplitude pattern across the duct, with m and n indicating the number of nodal pressure planes which are encountered in going across the duct in the y- and z-directions, respectively. This pressure pattern propagates in the x-direction, as specified by the factor $\exp(ik_x x)$, where k_x is determined below, and it is referred to as the (m, n) mode in the duct.

With a wave function of the form $\cos(k_y y)\cos(k_z z)\exp(ik_x x)$ inserted into Eq. (2), we find

$$k_x^2 + k_y^2 + k_z^2 = (\omega/v)^2$$

and it follows that for the (m, n) mode we have

$$\begin{aligned}k_x = k_{mn} &= [(\omega/v)^2 - k_m^2 - k_n^2]^{\frac{1}{2}} = (\omega/v)[1-(\omega_{mn}/\omega)^2]^{\frac{1}{2}} \qquad (5)\\ \omega_{mn} &= v(k_m^2 + k_n^2)^{\frac{1}{2}} = v[(m\pi/a)^2 + (n\pi/b)^2]^{\frac{1}{2}}\end{aligned}$$

The expression for the complex amplitude of a general wave field travelling in the positive x-direction is a linear combination of (m, n) modes

$$p = \sum p_{mn} = \sum A_{mn} \cos(m\pi y/a) \cos(n\pi z/b) \exp(ik_{mn}x) \tag{6}$$

where k_{mn} is given in Eq. (5). The sum extends over m and n from zero to infinity.

From the expression for the cut-off frequency ω_{mn} in Eq. (5) it follows that the cut-off wavelength for the $(m, 0)$ mode is $\lambda_{m0} = 2\pi v/\omega_{m0} = 2a/m$, i.e. an integer number of half wavelengths will 'fit' between the duct walls.

The propagation constant k_x for the (m, n) mode is real only if the frequency is above the cut-off frequency ω_{mn} for the mode; the wave then will propagate along the duct with constant amplitude. Below the cut-off frequency, on the other hand, k_x is imaginary, and the wave amplitude decays exponentially with x. The special case $m = 0$, $n = 0$ corresponds to the plane wave, which propagates at all frequencies. For all the other modes, often called 'higher order' modes, the duct acts like a high-pass filter.

The axial velocity distribution in the (m, n) mode is obtained from Eq. (3) and if the source at $x = 0$ has the same velocity distribution, only the (m, n) mode will be produced. For an arbitrary amplitude distribution at $x = 0$, several modes, both propagating and decaying, usually are generated.

The complex amplitude for the general wave field in Eq. (6) is a double sum of modes. The amplitude constants A_{mn} are determined so as to make the total axial velocity distribution in the duct match that of the source at the beginning of the duct. It is possible, of course, that the source is specified in terms of the pressure amplitude distribution, and the mode amplitudes then have to be chosen accordingly. The matching of the source field and the general expression in Eq. (6) is accomplished through a Fourier expansion of the source in terms of the duct modes.

Except for the plane wave, $(m, n) = (0, 0)$, the relation $\omega = \omega(k_x)$ between the frequency and the axial propagation constant is nonlinear, and in this regard the wave guide modes are dispersive, as discussed below.

Phase and group velocity. The wave pattern in a mode, defined by the amplitude distribution over the area of the duct, propagates in the positive x-direction with the phase velocity $v_{\text{p}} = \omega/k_x$. Having obtained the expression for k_x in Eq. (5), we obtain for the phase velocity

$$v_{\text{p}} = \omega/k_x = v/[1 - (\omega_{mn}/\omega)^2]^{\frac{1}{2}} \tag{7}$$

If we superimpose two (m, n) modes of slightly different frequencies we get beats, and the beat pattern propagates with the group velocity $v_{\text{g}} = \text{d}\omega/\text{d}k_x$ which follows by analogy with the discussion summarized in

Display 15.4.

Geometrical interpretation of phase velocity and group velocity in a wave guide.

$$k_x = \sqrt{(\omega/v)^2 - (m\pi/a)^2 - (n\pi/b)^2} = (\omega/v)\sqrt{1 - (\omega_{mn}/\omega)^2} \quad (1)$$

$$\omega_{mn} = v\sqrt{(m\pi/a)^2 + (n\pi/b)^2} \quad (2)$$

Phase velocity

$$v_p = \omega/k_x = v/\sqrt{1 - (\omega_{mn}/\omega)^2} \quad (3)$$

Group velocity

$$v_g = d\omega/dk_x = v\sqrt{1 - (\omega_{mn}/\omega)^2} \quad (4)$$

$$v_p v_g = v^2 \quad (5)$$

Special case, m = 0, n = 1

$$p_{01} = A\cos(\pi z/b)\cdot\exp(ik_{01}x) \quad (6)$$

$$\nu_{01} = \omega_{01}/2\pi = v/2b \quad (7)$$

$$\nu_{01}/\nu = v/2b\nu = \lambda/2b \quad (8)$$

$$k_{01} = (\omega/v)\sqrt{1 - (\lambda/2b)^2} = k\cdot\cos\phi = \omega/v_p \quad (9)$$

$$\sin\phi = \lambda/2b \quad k = \omega/v \quad (10)$$

$$v_p = v/\cos\phi \quad (11)$$

$$v_g = v\cos\phi \quad (12)$$

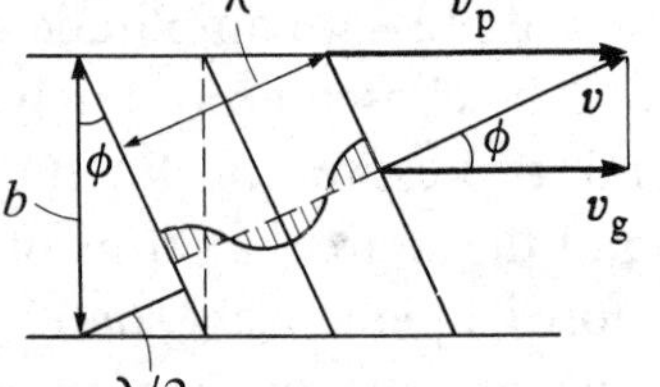

At cut-off, mode (0, 1):

$$\lambda_c = 2b \quad \phi = 90^\circ \quad (13)$$

For mode (0, n):

$$\sin\phi_n = n\lambda/2b \quad \lambda_c = 2b/n \quad (14)$$

Display 14.4. From the dispersion relation (5), we obtain

$$v_g = d\omega/dk_x = v[1-(\omega_{mn}/\omega)^2]^{\frac{1}{2}} \tag{8}$$

It is interesting to note that the phase velocity is always greater than, and the group velocity always smaller than, the free space wave speed v. Actually, the product of the two velocities is equal to the square of the wave speed

$$v_p v_g = v^2 \tag{9}$$

It is instructive to try to interpret v_p and v_g geometrically. For the purpose of illustration, we treat the special case of the (0, 1) mode, which has the complex pressure amplitude

$$p_{01} = A\cos(\pi z/b)\exp(ik_{01}x) \tag{10}$$

The cut-off frequency, obtained from Eq. (5), is $\nu_{01} = \omega_{01}/2\pi = v/2b$, and introducing the free space wavelength $\lambda = v\nu$, we obtain

$$k_x = k_{01} = (\omega/v)[1-(\omega_{01}/\omega)^2]^{\frac{1}{2}} = (\omega/v)[1-(\lambda/2b)^2]^{\frac{1}{2}} \tag{11}$$

With reference to the discussion summarized in Display 15.1, the (0, 1) mode can be considered to be a superposition of two intersecting plane waves, or, equivalently, as the field resulting from multiple reflections back and forth between the walls in the duct. Each of these plane waves has the propagation constant $k = \omega/v$ with a direction which makes an angle ϕ with the axis of the duct. In terms of this angle we can set $k_x = k_{01} = k\cos(\phi)$.

From the expression for k_{01}, it follows that $\sin(\phi) = \lambda/2b$, as shown in the figure in Display 15.4. The phase velocity v_p in Eq. (7) then can be interpreted as the trace velocity $v/\cos(\phi)$ along the duct axis of the plane wave and the group velocity $v_g = v\cos(\phi)$ as the component of the plane wave velocity along the axis of the duct, as shown.

For the $(0, n)$ wave we find in analogous manner, that the elementary plane waves involved travel in a direction which makes an angle ϕ_n with the x-axis, given by $\sin(\phi_n) = n\lambda/2b$.

Example. The cut-off phenomenon for a sound wave in a duct can be demonstrated in a simple manner as shown schematically in Display 15.5. A rectangular duct contains a sound source of two identical loudspeakers mounted side by side at one end of the duct. The speakers are driven by the same oscillator-amplifier and normally are connected in such a manner that they operate in phase. The wave produced in the duct is then dominated by the plane mode (0, 0). This wave propagates through the duct without attenuation at all frequencies.

If the leads between the amplifier and one of the speakers are inter-

Display 15.5.

Demonstration of the propagation and decay of the (0, 1) acoustic mode in a rectangular wave guide.

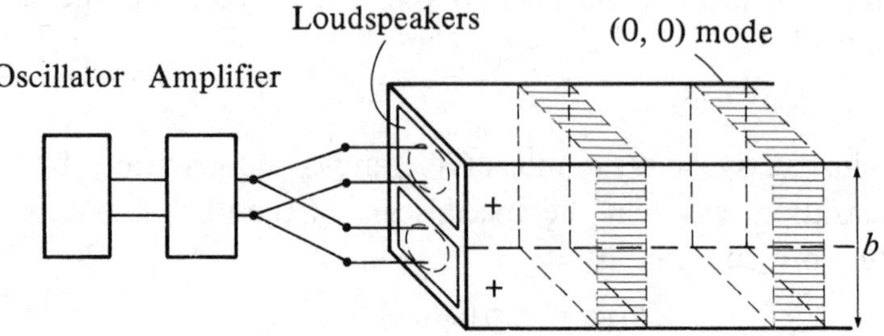

Speakers in phase: Plane wave propagation at all frequencies.

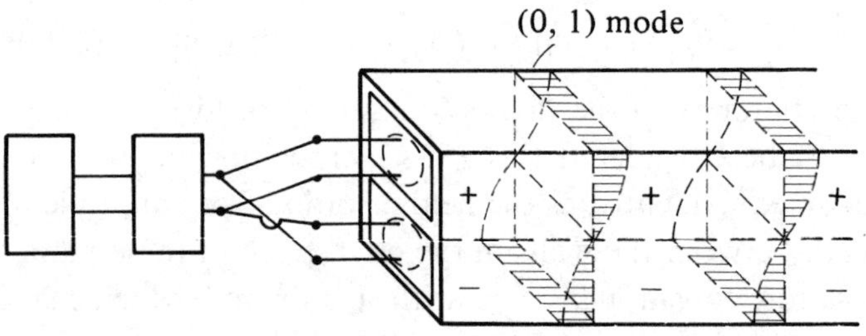

Speakers 180 degrees out of phase: (0, 1) mode generated.

$b = 25\,\text{cm}$

Cut-off frequency $\quad \nu_{01} = v/2b = 34\,000/2.25 = 684\,\text{Hz}$ (1)

Driving frequency $\nu = 400\,\text{Hz}$ (i.e. below cut-off).

$p_{01} \simeq \exp(-\beta x)$ (2)

$\beta = (\omega/v)\sqrt{(\nu_{01}/\nu)^2 - 1} = 1.39\,\omega/v \simeq 0.1\,\text{cm}^{-1}$ (3)

Driving frequency $\nu = 1000\,\text{Hz}$ (i.e. above cut-off).

$p_{01} \simeq \exp(\mathrm{i}k_{01}x)$ (4)

$k_{01} = (\omega/v)\sqrt{1 - (\nu_{01}/\nu)2} = 0.73\,\omega/v$ (5)

$v_\text{p} = 1.37\,v$ (6)

$v_\text{g} = 0.73\,v$ (7)

changed, as indicated, a phase shift of 180 degrees is introduced, and the speakers will move in opposition so that the average value of the axial velocity in the duct will be zero. This means that the plane wave will not be excited. Rather, the dominant mode will be the (0, 1) mode, with a pressure nodal plane along the center of the duct, as indicated.

In a particular case, the duct height was $b = 25$ cm, corresponding to a cut-off frequency of the (0, 1) mode of 684 Hz. With the speakers operating 180 degrees out of phase, the sound field in the duct will decay at frequencies below 684 Hz and propagate with negligible attenuation at frequencies above 684 Hz. By increasing the frequency through the cut-off value, a marked change in the acoustic output from the duct can readily be observed.

At frequencies below cut-off the amplitude at the exit end of the duct of the (0, 1) mode is reduced as a result of the exponential decay, and the sound that can be heard outside the duct is due in part to this decayed field and in part to the weak plane wave which is still present in the duct because of some unavoidable unequality of the speakers. Ordinarily the (0, 1) contribution dominates, however, and if one of the speakers is disconnected (so that the plane wave component generated by the remaining speaker is not cancelled by the second speaker), a substantial increase in the sound pressure is observed.

As a numerical example, we note that at 400 Hz, the sound pressure decreases by a factor of 1/e in a distance of 10 cm and with a duct length of 100 cm, i.e. 4 times the height, the (0, 1) mode has almost been 'filtered' out by the duct, which acts as a high-pass filter.

At a frequency of 1000 Hz, the (0, 1) mode propagates without attenuation, and a survey of the pressure amplitude across the duct with a microphone clearly shows the node at the center of the duct. (The node exists, of course, also below cut-off).

15.3 EM waves in a rectangular wave guide

With reference to Display 15.6 we consider a wave guide with conducting walls at $y = 0$, a_2 and $z = 0$, a_3, and for simplicity the conductivity is assumed to be infinite. For most metallic conductors this is a good approximation in calculating the field distribution within the wave guide. Under these conditions, the tangential component of the electric field must be zero at the walls. (Since this component is continuous across a boundary, a value different from zero would lead to an infinite current in the walls).

An important consequence of this boundary condition is that there can be no plane wave travelling in the axial direction of the duct. The reason

Display 15.6.

EM waves in a rectangular wave guide.

$$\nabla^2 E_y = (1/c^2)\partial^2 E_y/\partial t^2 \tag{1}$$

$$\partial^2 E_y/\partial x^2 + \partial^2 E_y/\partial y^2 + \partial^2 E_y/\partial z^2 + (\omega/c)^2 E = 0 \tag{2}$$

TE modes $E_x = 0$

$$\operatorname{div}\mathbf{D} = 0 \quad \partial E_y/\partial y + \partial E_z/\partial z = 0 \tag{3}$$

$$E_y = E_2 \cos(k_y y)\sin(k_z z)\cdot\exp(ik_x x) \tag{4}$$

$$E_z = E_3 \sin(k_y y)\cos(k_z z)\cdot\exp(ik_x x) \tag{5}$$

$$k_y \equiv k_m = m\pi/a_2 \quad k_z \equiv k_n = n\pi/a_3 \quad (n, m = 1, 2, \ldots) \tag{6}$$

$$k_x = \sqrt{(\omega/c)^2 - k_m^2 - k_n^2} \tag{7}$$

$$\text{From Eq. (3): } k_m E_2 = -k_n E_3 \tag{8}$$

$$\operatorname{curl}\mathbf{E} = -\partial\mathbf{B}/\partial t \tag{9}$$

$$H_x = A\cos(k_m y)\cos(k_n z)\exp(ik_x x) \tag{10}$$

$$H_y = B\sin(k_m y)\cos(k_n z)\exp(ik_x x) \tag{11}$$

$$H_z = C\cos(k_m y)\sin(k_n z)\exp(ik_x x) \tag{12}$$

$$\omega_{mn} = c\sqrt{k_m^2 + k_n^2} = c\sqrt{(m\pi/a_2)^2 + (n\pi/a_3)^2} \tag{13}$$

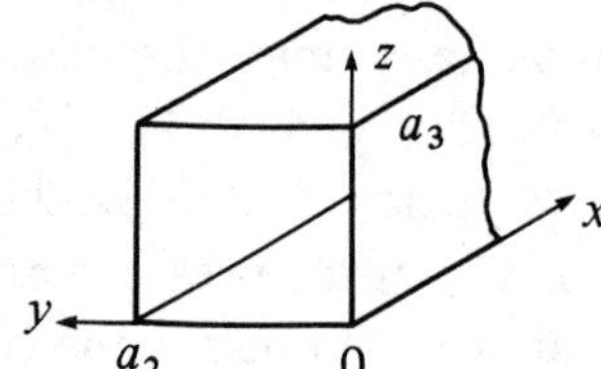

TM modes $H_x = 0$

$$\operatorname{div}\mathbf{B} = 0 \quad \partial H_y/\partial y + \partial H_z/\partial z = 0 \tag{14}$$

$$H_y = H_2 \sin(k_m y)\cos(k_n z)\cdot\exp(ik_x x) \tag{15}$$

$$H_z = H_3 \cos(k_m y)\sin(k_n z)\exp(ik_x x) \quad (k_m H_2 = -k_n H_3) \tag{16}$$

$$\operatorname{curl}\mathbf{H} = \partial\mathbf{D}/\partial t \tag{17}$$

$$E_x = P\sin(k_m y)\sin(k_n z)\cdot\exp(ik_x x) \tag{18}$$

$$E_y = Q\cos(k_m y)\sin(k_n z)\cdot\exp(ik_x x) \tag{19}$$

$$E_z = R\sin(k_m y)\cos(k_n z)\cdot\exp(ik_x x) \tag{20}$$

is that for such a wave the E-field would be constant across the area of the duct and perpendicular to the axis, and it would be impossible then to satisfy the condition of zero tangential component of this field at all the walls. This conclusion is not valid for a coaxial wave guide, however, as was demonstrated in Chapter 9, where we had a (locally) plane wave between the conductors with the E-field in the radial and the H-field in the circumferential direction.

Although there can be no plane EM wave travelling along the axis of the rectangular wave guide, we expect, as for the higher-order acoustic modes in the previous section, that an EM mode in the wave guide can be the superposition of plane waves travelling in an off-axis direction being reflected back and forth between the walls of the wave guide.

In a quantitative study of such wave guide modes, we have to construct a wave field which satisfies the field equations and the boundary conditions. As was shown in Chapter 9, the field components satisfy the wave equation. Thus, for the y-component of the E-field we have

$$\nabla^2 E_y = (1/c^2)\partial^2 E_y/\partial t^2 \tag{1}$$

with the corresponding complex amplitude equation

$$\partial^2 E_y/\partial x^2 + \partial^2 E_y/\partial y^2 + \partial^2 E_y/\partial z^2 + (\omega/c)^2 E_y = 0 \tag{2}$$

In general, the E- and H-field will have components along all the axes. It is useful, however, to consider separately the two important special cases when either the electric or the magnetic fields are in the transverse direction. The corresponding wave modes are usually referred to as TE (Transverse Electric) and TM (Transverse Magnetic) modes. The general wave field can be obtained as a superposition of these modes.

TE modes. As in the acoustic case, the x-axis is along the length of the wave guide and the walls are at $y = 0$, a_2, and $z = 0$, a_3, as already indicated. Since the x-component of the E-field is zero in a TE mode and since $\mathrm{div}\,(\mathbf{E}) = 0$, the y- and z-components of the field must be related so that

$$\partial E_y/\partial y + \partial E_z/\partial z = 0 \quad [\mathrm{div}\,(\mathbf{E}) = 0] \tag{3}$$

By analogy with the discussion in the previous section of the acoustic higher order modes in a rectangular duct, we find also that to satisfy the wave equation and the boundary conditions of zero tangential electric field at each wall, the electric field components must be of the form

$$E_y = E_2 \cos(k_y y) \sin(k_z z) \cdot \exp(ik_x x) \tag{4}$$

$$E_z = E_3 \sin(k_y y) \cos(k_z z) \cdot \exp(ik_x x) \tag{5}$$

$$k_y = k_m = m\pi/a_2 \quad k_z = k_n = n\pi/a \quad (n, m = 1, 2, \ldots) \tag{6}$$

$$k_m E_2 = -k_n E_3 \tag{7}$$

$$k_x = k_{mn} = [(\omega/c)^2 - k_m^2 - k_n^2]^{\frac{1}{2}} = (\omega/c)[1 - (\omega_{mn}/\omega)^2]^{\frac{1}{2}} \tag{8}$$

where Eq. (7) follows from Eq. (3) and Eq. (8) from Eq. (2), and the cut-off frequency ω_{mn} is given by

$$\omega_{mn}^2 = c^2[k_m^2 + k_n^2] = c^2[(m\pi/a_2)^2 + (n\pi/a_3)^2] \tag{9}$$

The corresponding magnetic field components are obtained from the Maxwell equation curl$(\mathbf{E}) = -\mu_0 \partial H/\partial t$,

$$H_x = A\cos(k_m y)\cos(k_n z)\cdot\exp(ik_x x) \tag{10}$$

$$H_y = B\sin(k_m y)\cos(k_n z)\cdot\exp(ik_x x) \tag{11}$$

$$H_z = C\cos(k_m y)\sin(k_n z)\cdot\exp(ik_x x) \tag{12}$$

Where $A = i(E_2/Z)(k_m^2 + k_n^2)/k_n k$, $B = (E_2/Z)k_x k_m/k_n k$, $C = (E_2/Z)(k_x/k)$, with $Z = \mu_0 c$ and $k = \omega/c$, to be checked in one of the problems.

As in the acoustic case, the mode (m, n) propagates without attenuation only if the frequency exceeds the cut-off value ω_{mn} in Eq. (9). It should be noted, that with $m = 0$, $n = 0$ the field is zero, indicating that the plane wave component is absent. For the acoustic (0, 0) mode, on the other hand, the cut-off frequency is zero, and propagation without decay occurs at all frequencies.

TM modes. In this case the x-component of the magnetic field is zero, and since div$(H) = 0$, the y- and z-components must be related as

$$\partial H_y/\partial y + \partial H_z/\partial z = 0 \quad [\text{div}(\mathbf{H}) = 0] \tag{13}$$

The field components are obtained in much the same way as for the TE mode, and we leave it for one of the problems to show that

$$H_y = H_2\sin(k_m y)\cos(k_n z)\cdot\exp(ik_x x) \tag{14}$$

$$H_z = H_3\cos(k_m y)\sin(k_n z)\cdot\exp(ik_x x) \tag{15}$$

$$k_m H_2 = -k_n H_3 \tag{16}$$

where Eq. (16) follows from Eq. (13).

The electric field components follow from the Maxwell equation curl$(\mathbf{H}) = \varepsilon_0 \partial\mathbf{E}/\partial t$

$$E_x = P\sin(k_m y)\sin(k_n z)\cdot\exp(ik_x x) \tag{17}$$

$$E_y = Q\cos(k_m y)\sin(k_n z)\cdot\exp(ik_x x) \tag{18}$$

$$E_z = R\sin(k_m y)\cos(k_n z)\cdot\exp(ik_x x) \tag{19}$$

where k_m, k_n, and k_x are the same as in Eqs. (6) and (8). It should be noted that the x-component of the electric field goes to zero at the duct walls as required by the boundary conditions. Again, we leave the calculation of the amplitude coefficients Q, P, and R (in terms of H) for one of the problems.

15.4 Normal modes in a rectangular cavity

It is apparent that normal modes in cavities play an important role in the study of the characteristics of various electromagnetic and acoustics resonators; the acoustics of lecture rooms and concert halls are particular examples. Less obvious is the basic role that normal modes play in our understanding of the spectrum of black body radiation, the specific heat of solids, and many other problems in statistical physics.

In our quantitative analysis of normal modes we can make use of results already obtained for wave propagation in a duct in the previous section although we shall now be concerned with the 'free' rather than the forced harmonic motion of the modes.

To convert the wave guide into a cavity we merely have to close both ends of the wave guide, and for the rectangular wave guide we do that by placing walls at $x = 0$ and $x = a_1$, as indicated in Display 15.7. Then, instead of having a travelling wave in the x-direction, as expressed by the factor $\exp(ik_x x)$, we now have a standing wave and a corresponding wave function.

We consider as an example sound waves in a rectangular cavity with rigid walls. In this case the standing wave function replacing $\exp(ik_x x)$ will be $\cos(l\pi/a_1)$, where $l = 0, 1, 2, \ldots$, and it follows from Eq. 2(6) that the complex amplitude functions for the normal modes in the cavity will be

$$p_{lmn} = A\cos(k_x x)\cos(k_y y)\cos(k_z z) \tag{1}$$

$$k_x a_1 = l\pi \quad k_y a_2 = m\pi \quad k_z a_3 = n\pi \tag{2}$$

Insertion of this wave function into the wave equation for the complex amplitude

$$\partial^2 p/\partial x^2 + \partial^2 p/\partial y^2 + \partial^2 p/\partial z^2 + (\omega/v)^2 = 0 \tag{3}$$

yields the following expression for the normal mode frequencies

$$\omega_{lmn} = v[(l\pi/a_1)^2 + (m\pi/a_2)^2 + (n\pi/a_3)^2]^{\frac{1}{2}} \tag{4}$$

where v is the wave speed.

Each combination of the integers l, m, n represents a characteristic frequency of oscillation of the field in the cavity. The corresponding spatial variation of the field is determined by k_x, k_y, and k_z, which can be interpreted as the components of a propagation vector k with the tip at the points k_x, k_y, k_z.

For example, the mode $l, 0, 0$, with the frequency $\omega_l = v\pi/a_1$, corresponds to a standing wave between the walls perpendicular to the x-axis with a

Display 15.7.

Normal modes in a rectangular cavity. Density of states.

$$p = A\cos(k_x x)\cos(k_y y)\cos(k_z z) \tag{1}$$

$$k_x a_1 = l\pi, \quad k_y a_2 = m\pi, \quad k_z a_3 = n\pi \tag{2}$$

$$k_x^2 + k_y^2 + k_z^2 = (\omega/v)^2 \equiv k^2 \quad k = \omega/v \tag{3}$$

$$\omega_{lmn} = v\sqrt{(l\pi/a_1)^2 + (m\pi/a_2)^2 + (n\pi/a_3)^2} \tag{4}$$

Density of modes

Volume in k-space per mode $= \pi^3/a_1a_2a_3 = \pi^3/V$ (5)

No. of modes with prop. const. $< k$ $\quad N(k) = (V/\pi^3)(1/8)4\pi k^3/3$ (6)

Modal density $\quad n(k) = \mathrm{d}N(k)/\mathrm{d}k = Vk^2/2\pi^2$ (7)

$$n(v) = \mathrm{d}N(v)/\mathrm{d}v = (V/v^3)4\pi v^2 \tag{8}$$

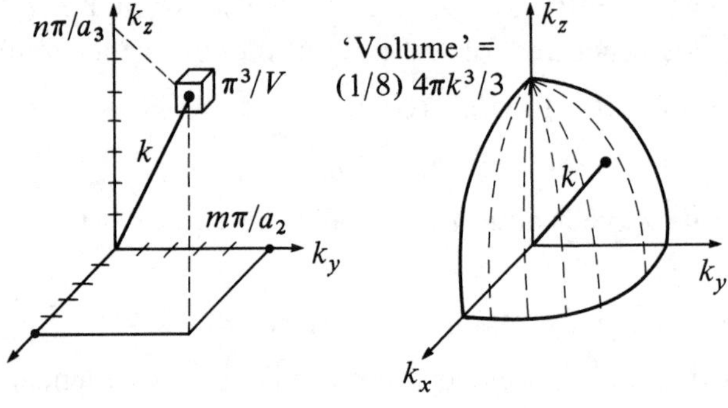

wavelength λ_{l00} such that $l\lambda_{l00}/2 = a_1$. Similarly, the modes $0, m, 0$ and $0, 0, n$ are standing waves between the walls perpendicular to the y- and z-axis, respectively.

A general mode l, m, n corresponds to a plane wave travelling in a direction of the propagation vector $\mathbf{k}_{lmn}$, and this direction is such, that after multiple reflections from the walls in the cavity the wave 'comes back on itself'.

The representative points of the modes, i.e. the tips of the propagation vectors $\mathbf{k}_{lmn}$, form a rectangular lattice in k-space, each cell in the lattice having the sides π/a_1, π/a_2, and π/a_3, and the volume $\pi^3/a_1a_2a_3 = \pi^3/V$, where $V = a_1a_2a_3$ is the total volume of the cavity. The corresponding average number of modes per unit volume in k-space is then V/π^3.

The volume of the lattice with propagation vectors with a magnitude smaller than $k = |k|$ is $(1/8)4\pi k^3/3$, i.e. 1/8 of the volume of a sphere of radius k. Consequently, the average number of modes within this volume is

$$N(k) = (V/\pi^3)(1/8)4\pi k^3/3 \tag{5}$$

A quantity which plays an important role in many problems is the number of modes $n(k)$ in a spherical shell of unit thickness in k-space. For sufficiently large values of k, the total number of modes $N(k)$ in Eq. (5) can be considered to be a continuous function of k, and it follows that

$$n(k)\,\mathrm{d}k = \mathrm{d}N(k) = (V/2\pi^2)k^2\,\mathrm{d}k \tag{6}$$

Since $\nu = \omega/2\pi = vk/2\pi$, we have $n(k)\,\mathrm{d}k = (V/2\pi^2)(2\pi\nu/v)^2(2\pi/v)\mathrm{d}\nu$ or

$$n(k)\,\mathrm{d}k = n(\nu)\,\mathrm{d}\nu = (V/v^3)4\pi\nu^2 \tag{7}$$

As indicated earlier, the volume of one cell in k-space is π^3/V so that the number of modes per unit volume in k-space, the 'density of states', is

$$n(k_x, k_y, k_z) = V/\pi^3 \tag{8}$$

Since the volume of a spherical shell of thickness $\mathrm{d}k$ is $(1/8)4\pi k^2\,\mathrm{d}k$, it follows that we can express $n(k)\,\mathrm{d}k$ as

$$n(k)\,\mathrm{d}k = n(k_x, k_y, k_z)(1/8)4\pi k^2\,\mathrm{d}k = (V/2\pi^2)k^2\,\mathrm{d}k \tag{9}$$

as before.

Problems

15.1 *Propagating and decaying waves.* The surface of an infinitely extended corrugated board in the yz-plane is defined by $x = A\cos(Ky)$, where $K = 2\pi/\Lambda$ and Λ is the period of the corrugation. The amplitude A is much smaller than Λ. The board is moving in the y-direction with the velocity U.

(a) What is the x-component of the velocity amplitude of an air layer with its equilibrium position in the yz-plane (on the positive x side)? Assume perfect slippage of the air at the board.
(b) Assuming a plane sound wave to be generated by the board, what will be the direction of propagation and the sound pressure amplitudes of the waves generated in the regions $x > 0$ and $x < 0$ if $U = 1.5\, v$, where v is the sound speed?
(c) What will be the 'wave drag' on the plate per m^2 if $A = 0.01\, \Lambda$? (Force per m^2 required to keep the board in motion).
(d) What is the sound field, pressure and fluid velocity, if $U = 0.5v$? If the xy-dependence of the complex amplitude is expressed as $\exp[ikx\cos(\theta) + ikx\sin(\theta)]$, what then is the angle θ? (include the possibility of a complex θ).

15.2 *Sound waves in a rectangular wave guide.* The cross section of a rectangular duct has the width 50 cm and height 38 cm. We wish to transmit a sound wave through the duct consisting only of a plane wave by means of a loudspeaker placed at the end of the duct. What is the highest frequency we can use in order that only the plane wave will be propagated? The sound speed is $v = 340$ m/sec.

15.3 *Energy flow in a higher order acoustic mode.* The pressure wave of the (m, n) mode in a rectangular wave guide has the complex amplitude $p_{mn} = A\cos(k_m y)\cos(k_n z)\exp(ik_x x)$, where y and z are the transverse coordinates and x is along the axis of the wave guide. $k_m = m\pi/a$, $k_n = n\pi/b$.

(a) Derive the complex amplitude for the axial fluid velocity.
(b) Derive the complex amplitude for the transverse velocity components.
(c) Calculate the acoustic power carried by the wave in the x-direction.

15.4 *Pressure release boundary.* Consider the propagation of sound waves in a rectangular water line. The line is coated with a soft rubber lining so that the sound pressure at the boundary can be considered to be zero. Derive the general expression for the acoustic pressure field in such a duct with a 'pressure release' boundary.

15.5 *Electromagnetic wave guide.* In a lecture demonstration, an EM wave was generated by a Klystron oscillator and transmitted between two parallel conducting plates, forming a wave guide. The wave travelled in the x-direction with the E-vector in the y-direction, parallel with the plates. The frequency of the Klystron was 10^{10} Hz. The transmitted signal amplitude was measured by a detector placed at the far end of the wave guide.

(a) As the separation between the plates was varied, the detected signal varied periodically with the plate separation. Determine the change in plate separation required to go from one maximum in the transmitted signal amplitude to the next.
(b) Assuming the E-field to be independent of y, derive the expressions for the E- and H-fields in the transmitted wave. The x-coordinate is in the direction of propagation parallel with the plates.

(c) What is the smallest plate separation for which a propagating wave can be transmitted between the plates?

(d) What happens if the separation is smaller than the 'cut-off' value obtained in (c)?

(e) Is there a similar cut-off if the E-field is perpendicular to the plates?

15.6 *Normal modes in a rectangular cavity.* One factor that influences the 'acoustics' of a room is the number of normal modes which are excited in a certain frequency interval. Consider a rectangular room with the dimensions 4m × 5m × 6m. Determine the number of modes in a third octave frequency band if the center frequency of the band is (a) 50 Hz, (b) 500 Hz, (c) 5000 Hz.

15.7 *Plane wave interpretation of higher order mode.* With reference to Display 15.5, the higher order mode involved can be interpreted as a plane wave being reflected back and forth between the upper and lower duct walls. Determine the direction of propagation of this wave with respect to the axis of the duct, when the frequency of the wave is 1000 Hz.

15.8 Derive the expressions for A, B, and C in Eqs. 4(10)–(12) and Q, P, and R in Eqs. 4(17)–(19).

16

Matter waves

After the discussion of wave characteristics in the last two chapters and diffraction and refraction in Chapters 11 and 12, it is appropriate at this point to say a few words about the wave nature of matter (de Broglie, 1924; Schroedinger, 1926). It was first demonstrated experimentally by electron diffraction (Davisson and Germer, 1927), and we shall devote this chapter to some of the basic concepts and relations involved.

16.1 Review

It is useful to start by reviewing some of the characteristics of waves which we have encountered so far.

First we recall that the space–time dependence of a one-dimensional harmonic wave has been described as

$$\Psi(x, t) = \mathrm{Re}\{\Psi(x)\exp(-\mathrm{i}\omega t)\} \qquad (1)$$

For a **travelling wave** we have

$$\Psi(x) = A\exp(\pm ikx) \quad A = A_0\exp(\mathrm{i}\alpha) \qquad (2)$$

where $\Psi(x)$ is the complex amplitude function, and ω the angular frequency, and the plus and minus sign refers to wave travel in the positive and negative x-direction, respectively. Quantity α is the phase angle of the complex amplitude at $x = 0$.

For a **standing wave** the complex amplitude function is

$$\Psi(x) = B\cos(kx - \gamma) \quad B = B_0\exp(\mathrm{i}\beta) \qquad (3)$$

where γ depends on the choice of the origin of x and β on the origin of t.

The relation between the angular frequency $\omega = 2\pi/T$ and the propagation constant $k = 2\pi/\lambda$ is determined by the dynamical properties of the medium that carries the wave. It is expressed by the **dispersion relation**

$\omega = \omega(k)$. For the acoustic and electromagnetic waves considered in Chapter 6, the dispersion relation is linear: $\omega = vk$, where v is the wave speed (phase velocity). In general, the dispersion relation is nonlinear, however, as for the periodic lattice, discussed in Chapter 14. The **phase velocity** v_p is then different from the **group velocity** v_g,

$$v_p = \omega/k \tag{4}$$
$$v_g = d\omega/dk$$

Illustrations of dispersive waves are given in Chapters 14, 18–20. For example, we shall find that the dispersion relation for bending waves on a thin plate is

$$\omega = Ck^2 \tag{5}$$

where C is a constant. In that particular case, the phase velocity is half the group velocity.

In a bounded region, in a cavity or between the supports of a string clamped at both ends, the wave is a superposition of waves travelling in different directions as a result of reflections from the boundaries.

In order for free (unforced) waves to exist under such conditions, only certain discrete values of the wavelength and the corresponding propagation constant can exist. For example, in the one-dimensional case, a wave function Ψ which has $\partial\Psi/\partial x = 0$ at $x = 0$ and $x = L$ has the complex amplitude $\Psi(x) = A\cos(kx)$ with $k = k_n = n\pi L$, where n is an integer. The normal mode functions are then $\Psi_n(x) = A\cos(k_n x/L)$ and the corresponding normal mode or eigenfrequencies are determined by the dispersion relation $\omega_n = \omega(k_n)$. For a linear dispersion relation we get $\omega_n = vk_n = vn\pi/L$. For bending waves on a plate, on the other hand, with the dispersion relation in Eq. (5), the frequencies are $\omega_n = Ck_n^2$.

The wave functions which correspond to the discrete values k_n, the eigenvalues, are often called the stationary states of the system. In practice, there is always some damping present, and a normal mode is not completely 'stationary'. Rather, the free oscillations decay in time, and this is expressed in terms of a complex value of the mode frequency, with the imaginary part expressing the inverse of a characteristic decay time, the 'life time' of the state.

As for the simple harmonic oscillator, forced harmonic motion of waves leads to resonance with infinite amplitude (in the absence of damping) at each of the normal mode frequencies.

Stationary states or resonances can be obtained even without reflecting boundaries perpendicular to the direction of propagation. As an illustration, let us start with a plane wave travelling in the x-direction between two

Display 16.1.

Review of wave characteristics.

Waves in bounded region. Normal modes

$\Psi(x) = A\exp(ikx) + B\exp(-ikx)$ (1)

Example If $\partial\Psi/\partial x = 0$ at $x = 0$ and $x = L$

$\Psi(x) = C\cos(kx)$ (2)

$k_nL = n\pi \quad k_n = n\pi/L$ (3)

Normal mode frequencies $\omega_n = \omega(k_n) = vk_n = (n\pi)v/L$ (4)

Normal mode functions (stationary states)

$\Psi_n(x) = C\cos(k_nx)$ (5)

Decaying mode $\omega = \omega_r + i\omega_i$ with $\omega_i < 0$

$\Psi(x,t) = \mathrm{Re}\{A(x)\cdot e^{-i\omega t}\} = e^{\omega_i t}\,\mathrm{Re}\{A(x)\cdot e^{-i\omega_r t}\} \quad (\omega_i < 0)$ (6)

$\tau = 1/|\omega_i|$ ('life time')

Periodic boundary condition

$2\pi r = n$ (7)

$k_n = 2\pi/\lambda_n = n$

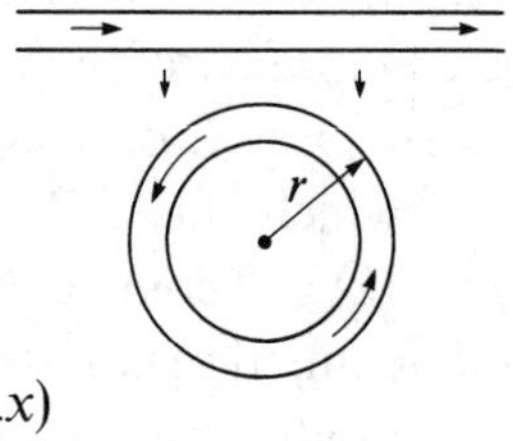

Spatial decay

$k = k_r + ik_i$ (8)

$\Psi(x) = A\cdot\exp(ikx) = e^{-k_ix}A\exp(ik_rx)$ (9)

Cut-off and wave decay

Example Mass–spring lattice

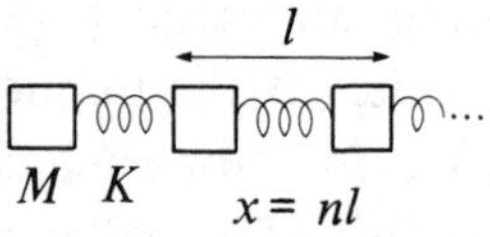

If $\omega > \omega_c = 2\sqrt{K/M}$ (10)

$k = (\pi/l) + ik_i$ (11)

$|\Psi(x)| = |\Psi(0)|\exp(-k_ix)$ (12)

Total reflection and wave decay

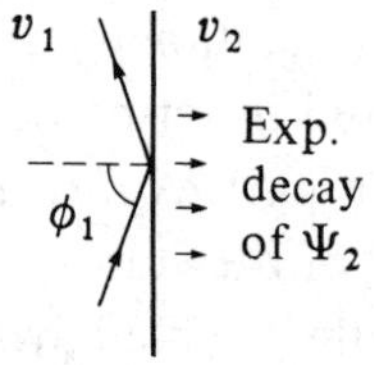

If $(v_2/v_1)\sin\phi_1 > 1$ (13)

$\Psi_2 \simeq \exp(-\sqrt{(v_2/v_1)^2\sin^2\phi_1 - 1}\cdot x)$ (14)

plane rigid boundaries, as indicated in Display 16.1. If the separation between the boundaries is small compared to the wavelength, the wave field will not be altered significantly if the planes are bent to form concentric cylinders, so that the wave travels in the annular region between the planes. A stationary state or resonance results if the length of the annulus is an integer number of wavelengths; both the value of the function and its slope will be the same after one round trip (periodic boundary conditions).

If the width of the annulus is not small compared to a wavelength there will be a radial variation $f(r)$ of the wave amplitude. The function $f(r)$ is expected to have maxima for values of r such that $2\pi r = n\lambda$.

In forced wave motion a one-dimensional wave will propagate without change in amplitude if there is no damping. If damping is present the wave amplitude will decay exponentially, accounted for by a complex value of k.

It is important to note, however, that we have encountered cases, in which exponential amplitude decay is obtained even without damping. One example referred to wave propagation on a mass–spring lattice (or the analogous LC transmission line) at a frequency above the cut-off frequency of the lattice. Other examples involved the total reflection at a boundary and the related decaying amplitude of a mode in a wave guide below the cut-off frequency of the mode. Later in this chapter we shall encounter an analogous decay for matter waves.

16.2 Free particle wave function

The interaction of electrons and other material particles with crystals has been found to give rise to angular distributions of scattered particles which are consistent with the assumption that the particles are diffracted as waves from the crystal 'grating' (Davisson and Germer, 1927). The wavelength to be assigned to the particle has been found to be $\lambda = h/p$ (de Broglie, 1924), inversely proportional to the momentum of the particle

$$\lambda = h/p \quad p = (h/2\pi)k = \hbar k \quad \hbar = h/2\pi \quad k = 2\pi/\lambda \tag{1}$$

$$h = 6.63 \times 10^{-34} \text{ joule sec}$$

where the constant of proportionality h is Planck's constant.

Furthermore, the works of Planck on black body radiation (1901) and of Einstein on the photo-electric effect (1905) support the hypothesis that the energy of radiation is quantized with the energy quantum proportional to the frequency,

$$E = h\nu \quad E = (h/2\pi)\omega = \hbar\omega \tag{2}$$

Again the constant of proportionality is Planck's constant.

The wave function which is assumed to be associated with the free motion

Display 16.2.

Wave function for free particle motion.

Wavelength of matter wave, free particle

$\lambda = h/p$ (1)

Planck's constant $h = 6.63 \times 10^{-34}$ joule sec

Energy–frequency relation

$E = h\nu$ (2)

Free particle wave function

$\Psi(x, t) = A\mathrm{e}^{\mathrm{i}kx - \mathrm{i}\omega t}$ (3)

$p = \hbar k \quad \hbar = h/2\pi$ (4)

$E = \hbar\omega$ (5)

Dispersion relation

$E = p^2/2m$ (6)

$\omega = \hbar k^2/2m$ (7)

$v_g = \mathrm{d}\omega/\mathrm{d}k = \hbar k/m = p/m$ (8)

Particle motion in arbitrary direction

$\Psi(\mathbf{r}, t) = A\exp^{\mathrm{i}\mathbf{k}\cdot\mathbf{r} - \mathrm{i}\omega t} \quad \mathbf{p} = \hbar\mathbf{k}$ (9)

Probability density $= |\Psi|^2 = \Psi\Psi^* = |A|^2$ (10)

(in the x-direction) of a particle is

$$\Psi(x,t) = A\exp(ikx - i\omega t)$$
$$k = p/\hbar \tag{3}$$
$$\omega = E/\hbar$$

To obtain the dispersion relation $\omega = \omega(k)$ we use the classical relation between energy and momentum for the free particle $E = p^2/2m$ and assume that the correspondence between p and k, and E and h, in Eq. (3) can be used, so that

$$\omega = (\hbar/2m)k^2 \tag{4}$$

We can now define a phase velocity and a group velocity of the matter wave

$$v_p = \omega/k = (\hbar/2m)k \tag{5}$$
$$v_g = d\omega/dk = (\hbar/m)k = p/m$$

It should be noted that v_g equals the classical velocity p/m of the particle and that $v_p = v_g/2$.

For a motion of the particle in an arbitrary direction, the wave function in Eq. (3) is replaced by

$$\Psi(r,t) = A\exp(i\mathbf{k}\cdot\mathbf{r} - i\omega t) \tag{6}$$
$$\mathbf{k} = \mathbf{p}/\hbar$$

with the propagation vector **k** being proportional to the momentum vector.

It should be noted that we have not taken the real part of the particle wave function. In wave mechanics, the entire complex wave function is used, and the physical meaning of the wave function is expressed in terms of the squared magnitude $|\Psi|^2 = \Psi\Psi^*$. This quantity is interpreted as the probability density, which, when multiplied by a volume element, gives the probability that the particle will be in this volume element at time t. Normalization of the wave function is obtained by the requirement that the integral of $\Psi\Psi^*$ over all space must equal 1.

16.3 The Uncertainty Principle

One might question the physical soundness of the representation of the free motion of a particle by means of a plane wave, since it extends over all space and, therefore, cannot describe the position of the particle to be described.

It should be noted though that the velocity of the particle is equal to the group velocity of the wave, and this suggests that a group of waves rather than a single plane wave should be used in the wave mechanical

Display 16.3.

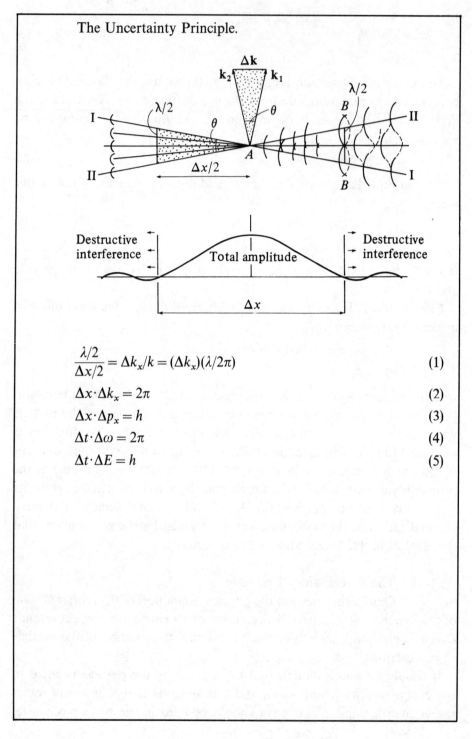

The Uncertainty Principle.

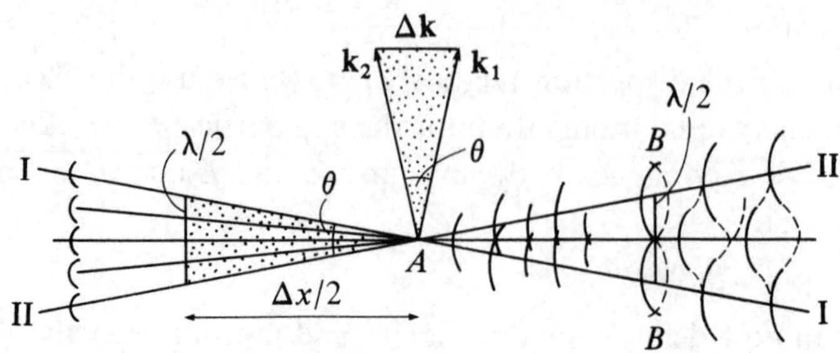

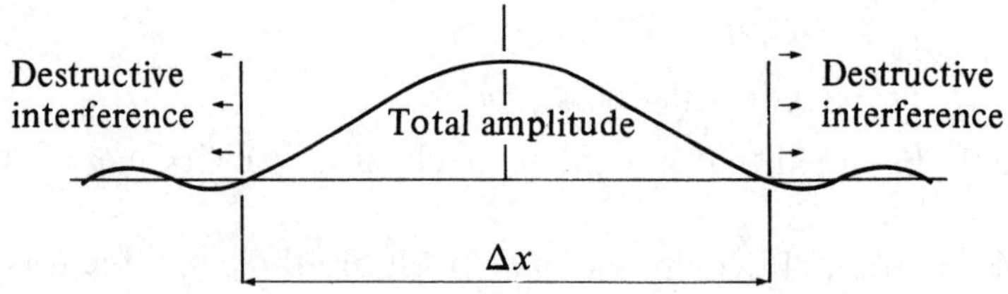

$$\frac{\lambda/2}{\Delta x/2} = \Delta k_x/k = (\Delta k_x)(\lambda/2\pi) \quad (1)$$

$$\Delta x \cdot \Delta k_x = 2\pi \quad (2)$$

$$\Delta x \cdot \Delta p_x = h \quad (3)$$

$$\Delta t \cdot \Delta \omega = 2\pi \quad (4)$$

$$\Delta t \cdot \Delta E = h \quad (5)$$

description of the location of the particle. This, indeed, turns out to be the case, and it is intimately related to an intrinsic characteristic of wave mechanics, expressed by the Uncertainty Principle. It says, in effect, that it is not possible to determine simultaneously the exact location and the exact momentum of a particle; if the momentum is determined exactly, the location is completely undetermined, and vice versa. To locate the particle to a certain region of space, an uncertainty in the momentum is required.

To see how this principle relates to the question we raised above, we note that for a single plane wave, the direction of propagation is determined exactly by the propagation vector **k** and the corresponding momentum **p**. According to the Uncertainty Principle, the location of the particle cannot be determined by the wave, which is consistent with the observation we made above.

To allow for an uncertainty in the direction of propagation we superimpose a group of plane waves with propagation vectors **k** which differ by an amount $\Delta\mathbf{k}$, as indicated in Display 16.3. We choose the amplitudes of these plane waves to be the same, and those constant phase surfaces which intersect at A are arranged to have the same phase. The outermost of these surfaces, corresponding to the propagation vectors **k** and $\mathbf{k} + \Delta\mathbf{k}$, are denoted by I and II. Since the complex amplitudes of all the waves in the range $\Delta\mathbf{k}$ are the same at A, the total wave amplitude will be a maximum at A and equal to the algebraic sum of the individual amplitudes.

As we go out from A along the x-axis in the region between I and II, the wave surfaces spread apart. In this region the phase surface for a wave in a certain direction will not go through A and its phase then will be different from that at A. The complex amplitudes from the various waves are then spread apart in the complex plane.

If we do not go too far from A, these amplitudes will all lie in the same half of the complex plane, so that their sum still will yield a substantial amplitude, although smaller than at A. At point B, on the other hand, the separation between I and II is half a wavelength, and the field contributions from I and II are 180 degrees out of phase and cancel each other. As we go beyond B, the complex amplitudes of the individual waves spread out over the entire complex plane, and the further we go, the more uniform will be the distribution of these amplitudes, and the total amplitude goes to zero.

16.4 Effect of potential. Barrier penetration

After the wave mechanical description of the free motion of a particle, we consider next the effect of the interaction with other particles, which we express in terms of a potential energy function $V(x)$. The total energy E of the particle is then the sum of the kinetic and the potential energy

$$E = p^2/2m + V(x) \tag{1}$$

The momentum of the particle depends on x according to

$$p = \sqrt{2m(E - V)} \tag{2}$$

and the propagation constant $k = p/\hbar$ and wavelength $\lambda = 2\pi/k$ of the corresponding matter wave are

$$k = (1/\hbar)\sqrt{2m(E - V)} \tag{3}$$
$$\lambda = h/p = h/\sqrt{2m(E - V)}$$

As an example, we study the wave mechanics of a particle under the influence of a potential in the form of a rectangular potential barrier, as indicated in Display 16.4. Outside the barrier, the potential is chosen to be zero; inside the barrier, between $x = 0$ and $x = L$, the potential is constant and equal to V_1.

A particle of energy E is incident on the barrier from the left. In classical mechanics, the barrier will act like a reflecting wall if E is smaller than the height V_1 of the barrier. If the barrier wall were not vertical, the particle would move up along the wall, slow down to rest, and then move back in the opposite direction.

To study this problem in wave mechanics, we need to determine the wave function throughout the region involved and then use the squared magnitude to determine the probability of finding the particle at a certain location.

We start by determining the propagation constant k from Eq. (3) inside and outside the barrier,

$$k = (1/\hbar)\sqrt{2mE} \qquad \text{(outside barrier)} \tag{4}$$
$$k = (1/\hbar)\sqrt{2m(E - V)} \qquad \text{(inside barrier)} \tag{5}$$

Outside the barrier the propagation constant is real. Inside the barrier, on the other hand, it will be imaginary if E is smaller than V_1,

$$k = \mathrm{i}(1/\hbar)\sqrt{2m(V_1 - E)} = \mathrm{i}\beta \tag{6}$$

and the function then will decay exponentially,

$$\Psi(x)/\Psi(0) = \exp(-\beta x)$$

Consequently, even though classical mechanics predicts reflection of the particle from the barrier, wave mechanics predicts a penetration into and through the barrier.

It is interesting in this context to recall that we encountered similar decaying wave fields in our discussion of total reflection at a boundary

Display 16.4.

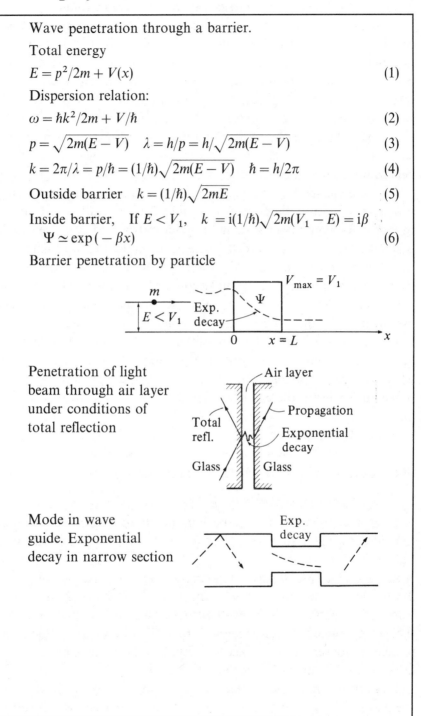

Wave penetration through a barrier.

Total energy

$$E = p^2/2m + V(x) \tag{1}$$

Dispersion relation:

$$\omega = \hbar k^2/2m + V/\hbar \tag{2}$$

$$p = \sqrt{2m(E - V)} \quad \lambda = h/p = h/\sqrt{2m(E - V)} \tag{3}$$

$$k = 2\pi/\lambda = p/\hbar = (1/\hbar)\sqrt{2m(E - V)} \quad \hbar = h/2\pi \tag{4}$$

Outside barrier $k = (1/\hbar)\sqrt{2mE}$ (5)

Inside barrier, If $E < V_1$, $k = \mathrm{i}(1/\hbar)\sqrt{2m(V_1 - E)} = \mathrm{i}\beta$

$$\Psi \simeq \exp(-\beta x) \tag{6}$$

Barrier penetration by particle

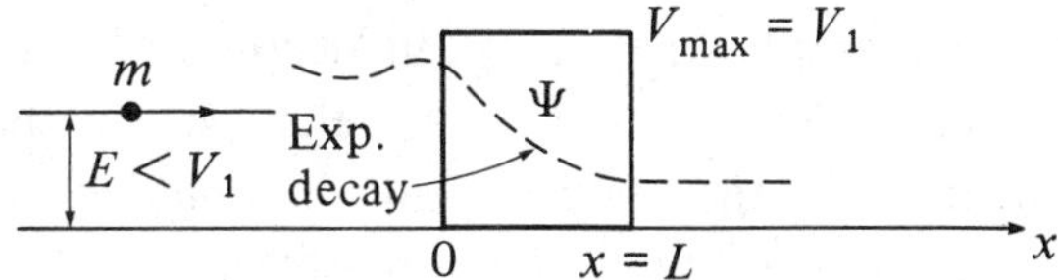

Penetration of light beam through air layer under conditions of total reflection

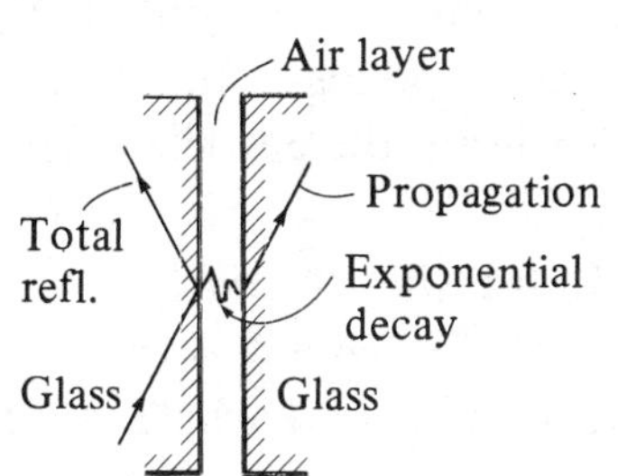

Mode in wave guide. Exponential decay in narrow section

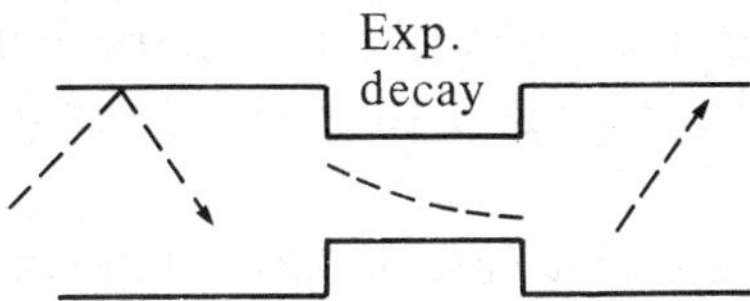

and in wave propagation in wave guides in Chapters 12 and 15, where analogues of the potential barrier penetration can be found.

As an example, we can get penetration of light across the air gap between two glass plates even under conditions of total reflection occurring in the first plate, as indicated schematically in the display. Although the wave decays in the air gap (since the wave velocity in the gap is larger than the trace velocity), it will generate a propagating wave in the second plate (since the trace velocity exceeds the wave velocity in the plate). It is noteworthy, that on the basis of geometrical optics, no such penetration is predicted, just as classical mechanics fails to account for particle penetration through a barrier.

We have included in the display another example, which refers to wave propagation in a rectangular wave guide. As we recall, a (1, 0) mode can propagate without attenuation in the guide only if the free space wavelength is smaller than twice the width of the wave guide. For wavelengths larger than this cut-off wavelength, the wave will decay exponentially.

In the example, the wave guide contains a constricted section with a width smaller than half a wavelength so that the wave will decay exponentially in this section. After having decayed, the wave will continue to propagate as it enters the downstream portion of the wave guide which has the same width as the upstream portion. The constricted part of the wave guide corresponds to the air gap between the glass plates and to the potential barrier in the particle case.

16.5 Refraction of matter waves

With $E = \hbar\omega$ and $p = \hbar k$, we can express the phase velocity $v_p = \omega/k$ of the matter wave of a particle in a potential V as

$$v_p = E/p = E/\sqrt{2m(E - V)} \tag{1}$$

Thus, the potential V can be considered to determine the index of refraction for the matter wave, and if V varies with position, the wave can be refracted.

In the simple case illustrated in Display 16.5, we have two regions 1 and 2, with the constant potentials V_1 and $V_2 > V_1$. This means, that the phase velocity in region 2 will be larger than in region 1, and as a plane wave crosses the plane boundary from region 1 to 2, it follows from Snell's law of refraction (conservation of trace velocity) that the angle of refraction will be larger than the angle of incidence, as shown.

This refraction of the matter wave is consistent with the trajectory of the particle. As an illustration let us consider a particle which slides up a

Display 16.5.

Refraction of matter waves.

Phase velocity of matter waves

$$E = \hbar\omega \tag{1}$$

$$p = \sqrt{2m(E - V)} = \hbar k \tag{2}$$

$$v_p = \omega/k = E/\sqrt{2m(E - V)} \tag{3}$$

Refraction of matter wave

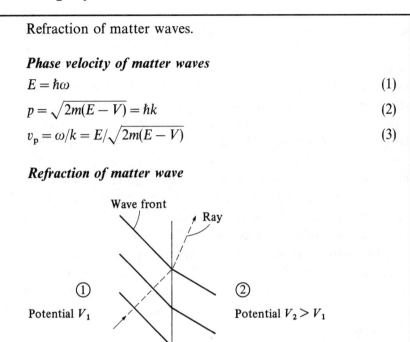

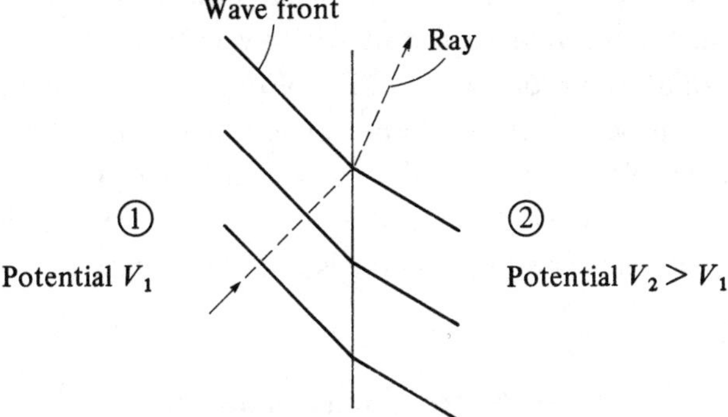

Particle moving uphill from one plane to another

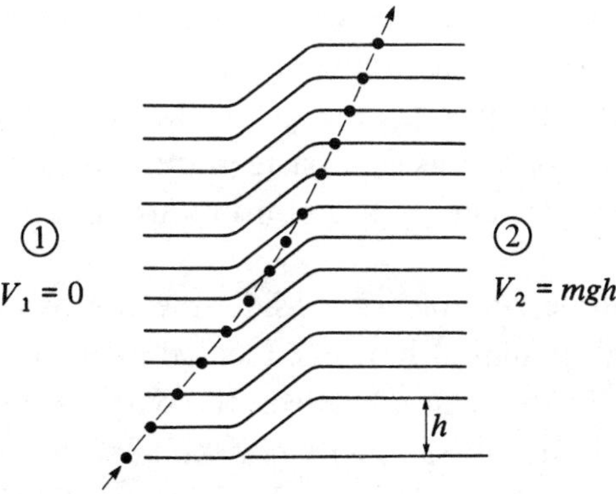

hill from one plane to another, i.e. from one gravitational potential to another. As a result, the component of the velocity normal to the hill will be reduced, whereas the tangential component will be unchanged. This means that the direction of the motion of the particle will be turned away from the normal direction as it moves up the hill, as indicated. We leave it for one of the problems to show that the change in direction is consistent with Snell's law, if the phase velocity of the matter wave is given by Eq. (1).

If the potential varies continuously with location, the wave fronts and the corresponding rays of the matter waves will change direction continuously in complete analogy with the refraction of sound waves in an inhomogenous atmosphere, as was discussed in Chapter 12.

It should be noted in this context that the channelling of waves along a circular path in the annular region in Display 16.2 can be thought of as a result of the radial variation of the index of refraction, which in the case of matter waves can be achieved by an appropriate radial variation of the potential. For example, the potential energy and hence the phase velocity of the matter wave of an electron in the field of a nucleus increases with distance from the nucleus, and this is consistent with closed (circular) orbits of the electron.

16.6 Stationary states. Schroedinger equation

As we have seen in previous chapters, the wave field in a bounded region is characterized by normal modes (stationary states) with a set of discrete values of the wavelength and corresponding normal mode frequencies.

This applies also to matter waves, describing the motion of a bound particle. The discrete values of the frequency are now replaced by a corresponding set of values of the particle energy; i.e. the energy is quantized. This feature of quantization of energy is a basic and important departure from classical mechanics, where energy is a continuous variable.

Particle in a 'box'. As a simple example we consider the motion of a particle between two infinitely high potential barriers, located at $x=0$ and $x=L$, as shown in Display 16.6. In the region between the barriers the potential is zero, and the particle is in free motion with constant speed, either in the positive or negative x-direction, as it bounces back and forth between the barriers. The corresponding matter wave then can be expressed as a superposition of free particle wave functions propagating in the positive and negative x-directions

$$\Psi(x, t) = A \exp(ikx - i\omega t) + B \exp(-ikx - i\omega t) \qquad (1)$$

Display 16.6.

Particle in a 'box'

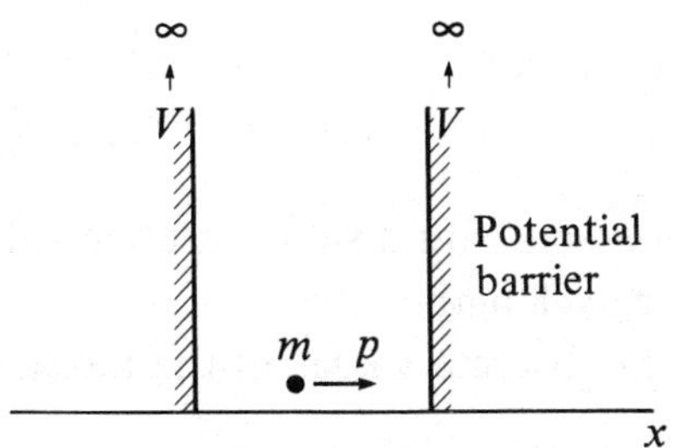

$$\Psi_+ = A \exp(ikx - i\omega t); \quad k = p/\hbar \quad \omega = E/\hbar \quad \hbar = h/2\pi \tag{1}$$

$$\Psi_- = B \exp(-ikx - i\omega t) \tag{2}$$

$$\Psi = C \sin(kx) e^{-i\omega t} \tag{3}$$

$$|\Psi|^2 = \Psi\Psi^* = |C|^2 \sin^2(kx) \tag{4}$$

$$kL = n\pi \quad k_n = n\pi/L \tag{5}$$

$$p_n = \hbar k_n \tag{6}$$

$$E_n = p^2/2m = (n\pi/L)^2(\hbar^2/2m) \tag{7}$$

For a barrier of finite height, as we have seen, there will be an exponentially decaying wave function within the barrier. With an infinitely high barrier, however, the decay rate will be infinite so that the wave function will be zero at the boundaries at $x = 0$ and $x = L$. To satisfy these conditions, we must have $B = -A$, so that the total wave function will be

$$\Psi(x, t) = C \sin(kx) \exp(-i\omega t) \tag{2}$$

$$k = k_n = n\pi/L \tag{3}$$

In other words, we obtain solutions only for a set of discrete values of the propagation constant k and the wavelength.

The corresponding values of the discrete values of the momentum and particle energy are

$$p_n = \hbar k_n \tag{4}$$

$$E_n = p_n^2/2m = (\hbar^2/2m)k_n^2 = (\hbar^2/2m)(n\pi/L)^2 \tag{5}$$

It should be realized that for ordinary macroscopic bodies, with a mass of the order of 1 kg and a box length L of the order of 1 m, the difference between two successive energy levels is small indeed, smaller than can be resolved experimentally. Consequently, the classical notion of energy as a continuous variable is justified from the standpoint of experiments.

Hydrogen atom. In the hydrogen atom an electron moves in the electric field of the nucleus, and the potential energy is proportional to $-1/r$, where r is the distance from the nucleus. In other words, the potential energy increases with r, and, according to Eq. 5(1), the same holds true for the phase velocity of the matter wave of the electron. As a result, the wave will be refracted, as indicated qualitatively in Display 16.7.

If the conditions are right, a ray will form a closed path, and if the length of such a path is an integer number of wavelengths, the wave comes back on itself in phase so that constructive interference results. This is the condition for a stationary state or resonance, in much the same way that it could be used to express the condition for a normal mode of a cavity or the annulus in Display 16.2. The condition for the stationary state can be written

$$\int \mathrm{d}s/\lambda = n \tag{6}$$

where ds is the length element and n an integer.

As an approximation, we use the plane wave relation between particle momentum and wavelength $\lambda = h/p$, so that the condition in Eq. (6) can be

Display 16.7.

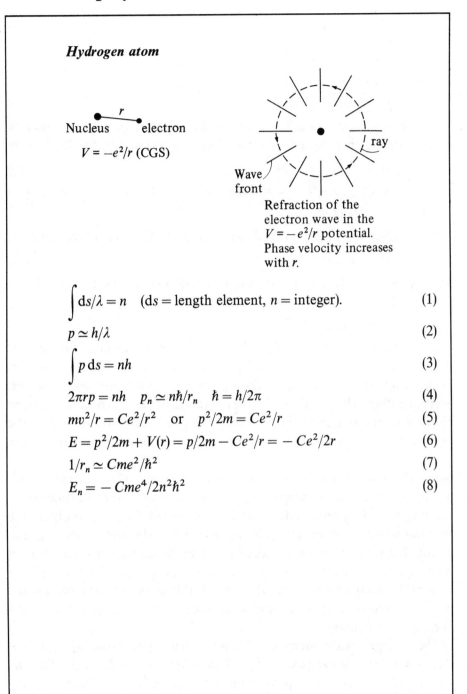

Hydrogen atom

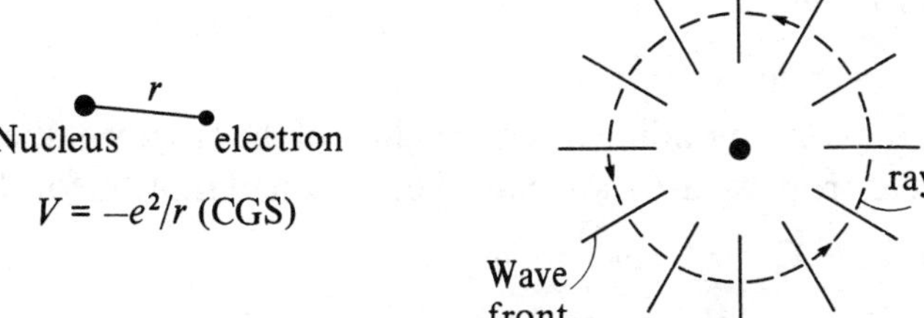

$V = -e^2/r$ (CGS)

Refraction of the electron wave in the $V = -e^2/r$ potential. Phase velocity increases with r.

$$\int \mathrm{d}s/\lambda = n \quad (\mathrm{d}s = \text{length element, } n = \text{integer}). \tag{1}$$

$$p \simeq h/\lambda \tag{2}$$

$$\int p\,\mathrm{d}s = nh \tag{3}$$

$$2\pi r p = nh \quad p_n \simeq n\hbar/r_n \quad \hbar = h/2\pi \tag{4}$$

$$mv^2/r = Ce^2/r^2 \quad \text{or} \quad p^2/2m = Ce^2/r \tag{5}$$

$$E = p^2/2m + V(r) = p/2m - Ce^2/r = -Ce^2/2r \tag{6}$$

$$1/r_n \simeq Cme^2/\hbar^2 \tag{7}$$

$$E_n = -Cme^4/2n^2\hbar^2 \tag{8}$$

written

$$\int p\,\mathrm{d}s = nh \tag{7}$$

For a circular orbit we obtain

$$2\pi rp = nh \tag{8}$$

$$p = p_n = n\hbar/r_n \tag{9}$$

To obtain the corresponding discrete values of the energy of the stationary states of the electron, we use the classical equation of motion for the electron

$$mv^2/r = Ce^2/r^2 \quad \text{or} \quad p^2/m = Ce^2/r \tag{10}$$

$$E = p^2/2m + V(r) = p^2/2m - Ce^2/r = -Ce^2/2r \tag{11}$$

$$C = 1/4\pi\varepsilon_0$$

Combining Eqs. (9) and (10) we can express the radius r_n of the nth orbit

$$1/r_n \simeq Cme^2/n^2\hbar^2 \tag{12}$$

and the corresponding expression for the energy of the nth state then follows from Eq. (11),

$$E_n \simeq -Cme^4/2n^2\hbar^2 \tag{13}$$

The approximate signs in Eqs. (12) and (13) reflect the fact that we have used the plane wave relation $\lambda = h/p$ between wavelength and momentum even though the waves are not plane. As the orbit radius increases, however, corresponding to large values of n, the curvature of the rays decrease and the plane wave approximation is expected to be good. Actually, for large n, the exact solution turns out to be in agreement with the result presented here.

Schroedinger equation. In problems with a potential V which varies with location, the solutions obtained by treating the wave functions locally as plane waves are approximate, as already indicated. In order to obtain the rigorous solutions we expect, as in the case of acoustic and electromagnetic waves, that we have to solve a wave equation for the matter waves. Such a wave equation was arrived at by Ervin Schroedinger in 1926. We shall not present the detailed reasoning which led to this equation but merely make use of plausibility arguments and a mnemonic for the construction of the Schroedinger equation.

The complex wave function of a plane harmonic wave $\Psi(x, y, z, t) = A\exp(\mathrm{i}\mathbf{k}\cdot\mathbf{r} - \mathrm{i}t) = A\exp[\mathrm{i}(k_x x + k_y y + k_z z) - \mathrm{i}\omega t]$ is such that differentiation with respect to x is equivalent to multiplying the wave function by $\mathrm{i}k_x$ and differentiation with respect to t is equivalent to multiplication by $-\mathrm{i}\omega$. For a matter wave, these factors become $\mathrm{i}p_x/\hbar$ and $-\mathrm{i}E/\hbar$, respectively.

Conversely, we can regard p and E as differential operators

$$p_x \rightarrow (\hbar/\mathrm{i})\partial/\partial x \tag{14}$$

$$E \rightarrow (\hbar/-\mathrm{i})\partial/\partial t \tag{15}$$

with corresponding expressions for p_y and p_z.

We now assume, that in the expression for the total energy

$$-E = (p_x^2 + p_y^2 + p_z^2)/2m + V(x, y, z) \tag{16}$$

we can use these operators acting on the wave functions to obtain

$$-(\hbar^2/2m)\nabla^2\Psi + V(x, y, z)\Psi = (\hbar/-\mathrm{i})\partial\Psi/\partial t \tag{17}$$

which is the time dependent Schroedinger equation for the wave function $\Psi(x, y, z, t)$.This equation is used in problems when the probability density $\Psi\Psi^*$ varies with time, such as in a description of a particle travelling in unbounded regions.

In a normal mode (stationary state) corresponding to a certain value of E, we expect the probability density to be time independent, and the corresponding spatial dependence of the wave function is obtained by putting $\Psi(x, y, z, t) = \Psi(x, y, z)\exp(-\mathrm{i}Et/\hbar)$. The time dependent Schroedinger equation (17) then reduces to the time independent Schroedinger equation for $\Psi(x, y, z)$

$$(\hbar^2/2m)\nabla^2\Psi + (E - V)\Psi = 0 \tag{18}$$

which is similar to the Helmholtz equation $\nabla^2 p + k^2 p = 0$, with a space dependent k.

Problems

16.1 *Wavelength.* The wavelength of soft X-rays is approximately 1 Ångström. To obtain the same matter wavelength of electrons (protons), what should be the velocity of the electrons (protons) and the corresponding energy in eV?

16.2 *Electron diffraction.* In the experiment by Davisson and Germer (1927), electrons were (Bragg) reflected (see Example 11E.2) by a nickel crystal. The electrons were accelerated through a potential drop of 54 V. Maximum intensity of the reflected electrons was observed at an angle of 50 degrees with respect to the direction of the incident electrons. Determine the wavelength of these electrons and the separation of the atomic planes in the nickel target.

16.3 *Uncertainty principle.* A wave function $\Psi(x)$ is defined as follows: $\Psi(x) = A$ for $-\Delta x/2 < x < \Delta x/2$ and zero elsewhere. Evaluate the Fourier Transform

$F(k) = \int \Psi(x) \exp(ikx)\,\mathrm{d}k$ (integration from $-\infty$ to $+\infty$). Introduce a width Δk of $F(k)$, say between the first zeroes of $F(k)$, and calculate the product $\Delta x \Delta k$ and relate the result to the Uncertainty Principle.

16.4 *Relativistic free particle.* In special relativity, the energy of a free particle is $E = \gamma mc^2$ and the momentum $p = \gamma mv$ where $\gamma = (1 - \beta^2)^{-\frac{1}{2}}$ and $\beta = v/c$ and m the restmass. Show that that $E^2 = p^2c^2 + m^2c^4$ and that the group velocity of the corresponding matter wave is the particle velocity v.

16.5 *Refraction.* In the demonstration referred to in Display 16.5, the height of the barrier is $H = 5$ cm. A particle of mass $m = 0.1$ kg moves up the barrier from the lower plane with an initial velocity of 1.5 m/sec and an angle of incidence of 30°. In what direction does the particle move on the upper plane? Compare the result with the angle of refraction for the corresponding matter wave.

16.6 *Energy levels.* Determine the energy levels (in electron volts) of an electron in an infinite potential well of width $L = 1$ Å ($= 10^{-8}$ cm).

PART 3

Special topics

17

Radiation and scattering

In Chapter 9 we discussed some aspects of the generation of electromagnetic waves, and we shall now do the same for sound waves. In addition, we treat some examples of the closely related problem of scattering, including the scattering of electromagnetic waves by electrons (Thomson and Rayleigh scattering).

17.1 Sound radiation from a pulsating sphere

As a prototype for a sound source with spherical symmetry, we use a pulsating sphere, as indicated in Display 17.1. The radius of the sphere is a, and the surface is oscillating in harmonic motion in the radial direction with the complex amplitude $u(a) = U$.

In the absence of reflecting boundaries, the sound pressure field produced by the sphere must have spherical symmetry and consist of an outgoing spherical wave. With reference to Chapter 10, the complex amplitude of the pressure is then

$$p(r) = (A/r)\exp(\mathrm{i}kr) \qquad k = \omega/v \tag{1}$$

The corresponding complex velocity amplitude $u(r)$ is obtained from the momentum equation $-\mathrm{i}\omega\rho u(r) = -\mathrm{d}p(r)/\mathrm{d}r$ (which follows from $\rho\partial u(r,t)/\partial t = -\partial p(r,t)/\partial r$). With $\mathrm{d}p/\mathrm{d}r = A(-1/r^2 + \mathrm{i}k/r)\exp(\mathrm{i}kr)$, we obtain

$$u(r) = (A/\rho vr)(1 + \mathrm{i}/kr)\exp(\mathrm{i}kr) \tag{2}$$

We note that the velocity contains two contributions, one of which varies as $1/r$ and one as $1/r^2$. These parts represent the **far field** and **near field**, respectively.

To determine the unknown amplitude constant A, we make use of the known complex velocity amplitude U at the surface of the sphere, and we

Display 17.1.

Sound radiation from a pulsating sphere. Monopole.

Surface velocity:

$u(a,t) = U\cos(\omega t)$ (1)

Complex amplitude: $u(a) = U$ (2)

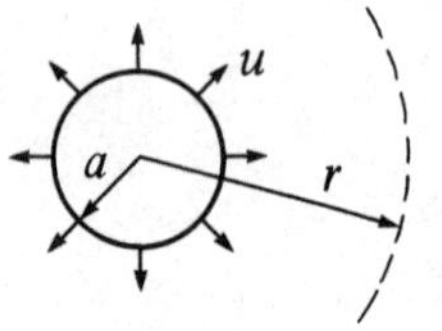

Pressure wave:

$p(r) = (A/r)\exp(ikr) \quad k = \omega/v$ (3)

Velocity wave:

$-i\omega\rho u = -dp/dr$ (4)

$u(r) = (A/\rho v r)[1 + i(1/kr)]\exp(ikr)$ (5)

Source condition:

$r = a \quad u(a) = U$ (6)

$A\exp(ika) = \rho v U a/(1 + i/ka)$ (7)

Complex pressure amplitude

$p(r) = \rho v U[ka/(1 + i/ka)](a/r)\exp[ik(r - a)]$ (8)

Radiation impedance

$z = p(a)/u(a) = \rho v[ka/(i + ka)] = \rho v(\theta + i\chi)$ (9)

$\theta = (ka)^2/[1 + (ka)^2]$ (10)

$\chi = -ka/[1 + (ka)^2]$ (11)

$z \simeq \rho v[(ka)^2 - ika] \quad ka \ll 1$ (12)

$z \simeq \rho v[1 - i/ka] \quad ka \gg 1$ (13)

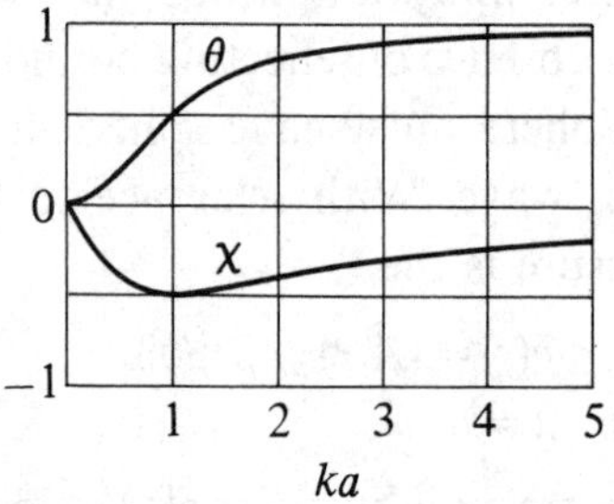

Monopole $(a \to 0)$

Source strength:

$q = \rho 4\pi a^2 U$ (14)

$p(r) = (-i\omega q/4\pi r)\exp(ikr)$ (15)

Power $= |\omega q|^2/8\pi\rho v$ (16)

obtain

$$A = [\rho v U a/(1 + \mathrm{i}/ka)] \exp(-\mathrm{i}ka) \tag{3}$$

so that the complex pressure amplitude in Eq. (3) becomes

$$p(r) = \rho v U[ka/(\mathrm{i} + ka)](a/r) \exp[\mathrm{i}k(r - a)] \tag{4}$$

Of particular interest is the complex pressure amplitude $p(a)$ at the surface of the sphere, and the ratio between $p(a)$ and $u(a) = U$, which is termed the **radiation impedance**

$$z = p(a)/u(a) = \rho v \cdot ka/(ka + \mathrm{i}) = \rho v(\theta + \mathrm{i}\chi) \tag{5}$$
$$\theta = (ka)^2/[1 + (ka)^2]$$
$$\chi = -ka/[1 + (ka)^2]$$

where the real part θ is the normalized **radiation resistance** and the imaginary part, the reactance. The resistance increases monotonically with $ka = 2\pi a/\lambda$ toward the value ρv, the plane wave resistance. The reactance, on the other hand, goes to zero as ka goes to infinity, and reaches a maximum value for $ka = 1$.

For small and large values of $ka = 2\pi a/\lambda$ (the ratio between the circumference of the sphere and the wavelength) the radiation impedance can be approximated by

$$z \simeq \rho v[(ka)^2 - \mathrm{i}ka] \quad ka \ll 1 \tag{6}$$

$$z \simeq \rho v[1 - \mathrm{i}/ka] \quad ka \gg 1 \tag{7}$$

At low frequencies the resistance is proportional to ω^2 and the reactance is proportional to ω. Therefore, at sufficiently low frequencies, the impedance is dominated by the reactance, which corresponds to an inertial mass load.

At high frequencies, the reactance goes to zero as $1/\omega$, and the resistance approaches the value characteristic of a plane wave, as already mentioned.

The spherically symmetrical field is often called a monopole field by analogy with the electrostatic field from a point charge. Similarly, with the radius of the sphere going to zero, the pulsating sphere becomes a point source, generally referred to as a **monopole**. The source strength of the monopole usually is expressed in terms of the amplitude of the oscillatory mass flow rate from the sphere, $q = 4\pi a^2 \rho U$, and in defining the monopole it is implied that this quantity remains finite as the radius of the sphere goes to zero. The complex pressure amplitude field for the monopole then becomes

$$p_0(r) = (-\mathrm{i}\omega q/4\pi r) \exp(\mathrm{i}kr)$$
$$q = 4\pi a^2 \rho u(a) \tag{8}$$

It should be noted that the pressure field is proportional to the radial acceleration amplitude $-i\omega u(a)$ of the surface of the sphere.

At large distances from the source, $kr \gg 1$, the relation between pressure and velocity is approximately the same as in a plane wave. In this far field region the pressure and velocity are in phase and $p \simeq \rho v u$.

Close to the source, in the near field, with $kr \ll 1$, the velocity lags behind the pressure by approximately 90 degrees. The pressure amplitude $p = zu \simeq -i\omega\rho a$ is then dominated by the reaction from the mass load of the surrounding fluid, and this load is not dependent on the sound speed and hence does not depend on the compressibility of the fluid. In the far field, on the other hand, the pressure is due solely to the compression of the fluid.

Radiated power. The acoustic power radiated by the monopole source is the integral of the normal component of the intensity over a surface which surrounds the source (compare analogous calculation of the radiated EM power in Chapter 9). We choose a spherical surface with the source at the center. Since the acoustic power is conserved, the integral of the acoustic intensity over the surface is independent of the radius r. If we choose r in the far field, we have $p = \rho v u$, and the intensity

$$I = \tfrac{1}{2}\mathrm{Re}\{pu^*\} = |p|^2/2\rho v.$$

With the expression for p given by Eq. (4) and with $q = 4\pi a^2 \rho U$, the power $P = 4\pi r^2 I$ becomes

$$P = \tfrac{1}{2}(4\pi a^2)|U|^2 \rho v (ka)^2/[1 + (ka)^2] = [|\omega q|^2/8\pi\rho v]/[1 + (ka)^2] \tag{9}$$

It is readily seen that if we carry out the corresponding calculation for $r = a$ we obtain the same result, and if we use the expression for the radiation resistance in Eq. (5), we can express the power as

$$P = \tfrac{1}{2}(4\pi a^2)|u(a)|^2 \rho v \theta \tag{10}$$

It follows from Eq. (9) that for a monopole, with $ka = 0$, the radiated power is

$$P = (\omega q)^2/8\pi\rho v \quad \text{(Monopole)} \tag{11}$$

As will be discussed in more detail later, the expression (8) can be used also as the average complex pressure amplitude (averaged over a spherical surface about the source) generated by an arbitrary source with dimensions small compared to the wavelength (often called an 'acoustically compact' source) with q being the amplitude of the total oscillator mass flow rate of the source.

17.2 Multipole sources

The monopole source, defined in the previous section, can be used as a building block in the construction of more complex radiators. As in electrostatics, we can combine monopoles of different signs to obtain multipole radiators, and we shall start with a discussion of the characteristics of the dipole source.

Dipole. A dipole consists of two closely spaced monopole sources of equal strength but 180 degrees out of phase, so that when the oscillator flow is outward in one monopole it is inward in the other. As shown in Display 17.2, the two monopoles A and B are placed on the x-axis a distance d apart, at x' and $x'-d$, respectively.

The complex amplitude p_a produced by A in the far field is chosen to lie on the real axis in the complex plane. The monopole B is 180 degrees out of phase, and the complex amplitude p_b at the observation point would be 180 degrees out of phase with p_a, and hence cancel it, were it not for the separation of A and B, which produces a time delay d/v between the observed signals from A and B. Consequently, at a point of observation on the x-axis to the right of the sources, p_b will have an additional phase lag of $(d/vT)2\pi = (d/\lambda)2\pi = kd$, where $k = 2\pi/\lambda$. If the point of observation is displaced from the x-axis by an angle θ, the corresponding additional phase lag will be $kd\cos(\theta)$ in the far field.

As a result of the additional phase difference between p_a and p_b, their sum will not be zero but will have a magnitude $kd|p_0|\cos(\theta)$, where $|p_0| = |p_a| = |p_b|$. The phase of this resulting pressure runs ahead of p_a by 90 degrees, as shown in the display.

In a more detailed analysis, we let the coordinates of A be x', $y'=0$, $z'=0$ and the point of observation x, y, z. According to Eq. 1(8), the complex amplitude from A is then

$$p_a = p_0(x, y, z; x', y', z') = (-\mathrm{i}\omega q/4\pi R)\exp(\mathrm{i}kR) \tag{1}$$

q = mass flow rate of monopole

$$R^2 = (x-x')^2 + (y-y')^2 + (z-z')^2$$

and the corresponding amplitude produced by B can be expressed as

$$p_b = -p_0(x, y, z; x'-d) = -[p_0(x, y, z) - (\partial p_0/\partial x')d]$$

From these expressions for p_a and p_b, we obtain for the total complex pressure amplitude

$$p_1 = p_a + p_b = (\partial p_0/\partial x')d = -(\partial p_0/\partial x)d = -(\partial p_0/\partial R)(\partial R/\partial x)d$$

Display 17.2.

Acoustic dipole radiation

Complex amplitude fields

Monopole:

$$p_0(x, y, z; x') = (-i\omega q/4\pi R)\exp(ikR) \tag{1}$$

$$q = 4\pi a^2 \rho U \tag{2}$$

$$R^2 = (x - x')^2 + y^2 + z^2 \tag{3}$$

$$p_0(x, y, z; x' - d) = -[p_0(x, y, z; x') - (\partial p_0/\partial x')d] \tag{4}$$

Dipole:

$$p_1 = (\partial p_0/\partial x')d = -(\partial p_0/\partial x)d \tag{5}$$

$$\partial R/\partial x = (x - x')/R = \cos(\theta) \tag{6}$$

$$p_1 = (-\omega^2 D/4\pi R v)(1 + i/kR)\cos(\theta) \quad D = qd \tag{7}$$

Far field intensity $(r \gg r')$

$$I = |p_1|^2/2\rho v = (\omega^2 D/4\pi R v)^2(1/2\rho v)\cos^2(\theta) \tag{8}$$

Radiated power

$$P = \int_0^\pi I 2\pi R \sin(\theta) R \mathrm{d}(\theta)$$

$$= (2\pi\omega^4 D^2/16\pi^2 v^2)\int \cos^2(\theta)\sin(\theta)\mathrm{d}\theta \tag{9}$$

$$P = (\omega^2 D)^2/12\pi\rho v^3 \tag{10}$$

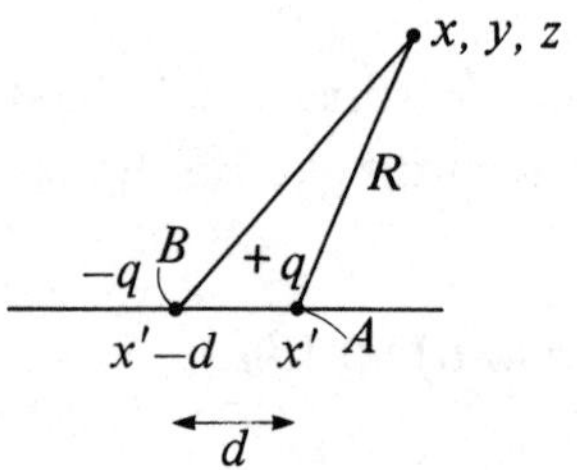

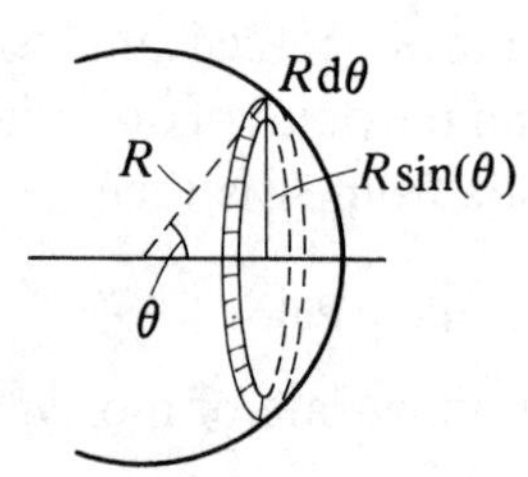

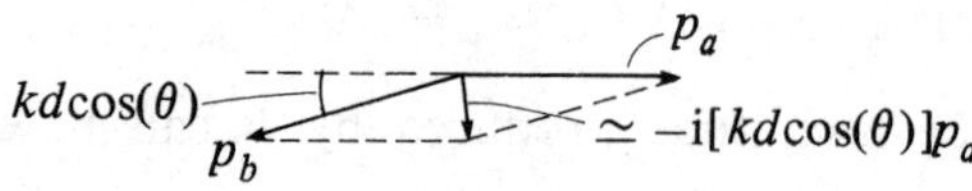

where $\partial R/\partial x = (x - x')/R = \cos(\theta)$ and θ is the angle between the x-axis and R, as shown.

With $\partial p_0/\partial R = (-\mathrm{i}\omega q/4\pi R)\exp(\mathrm{i}kR)(\mathrm{i}k - 1/R)$, we obtain for the total complex pressure amplitude

$$\begin{aligned} p_1 &= (-\mathrm{i}kdp_0)(1 + \mathrm{i}/kR)\cos(\theta) \\ &= (-\omega^2 qd/4\pi Rv)(1 + \mathrm{i}/kR)\cos(\theta)\exp(\mathrm{i}kR) \\ p_0 &= (-\mathrm{i}\omega q/4\pi R)\exp(\mathrm{i}kR) \end{aligned} \quad (2)$$

By definition, the two monopoles constitute a dipole in the limit of zero separation, provided that in this limit qd remains finite (i.e. q has to go to infinity as d goes to zero). The quantity $D = qd$ will be called the **dipole moment** of the source.

It should be noted that the complex pressure amplitude in the dipole field consists of two contributions varying as $1/R$ and $1/R^2$, representing the far field and near field, respectively. Note also that the pressure amplitude is a maximum in the direction along the dipole axis ($\theta = 0$) and zero perpendicular thereto.

The velocity field is obtained from the equation of motion, $\rho\partial\mathbf{u}/\partial t = -\operatorname{grad}(p)$, with the corresponding complex amplitude relation $-\mathrm{i}\omega\rho\mathbf{u} = -\operatorname{grad}(p)$.

In the far field we find $u = u_r = p/\rho v$, and the intensity becomes

$$\begin{aligned} I &= |p_1|^2/2\rho v = (\omega^2 D/4\pi Rv)^2(1/2\rho v)\cos^2(\theta) \\ D &= qd \end{aligned} \quad (3)$$

The intensity depends on the polar angle, and to determine the radiated power, we integrate the intensity over the surface of a sphere surrounding the dipole. We choose as the surface element a ring on the sphere with radius $R\sin(\theta)$ and width $R\mathrm{d}\theta$, and the area is $2\pi R^2\sin(\theta)\,\mathrm{d}\theta$. The power through this surface element is then $I(\theta)2\pi R^2\sin(\theta)\,\mathrm{d}\theta$, and the total power is obtained by integrating over θ from 0 to π,

$$\begin{aligned} P &= \int I(\theta)2\pi R^2\sin(\theta)\,\mathrm{d}\theta = (2\pi\omega^4 D^2/4\pi\rho v^3)\int\cos^2(\theta)\sin(\theta)\,\mathrm{d}\theta \\ &= (\omega^2 D)^2/12\pi\rho v^3 \end{aligned} \quad (4)$$

For comparison, we recall that the radiated power from the electromagnetic dipole (Eq. 9.5.(8)) is $(\omega^4 D_e^2)/12\pi\varepsilon_0 c^3$, where D_e is the electric dipole moment. We note that the dipole power is proportional to ω^4 for both the acoustic and EM dipole.

If we introduce the wave impedances $Z_e = \mu_0 c$ and $Z_a = \rho v$, the expressions for the electromagnetic and acoustic dipole power can be written $(k\omega D_e)^2 Z_e/12\pi$ and $(k\omega D)^2/12\pi Z_a$, respectively. This means that the quantities $(k\omega D_e)^2$ and $(k\omega D)^2$ have the dimensions of $[I]^2L^2$ and $[F]^2/L^2$, where I and F stand for electric current and mechanical force, respectively.

With the prototype of a monopole being a pulsating sphere, the dipole can be modelled as a small rigid sphere oscillating back and forth in a fluid. This motion produces an oscillatory flow through a fixed control surface surrounding the sphere which, at a given time, is outward in one direction and inward in the other. The net flow through the surface is zero. A free loudspeaker or oscillating piston produces a dipole field at low frequencies, as will be discussed in more detail in the next section.

Quadrupoles and higher multipoles. Two closely spaced dipoles of equal strength but with different directions form a quadrupole. If the dipoles, assumed to be aligned along the x-axis, are displaced with respect to each other in the x-direction, the quadrupole is called **longitudinal**; if the displacement is perpendicular to the x-axis, it is called **lateral**.

With reference to the introductory discussion of the dipole field, we recall that the complex pressure amplitude in the far field of a dipole pointing in the x-direction is obtained by multiplying the monopole field by $-ika_x\cos(\phi_x)$, where now we have denoted the relative displacement in the x-direction of the two monopoles by a_x, and the angular position of the observer with respect to the x-axis by ϕ_x.

In similar manner, the complex pressure amplitude of the longitudinal quadrupole field is obtained by multiplying the dipole field by $-ikb_x\cos(\phi_x)$, where b_x is the relative displacement in the x-direction of the two dipoles, which have opposite directions, as already indicated. We obtain then

$$p_{xx} = p_x(-ikb_x)\cos(\phi_x) = p_0(-k^2a_xb_x)\cos^2(\phi_x) \tag{5}$$

where p_0 and p_x refer to the monopole and dipole fields, respectively. Introducing the expression for p_0, we note that the amplitude of the quadrupole fields is proportional to $q_{xx} = qa_xb_x$, which is the **quadrupole moment** of the source. Actually, the definition of the quadrupole implies the limit $a_x, b_x \to 0$ as $q \to \infty$.

If, instead, the dipoles are displaced in the y-direction, the lateral quadrupole field is obtained by multiplying the dipole field by $-ikd_y\cos(\phi_y)$, where d_y is the displacement in the y-direction and ϕ_y the angular position of the observation point with respect to the y-axis, as indicated in the display, Thus,

$$p_{xy} = p_x(-ikb_y)\cos(\phi_y) = p_0(-k^2a_xb_y)\cos(\phi_x)\cos(\phi_y) \tag{6}$$

The corresponding quadrupole moment is now qa_xb_y.

Higher order multipoles are constructed in a similar manner, and it follows that the amplitude of the mth order multipole will contain an amplitude factor $(kd)^m p_0$, where we have used a characteristic length d to signify the relative displacement of the multipoles. It is important to realize that for low frequencies, $kd = 2\pi d/\lambda \ll 1$ (acoustically compact source), the far field amplitude of the multipole field is smaller than that of the monopole by a factor $(kd)^m$. Thus, if a source consists of a combination of multipoles, the far field at low frequencies is dominated by the multipole of lowest order.

17.3 Nonuniform spherical source. Spherical harmonics

The pulsating sphere in the previous section had a uniform surface amplitude distribution over the sphere. We now shall consider the radiation from a sphere with an arbitrary amplitude distribution.

The location on the surface of the sphere is expressed in terms of the azimuthal and polar angles ϕ and θ, respectively, as indicated in Display 17.3. We shall consider here only the important special case, when the velocity amplitude of the spherical surface depends only on the polar angle θ.

In our analysis we shall take advantage of the fact that the amplitude function can be decomposed in terms of a set of orthogonal functions of θ. This decomposition is analogous to the Fourier expansion of a function $f(x)$ in terms of sine and cosine functions (harmonic functions), familiar from the normal modes of a string, for example.

The corresponding functions in this spherical case are called **spherical harmonics** or **Legendre polynomials**. They can be thought of as the normal mode functions of a spherical membrane such as a balloon or a soap bubble and can be expressed mathematically in several different ways.

One possibility is to define the polynomials as the coefficients in a series expansion of the factor

$$1/R = [(x-a)^2 + y^2 + z^2]^{-\frac{1}{2}} = (1/r)[1 + (a/r)^2 - 2(a/r)(x/r)]^{-\frac{1}{2}}$$

in terms of a/r, where $r^2 = x^2 + y^2 + z^2$.

Quantity $1/R$ appears in the electrical potential from an electric point charge at $x = a$ or in the amplitude factor of the complex pressure amplitude from a monopole source at $x = a$. With the source at the origin of the coordinate system, the field is spherically symmetrical, but with the source off-center, an angular dependence is introduced through $x/r = \cos(\theta)$.

With $1/R = (1/r)F(a/r)$, we have, with $\alpha = a/r$,

$$F(\alpha) = [1 + \alpha^2 - 2\alpha\cos(\theta)]^{-\frac{1}{2}}$$

Display 17.3.

Sphere with nonuniform velocity distribution.
Legendre polynomials.

Complex velocity amplitude

At $r=a$: $u=u(\theta)$ (1)

Expansion in spherical harmonics (Legendre polynomials)

$$u(\theta)=\Sigma U_m P_m(\cos\theta) \quad m=0,1,2,\ldots \tag{2}$$

$$U_m=(1/M)\int_0^\pi u(\theta)P_m(\cos\theta)\sin(\theta)\mathrm{d}\theta \quad M=2/(2m+1) \tag{3}$$

$$U_0=\frac{1}{2}\int_0^\pi u(\theta)\sin(\theta)\mathrm{d}\theta=\frac{1}{4\pi}\int u(\theta)2\pi\sin(\theta)\,\mathrm{d}\theta=u_{\mathrm{av}} \tag{4}$$

Properties of Legendre polynomials

Possible definition of P_m

$$F(\alpha)=1/\sqrt{1+\alpha^2-2\alpha\cos(\theta)}=\Sigma P_m(\cos\theta)\alpha^m \quad m=0,1,2,\ldots \tag{5}$$

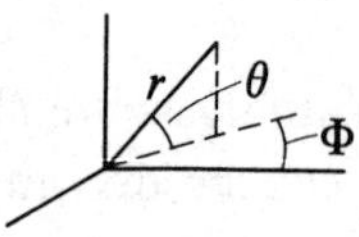

$$P_m=(1/m!)[\partial^m F/\partial\alpha^m]_{\alpha=0} \tag{6}$$

Can be written also as

$$P_m(\eta)=(1/2^m m!)\mathrm{d}^m(\eta^2-1)^m/\mathrm{d}\eta^m \quad \eta=\cos(\theta) \tag{7}$$

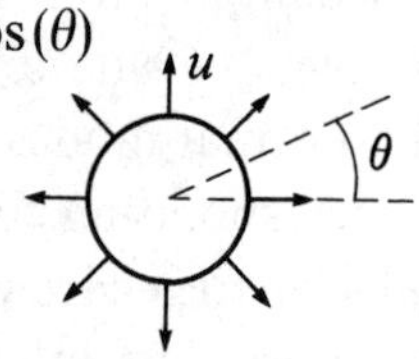

$$P_0(\eta)=1 \tag{8}$$

$$P_1(\eta)=\eta=\cos(\theta) \tag{9}$$

$$P_2(\eta)=\tfrac{1}{2}(3\eta^2-1)=\tfrac{1}{4}[3\cos(2\theta)+1] \tag{10}$$

$$P_3(\eta)=\tfrac{1}{2}(5\eta^3-3\eta)=(1/8)[5\cos(3\theta)+3\cos(\theta)] \tag{11}$$

Orthogonality

$$\int_{-1}^1 P_m(\eta)P_n(\eta)\mathrm{d}\eta$$

$$=\int_0^\pi P_m(\cos\theta)P_n(\cos\theta)\sin(\theta)\mathrm{d}\theta=0 \quad m\neq n \tag{12}$$

$$=2/(2m+1) \quad m=n \tag{13}$$

and this function is now expanded in a Taylor series in α

$$F(\alpha)=F(0)+(\partial F/\partial\alpha)\alpha+(1/2!)(\partial^2 F/\partial\alpha^2)\alpha^2+\cdots=\Sigma P_m\alpha^m \quad (1)$$

$$P_m=(1/m!)[\partial^m F/\partial\alpha^m]_{\alpha=0} \quad (2)$$

The expansion coefficients P_m, the Legendre polynomials, are functions of $\eta=\cos(\theta)$, and carrying out the differentiation in Eq. (2), we obtain

$$P_0(\eta)=1 \quad (3)$$
$$P_1(\eta)=\eta$$
$$P_2(\eta)=(\tfrac{1}{2})[3\eta^2-1]$$
$$\ldots$$
$$\eta=\cos(\theta)$$

It can be shown that the mth order Legendre polynomial can be written as

$$P_m(\eta)=\frac{1}{2^m m!}\frac{\mathrm{d}^m}{\mathrm{d}\eta^m}(\eta^2-1)^m \quad (4)$$
$$\eta=\cos(\theta)$$

where $m!=1\cdot 2\cdot 3\cdots m$ is the fractional.

The Legendre polynomials are orthogonal functions, and the orthogonality is stated here without proof,

$$\int_{-1}^{1} P_m(\eta)P_n(\eta)\mathrm{d}\eta$$
$$=\int_0^{\pi} P_m[\cos(\theta)]P_n[\cos(\theta)]\sin(\theta)\mathrm{d}\theta=0 \qquad n\neq m \quad (5)$$
$$=2/(2m+1) \qquad n=m$$

An arbitrary velocity distribution $u(\theta)$ on the sphere can be decomposed into spherical harmonics

$$u(\theta)=\Sigma U_m P_m[\cos(\theta)] \quad (6)$$

$$U_m=[(2m+1)/2]\int_0^{\pi} u(\theta)P_m[\cos(\theta)]2\pi\sin(\theta)\mathrm{d}\theta \quad (7)$$

$$U_0=\frac{1}{2}\int u(\theta)\sin(\theta)\mathrm{d}\theta=(1/4\pi)\int u(\theta)\sin(\theta)\,\mathrm{d}\theta=u_{\mathrm{av}} \quad (8)$$

It should be noted that the Legendre functions used here are not normalized, and the expression for the expansion coefficient U_m, therefore, contains the normalization constant $2/(2m+1)$.

The sound pressure contributed by the mth term in the series expansion for the surface velocity of the sphere has the same angular dependence $P_m(\cos\theta)$, and the sound pressure will be proportional to U_m.

The first term, corresponding to $m = 0$ and $P_0 = 1$, contributes a spherically symmetrical field, and is the only contribution in the case of a uniform velocity distribution. It is important to realize, however, that a symmetrical contribution, a monopole field, is obtained even if the velocity distribution is not uniform, as long as the average value of the velocity over the sphere is different from zero. This is expressed explicitly by Eq. (8), which shows that the monopole strength is determined by the average velocity.

The second term, corresponding to $m = 1$ and $P_1 = \cos(\theta)$, results from the velocity distribution which is produced by a sphere which oscillates back and forth in the $\theta = 0$ direction. It has a nodal circle at 90 degrees, along which the radial component of the velocity is zero. This angular dependence is characteristic of a dipole field, discussed in the previous section.

Similarly, the $m = 2$ term corresponds to a velocity distribution with two nodal circles, yielding an angular dependence of the sound pressure amplitude typical of a quadrupole field.

Far field amplitude at low frequencies. So far, our discussion of the sound pressure field from a sphere with an arbitrary surface velocity distribution has dealt only with the angular dependence of the field. A detailed analysis of the complete radial dependence of the amplitude will not be given here.

It follows from conservation of wave energy, however, that in the far field the amplitude varies as $1/r$ for each mode m and this is equivalent to a multipole field of order m. It was shown in the previous section, that the far field amplitude of this multipole is of the order of $(kd)^m p_0$, where p_0 is the monopole amplitude and d the characteristic distance between the monopoles which make up the multipole. In this case this distance corresponds to the sphere radius a.

Thus, if the wavelength is much larger than the radius of the sphere, $ka \ll 1$, the multipole amplitude decreases rapidly with m. Therefore, the main contribution to the far field at low frequencies is provided by the mode of lowest m, and if the source has an average velocity over the sphere different from zero, the monopole field will dominate ($m = 0$).

As an illustration, we shall comment on the performance of loudspeakers at low frequencies. With reference to Display 17.4, we regard a loudspeaker simply as an oscillating piston. If such a piston oscillates back and forth in free space, as indicated in Fig. (*d*), there will be no oscillatory net flow out through a control surface surrounding the piston. The reason is, of course, that the outflow on one side of the piston is cancelled by the inflow

Display 17.4.

(*a*) Pulsating sphere. (*b*) Oscillating sphere. (*c*) Loudspeaker in a sealed cabinet. (*d*) Free loudspeaker. (*e*), (*f*), (*g*). Four loudspeakers in a sealed cabinet with different relative phases.

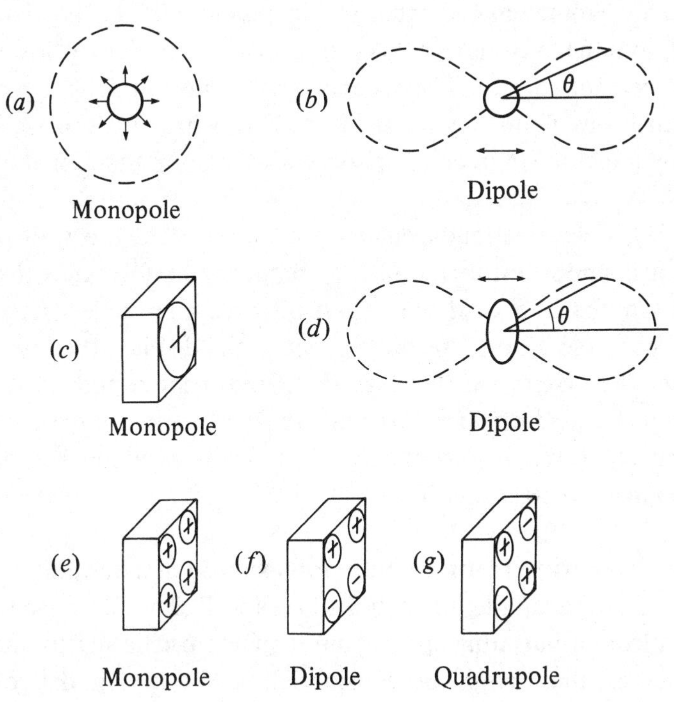

on the 'suction' side. The average velocity amplitude over the control surface and hence the monopole strength then will be zero. Thus, the leading low frequency contribution to the far field will be the dipole field, corresponding to $m = 1$.

On the other hand, if the piston is mounted in one of the walls of a sealed baffle cabinet, as indicated in Fig. (*c*), one side of the piston will be shielded from the outside, and this results in a net oscillatory flow amplitude over the control surface. The speaker system now will have a monopole field, with a low frequency far field pressure which is improved approximately by a factor λ/a in comparison with the amplitude for the unbaffled piston.

Similarly, if several loudspeakers are mounted in the wall of a baffle cabinet, it is important for good low frequency performance that they all be driven in phase to yield the largest possible monopole strength.

In the example of four speakers given in the display, the low frequency field from the system can be degraded from a monopole to a dipole or quadrupole far field by operating pairs of speakers in phase or 180 degrees out of phase. (Reversal of polarity of the leads to a speaker produces a phase change of 180 degrees).

17.4 Radiation from a continuous source distribution

In the examples in Sections 17.1 and 17.2, we discussed the sound radiation from a pulsating sphere and a sphere oscillating to and fro. For small spheres, these could be considered as physical models of acoustic monopole and dipole sources, respectively. Earlier, in Chapter 9, we considered the corresponding problem of electromagnetic radiation from an oscillating point charge yielding a dipole field, analogous to the acoustic field from the oscillating sphere.

We shall now consider the problem of the radiation from a continuous distribution of sources, and start with the acoustic field.

Acoustic radiation. The essential property of the pulsating sphere is that it produces a net time dependent mass flow through a control surface surrounding the sphere, and there is no net force on the fluid.

For the sphere oscillating to and fro, the situation is reversed; there is no net mass flow through the control surface, but a net rate of momentum transfer is produced. The corresponding force on the fluid can be interpreted as resulting from the acceleration of the 'induced mass' in the local flow field around the sphere. This induced mass equals half the mass of the fluid displaced by the sphere, which follows from the expression for the reaction force on the sphere which is discussed in Problem 17.3. In other words, the acoustic dipole field can be considered to be caused by a time dependent force applied to the fluid.

On the basis of these considerations, a continuous distribution of mass and momentum sources in the fluid are expected to represent a monopole and dipole source distribution, respectively.

In reality, a mass source distribution would require mass creation, which does not exist in a one component fluid. On the other hand, a distribution of sources of mass injection in the fluid can be described in terms of a source distribution, at least approximately. Rigorously, it should be expressed as a boundary effect.

A force distribution is not difficult to picture physically. It can be electrical in nature, and even gravity can be involved, since the gravitational force per unit volume will be time dependent if the fluid density is time dependent.

We shall express the rate of mass injection per unit volume by Q and the force per unit volume by F. The linearized equations for the conservation of mass and momentum are then

$$\partial\delta/\partial t + \rho\,\mathrm{div}\,(\mathbf{u}) = Q \tag{1}$$
$$\rho\partial u/\partial t = -\,\mathrm{grad}\,(p) + \mathbf{F} \tag{2}$$

where δ, p, and u are the perturbations in density, pressure, and fluid velocity, respectively, where $p = v^2\delta$ and v the speed of sound, as discussed in Chapter 10. Differentiating Eq. (1) with respect to t and taking the divergence of the Eq. (2), we can eliminate the velocity to obtain

$$\nabla^2 p - (1/v^2)\partial^2 p/\partial t^2 = -\,\partial Q/\partial t + \mathrm{div}\,(\mathbf{F}) = -\,S(r, t) \tag{3}$$

We can include in the force $\mathbf{F}$ also the effect of the pressure fluctuations $P(r, t)$, produced in the fluid as a result of (turbulent) or time dependent flow, which is not due to the compressibility of the fluid, but are of the nature of 'Bernoulli' forces. This contribution to the force is $-\,\mathrm{grad}\,(P)$, and the corresponding source term is then $-\,\mathrm{divgrad}\,(P) = -\,\nabla^2 P$. Thus the source function of the right hand side of Eq. (3), extended in this manner can be written

$$S(r, t) = \partial Q/\partial t - \mathrm{div}\,(\mathbf{F}) + \nabla^2 P \tag{4}$$

Before writing down the solution to the inhomogeneous wave equation (4), it is useful to recall the familiar inhomogeneous equation for the electrostatic potential. This equation follows from $\varepsilon\,\mathrm{div}\,(E) = \rho_c$ and $E = -\,\mathrm{grad}\,(\phi)$, and is

$$\nabla^2\phi = -\,\rho_c/\varepsilon \tag{5}$$

with the solution

$$\phi(r,t) = \int \rho_c(r',t)\,\mathrm{d}v'/4\pi\varepsilon|\mathbf{r}-\mathbf{r}'| \tag{6}$$

where $\mathrm{d}v'$ is the volume element of the source distribution.

The solution to the inhomogeneous wave equation is of the same form with the electric source function $\rho_c(r,t)/\varepsilon$ replaced by the source function $S(r,t')$, where t' is the retarded time $t-|\mathbf{r}-\mathbf{r}'|/v$, i.e.

$$p(r,t) = \int S(r',t')\,\mathrm{d}v'/4\pi|\mathbf{r}-\mathbf{r}'| \tag{7}$$

where r stands for the three coordinates x, y, z.

We now consider a Q- and **F**-distribution concentrated to such a small region at the origin that it can be considered to be a point source. Furthermore, we let the force be in the x-direction so that $\mathrm{div}(\mathbf{F}) = \partial F/\partial x$. The solution in Eq. (7) can then be written, for $r \gg r'$,

$$\begin{aligned} p(r,t) &= (1/4\pi r)\partial q(t')/\partial t - (1/4\pi r)\partial f(t')/\partial x \\ &= (1/4\pi r)\partial q(t')/\partial t + [\cos(\theta)/4\pi r v]\partial f(t')/\partial t \quad t' = t - r/v \end{aligned} \tag{8}$$

where $q = \int Q\mathrm{d}v'$ and $f = \int F\mathrm{d}v'$ and where we have used

$$\partial f/\partial x = -(1/v)(\partial f/\partial t)\partial r/\partial x \quad \text{and} \quad \partial r/\partial x = x/r = \cos(\theta)$$

The expressions for the pressure field from the point flow source q and the point force f at the origin are consistent with the results obtained for the pulsating and oscillating spheres in Sections 1 and 2.

The volume integral of the source term $-\mathrm{div}(\mathrm{grad}\,P)$, accounting for flow induced pressure fluctuations, can be expressed as a surface integral over the boundaries of the region considered where the integrand is the component of $\mathrm{grad}(P)$ normal to the boundary.

Electromagnetic radiation. In our discussion of the radiation from an oscillating point charge in Chapter 9, we noted that the radiation field could be expressed in terms of the time dependent dipole moment of the charge, and we expect that in the case of a continuous source distribution, the field can be expressed in terms of the dipole moment per unit volume.

If we denote the dipole moment per unit volume by P, the current density and the charge density can be expressed as

$$\mathbf{J} = \partial\mathbf{P}/\partial t \quad \rho_c = -\mathrm{div}(\mathbf{P}) \tag{9}$$

consistent with the continuity equation $\partial\rho_c/\partial t + \mathrm{div}(\mathbf{J}) = 0$.

The Maxwell equations can then be written

$$\mathrm{curl}\,(\mathbf{E}) + \mu\partial\mathbf{H}/\partial t = 0 \tag{10}$$

$$\mathrm{curl}\,(\mathbf{H}) - \varepsilon\partial\mathbf{E}/\partial t = \partial\mathbf{P}/\partial t \tag{11}$$

$$\mathrm{div}\,(\mathbf{H}) = 0 \tag{12}$$

$$\mathrm{div}\,(\mathbf{E}) = -(1/\varepsilon)\,\mathrm{div}\,(\mathbf{P}) \tag{13}$$

In this form, the 'source terms' in these equations are all expressed in terms of the dipole moment density **P**. The solution to these equations is facilitated, if we introduce a field vector **Π**, called the Hertz vector, in terms of which the fields are

$$\mathbf{E} = \mathrm{grad}\,(\mathrm{div}\,\Pi) - (1/c^2)\partial^2\Pi/\partial t^2 \quad \mathbf{H} = \varepsilon\,\mathrm{curl}\,(\partial\Pi/\partial t) \tag{14}$$

and the rectangular coordinates of this vector satisfy the wave equation

$$\nabla^2\Pi_i - (1/c^2)\partial^2\Pi_i/\partial t^2 = -(1/\varepsilon)P_i \tag{15}$$

where $c = (1/\varepsilon\mu)^{\frac{1}{2}}$ is the wave speed.

In other words, in terms of the Hertz vector, the problem of electromagnetic radiation is reduced to the solution of the ordinary wave equation with the dipole moment per unit volume as the source of the field. It should be emphasized, that P is the dipole moment of the free charge distribution. The free space solution to the wave equation is obtained in complete analogy with the solution for the acoustic field given above, and after having obtained the components of the Hertz vector, the electric and magnetic fields are obtained from Eq. (14).

17.5 Huygens' principle and scattering

Scattering can be regarded as radiation from a 'source' which is driven by an incident wave rather than by a local external generator. The results obtained earlier about acoustic and electromagnetic radiation, therefore, are directly applicable to the discussion in this section.

To illustrate the basic idea of scattering, let us consider first a plane harmonic wave travelling in the x-direction through a uniform fluid. As a result of the wave, a volume element of the fluid will be set into oscillatory motion in the direction of propagation of the wave, and it will be subject also to periodic compression and rarefaction. Thus, the 'response' of the fluid to the incident wave will involve both the inertial mass and the compressibility of the fluid. In the electromagnetic case, the corresponding quantities are the permeability and the permittivity of the medium describing variations in electron density and the manner in which the electrons are bound in matter.

Because of this response, a volume element of the fluid behaves in much the same way as a sphere which both pulsates and oscillates. From the standpoint of radiation theory, these motions will give rise to monopole and dipole radiation, discussed in the previous section. In the case of an incident plane wave travelling in a uniform fluid, all volume elements will represent sources of equal strength but with different phases, determined by the position of the element along the direction of propagation.

In a wave front, all volume elements oscillate in phase, and the total radiation field is the sum of the fields from the corresponding monopole and dipole sources. These fields are in phase in the forward direction and 180 degrees out of phase in the backwards direction, resulting in zero radiation backwards.

Ahead of the wave front, the fluid is at rest (not yet reached by the wave), and the propagation into this region can be interpreted as the result of the superposition of the 'wavelets' emitted by the monopoles and dipoles in the wave front. In other words, the incident field can be considered to be regenerated by sources which are driven by the incident wave. This interpretation of the propagation of a wave usually is called Huygens' principle.

The situation is different, however, if a foreign body is introduced into the fluid, in which case both the inertial mass and the compressibility of the fluid will be changed. This means that the strength of the 'Huygens' sources will be different from that of neighboring fluid elements. This results in an 'excess' or scattered wave contribution from the nonuniformity. If the size of the nonuniformity is small compared to a wavelength, the scattered monopole and dipole fields will be proportional to $\kappa_s - \kappa$ and $\rho_s - \rho$ respectively, where the subscript s refers to the scatterer and κ and ρ are the compressibility and mass density of the fluid.

Similar arguments can be used in the scattering of an electromagnetic wave. Here the sources of the scattered field are the electrons, free or bound, and the nonuniformities in the electron density and the manner in which the electrons are bound lead to scattering. These nonuniformities can be expressed macroscopically in terms of the free charge distribution and the magnetic permeability and the electric permittivity.

17.6 Acoustic scattering

As an example we shall consider the scattering of sound from a spherical inhomogeneity in a fluid. The radius of the sphere is much smaller than the wavelength of the incident sound, which we assume to be a plane harmonic wave. The sphere will be referred to as the scatterer or the target.

The incident wave will drive the sphere in a combined pulsating and translational harmonic motion, and the acoustic radiation from these motions will be the sum of a monopole and a dipole field, which constitutes the scattered sound.

The 'driving' pressure and corresponding force on the sphere generally is not known *a priori* and an exact calculation of this pressure can be complicated. If the scatterer is much smaller than the wavelength of the incident sound, however, it is a good approximation to assume that the driving pressure is the same as that in the incident wave. An analysis of scattering on this assumption is usually called the *Born approximation.*

From the definition of compressibility, it follows that if the pressure perturbation at the surface of the sphere is P, the change in volume of the sphere responsible for scattering, as explained above, will be $\mathrm{d}V = -V(\kappa_s - \kappa)P$ and the corresponding radial surface velocity u is given by $4\pi a^2 U = \mathrm{d}V/\mathrm{d}t = -V(\kappa_s - \kappa)\mathrm{d}P/\mathrm{d}t$. For harmonic time dependence, the corresponding relation between the complex amplitudes becomes

$$z = P/(-U) = \mathrm{i}(3/ka)[\kappa/(\kappa_s - \kappa)] \tag{1}$$

where we have introduced $\kappa = 1/\rho v^2$ for the fluid and $k = \omega/v$. In the Born approximation, the complex pressure amplitude P is taken to be the amplitude p of the incident wave at the location of the scatterer. Then, using the expression for the pressure field radiated from a pulsating sphere, we can express the scattered power in terms of the intensity of the incident wave. This monopole scattering will be discussed further below.

Dipole scattering from a sphere. In a similar manner, the dipole scattering can be analyzed. If the density ρ_s of the scatterer is infinite, so that it does not move, the scattered field (apart from sign) will be the same as the radiation field from an oscillating scatterer with the same velocity amplitude as that of the incident wave.

With a finite mass density ρ_s, however, the sphere will move, and the relative velocity of the fluid and the sphere will be reduced, being zero when $\rho_s = \rho$. Considering now the dipole scattering from a small sphere, the assumption in the Born approximation is that the pressure distribution over the sphere, and hence the force on it, will be the same regardless of the mass density of the sphere.

In the following discussion we denote by u, u_s and ρ, ρ_s the velocity and density of the fluid and target, respectively. Considering a sphere of unit volume, the force on the fluid sphere caused by the incident sound wave is $\rho \partial u/\partial t$. In the Born approximation the same force will be the driving force in the equation of motion for the target as given below.

The induced fluid mass, which for a sphere is $\rho/2$, is a result of the motion of the sphere *relative* to the fluid and the corresponding force on

Display 17.5.

Scattering from a spherical nonuniformity in compressibility in a fluid.

Compressibility of fluid: $\kappa = 1/\rho v^2$

Compressibility of scatterer: κ_s

Volume change responsible for scattering:

$dV/dt = 4\pi a^2 U = -V(dP/dt)(\kappa_s - \kappa) \quad V = 4\pi a^3/3$ (1)

Harmonic time dependence: $dP/dt = -i\omega P$

$P = 3U/i\omega a(\kappa_s - \kappa) = -i(3\rho v U/ka)[\kappa/(\kappa_s - \kappa)]$ (2)

Corresponding surface impedance of scatterer:

$z = P/(-U) = i(3\rho v/ka)[\kappa/(\kappa_s - \kappa)] = i\chi$ (3)

If losses present: $z = \rho v(\theta + i\chi)$ (4)

Incident compl. pressure amplitude $= p$ (5)

Scattered compl. pressure amplitude $= p'$ (6)

Total amplitude: $P = p + p'$ (7)

Radiation impedance: $z' = \rho v(\theta' + i\chi')$

$= p'/U \simeq [(ka)^2 - i(ka)]\rho v$ (8)

With $P = -zU$ and $p' = z'U$, Eq. (7) yields

$U = p/(z + z')$ (9)

Scattered pressure, far field:

$p'(r) = (a/r)z'U \exp(ikr)$ (10)

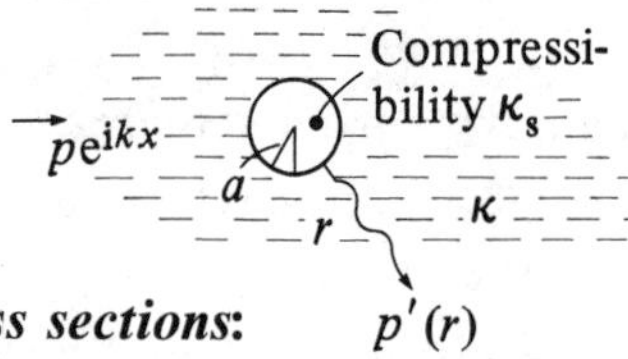

Scattering and absorption cross sections:

$\sigma_s = 4\pi a^2 \rho v \theta'/[(\theta + \theta')^2 + (\chi + \chi')^2]$ (11)

$\sigma_a = (\theta/\theta')\sigma_s$ (12)

$ka \ll 1$: $\sigma_s = 4\pi a^2 (ka)^2/\chi^2 = (4\pi a^2/9)(ka)^4[(\kappa_s - \kappa)/\kappa]^2$ (13)

Resonance: $\sigma_s = \lambda_0^2/\pi \quad (\sigma_a)_{max} = \lambda_0^2/4\pi$ (14)

Resonance frequency: $(\omega_0 a/v) = [3\kappa/(\kappa_s - \kappa)]^{1/2}$ (15)

the fluid is $(\rho/2)\partial(u_s - u)/\partial t$. The equation of motion of the sphere resulting from the interaction with the incident sound wave is then $\rho\partial u/\partial t = \rho_s\partial u_s/\partial t + (\rho/2)\partial(u_s - u)/\partial t$ from which follows $u_s/u = 3\rho/(\rho + 2\rho_s)$.

The relative velocity $u_s - u = (\rho - \rho_s)/(\rho + 2\rho_s)$. Thus, using the induced mass $\rho/2$ of the sphere, the force on the fluid becomes $(\rho/2)\partial(u_s - u)/\partial t$ and it is this force which is the cause of the scattered field. For a stationary sphere, corresponding to $\rho_s = \infty$ and $u_s = 0$, the force is $-(\rho/2)\partial u/\partial t$. We denote the corresponding scattered pressure by p_1. Since the scattered pressure is proportional to the force on the sphere, the scattered pressure by the target sphere becomes

$$p_s = p_1 \cdot 2(\rho_s - \rho)/(2\rho_s + \rho) \tag{2}$$

The scattered pressure field p_1 can be obtained directly from the result for the radiation field from the oscillating sphere, discussed in Section 2. For further details we refer to the problem section.

Monopole scattering from a sphere. We can improve on the Born approximation for monopole scattering from a small sphere, outlined above, by including the scattered field in the 'driving' pressure at the surface of the sphere. In carrying out this calculation, it is convenient to introduce the radiation impedance of the sphere as well as the impedance of the surface of the sphere, obtained as the ratio $P/(-U)$ in Eq. (1) (this impedance is defined as the ratio between the pressure and the inward velocity $-U$ of the surface of the sphere).

If there are no losses in the sphere, κ_s is real; losses can be accounted for, however, by making κ_s complex, and the impedance then contains also a real part, and we put

$$z = \rho v(\theta + i\chi) \tag{3}$$

Similarly, the radiation impedance z' is the ratio between the complex pressure amplitude p' and the radial velocity amplitude U in the scattered field at the surface of the sphere. This impedance is given in Eq. 17.1(5), which for small values of ka reduces to

$$z' = \rho v[(ka)^2 - ika] \tag{4}$$

The first term is the radiation resistance and the second term is the mass reactance, which results from the induced mass of the fluid around the sphere which equals three times the fluid mass displaced by the sphere. (Reaction force per unit area on the sphere is then $-i\omega\rho a U = -i\omega(M/4\pi a^2)U$, where $M = 3V\rho$ and $V = 4\pi a^3/3$ the volume of the sphere).

With the incident pressure amplitude at the surface of the sphere being p and the scattered p', we obtain $P = p + p' = -zU$, and with $p' = z'U$ it follows that

$$p = -(z + z')U = -[\theta + \theta' + i(\chi + \chi')]\rho v U \tag{5}$$

As indicated above, the radiation reactance represents a mass load and the sphere reactance χ, according to Eq. (1), represents a stiffness if $(\kappa_s - \kappa) > 0$. Consequently, the total reactance becomes zero, $\chi + \chi' = 0$, at a certain resonance frequency given by

$$(k_0 a)^2 = 3\kappa/(\kappa_s - \kappa) \quad k_0 = \omega_0/v \tag{6}$$

From Eq. (5) follows $U = -p/(z + z') = -p/[(\theta + \theta') + i(\chi + \chi')]\rho v$, and the scattered power can be expressed as $W' = 4\pi a^2(\rho v\theta')|U|^2/2$, or

$$W' = \{4\pi a^2\theta'/[(\theta + \theta')^2 + (\chi + \chi')^2]\}|p|^2/2\rho v \tag{7}$$

where $I = |p|^2/2\rho v$ is the intensity of the incident wave.

We note that the scattered power can be interpreted as the product of an area and the intensity of the incident wave. This area, which is called the **scattering cross section**, is

$$\sigma_s = 4\pi a^2\theta'/[(\theta + \theta')^2 + (\chi + \chi')^2] \tag{8}$$

In similar manner, the power absorbed by the target can be expressed as $4\pi a^2(\rho v\theta)|U|^2/2$, and the corresponding **absorption cross section** is

$$\sigma_a = 4\pi a^2\theta/[(\theta + \theta')^2 + (\chi + \chi')^2] \tag{9}$$

At resonance, the total reactance is zero, and the scattering cross section $4\pi a^2\theta'/(\theta + \theta')^2$. If the scatterer is loss free, $\theta = 0$, we obtain $\sigma_s = 4\pi a^2/\theta'$, or, with $\theta' = (ka)^2$,

$$\theta_s = \lambda_0^2/\pi \quad \text{(resonance)} \tag{10}$$

Assuming that we can vary the target resistance θ, we note that the resonance absorption cross section will have a maximum value when $\theta = \theta'$,

$$(\sigma_a)_{max} = 4\pi a^2/4\theta' = \lambda_0^2/4\pi \quad \text{(resonance)} \tag{11}$$

Under these conditions the scattering cross section has the same value as the absorption cross section.

At low frequencies, such that the target reactance $\chi = (3/ka)[\kappa/(\kappa_s - \kappa)]$ dominates (see Eq. (4)), the scattering cross section becomes

$$\sigma_s \simeq 4\pi a^2(ka)^2/\chi^2 = (4\pi a^2/9)(ka)^4[(\kappa_s - \kappa)/\kappa]^2 \quad (ka \ll 1) \tag{12}$$

It is interesting to note that σ_s is proportional to the fourth power of the frequency. For a rigid sphere $\kappa_s = 0$ and the scattering cross section will be $(4\pi a^2/9)(ka)^4$.

17.7 Electromagnetic scattering

A classic example of electromagnetic scattering, called **Thomson scattering**, involves a single free electron exposed to an incident plane

Display 17.6.

EM scattering by a charge. Thomson scattering.

Incident EM wave

$E_y = E_0 \cos(kx - \omega t) \quad H_z = H_0 \cos(kx - \omega t)$ (1)

$S = E_0 H_0/2 = \varepsilon_0 c E_0^2/2$ (2)

Induced motion of charge

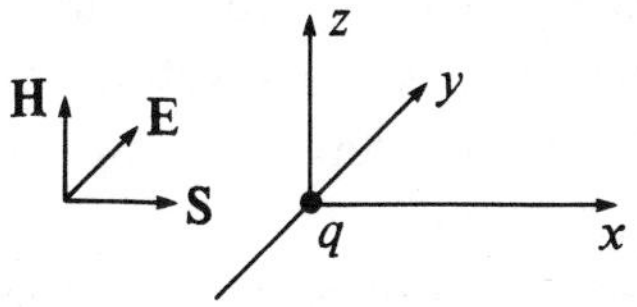

$m\mathrm{d}^2y/\mathrm{d}t^2 + Ky = qE_y$ (3)

Complex amplitude equation:

$m(-\omega^2 + \omega_0^2)y = qE$ (4)

$p = qy = (q^2/m)E_0/(\omega_0^2 - \omega^2)$ (5)

Scattered power

$P = (1/12\pi\varepsilon_0 c^3)\omega^4 |qy|^2$ (See Eq. 9.5(8)) (6)

$$P = (1/12\pi\varepsilon_0 c^3)[\omega^4/(\omega_0^2 - \omega^2)^2](q^2/m)^2 E_0^2$$
$$= (1/6\pi\varepsilon_0^2 c^4)[\omega^4/(\omega^2 - \omega^2)^2](q^2/m)^2 S \quad (7)$$

Electron scattering cross section $(q = -e)$

$\sigma = P/S = (2/3)\mathrm{A}\omega^4/(\omega_0^2 - \omega^2)^2$ (8)

$A = 4\pi a^2$ (9)

$a =$ classical electron radius $= e^2/4\pi\varepsilon_0 mc^2 \simeq 2.8 \times 10^{-17}\,\mathrm{m}^2$ (10)

Thomson scattering: $(\omega_0 = 0)$

$\sigma = 2A/3 \simeq 6.6 \times 10^{-33}\,\mathrm{m}^2$ (11)

Rayleigh scattering: $(\omega_0 \gg \omega)$

$\sigma = (2/3)A(\omega^4/\omega_0^4) = \mathrm{const.}\cdot\omega^4$ (12)

Electron charge: $e = 1.6 \times 10^{-19}$ Coul.
Electron mass: $m = 9.1 \times 10^{-31}$ kg
$4\pi\varepsilon_0 \simeq (1/9) \times 10^{-9}$ F/m
Speed of light: $c \simeq 3 \times 10^8$ m/sec.

harmonic wave. As a result of the force produced by the electric field in the wave, the charge will be driven in harmonic motion and will radiate (scatter) like a dipole source, as discussed in Chapter 9. When in motion, there will be a force on the charge also by the magnetic field in the wave, but this field is small compared to the electric force and will be neglected.

We analyze the somewhat more general problem of an elastically bound electric charge, so that it responds like a mass–spring oscillator. The result for a free electron is then obtained if the 'spring constant' in the oscillator is put equal to zero. The result obtained for the bound charge gives us an idea of the frequency dependence of the scattering by an antenna or an atomic electron.

The equilibrium position of the charge is at the origin in our coordinate system and the incident plane EM wave is travelling in the x-direction with the components

$$E_y = E_0 \cos(kx - \omega t) \tag{1}$$

$$H_z = H_0 \cos(kx - \omega t) \tag{2}$$

and the corresponding time average energy flux is given by the magnitude of the Poynting vector

$$S = E_0 H_0/2 = \varepsilon_0 c E_0^2/2 \tag{3}$$

If we neglect the damping produced by radiation, the equation of motion for the charge is

$$m\mathrm{d}^2y/\mathrm{d}t^2 + Ky = qE_y \tag{4}$$

and the corresponding complex amplitude of oscillation is

$$y = (q/m)E_y/(\omega_0^2 - \omega^2)$$
$$\omega_0^2 = K/m \tag{5}$$

With reference to Chapter 9, the amplitude in the radiated field from an electric charge is proportional to the acceleration, or to the second derivative of the dipole moment $p = qy$. The time average of the radiated power, according to Eq. 9.5(7) is

$$P = (1/12\pi\varepsilon_0 c^3)\omega^4 |p|^2 \tag{6}$$

The complex amplitude of the dipole moment is obtained from Eq. (5),

$$p = qy = (q^2/m)E_0/(\omega_0^2 - \omega^2) \tag{7}$$

and the radiated (scattered) power is seen to be proportional to E_0^2 and hence to the energy flux S in Eq. (3) of the incident wave

$$P = (1/6\pi\varepsilon_0^2 c^4)[\omega^4/(\omega_0^2 - \omega^2)^2](q^4/m^2)S \tag{8}$$

The corresponding scattering cross section of an electron, $|q| = e$, i.e. the ratio between scattered power and the incident energy flux, is

$$\sigma_s = P/S = (2/3)A\omega^4/(\omega_0^2 - \omega^2)^2 \quad (9)$$
$$A = 4\pi a^2$$
$$a = \text{classical electron radius} = e^2/4\pi\varepsilon_0 mc^2 = 2.8 \times 10^{-17}\,\text{m}^2$$

We have here introduced the characteristic area $A = 4\pi a^2$, where a is the classical radius of the electron.

The result for a free charge is obtained by putting the 'spring constant' K and hence $\omega_0 = 0$. In that case the scattering cross section becomes

$$\sigma_s = 2A/3 \simeq 6.6 \times 10^{-33}\,\text{m}^2 \quad (10)$$

which is **independent of frequency**. It is called the **Thomson cross section** (*re* Thomson, see Problem 1.20).

For a bound charge, such as an atomic electron, with a resonance frequency much higher than the incident wave frequency, the scattering cross section will be proportional to the **4th power** of the frequency,

$$\sigma_s = (2/3)A(\omega^4/\omega_0^4) = \text{const}\cdot\omega^4 \quad (11)$$

This result is essential in the explanation of the blue of the sky (**Rayleigh scattering**). The electrons in the air molecules scatter the incident white light, and the scattered power is predominantly in the direction perpendicular to the direction of the incident light, consistent with the angular dependence of dipole radiation. With the scattered intensity being proportional to the 4th power of the frequency, it is dominated by the blue part of the light spectrum, and this is the light we observe from the sky above as it is illuminated by light in the 'horizontal' direction.

Through the removal of the blue light through (lateral) scattering, the light transmitted through an air layer will be red, and this is the light we observe at sunrise and sunset, when the light has been transmitted through a comparatively thick air layer on its way to the observer.

Actually, part of the explanation of the laterally scattered light involves the density fluctuations in the atmosphere, which lead to a distribution of volume elements of scatterers which are uncorrelated so that their contributions to the total scattered power can be added, without regard to interference effects.

Problems

17.1 *Pulsating sphere.* With reference to Eq. 17.1(6), the radiation impedance of a small pulsating sphere is dominated by the reactance $-\mathrm{i}(\rho v)ka$ for $ka \ll 1$.

(a) Show that the total reaction of the surrounding fluid is the same as that of a fluid mass equal to three times the mass M of the fluid displaced by the volume of the sphere.
(b) In this same low frequency limit, determine the radial dependence of the velocity in the vicinity of the sphere and show that this velocity is the same as that obtained in an incompressible fluid. Show by direct integration of the kinetic energy in this flow (integration from a to infinity), that the total kinetic energy of this flow is $3Mu^2/2$, where u is the surface velocity of the sphere.

17.2 *Sound radiation.* The surface of a sphere of mean radius $a = 5$ cm oscillates in radial harmonic motion with the frequency 1000 Hz and with a uniform velocity amplitude 0.001 v, where v is the sound speed (348 m/sec). Neglecting sound absorption in the air, at what distance will the sound pressure be equal to the threshold of human hearing (0.0002 dyne/cm^2, rms)?

17.3 *Oscillating sphere.* A rigid sphere of radius a oscillates back and forth in harmonic motion, frequency ω, along the x-axis with the velocity amplitude U. For $\omega a/v \ll 1$, where v is the sound speed in the surrounding fluid, determine the pressure amplitude distribution over the surface of the sphere and calculate the corresponding reaction force on the fluid. Show that it is the same as the force required to oscillate a fluid mass equal to half of the fluid mass displaced by the sphere volume.

17.4 *Spherical harmonic.* A piston (loudspeaker) of area $A = 100$ sq cm is mounted in one side of a rectangular box (cabinet). The piston oscillates in harmonic motion at a frequency of 100 Hz with a displacement amplitude 0.01 cm.

(a) Determine the average sound pressure amplitude (rms value) over a spherical surface of radius $r = 10$ m (i.e. in the far field) with the loudspeaker in the center of the sphere.
(b) If the box is removed, so that the loudspeaker is free, estimate the change in the average rms pressure amplitude at $r = 10$ m and 100 Hz. In what direction is the sound pressure (1) a maximum? (2) a minimum?

17.5 *Scattering.* A sphere of radius $a = 1$ cm is submerged in water. The real part of the compressibility of the sphere is five times that of water but the mass density is the same as that of water. An incident plane sound wave is scattered by the sphere.

(a) At what frequency will the absorption cross section be a maximum?
(b) The absorption cross section in (a) depends on the internal damping in the sphere (expressed by the imaginary part of the compressibility). What is the maximum possible absorption cross section that can be obtained in (a) and what is the corresponding scattering cross section? The speed of sound in water is 1482 m/sec.

17.6 *Scattering of EM waves.* An electromagnetic plane wave propagates in the x-direction with the E-vector in the y-direction. The wave is scattered by a bound electron (atomic electron). Treating the electron as a harmonic oscillator with a resonance frequency ν_0, determine the scattering cross section of the electron if the incident frequency is

(a) $\nu = 0.5\nu_0$
(b) $\nu = 2\nu_0$
(c) In what direction is the scattered wave intensity (1) a maximum? (2) a minimum?

17.7 *Acoustic dipole scattering from a sphere.* With reference to the discussion of acoustic dipole scattering in Section 17.6 and dipole radiation in Section 17.2, calculate the scattered dipole field from a sphere if the density of the sphere is twice the density of the surrounding fluid. Assume $ka \ll 1$, where a is the radius and $k = 2\pi/\lambda$. Also calculate the scattering cross section.

18

Surface waves on a liquid

Surface waves on a liquid can be observed visually and studied experimentally with relatively simple means and are often used in (ripple tank) demonstrations of wave phenomena, such as interference, refraction, diffraction, and scattering. In shallow water, we shall find that the surface waves can be described by the standard field equations and wave equation with a linear dispersion relation, as discussed in Chapter 6.

Under more general conditions, the dispersion relation will be nonlinear, and if viscosity is included, the waves will be overdamped above a certain characteristic frequency.

As an application we shall show how (Brillouin) scattering of laser light from thermal fluctuations of a surface can be used for the measurement of surface tension and viscosity.

18.1 Waves on shallow water

With reference to Display 18.1, the undisturbed water depth is H and the perturbation caused by the wave is h. In equilibrium there is no motion, and the velocity induced by the wave is denoted by u. The vertical component is zero at the rigid bottom, at $z=0$, and in the shallow water approximation, the velocity is assumed to be horizontal and equal to u throughout the liquid layer. By 'shallow' water is meant, that the depth H is much smaller than the wavelength.

The perturbation in height gives rise to a perturbation in pressure

$$p=\rho g h \tag{1}$$

throughout the water layer, where ρ is the density and g the acceleration of gravity. This pressure will be a function of x and t, and the acceleration of the water will be proportional to the pressure gradient

$$\rho \partial u/\partial t = -\partial p/\partial x \tag{2}$$

Display 18.1.

Surface wave on shallow water.

Wave height $\quad h = h(x, t)$ (1)

Perturbation in pressure:

$p = \rho g h$ (2)

Perturbation in fluid velocity $\quad u = u(x, t)$ (3)

Equation of motion

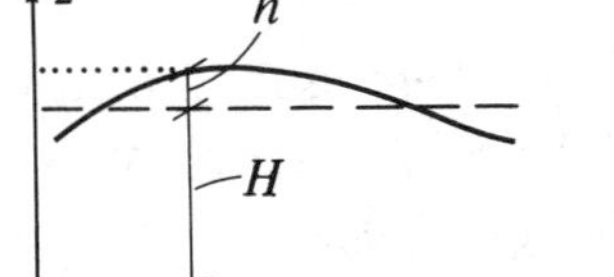

$\rho \partial u/\partial t = -\partial p/\partial x$ (4)

Conservation of mass

$\kappa \partial p/\partial t = -\partial u/\partial x$ (5)

$\kappa = 1/\rho g H$ (6)

Wave equation

$\partial^2 p/\partial x^2 = (1/v^2)\partial^2 p/\partial t^2$ (7)

$v = \sqrt{gH}$ (8)

Wave impedance $\quad (Z = p/u)$

$Z = \rho v$ (9)

Wave energy flux

$I = pu$ (10)

Wave power per unit width of wave front

Harmonic travelling wave $\quad h = h_0 \cos(kx - \omega t)$

$\langle HI \rangle = \tfrac{1}{2}(\rho g h_0^2)\sqrt{gH}$ (11)

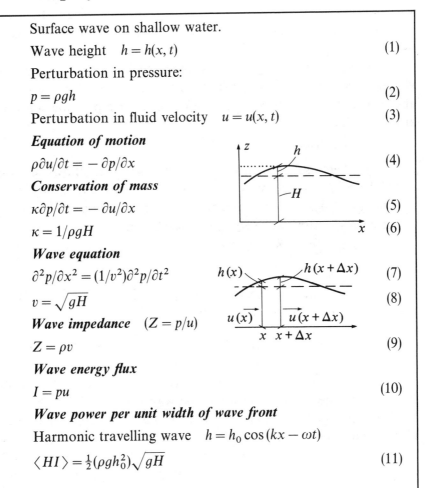

The compressibility of the liquid will be neglected and it follows that the variation in the mass $\rho(H+h)\Delta x$ per unit area in a slice of thickness Δx will be due to the variation of h with time, the rate of change being $\rho\partial h/\partial t = (1/g)\partial p/\partial t$.

The mass flow rate is $Q = \rho u(H+h)$ and the net influx into the control volume (see Display) of thickness Δx is $(-\partial Q/\partial x)\Delta x$. In the linear approximation $\partial Q/\partial x$ becomes $\rho H\partial u/\partial x$. The corresponding conservation of mass equation is then

$$(1/\rho gH)\partial p/\partial t = -\partial u/\partial x \tag{3}$$

We have here used as field variables p and u, and it follows that Eqs. (2) and (3) have the same form as that for a one-dimensional sound wave, with the quantity $1/\rho gH$ being analogous to compressibility.

The corresponding wave equation then becomes

$$\partial^2 p/\partial x^2 = (1/v^2)\partial^2 p/\partial t^2 \tag{4}$$
$$v = \sqrt{gH}$$

where the wave speed v is seen to be proportional to the square root of the water depth, independent of the density.

Wave energy density and flux. Since the field equations describing surface waves are completely analogous to the equations for sound waves, it follows that the expression for the wave impedance, wave energy density, and energy flux formally are the same as for sound waves.

Thus, the potential energy density is $\kappa p^2/2$, where $\kappa = 1/\rho gH$. The corresponding potential energy per unit area of a slice of unit thickness in the x-direction is obtained by multiplying $\kappa p^2/2$ by the depth H yielding $(1/2\rho g)p^2/2 = \rho gh^2/2$. It can be interpreted as the gravitational potential energy of a water column of height h above the equilibrium height H. The weight of this column is ρgh with the center of gravity a distance $h/2$ above the equilibrium height H.

The kinetic energy density is $\rho u^2/2$ and the total energy density is

$$E = \rho u^2/2 + \kappa p^2/2$$
$$\kappa = 1/\rho gH \tag{5}$$

The corresponding wave energy flux, or intensity, is

$$I = \rho u \tag{6}$$

Usually, we are interested in the energy carried by a wave per unit width of the wave front, and to obtain it, we have to multiply the intensity by the appropriate area. The magnitude of this area is simply the depth H, and the power per unit width of the wave is HI.

For a single travelling wave, we have $p = Zu$, where $Z = \rho v$ is the wave impedance. Then, for harmonic time dependence, the time average of HI is

$$\langle HI \rangle = H \langle pu \rangle = \tfrac{1}{2} H p_0 u_0 = \tfrac{1}{2}(\rho g h_0^2)\sqrt{gH} \tag{7}$$

where we have made use of $v = \sqrt{gH}$ and $p_0 = \rho g h_0$, with p_0 and h_0 being the amplitudes of pressure and wave height in the harmonic wave.

As an example, consider a travelling harmonic wave with an amplitude $h_0 = 0.5$ m on a layer of water with depth $H = 3$ m. The corresponding pressure amplitude is $p_0 = \rho g h_0$, and with $\rho = 1000$ kg/m and $g = 9.8$ m/s^2, we get $p = 4905$ N/m^2. The wave speed is $v = gH = 5.42$ m/s and the wave impedance $Z = 5420$ MKS. The energy flux is $I = p_0^2/2Z = 2220$ W/m, and the corresponding energy flow per unit width of the wave front is $HI = 6660$ W/m $= 6.6$ kW/m.

Tsunamis – giant ocean waves. Earthquakes, volcanic eruptions, and underwater landslides can generate surface waves on the ocean with wavelengths which exceed the ocean depth. The ocean then can be considered to be ‘shallow’, and the theory for wave motion in the last section can be applied.

A wavelength in excess of 1000 m and wave speeds of about 100 m/sec are not unusual. At such long wavelengths, a wave height of 1 m is barely perceptible on the ocean although the wave carries a huge amount of energy. For example, with a depth of 400 m and an amplitude of 1 m, the power is 310 kW per meter of wave front.

As the wave approaches the shore, the wave speed decreases as the depth decreases. On the basis of conservation of wave energy, it follows then from Eq. (7), that with decreasing depth H, the wave amplitude h must increase; actually it will vary inversely as the 1/4 power of H.

Consequently, the wave height in shallow water can be quite large, even if the wave height out on the deep ocean may be small, of the order of 1 m. Wave heights resulting from ocean waves of this kind have been reported to be 6–20 m in many cases, and the wave which followed the eruption of Krakotoa in 1883 gave rise to a 40 m high wave close to shore. It destroyed the coastal towns of Java and Sumatra, drowning 36 000 villagers in a matter of minutes.

More recently, the 1946 Alaskan earthquake produced a 30 m high wave, which demolished many building structures along the Alaskan coastline. About 5 hours later, the Alaskan tsunami had sped over 2000 miles across the Pacific ocean and struck the Hawaiian islands, causing damage in excess of 25 million dollars. The height of the wave at the islands was about 9 m.

Display 18.2.

Tsunamis—giant ocean waves.

A tsunami is not a single wave pulse but is more like a harmonic wave train. The crests are generally separated in time by intervals ranging from 15 to 90 minutes, and the entire wave train may last for hours. Often the largest crest is found in the third to eighth crest in the train.

After the 1946 tsunami, a Pacific Tsunami Warning System was set up. It has issued several warnings per year, and has been quite succesful in predicting giant coastal waves.

Channel waves. Tidal motion. The essential feature of shallow water waves is that the vertical component of the fluid velocity is negligible. Thus, regardless of the depth, a wave with predominantly horizontal motion of the fluid will propagate in the manner described in this section. For example, in channel flow, resulting from the removal of a barrier (such as a dam), which separates regions of different water level, the resulting wave has the shape of a step function, rounded at the corners, which propagates down the channel with a speed $v = \sqrt{Hg}$.

The penetration of tidal motion into rivers and channels leads to the same kind of wave motion, often referred to as a **bore**. If the channel depth or width decreases in the direction of wave travel, the amplitude will increase in much the same way as for tsunamis, described above. Although local variations in channel depth and width will affect the wave, it is well described by the theory of shallow water waves. For example, the tidal wave through the English Channel is known to travel with an average speed of about 34 m/s. The depth of the channel varies from 20 to 100 fathoms (1 fathom = 6 ft), and with an average depth of 60 fathoms, i.e. approximately 120 m, the wave speed calculated from $\sqrt{Hg}$ is 34 m/s, in good agreement with the observed value.

The largest variations in the range of the tide (difference between the maximum and minimum water levels) is usually caused by resonance of the water contained in bays. As an example it can be mentioned that the unusually large tidal range of about 50 ft in the Bay of Fundy (separating New Brunswick and Nova Scotia) is due to resonance of the bay.

The bay can be treated approximately as a 170 mile long rectangular channel with a water depth of 240 ft. The channel is 'closed' at the end, so that the fluid velocity (assumed horizontal and uniform in the shallow water approximation) will be zero at the end. The bay is analogous to an organ pipe, closed at one end, so that the fundamental resonance wavelength is four times the channel length.

The channel is driven by the tidal motion, which has a period T of 12 hours. The fundamental resonance frequency of the channel turns out to be quite close to the driving period, so that conditions for resonance excitation are nearly fulfilled. The vertical displacement of the water surface is analogous to the sound pressure in a pipe and will be a maximum at the closed end of the channel. We leave further details of the study of this motion as an excercise.

18.2 Surface tension

In the next section we shall consider the effect of surface tension on the waves on a liquid surface, and as a preparation, we devote this section to a review of the concept of surface tension.

It is an experimental fact that the interface between a liquid and a gas (or between two liquids that do not mix) behaves in much the same way as a stretched membrane. Thus, to increase the surface area of a liquid by dA requires an amount of work dW, which is proportional to the increase in the area,

$$dW = \sigma dA \tag{1}$$

The constant of proportionality, σ, is called the surface tension. It has the dimensions of force per unit length and, typically, is of the order of 20 dyne/cm. The surface tension for water is anomalously high, about 72 dyne/cm. The values for some other liquids are given in Display 18.3.

According to the definition, surface tension can be interpreted as the force per unit length of the surface, as indicated in the display, and it is sometimes useful to introduce a potential surface energy,

$$E = \sigma A \tag{2}$$

For an isolated liquid volume element there is a tendency for the liquid to seek a state of minimum surface energy, as will be illustrated below.

For a liquid film, such as a soap bubble, for example, both sides of the film make up the total surface area, and the corresponding force per unit length of the film will be 2σ.

Soap bubble. As an example, we consider the condition for equilibrium of a soap bubble. This condition is established by imagining the bubble cut in half and introducing the force of surface tension along the rim of the cut. If the radius of the bubble is r_0, this force has the magnitude $(2\pi r_0)2\sigma$. It must be balanced by the force produced by the pressure difference across the film. With the inside and outside pressures denoted by P_1 and P_2, this force is $(P_1 - P_2)\pi r_0^2$, and we obtain

$$(P_1 - P_2)\pi r_0^2 = (2\pi r_0)2\sigma$$
$$P_1 - P_2 = 4\sigma/r_0 \tag{3}$$

Display 18.3.

Surface tension.

Definition of surface tension

Surface area of liquid $= A$

Work required to increase A by $\mathrm{d}A$ $\quad \mathrm{d}W = \sigma \mathrm{d}A$ (1)

$\sigma =$ surface tension (force per unit length)

Surface (potential) energy $\quad E = \sigma A$ (2)

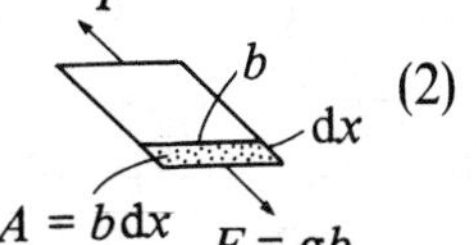

Liquid film

Total area $= 2A$ (two sides)

Force per unit length $= 2\sigma$ (3)

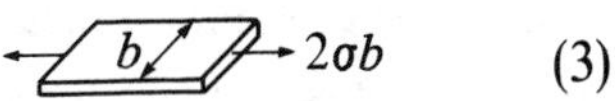

Soap bubble

Equilib. condition

$(P_1 - P_2)\pi r_0^2 = (2\pi r_0)2\sigma$ (4)

$P_1 - P_2 = 4\sigma / r_0$ (5)

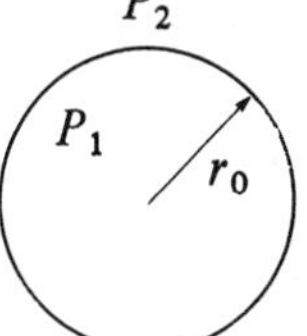

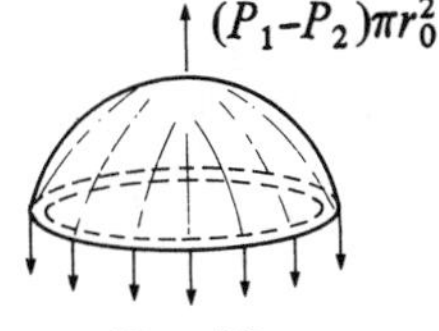

Some values of surface tension

Liquid	σ (dynes/cm)	In contact with:	Temp. (°C)
Acetic acid	27.6	air	20
Argon	13.2	vapor	−188
Benzene	28.9	air	20
Carbon tetrachloride	26.8	air	20
Ethyl alcohol	22.3	air	20
Helium	0.24	vapor	−270
n-Hexane	18.4	air	20
Mercury	470	air	20
Silver	800	air	970
Tin	526	hydrogen	253
Water	74	air	20

In a similar manner we can establish the equilibrium conditions for a gas bubble in the bulk of a liquid. The pressure and surface tension forces enter in the same manner in a study of bubble dynamics, which is of interest in the study of scattering of sound in liquids, for example.

Stability of a liquid column. As an illustration of the role of surface tension in the study of the stability of a liquid, we consider a cylindrical column of a liquid flowing out through a hole at the bottom of a container, as illustrated in Display 18.4. The velocity, and hence the diameter of the column, can be modulated by oscillating the container in a vertical harmonic motion. The resulting perturbation of the column will be wave-like with a wavelength which depends on the frequency of oscillation.

We can investigate the stability of the perturbed liquid cylinder by calculating the surface energy before and after the perturbation. If the energy decreases as a result of the perturbation, the cylinder is unstable for the type of perturbations considered.

The perturbation of the radius is assumed to be harmonic, $\delta \sin(2\pi x/\lambda)$ so that the total radius depends on the position x along the cylinder as

$$r(x) = \bar{r} + \delta \sin(2\pi x/\lambda) \tag{4}$$

where $\bar{r}$ is the mean value of the radius. This mean value is not equal to the unperturbed radius r_0 of the cylinder. The reason is due to the requirement that the unperturbed and perturbed liquid volumes V' and V be the same, and the relation between $\bar{r}$ and r_0 is obtained from $V = V'$, where the perturbed volume V' is

$$V' = \int_0^\lambda \pi r^2(x)\,\mathrm{d}x = \pi \int [r + \delta \sin(2\pi x/\lambda)]^2\,\mathrm{d}x = \pi\lambda(\bar{r}^2 + \delta^2/2) \tag{5}$$

The unperturbed volume is $V = \pi r_0^2 \lambda$, and the requirement $V = V'$ yields

$$\bar{r} = \sqrt{r_0^2 - \delta^2/2} \simeq r_0[1 - (\delta/2r_0)^2] \tag{6}$$

It follows that the mean radius of the perturbed cylinder is somewhat smaller than the unperturbed radius.

Having obtained the expression for the mean radius, we can now determine the perturbed surface area A' covering one wavelength of the cylinder

$$A' = \int_0^\lambda 2\pi r(x)\,\mathrm{d}s = \int 2\pi r(x)\sqrt{1 + (\mathrm{d}r/\mathrm{d}x)^2}\,\mathrm{d}x \tag{7}$$

and the corresponding surface energy is $E' = \sigma A'$.

With $\mathrm{d}r/\mathrm{d}x = 2\pi(\delta/\lambda)\cos(2\pi x/\lambda)$, and assuming $\delta/\lambda \ll 1$, we can approximate the square root in Eq. (7) by the first two terms in the power expansion

Display 18.4.

Instability of a liquid cylinder due to surface tension.

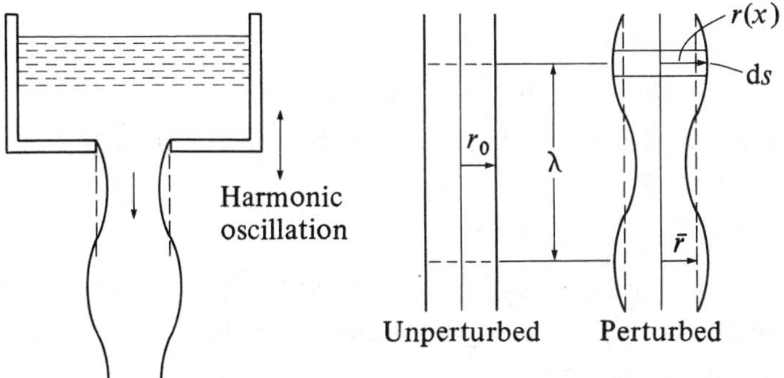

r_0 = radius of unperturbed liquid cylinder

$r(x)$ = radius of perturbed cylinder $= r + \delta \cdot \sin(2\pi x/\lambda)$ (1)

$\bar{r}$ = mean value of $r(x)$

λ = wave length of harmonic perturbation

Perturbation does not change liquid volume

Unperturbed $V = \pi r_0^2 \lambda$ (2)

Perturbed $V' = \int_0^\lambda \pi r^2(x)\,\mathrm{d}x = \pi \int [\bar{r} + \delta \cdot \sin(2\pi x)/\lambda)]^2\,\mathrm{d}x$

$= \pi\lambda(\bar{r}^2 + \delta^2/2)$ (3)

$V = V'$ yields $\bar{r} = \sqrt{r_0^2 - \delta^2/2} \simeq r_0[1 - (\delta/2r_0)^2]$ (4)

But it does change the surface area and the energy

Unperturbed $A = 2\pi r_0 \quad E = \sigma A$ (5)

Perturbed $A' = \int_0^\lambda 2\pi r(x)\,\mathrm{d}s$

$= \int 2\pi r(x)\sqrt{1 + (\mathrm{d}r/\mathrm{d}x)^2}\,\mathrm{d}x \quad E' = \sigma A'$ (6)

$\mathrm{d}r/\mathrm{d}x = 2\pi(\delta/\lambda)\cos(2\pi x/\lambda)$ (7)

$A' \simeq \int 2\pi r(x)[1 + (1/2)(\mathrm{d}r/\mathrm{d}x)^2]\,\mathrm{d}x = 2\pi\bar{r}\lambda[1 + (\pi\delta/\lambda)^2]$ (8)

$A' \simeq 2\pi r_0\lambda[1 - (\delta/2r_0)^2 + (\pi\delta/\lambda)^2]$ (9)

Condition for instability

$E' < E$ or $A' < A$

yields $\lambda > 2\pi r_0$ (10)

to obtain

$$A' \simeq \int 2\pi r(x)[1 + \tfrac{1}{2}(\mathrm{d}r/\mathrm{d}x)^2]\,\mathrm{d}x = 2\pi\bar{r}\lambda[1 + (\pi\delta/\lambda)^2]$$

and inserting the expression for $\bar{r}$ in Eq. (6), we get

$$A' \simeq 2\pi r_0\lambda[1 - (\delta/2r_0)^2 + (\pi\delta/\lambda)^2] \tag{8}$$

The condition for instability can now be expressed as $E' < E$ or $A' < A$, which corresponds to

$$\lambda > 2\pi r_0 \quad \text{(instability)} \tag{9}$$

In other words, the liquid cylinder is unstable for a harmonic perturbation in the radius if the wavelength of the perturbation is larger than the circumference of the cylinder.

In a more detailed analysis of the problem, the rate of growth of a perturbation can be calculated, and it can be shown that the maximum growth rate occurs at a wavelength which is approximately nine times the unperturbed radius.

As a historical note we mention that this problem of modulated liquid stream was investigated both theoretically and experimentally by none less than Niels Bohr (together with P.O. Pedersen). Under the conditions mentioned above, they found the stream to be highly unstable and to break up into drops after only a couple of stream diameters below the container.

Their study included also an investigation of the oscillation of a drop in free fall. A drop when formed is not completely spherical but rather resembles an ellipsoid, and, as a result, will oscillate about its spherical equilibrium shape as it falls. (The spherical shape has the minimum surface area for a given volume). The frequency of oscillation was measured by means of stroboscopic illumination and also calculated in terms of surface tension. From the measured frequency of oscillation, the surface tension of the liquid could then be determined.

It is interesting to note in this context, that the liquid drop model of the nucleus was proposed by Niels Bohr (1936), in which one part of the binding energy of a nucleus is interpreted in terms of the interaction between the necleons on the 'surface'of the nucleus, corresponding to the surface tension of a liquid drop.

18.3 Waves on deep water. Effect of surface tension

The discussion in the previous section is restricted to wavelengths much larger than the depth of the liquid involved. Under these conditions, the constraint of the horizontal bottom forces the velocity to be approximately parallel with the bottom and uniform throughout the liquid.

Display 18.5.

Surface wave on deep water. Gravity waves and ripples.

Surface displacement

$$h(x,t) = A\cos(kx - \omega t) \quad \text{at } z = 0 \tag{1}$$

Corresponding vertical velocity

$$v(z) = (\partial h/\partial t)\exp(kz) \quad z < 0 \tag{2}$$

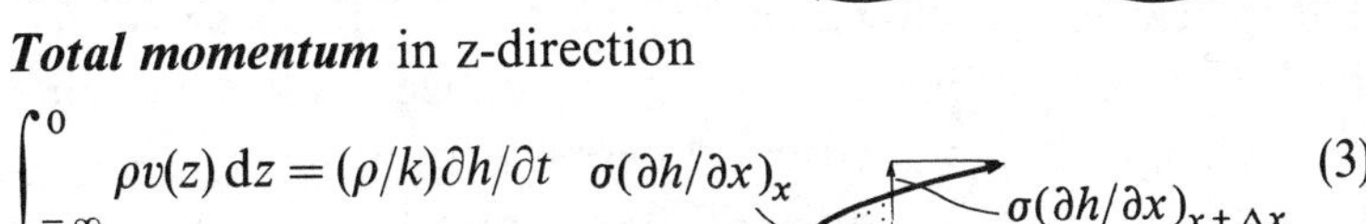

Total momentum in z-direction

$$\int_{-\infty}^{0} \rho v(z)\,\mathrm{d}z = (\rho/k)\partial h/\partial t \tag{3}$$

$\sigma(\partial h/\partial x)_x$ $\quad\sigma(\partial h/\partial x)_{x+\Delta x}$ $\quad(\rho g h)\Delta x$ $\quad\Delta x$

Forces in z-direction

$$\text{Gravity} \quad -(\rho g h)\Delta x \tag{4}$$

Surface tension:

$$\sigma(\partial h/\partial x)_{x+\Delta x} - \sigma(\partial h/\partial x)_x = \sigma(\partial^2 h/\partial x^2)\Delta x \tag{5}$$

Equation of motion

$$(\rho/k)\partial^2 h/\partial t^2 = -\rho g h + \sigma \partial^2 h/\partial x^2 \tag{6}$$

Harmonic time dependence:

$$\partial^2 h/\partial t^2 = -\omega^2 h \quad \partial^2 h/\partial x^2 = -k^2 h \tag{7}$$

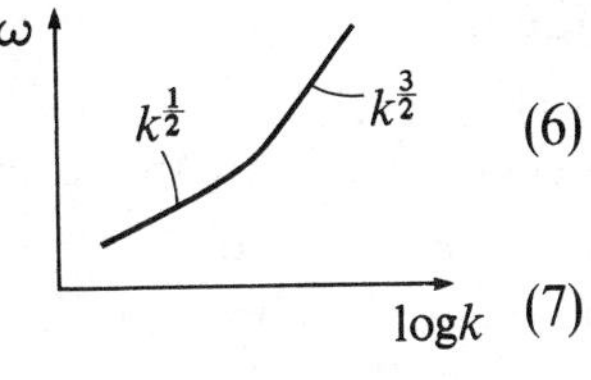

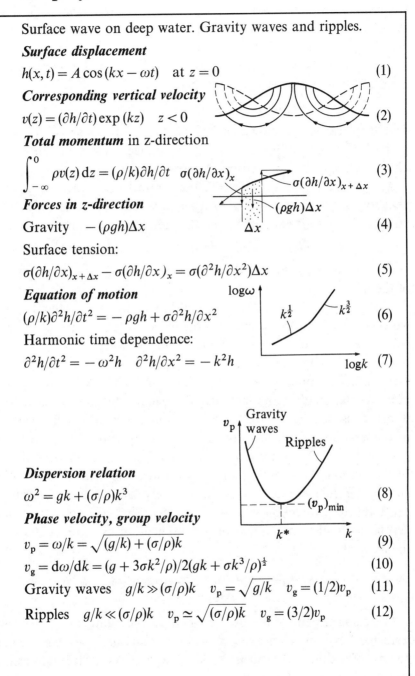

Dispersion relation

$$\omega^2 = gk + (\sigma/\rho)k^3 \tag{8}$$

Phase velocity, group velocity

$$v_p = \omega/k = \sqrt{(g/k) + (\sigma/\rho)k} \tag{9}$$

$$v_g = \mathrm{d}\omega/\mathrm{d}k = (g + 3\sigma k^2/\rho)/2(gk + \sigma k^3/\rho)^{\frac{1}{2}} \tag{10}$$

$$\text{Gravity waves} \quad g/k \gg (\sigma/\rho)k \quad v_p = \sqrt{g/k} \quad v_g = (1/2)v_p \tag{11}$$

$$\text{Ripples} \quad g/k \ll (\sigma/\rho)k \quad v_p \simeq \sqrt{(\sigma/\rho)k} \quad v_g = (3/2)v_p \tag{12}$$

As the depth of the liquid increases, the constraining effect of the bottom will be reduced, and the vertical component of the velocity must be included. We shall consider first the case when the depth is much larger than the wavelength, so that the effect of the bottom can be neglected.

Again, we consider a harmonic travelling wave with a vertical displacement of the liquid surface at $z = 0$ being

$$h(x, t) = A \cos (kx - \omega t) \tag{1}$$

This displacement is shown in Display 18.5 at times separated by half a period. The curved paths below the surface indicate qualitatively the direction of the fluid velocity at different depths.

At the liquid surface, at $z = 0$, the vertical component v of the fluid velocity is $v(0) = \partial h/\partial t$. This velocity component decreases with the depth, and the decrease is exponential so that for negative values of z we have

$$v(z) = (\partial h/\partial t) \exp (kz) \quad (z < 0) \tag{2}$$

where the decay constant is the same as the propagation constant $k = 2\pi/\lambda$ of the wave.

This decay of the velocity with depth is analogous to the decay which we encountered in the study of total reflection of a wave at a boundary when the trace velocity was smaller than the wave velocity. In this case the 'trace velocity' corresponds to the phase velocity ω/k of the surface wave, which is much smaller than the sound velocity in the liquid. (In a more formal analysis the fact that the decay constant is equal to k is a direct consequence of the wave equation, which in this case of an incompressible fluid reduces to the Laplace equation (Problem 18.9)).

A fluid element at a certain location x will oscillate in harmonic motion. In establishing the equation for the vertical component of the motion, we choose the fluid element to be an infinitely long vertical column of unit width (in the y-direction) and thickness Δx. Since the velocity varies with depth, the total z-component of the momentum has to be determined by the integral

$$\rho \int_{-\infty}^{0} v(z)\, \mathrm{d}z = \rho(\partial h/\partial t) \int_{-\infty}^{0} \exp (kz)\, \mathrm{d}z = (\rho/k)\partial h/\partial t \tag{3}$$

The displacement of the fluid element in the z-direction gives rise to a restoring force due to gravity with the z-component $-\rho g h \Delta x$, where the volume of the fluid element is Δx, since we have considered an element of unit width in the y-direction.

The force F_s of surface tension acts in the tangential direction of the surface. The vertical component of F_s at the right side of the element, at

$x + \Delta x$, is $F_s \sin(\theta)$, where θ is the angle of the surface tangent with the x-axis. For small displacement we have $\sin(\theta) \simeq \tan(\theta) = \partial h/\partial x$, and, accounting for the surface tension forces on the two sides of the element, the net force in the z-direction becomes

$$\sigma(\partial h/\partial x)_{x+\Delta x} - \sigma(\partial h/\partial x)_x = \sigma(\partial^2 h/\partial x^2)\Delta x \qquad (4)$$

With the total momentum given by Eq. (3), and the net force in the z-direction equal to $-\rho g h + \sigma \partial^2 h/\partial x^2$ per unit length along the x-axis, the equation of motion becomes

$$(\rho/k)\partial^2 h/\partial t^2 = -\rho g h + \sigma \partial^2 h/\partial x^2 \qquad (5)$$

For harmonic time dependence, this equation establishes the dispersion relation

$$\omega^2 = gk + (\sigma/\rho)k^3 \qquad (6)$$

from which follow the phase and group velocities

$$v_p = \omega/k = [(g/k) + (\sigma/\rho)k]^{\frac{1}{2}} \qquad (7)$$

$$v_g = d\omega/dk = \tfrac{1}{2}(g + 3\sigma k^2/\rho)(gk + \sigma k^3/\rho)^{-\frac{1}{2}} \qquad (8)$$

For long wavelengths, i.e. for small values of k, the force of gravity dominates, and we obtain

$$v_p \simeq \sqrt{g/k} \quad \text{(gravity waves)} \qquad (9)$$

$$v_g \simeq v_p/2$$

both v_p and v_g decreasing with increasing k (and frequency) and with the group velocity being smaller than the phase velocity.

For short wavelengths, on the other hand, we get

$$v_p \simeq \sqrt{(\sigma/\rho)k} \quad \text{(ripples, capillary waves)} \qquad (10)$$

$$v_g \simeq (3/2)v_p$$

with v_p and v_g increasing with k and with v_g larger than v_p.

With this dependence on k it follows that the phase velocity will have a minimum at a certain value of k. We leave it for one of the problems to determine the minimum phase velocity and the corresponding wavelength. For water these quantities are $(v_p)_{min} = 23$ cm/sec and the corresponding wavelength is $\lambda^* = 2\pi/k^* = 1.7$ cm.

The fact that there exist no surface waves with a phase velocity below the minimum value has interesting consequences. For example, in order for a surface wave to be generated by a uniform wind, the wind speed must exceed the minimum value of the phase velocity, 23 cm/sec for water.

Above the minimum, there are two possible wavelengths for each value of the phase velocity. Then, for a given wind speed above the minimum, two waves can be excited, one with $k < k^*$ and one with $k > k^*$, corresponding to long and short wavelengths. The long waves, the gravity waves, require a comparatively long time to build up, and a wind gust usually will excite only the short waves, the ripples or capillary waves.

The ripples make the surface of a body of water appear darker than the surrounding unrippled surface because the short wavelength irregularities of the surface result in the diffusion of the light reflected from the surface. The rapid motion of these dark regions over the water surface is a familiar occurrence.

Effect of water depth. The expression $v_p \simeq \sqrt{g/k}$ in Eq. (9) for the phase velocity of long waves on deep water goes to infinity as k goes to zero. This unrealistic result is due to the fact that the deep water assumption is no longer valid, since in the limit $k = 0$, the wavelength is infinite, and thus cannot be smaller than any finite value of the water depth. Rather, the limit of $k = 0$ must correspond to the shallow water approximation, with the wavelength large compared to H, and the phase velocity should be $v_p = \sqrt{Hg}$, as given by Eq. 1.(4).

It can be shown that we can account for the water depth in the dispersion relation simply by multiplying the right hand side of Eq. (6) by $\tanh(kH)$ to yield

$$\omega^2 = [gk + (\sigma/\rho)k^3] \tanh(kH) \tag{11}$$

In the shallow water limit, the factor $\tan(kH)$ can be approximated by kH, so that $\omega^2 \simeq gHk^2$ and $v_p \simeq \sqrt{gH}$, as it should.

In the other limit, $kH \to \infty$, we have $\tanh(kH) \simeq 1$, and we obtain $\omega^2 \simeq (\sigma/\rho)k^3$. In this limit, the dispersion relation is still unsatisfactory, since it gives an infinite phase velocity. This defect is due to our omission of viscosity, as will be discussed in the next section.

18.4 Effect of viscosity

As indicated above, the dispersion relation 3(11) leads to the unrealistic prediction of an infinite phase velocity as k goes to infinity. This defect can be traced to the omission of viscosity, which produces friction losses in the fluid which increase with k. These losses will reduce the phase velocity; in fact, the motion can become overdamped in much the same way as a harmonic oscillator can be overdamped. The phase velocity is then reduced to zero.

For a standing surface wave with a given wavelength, the frequency of oscillation, as determined from the dispersion relation, will be complex,

Display 18.6.

Dispersion relation for surface waves including the effect of viscosity. Maximum frequency and 'cut-off' wavelength.

Maximum frequency of surface waves

$\omega_m/2\pi = 0.0736 \cdot \sigma^2 \rho/\mu^3$ (1)

Corresponding wave length $\lambda_m = 4.91 \cdot \mu^2/\sigma\rho$ (2)

Critical wave length

$\lambda_c = 3.65 \cdot \mu^2/\sigma\rho$ (3)

$\lambda < \lambda_c$ overdamped (4)

Low frequencies

$\omega_i = -1/\tau = -2\mu k^2/\rho$ (5)

$h(t) \simeq \exp(-t/\tau)$ (6)

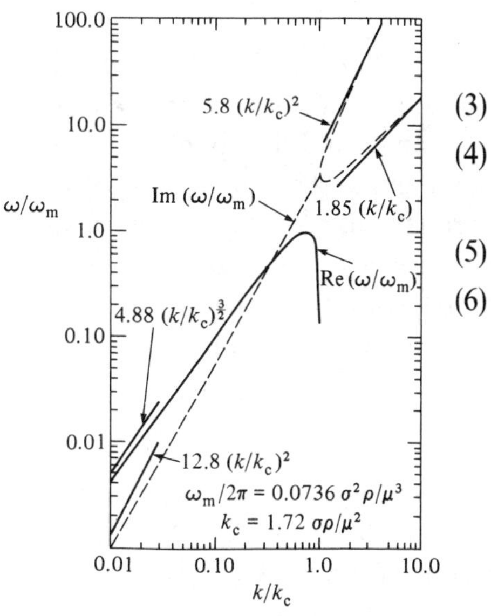

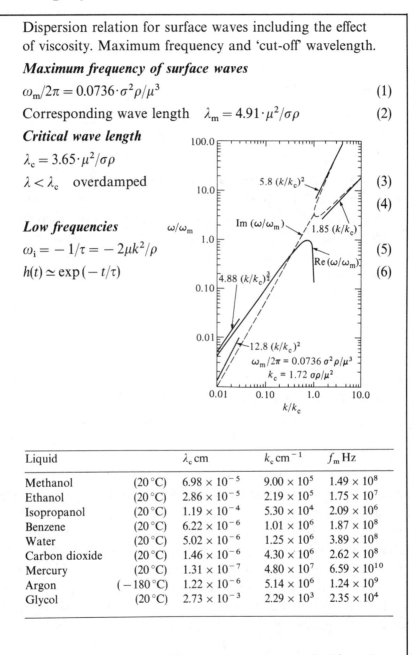

Liquid		λ_c cm	k_c cm^{-1}	f_m Hz
Methanol	(20 °C)	6.98×10^{-5}	9.00×10^{5}	1.49×10^{8}
Ethanol	(20 °C)	2.86×10^{-5}	2.19×10^{5}	1.75×10^{7}
Isopropanol	(20 °C)	1.19×10^{-4}	5.30×10^{4}	2.09×10^{6}
Benzene	(20 °C)	6.22×10^{-6}	1.01×10^{6}	1.87×10^{8}
Water	(20 °C)	5.02×10^{-6}	1.25×10^{6}	3.89×10^{8}
Carbon dioxide	(20 °C)	1.46×10^{-6}	4.30×10^{6}	2.62×10^{8}
Mercury	(20 °C)	1.31×10^{-7}	4.80×10^{7}	6.59×10^{10}
Argon	(−180 °C)	1.22×10^{-6}	5.14×10^{6}	1.24×10^{9}
Glycol	(20 °C)	2.73×10^{-3}	2.29×10^{3}	2.35×10^{4}

Reference: Robert H. Katyl and Uno Ingard, *Phys. Rev. Letters*, **20**, 248 (1968).

$\omega = \omega_r + i\omega_i$. With the time dependence expressed by the time factor $\exp(-i\omega t) = \exp(\omega_i t)\exp(-i\omega_r t)$, we note that a decay corresponds to a negative value of ω_i.

Unfortunately, the derivation of the complete dispersion relation is beyond the present scope, and we give only the final result, which can be presented in the form

$$(1 - i\omega\tau)^2 + (gk + \omega_0^2)\tau^2 - \sqrt{1 - i2\omega\tau} = 0 \qquad (1)$$

$$\omega = \omega_r + i\omega_i$$

$$\omega_0^2 = \sigma k^3/\rho \quad \tau = \rho/2\mu k^2$$

The numerically determined k-dependence ($k = 2\pi/\lambda$) of the frequency of oscillation ω_r and the attenuation constant $|\omega_i|$ are plotted in Display 18.6 as solid and dashed curves, respectively. Actually, both the frequency and the propagation constants are normalized with respect to certain characteristic values ω_m and k_c to be defined shortly.

In the long wavelength regime, $k \ll k_c$, the expressions for ω_r and ω_i reduce to (see Problem 18.8)

$$\omega_r^2 = gk + (\sigma/\rho)k^3 \qquad (2)$$

$$\omega_i = -1/\tau = -2\mu k^2/\rho \quad \mu = \text{shear viscosity coefficient} \qquad (3)$$

The expression for ω_r is consistent with the result obtained for the nonviscous liquid in Eq. 3.(6). (The range of k included in the graph does not extend far enough to low values of k to include the gravity wave regime, i.e. the $\sqrt{k}$-dependence of ω_r). We leave it for one of the problems to show that the long wavelength approximations in Eqs. (2) and (3) are consistent with the general dispersion relation in Eq. (1).

The numerical solution to Eq. (1) shows that, unlike the result for the nonviscous liquid, the frequency ω_r does not increase without limit with increasing k. Rather, the frequency reaches a maximum and then drops rapidly to zero at a certain value of k, which we have denoted by k_c. At this wavelength the motion of the surface is non-oscillatory and corresponds to critical damping of the oscillation. Above this value of k, the motion is overdamped and the frequency remains zero.

The numerically determined expressions for k_c and ω_m (with the corresponding value of k_m) are given by

$$\lambda_c = 2\pi/k_c = 3.65 \cdot \mu^2/\sigma\rho \qquad (4)$$

$$\nu_m = \omega_m/2\pi = 0.0736 \cdot \sigma^2\rho/\mu^3 \qquad (5)$$

$$\lambda_m = 2\pi/k_m = 4.91 \cdot \mu^2/\sigma\rho \qquad (6)$$

$$\sigma = \text{surface tension}, \mu = \text{shear viscosity}, \rho = \text{density}$$

It is notable that the maximum frequency is proportional to the square of the surface tension and inversely proportional to the third power of the shear viscosity. This strong viscosity dependence makes the maximum frequency for water approximately 10 000 times larger for water than for glycol.

The maximum frequency of oscillation ω_m and the smallest possible wavelength λ_c (corresponding to critical damping) are unique properties of a liquid, and they are listed for a few common liquids in the display.

The maximum frequency typically is of the order of 10^5–10^9 Hz and the corresponding wavelength is of the order of the wavelength of visible light. Therefore, such high frequency surface waves can be 'probed' by means of scattering of (laser) light from a liquid surface, as will be discussed in the next section.

The k-dependence of the decay constant, the magnitude of ω_i and the inverse of the characterstic decay time ('life time') of surface wave oscillations, is given by the dashed curve in the display.

At long wavelengths, the decay constant increases with the second power of k, as given by Eq. (3), and continues to increase monotonically with k. At the critical value k_c of k, the motion will be critically damped, and for larger values of k, overdamped. As for any harmonic oscillator, the overdamped regime is characterized by two values of the decay constant, corresponding to two different modes of decay. The degree of excitation of each of these modes depends on the manner in which the wave is excited.

From the long wavelength approximation for ω_i in Eq. (3) we conclude that gravity waves decay much slower than ripples. Correspondingly, the build-up time of gravity waves is longer than for ripples. This explains why a wind gust of relatively short duration can excite ripples on a body of water but not the gravity waves.

In the other regime, $k \gg k_c$, the two branches of the decay constant approach asymptotically the values

$$|\omega_i| \simeq 5.8(k/k_c)^2\omega_n \tag{7}$$

$$|\omega_i| \simeq 1.85(k/k_c)\omega_m \quad (k \gg k_c) \tag{8}$$

k_c, ω_m, see Eqs. (4), (5)

18.5 Thermal ripples and Brillouin scattering

As discussed in Chapters 5 and 14, the motion of a system of coupled oscillators can be expressed as a superposition of the normal modes of the system. This decomposition can be applied also to thermal motion, as in the Debye theory of specific heat of a solid where the thermal motion is expressed in terms of the acoustic modes of the solid. Each mode can be regarded as a harmonic oscillator, and the thermal energy is the sum of the oscillator energies. The highest modal frequency corresponds to a wave-length equal to twice the intermolecular distance (see Chapter 14).

The thermal motion of a liquid surface can be treated in a similar manner as a superposition of the normal modes of the surface waves of the liquid containing all frequencies up to a limit set by the intermolecular distance.

These thermally excited surface oscillations produce a thermal 'roughness' of the surface. Indeed, a close examination reveals that the light reflected from a liquid surface in equilibrium contains not only the ordinary specular reflection (discussed in Chapter 12), but also a contribution which is diffusely scattered over an angular range surrounding the specularly reflected light, as indicated schematically in Display 18.7. Although weak, this scattered light can be detected by means of a sensitive photomultiplier and associated instrumentation.

As we shall see, the light scattered in a certain direction carries information about the wavelength, frequency, and decay time of the surface waves responsible for the scattering in that direction. Extraction of this information makes it possible to determine the dispersion relation, from which the surface tension and viscosity can be determined.

Each standing wave can be regarded as a superposition of two waves travelling in opposite directions. Each of these waves acts like a travelling grating from which the incident light wave is scattered. For a given angle of incidence, the scattered light will have a pronounced maximum in a direction established by the condition for constructive interference, as discussed in Chapter 11 and later in this section.

Thus, in the scattering of light from thermal ripples, the scattering angle, which is measured, corresponds to a particular **wavelength** of the surface waves, which dominate the contribution to the scattered light at that angle.

The relation between the scattering angle and the wavelength is determined by the geometrical condition for constructive interference, which is not altered by the oscillatory motion in the wave. The effect of the oscillation is to produce a shift in the frequency of the scattered light by an amount equal to the frequency ω of the surface wave. Actually, the scattered light contains one upshifted and one downshifted component, as indicated in the display. They can be interpreted as Doppler shifts resulting from the reflection of light from an oscillating surface.

In other words, with the incident light containing only a single frequency line, as produced by a laser, the scattered light will consist of two lines

Display 18.7.

Light scattering by thermal ripples on a liquid surface. (Brillouin scattering).

Light scattering by thermal 'roughness' of liquid surface

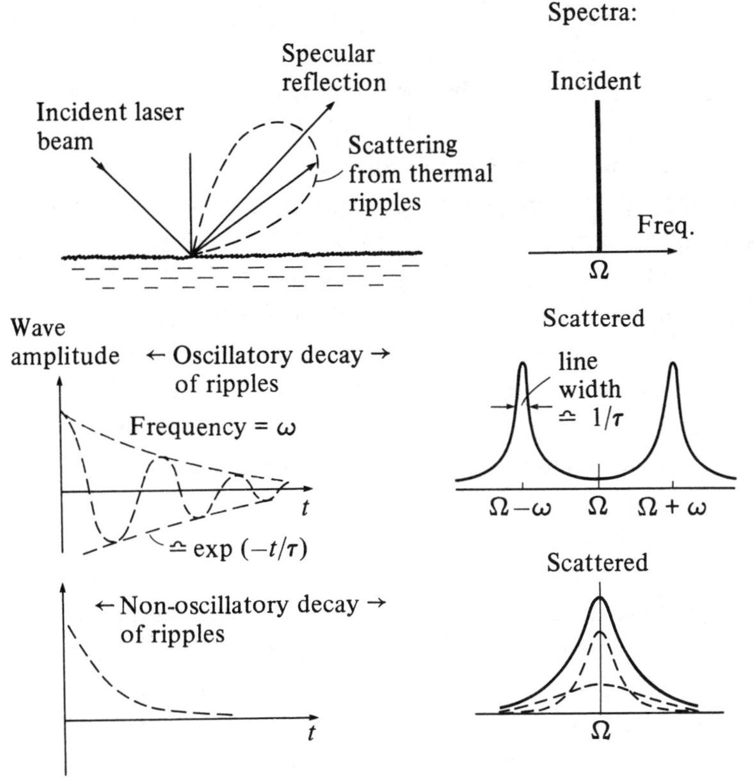

Surface tension and viscosity

Can be determined from measurement of scattering angle, frequency shift and line width

separated by 2ω. A spectrum analysis of the scattered light yields the frequency shift and hence the **frequency** of oscillation of the surface wave.

The thermal motion of a surface wave mode is similar to the motion of an oscillator driven by a random force. Such a force has a broad band frequency spectrum and the corresponding spectrum of the displacement function of the oscillator will not be a harmonic function, corresponding to a sharp line spectrum, but rather a continuous band with a peak at the resonance frequency of the oscillator. The width of the band is determined by the decay constant of the oscillator. (For comparison, see the frequency spectrum of the free damped oscillation of an oscillator discussed in Section 3.7 and summarized in Display 3.13).

Accordingly, the frequency spectrum of a surface wave mode of a given wavelength will not be a line, as implied earlier, but a band of frequencies centered at ω. The scattered spectral lines $\Omega-\omega$ and $\Omega+\omega$, referred to above, will have the same width as the spectrum of the surface wave, and from the measurement of this width, we can determine the **decay constant** (i.e. the imaginary part of the frequency) of the mode.

By repeating the measurement of wavelength, frequency, and decay rate at different scattering angles (i.e. different wavelengths), we can determine experimentally the dispersion relation for the surface waves and by comparison with the theoretical result (Eqs. 4(1) or 4(2)–(3)) makes it possible to determine the surface tension and the shear viscosity.

A complete quantitative analysis of the scattering process, as described qualitatively above, will not be given here. We limit ourselves to elementary aspects of the problem and start with the relation between the wavelength and the scattering angle. This relation can be established from a conventional analysis of diffraction, along the lines of Chapter 11. It is convenient and instructive to use another approach, however, in which we regard the scattering process as an inelastic collision between a photon and a surface wave phonon or 'ripplon'.

Like the photon, the ripplon is a quasi-particle, which represents a quantum of oscillator energy. With the frequency of the incident light denoted by Ω and the propagation constant by $\mathbf{K}_i$, the energy and momentum of the corresponding photon are $\hbar\Omega$ and $\hbar\mathbf{K}_i$, respectively, where $2\pi\hbar$ is the Planck constant. Similarly, if the angular frequency and propagation constant of the surface wave are ω_r and $\mathbf{k}$, the energy and momentum of the ripplon are $\hbar\omega_r$ and $\hbar\mathbf{k}$.

In Display 18.8 we have placed the liquid surface in the xy-plane and the incident photon in the xz-plane, which means that the component $\mathbf{K}$ of $\mathbf{K}_i$ on

Display 18.8.

Light scattering from thermal ripples. Relation for the determination of the wavelength of the ripples which are responsible for the dominant contribution to the light scattered in the direction defined by the angles θ', ϕ' when the angle of incidence is θ.

Liquid surface is in the xy-plane.

θ = angle of incidence.

θ', ϕ' = polar and azimuthal angles of direction of scattered wave.

$\mathbf{K}_i$ = prop. vector, incident wave $\mathbf{K}$ = component in xy-plane
$K_i = 2\pi/\Lambda_i$

$\mathbf{K}_s$ = prop. vector, scattered wave $\mathbf{K}'$ = component in xy-plane.

$\mathbf{k}$ = prop. vector, surface wave $k = 2\pi/\lambda$

$(k/K_i)^2 = (\Lambda_i/\lambda)^2 = \sin^2(\theta) + \sin^2(\theta') - 2\sin(\theta)\sin(\theta')\cos(\phi')$

If $\mathbf{K}_i$ and $\mathbf{K}_s$ in the same plane, $\phi' = 0$

$k/K_i = |\sin(\theta) - \sin(\theta')|$

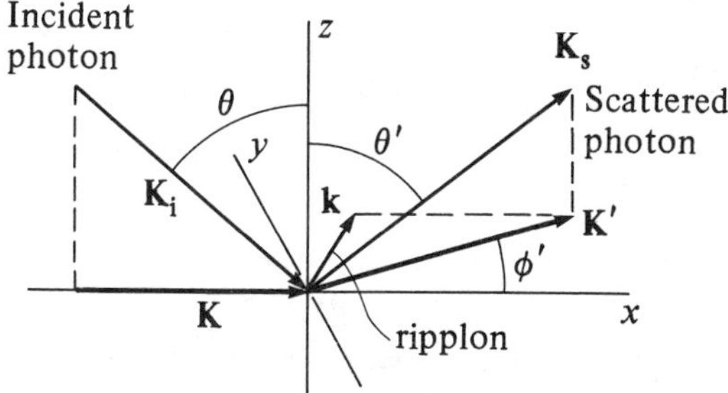

the xy-plane is along the x-axis, where $K = K_i \sin(\theta)$ and θ the angle of incidence.

The direction of the scattered photon is specified by the polar and azimuthal angles θ' and ϕ', as shown. The propagation vector is denoted by $\mathbf{K}_s$ and the component on the xy-plane is denoted by $\mathbf{K}'$.

Treating the interaction of the light wave with the surface wave as an inelastic collision between a photon and a ripplon, conservation of momentum along the liquid surface requires that $\mathbf{K} + \mathbf{k} = \mathbf{K}'$ or $k^2 = K'^2 + K^2 - 2\mathbf{K}\cdot\mathbf{K}'$. With $K = K_i \sin(\theta)$, $K' = K_s \sin(\theta')$, $\mathbf{K}\cdot\mathbf{K}' = K_i K_s \sin(\theta) \sin(\theta') \cos(\phi')$ and, since $K_i \simeq K_s$ (see discussion below), we get

$$(k/K_i)^2 = \sin^2(\theta) + \sin^2(\theta') - 2\sin(\theta)\sin(\theta')\cos(\phi') \qquad (1)$$

This relation yields the particular wavelength $\lambda = 2\pi/k$ of the ripples which contribute to the scattering in the direction (θ', ϕ') for a given angle of incidence θ and a given wavelength $2\pi/K_i$ of the incident light.

If the scattered light is observed in the plane of the incident light, so that $\phi' = 0$, Eq. (1) simplifies to

$$k/K_i = |\sin(\theta) - \sin(\theta')| \qquad (2)$$

If in the inelastic photon–ripplon collision, the incident photon absorbs a ripplon, the energy of the photon will be increased and the opposite holds true if the photon creates a ripplon. Conservation of energy in these processes requires that the frequency of the scattering photon becomes

$$\Omega' = \Omega \pm \omega_r \qquad (3)$$

where the plus sign corresponds to absorption and the minus sign to the creation of a ripplon. We have $K_i = \Omega/c$ and $K_s = \Omega'/c$, and since $\Omega \gg \omega_r$, it follows that the approximation $K_i \simeq K_s$ used above in the conservation of momentum equation is justified.

Although the scattered frequency differs only slightly from the incident, it is possible to measure the shift $\omega_r = \Omega' - \Omega$, i.e. the frequency of the ripples involved, as will be discussed below.

As indicated earlier, the spectrum of the oscillatory decay of a ripple is not a single frequency but rather a band with a width related to the decay rate and the two lines in the spectrum of the scattered light, centered at $\Omega + \omega_r$ and $\Omega - \omega_r$, are not sharp. It can be shown (compare Display 3.13 and Display 18.7) that the scattered intensity consists of two 'Lorentzian' lines, which means that the frequency dependence is

$$I' = A[\omega_i/D_- + \omega_i/D_+]$$
$$D_- = \omega_i^2 + [\Omega' - (\Omega - \omega_r)]^2 \quad D_+ = \omega_i^2 + [\Omega' - (\Omega + \omega_r)]^2 \qquad (4)$$

The two lines have their maxima at $\Omega' = \Omega - \omega_r$ and $\Omega' = \Omega + \omega_r$ with the value $|A/\omega_i|$. The constant A is proportional to the intensity of the incident light and can be shown to be proportional to the square of the incident frequency. Measurement of the locations of the two lines determines the ripple frequency, i.e. the real part ω_r, and the bandwidth of the lines establishes the decay constant, i.e. the imaginary part ω_i. With the wavelength already established from the known scattering angle, we have the quantities required to determine surface tension and viscosity by comparison with the theoretical dispersion relation. For sufficiently small values of k/k_c, the dispersion relation reduces to Eqs. 4(2)–(3), and the comparison becomes particularly simple. The surface tension is then obtained directly from the frequency shift and the viscosity from the line width.

The technique of using light scattering as a diagnostic tool in this manner was used initially in the study of compressional waves in liquids and transparent solids, and in that context, it is usually referred to as **Brillouin scattering.**

Comments on experimental technique. Before the advent of the laser, it was not possible to produce a sufficiently narrow frequency spectrum of the incident light to make possible the separation of the lines in the scattered doublet. With laser light, this problem has been eliminated, and extensive studies of compressional waves in liquids and solids have been made using Brillouin scattering.

The standard spectroscopic tool in such measurements has been the Fabry–Perot interferometer. Typically, the bandwidth of such a spectrometer is of the order of 200 MHz for measurements of frequencies about 10 GHz. The phonon frequencies in the bulk of a liquid or solid considered in Brillouin scattering experiments typically are in the range 1–10 GHz, and the width of the scattered lines of the order of 500 MHz. Under such conditions, the resolution of a Fabry–Perot interferometer is sufficient to resolve the 'Brillouin doublet' and hence to measure the frequency shift in the light scattered by such phonons.

For ripples on a liquid surface, however, the frequencies involved are considerably lower, and the resolution of the Fabry–Perot interferometer generally is not sufficient to resolve the doublet in the scattered light.

Better resolution can be obtained by means of **heterodyne spectroscopy**. In this method, the scattered light is detected by a photo-multiplier and mixed with a portion of the incident light or the specularly reflected light. Because of the square law response of this detector, the output current will be proportional to the square of the total electric field at the detector and,

Display 18.9.

Downshifted spectral lines of light scattered from thermal ripples on Methanol at three different wavelengths. Instrumental bandwidth = 3.4 kHz.

Spectrum of scattered light.

$I = A(\omega_i/D_- + \omega_i/D_+)$

$D_- = \omega_i^2 + [\Omega' - (\Omega - \omega_r)]^2 \quad D_+ = \omega_i^2 + [\Omega' - (\Omega + \omega_r)]^2$

Ω' = angular freq. of scattered light

Ω = angular freq. of incident light

ω_r = angular freq. of surface wave

Example

Downshifted spectra through heterodyne spectroscopy. (difference frequency $\Omega' - \Omega$ output spectrum from photomultiplier resulting from superposition of scattered light (Ω') and unshifted specularly reflected light (Ω)).

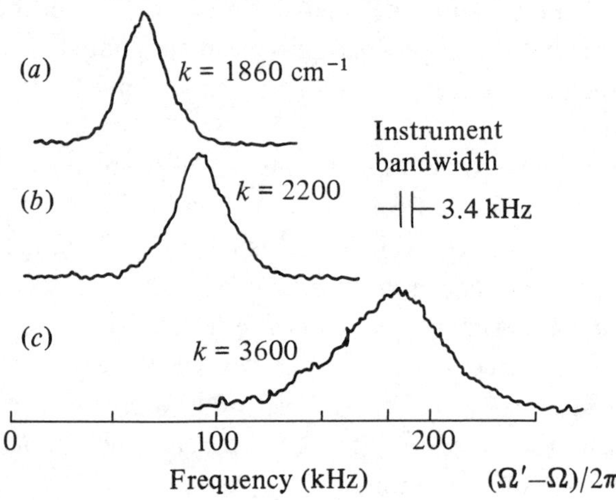

From Robert H. Katyl and Uno Ingard, *Phys. Rev. Letters*, **20**, 248 (1968).

therefore, will contain both the sum and difference frequencies of the scattered and incident (unshifted) light. The difference frequency, as we have seen, is equal to the frequency of the ripples, and this low frequency signal can be frequency analyzed with a conventional electronic spectrum analyzer.

Typical spectral 'lines' obtained in this manner are shown in Display 18.9; they were obtained in scattering from thermal ripples on methanol at three different scattering angles. An increase of the scattering angle corresponds to a decrease in the ripple wavelengths, and an increase in the width of the spectral lines.

If the wavelength is shorter than λ_c, the wave is overdamped, and there will be no average frequency shift of the scattered light, only a line broadening. As we recall, there are still two modes of oscillations with different decay rates, and, correspondingly, the scattered line is a superposition of two Lorentzian lines of different width. In thermal equilibrium, the energy of oscillation is the same for both, and through careful measurements of the line shape, it is possible to separate the two lines, and to determine the corresponding decay constants and the coefficients of surface tension and viscosity.

18.6 Liquid–liquid interface. Critical phenomenon

The light scattering technique described above lends itself equally well to the study of the thermal fluctuation of the interface between two liquids which do not mix. As an example, we describe here the result of scattering experiments involving the interface between methanol and hexane. These liquids are known to be immiscible only at temperatures below a critical temperature of 36.5 °C, and it is of interest to study the temperature dependence of the surface tension in the vicinity of the critical temperature.

We shall start with a discussion of the dispersion relation for surface waves on the liquid–liquid interface. The densities of the two liquids will be denoted by ρ_1 and $\rho_2 > \rho_1$.

In the absence of viscosity, the derivation of the dispersion relation is completely analogous to that for the free liquid surface. Thus, in the equation of motion 3(5) for the free surface, only the inertia term on the left hand side and the gravitational restoring force on the right hand side need to be modified. The density ρ in the inertia term has to be replaced by $\rho_1 + \rho_2$ and the restoring force ρgh by $(\rho_2 - \rho_1)gh$. (The force $\rho_2 gh$, which applies in the case of a free surface, must be reduced by the buoyancy force $\rho_1 gh$). The dispersion relation 3(6) for the free liquid surface then will be modified to

$$\omega^2 = [gk(\rho_2 - \rho_1) + \sigma k^3]/(\rho_2 + \rho_1) \tag{1}$$

with corresponding expressions for the phase- and group-velocities.

Display 18.10.

Dispersion relation for surface waves on an interface between two liquids. Light scattering experiment.

Liquid–liquid interface

Dispersion relation

$$(1 - i\omega\tau)^2 + (\omega_1\tau)^2 - a\sqrt{1 - i2\omega\tau a/b} - (1-a)\sqrt{1 - i2\omega\tau(1-b)/(1-a)} = 0, \text{ where}$$

$\tau = (\rho_1 + \rho_2)/2(\mu_1 + \mu_2)k^2$

$\omega_1^2 = [gk(\rho_2 - \rho_1) + \sigma k^3]/(\rho_2 + \rho_1)$

$a = \mu_1/(\mu_1 + \mu_2) \quad b = \rho_1/(\rho_2 + \rho_1)$

$\mu_1, \mu_2 =$ coefficients of viscosity

Long wavelength limit $k \ll (\rho_1 + \rho_2)\sigma/(\mu_1 + \mu_2)^2$

$\omega_r^2 \simeq [kg(\rho_2 - \rho_1) + \sigma k^3]/(\rho_2 + \rho_1)$

$\omega_i \simeq -2(\mu_1 + \mu_2)k^2/(\rho_1 + \rho_2)$

Example

Measured temperature dependence of the interfacial surface tension between Methanol and Hexane. Surface tension approaches zero at the critical mixing temperature 36.5 C.

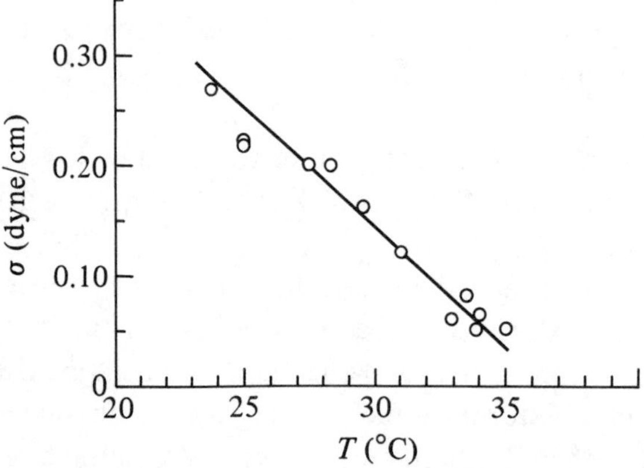

From Robert H. Katyl and K.U. Ingard, Light Scattering from Thermal Fluctuations of a Liquid Surface, Ch. 6, pp. 70–87 in *In Honor of P.M. Morse*, Editors Herman Feshbach and K. Uno Ingard, The M.I.T. Press, Cambridge, Mass. (1969).

As before, the first term corresponds to gravity waves and the second to ripples (capillary waves). It should be noted that the frequency of oscillation and the corresponding wave velocity for gravity waves goes to zero as the density difference goes to zero. This can be utilized for demonstrating waves in 'slow motion'.

As before, the derivation of the complete dispersion relation when viscosity is included will not be given here. The result is

$$(1-\mathrm{i}\omega\tau)^2+(\omega_1\tau)^2-a\sqrt{1-\mathrm{i}2\omega\tau a/b}$$
$$-(1-a)\sqrt{1-\mathrm{i}2\omega\tau(1-b)/(1-a)}=0$$

where $\tau=(\rho_1+\rho_2)/2(\mu_1+\mu_2)k_2$

$$\omega_1^2=[gk(\rho_2-\rho_1)+\sigma k^3]/(\rho_2+\rho_1) \quad \sigma=\text{surface tension}$$
$$a=\mu_1/(\mu_1+\mu_2) \quad b=\rho_1/(\rho_1+\rho_2)$$
$$\mu_1,\mu_2=\text{coefficients of viscosity} \tag{2}$$

which reduces to Eq. 4(1) for a free surface when $a=b=0$.

The measurement of the coefficients of the interfacial surface tension and viscosity by means of laser light scattering can be made in the same manner as described in Section 5. The only difference is that the separation of the lines in the scattered light generally is smaller than for a free surface. The scattered lines are then apt to overlap so that the line separation and the line width cannot be determined independently. Instead, the coefficients of interfacial surface tension and viscosity have to be determined by a more detailed comparison between the measured and the calculated total spectrum of the scattered light.

Results obtained in this manner of the temperature dependence of the surface tension of the interface between methanol and *n*-hexane is shown in Display 18.10. It is interesting to note that the surface tension decreases to zero as the temperature approaches the critical mixing temperature (36.5 °C). It is also noteworthy that the method allows measurement of values of surface tension as low as 0.01 dyne/cm.

Problems

18.1 *Channel wave.* Consider a surface wave in a channel, depth H, in which the velocity u of the fluid is uniform. If the wave amplitude is A, what is the fluid velocity u?

18.2 *Minimum phase velocity.* From Eq. 3(7), show that the phase velocity has a minimum value for a certain wavelength and derive expressions for this minimum value and the corresponding wavelength. Determine the numerical values for water.

18.3 *Wave energy.* Consider a surface wave on deep water with the surface displacement $A\cos(kx-\omega t)$, where $k=2\pi/\lambda$.

(a) With the time average value of the kinetic energy in the wave set equal to the time average potential energy, determine the time average total wave energy per square kilometer if $A=2\,\mathrm{m}$ and $\lambda=20\,\mathrm{m}$.
(b) If the wind over the water surface has the same velocity as the phase velocity of the wave, determine the thickness of the air layer such that the kinetic energy of the air equals the total wave energy.

18.4 *Oscillations of a drop.* A liquid drop in free fall oscillates in harmonic motion about its spherical equilibrium shape. From dimensional considerations, determine how the period of oscillation depends on the surface tension σ, the liquid density ρ, and the radius R.

18.5 *Maximum frequency of surface waves.* Determine whether the expressions in Eq. 4(1) and 4(2) for the maximum frequency of surface waves and the corresponding smallest wavelength are dimensionally correct.

18.6 *Wave from a stone thrown into water.* A stone thrown into a lake enters the water at $t=0$ at the origin $r=0$. The resulting impulsive deformation of the water surface generates a wave on the water surface. This wave can be considered to be a superposition of waves with all possible wavelengths, each wave component travelling with a different wave speed because of dispersion.

Explain why the dominant wavelength observed at a distance r from the origin at time t is the wavelength which corresponds to a group velocity equal to r/t. Express this wavelength in terms of r and t, assuming that only the gravity waves are involved.

18.7 *Refraction.* At a distance from the shore where the water depth is 5 m, waves are incident on the shore at an angle of incidence of 45 degrees. What is the angle of incidence closer to the shore where the depth is 2 m? Assume the depth small compared to the wavelength.

18.8 *Dispersion relation (including viscosity).* Show that the long wavelength approximations in Eqs. 4(2)–(3) for the real and imaginary parts of the complex frequency of surface wave oscillation are consistent with the general dispersion relation in Eq. 4(1).

18.9 *Velocity distribution in a surface wave.* Derive Eq. 18.3(2) by solving Laplace's equation for harmonic surface wave motion.

19

Plasma oscillations and hydromagnetic waves

In recent years, fusion research has motivated considerable interest in the physics of plasmas (ionized gases), and several different types of wave motion have been discovered and studied in detail both theoretically and experimentally. In this chapter we shall discuss two important examples, longitudinal oscillations of the electrons in a plasma, and transverse waves (Alfven waves) in a conducting fluid in a magnetic field.

19.1 **The plasma state**

In order of increasing energy, matter in bulk occurs in the form of a solid or a fluid; the latter including the liquid, gaseous, and plasma state. A plasma is an ionized gas with an equal number N_0 of electrons and ions per unit volume in the unperturbed state, so that the plasma is electrically neutral on the average. It is necessary that the thermal kinetic energy of the electrons be large enough to prevent recombination of the electrons and ions and a 'collapse' of the plasma. This condition can be expressed approximately as

$$k_B T \gg Ce^2/l \quad C = 1/4\pi\varepsilon_0 \simeq 9 \times 10^9 \tag{1}$$

$$e = 1.59 \times 10^{-19}\,\text{Coul.} = 4.8 \times 10^{-10}\,\text{ese}$$

$$k_B = \text{Boltzmann constant} = 1.38 \times 10^{-23}\,\text{joule/K}$$

$$l^3 = 1/N_0$$

where l is the average distance between the electrons.

A plasma can be produced and maintained by means of a static or a high frequency electric field. Thus, in a glow discharge in a gas at relatively low pressure, the plasma state is approximately realized in the 'positive column' of the discharge. The degree of ionization (fraction of ions in the gas) in this column is usually less than 1%. In a discharge, in which the gas pressure is 1 mm Hg corresponding to 10^{16} neutrals per cm^3, the number of ions (and electrons) is then about 10^{14} per cm^3. Higher degrees of ionization are found in arc discharges and, of course, in various types of plasma machines used in fusion research.

In the ionosphere, the ionization is caused by the Sun's radiation, and the electron density shows a diurnal variation. The degree of ionization is only about 1 part in a million. Such a weakly ionized gas can be maintained in the laboratory by means of a static electric field, which typically is of the order of a few volts per centimeter. The electrons and ions are caused to drift through the gas with a drift speed determined by the 'drag' resulting from the collisions with the neutral particles.

The typical drift speed for the electrons is about 10^6 cm/sec and that of ions, 10^3 cm/sec. The corresponding kinetic energy is small compared to the energy transferred to the charged particles from the electric field, and it follows that almost all of the energy goes into the random motion of the electrons as they find their way through the neutral background gas, bouncing between the gas molecules in an irregular manner until they reach electrodes or recombine with ions. The lost electrons are replaced through new ionization, so that the electron density remains constant in steady state. The temperature corresponding to the random motion of the electrons is of the order of 10^4 K, the kinetic energy being approximately 1 eV.

Some plasma characteristics. Before starting our study of wave motion, it is instructive to summarize some characteristics, lengths and times, which play an important role in the description of a plasma.

One of the most significant properties of a plasma is the **period of plasma oscillations**. These oscillations occur as a result of the electrostatic restoring force which is produced locally when electrons are separated from the ions as a result of a perturbation.

With reference to Display 19.1, we consider the perturbation of the number of electrons contained within a sphere of radius r resulting from a small radial displacement of the electrons located in a surface layer of the sphere. With the radial displacement of the electrons denoted by ξ, the number of electrons which leave the sphere of radius r is $4\pi r^2 \xi N_0$, where N_0 is the number of electrons per unit volume. This results in a positive net charge within the sphere equal to $4\pi r^2 \xi N_0 e$.

The electric field produced by this charge is then $N_0 e\xi/\varepsilon_0$ and this results in a radial force on a displaced electron $F_r = -N_0 e^2 \xi/\varepsilon_0$. This is a linear restoring force of the form $-K\xi$, where $K = e^2 N_0/\varepsilon_0$ is the

Display 19.1.

Characteristic lengths and times in a plasma.

Ionized gas, plasma

$k_B T > Ce^2/l \quad C = 1/4\pi\varepsilon_0 \quad l^3 = 1/N_0$ (1)

$e = 1.59 \times 10^{-19}$ Coul. $\quad k_B = 1.38 \times 10^{-23}$ joule/°K

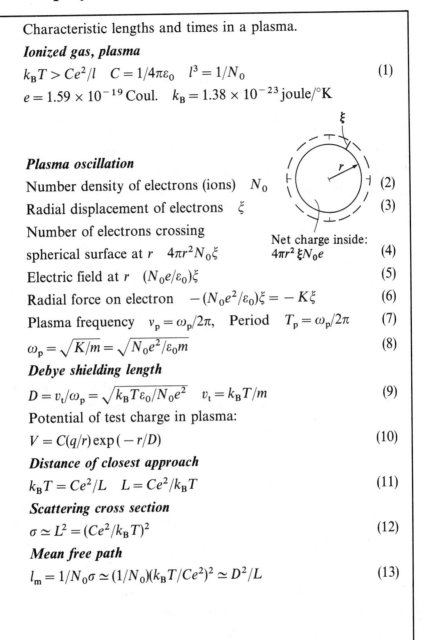

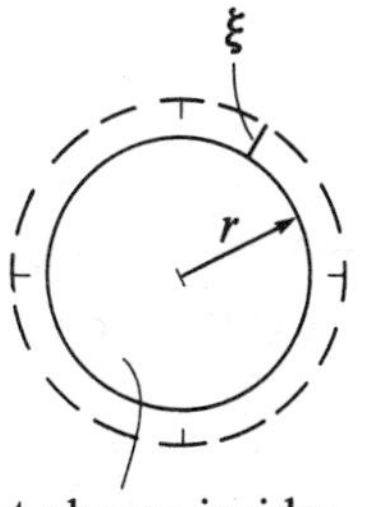

Plasma oscillation

Number density of electrons (ions) $\quad N_0$ (2)

Radial displacement of electrons $\quad \xi$ (3)

Number of electrons crossing spherical surface at $r \quad 4\pi r^2 N_0 \xi$ (4)

Net charge inside: $4\pi r^2 \xi N_0 e$

Electric field at $r \quad (N_0 e/\varepsilon_0)\xi$ (5)

Radial force on electron $\quad -(N_0 e^2/\varepsilon_0)\xi = -K\xi$ (6)

Plasma frequency $\quad \nu_p = \omega_p/2\pi$, $\quad$ Period $\quad T_p = \omega_p/2\pi$ (7)

$\omega_p = \sqrt{K/m} = \sqrt{N_0 e^2/\varepsilon_0 m}$ (8)

Debye shielding length

$D = v_t/\omega_p = \sqrt{k_B T \varepsilon_0/N_0 e^2} \quad v_t = k_B T/m$ (9)

Potential of test charge in plasma:

$V = C(q/r)\exp(-r/D)$ (10)

Distance of closest approach

$k_B T = Ce^2/L \quad L = Ce^2/k_B T$ (11)

Scattering cross section

$\sigma \simeq L^2 = (Ce^2/k_B T)^2$ (12)

Mean free path

$l_m = 1/N_0\sigma \simeq (1/N_0)(k_B T/Ce^2)^2 \simeq D^2/L$ (13)

equivalent spring constant. The resulting motion of the electron will be harmonic with the frequency

$$\omega_p = \sqrt{K/m} = \sqrt{N_0 e^2/\varepsilon_0 m} \qquad (2)$$
$$T_p = 2\pi/\omega_p$$

This harmonic motion is called a plasma oscillation and $\omega_p/2\pi$ is the **plasma frequency** and T_p the period. For a laboratory plasma with an electron density 10^{14} per cm^3, the plasma frequency becomes approximately 10^{11} Hz.

Having introduced the period of plasma oscillations, we can construct a characteristic length by multiplying by the thermal speed of the electron $v_t \simeq \sqrt{k_B T/m}$ to obtain a length, defined by

$$D = v_t/\omega_p = \sqrt{k_B T \varepsilon_0 / N_0 e^2} \qquad (3)$$

This length is called the **Debye shielding length**. The physical significance of this length refers to the 'shielding' of the field from a test charge q in the plasma, the shielding resulting from the swarm of electrons surrounding the charge. The electric potential from this test charge will be shown to be

$$V(r) = (q/4\pi\varepsilon_0 r)\exp(-r/D) \qquad (4)$$

where the factor $\exp(-r/D)$ expresses the shielding effect of the electrons. The characteristic range of the electric field is the Debye shielding length D.

Another characteristic length is the **distance of closest approach** in a head-on collision of two electrons, when each electron has the thermal energy $k_B T/2$. The energy $k_B T$ is then converted into electrical potential energy so that $kT = e^2/4\pi\varepsilon_0 L$, so that

$$L = Ce^2/k_B T \quad C = 1/4\pi\varepsilon_0 \qquad (5)$$

The distance of closest approach, L, is sometimes called the Landau length.

This length can be taken as a measure of the 'size' of an electron as far as collisions are concerned, and the corresponding area can be taken as the **scattering cross section**

$$\sigma \simeq L^2 = (Ce^2/k_B T)^2 \qquad (6)$$

It should be noted that this 'Coulomb cross section' decreases with increasing temperature, and the **mean free path**

$$l_m = 1/N_0\sigma \simeq (1/N_0)(4\pi\varepsilon_0 k_B T/e^2)^2 \simeq D^2/L \qquad (7)$$

increases with temperature.

After having introduced the characteristic lengths l, L, and D (Eqs. (1), (3), (5)), we note that the energy condition $k_B T \gg Ce^2/l$ in Eq. (1) can be expressed as $l \gg L$.

Furthermore, it follows that $D = l\sqrt{l/4\pi L}$ and that the volume of the Debye sphere can be written

$$V = (4\pi/3)D^3 = (\sqrt{4\pi}/3)l^3(l/L)^{3/2} = (\sqrt{4\pi}/3)(l/L)^{3/2}/N_0 \tag{8}$$

In other words, the number of particles in a Debye sphere can be written $N_0 V \simeq (l/L)^{3/2}$, and the condition $k_B T \gg Ce^2/l$ or, equivalently, $l \gg L$, implies that in a plasma the number of particles within a Debye sphere must be much greater than 1.

19.2 Longitudinal waves. (Plasma oscillations)

We have already determined the frequency of plasma oscillations, and we shall now study these oscillations in more detail. The electrons will be treated as a continuous charge distribution and the positive ions are represented as an immobile, uniform background charge. On the basis of these assumptions, we can analyze longitudinal waves in the electron gas in much the same way as for the sound waves in a neutral fluid. The only significant difference is the occurrence of an electric force caused by the organized or 'collective' motion of the electrons. As before, the unperturbed pressure in the gas is a result of the random thermal motion.

We shall be concerned with small amplitude oscillations, and the equations of motion will be linearized. The corresponding perturbations will be denoted by a subscript 1, and the unperturbed quantities will have the subscript 0. In addition to the number density N and mass density Nm of the electrons, we need to introduce also the charge density $-Ne$ and the electric field $\mathbf{E}$. The corresponding electric force per unit volume, $\mathbf{F} = -Ne\mathbf{E}$, generally contains two contributions to the first order perturbation $\mathbf{F}_1 = -N_0 e\mathbf{E}_1 - N_1 e\mathbf{E}_0$. In the case considered here, the external electric field $\mathbf{E}_0$ is zero so that

$$\mathbf{F}_1 = -N_0 e\mathbf{E}_1 \tag{1}$$

In terms of these quantities, the linearized momentum equation for the electrons is

$$\rho_0 \partial\mathbf{u}_1/\partial t = -\operatorname{grad}(p_1) - N_0 e\mathbf{E}_1 \tag{2}$$

and the equation for mass conservation is

$$\partial\rho_1/\partial t + \rho_0 \operatorname{div}(\mathbf{u}_1) = 0 \tag{3}$$

The perturbation in the electric field is related to the perturbation in the charge density through the Maxwell equation

$$\varepsilon_0 \operatorname{div}(\mathbf{E}_1) = -N_1 e \tag{4}$$

Finally, the relation between the perturbation in pressure and mass

Display 19.2.

Longitudinal waves in a plasma. (Electron plasma oscillation)
m = electron mass
N = number density of electrons
$\mathbf{E}$ = electric field
u = velocity
p = pressure
$\rho = Nm$ mass density
$1/\rho v^2$ = electron gas compressibility
Unperturbed values – subscript 0
Perturbed values – subscript 1
Force on electrons per unit volume

$$\mathbf{F} = -Ne\mathbf{E} \tag{1}$$

$$\mathbf{F}_1 = -N_0 e\mathbf{E}_1 \quad (\mathbf{E}_0 = 0) \tag{2}$$

Linearized equations

$$\partial\rho_1/\partial t + \rho_0 \operatorname{div}(\mathbf{u}_1) = 0 \tag{3}$$

$$\rho_0 \partial\mathbf{u}_1/\partial t = -\operatorname{grad}(p_1) - N_0 e\mathbf{E}_1 \tag{4}$$

$$\varepsilon_0 \operatorname{div}(\mathbf{E}_1) = -N_1 e \tag{5}$$

$$p_1 = v^2 \rho_1 \tag{6}$$

Klein–Gordon equation

$$\nabla^2 \rho_1 - (1/v^2)\partial^2\rho_1/\partial t^2 - (\omega_p/v)^2 \rho_1 = 0 \tag{7}$$

$$\omega_p^2 = N_0 e^2/m\varepsilon_0 \tag{8}$$

Dispersion relation

$$\nabla^2 \rho_1 + [(\omega^2 - \omega_p^2)/v^2]\rho_1 = 0$$

(Eq. (7), harmonic time dependence) (9)

$$\omega^2 = \omega_p^2 + v^2 k^2 \tag{10}$$

$$k = (\omega/v)[1 - (\omega_p/\omega)^2]^{\frac{1}{2}} \tag{11}$$

$$\omega \ll \omega_p \quad \rho_1(x) = \rho_1(0)\exp(-x/D) \quad D = v/\omega_p \tag{12}$$

Phase velocity and group velocity

$$v_p = \omega/k = v[1 - (\omega_p/\omega)^2]^{-\frac{1}{2}} \tag{13}$$

$$v_g = \mathrm{d}\omega/\mathrm{d}k = v[1 - (\omega_p/\omega)^2]^{\frac{1}{2}} \tag{14}$$

density can be expressed in the same way as for the neutral gas

$$p_1 = v^2 \rho_1 \tag{5}$$

For ordinary sound waves in a neutral gas, v turned out to be the wave speed. As we shall find, this is still true here for small values of the electron density. In any event, v is a characteristic thermal speed of the electrons.

By differentiating Eq. (3) with respect to t and taking the divergence of Eq. (2), we can eliminate the velocity u to obtain the equation for the density perturbation

$$\nabla^2 \rho_1 - (\omega_p/v)^2 \rho_1 = (1/v^2)\partial^2 \rho_1/\partial t^2 \tag{6}$$

where we have introduced the plasma frequency

$$\omega_p = \sqrt{N_0 e^2/m\varepsilon_0} \tag{7}$$

discussed in Section 1.

Eq. (6) has the same form as the Klein–Gordon equation, familiar to physicists as the relativistic version of the Schroedinger equation. For harmonic time dependence, the equation for the complex density amplitude becomes

$$\nabla^2 \rho_1 + [(\omega^2 - \omega_p^2)/v^2]\rho_1 = 0 \tag{8}$$

For a wave travelling in the positive x-direction, $\rho_1 = A \exp(ikx)$, for which Eq. (8) yields the **dispersion relation**

$$\omega^2 = \omega_p^2 + v^2 k^2 = \omega_p^2[1 + (kD)^2] \quad D = \text{Debye length} \tag{9}$$

i.e. the condition imposed by the wave equation on the propagation constant k in order for the harmonic wave to be a solution.

As the plasma frequency goes to zero, the dispersion relation approaches the familiar linear form for ordinary sound waves, $\omega = vk$, with v being the phase velocity. In the other extreme of a 'cold' plasma, with negligible thermal speed, a harmonic motion can exist throughout the plasma only if the frequency equals the plasma frequency.

Under more general conditions, the propagation constant corresponding to an imposed frequency ω is

$$k = (\omega/v)\sqrt{1 - (\omega_p/\omega)^2} \tag{10}$$

The propagation constant is real only if the frequency exceeds the plasma frequency. Below this frequency, k becomes imaginary, and the wave decays exponentially with x. If the frequency is much lower than the plasma frequency, the exponential decay of the amplitude can be approximated by

$$\rho_1(x) \simeq \rho_1(0)\exp(-\omega_p x/v) = \rho_1(0)\exp(-x/D) \quad \omega \ll \omega_p \tag{11}$$

where D is the Debye shielding length, defined in Eq. 1(3).

Since the plasma is electrically neutral, the field from a positive test charge q introduced into the plasma would yield the same potential $q/4\pi\varepsilon_0 r$ as in free space if the plasma were not disturbed by q. The charge does disturb the plasma, however; the electron density in the vicinity of q will be larger (and the ion density smaller) then in the rest of the plasma because of the field of q, and this results in a net negative space charge around q. This space charge makes the net charge inside a sphere of radius r less than q, and this results in a reduction of the field, as expressed by $\exp(-r/D)$.

The shielding increases with decreasing electron temperature (velocity), and a reduction of the temperature eventually results in electrons being captured by q to form a neutral particle, in which case the shielding is complete outside the 'radius' of the particle. This is consistent with the fact that D increases with T.

Returning to the dispersion relation (9), we obtain the following expressions for the **phase velocity** and the **group velocity** of the electron plasma oscillation

$$v_\mathrm{p} = \omega/k = v[1-(\omega_\mathrm{p}/\omega)^2]^{-\frac{1}{2}} \tag{12}$$

$$v_\mathrm{g} = \mathrm{d}\omega/\mathrm{d}k = v[1-(\omega_\mathrm{p}/\omega)^2]^{\frac{1}{2}} \tag{13}$$

$$v_\mathrm{p} v_\mathrm{g} = v^2 \tag{14}$$

As indicated earlier, these wave velocities approach the characteristic value v when the plasma frequency is much smaller than the driving frequency.

Energy considerations. As a preliminary to the derivation of expressions for energy density and flux in electron plasma oscillations, we shall express the Maxwell equation, 2(4),

$$\varepsilon_0 \operatorname{div}(\mathbf{E}_1) = -N_1 e = -\rho_1 e/m$$

in a slightly different form. Differentiation with respect to t yields $\varepsilon_0 \operatorname{div}(\partial \mathbf{E}_1/\partial t) = -(e/m)\partial\rho_1/\partial t$, and by making use of the conservation of mass equation 2(3), $\partial\rho_1/\partial t + \rho_0 \operatorname{div}(\mathbf{u}_1) = 0$, we obtain

$$\varepsilon_0 \partial \mathbf{E}_1/\partial t = N_0 e \mathbf{u}_1 \tag{15}$$

This equation together with the equations (2) and (3) for mass and momentum balance

$$(1/v^2)\partial p_1/\partial t + \rho_0 \operatorname{div}(\mathbf{u}_1) = 0 \quad (p_1 = v^2 \rho_1) \tag{16}$$

$$\rho_0 \partial \mathbf{u}_1/\partial t = -\operatorname{grad}(p_1) - N_0 e \mathbf{E}_1 \tag{17}$$

will now be used for the derivation of an equation for the wave energy in much the same way as we did for acoustic and electromagnetic waves in Chapters 9 and 10.

Thus, we multiply Eq. (16) by p_1, and Eqs. (15) and (17) by $\mathbf{E}_1$ and $\mathbf{u}_1$, respectively (scalar multiplication), to obtain

$$\kappa\partial(p_1^2/2)/\partial t + \rho_1 \operatorname{div}(\mathbf{u}_1) = 0 \tag{18}$$

$$\rho_0\partial(u_1^2/2)/\partial t + \mathbf{u}_1\cdot\operatorname{grad}(p_1) = -N_0 e\mathbf{u}_1\cdot\mathbf{E}_1 \tag{19}$$

$$\partial(\varepsilon_0 E_1^2/2)/\partial t = N_0 e\mathbf{E}_1\cdot\mathbf{u}_1 \tag{20}$$

Addition of these equations leads to

$$\partial W/\partial t + \mathbf{u}_1\cdot\operatorname{grad}(p_1) + p_1 \operatorname{div}(u_1) = 0$$

and with $\mathbf{u}_1\cdot\operatorname{grad}(p_1) + p_1 \operatorname{div}(\mathbf{u}_1) = \operatorname{div}(p_1\mathbf{u}_1)$ we obtain the energy equation

$$\partial W/\partial t + \operatorname{div}(\mathbf{I}) = 0 \tag{21}$$

$$\mathbf{I} = p_1 u_1 \quad W = \kappa p_1^2/2 + \rho_0 u_1^2/2 + \varepsilon_0 E_1^2/2$$

This is the standard form of an energy conservation equation in which W is the energy density and I the energy flux.

For a harmonic plane wave travelling in the positive x-direction, $p_1(x,t) = p_{10}\cos(kx-\omega t)$, with similar expressions for the other field variables; the relations between the complex amplitudes $u_1(x,\omega)$, $p_1(x,\omega)$, and $E_1(x,\omega)$, become

$$u_1 = (\rho_1/\rho_0)(\omega/k) = (\rho_1/\rho_0)v_p = p_1/\rho_0 v_g \tag{22}$$

$$\varepsilon_0 E_1 = N_0 e(u_1/-\mathrm{i}\omega) = (\xi_1/l)(e/l^2) \quad \xi_1 = u_1/-\mathrm{i}\omega \quad l^3 = 1/N_0 \tag{23}$$

where we have introduced the displacement amplitude $\xi_1 = u_1/(-\mathrm{i}\omega)$. It should be noted that the **wave impedance** is $Z = p_1/u_1 = \rho_0 v_g$, containing the group velocity rather than the phase velocity as a factor.

From these relations we can express the potential energy contributions in terms of the kinetic energy,

$$\varepsilon_0 E_{10}^2/2 = (\omega_p/\omega)^2(\rho_0 u_{10}^2/2) \tag{24}$$

$$\kappa p_{10}^2/2 = (\rho_0^2 v_g^2 u_{10}^2)/2\rho_0 v^2 = (\rho_0 u_{10}^2/2)[1-(\omega_p/\omega)^2] \tag{25}$$

and it follows that the total potential energy density, thermal and electrostatic, is equal to the kinetic energy density. Consequently, the time-average of the total energy density can be expressed as $\langle W\rangle = \frac{1}{2}\rho_0 u_{10}^2$, and the time-average of the energy flux is then

$$\langle I\rangle = \tfrac{1}{2}(p_{10}u_{10}) = \tfrac{1}{2}(\rho_0 v_g u_{10}^2) = \langle W\rangle v_g \tag{26}$$

It is interesting to see that the flux is the product of the energy density and the group velocity. This calculation should be compared with the calculation of the power carried by a harmonic wave on a periodic lattice in Chapter 14.

Display 19.3.

Wave energy density and energy flux in a longitudinal wave in a plasma.

Linearized equations

$$(1/v^2)\partial p_1/\partial t + \rho_0 \operatorname{div}(\mathbf{u}_1) = 0 \tag{1}$$

$$\rho_0 \partial \mathbf{u}_1/\partial t = -\operatorname{grad}(p_1) - N_0 e \mathbf{E}_1 \tag{2}$$

$$\varepsilon_0 \partial \mathbf{E}_1/\partial t = N_0 e \mathbf{u}_1 \quad [\varepsilon_0 \operatorname{div}(E_1) = -(\rho_1/m)e] \tag{3}$$

$$p_1 = v^2 \rho_1 \tag{4}$$

Energy conservation equation

$$\partial W/\partial t + \operatorname{div}(p_1 \mathbf{u}_1) = 0 \tag{5}$$

Wave energy density:

$$W = \kappa(p_1^2/2) + \rho_0(u_1^2/2) + \varepsilon_0 E_1^2/2 \tag{6}$$

Energy flux:

$$I = p_1 u_1 \tag{7}$$

Harmonic travelling wave

$$p_1(x,t) = p_{10} \cos(kx - \omega t)$$

$$u_1(x,t) = u_{10} \cos(kx - \omega t), \text{ etc.} \tag{8}$$

$$p_{10} = \rho_0 v_g u_{10} \tag{9}$$

$$\langle W \rangle = \tfrac{1}{2}\rho_0 u_{10}^2 \tag{10}$$

$$\langle I \rangle = \tfrac{1}{2}(p_{10} u_{10}) = \langle W \rangle \cdot v_g \tag{11}$$

19.3 Transverse waves. (Hydromagnetic or Alfvén waves)

As another example of wave motion in a plasma we consider the wave mode resulting from the interaction of a conducting fluid with an external magnetic field. It is a transverse wave, which travels along the magnetic field. Predicted by Hannes Alfvén in 1942, this wave has been observed in numerous experiments both in the laboratory and in space. Although most of these experiments have dealt with plasmas, the Alfvén wave can exist in any conducting fluid, such as mercury, liquid sodium, etc.

In order to explain qualitatively the mechanism of this wave mode, we consider a sheet-like element in a conducting fluid in a constant magnetic field, as indicated in Display 19.4. The magnetic field is in the z-direction and the sheet element is parallel with the xy-plane.

If the element is set in motion in the y-direction (into the plane of the paper), an electromotive force in the x-direction is developed in the element (conductor moving across a magnetic field), a current will be produced in the surrounding fluid by the electromotive force.

A portion of this current goes through the adjacent sheet element above the first. The current is in the x-direction, and through the interaction with the magnetic field, a force in the y-direction is produced on the element. The resulting motion will generate a new electromotive force, which, in turn, results in a current flowing through the adjacent third element above the second.

This process continues, and due to the intertial mass of the fluid, the elements will not respond instantaneously, but the motion will be delayed. Thus, the initial displacement of the first sheet element results in a displacement wave which travels in the z-direction with a certain wave speed. (There will be a corresponding wave travelling in the negative z-direction). The wave speed is expected to increase with the magnetic field strength and decrease with increasing mass density of the fluid.

To simplify the mathematical analysis of the motion, we let the sheet elements extend over the entire plane. The motion is still in the y-direction, but the field variables will be independent of the y-coordinate.

The induced electromotive force is in the x-direction and the same applies to the corresponding electric current. The magnetic field perturbation produced by this field will be in the y-direction. The time dependence of this field results in an induced electric field in the x-direction. These relations are indicated schematically in the display.

The unperturbed fluid is assumed to be at rest, so that the only motion involved is expressed by the velocity perturbation u_y. Similarly, the electric current is expressed by the perturbation in current density I_x, and the

Display 19.4.

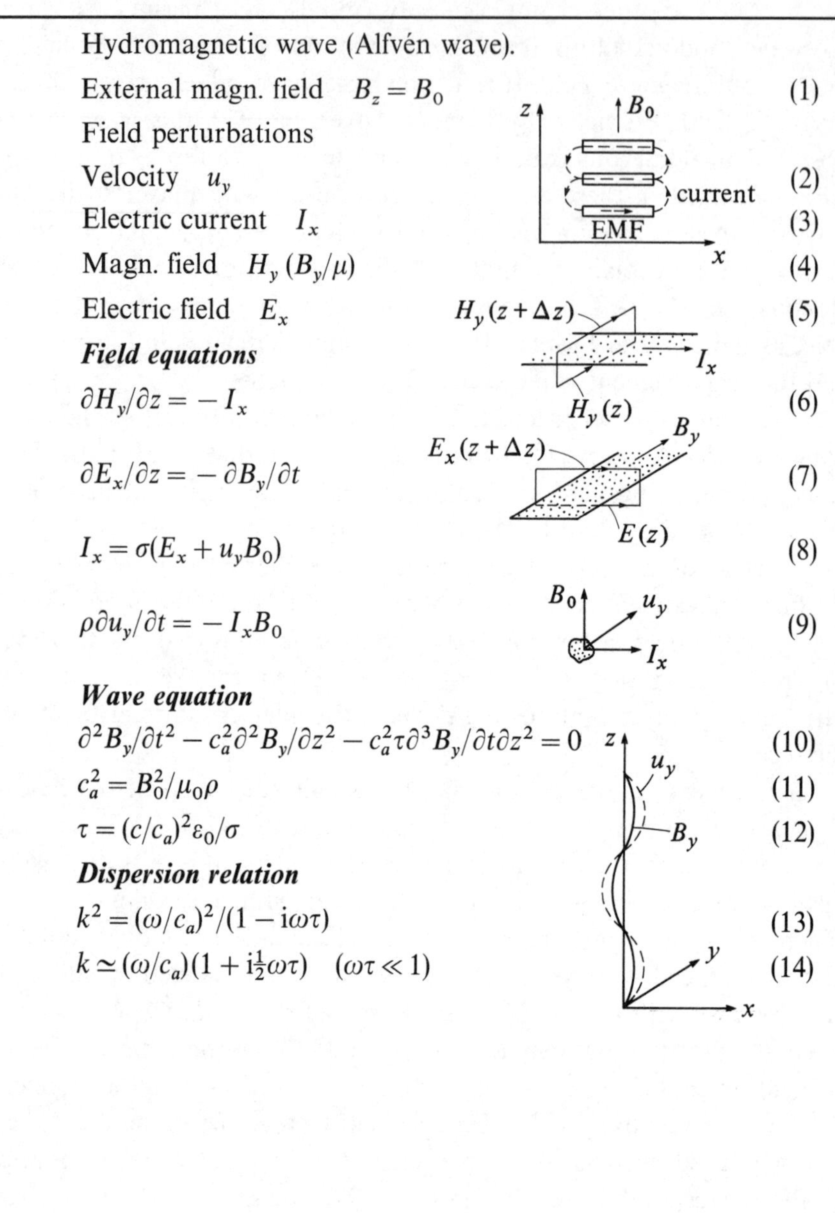

Hydromagnetic wave (Alfvén wave).

External magn. field $B_z = B_0$ (1)

Field perturbations

Velocity u_y (2)

Electric current I_x (3)

Magn. field H_y (B_y/μ) (4)

Electric field E_x (5)

Field equations

$\partial H_y/\partial z = -I_x$ (6)

$\partial E_x/\partial z = -\partial B_y/\partial t$ (7)

$I_x = \sigma(E_x + u_y B_0)$ (8)

$\rho \partial u_y/\partial t = -I_x B_0$ (9)

Wave equation

$\partial^2 B_y/\partial t^2 - c_a^2 \partial^2 B_y/\partial z^2 - c_a^2 \tau \partial^3 B_y/\partial t \partial z^2 = 0$ (10)

$c_a^2 = B_0^2/\mu_0 \rho$ (11)

$\tau = (c/c_a)^2 \varepsilon_0/\sigma$ (12)

Dispersion relation

$k^2 = (\omega/c_a)^2/(1 - \mathrm{i}\omega\tau)$ (13)

$k \simeq (\omega/c_a)(1 + \mathrm{i}\tfrac{1}{2}\omega\tau) \quad (\omega\tau \ll 1)$ (14)

perturbations in the magnetic and electric fields are H_y ($B_y = \mu H_y$) and E_x. The constant magnetic field in the z-direction is expressed by the induction B_0.

The field variables are related through the Maxwell equations

$$\mathrm{curl}\,(\mathbf{H}) = \mathbf{I} + \varepsilon \partial \mathbf{E}/\partial t \tag{1}$$

$$\mathrm{curl}\,(\mathbf{E}) = -\,\partial \mathbf{B}/\partial t \tag{2}$$

and the equation for the conduction current

$$\mathbf{I} = \sigma(\mathbf{E} + \mathbf{u} \times \mathbf{B}) \tag{3}$$

The dynamics of the fluid is expressed by the equation of motion

$$\rho \partial \mathbf{u}/\partial t = \mathbf{I} \times \mathbf{B} - \mathrm{grad}\,(p) \simeq \mathbf{I} \times \mathbf{B} \tag{4}$$

The relationship between the current density and the electric field, expressed in terms of the conductivity of the fluid, implies a reference frame which moves with the fluid. In this frame, in addition to the electric field **E**, there is an equivalent electric field $\mathbf{u} \times \mathbf{B}$ which corresponds to the induced electromotive force (related to the Lorentz force on a moving charge). This explains the form of Eq. (3).

In regard to Eq. (4), we note that for an incompressible fluid, the pressure gradient is zero in the linear approximation, and we shall limit our study to that case. We note also, that in the linear approximation, the driving force density $\mathbf{I} \times \mathbf{B}$ contains only the term $I_x B_0$.

Because of the simple model chosen in our study, all field variables are independent of y, the velocity and magnetic field components have only y-components, and the electric current and field have only x-components.

The component forms of Eqs. (1)–(4) are then:

$$\partial H_y/\partial z = -\,I_x \tag{5}$$

$$\partial E_x/\partial z = -\,\partial B_y/\partial t \tag{6}$$

$$I_x = \sigma(E_x + u_y B_0) \tag{7}$$

$$\rho \partial u_y/\partial t = -\,I_x B_0 \tag{8}$$

We can express all the field variables in terms of B_y and obtain an equation for it. To do that we differentiate Eq. (5) with respect to z, Eq. (6) with respect to t, and Eq. (7) with respect to both z and t and then combine it with Eq. (8). This results in the wave equation

$$\partial^2 B_y/\partial t^2 - c_a^2 \partial^2 B_y/\partial z^2 - (1/\sigma\mu_0)\partial^3 B_y/\partial t\,\partial z^2 = 0 \tag{9}$$

$$c_a^2 = B_0^2/\mu_0 \rho \tag{10}$$

With $\mu \simeq \mu_0$, we can express $1/\sigma\mu$ as $c^2(\varepsilon_0/\sigma)$.

In the limiting case of infinite conductivity, this equation reduces to the ordinary wave equation

$$\partial^2 B_y/\partial t^2 = c_a^2 \partial^2 B_y/\partial z^2 \tag{11}$$

The wave propagates in the z-direction with the speed c_a, with the velocity and the magnetic field perturbations directed in the y-direction; i.e. the wave is transverse.

We note that the expression for the wave speed, $B_0/\sqrt{\rho\mu}$ is the same as for the transverse wave on a string with tension μH_0^2 and mass per unit length equal to the density ρ of the fluid. Actually, this analogy has a real significance, because the lines of the total magnetic field (external plus perturbation) are 'frozen' into the fluid and oscillate in phase with the fluid. This follows from the fact that the perturbation B_y is proportional to the velocity u_y, as indicated schematically in the display, and we leave if for one of the problems to calculate the ratio between B_y and u_y. Furthermore, the magnetic stress (force per unit area) along the field lines is known to be μH_0^2 and a fluid 'string' with this tension has unit area and hence the mass per unit length.

The conductivity, contained in the third term in Eq. (9), leads to wave attenuation, corresponding to the energy loss due to ohmic heating. This is readily seen in the case of harmonic time dependence. Thus, with the time dependence $\exp(-\mathrm{i}\omega t)$, and the plane wave complex amplitude $B_y = A\exp(\mathrm{i}kx)$ inserted into Eq. (9), we obtain the dispersion relation

$$k^2 = (\omega/c_a)^2/(1 - \mathrm{i}\omega\tau) \tag{12}$$

$$\tau = (c/c_a)^2(\varepsilon_0/\sigma) \tag{13}$$

For large values of the conductivity, we obtain the following approximation for the complex propagation constant

$$k = k_r + \mathrm{i}k_i \simeq (\omega/c_a) + \mathrm{i}\omega^2\tau/2c_a \quad \omega\tau \ll 1 \tag{14}$$

which shows that the attenuation constant k_i, and the attenuation per unit length, is proportional to the square of the frequency.

The magnetic field lines are 'frozen' into the fluid only in the limiting case of infinite conductivity. In reality, there will be some 'slippage' of the field lines with respect to the fluid, which increases with decreasing conductivity. It becomes particularly significant, as does the attenuation, at high frequencies; for which the period of oscillation becomes smaller than the characteristic relaxation time τ in Eq. (13).

Effect of compressibility. If the fluid is compressible and if we have no magnetic field, wave motion in the fluid is limited to the ordinary sound wave. Even if a B-field exists, it does not affect the sound wave if the direction of propagation is in the direction of the field, and if a sound wave is launched normal to the B-field, the wave remains a longitudinal wave.

In general the wave motion can be classified in terms of the direction of the fluid velocity in relation to the plane defined by the propagation vector $\mathbf{k}$ of the wave and the magnetic field $\mathbf{B}_0$. If the fluid velocity is perpendicular to this plane, the wave is purely transverse, an Alfven wave, with a wave speed determined by the component of the magnetic field in the direction of propagation. On the other hand, if the fluid velocity is parallel with the plane, the wave generally contains both a longitudinal and a transverse component.

Problems

19.1 *Hydrogen plasma.* (a) What is the electron plasma frequency in a fully ionized hydrogen plasma with an electron density of 10^{16} electrons per cm^3?
(b) If the plasma is confined in a magnetic field of 10 000 gauss, what is the speed of the Alfven wave?

19.2 *Energy flux in an Alfvén wave.* Derive an energy equation for an Alfvén wave, analogous to the energy equation for plasma oscillations, summarized in Display 19.3.

19.3 *Hydromagnetic wave.* (a) Determine the ratio between the complex amplitudes of fluid velocity and magnetic field for a harmonic hydromagnetic wave travelling in the z-direction, as described in Display 19.4. (b) The wave is generated by a conducting plate in the xy-plane, oscillating in harmonic motion in the y-direction with the complex amplitude U. Calculate the force per unit area on the plate required to generate the wave. What is the time average power per unit area required to drive the plane? Let the conductivity be infinite.

19.4 *Dispersion relation for hydromagnetic waves.* With reference to the dispersion relation 3.(12) for hydromagnetic waves, derive the general expression for the phase velocity and the attenuation constant, accounting for the effect of a finite conductivity.

19.5 *Plasma frequency in a solid.* Regard the conduction electrons in a solid conductor as an electron gas with an electron density of the order of 10^{24} electrons per cm^3 (one electron per atomic site). Can plasma oscillations in this gas be excited by the thermal motion of the atoms in the solid?

19.6 *Spherical plasma wave.* In the discussion of spherical plasma oscillations, collisions were neglected. If collisions are included, there will be a radial propagating plasma wave which carries energy from the source.

Suppose the radial oscillatory motion of the electrons is produced by a pair of concentric grids between which an electric field is maintained at the angular frequency ω. As a result, the radial velocity amplitude of the electrons will be U at $r = a$. Determine the corresponding field variables at r and the radiated power.

20

Waves in solids

A compressional wave in a thin solid rod is analogous to the wave in a fluid column, as was mentioned in Chapter 6, but under more general conditions, there are essential differences between waves in fluids and solids.

The reason is that a solid has a stiffness not only in deformations which involve a volume change but also in deformations (shear) which do not. Therefore, a solid can support not only longitudinal but also transverse waves and combinations of both. In a fluid, on the other hand, the excitation of a transverse or shear motion leads to diffusion (of vorticity) rather than wave propagation. An exception is the hydromagnetic wave, in which an external magnetic field provides the rigidity of the fluid in shear, as discussed in the previous chapter.

Another characteristic of waves in solids is that conversion of longitudinal to transverse waves, and vice versa, generally occurs in reflection and scattering. An extensive treatment of these and related aspects of waves in solids is beyond the present scope, and only some of the simplest but yet important examples will be considered.

Thus, we begin with the elements of longitudinal and shear waves and a discussion of bending waves on a thin plate. This is followed by a discussion of strain and stress and some general aspects of wave propagation in isotropic solids.

20.1 Longitudinal waves in a rod

We start by considering longitudinal waves in a uniform rod, which was referred to briefly in Section 6.5. The thickness of the rod is assumed small compared to a wavelength to justify the assumption of a uniform axial deformation.

In the description of fluid motion (see Section 10.1), we used the velocity as a field variable with the space–time coordinates as independent variables. In this Eulerian description, as we recall, the velocity at a certain position and time refers to the fluid element which happens to be at that position at that time. At a later time at the same position, the velocity refers to a different fluid element.

In a solid, on the other hand, the Lagrangian description of motion is generally used, in which the motion of a particular element of the solid is followed. The displacement of the element from its equilibrium position is then used as the field variable; it is a function of time and the coordinates of the equilibrium position. These coordinates specify the particular element considered. (In principle, any other parameter or set of parameters which identifies or 'labels' the element can be used as independent variables).

Strain and stress. Young's modulus. With reference to Display 20.1, we consider the element which in equilibrium is located between x and $x + \Delta x$. The displacement from the equilibrium position x is expressed by $\xi(x, t)$ so that a particle at x will be displaced to $x' = x + \xi(x, t)$. Similarly, a particle initially at $x + \Delta x$ will be displaced to $x + \Delta x + \xi(x + \Delta x, t) = x + \Delta x + \xi(x, t) + (\partial\xi/\partial x)\Delta x = x' + \Delta x'$, where $\Delta x' = \Delta x(1 + \partial\xi/\partial x)$. This last relation between the thicknesses of the elements in the initial and deformed states can be obtained directly, of course, from $x' = x + \xi(x, t)$ as $\Delta x' = \Delta x + (\partial\xi/\partial x)\Delta x$.

The **strain** $\boldsymbol{\varepsilon}$ of the element is defined as the relative change in length of the element under consideration

$$\varepsilon = (\Delta x' - \Delta x)/\Delta x = \partial\xi/\partial x \tag{1}$$

Under more general conditions, this strain is denoted by ε_{xx} to indicate that it involves the x-variation of the x-component of a displacement.

The corresponding **stress** $\boldsymbol{\sigma_{xx}}$ is the force in the x-direction per unit area on the surface (of the volume element considered) which has its outward normal in the positive x-direction.

Thus, the stress on the right hand side, at $x + \Delta x$, of the element is $\sigma_{xx}(x + \Delta x, t)$, pulling in the positive x-direction. On the left hand side of the element, the outward normal points in the negative x-direction, and the stress $\sigma_{xx}(x, t)$ acts on the portion of the solid to the left of x, the normal of its surface pointing in the positive x-direction. The stress on the element is then the reaction $-\sigma_{xx}(x, t)$.

For small deformations, the stress is known from experiments to be proportional to the strain. The constant of proportionality however, depends on whether the side walls of the bar are free or constrained. If the

Display 20.1.

Longitudinal wave motion in a thin rod.

Equilibrium position x (1)

Displaced position x' (2)

Displacement function $\xi(x, t)$ (3)

$x' = x + \xi(x, t)$ (4)

$\Delta x' = \Delta x + (\partial\xi/\partial x)\Delta x$ (5)

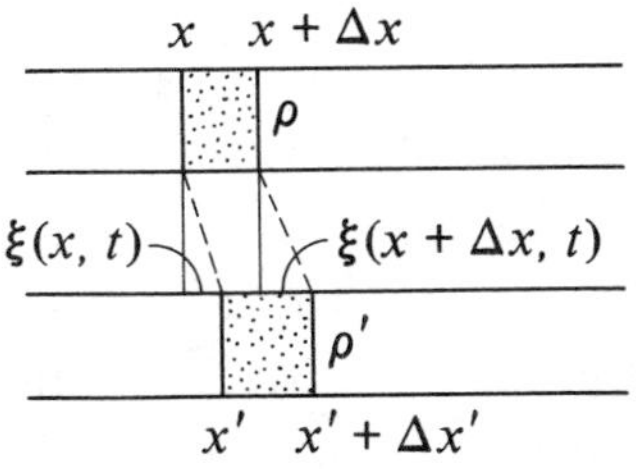

Strain

$\varepsilon = (\Delta x' - \Delta x)/\Delta x = \partial\xi/\partial x$ (6)

Stress

$\sigma = E\varepsilon = E\partial\xi/\partial x$ (7)

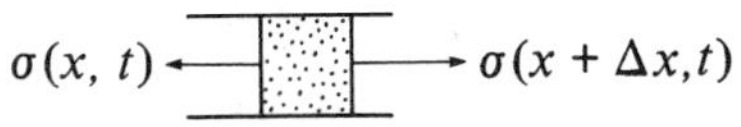

$E =$ Young's modulus

Conservation of mass:

$\rho'\Delta x' = \rho\Delta x$ (8)

$\rho' = \rho/(1 + \partial\xi/\partial x)$ (9)

Eq. of motion. Wave equation

$(\rho\Delta x)\partial^2\xi/\partial t^2 = \sigma(x + \Delta x) - \sigma(x) = (\partial\sigma/\partial x)\Delta x$ (10)

$\rho\partial^2\xi/\partial t^2 = E\partial^2\xi/\partial x^2$

$\partial^2\xi/\partial t^2 = v^2\partial^2\xi/\partial x^2 \quad v = \sqrt{E/\rho}$ (11)

Young's modulus, density and longitudinal wave speed

	$E(\mathrm{N/m^2})$	$\rho(\mathrm{kg/m^3})$	$v(\mathrm{m/sec})$
Aluminum, cast	$5.6–7.7 \times 10^{10}$		
Aluminium, rolled	6.8–7.0	2.7×10^3	5100
Brass, rolled	9.02	8.4	3300
Copper, rolled	12.1–12.9	8.5	3800
Iron, cast	8.4–9.8		
Iron, wrought	18.3–20.4	7.8	5100
Lead, rolled	1.5–1.7	11.4	1100
Steel, C 0.38	20.0	7.8	5100
Tungsten, drawn	35.5	14.0	5000

side walls are free, as will be assumed in the present case, an axial strain results in a lateral contraction or expansion of the rod, depending on the sign of the strain, and the required axial stress will be smaller than if the side walls had been constrained. The corresponding constant of proportionality is called **Young's modulus *E***

$$\sigma = E\varepsilon = E\partial\xi/\partial x \tag{2}$$

This definition will be discussed in more detail later in this chapter, when we study the relationship between various elastic constants used in the description of a solid. In that context the reason for the assumption of a thin bar will be illuminated. If the thickness is not small compared to the wavelength, the analysis given here has to be modified.

For simple crystal lattices, it is possible to calculate the elastic constants from first principles in terms of intermolecular forces, but in general, the elastic constants should be regarded as experimentally determined quantities. Some values of E are listed in the display.

It should be noted that the strain is positive when the element under consideration is stretched. Correspondingly, the stress is positive when the force on each end of the element represents a pull. In a fluid, as we recall, the stress is expressed in terms of the pressure, with the corresponding force directed inward on each end of the element. In other words, a positive pressure corresponds to a negative stress.

Wave equation. The net force in the x-direction on the element between x and $x + \Delta x$ is the sum of the stresses on the two sides of the element. As we have seen above, the stress on the right hand side is $\sigma(x + \Delta x, t)$ and the stress on the left hand side is $-\sigma(x, t)$. Consequently, the net force on the element will be $\sigma(x + \Delta x, t) - \sigma(x, t) = (\partial\sigma/\partial x)\Delta x = E(\partial^2\xi/\partial x^2)\Delta x$.

The mass of the element involved is $\rho\Delta x$, and the equation of motion becomes

$$\partial^2\xi/\partial t^2 = v^2\partial^2\xi/\partial x^2 \tag{3}$$

$$v = \sqrt{E/\rho}$$

which is the wave equation for the displacement.

Having obtained ξ, we can determine the corresponding variation in the density of the material from conservation of mass

$$\rho\Delta x = \rho'\Delta x'$$

$$\rho' = \rho/(1 + \partial\xi/\partial x) \tag{4}$$

where, as before, $\Delta x'$ is the length of the element in the displaced (and deformed) state and ρ' the corresponding density.

The wave speed v for steel is approximately 5000 m/sec, nearly 15 times larger than the sound speed in air. For cork, the wave speed typically is 1000 m/sec and for vulcanized rubber as low as 100 m/sec. Wood is anisotropic, and the wave speed can be considerably different along and across the fibers. For ash, the wave speeds along and across the fibers are about 4700 and 1400 m/sec.

20.2 Shear waves

In the longitudinal wave motion, the layers at x and $x + \Delta x$ were displaced in the x-direction, and the resulting strain caused a change in volume and density. Therefore, it is often referred to as a compressional or dilatational strain.

We now consider displacements of the two layers at x and $x + \Delta x$ in the y-direction. This lateral displacement is denoted by η, and we are only interested in the case when η varies with x (otherwise we have a mere translation of the body as a whole in the y-direction). The distance between the layers remains the same, and there will be no volume and density change. As a result of such a shear deformation, an initially square element will be deformed into a rhombus, as shown in Display 20.2.

The corresponding **shear strain** ε_{yx} is defined in terms of the relative displacements of the two layers

$$\varepsilon_{yx} = [\eta(x + \Delta x) - \eta(x)]/\Delta x = \partial\eta/\partial x \tag{1}$$

in complete analogy with the definition of the axial strain.

For small displacements, the shear strain can be thought of as an angular deflection ϕ with the respect to the x-axis, as indicated. The shear strain is positive when the angle is positive, i.e. when the displacement at $x + \Delta x$ is larger than at x.

To define the **shear stress** σ_{yx}, we proceed by analogy with the definition of the axial stress. Thus we consider a plane perpendicular to the x-axis (and the rod), and define the shear stress σ_{yx} as the force in the y-direction acting per unit area of the surface which has its outward normal pointing in the positive x-direction. Thus, on the right hand side (at $x + \Delta x$) of the element under consideration, the shear stress $\sigma_{yx}(x + \Delta x, t)$ is the force per unit area in the y-direction. On the left hand side, the outward normal to the surface points in the negative x-direction and the stress on the element is $-\sigma_{yx}(x, t)$. (The stress $\sigma_{yx}(x, t)$ acts on the surface of the portion of the solid to the left of x).

For small deformations, the stress is found to be proportional to the

$$\sigma_{yx} = G\varepsilon_{yx} \tag{2}$$

Display 20.2.

Shear wave.

Lateral displacement $\eta(x, t)$ (1)

Shear strain:

$$\varepsilon_{yx} = [\eta(x+\Delta x) - \eta(x)]/\Delta x = \partial\eta/\partial x \simeq \phi \tag{2}$$

Shear stress:

$$\sigma_{yx} = G\varepsilon_{yx} = G\partial\eta/\partial x \tag{3}$$

G = shear modulus

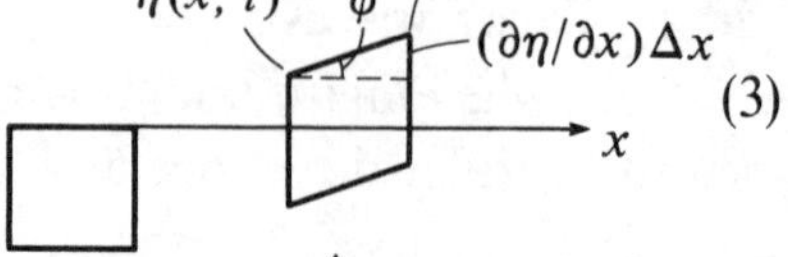

Eq. of motion:

$$(\rho\Delta x)\partial^2\eta/\partial t^2 = (\sigma_{yx})_{x+\Delta x} - (\sigma_{yx})_x = (\partial\sigma_{yx}/\partial x)\Delta x$$
$$= G(\partial^2\eta/\partial x^2)\Delta x \tag{4}$$

Wave equation:

$$\partial^2\eta/\partial t^2 = v_t^2\partial^2\eta/\partial x^2 \tag{5}$$

$$v_t = \sqrt{G/\rho} \tag{6}$$

$(\sigma_{yx})_{x+\Delta x}$

$(\sigma_{yx})_x$

mass $\rho\Delta x$

Shear modulus (modulus of rigidity), density and shear wave speed

	$G(\text{N/m}^2)$	$\rho(\text{kg/m}^3)$	v_t(m/sec)
Aluminum, rolled	2.37×10^{10}	2.71×10^3	3200
Brass, rolled	3.53	8.4	2000
Copper, rolled.	4.24	8.5	2200
Lead, rolled.	0.54	11.4	688
Steel, C 0.38	8.11	7.8	3220
Tungsten, drawn	14.8	14.0	3250

strain, where the constant of proportionality G is called the **shear modulus**. Some values of G are listed in the display.

Returning to the element between x and $x + \Delta x$, the net force on it in the y-direction is

$$\sigma_{yx}(x + \Delta x, t) - \sigma_{yx}(x, t) = (\partial \sigma_{yx}/\partial x)\Delta x = G(\partial^2 \eta/\partial x^2)\Delta x$$

where we have made use of $\varepsilon_{yx} = \partial \eta/\partial t$ in Eq. (1).

With the mass of the element being $\rho \Delta x$, the equation of motion then becomes

$$\partial^2 \eta/\partial t^2 = v_t^2 \partial^2 \eta/\partial x^2 \quad (3)$$

$$v_t = \sqrt{G/\rho} \quad (4)$$

This shows that the transverse displacement propagates as a wave with the wave speed v_t. For steel we have $v_t = 3225$ m/sec, about 65% of the longitudinal wave speed. We shall prove later that for any isotropic solid, the shear wave speed is always smaller than the longitudinal, a result which is consistent with the results obtained in our study of wave motion on a spring.

20.3 Wave reflection. Energy considerations

If we use the velocity rather than the displacement as a field variable, the longitudinal and transverse wave motion can be described in terms of field equations of the 'standard' form of Chapter 6.

Thus, for the longitudinal motion, with $u_x = \partial \xi/\partial t$, the equation of motion is

$$\rho \partial u_x/\partial t = \partial \sigma_{xx}/\partial x \quad (1)$$

Furthermore, by differentiating the stress–strain relation $\sigma_{xx} = E\varepsilon_{xx} = E\partial \xi/\partial x$, with respect to time we obtain

$$(1/E)\partial \sigma_{xx}/\partial t = \partial u_x/\partial x \quad (2)$$

Analogous equations are obtained from the transverse motion, with E replaced by G and u_x by $u_t = \partial \eta/\partial t$.

On comparison with the standard field equations $\rho \partial u/\partial t = -\partial F/\partial x$ and $\kappa \partial F/\partial t = -\partial u/\partial x$, we note that Eqs. (1) and (2) are of the standard form, with the generalized force represented by $-\sigma_{xx}$ and $-\sigma_{yx}$ and the compressibility κ represented by $1/E$ and $1/G$, respectively.

It follows that with these substitutions, the results obtained for wave impedance, reflection, transmission, and energy flux, can be applied directly to the longitudinal and transverse waves discussed in the last two sections.

20.4 Bending waves on a plate

A special type of wave motion of considerable practical importance is the bending wave on a plate. Although not as simple as the examples discussed in Sections 1 and 2, the bending wave can be analyzed by applying the simple concepts and results obtained in these sections.

We assume the plate to extend over the entire xy-plane in the y-direction, and we make the motion independent of y. The thickness of the plate is h, and it will be assumed small compared to the wavelength. In equilibrium the plate is parallel with the xy-plane.

In a bent plate, the concave side of the plate is compressed and convex side stretched. The mid-plane remains unstrained and is often called the neutral layer of the plate. The average axial stress over the cross sectional area of the plate is zero, and the stresses involved result from a bending moment and a shear force. The latter is necessary to account for a lateral displacement ζ of the plate in the z-direction. The x-dependence of ζ results in a slope of the neutral layer of the plate which for small displacement corresponds to an angle of inclination

$$\theta = \partial\zeta/\partial x \tag{1}$$

as indicated in Display 20.3.

In deriving the equation for the space–time dependence of the displacement of the plate, we consider a plate element between $x - \Delta x$ and x (we could equally well have chosen x and $x + \Delta x$) of unit width (in the y-direction) so that the area perpendicular to the x-axis numerically is equal to the thickness h of the plate. At each end of the plate element, there will be a bending moment M and a shear force $\sigma_{zx} = f$, as indicated.

The shear stress was defined in the previous section, and we use the same sign convention, the shear stress being the rate of transfer of z-momentum from right to left across a reference plane. Similarly, we define the bending moment M as the rate of transfer of angular momentum (positive in the counter clockwise direction) across the plane.

With these definitions, the shear force and the bending moment on the right hand side of the element (at x) will be $f(x,t)$ and $M(x,t)$. At the left end of the plate element, at $x - \Delta x$, the quantities $f(x - \Delta x, t)$ and $M(x - \Delta x, t)$, by definition, act on the plate to the left of the element, and the corresponding reactions on the element are $-f(x - \Delta x, t)$ and $-M(x - \Delta x, t)$.

The net force on the element in the z-direction will be $f(x) - f(x - \Delta x) = (\partial f/\partial x)\Delta x$, and with the mass density of the plate denoted by ρ, the equation of motion is

$$\rho h \partial^2\zeta/\partial t^2 = \partial f/\partial x \tag{2}$$

Display 20.3.

Bending wave on a thin plate.

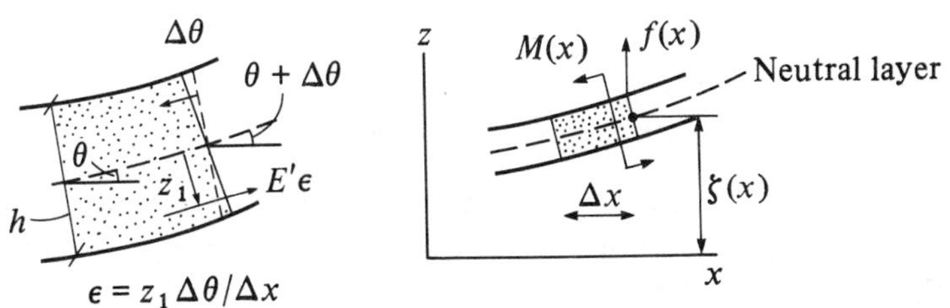

$\zeta(x, t)$ = displacement from equilibrium in z-direction (1)

f = shear force. M = bending moment (2)

$\theta \simeq \partial\zeta/\partial x$ (3)

Eqs. of motion for element:

Translational $(\rho h \Delta x)\partial^2\zeta/\partial t^2 = f(x) - f(x - \Delta x)$
$= (\partial f/\partial x)\Delta x$ (4)

Rotational $f\Delta x + (\partial M/\partial x)\Delta x = \rho I \Delta x \partial^2\theta/\partial t^2$ (5)

$\rho I = \rho \int z_1^2 \, \mathrm{d}z_1 = \rho h^3/12$ (6)

Strain $\varepsilon = z_1 \partial\theta/\partial x$ (7)

$M = E' \int \varepsilon z_1 \, \mathrm{d}z_1 = E'(\partial\theta/\partial x) \int z_1^2 \, \mathrm{d}z_1 = E' I \partial\theta/\partial x$ (8)

$E' = E/(1 - \sigma^2)$ See problem section. (9)

σ = Poisson ratio.

$\partial^2\zeta/\partial t^2 = -v^2(I/h)\partial^4\zeta/\partial x^4 + (I/h)\partial^4\zeta/\partial t^2 \partial x^2$ (10)

$v^2 = E/\rho(1 - \sigma^2)$ (11)

Harmonic wave $\exp(\mathrm{i}kx - \mathrm{i}\omega t)$

Dispersion relation:

$\omega^2[1 + (kh)^2/12] = (h^2/12)v^2 k^4$ (12)

$\omega \simeq (h/\sqrt{12})vk^2 \quad (kh \ll 1)$ (13)

$v_\mathrm{p} = \omega/k = (kh/\sqrt{12})v \quad v_\mathrm{g} = \mathrm{d}\omega/\mathrm{d}k = 2v_\mathrm{p}$ (14)

In addition to the translational motion of the element, we have to consider also the rotational motion, expressed in terms of the angle θ. The total torque on the element will be $M(x,t) - M(x-\Delta x, t) + f\Delta x = (\partial M/\partial x)\Delta x + f(\Delta x)$. The moment of inertia $\rho I \Delta x$ of our element about the center of mass axis is the same as for a thin rod of length h, i.e. $I\Delta x = (h^3/12)\Delta x$, and we obtain

$$\rho I \partial^2\theta/\partial t^2 = \partial M/\partial x + f \quad (I = h^3/12) \tag{3}$$

Differentiating Eq. (3) with respect to x, we can express $\partial f/\partial x$ in terms of ζ from Eq. (2), and from Eq. (1) we obtain θ in terms of ζ. Combining Eqs. (1)–(3) we get

$$\rho h \partial^2\zeta/\partial t^2 - \rho I \partial^4\zeta/\partial t^2 \partial x^2 = -\partial M/\partial x \tag{4}$$

It remains to express the moment M in terms of ζ. We note that a relative angular deflection $\Delta\theta$ of the two ends of the element results in a stretching $z_1\Delta\theta$ of a layer a distance z_1 below the neutral layer. Since the unperturbed length of the layer is Δx, the corresponding strain is $\varepsilon_{xx} = z_1\Delta\theta/\Delta x \simeq z_1\partial\theta/\partial x = z_1^2\partial^2\zeta/\partial x^2$.

With the elastic constant denoted by E', the stress $\sigma_{xx} = E'\varepsilon_{xx} = E'z_1^2\partial^2\zeta/\partial x^2$ is positive below the neutral layer and negative above, and the bending moment M is obtained by integrating the moment of this stress, $\sigma_{xx}z = E'z_1^2\partial^2\zeta/\partial x^2$, over the cross sectional area of the plate

$$M = \int \sigma_{xx} z_1 \, dz_1 = E' I \partial^2\zeta/\partial x^2 \tag{5}$$

$$I = \int z^2 \, dz_1 = h^3/12$$

where the integration is from $-h/2$ to $h/2$.

For a plate, only the horizontal surfaces of the elementary volume considered are free; the surfaces perpendicular to the y-axis are not. They are connected with the remainder of the (infinitely extended) plate. This constraint prevents a free contraction of the element in the y-direction as a result of a strain in the x-direction, and this makes the modulus of elasticity E' somewhat larger than the Young's modulus E, which referred to a (thin) rod, in which all the surfaces were free. In our plate, only two surfaces are free, and it can be shown that

$$E' = E/(1-\sigma^2) \tag{6}$$

where σ is another elastic constant, usually called the Poisson ratio (see Section 7).

Having expressed M in terms of ζ through Eq. (5), we can complete

Eq. (4) to obtain

$$\partial^2\zeta/\partial t^2 = -v^2(I/h)\partial^4\zeta/\partial x^4 + (I/h)\partial^4\zeta/\partial t^2\partial x^2 \tag{7}$$

$$v^2 = E/\rho(1-\sigma^2) \tag{8}$$

Unlike the ordinary wave equation for longitudinal and transverse waves, this is a fourth order equation with respect to x. As a result, the corresponding dispersion relation becomes non-linear. The dispersion relation is the condition Eq. (7) imposes on ω and k in a harmonic wave, $\zeta(x,t) = A\cos(kx - \omega t)$, in order for the wave to be a solution,

$$\omega^2[1 + (kh')^2] = (kh')^2(vk)^2 \tag{9}$$

$$\omega \simeq (kh')vk \quad (kh \ll 1) \tag{10}$$

$$h' = \sqrt{I/h} = h/\sqrt{12}$$

The dispersion relation has the same form as for the matter wave of a free particle, as discussed in Chapter 16. The phase and group velocities are

$$v_p = \omega/k = (kh')v = \sqrt{h'v}\sqrt{\omega} \tag{11}$$

$$v_g = d\omega/dk = 2v_p \tag{12}$$

$$h' = h/\sqrt{12} \quad v^2 = E/\rho(1-\sigma^2)$$

The fact that the bending wave speed increases with frequency has many interesting consequences. For example, the radiation of sound from a bending wave on a plate will be efficient only at sufficiently high frequency at which the bending wave speed exceeds the speed of sound in the surrounding fluid. This follows from our previous studies of the radiation from a travelling wave along a boundary in connection with refraction, where we found that radiation occurs only if the 'trace velocity' along the boundary is larger than the wave speed in the surrounding fluid. In this case the phase velocity of the bending wave takes the place of the trace velocity, and the plane wave radiated from the plate will travel in a direction such that the trace velocity of the radiated wave equals the bending wave speed.

Similarly, a plane wave incident on an infinite plate at an angle for which the trace velocity matches the bending wave speed, will be transmitted unimpeded through the plate, if there is no internal damping in the plate. Such an angle exists only at frequencies for which the bending wave speed exceeds the sound speed in the surrounding fluid. The frequency at which $v_p = c$ is generally called the **critical frequency**, and it follows from Eq. (11) that this frequency is

$$\omega_c = c^2/vh' \tag{13}$$

where c is the sound speed in the surrounding fluid and $h' = h/\sqrt{12}$.

If the angle of incidence of a plane wave is ϕ, the trace velocity is $c/\sin(\phi)$ which equals the bending wave speed at the coincidence angle

$$\sin(\phi_c) = c/v_p = \sqrt{\omega_c/\omega} \tag{14}$$

where ω_c is the critical frequency in Eq. (13).

Example. For glass, we have $E \simeq 6 \times 10^{11}$ dyne/cm^2, $\rho = 2.5$ g/cm^3, and $\sigma \simeq 0.25$. The characteristic speed v is then

$$v = \sqrt{E/\rho(1-\sigma^2)} = 5046 \text{ m/sec.}$$

If the thickness of the glass plate is h cm, the critical frequency in Hz in air (sound speed $c = 340$ m/sec) is

$$\nu_c = \omega_c/2\pi = (1/2\pi)c^2/vh' \simeq 1264/h \text{ Hz} \quad (h \text{ in cm})$$

For a glass plate with a thickness of 0.5 cm, the critical frequency becomes $\nu_c \simeq 2528$ Hz. Then, a plane sound wave with a frequency of 4000 Hz will be transmitted through the glass plate almost unimpeded at an angle of incidence

$$\phi_c = \arcsin(\sqrt{2528/4000}) = 52.7 \text{ degrees.}$$

The result for a steel plate is almost the same, since the characteristic speed v is approximately 5000 m/sec for both glass and steel.

20.5 Displacement, deformation, and rotation

The examples we have dealt with so far were simple enough so that the stresses and strains involved could be expressed in terms of single component of the displacement vector and its x-derivative. In general, all components and their derivatives are involved.

To discuss this general case we refer to Display 20.4 and carry out an analysis of the displacements of adjacent points A and B analogous to that in Section 1. As we recall, in the one-dimensional case, the points A and B were initially at x and $x + \mathrm{d}x$.

With the displacement of A denoted by $\xi(x, t)$, the displacement of B was $\xi(x + \mathrm{d}x, t)$ so that the new coordinates for A and B became $x' = x + \xi(x,t)$ and $x' + \mathrm{d}x' = x + \mathrm{d}x + \xi(x + \mathrm{d}x, t) = x + \mathrm{d}x + \xi(x, t) + (\partial\xi/\partial x)\mathrm{d}x = x' + (1 + \partial\xi/\partial x)\mathrm{d}x$.

The separation of A and B after the displacement was then $\mathrm{d}x' = \mathrm{d}x(1 + \partial\xi/\partial x)$, and the change in the relative positions was expressed by

$$(\mathrm{d}x' - \mathrm{d}x) = (\partial\xi/\partial x)\mathrm{d}x. \tag{1}$$

By analogy, we now specify the locations of A and B before the deformation by the position vectors $\mathbf{r}$ and $\mathbf{r} + \mathrm{d}\mathbf{r}$ with the coordinates x, y, z,

Display 20.4.

The strain tensor.

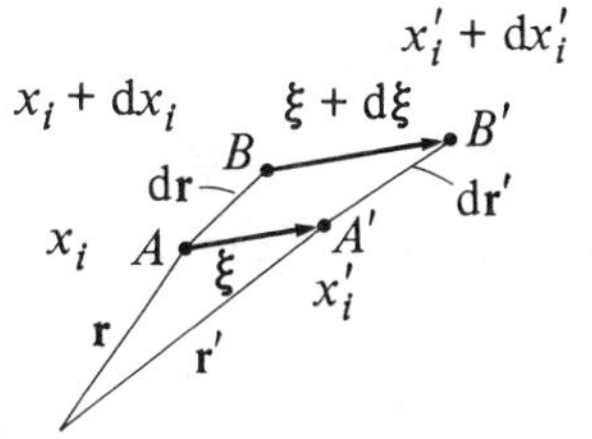

Initial positions of A and B

$A \quad \mathbf{r}$, components x_i (1)

$B \quad \mathbf{r} + \mathrm{d}\mathbf{r}$, components $x_i + \mathrm{d}x_i$ (2)

Displacement

of $A \quad \boldsymbol{\xi}(r, t)$, components $\xi_i(\mathbf{r}, t)$ (3)

of $B \quad \boldsymbol{\xi}(\mathbf{r} + \mathrm{d}\mathbf{r}, t)$, components $\xi_i(\mathbf{r} + \mathrm{d}\mathbf{r}, t)$

$= \xi_i(\mathbf{r}, t) + (\partial\xi_i/\partial x_k)\mathrm{d}k_k$ (sum over k) (4)

Positions after displacement

of $A \quad x_i' = x_i + \xi_i(\mathbf{r}, t)$ (5)

of $B \quad x_i' + \mathrm{d}x_i' = x_i + \mathrm{d}x_i + \xi_i(\mathbf{r} + \mathrm{d}\mathbf{r}, t)$

$= x_i + \mathrm{d}x_i + \xi_i(\mathbf{r}) + \xi_{ik}\,\mathrm{d}x_k \quad \xi_{ik} = \partial\xi_i/\partial x_k$ (6)

$\mathrm{d}x_i' - \mathrm{d}x_i = \xi_{ik}\mathrm{d}x_k$ (sum over k)

Rotation

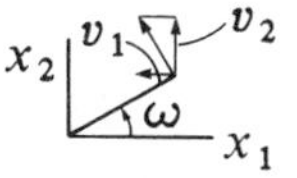

$v_2 = \omega x_1 \quad v_1 = -\omega x_2$

$\omega = \frac{1}{2}(\partial v_2/\partial x_1 - \partial v_1/\partial x_2)$ (7)

Displacement tensor $\quad \xi_{ik} = \partial\xi_i/\partial x_k$ (8)

Rotation tensor $\quad \phi_{ik} = \frac{1}{2}(\partial\xi_i/\partial x_k - \partial\xi_k/\partial x_i)$ (9)

Strain tensor $\quad \varepsilon_{ik} = \xi_{ik} - \phi_{ik} = \frac{1}{2}(\partial\xi_i/\partial x_k + \partial\xi_k/\partial x_i)$ (10)

and $x + \mathrm{d}x$, $y + \mathrm{d}y$, $z + \mathrm{d}z$. The writing lends itself better to simplification if we denote the coordinates x, y, z, of $\mathbf{r}$ by x_1, x_2, x_3 or simply by $x_i (i = 1, 2, 3)$.

As a result of the displacement ξ of A with the coordinates $\xi_i(x, y, z, t)$, the new coordinates of A will be

$$x_i' = x_i + \xi_i(\mathbf{r}, t) \tag{2}$$

and the new coordinates of B

$$\begin{aligned} x_i' + \mathrm{d}x_i' &= x_i + \mathrm{d}x_i + \xi_i(\mathbf{r} + \mathrm{d}\mathbf{r}, t) = x_i + \mathrm{d}x_i + \xi_i(\bar{r}) \\ &+ \sum_k (\partial \xi_i / \partial x_k) \mathrm{d}x_k = x_i' + \mathrm{d}x_i + \sum_k (\partial \xi_i / \partial x_k) \mathrm{d}x_k \end{aligned} \tag{3}$$

where, as before, $\mathbf{r}$ stands for the coordinates x_k with $k = 1, 2, 3$.

It follows that the components of the separation vector between A and B after the displacement is

$$\mathrm{d}x_i' = \mathrm{d}x_i + \sum_k (\partial \xi_i / \partial x_k) \mathrm{d}x_k \tag{4}$$

The change in the relative positions of A and B, which corresponds to Eq. (1), is then

$$\mathrm{d}x_i' - \mathrm{d}x_i = \sum_k (\partial \xi_i / \partial x_k) \mathrm{d}x_k = \sum_k \xi_{ik} \mathrm{d}x_k \tag{5}$$

where we have introduced the quantity

$$\xi_{ik} = \partial \xi_i / \partial x_k \tag{6}$$

With ξ_i being a function of all three of the initial coordinates, the change in the relative position contains three contributions. Since we often have to sum over these three contributions, we shall not always write out the sum symbol Σ, but infer summation over the repeated indices in the product $\xi_{ik} \mathrm{d}x_k$ (summation convention).

The quantity ξ_{ik} is called the **displacement tensor**. Unlike a scalar, which is defined by a single number, or an ordinary three dimensional vector, which has three components, the displacement tensor is a 3×3 array of numbers corresponding to the values $1, 2, 3$ for i and k.

The strain tensor. A mere translation of the body as a whole clearly will not result in a change in the relative positions of A and B, i.e. $\mathbf{dr}' = \mathbf{dr}$. A rotation of the body, on the other hand, does result in a change of $\mathbf{dr}' - \mathbf{dr}$, and does contribute to the displacement tensor ξ_{ik}. No deformation of the body is involved, however; the distance between A and B remains the same.

Since stresses in the body are related to deformations, it is of interest to extract from the displacement tensor the contribution which results from a pure rotation. To do that we first have to obtain the tensor description of a rotational displacement, and we start by considering the familiar rigid body rotation.

If the body rotates about the x_3-axis, the velocity components of a particle with the coordinates x_1, x_2 are

$$v_1 = -\omega x_2 \quad v_2 = \omega x_1 \tag{7}$$

Regarding the velocity components as functions of x_1, x_2 we can express ω as

$$\omega = \partial v_2/\partial x_1 = -\partial v_1/\partial x_2 = \tfrac{1}{2}(\partial v_2/\partial x_1 - \partial v_1/\partial x_2) \tag{8}$$

By multiplying this expression by the time interval dt, we obtain a corresponding tensor description of an angular displacement. The tensor

$$\phi_{ik} = \tfrac{1}{2}(\partial \xi_i/\partial x_k - \partial \xi_k/\partial x_i) \tag{9}$$

is of this form and will be called the **rotation tensor.**

If we now subtract this tensor from the displacement tensor, we obtain the deformation or **strain tensor**

$$\varepsilon_{ik} = \xi_{ik} - \phi_{ik} = \tfrac{1}{2}(\partial \xi_i/\partial x_k + \partial \xi_k/\partial x_i) \tag{10}$$

Unlike the rotation tensor, it is symmetrical with respect to an interchange of i and k.

Example. It is instructive to illustrate with a simple example the geometrical interpretation of Eq. (10), that a displacement (tensor) is the sum of a rotation (tensor) and a deformation or strain (tensor).

Thus, with reference to Display 20.5, we consider a displacement which brings the corners A, B, C, D of a square into the positions A', B', C', D' in the x_1, x_2-plane. The corresponding displacement tensor has only one component, namely $\xi_{12} = \partial \xi_1/\partial x_2$.

We note that the initial separation of A and B has only the component dx_2. Applying Eq. (5) for the change in the relative positions of A and B, we find that the components of the new separation vector d**r**′ between A and B are

$$dx_1' = dx_1 + \xi_{12} dx_2 = (\partial \xi_1/\partial x_2) dx_2$$

and

$$dx_2' = dx_2$$

The same result is obtained for the separation DC. The displacement can be described in terms of the change in the directions of AB and DC which for small displacements, are given by the angle ξ_{12} with the x_2-axis.

The corresponding rotation tensor has the components $\phi_{12} = \tfrac{1}{2}\xi_{12}$ and $\phi_{21} = -\phi_{12}$. Again, the change in the components of the separation vectors are obtained from Eq. (5) with ξ_{ik} replaced by ϕ_{ik}. Thus for AB we obtain

$$dx_1' = \phi_{12} dx_2 = \tfrac{1}{2}\xi_{12} dx_2$$

Display 20.5.

Relations between tensors. Example.

Displacement tensor $\xi_{ik} = \partial\xi_i/\partial x_k$ (1)

Rotation tensor $\phi_{ik} = \frac{1}{2}(\xi_{ik} - \xi_{ki})$ (2)

Strain tensor $\varepsilon_{ik} = \xi_{ik} - \phi_{ik} = \frac{1}{2}(\xi_{ik} + \xi_{ki})$ (3)

Example

$\xi_{12} = \varepsilon_{12} + \phi_{12}$ (4)

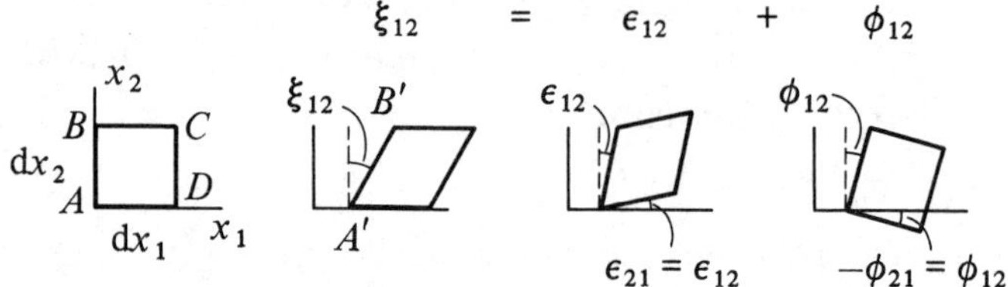

Line and volume element

$(\mathrm{d}l)^2 = (\mathrm{d}x_1)^2 + (\mathrm{d}x_2)^2 + (\mathrm{d}x_3)^2 = (\mathrm{d}x_i)^2$ (sum over i implied) (5)

$(\mathrm{d}l')^2 = (\mathrm{d}x_i')^2 = (\mathrm{d}x_i)^2 + [\Sigma(\partial\xi_i/\partial x_k)]^2 + 2\Sigma(\partial\xi_i/\partial x_k)\mathrm{d}x_i\mathrm{d}x_k$ (6)

$\simeq (\mathrm{d}l)^2 + 2(\partial\xi_i/\partial x_k)\mathrm{d}x_i\mathrm{d}x_k$ (sum over i, k) (7)

$(\mathrm{d}l' - \mathrm{d}l)/\mathrm{d}l \simeq (\partial\xi_i/\partial x_k)\mathrm{d}x_i\mathrm{d}x_k/(\mathrm{d}l)^2 = \xi_{ik}\gamma_i\gamma_k = \varepsilon_{ik}\gamma_i\gamma_k$ (8)

$\gamma_i = \mathrm{d}x_i/\mathrm{d}l$ (9)

$(V' - V)/V = \varepsilon_{11} + \varepsilon_{22} + \varepsilon_{33} = \varepsilon_{nn}$ (sum over n) (10)

and for AC

$$\mathrm{d}x_2' = \phi_{21}\,\mathrm{d}x_1 = -\phi_{12}\,\mathrm{d}x_1$$

with analogous results from DC and BC. For small deformations this operation corresponds to a rigid body rotation by an angle $\frac{1}{2}\xi_{12}$, as indicated in the display.

It remains to determine the displacements which correspond to the strain tensor $\varepsilon_{ik} = \xi_{ik} - \phi_{ik}$. It has the components $\varepsilon_{12} = \varepsilon_{21} = \frac{1}{2}\xi_{12}$. With ξ_{ik} in Eq. (5) replaced by ε_{ik}, we find in the same manner as above that the effect of ε_{ik} is to turn AB and AD in opposite directions by an angle $\frac{1}{2}\xi_{12}$, with similar results for DC and BC. This can be seen geometrically by 'subtracting' the displacements produced by ξ_{ik} and ϕ_{ik}, shown in the display.

The strain represents a deformation of the material, free from a superimposed rigid body rotation, and the stresses involved will be a linear function of the components of the strain tensor, as will be discussed later.

Change in line element and volume. To illustrate further the use of the strain tensor, we calculate the relative change in the magnitude $\mathrm{d}l$ of the displacement vector $\mathbf{dr}$ between A and B. The components of $\mathbf{dr}$ are $\mathrm{d}x_i$, and we have

$$(\mathrm{d}l)^2 = (\mathrm{d}x_1)^2 + (\mathrm{d}x_2)^2 + (\mathrm{d}x_3)^2 = (\mathrm{d}x_i)^2 \quad \text{(sum over } i) \tag{11}$$

After a deformation the corresponding separation between the new positions of A and B is $\mathrm{d}l'$, which is expressed in terms of $\mathrm{d}x_i'$ in similar manner,

$$(\mathrm{d}l')^2 = (\mathrm{d}x_i')^2 = [\mathrm{d}x_i + \xi_{ik}\,\mathrm{d}x_k]^2$$
$$= (\mathrm{d}l)^2 + 2\xi_{ik}\,\mathrm{d}x_i\,\mathrm{d}x_k + (\xi_{ik}\,\mathrm{d}x_k)^2 \tag{12}$$
$$\simeq (\mathrm{d}l)^2 + 2\xi_{ik}\,\mathrm{d}x_i\,\mathrm{d}x_k \quad \text{(sum over } i, k) \tag{13}$$

where we have used Eq. (5). The third term in Eq. (12) is of second order in the field variable, and since we limit ourselves here to small displacements, it can be neglected. Similarly we can use the approximation $[(\mathrm{d}l)^2 + 2\xi_{ik}\,\mathrm{d}x_i\,\mathrm{d}x_k]^{\frac{1}{2}} \simeq \mathrm{d}l[1 + \xi_{ik}\,\mathrm{d}x_i\,\mathrm{d}x_k/(\mathrm{d}l)^2]$ to obtain

$$(\mathrm{d}l' - \mathrm{d}l)/\mathrm{d}l = \xi_{ik}\,\mathrm{d}x_i\,\mathrm{d}x_k/(\mathrm{d}l)^2 = \xi_{ik}\gamma_i\gamma_k = \varepsilon_{ik}\gamma_i\gamma_k \quad \text{(sum over } i, k)$$
$$\gamma_i = \mathrm{d}x_i/\mathrm{d}l \tag{14}$$

In other words, the relative change in the distance between A and B is a quadratic form in terms of the direction cosines $\gamma_i = \mathrm{d}x_i/\mathrm{d}l$ of the line vector $\mathrm{d}r$ from A to B. The coefficients in this quadratic form are the components of the displacement tensor ξ_{ik} (or strain tensor ε_{ik}).

By expressing this tensor as the sum of the strain and rotation tensor, we see explicitly that the rotation tensor does not contribute to the relative change in the distance between A and B. Since the rotation tensor is anti-symmetric with respect to an interchange of i and k, $\phi_{ik} = -\phi_{ki}$, the sum over i and k in Eq. (14) makes the terms containing the elements of the rotation tensor cancel each other and the remaining contribution is due to the strain only, as indicated in Eq. (14).

To derive the important expression for the relative change in the volume we consider an element with the sides dx_1, dx_2, dx_3 so that the volume before deformation is $V = dx_1\, dx_2\, dx_3$. After the deformation the sides of the element generally change in both length and direction and will be described by the vectors **a**, **b**, and **c**, respectively. According to Eq. (5), the components of these vectors are $a_1 = dx_1 + \xi_{11}\, dx_1$, $a_2 = \xi_{21}\, dx_1$, $a_3 = \xi_{31}\, dx_1$, with similar expressions for the components of **b** and **c**.

The new volume is obtained from $V' = (\mathbf{a} \times \mathbf{b})\cdot\mathbf{c}$, and we express it in terms of the vector component form and neglect terms of second and higher order in ξ_{ik}, we obtain

$$(V' - V)/V = \xi_{11} + \xi_{22} + \xi_{33} = \varepsilon_{11} + \varepsilon_{22} + \varepsilon_{33} = \varepsilon_{nn} \tag{15}$$

where summation over n is implied in the last expression. We note that the relative volume change is the sum of the diagonal terms of the strain tensor and that the rotation tensor does not contribute since $\phi_{11} = \phi_{22} = \phi_{33} = 0$. The sum ε_{nn} is called the **trace** of the tensor.

20.6 Stress–strain relation

In Sections 1 and 2 we have already introduced the purely axial and transverse stresses, defined in Eqs. 1(2) and 2(2). In general each surface of a volume element with the sides dx_1, dx_2, dx_3 is acted on by both an axial and a transverse stress component. By definition, the **stress component** σ_{ik} is the i-component of the force acting on the surface which has its outward normal pointing in the positive k-direction.

For small deformations, each stress component is assumed to be a linear function of the strain components. As we recall, the strain tensor is the part of the displacement tensor which contains a deformation of the material, and thus is related to the stress. With 9 strain components, such a linear relation will contain 9 elastic constants for each of the 9 stress components, i.e. a total of 81 constants. Fortunately, they are not all independent, however; various symmetry conditions reduce the number to 21, and for an isotropic solid the symmetry is so high, that the number of independent elastic constants is reduced to 2.

Display 20.6.

Stress–strain relation.

Strain tensor

$\varepsilon_{ik} = (1/2)(\partial\xi_i/\partial x_k + \partial\xi_k/\partial x_i)$ (1)

Volume change:

$\Delta V/V = \varepsilon_{11} + \varepsilon_{22} + \varepsilon_{33} = \varepsilon_{nn}$ (sum over n) (2)

$\varepsilon_{ik} = C_{ik} + S_{ik}$ (3)

$C_{ik} = (1/3)\delta_{ik}\varepsilon_{nn}$ (compression) (4)

$S_{ik} = \varepsilon_{ik} - C_{ik} = \varepsilon_{ik} - (1/3)\delta_{ik}\varepsilon_{nn}$ (shear) (5)

$\delta_{ik} = 1$ if $i = k$, $\delta_{ik} = 0$ if $i \neq k$ (6)

Note $S_{nn} = 0$ $C_{nn} = \varepsilon_{nn} = \Delta V/V$ (sum over n) (7)

Stress tensor

For uniform compression, $\varepsilon_{11} = \varepsilon_{22} = \varepsilon_{33}$, the stress is (8)

$\sigma_{11} = (1/\kappa)\Delta V/V = K(\Delta V/V)$ (9)

κ = compressibility $K = 1/\kappa$ = elastic bulk modulus (10)

For pure shear, $\varepsilon_{ik} \rightarrow \partial\xi_1/\partial x_2$, the stress is (11)

$\sigma_{12} = G\partial\xi_1/\partial x_2$ G = shear modulus (modulus of rigidity) (12)

General stress–strain relation:

$\sigma_{ik} = 3KC_{ik} + 2GS_{ik} = K\delta_{ik}\varepsilon_{nn} + 2G[\varepsilon_{ik} - (1/3)\delta_{ik}\varepsilon_{nn}]$ (13)

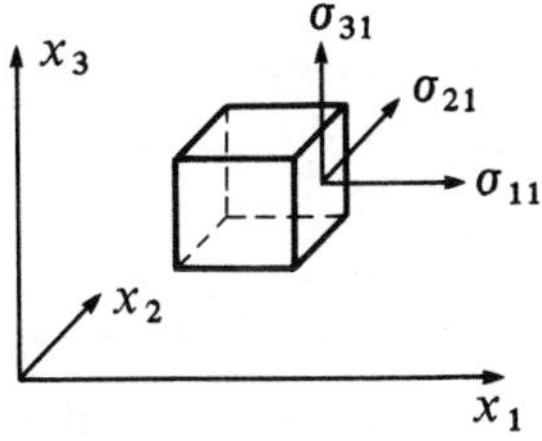

In the special deformations considered in Sections 1 and 2, we have already encountered two independent elastic constants, Y and G. The first involved a volume change and the other did not.

In the general deformation considered here, we now seek the elastic constant which corresponds to a pure volume change, and one which expresses the stresses in a deformation with no volume change.

As we recall, the components of the strain tensor are

$$\varepsilon_{ik} = \tfrac{1}{2}(\partial\xi_i/\partial x_k + \partial\xi_k/\partial x_i) \tag{1}$$

and the relative volume change in a deformation is

$$\Delta V/V = \varepsilon_{11} + \varepsilon_{22} + \varepsilon_{33} = \varepsilon_{nn} \tag{2}$$

We now wish to split up the strain tensor into two parts,

$$\varepsilon_{ik} = C_{ik} + S_{ik} \tag{3}$$

in which C_{ik} is to represent a pure volume change and S_{ik} a deformation without a volume change.

We have already shown that the sum of the diagonal terms of the strain tensor is the relative volume change. Therefore, in constructing the 'compression tensor' C_{ik}, we require it to have the following properties. First, the sum of its diagonal terms should be ε_{nn}, the relative change in volume, and second, the remaining elements should be zero to guarantee C_{ik} describes nothing but a volume change.

The first requirement is fulfilled if we put $C_{11} = C_{22} = C_{33} = \varepsilon_{nn}/3$, and to satisfy the second requirement we make C_{ik} proportional to the particular tensor δ_{ik} (Kronecker delta) with $\delta_{11} = \delta_{22} = \delta_{33} = 1$ and $\delta_{ik} = 0$ for $i \neq k$. Thus,

$$C_{ik} = (\varepsilon_{nn}/3)\delta_{ik} \tag{4}$$

The second deformation tensor, S_{ik}, is obtained as the difference between ε_{ik} and C_{ik}, i.e.

$$S_{ik} = \varepsilon_{ik} - C_{ik} = \varepsilon_{ik} - (\varepsilon_{nn}/3)\delta_{ik} \tag{5}$$

It follows that $S_{11} = \varepsilon_{11} - (\varepsilon_{11} + \varepsilon_{22} + \varepsilon_{33})/3$, with similar expressions for S_{22} and S_{33}. We note that $S_{11} + S_{22} + S_{33} = 0$, as it should, and that $S_{ik} = \varepsilon_{ik}$ for $i \neq k$.

Compressibility and bulk modulus. We are now prepared to define elastic constants, and we start with the compressibility κ. It is defined operationally as the relative change in the volume of a cube per unit stress when all sides are acted on by equal compressional forces per unit area, i.e. equal stresses

$\sigma_{11} = \sigma_{22} = \sigma_{33}$,

$$\kappa = (1/\sigma_{11})\Delta V/V \tag{6}$$

or

$$\sigma_{11} = K\Delta V/V \tag{7}$$

where $K = 1/\kappa$ is the **bulk modulus.**

Shear modulus. As already explained in Section 2, the shear modulus is based on the particular deformation shown in Display 20.5, represented by the displacement tensor $\xi_{12} = \partial\xi_1/\partial x_2$. The stress σ_{12} is the force in the x_1-direction acting per unit area on the top surface of the element. The shear modulus G is then defined by the relation

$$\sigma_{12} = G(\xi_{12}) = G\partial\xi_1/\partial x_2 \tag{8}$$

General stress–strain relation. On the premise that the general expression for the stress σ_{ik} be a linear function $\sigma_{ik} = a_1 C_{ik} + a_2 S_{ik}$ of the deformation tensors C_{ik} and S_{ik}, we have to adjust the constants a_1 and a_2 so that this expression reduces to those involved in the special deformations used in the definition of the bulk and shear moduli in Eqs. (5) and (6). The appropriate values of a_1 and a_2 are seen to be $a_1 = 3K$ and $a_2 = 2G$, so that

$$\sigma_{ik} = 3KC_{ik} + 2GS_{ik} = K\varepsilon_{nn}\delta_{ik} + 2G[\varepsilon_{ik} - (\varepsilon_{nn}/3)\delta_{ik}] \tag{9}$$

To compare this result with Eq. (6), we put $i = k = 1$ and make $\varepsilon_{11} = \varepsilon_{22} = \varepsilon_{33}$ so that $\varepsilon_{nn} = 3\varepsilon_{11}$. The second term in Eq. (9) then becomes zero and we obtain $\sigma_{11} = K\varepsilon_{nn}$ as it should. Similarly, to compare with Eq. (8), we put $i = 1$ and $k = 2$. The first term in Eq. (9) then vanishes, and the second term reduces to $G\xi_{12}$, as it should.

20.7 Relations between elastic constants

With reference to Section 1, Young's modulus is defined as the ratio between the axial stress and the axial strain in a free rod

$$E = \sigma_{11}/\varepsilon_{11} \tag{1}$$

Since the side surfaces of the rod are free, the lateral stress components are zero, $\sigma_{22} = \sigma_{33} = 0$.

In addition to the axial strain there will be lateral strains $\varepsilon_{22} = \varepsilon_{33}$. The ratio

$$\sigma = -\varepsilon_{22}/\varepsilon_{11} \tag{2}$$

is an elastic constant called the **Poisson ratio**.

It is of interest to determine the relation between the elastic constants E

Display 20.7.

Relations between elastic constants.

Free bar ($\sigma_{22} = \sigma_{33} = 0$)

Definition of Young's modulus E:

$\sigma_{11} = E\partial\xi_1/\partial x_1 = E\varepsilon_{11} \quad \sigma_{22} = \sigma_{33} = 0$ (1)

Definition of Poisson ratio σ:

$\sigma = -\varepsilon_{22}/\varepsilon_{11} \quad (\varepsilon_{22} = \varepsilon_{33})$ (2)

$\sigma_{22} = 0$ σ_{11} $\sigma_{33} = 0$

General stress–strain relation:

$\sigma_{ik} = K\delta_{ik}\varepsilon_{nn} + 2G[\varepsilon_{ik} - (1/3)\delta_{ik}\varepsilon_{nn}]$ K = bulk modulus (3)

$\sigma_{11} = K\varepsilon_{nn} + 2G(\varepsilon_{11} - \varepsilon_{nn}/3)$ G = shear modulus (4)

$\sigma_{22} = K\varepsilon_{nn} + 2G(\varepsilon_{22} - \varepsilon_{nn}/3)$ (5)

$\sigma_{33} = K\varepsilon_{nn} + 2G(\varepsilon_{33} - \varepsilon_{nn}/3)$ (6)

Application, free bar

With $\sigma_{22} = \sigma_{33} = 0$, addition of Eqs. (3), (4), (5) gives (note $\varepsilon_{nn} = \varepsilon_{11} + \varepsilon_{22} + \varepsilon_{33}$) (7)

$\sigma_{11} = 3K\varepsilon_{nn} \quad \varepsilon_{nn} = \sigma_{11}/3K$ (8)

Insertion into Eq. (4):

$\sigma_{11} = [9GK/(3K + G)]\varepsilon_{11}$, i.e. (9)

$E = 9GK/(3K + G)$ (10)

With $\varepsilon_{nn} = \varepsilon_{11} + 2\varepsilon_{22}$, Eqs. (8) and (9) give (11)

$\sigma = -\varepsilon_{22}/\varepsilon_{11} = (3K - 2G)/(6K + 2G)$ (12)

$\sigma = (E/2G) - 1$ (13)

$K = E/3(1 - 2\sigma)$ (14)

$G = E/2(1 + \sigma)$ (15)

Steel $E = 20 \times 10^{10}$, $G = 8.11 \times 10^{10}$ N/m^2.
Gives $\sigma \simeq 0.23$

and σ, the bulk modulus K, and the shear modulus G, introduced in the last section. To do that we apply the general stress–strain relation in Eq. 6(9) to the free rod problem and obtain

$$\sigma_{11} = K\varepsilon_{nn} + 2G(\varepsilon_{11} - \varepsilon_{nn}/3) \tag{3}$$

$$\sigma_{22} = K\varepsilon_{nn} + 2G(\varepsilon_{22} - \varepsilon_{nn}/3) \tag{4}$$

$$\sigma_{33} = K\varepsilon_{nn} + 2G(\varepsilon_{33} - \varepsilon_{nn}/3) \tag{5}$$

With $\sigma_{22} = \sigma_{33} = 0$ and $\sigma_{11} = E\varepsilon_{11}$, addition of Eqs. (3)(4) and (5) yields

$$\sigma_{11} = 3K\varepsilon_{nn} = E\varepsilon_{11} \tag{6}$$

and combining this expression with $\varepsilon_{nn} = \varepsilon_{11} + \varepsilon_{22} + \varepsilon_{33} = (1 - 2\sigma)\varepsilon_{11}$, we obtain

$$\sigma = \tfrac{1}{2} - (E/6K) \tag{7}$$

which shows that the Poisson ratio is always less than $\frac{1}{2}$.

Accounting for Eq. (3), we find the additional relations

$$E = 9GK/(3K + G) \tag{8}$$

$$\sigma = (3K - 2G)/(6K + 2G) \tag{9}$$

which are often written in the form

$$K = E/3(1 - 2\sigma) \tag{10}$$

$$G = E/2(1 + \sigma) \tag{11}$$

For steel, $E = 2 \times 10^{11}$ and $G = 8.1 \times 10^{10}\ \mathrm{N/m^2}$, and it follows from Eq. (11) that $\sigma = 0.23$.

20.8 Longitudinal and transverse waves

We have established that the component σ_{ik} of the stress tensor is the force component along the ith coordinate axis per unit area of the surface element which has it outward normal pointing in the positive k-direction. In Display 20.8, on the volume element under consideration, we have shown the three stress components on the surface which is perpendicular to the x_1-axis at location x_1. There are similar sets of stresses on the surfaces perpendicular to the x_2- and x_3-axes at locations x_2 and x_3.

On each of the surfaces at x_1–dx_1, x_2–dx_2, and x_3–dx_3, the stresses are in opposite directions from those at x_1, x_2, and x_3, respectively and the magnitudes are different in accordance with the variation of the stress with location.

The net force resulting from the stresses on each pair of opposite surfaces then can be expressed in terms of the spatial derivative of the stress. We have carried out similar discussions several times in previous chapters, and

Display 20.8.

Wave motion in a solid.

$$\rho\partial^2\xi_i/\partial t^2 = \partial\sigma_{ik}/\partial x_k \quad \text{(sum over } k\text{)} \tag{1}$$

$$\sigma_{ik} = K\delta_{ik}\varepsilon_{nn} + 2G[\varepsilon_{ik} - (1/3)\delta_{ik}\varepsilon_{nn}] \tag{2}$$

$$\varepsilon_{ik} = (1/2)(\partial\xi_i/\partial x_k + \partial\xi_k/\partial x_i) \tag{3}$$

$$\varepsilon_{nn} = \partial\xi_1/\partial x_1 + \partial\xi_2/\partial x_2 + \partial\xi_3/\partial x_3 \tag{4}$$

Longitudinal motion $\quad \xi_i \to \xi_1 \quad \partial/\partial x_k \to \partial/\partial x_1$ (5)

$$\rho\partial^2\xi_1/\partial t^2 = \partial\sigma_{11}/\partial x_1 \tag{6}$$

$$\sigma_{11} = (K + 4G/3)\partial\xi_1/\partial x_1 \tag{7}$$

$$\partial^2\xi_1/\partial t^2 = v_l^2\partial^2\xi_1/\partial x_1^2 \tag{8}$$

$$v_l = \sqrt{(K + 4G/3)/\rho} = \sqrt{[(1-\sigma)/(1-2\sigma)(1+\sigma)]\,E/\rho}$$

(longit. wave speed) (9)

Transverse motion $\quad \xi_i \to \xi_2, \quad \partial/\partial x_k \to \partial/\partial x_1$ (10)

$$\rho\partial^2\xi_2/\partial t^2 = \partial\sigma_{21}/\partial x_1 \tag{11}$$

$$\sigma_{21} = G\partial\xi_2/\partial x_1 \tag{12}$$

$$\partial^2\xi_2/\partial t^2 = v_t^2\partial^2\xi^2/\partial x_1^2 \tag{13}$$

$$v_t = \sqrt{G/\rho} = \sqrt{[1/2(1+\sigma)]E/\rho} \quad \text{(transverse wave speed)} \tag{14}$$

With $\sigma = (3K - 2G)/(6K + 2G)$ (See Eq. 21.5(11)) (15)

$$v_t/v_l = \sqrt{(1-2\sigma)/2(1-\sigma)} \tag{16}$$

General wave motion

$$\boldsymbol{\xi} = \boldsymbol{\xi}_l + \boldsymbol{\xi}_t \tag{17}$$

$$\boldsymbol{\xi}_l = \text{grad}\,\Phi \quad \boldsymbol{\xi}_t = \text{curl}\,\mathbf{A} \tag{18}$$

$$\rho\partial^2\boldsymbol{\xi}_l/\partial t^2 = v_l^2\nabla^2\boldsymbol{\xi}_l \tag{19}$$

$$\rho\partial^2\boldsymbol{\xi}_t/\partial t^2 = v_t^2\nabla^2\boldsymbol{\xi}_t \tag{20}$$

there should be no need at this point to write out all the components of the net force in detail. It is sufficient to do it for one of the components for one of the surface pairs. Thus, the contribution to the force in the x_2-direction from the surfaces perpendicular to the x_1-axis is given by

$$[\sigma_{21}(x_1, t) - \sigma_{21}(x_1 - \mathrm{d}x_1, t)]\,\mathrm{d}x_1(\mathrm{d}x_2\,\mathrm{d}x_3) = (\partial\sigma_{21}/\partial x_1)\Delta V$$

where $V = \mathrm{d}x_1\,\mathrm{d}x_2\,\mathrm{d}x_3$ is the volume element. The contributions from the other surface pairs are obtained in an analogous manner.

Adding these contributions, corresponding to summation over $k = 1, 2, 3$, we obtain the net force component in the direction of the ith coordinate axis $\Sigma\partial\sigma_{ik}/\partial x_k$ per unit volume.

The equation of motion for the ith component of the displacement is then

$$\rho\partial^2\xi_i/\partial t^2 = \partial\sigma_{ik}/\partial x_k \quad \text{(sum over } k = 1, 2, 3) \tag{1}$$

where the ik-component of the stress, according to Eq. 6(9), is

$$\sigma_{ik} = K\delta_{ik}\varepsilon_{nn} + 2G[\varepsilon_{ik} - (1/3)\delta_{ik}\varepsilon_{nn}] \tag{2}$$

$$\varepsilon_{ik} = \tfrac{1}{2}(\partial\xi_i/\partial x_k + \partial\xi_k/\partial x_i)$$

$$\varepsilon_{nn} = \varepsilon_{11} + \varepsilon_{22} + \varepsilon_{33}$$

where K is the bulk modulus and G the shear modulus.

We shall analyze only two special motions, corresponding to longitudinal and transverse waves. In both cases we choose the displacements to depend only on the x_1-coordinate but the directions of the displacements are different, along the x_1- and x_2-axes, respectively.

Longitudinal wave. With the direction of propagation and displacement both in the x_1-direction, Eq. (1) reduces to

$$\rho\partial^2\xi_1/\partial t^2 = \partial\sigma_{11}/\partial x_1 \tag{3}$$

With the stress, obtained from Eq. 6(9), being

$$\sigma_{11} = (K + 4G/3)\partial\xi_1/\partial x_1 \tag{4}$$

the equation of motion becomes the ordinary wave equation

$$\partial^2\xi_1/\partial t^2 = v_l^2\partial^2\xi_1/\partial x^2 \tag{5}$$

$$v_l^2 = (K + 4G/3)/\rho = (E/\rho)(1 - \sigma)/[(1 - 2\sigma)(1 + \sigma)] \tag{6}$$

where v_l is the wave speed for the longitudinal one-dimensional wave in the solid. In obtaining the last expression for this wave speed in Eq. (6), we have made use of the relations in Section 7.

It should be noted that this wave speed, which refers to one-dimensional longitudinal wave propagation in an infinitely extended solid, is somewhat larger than the wave speed $\sqrt{E/\rho}$ for the wave on a thin rod. The basic difference between the two waves is that in the thin rod a lateral displacement is present. The lateral motion increases the kinetic energy in the wave per unit length, which, in effect corresponds to a larger inertial mass of the system and hence a lower wave speed.

The effect is rather small, however. For steel, with a Poisson ratio of $\sigma \simeq 0.23$ the ratio between the two speeds is 1.08.

Transverse wave. In the case of transverse motion, with the displacement in the x_2-direction, the equation of motion becomes

$$\rho \partial^2 \xi_2/\partial t^2 = \partial \sigma_{21}/\partial x_1 \tag{7}$$

where, as before, the stress is obtained from Eq. 6(9),

$$\sigma_{21} = G \partial \xi_2/\partial x_1 \tag{8}$$

Eq. (7) then becomes the wave equation

$$\partial^2 \xi_2/\partial t^2 = v_t^2 \partial^2 \xi_2/\partial x_1^2 \tag{9}$$

$$v_t^2 = G/\rho = (E/\rho)/[2(1+\sigma)] \tag{10}$$

where the transverse wave speed is always smaller than the longitudinal. The ratio between the two can be expressed as

$$v_t/v_l = [(1-2\sigma)/3(1-\sigma)]^{\frac{1}{2}} \tag{11}$$

With $\sigma = 0.23$, we get $v_t/v_l = 0.59$.

General wave motion. The general solution to Eq. (1) with the associated stress in Eq. (2), can be constructed from solutions of the ordinary wave equation. This follows from the observation that Eq. (1) can be satisfied if the displacement field is decomposed into two parts, $\xi = \xi_l + \xi_t$, where

$$\xi_l = \text{grad}\,(\phi) \tag{12}$$

$$\xi_t = \text{curl}\,(\mathbf{A}) \tag{13}$$

It follows from Eqs. (1) and (2), that with this decomposition, we find that ξ_l and ξ_t satisfy the wave equations

$$\partial^2 \xi_l/\partial t^2 = v_l^2 \nabla^2 \xi_l \tag{14}$$

$$\partial^2 \xi_t/\partial t^2 = v_t^2 \nabla^2 \xi_t \tag{15}$$

The first part of the displacement is the gradient of a scalar potential ϕ, and the second is the curl of a vector $\mathbf{A}$. The scalar potential accounts for the volume change, dilatation, and it can readily be shown that the relative change in volume, as expressed by ε_{nn}, satisfies Eq. (14).

20.9 Wave conversion

If a longitudinal or transverse wave is reflected at a boundary or scattered by an object or inhomogeneity in the solid, the reflected, refracted, and scattered waves generally will be mixtures of longitudinal and transverse waves. In other words, wave conversion generally occurs in these interactions, and this conversion is strictly a linear effect.

In our study of refraction and scattering in a fluid, we did not encounter this phenomenon, only longitudinal waves were involved. It should be remembered, however, that this study dealt with an ideal fluid without viscosity. In reality, the effect of viscosity leads to excitation of a shear motion at a (plane) boundary (by an obliquely incident longitudinal wave). This shear motion decays exponentially with distance from the boundary, however, and generally has only a small effect on reflection and transmission coefficients.

In a solid, the situation is quite different because of the rigidity of the solid. The excitation of a shear motion at a boundary by an incident longitudinal wave no longer decays but propagates and will appear together with the 'regular' longitudinal components in the reflected, refracted, or scattered waves.

The quantitative analysis of this wave conversion or coupling at a plane boundary is similar to the analysis which we have carried out in Chapter 12. The difference is, that the boundary conditions of continuity of stress and displacement now involve more components, and more than one wave type is required to fulfill these conditions.

We shall not carry out such an analysis here but merely illustrate some examples of wave conversion at a plane boundary.

Thus, with reference to Display 20.9, we consider in the first figure an incident longitudinal wave in the vertical plane incident on a horizontal boundary. It has a component of displacement parallel with the boundary, and this gives rise to the excitation of a reflected and a refracted transverse wave, denoted by T' and T''. The displacement in these transverse waves is in the vertical plane which is said to be the plane of 'polarization' of the waves. The total reflected and refracted wave fields are superpositions of longitudinal and transverse waves.

The angles of reflection and refraction of the various waves are determined by the requirement that the trace velocities along the boundary of all waves must be the same. Since the longitudinal wave speed is always larger than the transverse, it follows that the angles of reflection and refraction of the longitudinal waves are larger than the corresponding angles for the transverse waves, as indicated.

Display 20.9.

Reflection and transmission of elastic waves at a plane boundary. Wave conversion.

Incident wave:
Longitudinal L
Reflected waves:
Longitudinal, L'
Transverse, T', vertical plane
Transmitted waves:
Longitudinal, L''
Transverse, T'', vertical plane

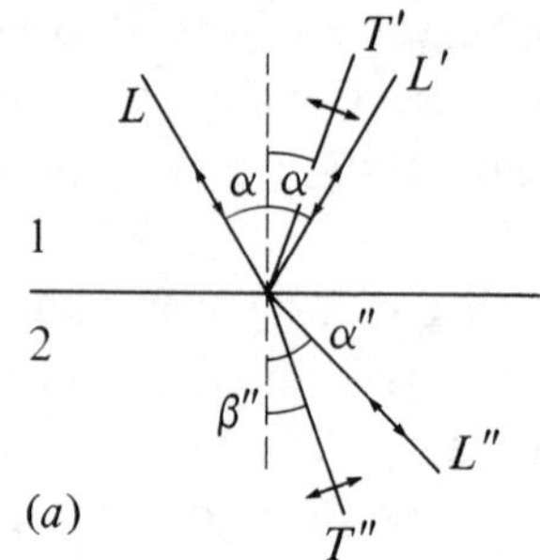

(*a*)

Incident wave:
Transverse, T, vertical plane
Reflected waves:
Transverse, T', vertical plane
Longitudinal, L'
Transmitted waves:
Transverse, T'', vertical plane
Longitudinal, L''

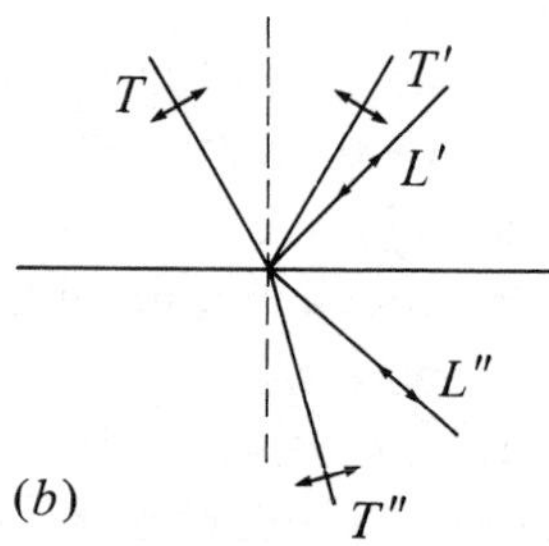

(*b*)

Incident wave:
Transverse, T, horizontal plane
Reflected wave:
Transverse, T', horizontal plane
Transmitted wave:
Transverse, T'', horizontal plane

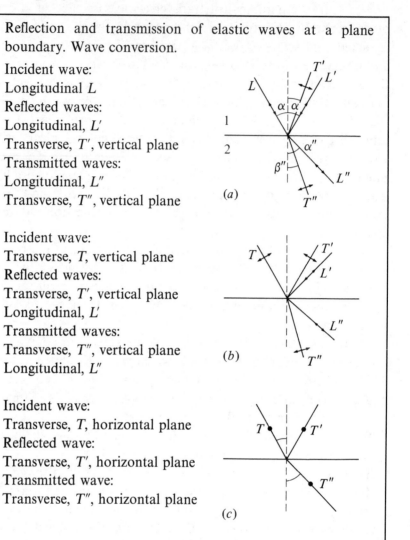

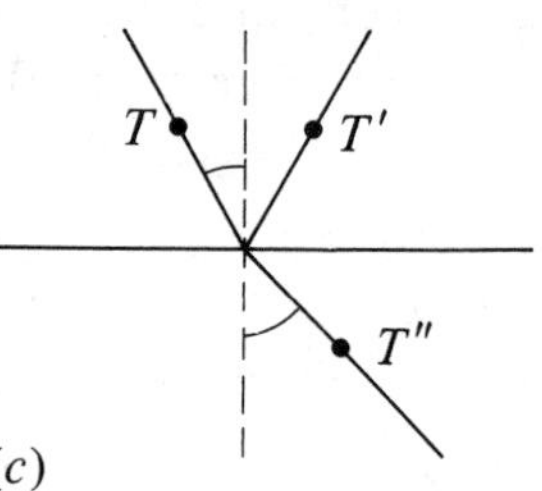

(*c*)

In the second figure, the incident wave is transverse, polarized in the vertical plane. It has a displacement component in the direction normal to the plane, and this leads to excitation of longitudinal waves which will be superimposed on the transverse waves in the reflected and refracted fields.

In the last figure, the incident wave is transverse with the plane of polarization horizontal. In this case there is no displacement component normal to the boundary, and the reflected and refracted waves are transverse, horizontally polarized.

20.10 The Rayleigh wave

The examples in the last section indicated that an elastic wave field generally is a superposition of longitudinal and transverse waves. The resulting total displacement field can be quite complex and depends not only on the relative amplitudes of the two wave types but also on the directions of propagation, the plane of polarization and the phase relationship between the waves.

A further complication is the inhomogeneity and anisotropy of the solid, which is often encountered in practice, as in the case of wave propagation in the earth. The mathematical analysis of wave motion then becomes quite complex, and, as in seismology, one has to resort to numerical analysis with the aid of powerful computers. Numerical analysis, however, generally does not provide the same kind of insight that can be derived from closed form mathematical solutions, and, as in other fields of science and engineering, it is useful to invent simple model problems which can be solved exactly and which contain the essence of the phenomena to be studied.

One such problem, first analyzed by Lord Rayleigh, concerns the wave motion confined to the surface of a semi-infinite homogeneous solid. It is analogous to the surface wave on a liquid, also studied extensively by Lord Rayleigh.

To analyze this problem, we decompose the total displacement field into a longitudinal and a transverse part, $\xi = \xi_l + \xi_t$, and each of these components satisfy the ordinary wave equation

$$\partial^2 \xi_m/\partial t^2 = v_m^2(\partial^2 \xi_m/\partial x^2 + \partial^2 \xi_m/\partial z^2) \quad (1)$$

where $m = \mathrm{l}$ and $m = \mathrm{t}$ yield the equations for the longitudinal and transverse displacement component, respectively.

With the free boundary of the solid placed in the xy-plane, the range of z will be from zero to negative infinity defining the half space occupied by the solid. (See Display 20.10).

Display 20.10

Surface wave on a solid. (Rayleigh wave).

Total displacement:

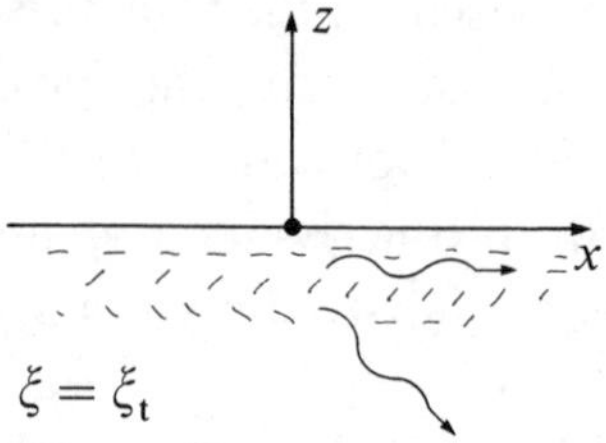

$$\xi = \xi_l + \xi_t \quad (1)$$

Each component satisfies

$$\partial^2\xi/\partial t^2 = v^2 \partial^2 \xi/\partial x^2 \quad (2)$$

$$v = v_l \quad \text{for} \quad \xi = \xi_l, \quad v = v_t \quad \text{for} \quad \xi = \xi_t \quad (3)$$

$$\xi = f(z)\exp(ikx) \quad [\text{time factor } \exp(-i\omega t)] \quad (4)$$

$$d^2f/dz^2 = [k^2 - (\omega/v)^2]f \quad (5)$$

$$f(z) = C\cdot\exp[\sqrt{k^2-(\omega/v)^2}\cdot z] \quad (z<0) \quad (6)$$

Boundary conditions at $z = 0$:

$\sigma_{xz} = \sigma_{yz} = \sigma_{zz} = 0$ (and a fair amount of algebra) (7)

lead to dispersion relation

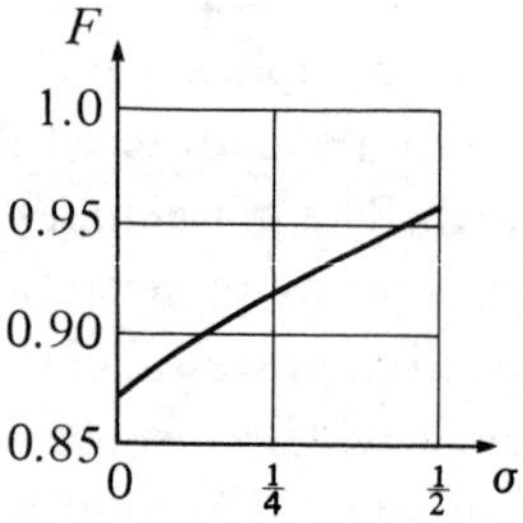

$\omega = v_t k \cdot F(\sigma)$ (σ = Poisson ratio) (8)

$$v_R = \omega/k = F(\sigma)\cdot v_t \quad (9)$$

'Penetration depth':

$$f(z) = C\exp[kz\sqrt{1-(v_t/v)^2F^2}] \quad (10)$$

$|f(z)/f(0)| = 1/e$ for

$z = -1/k\sqrt{1-(v_t/v)^2F^2} = -1/\alpha$ ($\alpha = \alpha_l$ for $v = v_l$, $\alpha = \alpha_t$ for $v = v_t$) (11)

For transverse wave

$z \simeq -0.3\lambda$ ($\sigma = 0$) and -0.5λ ($\sigma = 0.5$) (12)

Displacement field (see problem section):

$$\xi_{l,x} = Ae^{k\beta_l z}e^{ikx} \quad (13)$$

$$\xi_{l,x} = -i\beta_l\xi_{l,x} \quad (14)$$

$$\xi_{t,x} = Be^{k\beta_t z}e^{ikx} \quad (15)$$

$$\xi_{t,z} = -(i/\beta_t)\xi_{t,x} \quad (16)$$

where $\beta_l = \sqrt{1-(v_R/v_l)^2}$ and $\beta_t = \sqrt{1-(v_R/v_t)^2}$.

Reference: The surface wave speed function $F(\sigma)$ is from L. Knopoff, *Bulletin of the Seismological Society of America*, (1952), 307–308.

For harmonic time dependence, the wave equation for the complex amplitude of each component of ξ_l and ξ_t is of the form

$$d^2\xi_n/dx^2 + d^2\xi_n/dz^2 + (\omega/v_n)^2\xi_n = 0 \quad n = \mathrm{lx, lz} \quad \text{or} \quad \mathrm{tx, tz} \tag{2}$$

We are interested in a wave travelling in the positive x-direction, the components of the displacement will be of the form

$$\xi_n = f_n(z)\exp(ikx) \tag{3}$$

where k is the propagation constant in the x-direction. Since the trace velocities of both the longitudinal and the transverse waves must be the same, it follows also that the propagation constants in the x-direction will be the same and equal to k for both longitudinal and transverse x- and z-components of the displacement.

Inserting this form of the solution in Eq. (2), we obtain the differential equation for the amplitude function $f_n(z)$

$$d^2 f_n/dz^2 = [k^2 - (\omega/v_n)^2]f_n \tag{4}$$

where, as before, label n stands for the components of longitudinal and transverse displacement functions. The solution is of the form

$$f_n(z) = C_n \exp[z\sqrt{k^2 - (\omega/v_n)^2}] \quad (z < 0) \tag{5}$$

Actually, there is a second solution corresponding to a negative sign in the exponent. Since z is negative, however, and the requirement that the function be finite within the solid, we cannot use the negative sign, since the function would go to infinity when z goes to negative infinity.

The total displacement is a linear combination of the longitudinal and transverse parts,

$$\xi = \mathrm{A}\xi_l + \mathrm{B}\xi_t \tag{6}$$

and the ratio between the coefficients A and B in this linear combination is determined by the boundary conditions, as we shall see shortly.

From Eq. (6) we can determine the x- and z-components of the displacement as functions of x and z, once ξ_l and ξ_t have been found. From the spatial derivatives of these displacements, we can determine the components of the strain and the stress. Since the boundary at $z = 0$ is free, the stress on a surface element in the boundary must be zero, i.e.

$$\sigma_{xz} = \sigma_{yz} = \sigma_{zz} = 0 \tag{7}$$

The condition $\sigma_{yz} = 0$ is automatically satisfied, and from the other two conditions, we can determine the ratio between a_l and a_t and obtain the expression for the unknown propagation constant k in terms of the frequency ω, i.e. the dispersion relation. After some algebra, left for a problem, (Problem 20.8), the dispersion relation can be expressed as

$$\omega = v_t F(\sigma) k \tag{8}$$

where F is a function of the Poisson ratio (or of v_t/v_l) and it is plotted in the display. Furthermore, if we introduce the quantity

$$\beta_n = [1 - (v_t/v_n)^2 F^2]^{\frac{1}{2}} \quad (n = l, t) \tag{9}$$

we find the following expressions for the components of the displacement field

$$\xi_{l,x} = A e^{k\beta_l z} e^{ikx} \tag{10}$$

$$\xi_{l,z} = i\beta_l \xi_{l,x} \tag{11}$$

$$\xi_{t,x} = B e^{k\beta_t z} e^{ikx} \tag{12}$$

$$\xi_{t,z} = -(i/\beta_t)\xi_{t,x} \tag{13}$$

where $B/A = -2\beta_l\beta_t/(1 + \beta_t^2)$

This wave field describes the surface wave mode of the solid, known as the Rayleigh wave. Its dispersion relation is linear, with the phase velocity

$$v_R = \omega/k = F(\sigma)v_t. \tag{14}$$

The constant $F(\sigma)$ is a function of the Poisson ratio, which in turn is determined by the ratio between the transverse and longitudinal wave speeds. This function varies between 0.87 and 0.96 as the Poisson ratio goes from 0 to 0.5, as shown graphically in the display. This means that the speed of the Rayleigh wave is slightly smaller than the transverse wave speed and it is independent of frequency.

The wave amplitude decreases exponentially with depth, with the decay constant for the longitudinal and transverse wave components being $\alpha_l = k\beta_l$ and $\alpha_t = k\beta_t$. The inverse of α defines a 'penetration depth' at which the amplitude is smaller than the amplitude at the surface by a factor of 1/e. The depth typically is of the order of $0{\cdot}4\lambda$, where $\lambda = 2\pi/k$ is the wavelength of the Rayleigh wave.

The calculation of the elastic wave field which is produced by specific sources on the surface or within the solid, such as used frequently in seismology, is a considerably more complicated problem, far beyond the present scope.

The fields in some idealized cases are known. For example, in the field produced by a vibrating circular plate on the surface, 67% of the energy has been found to go into the Rayleigh wave, 26% into the purely transverse bulk wave, and 7% into the longitudinal bulk wave.

The Rayleigh wave in this case has cylindrical symmetry, and its amplitude decreases slower with the distance from the source than the spherical bulk wave. Thus, far from the source, the Rayleigh wave will be dominant close to the surface, and it is known to be the most destructive in earthquakes.

Among other problems which have been studied extensively can be mentioned wave propagation in horizontally layered half spaces, which are of particular interest in seismology.

20.11 Seismology

Seismology involves not only the study of earthquakes; an important part of the field deals with the use of elastic waves as a diagnostic tool in the study of soil characteristics and the composition of the Earth. Waves are then generated by one or more sources on the surface or in the ground, and one or more receivers are used to detect the waves at various locations. The received signals are then compared with those predicted on the basis of different models of the ground.

In physics a similar approach is used in the study of atomic and nuclear structure with waves and particles used as 'probes', and matter in bulk is studied in similar manner. An example was given in Chapter 18, where we discussed the measurement of surface tension and viscosity of a liquid from studies of light scattering.

Seismology is an extensive field, and we shall present here only some very simple examples, illustrated in Display 20.11. In the first example, Fig. (*a*), is shown a source and some receivers placed on the ground surface. The source produces an impulse on the ground, which generates several types of waves, purely longitudinal and transverse, which travel out in the hemisphere below the source, and a Rayleigh wave, which travels along the ground surface.

These wave pulses arrive at the receivers and produce output traces as functions of time, similar to those illustrated in Display 6.4. Each of the traces contains the longitudinal, the transverse, and the Rayleigh wave, in that order. Apart from a decrease in amplitude, the traces at the three locations have approximately the same form except for a difference in the time of arrival. The lines connecting the pulses are the 'wave lines', discussed in Chapter 6, and by measuring the slopes of these lines, the corresponding wave speeds can be determined. In seismology, the waves, in order of their arrival, are often denoted by the letters P ('primary'), S ('secondary'), and R ('Rayleigh').

In the second example, illustrated in Fig. (*b*), we have a layer of depth D above a uniform half space with a different wave speed. The wave pulses

Display 20.11.

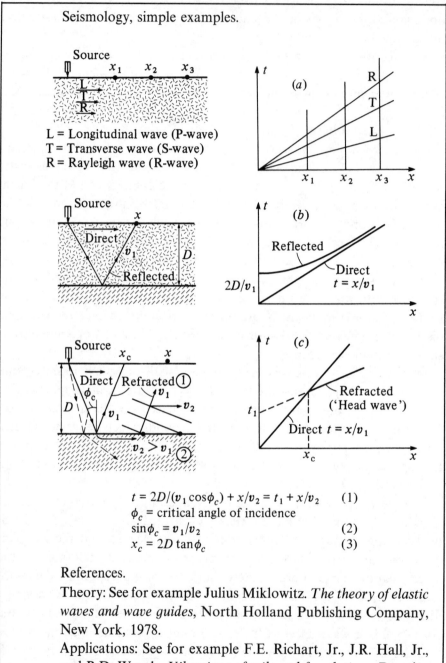

Seismology, simple examples.

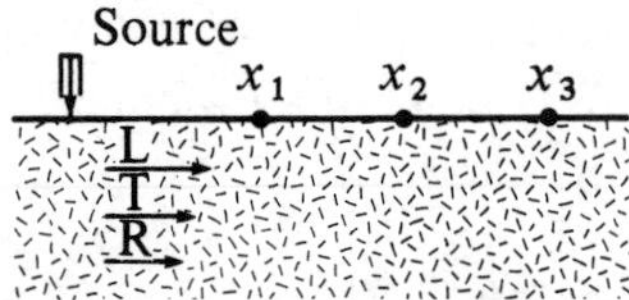

L = Longitudinal wave (P-wave)
T = Transverse wave (S-wave)
R = Rayleigh wave (R-wave)

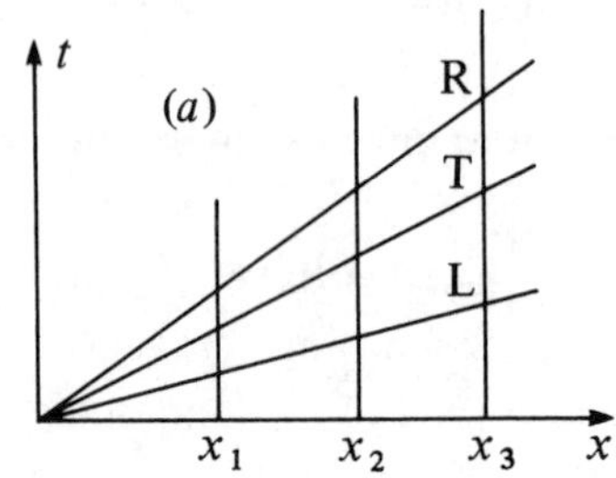

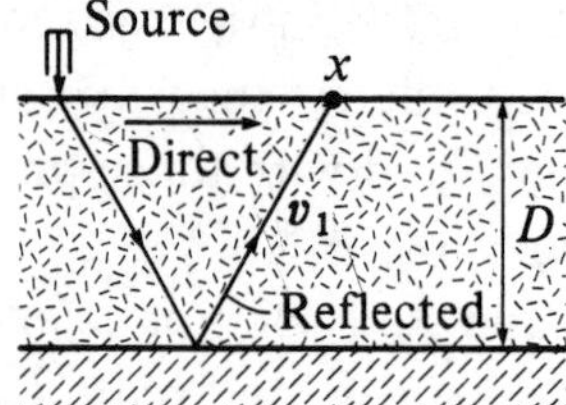

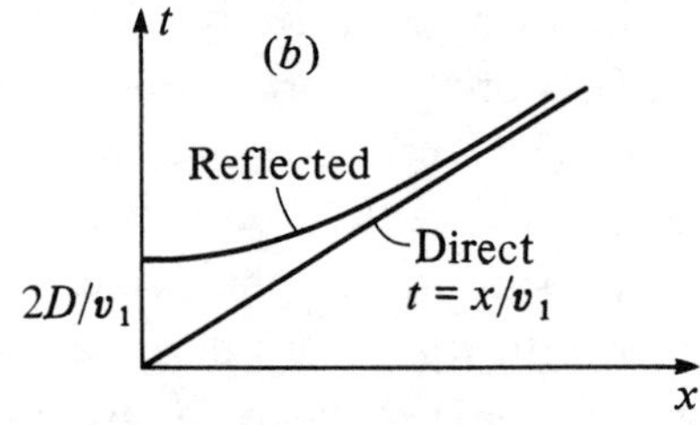

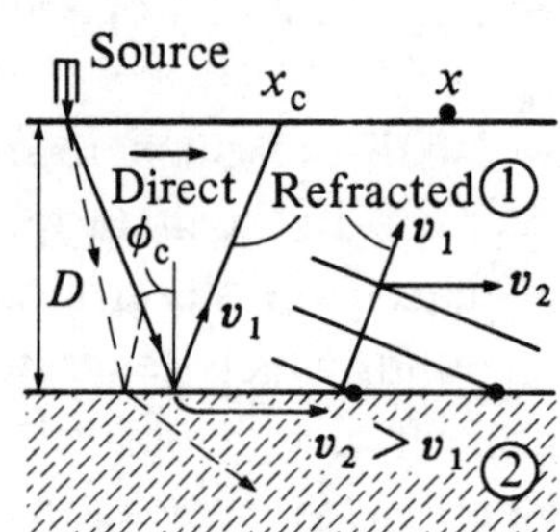

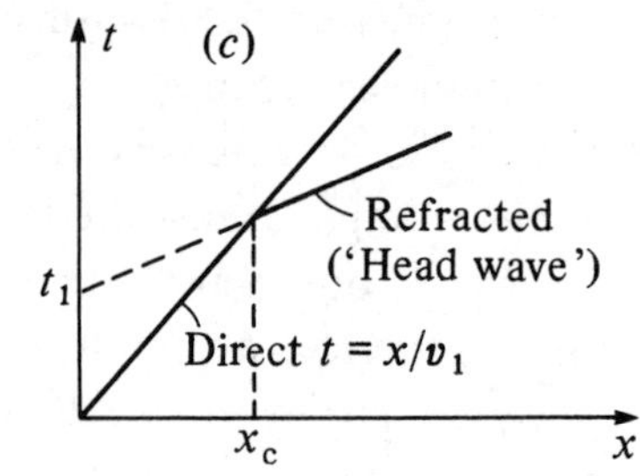

$$t = 2D/(v_1 \cos\phi_c) + x/v_2 = t_1 + x/v_2 \qquad (1)$$

ϕ_c = critical angle of incidence

$$\sin\phi_c = v_1/v_2 \qquad (2)$$

$$x_c = 2D \tan\phi_c \qquad (3)$$

References.

Theory: See for example Julius Miklowitz. *The theory of elastic waves and wave guides*, North Holland Publishing Company, New York, 1978.

Applications: See for example F.E. Richart, Jr., J.R. Hall, Jr., and R.D. Woods, *Vibrations of soils and foundations*, Prentice Hall, Inc., 1970.

arriving at the receivers now contain waves reflected at the bottom of the layer in addition to the direct waves.

Considering now only the 'first arrivals', i.e. the longitudinal wave pulses, we obtain wave lines as shown. The straight line corresponds to the direct wave, and the inverse of the slope of this line is the longitudinal wave speed v in the layer.

The other line is curved and approaches the first line asymptotically for large values of x (source–receiver distance). It intercepts the t-axis at a time $t' = 2D/v$, which is the time of travel of the longitudinal wave down to the layer and back. It follows that a determination of this intercept and the asymptotic slope yields both the wave speed and the depth D of the layer.

In the third example, Fig. (c), we have a layer (region 1) with a longitudinal wave speed v_1, which is smaller than the wave speed v_2 in the underlying half space (region 2). Under such conditions, as discussed in Chapter 12, total reflection occurs at the bottom of the layer at the critical angle of incidence ϕ_c, and the refracted wave will travel along the boundary, corresponding to an angle of refraction of 90 degrees.

The wave fronts of the totally reflected wave can be considered to be generated by the refracted wave as it travels along the boundary with a speed v_2, which is larger than the speed v_1 in the layer. We shall be interested only in the 'first arrival', and the corresponding reflected wave then travels in the direction inclined an angle ϕ_c with the vertical, as shown. (In seismology, this wave is sometimes called the 'head wave').

For values of x below a certain critical distance

$$x_c = 2D\tan(\phi_c) \tag{1}$$

where D is the layer depth and ϕ_c the critical angle, the first arrival is the direct wave in the layer, and the wave line is a straight lines with a slope equal to the inverse of the wave speed v_1 in the layer.

Beyond the critical distance, the first arrival is the head wave, corresponding to the critical angle of incidence, and the time of arrival of this wave versus the distance x is

$$t = 2D/v_1 \cos(\phi_c) + x/v_2 = t_1 + x/v_2 \tag{2}$$

The slope of this line is the inverse of the longitudinal wave speed v_2 in region 2. The intercept with the t-axis, $t_1 = (2D/v_1)\cos(\phi_c)$, contains the velocity v_1, determined from the wave line for the direct wave, the critical angle ϕ_c, determined from the measured x_c, and the depth D.

Thus, from an experimental determination of the wave lines shown in Fig. (c), it is possible to determine not only the depth of the layer but also the wave speeds in the layer and the underlying region.

In modern seismology and geophysics, the experimental techniques and data processing have advanced rapidly, and it is now possible to process data from receiver arrays on the ground so as to yield computer generated three-dimensional maps of the wave speed and wave impedance of the soil versus depth.

Problems

20.1 *Stress–strain.* (a) A vertical steel rod of circular cross section has an area of $2\,\text{cm}^2$. It carries a load of 10 000 kg at its lower end. What is the relative change of the length of the rod produced by this load (b) The rod, is also subjected to a torque such that the shear stress at $r = a$ is $10^9\,\text{N/m}^2$. Determine the angular deflection (in degrees) of an initially axial line on the surface of the rod.

20.2 *Longitudinal wave in a rod.* Derive the expression for the energy density and energy flux in a longitudinal wave in a rod with density ρ and Young's modulus E.

20.3 *Bending wave on a plate.* Consider a harmonic bending wave on a 3 mm thick window pane. (a) Determine the critical frequency at which the phase velocity of this wave is equal to the sound speed in the surrounding air. (b) What is the phase velocity at twice this frequency, and determine the direction of wave travel of the sound emitted from the plate at this frequency. (c) What is the nature of the sound field at frequencies below the critical frequency? (Regard the plate as infinite. Young's modulus for glass $E = 6 \times 18^{11}\,\text{dyne/cm}^2$ and the density is $\rho = 2.5\,\text{g/cm}^2$).

20.4 *Dispersion of bending waves.* A steel plate (thickness 2 mm) is given an impulse along a line $x = 0$ so that a bending wave is generated. The pulse can be Fourier decomposed in terms of harmonic wave components of the form $\cos(kx - \omega t)$ involving all possible wavelengths. An observer located at x is able to measure the dominant wavelength there at time t.

(a) Explain why the dominant wavelength is determined by the condition $d\phi/dk = 0$, where $\phi(k) = kx - \omega(k)t$ is the phase function.
(b) Calculate the wavelength in terms of x, t, and the plate parameters.

20.5 *Strain tensor.* The displacement vector in a solid has the components $\xi_1 = \alpha x_2$, $\xi_2 = \beta x_2$, and $\xi_3 = \gamma x_2$.

(a) Calculate the components of the displacement tensor, the strain tensor, and the rotation tensor.
(b) The components of a line element before the deformation are $\Delta x_1 = \Delta x_2 = \Delta x_3 = a$. What will be the change in length of this element as a result of the deformation?

(c) What will be the relative change in the volume of an initially cubical element as a result of the deformation?

20.6 *Elastic constants.* Prove the relations 28.7(7)–(11).

20.7 *Longitudinal wave speed in a plate.* Consider a thin plate (infinite extent) in the xy-plane. A uniform (longitudinal) stress σ_x is applied in the x-direction. Show that the corresponding strain ε_x is given by $\sigma_x/\varepsilon_x = E/(1-\sigma^2)$, where σ is the Poisson ratio. Determine also the corresponding longitudinal wave speed and compare the result with the speed of a longitudinal wave on a thin rod and in the bulk of an extended solid.

20.8 *Rayleigh wave.* Derive Eqs. 20.10(11)–(13) for the displacement field in a Rayleigh wave.

21

Feedback oscillations

In the free oscillator considered in Chapter 4, the amplitude decreases with time as a result of friction. Under certain conditions, however, the motion of an oscillator will generate a disturbance, which is fed back to the oscillator to act as a driving force. Such a 'feedback' can cause the amplitude to grow rather than decay. The energy of oscillation is then typically supplied by a steady 'force', an electric current or fluid flow or heat, which is modulated by the oscillator. This is the case in various mechanical clocks, electronic oscillators, musical wind instruments, bowed string instruments, the human voice, and various kinds of whistles, control valves and pipes. The last two examples have received considerable attention in recent years in nuclear power plants. The basic mechanisms involved in some of these 'self-sustained' oscillations will be illustrated in this chapter in terms of some specific examples.

21.1 The frequency equation

The whistling of a sound system in a lecture room is a common occurrence. The amplitude of the resulting shrill sound usually reaches levels, which far exceed the intended level of the amplified voice.

The situation is illustrated in Display 21.1. The oscillator involved is a loudspeaker, which is driven by the amplified voltage from a microphone M. The microphone in turn is driven by the sound from the lecturer with the complex amplitude denoted by p'. There is also amplitude contribution p'' from the loudspeaker so that the total complex pressure amplitude at M is

$$p = p' + p'' \qquad (1)$$

The complex amplitude of the corresponding output voltage from the microphone is denoted by $V = C_1 p$, where C_1 depends on the characteristics

Display 21.1.

Frequency equation for feedback oscillations.

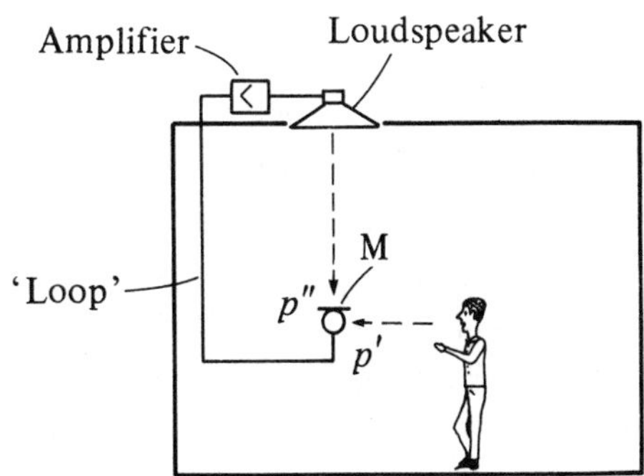

Complex amplitudes

Driving pressure p' (1)

Feedback pressure p'' (2)

Total pressure at microphone $p = p' + p''$ (3)

'Loop gain' $C(\omega)$ (4)

$p'' = C(\omega)p = C(\omega)(p' + p'')$ (5)

$p'' = p'/(1 - C)$ (6)

Condition for free oscillations

Frequency equation:

$C(\omega) = 1$ (7)

of the microphone. The amplified voltage amplitude $V_1 = C_2 V$ drives the loudspeaker which produces the pressure $p'' = C_3 V_1$, where C_3 depends on the characteristics of the loudspeaker and the transmission path from the loudspeaker to M. It follows that $p'' = C_1 C_2 C_3 p = C(\omega)p = C(\omega)(p' + p'')$ or

$$p'' = p'/[1 - C(\omega)] \tag{2}$$

where $C(\omega) = C_1 C_2 C_3$ is a function of frequency determined by the characteristics of the elements in the feedback 'loop'.

This equation is analogous to the frequency response equation for an oscillator, $I(\omega) = V(\omega)/Z(\omega)$, where I and V are the complex amplitudes of current and voltage and $Z(\omega) = R - \mathrm{i}(\omega L - 1/\omega C)$ the impedance of the circuit. We recall, that $Z(\omega) = 0$ is the frequency equation for free oscillations expressing the conditions under which there can be a solution for I even if V is zero.

The solution to $Z(\omega) = 0$ is a complex frequency of the form $\omega = -\mathrm{i}\gamma + \omega'$, where $\gamma = R/2L$, indicating that the amplitude of free oscillations decreases exponentially with time, as expressed by the factor $\exp(-\gamma t)$. The angular frequency of oscillation is $\omega' = \sqrt{\omega_0^2 - \gamma^2}$, where $\omega_0^2 = 1/LC$. In **forced** harmonic oscillations the maximum amplitude $I_m = V/R$ is obtained when the driving frequency is equal to ω_0.

The important difference in the present case is that the frequency equation $C(\omega) = 1$ can have a solution $\omega = \omega_r + \mathrm{i}\omega_i$ with a positive value of ω_i, so that instead of a decay we get an exponential growth of the amplitude. Thus, an infinitesimal disturbance, such as a short pulse, can trigger the system into self-sustained oscillations. As before, the period and frequency of oscillation are determined by the real part of ω, which depends on the characteristics of the system, and not by the frequency of an external force.

With a positive imaginary part (negative γ) of the complex frequency, the amplitude will grow exponentially with time, but it is clear that such a growth cannot go on indefinitely. In reality, the amplitude levels off after a comparatively short time and then remains at some final steady state value. One reason for this reduction of the growth rate can be an increase of the damping resistance with amplitude. When this resistance becomes equal to the magnitude of the negative resistance that is due to the feedback, the net growth rate will be zero.

In our example the constant $C(\omega)$ in Eq. (2) is considered to be the product of factor E, which contains the characteristics of the electrical components, and a 'geometrical' factor G, which describes the change of amplitude and phase of the sound wave along the path of propagation. In the particular case of acoustic feedback in a room, the latter factor is difficult to calculate because of multiple reflections of the sound from the walls.

Example. In order to make such a calculation simple, we replace the room by a tube, as illustrated in Display 21.2. The loudspeaker is mounted at the beginning of the tube, at $x = 0$, and a microphone is inside the tube at location x. The microphone output terminals are connected to a loudspeaker via an amplifier.

We account for the electrical characteristics of the microphone, amplifier, and the loudspeaker by the factor E, as mentioned above, so that $p(0) = E \cdot p(x)$, where $p(0)$ and $p(x)$ are the complex amplitudes of the sound pressures at the loudspeaker and the microphone, respectively. If the tube is long enough so that reflections can be ignored, the relation between the complex amplitudes at x and 0 is simply $p(x) = p(0)\exp(ikx) = G \cdot p(0)$. The propagation constant $k = \omega/v$, where ω is the unknown (complex) frequency of oscillation and v the sound speed. The quantity C in Eq. (2) is then $C = EG$, and the condition for free oscillations of the system is expressed by

$$C(\omega) = E(\omega)G(\omega) = E\exp(i\omega x/v) = 1 \tag{3}$$

For simplicity we assume E to be real, and to solve Eq. (3), we put $\omega = \omega_r + i\omega_i$ and replace $1/E$ by $(1/E)\exp(in2\pi)$ to obtain

$$\exp(-\omega_i x/v) \cdot \exp(i\omega_r/v) = \frac{1}{E} e^{in2\pi} \tag{4}$$

which leads to

$$\omega_i = (v/x)\log(E) \tag{5}$$

$$\omega_r x/v = n2\pi \quad \omega_r/2\pi = n(v/x) \quad x = n\lambda \tag{6}$$

The time dependence of the amplitudes is expressed by the factor $\exp(-i\omega t) = \exp(\omega_i t)\exp(i\omega_r t)$. If E is larger than unity, the imaginary part of the frequency is positive, and the amplitude will grow exponentially with time; the system is unstable at all locations of the microphone.

As shown in Eq. (6), the real part of the frequency is such that the length x is an integer number of wavelengths. In other words, the frequency of oscillation can be varied by moving the microphone along the axis. It should be noted, that the amplitude growth in one period is independent of x, $\omega_i T = 2\pi(\omega_i/\omega_r) = (1/n)\log(E)$.

If E is less than unity, the imaginary part of the frequency is negative and the system is stable.

If the tube has finite length and is terminated by a rigid wall, we will get a standing rather than a travelling wave in the tube. The phase difference between the pressures at locations half a wavelength apart is then 180 degrees. This means that if the system is unstable with the microphone placed at one of these locations it will be stable at the other. Thus, as the microphone is moved along the tube, regions of stability and instability

Display 21.2.

Feedback oscillations in a tube.

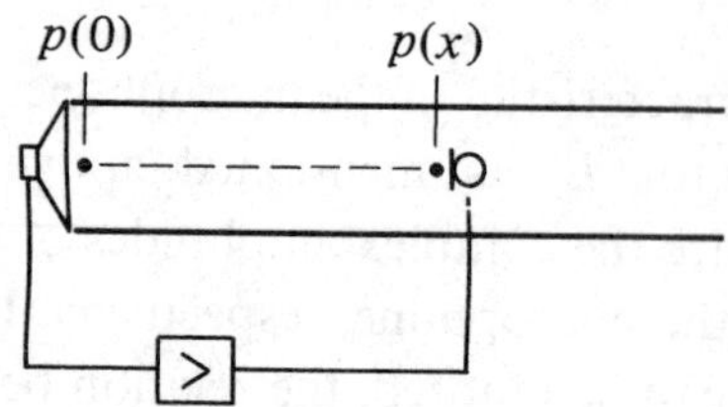

Complex amplitudes

$$p(0) = Ep(x) \quad (1)$$

$$p(x) = G(\omega)p(0) \quad (2)$$

$$G(\omega) = \exp(\mathrm{i}\omega x/v) \quad (3)$$

Frequency equation for free oscillations

$$C(\omega) = E(\omega)G(\omega) = 1 \quad (4)$$

$$\exp(\mathrm{i}\omega x/v) = 1/E \quad (5)$$

Solution

Consider the case when E = constant

Let $\omega = \omega_r + \mathrm{i}\omega_i$ (6)

$$\exp(-\omega_i x/v)\exp(\mathrm{i}\omega_r x/v) = (1/E)\exp(\mathrm{i}n2\pi) \quad (7)$$

$$\omega_i = (v/x)\log(E) \quad (8)$$

$$\omega_r x/v = n2\pi \quad \omega_r/2\pi = n(v/x) \quad x = n\lambda \quad (9)$$

$$\omega_i T = (1/n)\log(E) \quad (10)$$

will be encountered in much the same way as in a room. We leave further details for the problem section.

21.2 Valve instability

The sound produced in a harmonica and other wind instruments is a result of a feedback instability, in which the energy of oscillation is drawn from a steady stream of air. As far as the oscillations are concerned, the reed in a harmonica acts in much the same way as a control valve, illustrated schematically in Display 21.3. This type of valve, often referred to as a plug valve, is used for the control of fluid flow such as the steam in a power plant.

The regulating characteristic of such a valve is expressed by the relationship between the mass flow rate Q through the valve and the 'lift' y of the valve, with the upstream and downstream pressures P_1 and P_2 as parameters. For given values of the pressures, the regulating characteristic $Q(y)$ depends on the geometry of the valve, and the slope $\alpha = \partial Q/\partial y$ usually has a maximum for some value of y different from zero and approaches zero asymptotically with increasing y.

The valve plug is supported by a structural element, which is part of the control system. Like the harmonium reed, this system has one or more modes of motion, and in the vicinity of a mode frequency, the system can be described dynamically as a damped mass–spring oscillator.

In a power plant, the valve controls the steam from the steam generator (reactor, for example), and the pipe after the valve carries the steam to a plenum chamber, from which it is guided to the turbines. Thus, the pipe section between the valve and the plenum chamber acts like an organ pipe with its characteristic modes and frequencies. The instability can occur as a result of the interaction between these modes and the valve oscillator.

If the valve plug oscillates, the mass flow into the pipe will be modulated at the frequency of oscillation, and this periodic flow will excite one or more modes in the pipe. As a result, a reaction force will be produced on the valve plug, and under certain conditions, the power fed back into the valve by the reaction force will be positive and larger than the power dissipated through friction. The system is then unstable and self-sustained oscillations result.

Complex amplitudes and basic equations. We start the quantitative analysis of this oscillation with a discussion of the variables involved and the basic relations between them.

The static mass flow characteristic of the valve is expressed by the mass

Display 21.3.

Basic equations for the analysis of valve instability.

Valve characteristic

Mass flow rate $Q = Q(y, P_1, P_2)$ (1)

Complex amplitudes

Valve displacement and velocity $\delta y = \eta \quad u' = -\mathrm{i}\omega\eta$ (2)

Pressures $p_1 = \delta P_1 \quad p_2 = \delta P_2$ (3)

Mass flow rate $\delta Q = q = (\partial Q/\partial y)\eta + (\partial Q/\partial P_1)p_1 + (\partial Q/\partial P_2)p_2$ (4)

In this analysis $q \simeq \alpha\eta \quad \alpha = \partial Q/\partial y$ (5)

Fluid velocity u (6)

Basic equations

Conservation of mass:

$q = \alpha\eta = \alpha u'/(-\mathrm{i}\omega) = \rho A(u + u')$ (7)

Eqs. of motion:

$p = zu = z'u'$ (8)

$z' = -\mathrm{i}\omega m' + r' + \mathrm{i}k'/\omega$ (9)

$z = r + \mathrm{i}x = [\theta_t - \mathrm{i}\tan(kL)]/[1 - \mathrm{i}\theta_t \tan(kL)]$ (10)

$k = \omega/v \quad \theta_t = r_t/\rho v$ (11)

Frequency equation

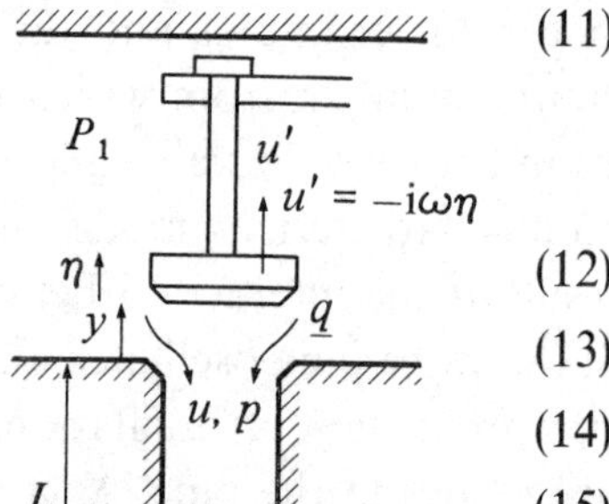

From Eqs. (7) and (8)

$z' + z - \mathrm{i}(\alpha/\rho A\omega) = 0$ (12)

$\omega = \omega_r + \mathrm{i}\omega_i$ (13)

$\omega_i < 0$ stable (14)

$\omega_i > 0$ unstable (15)

$\omega_i = 0$ neutrally stable (16)

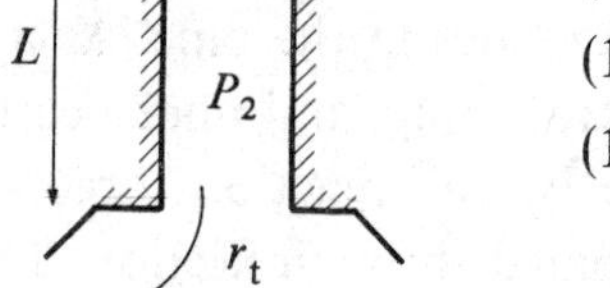

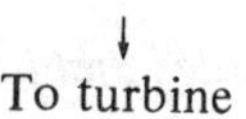

flow rate function $Q = Q(y, P_1, P_2)$, where y is the 'lift' of the valve and P_1 and P_2 the static pressures on the upstream and downstream sides of the valve.

The displacement of the valve plug from the equilibrium position y is denoted by η, and the perturbations in P_1 and P_2 by p_1 and p_2. The corresponding variation in Q can then be expressed as

$$\delta Q = (\partial Q/\partial y)\eta + (\partial Q/\partial P_1)p_1 + (\partial Q/\partial P_2)p_2 = \alpha\eta + \beta_1 p_1 + \beta_2 p_2 \qquad (1)$$

where $\alpha = \partial Q/\partial y$, $\beta_1 = \partial Q/\partial P_1$, and $\beta_2 = \partial Q/\partial P_2$.

In the analysis presented here, we shall consider only the variation of Q due to η. This is the contribution which can lead to an instability, which is the effect of particular interest. The variations of Q due to the β-terms can be shown to lead to damping. Their effects can be incorporated in the friction damping parameter of the valve, to be introduced later.

The variation in the mass flow rate gives rise to a fluctuation ρAu of the mass flow rate into the pipe, where ρ is the fluid mass density, u the velocity perturbation of the fluid at the entrance to the pipe, and A the pipe area. If the valve plug were fixed, conservation of mass would give $\rho Au = \delta Q$, if we neglect the effect of compressibility of the fluid in the small volume between the entrance and the exit of the valve. To account for the motion of the valve, velocity u', we have to modify this relation to $\delta Q = \rho A(u + u')$, where we have assumed the valve area to be the same as the pipe area A.

With the downstream side of the valve terminated by a pipe and the upstream side connected to a large open fluid volume, the perturbation in pressure on the downstream side will have the dominant influence in this case, and we shall consider only the effect of this perturbation, denoted by p.

We are interested in a time dependence expressed by the time factor $\exp(-\mathrm{i}\omega t)$, and in the continued analysis we regard the variables introduced above as complex amplitudes. The relations $\delta Q = \alpha\eta = \rho A(u + u')$ can then be written,

$$\mathrm{i}(\alpha/\rho A\omega) = 1 + u/u' \qquad (2)$$

where we have used $u' = -\mathrm{i}\omega\eta$.

Furthermore, the relation between the pressure fluctuation p at the pipe entrance and the velocities u and u' can be expressed as

$$p = zu = z'u' \qquad (3)$$

where z and z' are the impedances per unit area of the fluid column and the valve, respectively.

Frequency equation. Combining Eqs. (2) and (3) we obtain

$$z' + z - \mathrm{i}(\alpha/\rho A\omega)z = 0 \tag{4}$$

which is the frequency equation for the free oscillations of the system. To solve this equation, we must have explicit expressions for the frequency dependence of the impedances z' and z.

The valve is treated as a simple harmonic oscillator so that the impedance per unit area of the valve plug is

$$z' = r' - \mathrm{i}\omega m' + \mathrm{i}k'/\omega \tag{5}$$

where $r' = R/A$, $m' = M/A$, $k' = K/A$, and A is the area of the valve assumed to be the same as the pipe area.

In calculating the impedance z of the fluid column in the pipe, we include an acoustic termination resistance r_t to account for the losses at the end of the duct. This resistance is known to be approximately $r_t = \rho U$, where U is the velocity of the mean flow in the pipe. The corresponding normalized resistance

$$\theta_t = r_t/\rho v \simeq U/v = M$$

is then approximately equal to the Mach number of the mean flow.

It is left for one of the problems to show that the acoustic impedance of the fluid column in the pipe is

$$z/\rho v = (1/\rho v)(r + \mathrm{i}x) = [\theta_t - \mathrm{i}\tan(kL)]/[1 - \mathrm{i}\theta_t \tan(kL)] \tag{6}$$

$$k = \omega/v \quad v = \text{wave speed}$$

$$\theta_t = r_t/\rho v \simeq U/v = M$$

Having obtained the expressions for z' and z, we now have to solve Eq. (3) for the complex frequency $\omega = \omega_r + \mathrm{i}\omega_i$ in much the same manner as in the example in Display 21.2. In the present case, however, the equation is much more complicated, and has to be solved numerically in the general case.

With the time dependence expressed by $\exp(-\mathrm{i}\omega t)$, this factor becomes $\exp(\omega_i t)\exp(-\mathrm{i}\omega_r t)$. In other words, if the imaginary part of the frequency is found to be positive, the amplitude of the system will grow exponentially with time; i.e. the system will be unstable. A negative value of ω_i yields exponential decay, familiar from Chapter 4.

To determine the complex frequency, we have to solve Eq. (4) numerically. Such a study will not be made here. Rather, we shall limit the analysis to the conditions for neutral stability corresponding to $\omega_i = 0$.

Stability criterion. With $\omega_i = 0$ we have $\omega = \omega_r$, and the propagation constant $k = \omega_r L/v$ in Eq. (6) for the pipe impedance is real. The real and imaginary parts of this impedance can then be expressed as

$$r/\rho v = \theta_t[1 + \tan^2(kL)]/[1 + \theta_t^2 \tan^2(kL)] \qquad (7)$$

$$x/\rho v = -(1 - \theta_t^2)\tan(kL)/[1 + \theta_t^2 \tan^2(kL)] \quad (k = \omega_r/v, \omega_i = 0) \qquad (8)$$

Eq. (4) now reduces to

$$[r' + r + r_f] + i[-\omega m' + x + (k' - k_f)/\omega] = 0$$

$$r_f = (\alpha/\rho A)(x/\omega) \quad k_f = (\alpha/\rho A)r \quad (\omega_i = 0) \qquad (9)$$

which can be regarded as the frequency equation for an oscillator with an impedance $z' + z$ augmented by a feedback resistance r_f and a reactance $-k_f/\omega$, which, in effect, corresponds to a reduction in the spring constant. The quantities r_f and k_f are caused by the feedback and are proportional to the coupling constant $\alpha = \partial Q/\partial y$. Of primary importance is the fact that r_f can be negative, since the reactance x is negative if the pipe reactance represents a mass load.

Both the real and imaginary parts of the frequency equation (8) must be zero, i.e.

$$r' + r + r_f = 0 \qquad (10)$$

$$-\omega m' + x + (k' - k_f)/\omega = 0 \qquad (11)$$

where r, x, r_f, and k_f, as given in Eqs. (7), (8), and (9), are functions of the frequency.

In order for the amplitude to remain constant, it is necessary, of course, that the feedback resistance be negative so that it will cancel the resistances presented by the valve and the fluid column. This condition is expressed by Eq. (10) as part of the solution to the frequency equation. A negative value of r_f requires that x be negative, and this, in turn, means that the pipe reactance must be mass-like.

A negative value of x in Eq. (11) adds to the inertial mass reactance of the valve and since the spring constant is reduced by k_f, it follows that the frequency will be smaller than that of the free uncoupled valve oscillator.

By combining Eqs. (10) and (11), we can calculate the frequency of oscillation in terms of the system parameters. The equations determine not only the frequency, however, but impose also a relation between the system parameters, which must be fulfilled in order that the system be neutrally stable. This relation is obtained by introducing the calculated frequency in either of the two equations.

In discussing Eqs. (10) and (11) it is convenient to introduce dimensionless system parameters. They can be chosen in many different ways and among the options can be mentioned the following.

One quantity, to be called the **instability parameter**, is $\gamma = \alpha/\rho A\omega_0 = \alpha'/A\omega_0$, where $\alpha' = \alpha/\rho = \partial W/\partial y$, $W = Q/\rho$ the volume flow rate through the valve, and $\omega_0 = \sqrt{K/M}$ the angular frequency of the free uncoupled oscillations of the valve.

The length L of the pipe can be expressed in terms of $k_0 L = 2\pi L/\lambda_0 = \omega_0 L/v$. It should be pointed out, that L is the 'dynamic' length of the fluid column which is slightly larger than the physical length because of an 'end correction', which is of the order of the diameter of the pipe. The mass M of the valve plug can be expressed in terms of $\omega_0 M/\rho v$ or in terms of the ratio $M/A\rho L$, where $A\rho L$ is the mass of the fluid in the pipe. Finally, the resistances in the pipe and in the valve are expressed in terms of $\theta = r/\rho v$ and $\theta' = r'/\rho v$.

Eqs. (10) and (11) yield the frequency of oscillation and determine the value of the instability parameter γ required to make the feedback resistance overcome the friction in the system and make it neutrally stable. If we plot this value of γ as a function of $k_0 L$ for given values of θ_t and θ', we obtain a stability diagram or 'contour'. If the 'operating point' (γ, k_0, L) of the system lies on the contour, the system is neutrally stable, but if the point lies above (below) the contour, the system is unstable (stable). We shall not compute the entire contour but merely indicate some general characteristics by analyzing the special cases $k_0 L \ll 1$, $k_0 L \simeq (2n-1)\pi/2$, and $k_0 L \simeq n\pi$, where $n = 1, 2, \ldots$.

In the first, we let $k_0 L \ll 1$. The expressions for the pipe resistance and reactance in Eqs. (7) and (8) then reduce to

$$r/\rho v \simeq \theta_t$$
$$x/\rho v \simeq -(1-\theta_t^2)kL$$

and in terms of the parameters defined above, Eq. (10) can be written

$$\gamma = (1/k_0 L)(\theta_t + \theta')/(1-\theta_t^2), \quad (k_0 L \ll L)$$
$$\gamma = \alpha/\rho A\omega_0, \quad \theta' = r'/\rho v \tag{12}$$

Similarly, Eq. (11) gives the frequency of oscillation

$$(\omega/\omega_0)^2 = [1 - \gamma(\rho v/\omega_0 m')\theta_r]/[1 + (\rho AL/M)(1-\theta_t^2)] \tag{13}$$

In the majority of cases we have $\rho v \ll \omega_0 m'$ and $\rho AL \ll M$, and it follows then that the frequency of oscillation will be only slightly less than the frequency of uncoupled oscillations of the valve.

After having considered the region $k_0 L \ll 1$ of the stability diagram, we turn to the case when the frequency of oscillation coincides with the frequency at which the reactance x has its maximum negative value. According to Eq. (8), this occurs when $\tan(kL) = 1/\theta_t$, and the corresponding value of x is $-(1-\theta_t^2)/2\theta_t$. Generally θ_t lies in the range 0.05–0.1

and the value of kL corresponding to $\tan(kL) = 1/\theta_t$ is somewhat smaller than $(2n-1)\pi/2$, where n is an integer.

For a heavy valve, as encountered in a nuclear power plant, for example, the frequency of coupled oscillations is only slightly lower than the natural frequency ω_0 of the valve, and in this discussion we shall put $kL \simeq k_0 L$. According to Eq. (9), the maximum negative resistance is then $r_f = -(\alpha/\rho A)(1-\theta_t^2)/2\theta_t\omega_0$. The corresponding value of the input resistance of the pipe is $r = \rho v(1+\theta_t^2)/2\theta_t$ (see Eq. (7)). With these values and the normalized parameters defined earlier introduced into Eq. (10), we obtain

$$\gamma = 2\theta_t(\theta' + \theta_t)/(1-\theta_t^2) \quad (\omega \simeq \omega_0) \tag{14}$$

Thus, in this case of a heavy valve, for which the frequency of feedback oscillations is approximately equal to the frequency of the uncoupled valve, the value of γ for neutral stability will have minima for the values $L/\lambda_0 \approx (2n-1)/4$ considered here; the risk for instability is then greatest, as indicated in Display 21.8.

For a value $k_0 L \simeq n\pi$, which corresponds to a pipe length equal to an integer number of half wavelengths, we can replace $\tan(kL)$ by $(\omega - \omega_n)L/v$, where $\omega_n = n\pi v/L$. It follows then from Eqs. (7) and (8), that the expressions for r and x will have the same form as in the first case, $k_0 L \ll 1$, if we replace kL, by $(\omega - \omega_n)L/v$ in the approximate expressions for r and x used in Eq. (12); the risk for instability is at a minimum.

Then, if, as before, we approximate the factor $1/\omega$ by $1/\omega_0$ in the expression for r_f in Eq. (9), we obtain the same relation between γ and $k_0 L$ as given in Eq. (12) with $(k_0 L)$ replaced by $(\omega_0 - \omega_n)L/v$. Actually, the stability diagram can be considered to be a superposition of curves of approximately the same shape but displaced with respect to one another by an amount $k_0 L = \pi$. Each curve corresponds to a particular wave mode in the pipe. In order to assure that all modes are stable, the operating point $(\gamma, k_0 L)$ of the system must lie below the contours for all the modes.

Lateral oscillations. In addition to the axial feedback oscillations of the valve considered above, there can occur also lateral oscillations resulting from the interaction of lateral modes of oscillation of the valve (bending modes of the valve stem) and higher order acoustic modes in the pipe.

A lateral oscillation of the valve produces a non-uniformity of the flow entering the pipe, oscillating back and forth across the pipe. The net mass flow into the duct generally is not substantially affected by this motion, and the axial plane acoustic modes of the fluid column in the pipe will not be strongly excited. Instead, the flow oscillations couple to one or more

Display 21.4.

Stability diagram for axial valve oscillations.

Frequency equation $z' + z - i(\alpha/\rho A\omega)z = 0$ (1)

$$z' = -i\omega m' + r' + ik'/\omega \quad (2)$$

$$z = (\rho v)(r + ix)[\theta_t - i\tan(kL)]/[1 - i\theta_t \tan(kL)] \quad (3)$$

$$k = \omega/v \quad \omega = \omega_r + i\omega_i \quad z = r + ix \quad (4)$$

Stability criterion

Neutral stability $\omega_i = 0$ (5)

Feedback resistance and reactance:

$$r_f = (\alpha/\rho A)(x/\omega) \quad x_f = ik_f/\omega = (\alpha/\rho A)(r/\omega) \quad (6)$$

From Eq. (1)

$$r' + r + r_f = 0 \quad (7)$$

$$-\omega m' + x + (k' - k_f)/\omega = 0 \quad (8)$$

Stability diagram

Let $\omega_0 = \sqrt{k'/m'} = \sqrt{K/M}$ (9)

$\gamma = \alpha/\rho A\omega_0$ (Instability parameter) (10)

Heavy valve $(\omega \simeq \omega_0)$

$$k_0L \ll 1 \quad \gamma \simeq (1/k_0L)(\theta_t + \theta')/(1 - \theta_t^2) \quad \theta' = r'/\rho v \quad \theta_t = r_t/\rho v \quad (11)$$

$$k_0L \simeq (2n-1)\pi/2 \quad \gamma \simeq 2\theta_t(\theta' + \theta_t)/(1 - \theta_t^2) \quad (12)$$

$$k_0L \simeq n\pi \quad \gamma \simeq [1/(k_0 - k_n)L](\theta_t + \theta')/(1 - \theta_t^2) \quad k_nL = n\pi \quad (13)$$

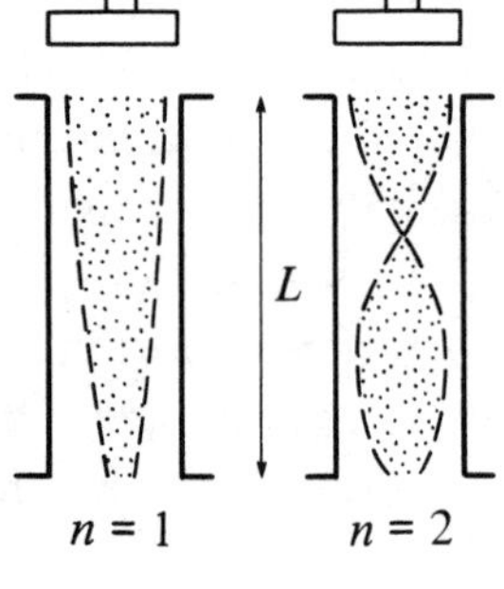

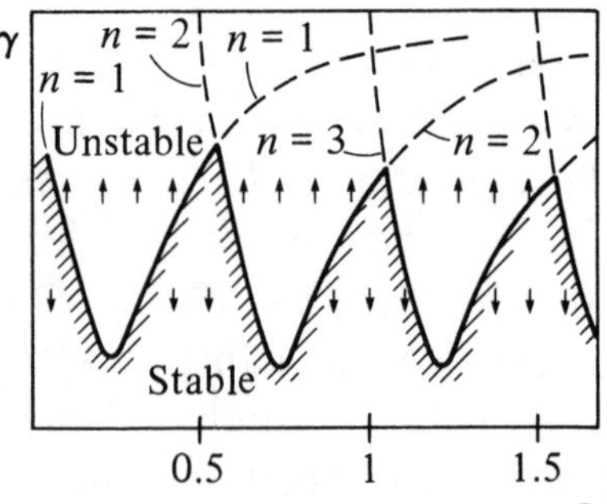

higher-order acoustic modes in the pipe. For the lowest of these modes, the pressure amplitudes at two diametrically opposite locations are out of phase, and a reaction torque will be developed on the valve.

The analysis of the lateral feedback oscillations is analogous to the analysis of the axial oscillations and it can be shown that in order for this torque to result in self-sustained lateral oscillations of the valve, it is necessary that the frequency of oscillation be slightly lower than the cut-off frequency (lateral resonance frequency) of the mode. The corresponding half wavelength is smaller than the diameter of the pipe, and the corresponding frequency often is in the kilocycle range, i.e. considerably higher than the frequency of axial oscillations, which usually is below 100 Hz.

Steady state oscillations and structural fatigue. As indicated earlier, the exponential amplitude growth of an unstable oscillator will last for a comparatively short time, during which the amplitude is still small enough to keep the system response linear. As the amplitude grows, however, nonlinear effects become important. There can be a nonlinear damping, for example, or the spring constant of the valve oscillator can be nonlinear, so that the system detunes itself as the amplitude increases. Both these effects will limit the amplitude growth, and the system settles down in a steady state oscillation with a certain constant amplitude.

When the power carried by the fluid flow is large and the damping of the valve system is small, the steady state amplitude of the self-sustained oscillation can be quite large. In fact, the corresponding oscillating elastic stresses can become so large that mechanical 'fatigue' failure results. (Compare the breaking of a metal wire through repeated bending back and forth).

In order for fatigue failure to occur, the maximum stress produced by the oscillation must exceed a certain critical value, which varies from one material to the next. For a given value of the stress larger than the critical value, the 'life' time of the structure is expressed in terms of the number of cycles of oscillation that the structure can endure. Therefore, for a given maximum stress, the time until failure occurs is shorter the higher the frequency of oscillation. It should be realized, however, that if the system can oscillate in several unstable modes, the lowest mode usually has the smallest damping and the largest amplitude and stress.

21.3 Instability of a vortex sheet

Vortex line and vortex sheet. In a line vortex in a fluid, the fluid elements move in circular paths around a straight line, and the speed varies

Display 21.5.

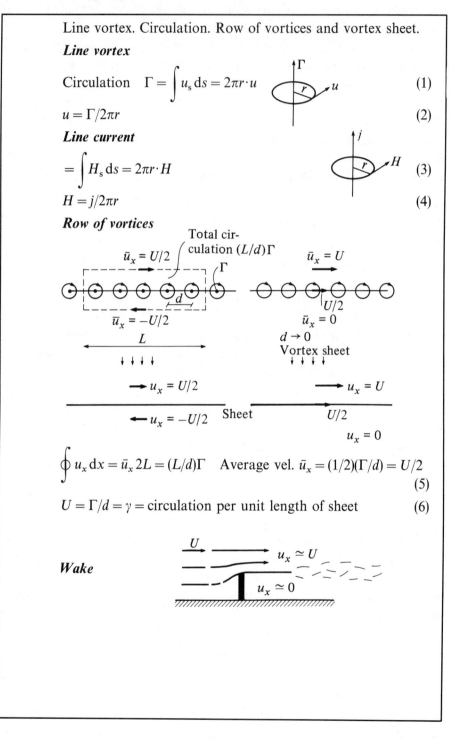

Line vortex. Circulation. Row of vortices and vortex sheet.

Line vortex

Circulation $\Gamma = \int u_s \, ds = 2\pi r \cdot u$ (1)

$u = \Gamma/2\pi r$ (2)

Line current

$= \int H_s \, ds = 2\pi r \cdot H$ (3)

$H = j/2\pi r$ (4)

Row of vortices

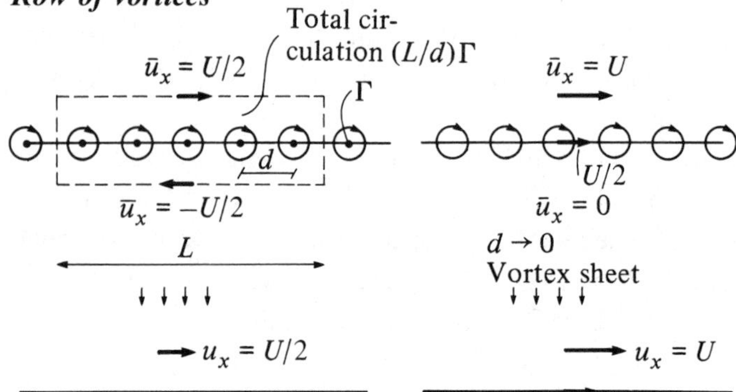

$\oint u_x \, dx = \bar{u}_x 2L = (L/d)\Gamma$ Average vel. $\bar{u}_x = (1/2)(\Gamma/d) = U/2$ (5)

$U = \Gamma/d = \gamma =$ circulation per unit length of sheet (6)

Wake

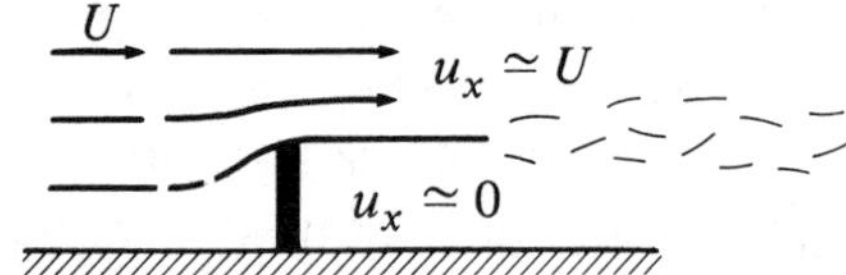

as $1/r$ with the distance r from the line. This means that the line integral of the velocity along a path surrounding the line is independent of r; the constant value of the integral is called the **circulation**, and we shall denote it by Γ.

The r-dependence of the velocity u is the same as for the magnetic field H produced by an electric line current j, and with u being analogous to H the circulation Γ corresponds to the current j.

A tornado can be described approximately as a line vortex. It is important to realize that a vortex is carried along with the mean flow velocity perpendicular to the vortex line.

Of particular interest is a row of equidistant vortex lines, as shown in Display 21.5. The total velocity field is obtained by adding the velocity contributions in the circular motions about the individual vortex lines. As the separation between the lines goes to zero, the configuration becomes a vortex sheet in much the same way as electric currents form a current sheet. The resulting velocities in the regions above and below the sheet then will be uniform and opposite in direction, parallel with the vortex sheet. Again, it is instructive to compare with the magnetic field produced by a current sheet.

We have denoted the flow velocities above and below the sheet by $U/2$ and $-U/2$ and the vortex lines are at rest. If we now superimpose a uniform flow with a velocity $U/2$, the resulting velocities above and below the sheet will be U and 0 and the vortex lines will be moving with the velocity $U/2$.

The flow field thus obtained, with a plane vortex sheet separating a region at rest and one in motion, is an idealization, but it can be used as an approximate description of the 'wake' behind a barrier in a moving fluid, as indicated. It is an experimental fact that a flow separation of this kind occurs behind a blunt body when the incident flow velocity exceeds a critical value. The resulting vortex sheet is unstable, however, and starts to break up and the motion eventually becomes chaotic.

Instability of a row of vortex lines. The instability of the vortex sheet can be understood qualitatively from the motion of an infinite row of discrete vortex lines. The lines are in the xy-plane, parallel with the y-axis. The locations of the lines along the x-axis are $\pm nd$, where d is the separation between two adjacent lines and $n = 0, 1, 2, \ldots$.

The flow velocity contributions by the individual vortices at a particular vortex line, say at $x = 0$, add up to zero, and a vortex line cannot be moved out of the line through the action of the others in this configuration.

If, as results of a perturbation (fluctuation), the line is displaced out of the plane in the z-direction, however, the flow field from the other vortices will make the line move in the x-direction, and a displcement in the

Display 21.6

Instability of row of vortices and vortex sheet.

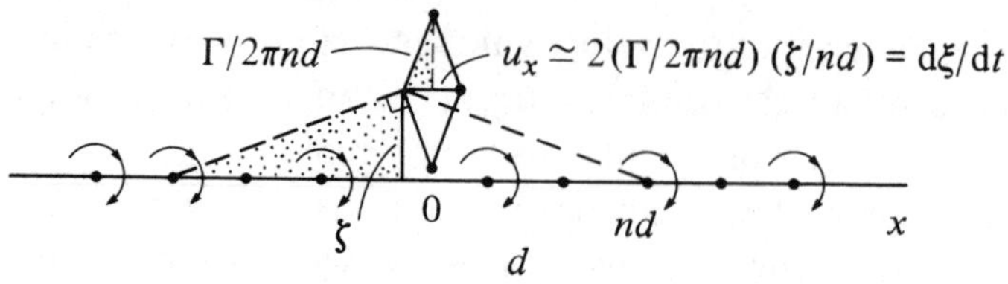

Initial displacement $\zeta\,(\zeta \ll d)$ of line $n=0$ in the z-direction results in a convective motion in the x-direction

$$\mathrm{d}\xi/\mathrm{d}t = \beta\zeta \tag{1}$$

$$\beta = (\Gamma/\pi d^2)\sum_1^\infty(1/n^2) = \pi\Gamma/6d^2 \quad \left[\sum_1^\infty(1/n^2) = \pi^2/6\right] \tag{2}$$

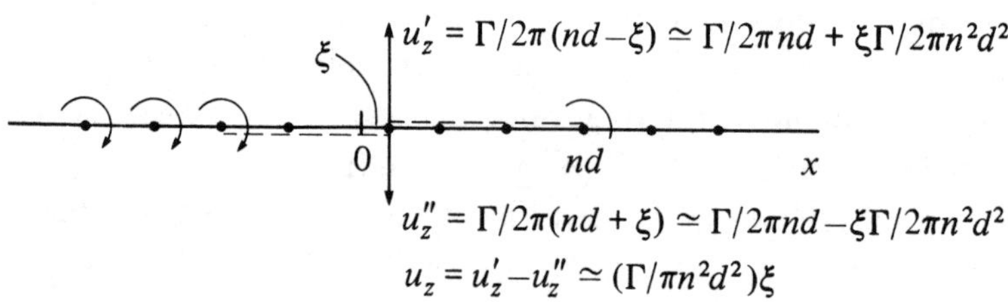

$u_z' = \Gamma/2\pi(nd-\xi) \simeq \Gamma/2\pi nd + \xi\Gamma/2\pi n^2d^2$
ξ
0
nd
x
$u_z'' = \Gamma/2\pi(nd+\xi) \simeq \Gamma/2\pi nd - \xi\Gamma/2\pi n^2d^2$
$u_z = u_z' - u_z'' \simeq (\Gamma/\pi n^2d^2)\xi$

Initial displacement $\xi(\xi \ll d)$ of line $n=0$ in the x-direction results in a convective motion in the z-direction

$$\mathrm{d}\zeta/\mathrm{d}t = \beta\xi \tag{3}$$

For simult. small displ. ξ, ζ, Eqs. (1) and (3) are still valid.

$$\mathrm{d}^2\xi/\mathrm{d}t^2 - \beta^2\xi = 0 \tag{4}$$

$$\xi = A\exp(\beta t) + B\exp(-\beta t) \quad \xi = \xi_0 \quad \text{at} \quad t=0 \tag{5}$$

$$\zeta = A\exp(\beta t) - B\exp(-\beta t) \quad \zeta = \zeta_0 \quad \text{at} \quad t=0 \tag{6}$$

$$A = (\xi_0 + \zeta_0)/2 \quad B = (\xi_0 - \zeta_0)/2 \tag{7}$$

x-direction, in turn, results in a motion in the z-direction. Thus, an initial displacement of a line out of the row will grow, and the vortex row and a corresponding vortex sheet is unstable.

To formulate these observations quantitatively, we proceed as shown in Display 21.6. Thus, we consider first a small displacement ζ in the z-direction of the vortex line which is initially at $x = 0$. The total velocity at the line in the new location contributed by the vortices at $x = nd$ and $x = -nd$ no longer is zero but has a component in the x-direction $2(\Gamma/2\pi nd)(\zeta/nd)$, where we have approximated the tangent for the angle between the velocity from each of the vortex lines at $x \pm nd$ and the x-axis by ζ/nd. The sum of the velocity contributions from all the vortices in the row then is

$$u_x = \mathrm{d}\xi/\mathrm{d}t = \beta\zeta \tag{1}$$

$$\beta = (\Gamma/\pi d^2)\sum_1^\infty (1/n^2) = \Gamma\pi/6d^2 \quad \left[\sum_1^\infty (1/n^2) = \pi^2/6\right] \tag{2}$$

Similarly, a small displacement $\xi(\xi \ll d)$ in the x-direction leads to a velocity contributions $\Gamma/2\pi(nd - \xi)$ and $-\Gamma/2\pi(nd + \xi)$ in the z-direction from the vortex lines at $x = nd$ and $x = -nd$, respectively. The sum of these contributions, to first order in ξ, is $(\xi/\pi n^2 d^2)\Gamma$, and, accounting for all the vortex lines in the row, we obtain

$$u_z = \mathrm{d}\zeta/\mathrm{d}t = \beta\xi \tag{3}$$

where β is given in Eq. (2).

Thus, an arbitrary small displacement with the components (ξ, ζ) of the central vortex line gives rise to a velocity with the components given by Eqs. (1) and (2). By eliminating one of the variables by time differentiation, we obtain

$$\mathrm{d}^2\xi/\mathrm{d}t^2 - \beta^2\xi = 0 \tag{4}$$

The general solutions for ξ and the corresponding displacement ζ are

$$\xi = A\exp(\beta t) + B\exp(-\beta t) \tag{5}$$

$$\zeta = A\exp(\beta t) - B\exp(-\beta t) \tag{6}$$

If the displacements at $t = 0$ are $\xi = \xi_0$ and $\zeta = \zeta_0$ the constants A and B are $A = (\xi_0 + \zeta_0)/2$ and $B = (\xi_0 - \zeta_0)/2$.

The displacements grow exponentially with time which means that the row of vortices and the corresponding vortex sheet are unstable. Eventually, the displacements become so large that the linear approximation breaks down, and the Eqs. (5) and (6) no longer are valid. Furthermore, other vortex lines will be gradually brought from their equilibrium positions and set in motion, which quickly becomes quite complicated and eventually chaotic. To follow the evolution of the vortex lines requires numerical analysis.

21.4 Whistles and wind instruments

In the feedback oscillations of the valve, it was essential that the valve plug participated in the oscillatory motion. As a result of this motion, the fluid flow was modulated and the corresponding reaction force on the valve made it unstable. The tone generation in reed type instruments and in the vocal tract are related to this type of instability.

There is another type of oscillation, such as in a flute, where it is not essential that the mechanical structure participates in the motion. Rather, the instability is caused by the creation of vortex sheets as the flow interacts with (rigid) structural elements. As in the case of valve oscillations, a detailed analysis goes beyond the scope of this book, and we shall consider only some elementary aspects of the subjects.

First, it should be mentioned that the row of vortex lines in the previous section (Display 21.5) is unstable for a more general type of displacement than that which we dealt with in the last section. In particular, a spatially harmonic initial displacement of the line of vortices or a vortex sheet is unstable, and the amplitude will have an exponential initial growth with time over a wide range of wavelengths.

If the motion of the vortex sheet is subject to boundary conditions, resulting from the interaction of the sheet with downstream obstacles (or in some cases with another vortex sheet), certain discrete wavelengths and corresponding modes of oscillation compatible with the boundary conditions will be defined. The wavelength of a mode is determined by some characteristic length between the boundaries. The corresponding frequency of oscillation then follows from dispersion relation for the wave motion on the vortex sheet.

The mode selection can be thought of qualitatively as a feedback process, in which the interaction of the sheet with a downstream boundary results in a feedback signal to the upstream end of the vortex sheet which promotes a wave disturbance which satisfies the boundary conditions. If the feedback signal is assumed to be a sound wave, travelling the distance d from the downstream to the upstream end of the sheet, the round trip time in this feedback loop will be

$$T = d/V + d/v \approx d/V \quad (v \gg V) \tag{1}$$

Display 21.7.

Excitation of acoustic modes in cavity by grazing flow.

Flow velocity above vortex sheet $= U$ Sound speed $= v$

Wave speed on sheet $V = \beta U \quad \beta \simeq 0.5$ (1)

Round trip time in loop:

$T \simeq d/V + d/v$ (2)

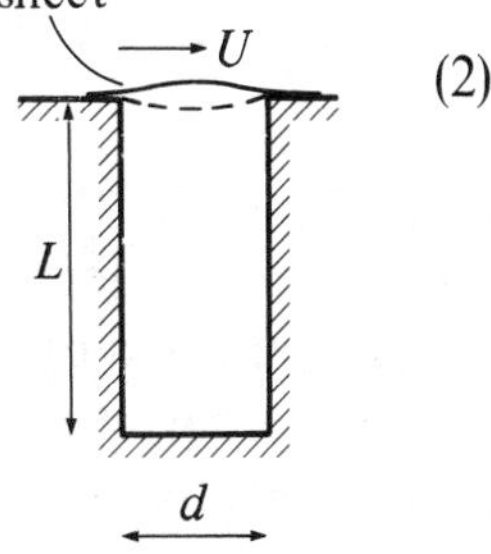

Frequency of mth mode of vortex sheet

$\nu_m \simeq m/T = mVv/d(V+v) = m(U/d)\cdot\beta/(1+\beta M) \quad m = 1, 2, \ldots$ (3)

$M = U/v =$ Mach number of incident flow. (4)

Frequency of nth acoustic mode in cavity

$(2n-1)\lambda_n/4 = L' \simeq L + 0.3d$ (5)

$\nu'_n = v/\lambda_n = (2n-1)v/4L' \quad n = 1, 2, 3, \ldots$ (6)

Necessary condition for acoustic feedback oscillation $\nu_m \simeq \nu'_n$

Corresponding value of U (from Eqs. (3), (5), and (6)):

$$M = U/v \simeq (2n-1)(d/\beta)/[4mL' - (2n-1)d]$$
$$\simeq (d/L')(2n-1)/4m\beta \quad (7)$$

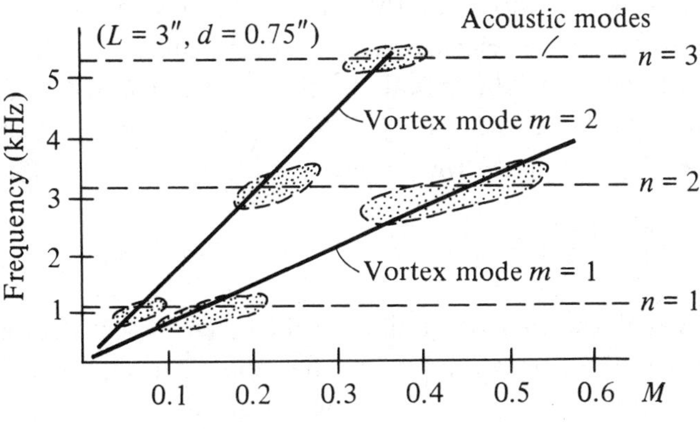

where V is the convection speed of the wave on the sheet and v the sound speed.

To sustain the oscillations, i.e. to keep the feed back signal 'synchronized' with a mode, the round trip must equal an integer number of periods of oscillation of the mode, $T = mT_m$, so that the frequency of the mth mode will be

$$\nu_m = m/T \approx mV/d = m\beta U/d \quad (m = 1, 2, \ldots) \tag{2}$$

In what follows, we shall assume the speed V to be independent of frequency and proportional to the unperturbed incident flow speed U, $V = \beta U$. In the idealized case discussed in Display 21.5, we had $\beta = 0.5$ which is of the right order of magnitude in most other situations.

Each of the vortex modes can be regarded as an oscillator which can interact with acoustic modes of adjacent cavities similar to the interaction between the valve plug and the acoustic pipe modes discussed in Section 21.2. As in many musical wind instruments, such an interaction can lead to selfsustained oscillations of the acoustic (and vortex) modes.

Example. Flow excitation of a bottle. As an example we consider the excitation of a cavity or bottle by grazing flow, as illustrated in Display 21.7. The cavity is assumed to be two-dimensional and the flow is perpendicular to the cavity edges. The flow over the cavity is modelled as a shear layer with the free stream velocity being U.

The vortex mode wavelength is determined by the width d of the cavity, and the modal frequencies are given by Eq. 2. Each mode can interact with the (axial) acoustic modes in the cavity, which has the depth L.

In the nth acoustic mode of the cavity, an odd number of quarter wavelengths is equal to the effective depth L' of the cavity, which is known to be somewhat larger than the actual length L, $L' = L + \delta$, where $\delta \simeq 0.3\,d$ is the 'end correction' of the cavity. The corresponding acoustic mode frequencies are, with v being the sound speed,

$$\nu_n = (2n - 1)v/4L' \quad n = 1, 2, \ldots \tag{3}$$
$$L' \simeq L + 0.3d$$

By analogy with the valve instability problem, we expect strong coupling to occur between the vortex modes and the acoustic modes when the mode frequencies are about the same. The corresponding flow velocity, as obtained from Eqs. (1) and (2), is then given by

$$M = U/v \simeq (2n-1)(d/\beta)/[4mL' - (2n-1)d] \simeq (d/L')(2n-1)/4m\beta \quad (4)$$

$$\beta = V/U \simeq 0.5 \quad m, n = 1, 2, \ldots, \quad L' \simeq L + 0.3d$$

The graph in the display refers to an experiment with a cavity with $d = 1.9$ cm and $L = 7.6$ cm mounted in the wall of a rectangular duct. The effective depth of the cavity was approximately $L' \simeq 8.1$ cm and the first three acoustic mode frequencies, corresponding to $n = 1, 2$, and 3, were 1040, 3120, and 5200 Hz, shown as dashed horizontal lines, i.e. independent of the incident flow velocity.

The natural frequencies of the vortex oscillations, on the other hand, increase with the flow velocity, as shown by the solid lines for the first two modes, $m = 1, 2$. Self-sustained oscillations of the cavity are expected in the vicinity of the intersections of the two sets of lines, and this indeed, was confirmed by the experiments.

Frequently, more than one mode occurred simultaneously. Thus, for a Mach number $M = 0.2$, two pronounced oscillations with about the same amplitude were obtained for the modes identified by $m = 1, n = 1$ and $m = 2$, $n = 2$. At a somewhat lower Mach number, the former oscillation dominated, and at a somewhat higher Mach number, the latter.

Other systems involving the coupling between a vortex oscillation and the acoustic modes of an air column, such as the organ pipe and the ordinary mouth whistle, can be treated in a similar manner.

Oscillations of the wake behind a cylinder. Kármán vortex street. The wake behind a cylinder or some other blunt body in a fluid flow is known to oscillate in periodic motion in the transverse direction over a wide velocity range of the incident flow. A transverse momentum flux in the fluid and a corresponding reaction force on the cylinder are consequences of this motion. As indicated in Display 21.8, the oscillations give rise to a set of discrete line vortices, known as the **Kármán vortex street.**

This phenomenon is a result of the formation of two shear layers on the two sides of the cylinder and the interaction between these layers. A single layer, as we have seen in the previous section, is unstable and breaks up and eventually becomes chaotic. With two layers interacting, however, a periodic rather than a chaotic motion results, and the period of oscillation will be of the order of the characteristic frequency U/d. Actually, the frequency is known to be

$$v \simeq S(U/d) \quad (5)$$

Display 21.8.

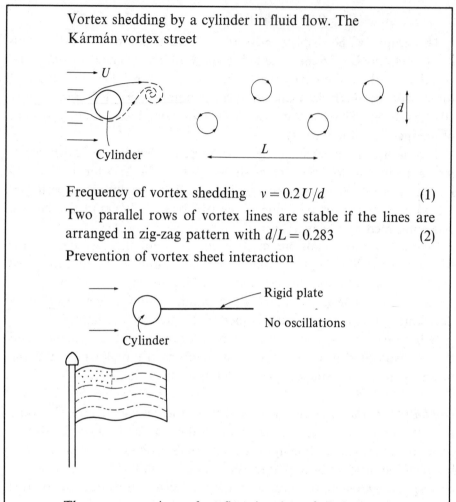

Vortex shedding by a cylinder in fluid flow. The Kármán vortex street

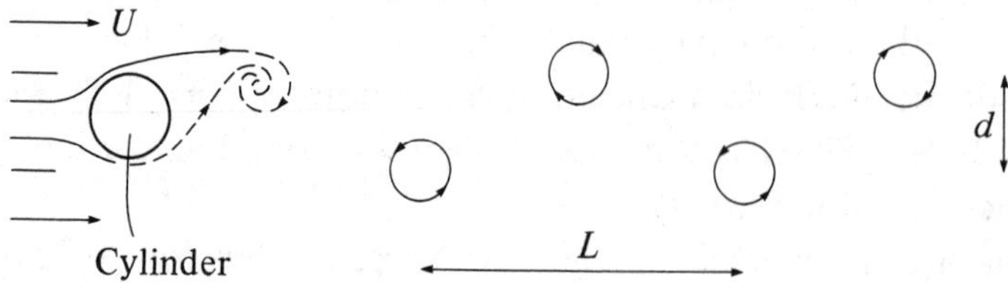

Frequency of vortex shedding $\nu = 0.2\,U/d$ (1)

Two parallel rows of vortex lines are stable if the lines are arranged in zig-zag pattern with $d/L = 0.283$ (2)

Prevention of vortex sheet interaction

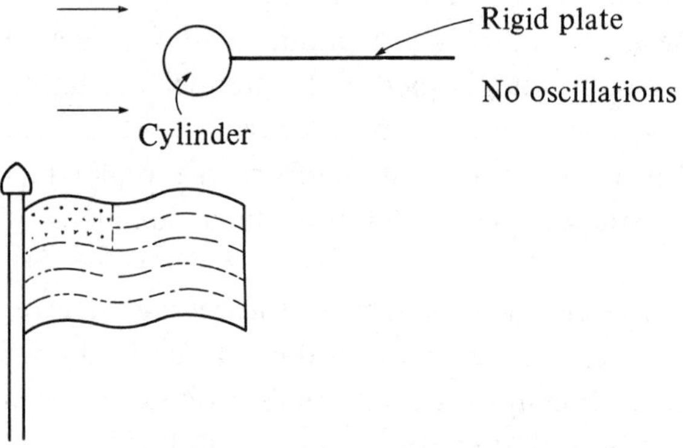

The wave motion of a flag in the wind is induced, at least in part, by the vortex shedding by the flag pole

where the dimensionless number S, usually called the Strouhal number, is approximately constant and equal to 0.2 over a wide range of velocities.

If the two shear layers are prevented from interacting by means of a rigid plate, as indicated, the periodic oscillations will be suppressed. On the other hand, if the plate is replaced by a thin flexible cloth, like a flag, it will be set in oscillations by the flow.

The oscillatory flow behind a cylinder rolls up into vortices, which arrange themselves in zig-zag pattern. It should be mentioned in this context, that, unlike a single row, a double row of line vortices is stable if the vortices are arranged in a zig-zag pattern with a separation d between the rows and an axial distance L between the vortex lines in the axial direction such that $d/L = 0.238$. Any other configuration is unstable. This result was established by Kármán and by Heisenberg.

The frequency of vortex shedding generally falls in the audible range, and the whining of telephone wires in the wind is due to this kind of oscillation. It is also the mechanism of tone generation in the aeolian harp, and the oscillation often is called the aeolian tone.

The sound generated by the wake oscillations usually is quite weak, however. The main reason is the experimental fact that the oscillations generally are not in phase over the entire length of the cylinder but only over 'patches', which typically are of the order of five diameters in length. Each such patch can be regarded as an oscillator, and the entire wake motion is composed of such oscillators with their phases distributed randomly along the cylinder.

The sound can be considerably enhanced, however, if the wake is coupled to a resonator with a resonance frequency equal to the vortex shedding frequency. The resonator can be an air column in a cavity or it can be a mechanical structure, usually the cylinder itself. The resonator stimulates the elementary vortex oscillators to move in phase in much the same way as a standing light wave in an optical laser resonator stimulates the atoms to emit in phase.

An example of considerable practical importance is illustrated in Display 21.9, where a duct with flow contains a cylinder mounted perpendicular to the duct axis. The periodic wake behind the cylinder interacts with the standing acoustic waves in the transverse direction and the strongest interaction occurs at the cut-off frequency of the mode. In a rectangular duct, the cut-off frequency of the $(n, 0)$ mode is

$$\nu_{n,0} = nv/2D \quad n = 1, 2, \ldots \tag{6}$$

where v is the sound speed and D the separation of the walls in the duct

Display 21.9.

Acoustically stimulated vortex shedding in a rectangular duct.
Vortex shedding by cylinder in rectangular duct.

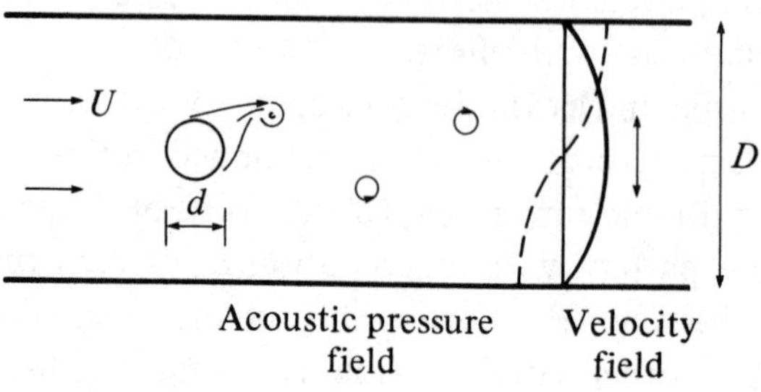

Vortex shedding frequency $\nu = 0.2U/d$ (1)

Frequency of nth acoustic mode $\nu'_n = nv/2D \quad n = 1, 2, \ldots$ (2)

Resonance if $\nu = \nu'_n$ (3)

Corresponding flow speed $0.2 \cdot U/d = n \cdot v/2D$

$M = U/v = n \cdot 2.5\, d/D$ (4)

in the direction perpendicular to the cylinder. (To account for the flow velocity on this frequency we have to multiply by $\sqrt{1-M^2}$, where $M = U/v$, but the corresponding change in the frequency is usually insignificant).

This frequency will equal the vortex shedding frequency in Eq. (1), $0.2U/d = nv/2D$, at the flow velocity

$$M = U/v = 2.5n \cdot d/D \tag{7}$$

where d is the diameter of the cylinder and D the duct width. A similar stimulated emission of sound through periodic vortex shedding occurs also from radial cylinders through the interaction with circumferential acoustic modes.

Instead of an acoustic mode, the cylinder itself can provide the feedback to stimulate the vortex shedding by oscillating in the transverse direction in its fundamental mode with a frequency equal to the vortex shedding frequency. In either case, the amplitudes of the oscillation will be very large, and as a result mechanical fatigue failure of duct walls and cylinders have occurred in heat exchangers and similar installations. The simplest way of preventing these instablities is to avoid coincidence of the frequencies of the vortex shedding and the acoustic and/or mechanical resonances.

21.5 Lasers

In the vortex shedding oscillators involving one or more cylinders in a flow, there is a 'spontaneous' transition of the fluid from the state of uniform motion to a state involving vortices, and this transition is associated with the emission of sound. The vortices along a cylinder or from one cylinder to the next and the corresponding sound field contributions are uncorrelated, however, and the resulting emitted acoustic power is comparatively small.

On the other hand, with the oscillators placed in the field of a resonator, driven at resonance by the spontaneous sound, this feedback tends to synchronize the vortex generation and correlate the elementary sound field contributions thus increasing the intensity of the resulting total field.

In a laser, the elementary oscillators involved are atoms, with an abnormally high proportion of atoms in an excited state. This 'inverted' non-equilibrium distribution is maintained by an external power source.

There is a spontaneous transition from the excited state to a lower lying state, and in the process, light is emitted; the transitions and light emissions from the various atoms are uncorrelated, however.

With the atoms placed in the field of an optical resonator (cavity between two mirrors) driven at resonance by the emitted light, the feedback field produces a stimulated emission of light such that the contributions from the various atoms are correlated. The resulting light intensity will be increased by a factor of the order of the number of oscillators involved. (The intensity produced by N uncorrelated oscillators is proportional to N. For N oscillators operating in phase, the total electric field will be proportional to N and the intensity proportional to N^2).

21.6 Heat maintained oscillations. The Rijke tube

A time independent heat source, like a time independent flow, can lead to feedback oscillations. For example, the thermal expansion of a body can alter the heat transfer to it and thus give rise to a modulation of the heat transfer, which can give rise to oscillations, such as the rocking motion of a kettle. In combustion chambers or in chemical gas reactions in general, the rate of heat generation depends on the gas density and the temperature, and a density variation produced through acoustic feedback can lead to feedback instability and oscillations involving one or several acoustic normal modes of the combustion chamber or gas container.

A simple and amusing demonstration of a heat maintained feedback oscillation is illustrated in Display 21.10. The device involved is simply a vertical steel tube, typically 1–2 m long and with a diameter of about 10 cm, which contains a wire mesh screen in the lower half of the tube.

The screen is heated with a Bunsen burner, and after removal of the burner (to avoid distortion of the air flow in the tube), a strong tone is emitted at the frequency of the fundamental mode of the air column in the tube. As the screen cools, the intensity of the tone decreases. It is observed, that a tone is produced only if the screen is located in the **lower half** of the tube. Although a detailed quantitative explanation of the oscillations in the tube (usually called the Rijke tube) is rather complicated, we can understand this effect qualitatively as follows.

The heating of the air in the tube by the screen creates a convected air flow up through the tube. Superimposed on this steady flow is the oscillatory gas flow in the sound field in the tube, which can be considered to be initiated by an unavoidable fluctuation. In the fundamental acoustic mode, the oscillatory flow goes in and out of the ends of the tube in counter motion, the velocity at the center of the tube being zero. When the flow is inwards (outwards), the sound pressure p in the tube increases (decreases) with time.

At a time when the flow is inwards, i.e. when the sound pressure increases with time, the oscillatory flow velocity u is in the **same** direction as the convection flow velocity U in the lower half of the tube. Thus, with the

Display 21.10.

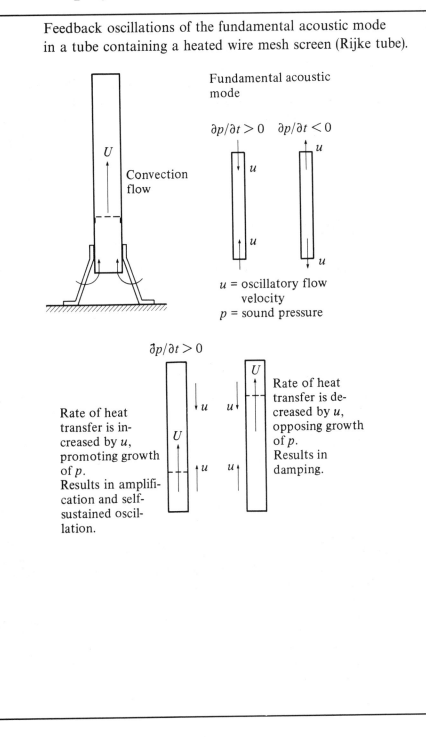

Feedback oscillations of the fundamental acoustic mode in a tube containing a heated wire mesh screen (Rijke tube).

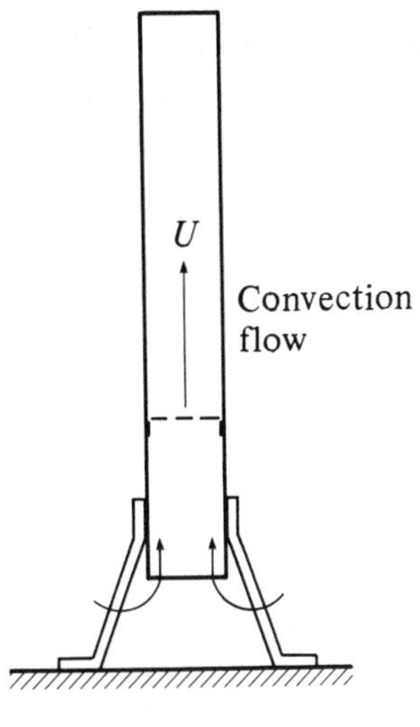

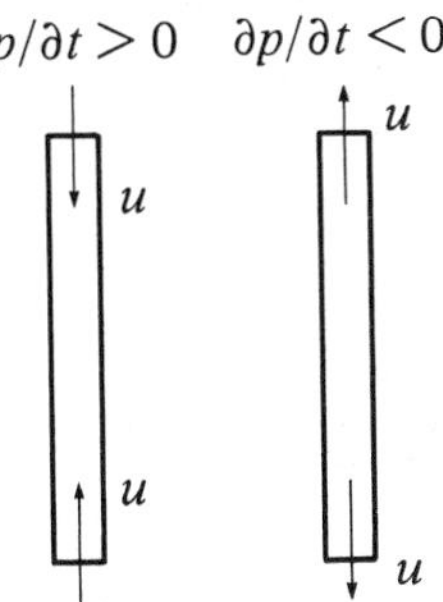

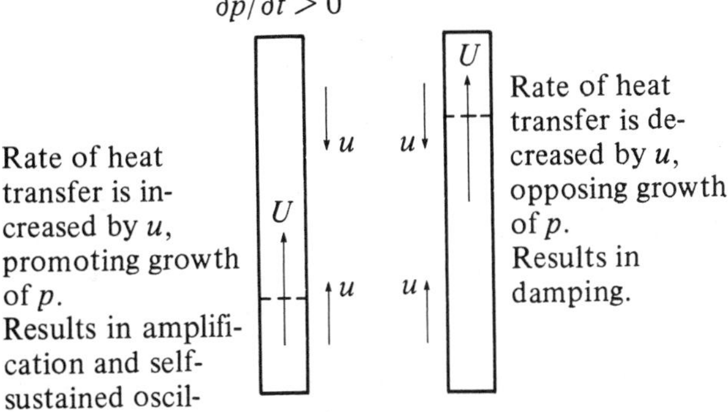

heated screen located in the lower half of the tube, the rate of heat transfer to the gas from the screen then will be increased through the cooling action of the air motion in the sound field. This tends to increase the pressure in the tube, i.e. promote the growth in sound pressure already under way, and a feedback oscillation can occur.

In the upper half of the tube, on the other hand, the inflow **opposes** the convection flow, and with the screen located in this half, the rate of heat transfer to the gas is reduced through the action of the acoustic mode. This tends to decrease the pressure in the tube, in opposition to the existing rate of increase, and the acoustic feedback in this case leads to an additional damping of the acoustic mode.

Examples

E21.1 *Vortex flow meter.* As discussed in Section 21.3, a cylinder, or any other blunt object, placed in a fluid stream perpendicular to the flow, produces an oscillatory wake behind the object (Karman vortex street). The frequency of oscillation can be expressed as $v = SU/d$, where S is a dimensionless number, U the flow velocity and d the cylinder diameter (or width of the wake). The Strouhal number depends on the Reynolds number, $R = Ud/\eta'$, where η' is the kinematic viscosity of the fluid involved. (For air at atmospheric pressure and room temperature $\eta' \simeq 0.15$ CGS units and it is approximately equal to the product of the mean free path and the mean thermal molecular speed). It is known from experiments that $S \simeq 0.2$ over a wide range of Reynolds numbers, from 100 to 10^7.

This relation between frequency and flow velocity can be used for the determination of the flow velocity from measurement of the frequency. Since the flow velocity across a duct at large Reynolds numbers is approximately uniform, a measurement of U determines also the volume flow rate AU through a duct of area A.

The method of measurement is based on the oscillatory stress produced in the cylinder. This stress (or corresponding strain) can be determined by means of a strain gauge or piezoelectric transducer. To avoid having the transducer in contact with the fluid, it can be mounted on a portion of the vortex shedder which is made to protrude outside the duct wall. An instrument of this kind has the advantage of being simple, rugged, and reliable, requiring little or no maintenance.

The material in the vortex shedder can be stainless steel, and it can be used over a wide range of temperatures (typically from -40 to 400 °C). The transducer output increases approximately as the square of the flow velocity, and in order to get a sufficiently large transducer output signal, the flow velocity must not be too small. The typical Reynolds number range of an instrument of this kind is 5×10^3–7×10^6.

Problems

21.1 *Instability in a tube.* (a) With reference to Display 21.1, determine the rate of amplitude growth and the frequency of oscillation if the 'loop gain' $E = 5$ and $x = 2$m. (b) Derive the equation for the determination of the complex frequency of feedback oscillations when the tube in (a) is terminated by a rigid wall at a distance $L = 5$ m from the loudspeaker.

21.2 *Flow meter.* Consider a tube carrying a mean flow with a velocity U. A sound source is mounted on the side wall of the tube and emits sound at a frequency ω in the upstream and downstream directions, as indicated schematically in the figure. The microphones M_1 and M_2 are mounted a distance L from the sound source and are connected via an amplifier to the sound source, as shown. Neglect reflections in the tube.

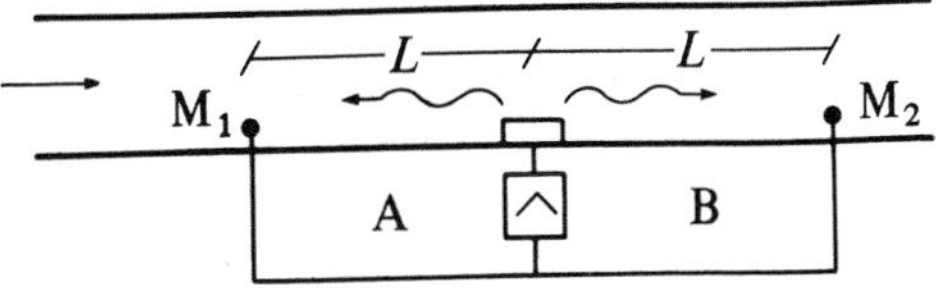

(a) What is the lowest frequency of feedback oscillations in each of the loops A and B shown in the figure?
(b) Show that the beat frequency resulting from the superposition of the electric currents from these oscillations is proportional to the mean flow U.

21.3 *Valve instability; acoustic feedback from pipe.* Prove the expression for the pipe impedance in Eq. 21.2(6) and the corresponding expressions for the real and imaginary parts in Eqs. (7) and (8).

21.4 *Whistle.* A bottle of depth 30 cm and with a neck diameter of 2 cm is moved through the air. Estimate the lowest velocity of the bottle at which flow excited whistling of the bottle can occur.

21.5 *Kármán vortex oscillations.* A heat exchanger consists of a rectangular duct through which hot air is flowing. Water pipes are mounted in the duct perpendicular to the duct axis and parallel with the short side walls in the duct. The separation of these walls is 4 m and the diameter of the water pipes is 5 cm. At what air flow speed do you expect unstable acoustically driven feedback oscillations to occur, if the temperature of the air is 500 K? Suggest a simple method of eliminating the instability at this speed.

Index